W9-ARZ-745

ELECTRONIC DEVICES AND CIRCUIT THEORY

Robert Boylestad

QUEENSBOROUGH COMMUNITY COLLEGE
ASSOCIATE PROFESSOR

Louis Nashelsky

QUEENSBOROUGH COMMUNITY COLLEGE
ASSOCIATE PROFESSOR

PRENTICE-HALL, INC., ENGLEWOOD CLIFFS, N.J.

ELECTRONIC DEVICES AND CIRCUIT THEORY

ELECTRONIC DEVICES AND CIRCUIT THEORY

Robert Boylestad and Louis Nashelsky

10 9 8 7

ISBN: 0-13-250-548-7

PRINTED IN THE UNITED STATES OF AMERICA
LIBRARY OF CONGRESS CARD CATALOG NUMBER: 78-151394

Prentice-Hall International, Inc., London
Prentice-Hall of Australia, Pty. Ltd., Sydney
Prentice-Hall of Canada, Ltd., Toronto
Prentice-Hall of India Private Limited, New Delhi
Prentice-Hall of Japan, Inc., Tokyo

Dedicated to:
Else Marie, Eric, Alison, and Stacey
and to
Katrin, Kira, and Larren

Contents

DIODE RECTIFIERS AND FILTERS 65

3

TRANSISTORS AND VACUUM TUBES 116

4

DC BIASING 152

5

FIELD EFFECT TRANSISTORS (FETs) 305

7

MULTISTAGE SYSTEMS, DECIBELS (dB), AND FREQUENCY CONSIDERATIONS 350

LARGE-SIGNAL AMPLIFIERS 419

PNPN AND OTHER DEVICES 459

10

INTEGRATED CIRCUITS (ICs) 497

11

DIFFERENTIAL AND OPERATIONAL AMPLIFIERS 520

12

FEEDBACK AMPLIFIERS AND OSCILLATOR CIRCUITS 572

13

PULSE AND DIGITAL CIRCUITS 642

14

REGULATORS AND MISCELLANEOUS CIRCUIT APPLICATIONS 683

15

CATHODE RAY OSCILLOSCOPE 708

APPENDICES

Preface

This text is designed primarily for use in a two-semester or three-trimester sequence in the basic electronics area. It is expected that the student has taken a course in dc circuit analysis and has either taken or is taking a course in ac circuit analysis. This text only requires a mathematical background similar to that required for the ac circuit analysis course.

In an effort to aid the student the text contains extensive examples that stress the main points of each chapter. There is also a generous number of illustrations to help the student through the new concepts and techniques introduced. Important conclusions are emphasized by boxed equations or boldface answers to make the student aware of the essential points covered.

The text is the result of a two-semester electronics course sequence which both authors were actively involved in teaching over a period of seven years. However, the fifteen chapters actually contain more material than can be covered in two 15-week semesters (or three 10-week trimesters). This preface will provide some guidance on how the authors feel the material can be organized.

Essentially, the first six chapters provide the basic background to electronic devices—including construction, biasing, and operation as single stages. The material in these chapters can be included in the first semester with the option left to the teacher of stressing some areas more than others, or some not at all. The course would begin with the theory and operation of two-terminal devices, stressing semiconductor diodes. Since the theory course is usually taught in conjunction with a laboratory course the material has been organized with regard to providing practical circuit examples which can be operated in the lab. In this regard a laboratory manual which follows closely the material in the present text has been prepared by G. Kousourou and is available from Prentice-Hall.

Chapter 2 on diode rectifiers and filters provides some practical examples of diode application to the basic electronic area of power supplies. Other texts generally place this material at the end of the text. Our own experience shows that this practical study serves as an interlude between the chapters on basic device theory and provides some valuable lab experiments.

Chapter 3 covers the transistor device, its construction, and theory of operation. As mentioned earlier it is possible to follow Chapter 1 with this chapter. The operation of the transistor is presented both mathematically and graphically. The amplifying action of the transistor is defined and demonstrated.

It is the authors' experience that the student can better comprehend the operation of the transistor device if, initially, the dc bias and ac operation are treated separately. Thus, Chapter 4 deals only with the dc bias of the transistor (and tube). This is done for common-emitter, common-bias, and common-collector (emitter follower) configurations for a variety of bias circuit types. Numerous examples help demonstrate the theory presented. Also, some design problems are included to provide a well rounded treatment. The chapter is quite extensive and only part of it may be covered if sufficient time isn't available. Emphasis is placed on the transistor device rather than tube circuits, as it should be in the present electronic field.

Chapter 5 is one of the most important in the basic coverage area and should be given sufficient time in any course. The development of the transistor ac equivalent circuit model is covered in detail followed by analysis of the ac operation of the full small-signal circuit. The treatment in this chapter (as in Chapter 4) is essentially mathematical. However, the mathematics are kept short and direct, with a generous number of examples provided so that students will be able to follow the ideas presented. The hybrid equivalent circuit of the transistor is comprehensively presented and then the usual engineering simplifications are included in ac analysis to provide a more practically meaningful treatment. Also, an approximate equivalent circuit technique is presented which considerably simplifies analysis of the various circuit configurations found in later chapters.

If possible the material on the field effect transistor (FET) should also be covered in the first semester of electronics. After having presented and developed the concepts of dc bias and ac analysis of the transistor (and tube), Chapter 6 then covers a number of practical FET circuits. We had considered including the FET dc biasing in Chapter 4 and ac analysis in Chapter 5. It was our feeling from classroom experience that this would require spending too much time on each topic and the FET would appear to be a minor device to the student. By covering the FET in a separate chapter, its significance is stressed and its operation can be properly presented.

Chapter 7 would be the first topic in the second semester and covers the operation of multistage transistor, FET (and tube) circuits. Stage loading, overall gain calculations, and use of decibels are all covered in this important chapter. A number of examples help emphasize the main points of the chapter.

Chapter 8 covers the operation of power transistors in a few basic power amplifier circuits. Most important is the operation of the push-pull circuit. Transistor push-pull circuits containing a transformer as well as transformerless circuits are covered.

Chapter 9 is a "catch-all" of a number of PNPN devices—covering their construction, operation, and circuit applications. It can be covered quickly or even passed over, if desired, without loss of continuity.

Chapter 10 is a short treatment of the fabrication and construction of integrated circuits (IC) and can be assigned mainly as student reading.

Chapter 11 provides coverage of two very important topics and should be considered essential to the second semester coverage. Due to the popularity of linear IC units, both the differential and operation amplifier are now regarded as basic units and may soon be considered as significant as the transistor in electronic circuits. A comprehensive treatment is accordingly given each topic as well as examples and practical applications.

Chapter 12 on feedback amplifiers and oscillators should be covered at least partially in the second semester. The material can also be deferred to a third electronics course on communications if desired. The chapter is extensive and need not be fully covered if time is limited.

Chapter 13 on digital circuits provides a good survey of important digital circuits. It is so important in the present electronics field to know this area well. If no course devoted exclusively to computer circuits and logic is taught in your curriculum, then the material of this chapter should be closely covered. It is the authors' feeling that digital circuits are, today, very much a part of basic electronics.

Chapter 14 provides coverage of voltage regulators which should be covered in the second semester. The miscellaneous circuits provided are for the student's own practical study and can be used to stimulate or motivate his interest.

Chapter 15 can be integrated anywhere in the two semesters. In our own courses we cover this material first in order to introduce the student to the basic electronic measuring device—the cathode ray oscilloscope (CRO). The fundamental operation and use of the CRO is stressed in this chapter. Again generous examples help emphasize the main points covered. The operation and measurements using the CRO is so important that it is strongly urged that this material, although appearing at the end of the book, be stressed in classroom teaching.

To improve the use of this text by both student and instructor there are numerous practical examples in most chapters. Problems at the end of these chapters are keyed to the particular section in which the problems are covered.

We wish to thank Professors Aidala and Katz of the ET department at Queensborough Community College for their continued help and encouragement over the years. They have provided us with both courses and atmosphere conducive to the best in learning and teaching. We also wish to thank Professor Kosow of Staten Island Community College for his valuable assistance with the original manuscript and for his numerous suggestions and criticisms.

We thank both Sylvia Neiman and Doris Topel, department secretaries at Queensborough Community College, for their excellent typing of the manuscript. Finally we wish to thank each other for a remarkably pleasant and rewarding collaboration.

ROBERT BOYLESTAD / LOUIS NASHELSKY

Bayside, N.Y.

ELECTRONIC DEVICES AND CIRCUIT THEORY

1

Two-Terminal Devices

1.1 INTRODUCTION

This chapter is devoted primarily to the introduction of a very important two-terminal device, the *diode*. The diode is one of the fundamental building blocks of the wide variety of electronic circuits in use today. It is essential to the operation of such representative systems as rectifiers, doublers, limiters, clampers, clippers, modulators and demodulators, waveforming circuits, and frequency converters. The diode is available in many different sizes and shapes with varying modes of operation. The *vacuum diode*, *gas diode*, and *semiconductor diode* will all be considered in this chapter. The last sections will contain a brief description of the *Zener*, *varicap*, *tunnel*, *photoelectric*, and *silicon power diodes*. Also to be covered is a temperature-sensitive resistor, the *thermistor*. The silicon-controlled rectifier (SCR) will be discussed in Chapter 9.

The first diode, called Fleming's valve, was developed by J. Ambrose Fleming in 1902. Its basic construction consisted of two elements, a filament and metallic plate in an evacuated glass envelope, similar in many respects to the modern high-vacuum diode. It was not until the early 1930s that a radically new type of diode became increasingly important, the semiconductor diode. This solid-state device, much smaller than the vacuum diode with characteristics closer to the ideal switching characteristics, led the way to the development of the transistor amplifier (Chapter 3) by J. Bardeen and W. Brattain of Bell Laboratories in 1948.

In recent years, emphasis has been almost completely on the development of the semiconductor diode. Except for very high frequencies or high-power applications the solid-state diode seems destined to eliminate the vacuum diode from the competitive market. It would appear that the ever-increasing interest in the solid-state area *may* eventually lead to the "complete" elimination of the vacuum-tube diode from future design and development considerations.

Before considering the basic theory of operation of the vacuum, gas, and semiconductor diodes, the *ideal* diode is presented to introduce basic diode action and establish a basis for later comparison with actual diode characteristics.

1.2 IDEAL DIODE

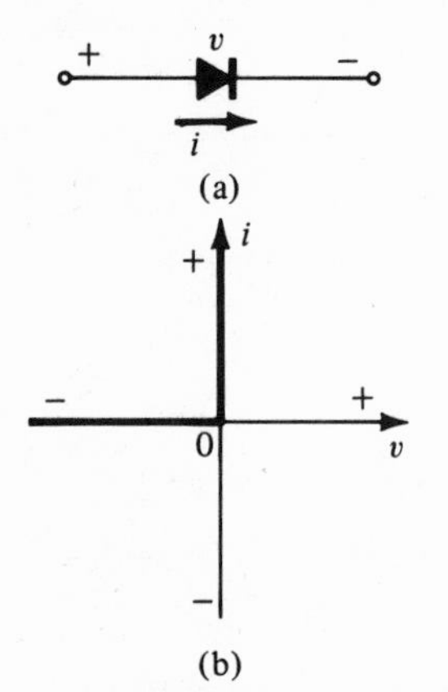

FIG. 1.1
Ideal diode. (a) symbol; (b) characteristics.

The ideal diode is a *two-terminal* device having the symbol and characteristics shown in Fig. 1.1a and b, respectively.

In the description of the elements to follow, it is critical that the various *letter symbols*, *voltage polarities*, and *current directions* be defined. If the polarity of the applied voltage is consistent with that shown in Fig. 1.la, the portion of the characteristics to be considered in Fig. 1.1b is to the right of the vertical axis. If a reverse voltage is applied then the characteristics to the left are pertinent. If the current through the diode has the direction indicated in Fig. 1.1a the portion of the characteristics to be considered are above the horizontal axis, while a reversal in direction would require the use of the characteristics below the axis. For all device characteristics to appear in this text the *ordinate* will be the *current axis*, while the *abscissa* will be the *voltage axis*.

One of the important parameters for the diode is the resistance at the point or region of operätion. If we consider the region defined by the direction of i and polarity of v in Fig. 1.1a (upper right quadrant of Fig. 1.1b) we shall find that the value of the forward resistance, R_f, as defined by Ohm's law is

$$R_f = \frac{V_f}{I_f} = \frac{0}{2, 3\text{ mA}, \ldots, \text{ or any positive value}} = 0\Omega$$

where V_f is the forward voltage across the diode and I_f is the forward current through the diode. *The ideal diode, therefore, is a short circuit for the forward region of conduction* ($i_f \neq 0$).

If we now consider the region of negatively applied potential (third quadrant) of Fig. 1.1b,

$$R_r = \frac{V_r}{I_r} = \frac{-5, -20, \text{ or any reverse bias potential}}{0}$$
$$= \text{very large number, which for our purposes we shall consider to be infinite } (\infty)$$

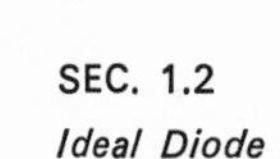

where V_r is the reverse voltage across the diode and I_r is the reverse current in the diode. *The ideal diode, therefore, is an open circuit in the region of nonconduction* ($i_r = 0$).

In general, it is relatively simple to determine whether a diode is in the region of conduction or nonconduction by simply noting the direction of the current i. For conventional flow (opposite to that of electron flow), if the resultant diode current has the same direction as the arrowhead of the diode symbol, the diode is operating in the conducting region.

As an introductory example of one practical application of the diode let us consider the process of rectification by which an alternating voltage having zero average value is converted to one having a dc or average value greater than zero. The circuit required is shown in Fig. 1.2 with an ideal diode.

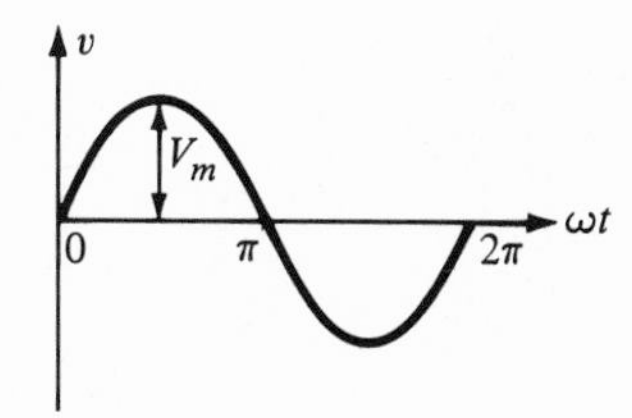

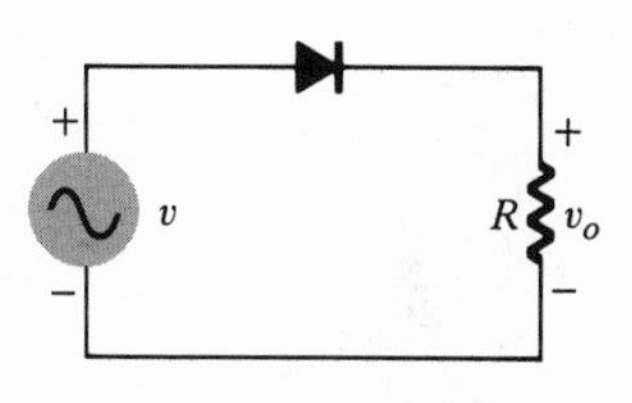

FIG. 1.2
Basic rectifying circuit.

For the region defined by $0 \rightarrow \pi$ of the sinusoidal input voltage v, the polarity of the voltage drop across the diode, would be such that the short-circuit representation would result and the circuit would appear as shown in Fig. 1.3a. For the region $\pi \rightarrow 2\pi$ the open-circuit representation would be applicable and the circuit would appear as shown in Fig. 1.3b.

For future reference note the polarities of the input v for each circuit. For sinusoidal inputs the polarity indicated will be for the positive portion of the sinusoidal waveform as shown in Fig. 1.2.

For the situation shown in Fig. 1.3a the output voltage, v_o, will appear exactly the same as the input voltage v, as long as the diode is forward biased. In Fig. 1.3b, because of the open-circuit representation of the ideal diode, the output voltage v_o equals zero from π to 2π of the impressed voltage v. The complete resultant output wave-

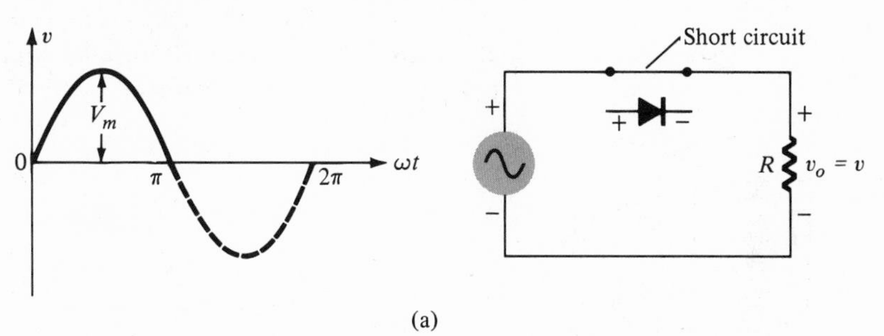

(a)

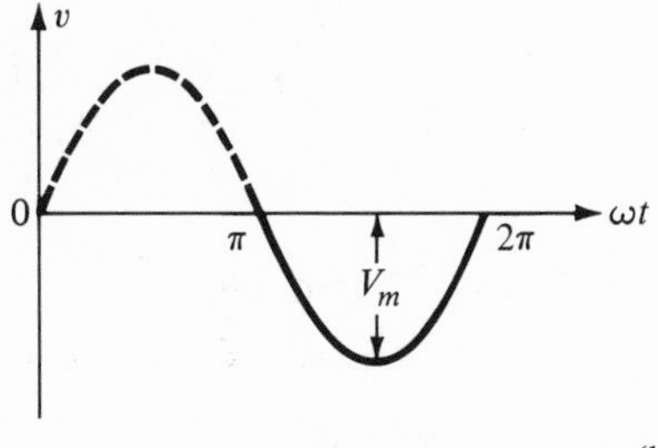

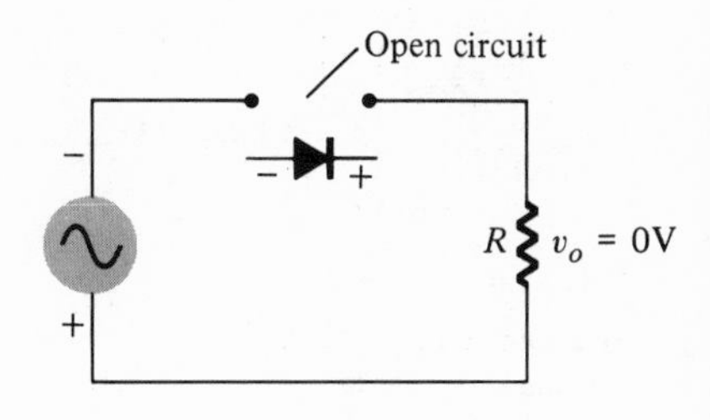

(b)

FIG. 1.3
Rectifying action of the circuit of Fig. 1.2.

form is shown in Fig. 1.3c for the entire sinusoidal input. For each cycle of the input voltage, v, the waveform of v_o, will repeat itself so that each waveform has the same frequency. A closer examination of the various figures will also reveal that the impressed *emf* v and v_o are *in phase;* that is, the positive pulse of each appears during the same time period. Phase relationships will become increasingly important when we consider the transistor and vacuum-tube amplifier.

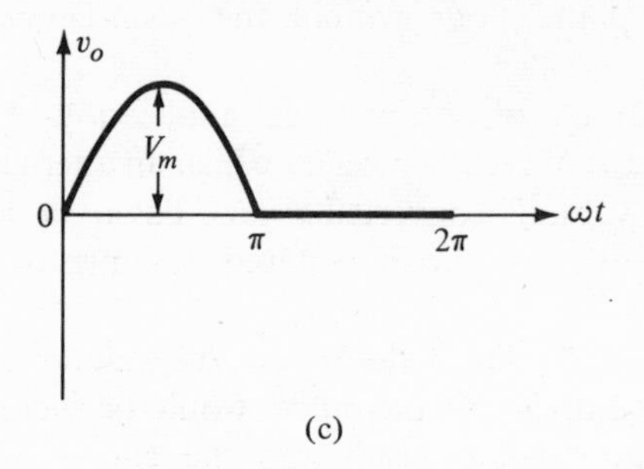

FIG. 1.3

Rectifying action of the circuit of Fig. 1.2.

1.3 VACUUM-TUBE DIODE

The basic vacuum diode consists of a *cathode* and an *anode* (metallic plate) in an *evacuated tube* in the relative positions shown in Fig. 1.4. The *cathode* may be heated either directly or indirectly. Various types of cathodes are shown in Fig. 1.5.

The cathode provides a large number of "free" electrons in the region between the cathode and the plate. In either the directly or indirectly heated type this is accomplished by bringing the cathode to an electron emission temperature by applying a specified heater potential across the heater filament terminals. The applied potential will

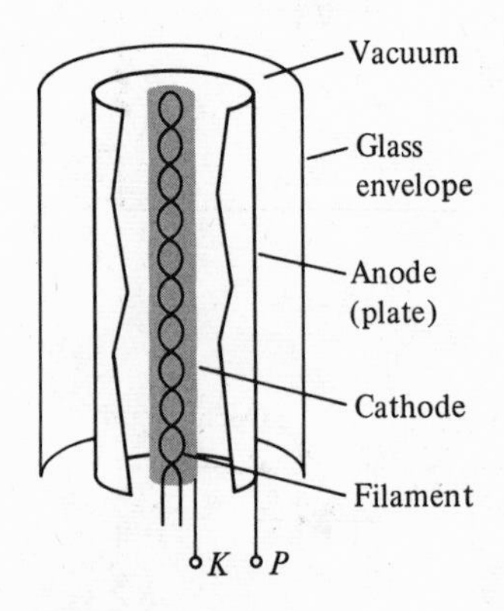

FIG. 1.4

Basic construction of the vacuum-tube diode.

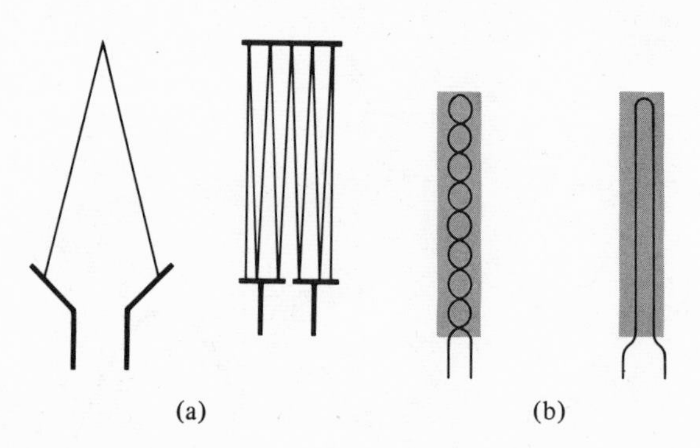

FIG. 1.5

Cathodes: (a) directly heated; (b) indirectly heated.

develop a current I through the filament, resulting in an I^2R heating loss much like the heating element of a toaster. As the temperature of the filament increases, the thermal agitation of the electrons will increase to a point where the electrons have sufficient kinetic energy to leave the surface of the cathode and assume their "free" state. This type of emission is referred to as *thermionic* emission. The tube is evacuated to prevent the cathode from "burning up" due to the oxygen in the air and to increase the mobility of the electrons.

In the directly heated type (the filament is the cathode) the electrons are emitted directly by the filament material. In the indirectly heated type, the electrons are emitted from a surface not directly connected to the filament (or heater) that is brought to emission temperature by the radiating heat of the filament. The indirectly heated type is the more commonly used of the two. If a 60-cycle ac potential were applied to the directly heated type the number of thermionically generated "free" electrons would vary at each instant of time, since the current through the filament is determined by the instantaneous value of the applied signal. This varying emission of electrons may result in a 60-cycle hum at the output of the system. This undesirable effect is negligible is the indirectly heated type. A second advantage of the indirectly heated type is that the entire cathode is at a relatively fixed potential, whereas the potential of the directly heated type will vary from point to point along the filament.

The graphic symbols for vacuum-tube diodes with directly or indirectly heated cathodes are shown in Fig. 1.6a. There are no special markings on a tube to indicate which pins are connected to the plate, cathode, or filament. A tube manual must be consulted for the pin connections as indicated in Fig. 1.6 for the directly heated and indirectly heated types. The first number of the tube type indicates the rms voltage to be applied to the filament, while the remaining numbers and letters refer to a particular production series.

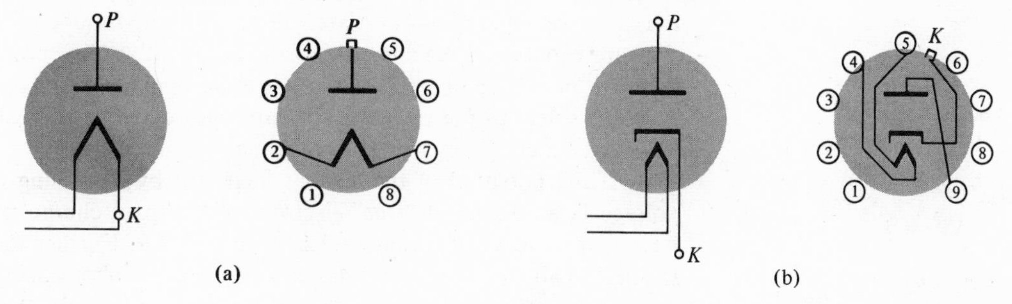

FIG. 1.6
Vacuum-tube diode symbols and pin connections: (a) directly heated; (b) indirectly heated.

Let us now examine the basic operation of a vacuum-tube diode using the circuit of Fig. 1.7. Initially the ac heater voltage of 6.3 V rms (typical value) is applied to the filament with the input voltage V set at 0 V. As the temperature of the filament rises an increasing number of electrons will be emitted by the cathode as shown by Fig. 1.8.

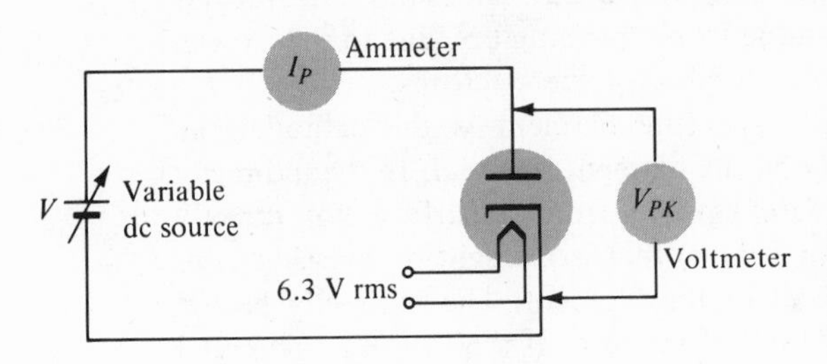

FIG. 1.7
Circuit for determining vacuum-tube diode characteristics.

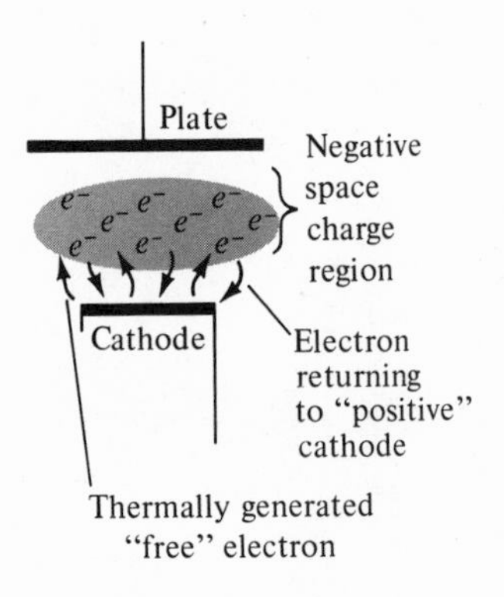

FIG. 1.8
Negative space charge region.

Consider, however, that as the negatively charged electrons leave the surface of the cathode a net positive charge will result on the cathode. This net positive charge will in turn attract the "free" electrons back to the cathode. At the same instant, however, additional electrons are being liberated, resulting in a continual transfer of electrons from the cathode to the region between the plate and cathode. When the full heating effect of the filament has been reached an equilibrium condition will exist such that the number of electrons leaving the cathode will be equal to the number returning. Once equilibrium is established, there will exist at any instant of time a number of "free" electrons between the plate and cathode. This region or cloud of "free" electrons is referred to as the region of *negative space charge* (Fig. 1.8). The number of electrons in this region can be increased by raising the temperature of the filament by increasing the filament voltage. For the diode of Fig. 1.8, however, increasing the filament voltage above the rated 6.3 V rms may result in permanent damage to the tube.

If the input voltage V remains set at 0 V the plate of the tube will remain at zero potential and (ideally) have no effect on the electrons of the space charge region. There are a relatively small number of electrons that are capable of reaching the plate with zero applied potential but this flow of charge or current is never more than a few microamperes. It is due solely to the electrons that are released through heating with sufficient kinetic energy to reach the plate of the diode.

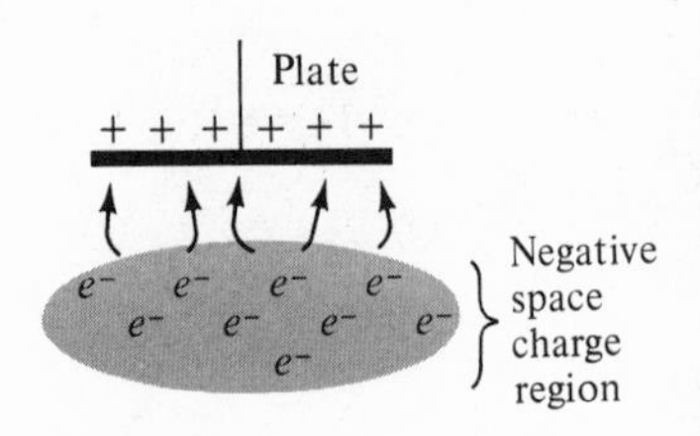

FIG. 1.9
The effect of a positive plate on the electrons of the negative space charge region.

If the potential of the plate is increased by increasing the input voltage V, a number of "free" electrons of the space charge region will be attracted to the plate as indicated in Fig. 1.9. Further increase in the input voltage V will result in more and more electrons being attracted to the plate, resulting in an exponential increase in the flow of charge or current as shown in Fig. 1.10a. The maximum current is not limited by the applied potential but by the temperature of the cathode material. Once the rate of cathode emission is equal to the rate of absorption by the plate the current will level off and not increase with further increase in applied voltage. The maximum current can only be increased by increasing the filament temperature. The maximum current for a particular filament temperature is referred to as the *saturation current.*

If the polarity of the applied voltage V of Fig. 1.7 is reversed, resulting in a negative potential on the plate, the negatively charged electrons will be repelled and the resulting flow of charge will be zero. This is shown quite clearly by the horizontal characteristics of Fig. 1.10a for negatively applied plate potentials. It should also be quite obvious from Fig. 1.10a that the vacuum-tube diode is closely related to the ideal diode in the reverse-bias region ($R_r = \infty\ \Omega$) but is far from ideal in the forward-bias region. In this region, the resistance R_f is not zero or fixed in magnitude but varies from point to point along the curve. The symbol for the vacuum-tube diode is shown in its most common form in Fig. 1.10b with the defined polarities and direction for Fig. 1.10a. Note that the direction of the current (region of conduction) is from plate to cathode as defined by conventional flow. The exponential curve of Fig. 1.10a is called a nonlinear curve since a change in voltage (or current) will not result in an equal change in current (or voltage). This is shown in Fig. 1.10a. The resultant change in i for Δv_2 is almost twice the change for i for Δv_1 although $\Delta v_1 = \Delta v_2$. For a linear curve, such as for a fixed resistance, the per cent of change in one quantity, v or i, will result in a corresponding per cent of change in the other.

The calculation of the resistance of the vacuum-tube diode at various points along its characteristic curve and the analysis of circuits using the vacuum-tube diode will be considered in greater depth in Section 1.6.

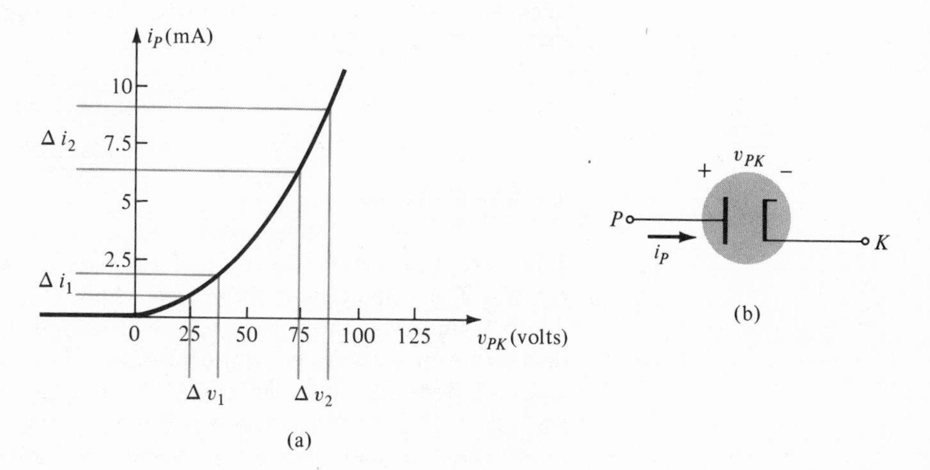

FIG. 1.10
Vacuum-tube diode: (a) plate characteristics; (b) symbol.

(a)
35W4
Half-wave rectifier
7-pin miniature type
PIV = 330 V
Peak I_P = 600 mA
Heater voltage = 35 V

(b)
5U4-GB
Full-wave rectifier
PIV = 1550 V
Peak I_P = 1 ampere
Filament voltage = 5 V

(Courtesy Radio Corporation of America)

In Fig. 1.11 various types of vacuum diodes are shown with pertinent data. Note the ceramic tube, which was designed to withstand extreme shock and high temperatures.

1.4 GAS DIODES

There are, in general, two types of gas diodes: the *hot* and the *cold cathode*. The graphic symbol for each is shown in Fig. 1.12. The solid dot is included in the tube envelope to denote the presence of gas (mercury vapor, argon, helium, or neon) in the tube. Each has a plate and a cathode in a glass envelope. The hot cathode has a filament, while the cold cathode employs no heating element. Although the mode of operation of the cold and hot cathode diodes is quite different, the resulting operating characteristics are somewhat similar, as indicated by Figs. 1.13 and 1.14.

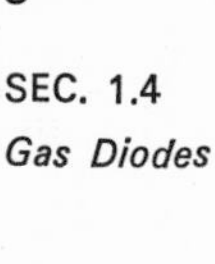

(c)
3CN3-B
High voltage rectifier
PIV = 30 KV
PEAK I_P = 110 mA
Heater voltage = 3.65 V

(d)
6H6
Twin diode
PIV = 465 V
Peak I_P = 53 mA
Heater voltage = 6.3 V

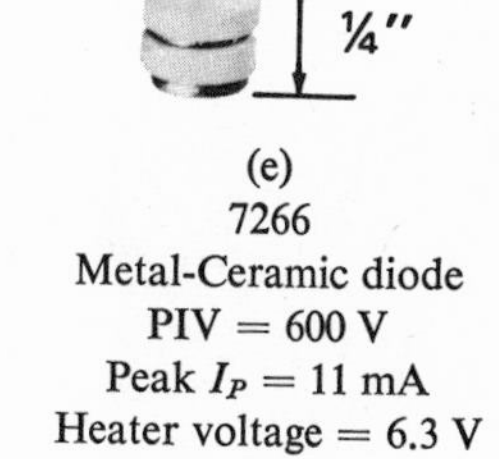

(e)
7266
Metal-Ceramic diode
PIV = 600 V
Peak I_P = 11 mA
Heater voltage = 6.3 V

(Courtesy General Electric Company)

FIG. 1.11
Various types of vacuum-tube diodes.

The characteristics of the hot cathode gas diode are quite similar to those of the vacuum diode in the region of *low* plate potentials (Fig. 1.13).

The effect of the gas is negligible in this region. Eventually, however, as the plate potential increases the electrons liberated by thermionic emission will, through their attraction for the plate, develop sufficient kinetic energy ($\frac{1}{2}\, mv^2$) to release electrons from the gas molecules through collision with the gas molecules. This process of removing negatively charged electrons from the gas molecule and forming positive ions is called *ionization*. Each gas has a different *ionization potential*. For mercury vapor it is approximately 10.4 V; for argon, 15.7 V; for neon, 21.5 V; and for helium, 24.0 V. Once ionization has occurred, there are an increasing number of "free" electrons and positive ions in the region surrounding the plate and cathode. The positive ions will reduce the retarding effect of the negative space charge on the negatively charged electrons so that a larger number of "free" electrons can travel relatively unaffected to the plate of the tube. The resulting flow of charge or current is then limited only by the external circuit or

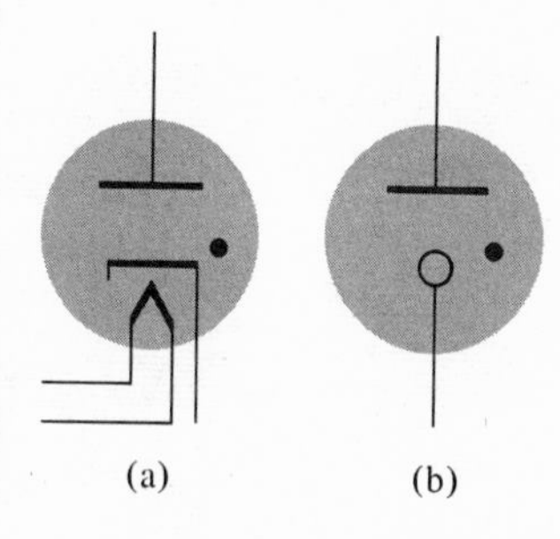

FIG. 1.12
Gas diodes: (a) hot cathode; (b) cold cathode.

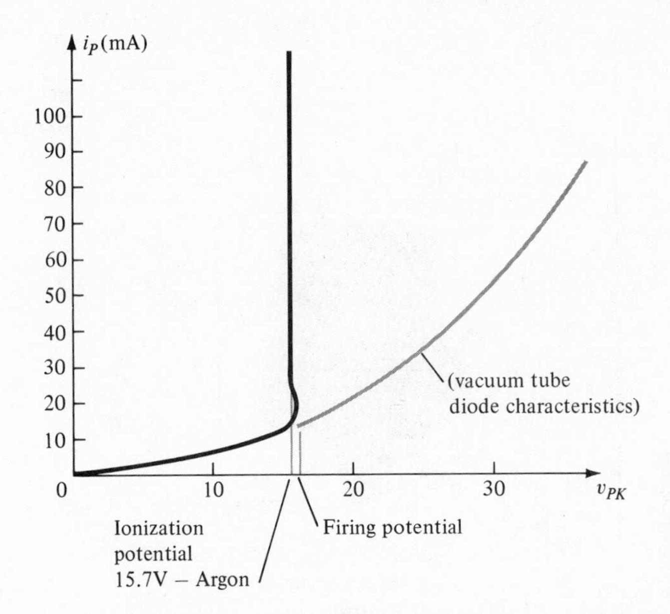

FIG. 1.13
Hot cathode gas diode characteristics.

current rating of the tube. Note in Fig. 1.13 that the potential necessary to induce ionization is slightly greater than the ionization potential. This *firing potential* is usually a few tenths of a volt higher than the ionization potential. Note also the almost vertical rise of current of the characteristic after the firing potential has been reached. Forgetting for a moment the shift in the characteristics from the vertical axis by an amount equal to the ionization potential, consider how closely the characteristics match those of the ideal diode.

The mode of operation of the cold cathode diode is completely dependent on the fact that there will be a number of "free" electrons

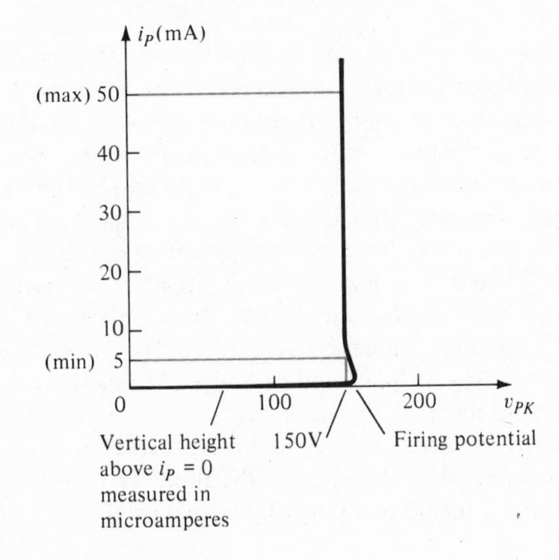

FIG. 1.14
Cold cathode gas diode characteristics (*OD*3).

and positive ions within any gaseous region due to natural causes. These causes include, among others, the energy contained by light in the form of photons and the internal thermal energy of the gas molecules. Since there is no heating element in the cold cathode diode there is no negative space charge to contend with between the plate and the cathode. As the plate potential increases from 0 V, the plate current will increase to a maximum as determined by the free electrons due to natural causes. The magnitude of this current is limited to microamperes. To increase the charge flow above this level the plate potential must be increased to the point where the "free" electrons of natural causes can induce ionization through collisions with the gas molecules. The resulting flow of charge due to ionization by these relatively few "free" electrons is still not sufficiently high, but the resulting positive ions will migrate and collide with the cathode. Their heavy mass and correspondingly high kinetic energy will impart sufficient thermal energy to the surface of the cathode to result in thermionic emission and also to release electrons purely through their collisions with the cathode molecules, resulting in a number of electrons developing the necessary velocity to leave the surface of the cathode. The flow of charge (electrons) to the plate can now rise to a value limited only by the external circuit or the current rating of the tube. This is demonstrated by the vertical rise in the characteristics at 150 V in Fig. 1.14. The breakdown potential of a gas tube can be controlled by the type of gas employed, the spacing between the electrodes, and the gas pressure within the tube. The firing potential of the cold cathode diode is, in general, a few volts higher than the ionization potential, whereas it was only a few tenths of a volt for the hot cathode gas diode. Both the hot and cold cathode diodes can be turned "off" by lowering the plate to cathode potential below the ionization potential or breaking the path of conduction.

Gas diodes are most commonly used in voltage regulation, rectifiers, voltage reference circuits, and flourescent lamps. The cold cathode is also used in the construction of neon signs. In the region of ionization the sign will glow with a color determined by the gas employed. As a matter of interest, neon gas will glow red-orange or helium-yellow, while mercury vapor with a mixture of argon and neon will be blue.

EXAMPLE 1.1 The following circuit is a simple voltage regulator employing a hot cathode gas tube. The function of the regulator is to maintain a fixed potential across the load even though the source voltage may vary as shown in Fig. 1.15a. The gas-tube characteristics have been idealized as shown in the same figure.

Solution: At the instant $v_i = 14$ V, the circuit can be redrawn as in Fig. 1.16. Removing the gas tube for a moment and applying the voltage divider rule will result in the following for V_L:

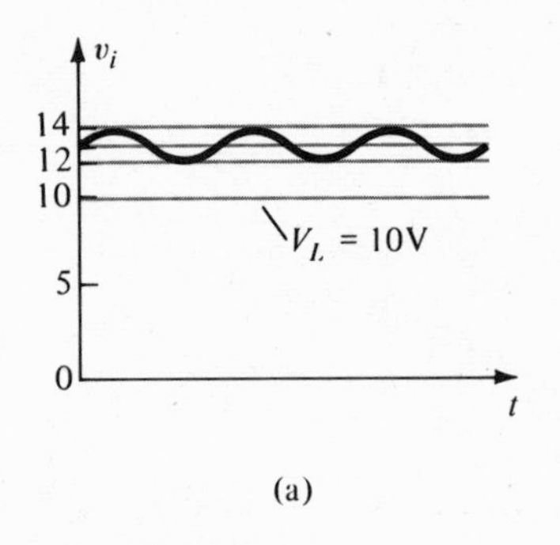

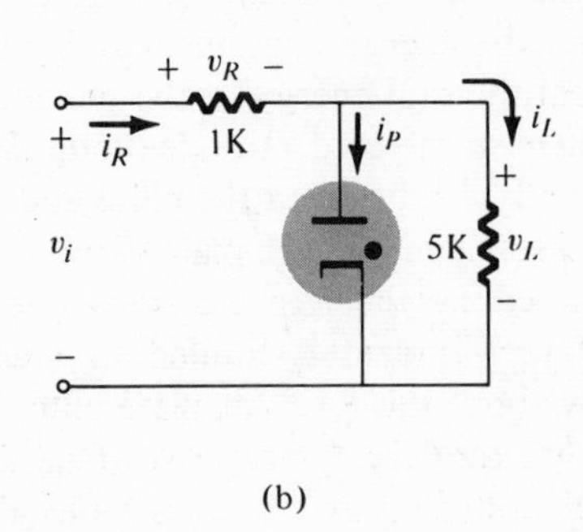

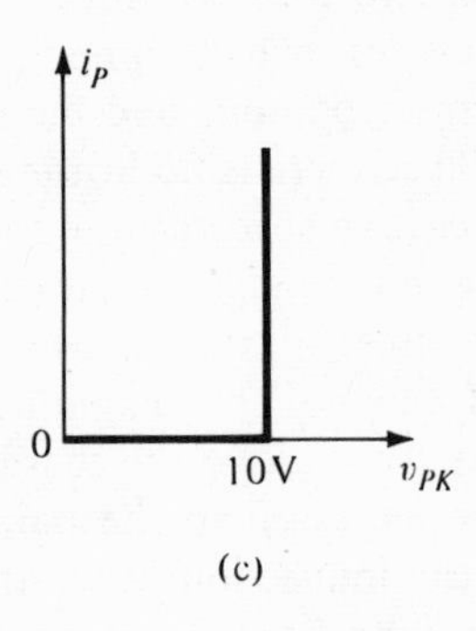

FIG. 1.15

Gas-tube voltage regulator: (a) input; (b) circuit; (c) gas-tube characteristics.

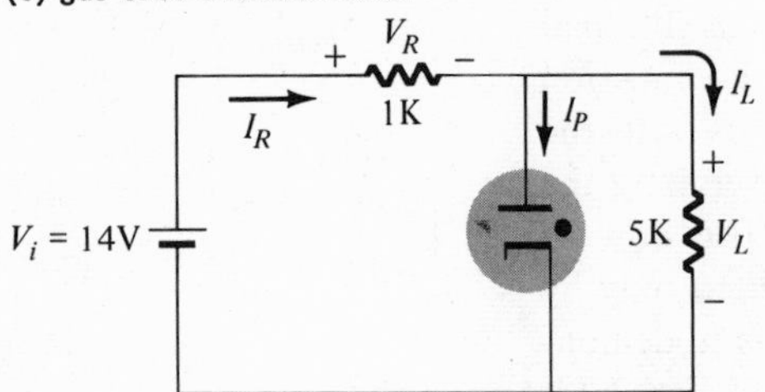

FIG. 1.16

Circuit of Fig. 1.15 at instant v_i = 14 V.

$$V_L = \frac{5\text{ K}(14)}{5\text{ K} + 1\text{ K}} = \frac{70}{6} \cong 11.7\text{ V}$$

Since 11.7 V > 10 V the gas tube has fired and the potential drop across it is fixed at 10 V. Therefore,

$$V_L = 10\text{ V} \qquad \text{and} \qquad I_L = \frac{10}{5\text{ K}} = 2\text{ mA}$$

$$V_R = 14 - 10 = 4\text{ V} \qquad \text{and} \qquad I_R = \frac{4}{1\text{ K}} = 4\text{ mA}$$

resulting in

$$I_{PK} = 4 - 2 = 2\text{ mA}$$

At $v_i = 13$ V:

$$V_L = \frac{5\text{ K}(13)}{5\text{ K} + 1\text{ K}} = \frac{65}{6} \cong 10.8\text{ V} > 10\text{ V}$$

so that the tube remains in the "on" condition and

$$V_L = 10\text{ V}, \qquad I_L = 2\text{ mA}$$

but V_R is now

$$13 - 10 = 3\text{ V}$$

and

$$I_R = \frac{3}{1\text{ K}} = 3\text{ mA}$$

with

$$I_{PK} = 3 - 2 = 1 \text{ mA}$$

At $v_i = 12$ V:

$$V_L = \frac{5\text{ K}(12)}{5\text{ K} + 1\text{ K}} = \frac{60}{6} = 10 \text{ V}$$

so that the tube is in the transition state, but

$$V_L = 10 \text{ V}, \qquad I_L = 2 \text{ mA}$$

and

$$V_R = 12 - 2 = 2 \text{ V}$$

with

$$I_R = \frac{2}{1\text{ K}} = 2 \text{ mA}$$

and I_{PK} theoretically $= 0$.

The output voltage, V_L, therefore, is fixed at 10 V, even though v_i varies from 12 to 14 V. Voltage regulators will be examined in greater depth in Chapter 14.

(Courtesy Radio Corporation of America)

FIG. 1.17
OA2 Cold-cathode gas diode.

A typical cold-cathode gas diode appears in Fig. 1.17.A tube manual must be consulted for the proper connections using this type of tube.

1.5 SEMICONDUCTOR DIODES

As the name implies, the semiconductor diode is constructed of *semiconductor materials.* Semiconductors are neither good conductors, such as copper or aluminum, nor good insulators, such as bakelite or rubber, but belong to a class of materials between the two. There are approximately 10^{28} free electrons/m^3 for conductors and 10^7 free electrons/m^3 for insulators, at room temperature, with the number of free electrons per unit volume for semiconductors varying within this range. In general, semiconductor materials also have *negative* temperature coefficients, indicating that the resistance of the material decreases as temperature increases. The reverse occurs for most metallic conductors.

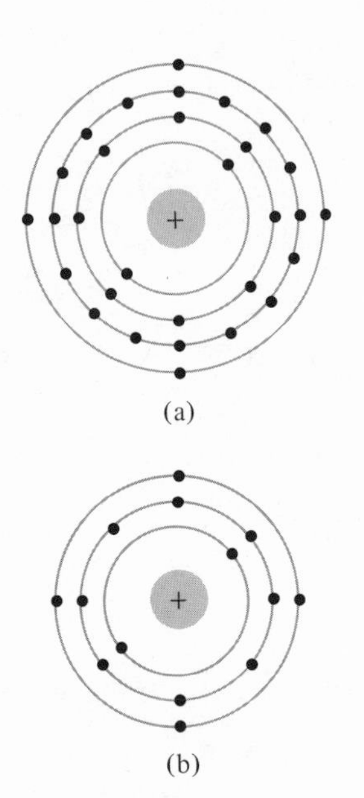

FIG. 1.18
Atomic structure: (a) germanium; (b) silicon.

To appreciate fully the behavior of the semiconductor diode a very brief and limited introduction to solid-state physics is necessary. We shall begin by considering the basic structure of the *atom.* The atom is composed of three basic particles; the *electron, proton,* and *neutron.* In the atomic lattice, the neutrons and protons form the *nucleus,* while the electrons revolve around the nucleus in a fixed *orbit.* The atomic structures of the two most commonly used semiconductors, *germanium* and *silicon,* are shown in Fig. 1.18.

As indicated by Fig. 1.18a, the germanium atom has 32 orbiting electrons, while silicon has 14 orbiting electrons. In each case, there are four electrons in the outermost (valence) shell. The potential (ionization potential) required to remove any one of these four valence electrons is lower than that required for any other electron in the structure. In a pure germanium or silicon crystal these four valence electrons are bonded to four adjoining atoms, as shown in Fig. 1.19 for germanium.

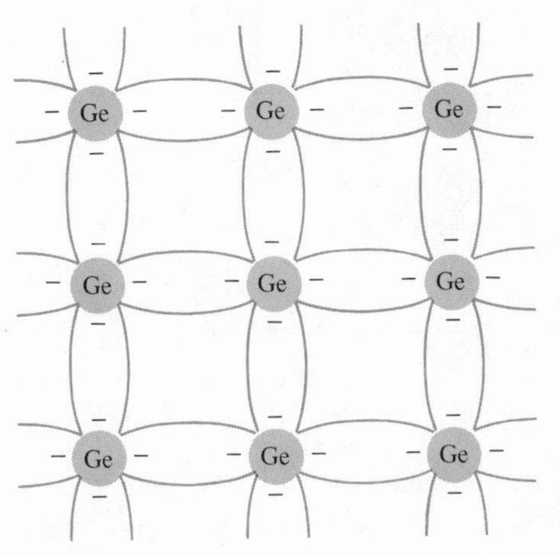

FIG. 1.19
Covalent bonding of the germanium atom.

This type of bonding, formed by *sharing* electrons, is called *covalent bonding*. Although the covalent bond will result in a stronger bond between the valence electrons and their parent atom, it is still possible for the valence electrons to absorb sufficient kinetic energy from natural causes to break the covalent bond and assume the "free" state. These natural causes include effects such as light energy in the form of photons and thermal energy from the surrounding medium. At room temperature there is approximately one "free" electron per 5×10^{10} germanium atoms. These "free" electrons, due only to natural causes, are called *intrinsic* carriers.

n- and *p*-Type Materials

The semiconductor diode is formed using two types of materials, the *n-type* and the *p-type*. Both are formed by adding a predetermined number of impurity atoms into a germanium or silicon base. The *n*-type is created by adding those impurity elements that have *five* valence electrons, such as *antimony*, *arsenic*, and *phosphorus*. The effect of such impurity elements is indicated in Fig. 1.20 (using antimony as

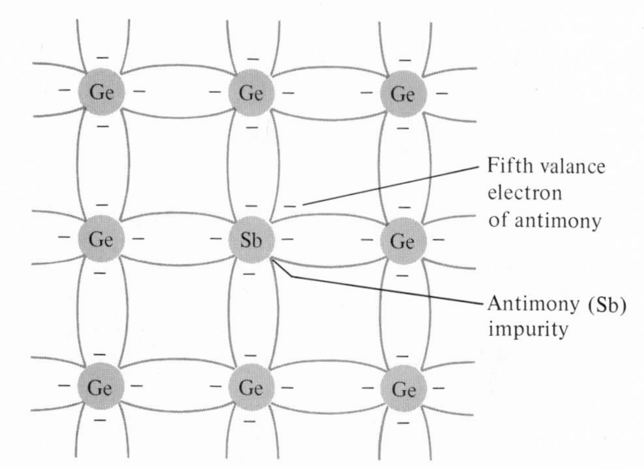

FIG. 1.20
Antimony impurity in *n*-type material.

the impurity in a germanium base). Note that the four covalent bonds are still present. There is, however, an additional fifth electron due to the impurity atom, which is *unassociated* with any particular covalent bond. This remaining electron, loosely bound to its parent (antimony) atom is relatively free to move within the newly formed *n*-type material. Since the inserted impurity atom has donated a relatively "free" electron to the structure, impurities with five valence electrons are called *donor* atoms. It is important to realize that even though a large number of "free" carriers have been established in the *n*-type material it is still electrically *neutral* since ideally the number of positive charged protons in the nuclei is still equal to the number of "free" and orbiting negatively charged electrons in the structure. The process of adding impurities in the manner described to establish a large number of free carriers is called *doping*.

The *p*-type material is formed by doping a pure germanium or silicon crystal with impurity atoms having *three* valence electrons. The elements most frequently used for this purpose include *boron*, *gallium*, and *indium*. The effect of one of these elements, boron, on a base of germanium, is indicated in Fig. 1.21.

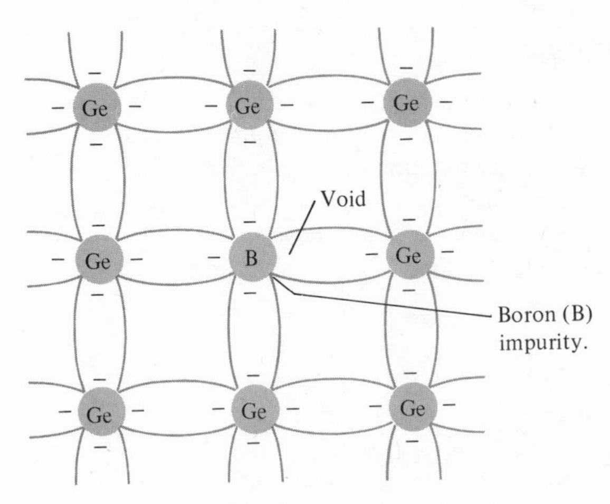

FIG. 1.21
Boron impurity in *p*-type material.

Note that there is now an insufficient number of electrons to complete the covalent bonds of the newly formed lattice. The resulting vacancy is called a *hole* and is represented by a small circle or positive sign due to the absence of a negative charge. Since the resulting vacancy will readily accept a "free" electron, the impurities added are called *acceptor* atoms. The resulting *p*-type material is electrically neutral for the same reasons as for the *n*-type material.

The effect of the hole on conduction is shown in Fig. 1.22. If a valence electron gains sufficient kinetic energy to break its covalent bond and fills the void created by a hole, then a vacancy, or hole, will be created in the covalent bond that released the electron. There is therefore a transfer of holes as shown in Fig. 1.22. As pointed out earlier the direction of flow to be used in this text is that of *conventional* flow, which is indicated by the direction of hole flow.

The graphic representation of the *n*- and *p*-type material is shown in Fig. 1.23. Note the larger number of electrons in the *n*-type material and holes in the *p*-type material. These charge carriers for obvious reasons are called the *majority* carriers. The positive ions of the *n*-type and the negative ions of the *p*-type are the donor and acceptor ions, respectively. The holes and negative ions in the *n*-type material

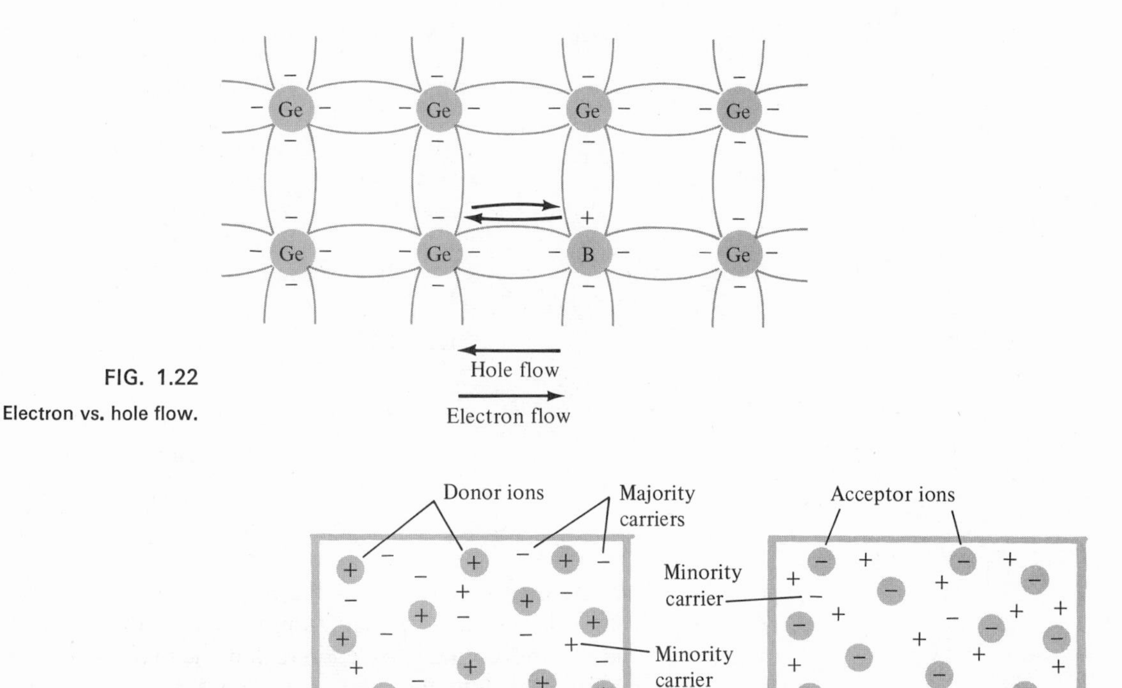

FIG. 1.22
Electron vs. hole flow.

FIG. 1.23
(a) *n*-type material; (b) *p*-type material.

and electrons and positive ions in the *p*-type material are due purely to those impurities that exist in every material not through design but because an absolutely *pure* germanium or silicon crystal cannot be obtained. Since carriers of the type just mentioned are very few compared to that of the majority carriers, they are called *minority* carriers.

p-n Junction

The semiconductor diode is formed by "joining" an *n*- and *p*-type material as shown in Fig. 1.24, using techniques to be described below. At the instant the two materials are "joined" the electrons and holes in the region of the junction will combine resulting in a lack of carriers in the region near the junction. This region of uncovered positive and negative ions is called the *depletion* region due to the depletion of carriers in this region.

The minority carriers in the *n*-type material that find themselves within the depletion region will pass directly into the *p*-type material. The closer the minority carrier is to the junction the greater the attraction for the layer of negative ions and the less the opposition of the

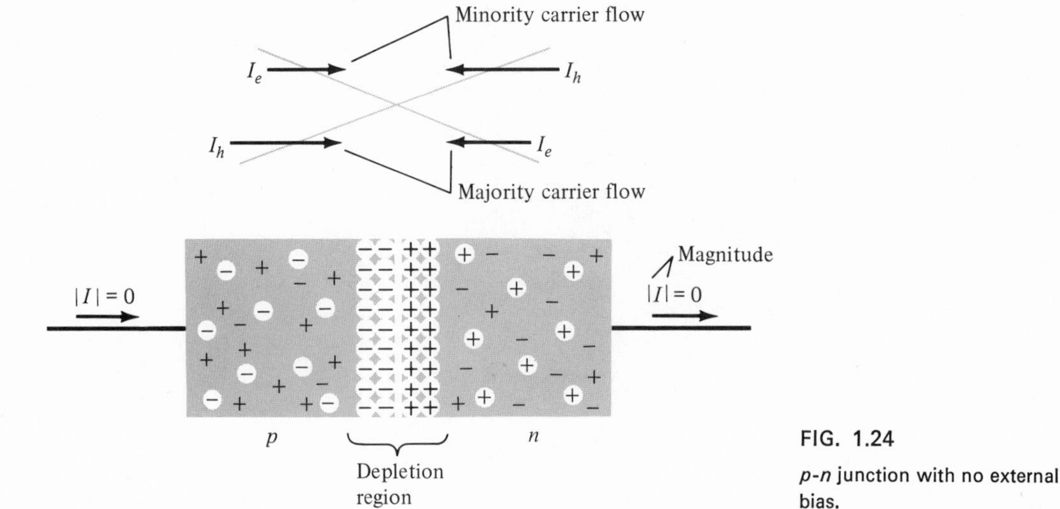

FIG. 1.24
p-n junction with no external bias.

positive ions in the depletion region of the *n*-type material. For the purposes of future discussions we shall assume that all the minority carriers that find themselves in the depletion region due to their random motion will pass directly into the *p*-type material. Similar discussion can be applied to the minority carriers (electrons) of the *p*-type material. This carrier flow has been indicated in Fig. 1.24 for the minority carriers of each material.

The majority carriers in the *n*-type material must overcome the attractive forces of the layer of positive ions in the *n*-type material and the shield of negative ions in the *p*-type material in order to migrate into the neutral region of the *p*-type material. The number of majority carriers is so large in the *n*-type material, however, that there will be invariably a small number of majority carriers with sufficient kinetic energy to pass through the depletion region into the *p*-type material. Again, the same type of discussion can be applied to the majority carriers of the *p*-type material. The resulting flow due to the majority carriers is also shown in Fig. 1.24.

A close examination of Fig. 1.24 will reveal that the relative magnitudes of the flow vectors are such that the net flow in either direction is zero. This cancellation of vectors has been indicated by crossed lines. The length of the vector representing hole flow has been drawn longer than that for electron flow to demonstrate that the magnitude of each need not be the same for cancellation and that the doping levels for each material may result in an unequal carrier flow of holes and electrons. In summary, *the net flow of charge in any one direction with no applied emf is zero.*

If an external potential of V volts is applied across the *p-n* junction as shown in Fig. 1.25, the number of uncovered negative ions in

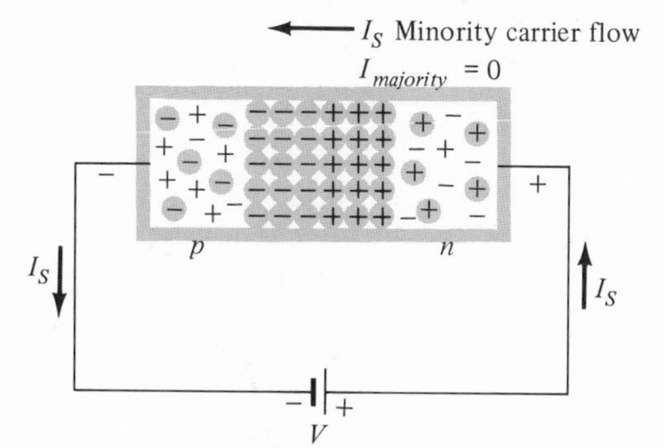

FIG. 1.25
Reverse-biased *p-n* junction

the depletion region of the *n*-type material will increase due to the large number of "free" electrons drawn to the positive potential of the applied *emf.* For similar reasons, the number of uncovered negative ions will increase in the *p*-type material. The net effect, therefore, is a widening of the depletion region. This widening of the depletion region will establish too great a barrier for the majority carriers to overcome, effectively reducing the majority carrier flow to zero (Fig. 1.25).

The number of minority carriers, however, that find themselves entering the depletion region will not change with reverse bias, resulting in minority carrier flow vectors of the same magnitude indicated in Fig. 1.24 with no applied *emf.* The current that exists under reverse-bias conditions is called the *reverse saturation current* and is represented by the subscript *s*. It is seldom more than a few microamperes in magnitude.

The effect of a *forward bias* on the *p-n* junction is indicated in Fig. 1.26. Note that the minority carrier flow has not changed in magnitude, but the reduction in the width of the depletion region has resulted in a heavy majority carrier flow across the junction. The magnitude of

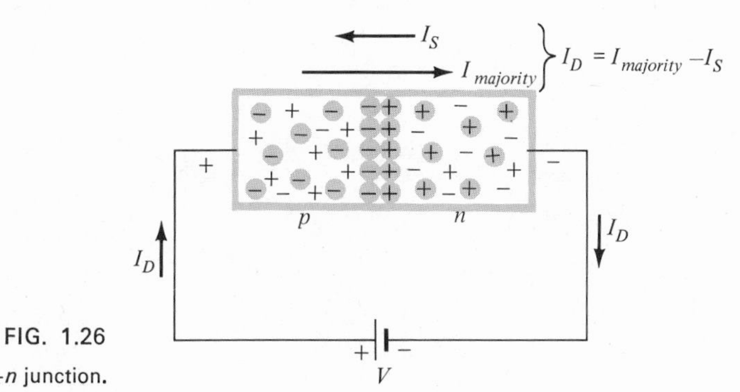

FIG. 1.26
Forward-biased *p-n* junction.

the majority carrier flow will increase exponentially with increasing forward bias as indicated in Fig. 1.27.

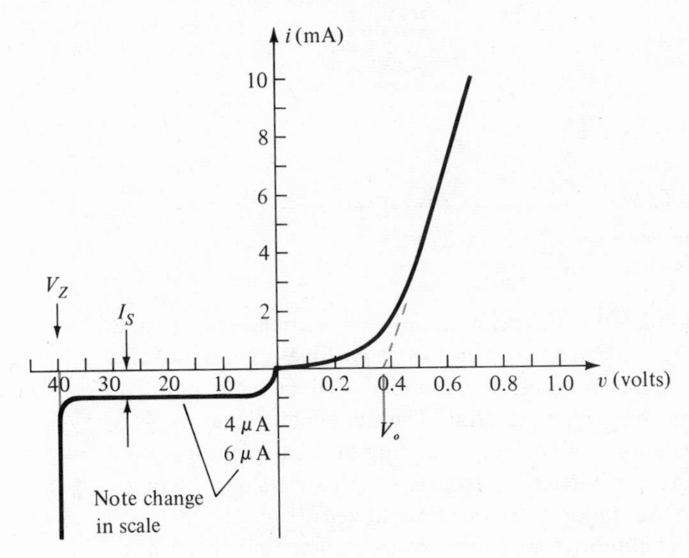

FIG. 1.27
Semiconductor diode characteristics.

It can be demonstrated through the use of solid-state physics that the diode current in the forward- and reverse-bias region can be mathematically related to temperature (T) and applied bias (V) in the following manner:

$$\boxed{I = I_s(e^{11,600V/T_K} - 1)} \tag{1-1}$$

where I_s is the reverse saturation current and T_K is degrees Kelvin; $T_K = T_C + 273°$.

The validity of the above equation can be demonstrated by considering $V = 0$ and $V = -0.2$ V. At $V = 0$ V Eq. (1-1) becomes

$$I = I_s(e^0 - 1) = I_s(1 - 1) = 0$$

as indicated in Fig. 1.27. For $V = -0.2$ V, at room temperature (25°C),

$$T_K = 25° + 273° = 298°$$

and

$$e^{11,600V/T} = e^{-11,600(0.2)/298} = e^{-7.75} \cong 0$$

so that

$$I = I_s(e^{11,600V/T} - 1) = I_s(0 - 1) = -I_s$$

the saturation current, as indicated in Fig. 1.27. As demonstrated by Eq. (1.1), temperature can have a marked effect on the current through a diode. It has been found experimentally *that the reverse saturation current I_s will almost double in magnitude for every* 10°C *change in temperature.*

Zener Region

Note the sharp change in the characteristics of Fig. 1.27 at the reverse bias potential V_Z (the subscript Z refers to the name Zener). Consider also the similarities between this region (in the negative sense) and that obtained for the hot and cold cathode diode. In fact, in a later section we will find that the (almost) ideal characteristics of this region are put to full use on the design of *Zener diodes.* This constant-voltage effect is induced by a high reverse-bias voltage across the diode. When the applied reverse potential becomes more and more negative, a point is eventually reached where the minority carriers have sufficient velocity to liberate additional carriers through ionization. These additional carriers can then aid the ionization process to the point where a high avalanche current is established. In addition to the ionization or avalanche process, many valence electrons will have a sufficiently strong attraction for the applied potential to break the covalent bond and assume the "free" state. The latter is called the *Zener* effect. If the reverse bias potential is brought below the Zener value the characteristics will trace the same curve in reverse. It is this latter characteristic that allows us to use the Zener region in the design of Zener diodes (Section 1.12).

The Zener region of the semiconductur diode described must be avoided if the response of a system is not to be completely altered by the sharp change in characteristics in this reverse-voltage region. The maximum reverse-bias potential that can be applied before entering this region is called the *peak inverse voltage*, referred to simply as the PIV rating.

Silicon and Germanium Characteristics

Silicon diodes have, in general, higher PIV and current ratings and wider temperature ranges than germanium diodes. PIV ratings for silicon can be in the neighborhood of a 1000 V, whereas the maximum value for germanium is closer to 400 V. Silicon can be used for applications in which the temperature may rise to about 200°C (400°F), whereas germanium has a much lower maximum rating (100°F). The disadvantage of silicon, however, as compared to germanium, is the higher forward-bias voltage required to reach the region of upward swing. It is typically in the order of magnitude of 0.6 *V for silicon* and

0.35 *V for germanium.* The closer the upward swing is to the vertical axis, the more closely we approach the ideal diode characteristic.

Semiconductor Diode Fabrication

Semiconductor diodes are normally one of the four following types: grown junction, alloy, diffused, or point-contact.

The first step in the manufacture of any semiconductor device is to obtain semiconductor materials, such as germanium or silicon, of the desired purity level. Impurity levels of *less* than *one* part in *one billion* (1 in 1,000,000,000) are required for most semiconductor fabrications today. The basic processes involved in the production of semiconductor materials with this low level of impurities are indicated in Fig. 1.28.

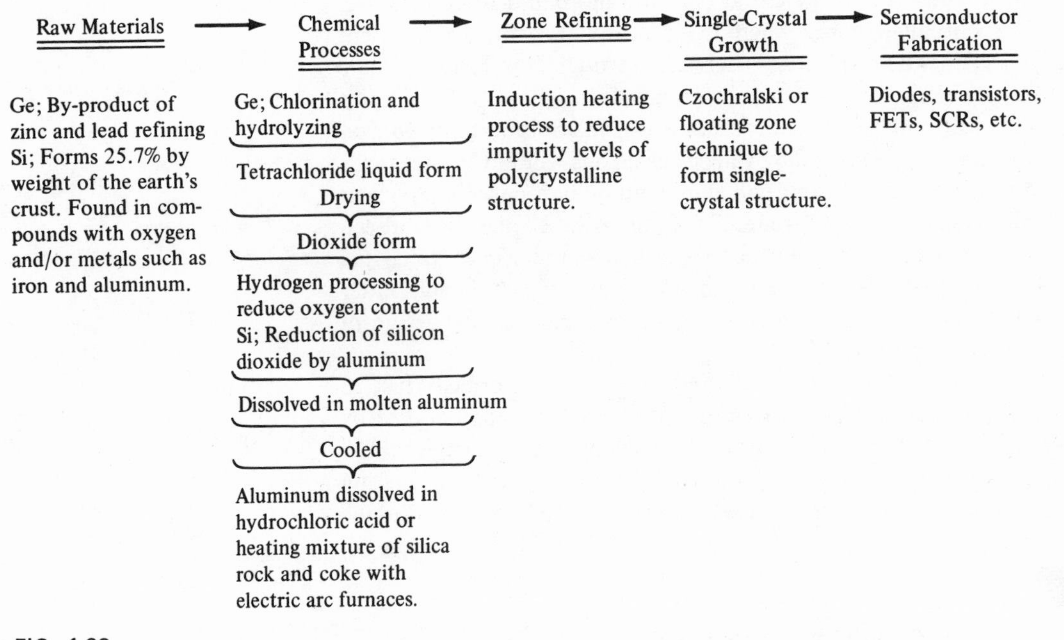

FIG. 1.28
Sequence of events leading to semiconductor fabrication.

As indicated in Fig. 1.28, the raw materials are first subjected to a series of chemical reactions and a zone refining process to form a polycrystalline crystal of the desired purity level. The atoms of a polycrystalline crystal are haphazardly arranged, while in the single crystal desired, the atoms are arranged in a symmetrical, uniform, geometrical lattice structure.

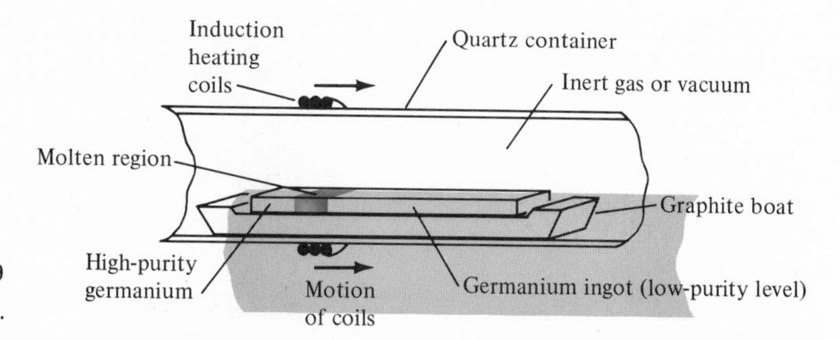

FIG. 1.29
Zone refining process.

Zone refining apparatus is shown in Fig. 1.29. It consists of a graphite or quartz boat for minimum contamination, a quartz container, and a set of RF (radio-frequency) induction coils. Either the coils or boat must be moveable along the length of the quartz container. The same result will be obtained in either case, although moving coils are discussed here since it appears to be the more popular method. The interior of the quartz container is filled with either an inert (little or no chemical reaction) gas, or vacuum, to reduce further the chance of contamination. In the zone refining process, a bar of germanium is placed in the boat with the coils at one end of the bar as shown in Fig. 1.29. The radio-frequency signal is then applied to the coil, which will induce a flow of charge (eddy currents) in the germanium ingot. The magnitude of these currents is increased until sufficient heat is developed to melt that region of the semiconductor material. The impurities in the ingot will have a lower melting temperature and therefore will be in a more liquid state than the surrounding semiconductor material. If the induction coils of Fig. 1.29 are now slowly moved to the right to induce melting in the neighboring region the "more fluidic" impurities will "follow" the molten region. The net result is that a large percentage of the impurities will appear at the right end of the ingot when the induction coils have reached this end. This end piece of impurities can then be cut off and the entire process repeated until the desired purity level is reached.

The final operation before semiconductor fabrication can take place is the formation of a single crystal of germanium or silicon. This can be accomplished using either the *Czochralski* or the *floating zone* technique, the latter being the more recently devised. The apparatus employed in the Czochralski technique is shown in Fig. 1.30a. The polycrystalline material is first transformed to the molten state by the RF induction coils. A single crystal "seed" of the desired impurity level is then immersed in the molten germanium and gradually withdrawn while the shaft holding the seed is slowly turning. As the "seed" is withdrawn, a single-crystal germanium lattice structure will grow on the "seed" as shown in Fig. 1.30a. The resulting single-crystal ingot

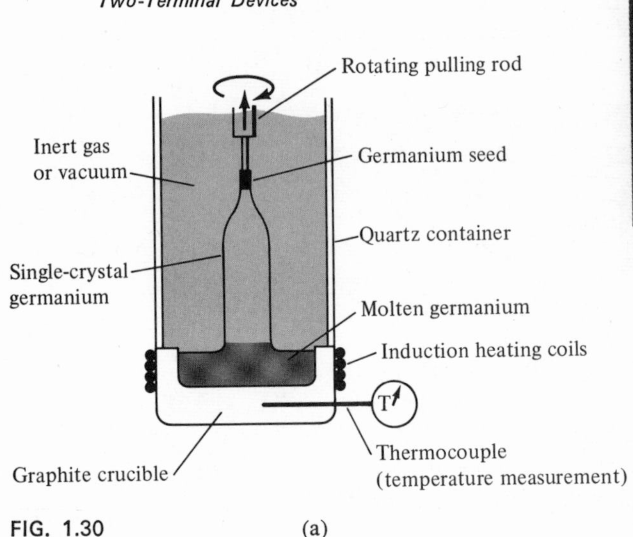

(a)

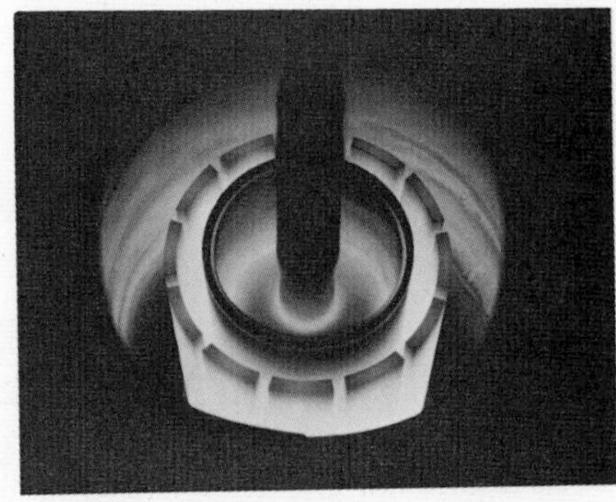

(Courtesy Texas Instruments Incorporated)

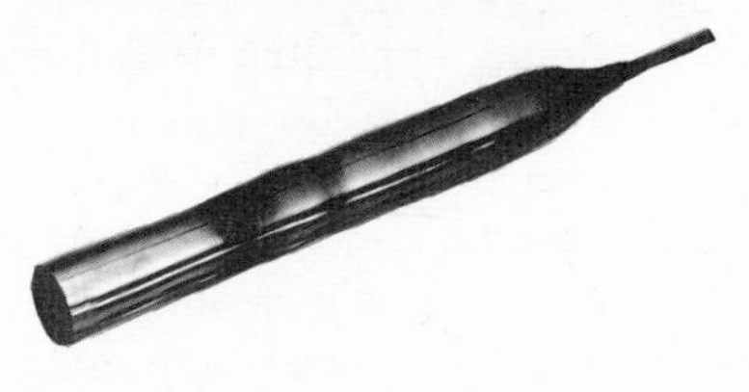

(Courtesy Motorola Incorporated)

(b)

FIG. 1.30
Czochralski technique.

can be as large as 7–10 inches in length and 1–3 inches in diameter (Fig. 1.30).

The floating zone technique eliminates the need for having both a zone refining and single-crystal forming process. Both can be accomplished at the same time using this technique. A second advantage of this method is the absence of the graphite or quartz boat, which often introduces impurities into the germanium or silicon ingot. Two clamps hold the bar of germanium or a silicon in the vertical position within a set of movable RF induction coils as shown in Fig. 1.31. A small single-crystal "seed" of the desired impurity level is deposited at the lower end of the bar and heated with the germanium bar until the

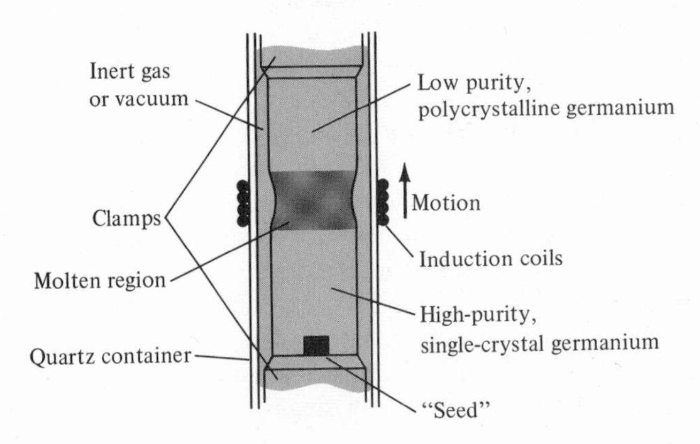

FIG. 1.31
Floating zone technique.

molten state is reached. The induction coils are then slowly moved up the germanium or silicon ingot while the bar is slowly rotating. As before, the impurities follow the molten state resulting in an improved impurity level single-crystal germanium lattice below the molten zone. Through proper control of the process, there will always be sufficient surface tension present in the semiconductor material to ensure that the ingot does not rupture in the molten zone.

The single-crystal structure produced can then be cut into wafers sometimes as thin as $\frac{1}{1000}$ (or 0.001) of an inch ($\cong \frac{1}{5}$ the thickness of this paper). This cutting process can be accomplished using the setup of Fig. 1.32a or b. In Fig. 1.32a, tungsten wires (0.001 in. in diameter)

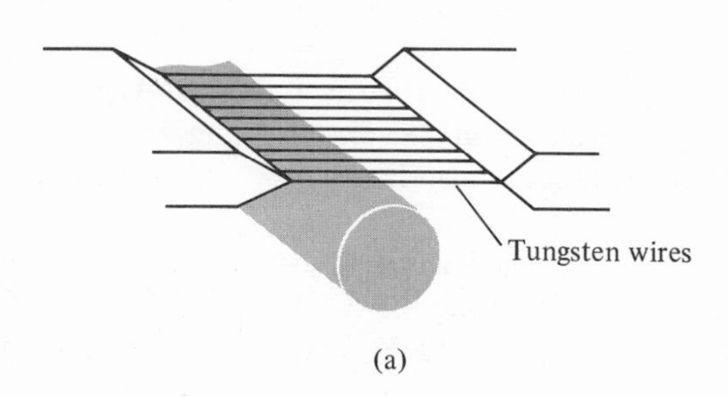

(a)

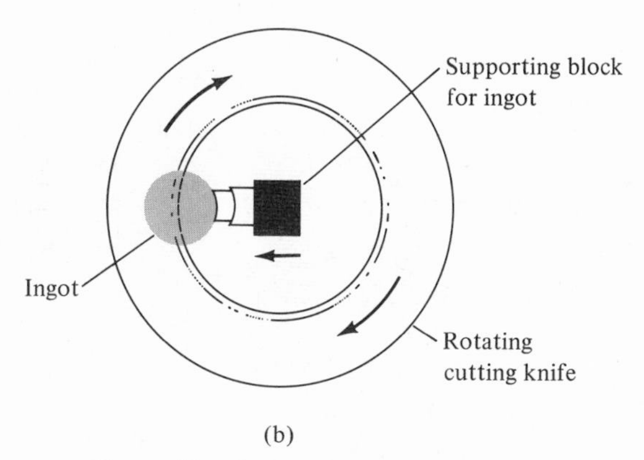

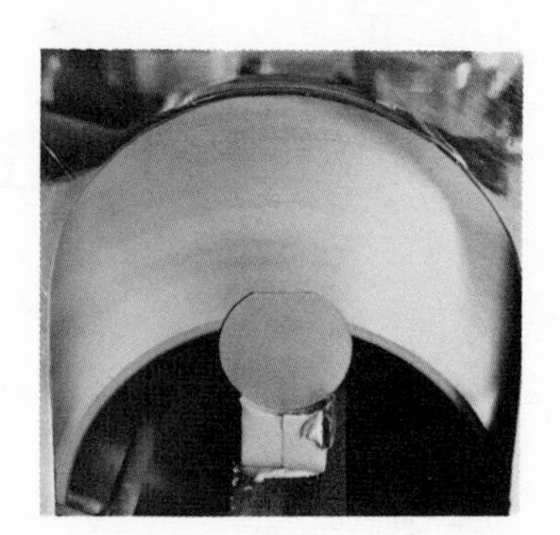

(b)

(Courtesy Texas Instruments Incorporated)

FIG. 1.32
Slicing the single-crystal ingot into wafers.

with abrasive deposited surfaces are connected to supporting blocks at the proper spacing and then the entire system is moved back and forth as a saw. The system of Fig. 1.32b is self-explanatory. We shall now consider the four basic processes most commonly used in the manufacture of semiconductor diodes.

Diodes of this type are formed during the Czochralski *crystal pulling* process. Impurities of *p*- and *n*-type can be alternately added to the molten semiconductor material in the crucible resulting in a *p-n* junction as indicated in Fig. 1.33 when the crystal is pulled. After slicing, the large-area device can then be cut into a large number (sometimes thousands) of smaller-area semiconductor diodes. The area of grown junction diodes is sufficiently large to handle high currents (and therefore have high power ratings). The large area, however, will introduce undesired junction capacitive effects.

ALLOY

The alloy process will result in a junction-type semiconductor diode that will also have a high current rating and large PIV rating. The junction capacitance is also large, however, due to the large junction area.

The *p-n* junction is formed by first placing a *p*-type impurity on an *n*-type substrate and heating the two until liquefaction occurs where the two materials meet (Fig. 1.34). An alloy will result that, when

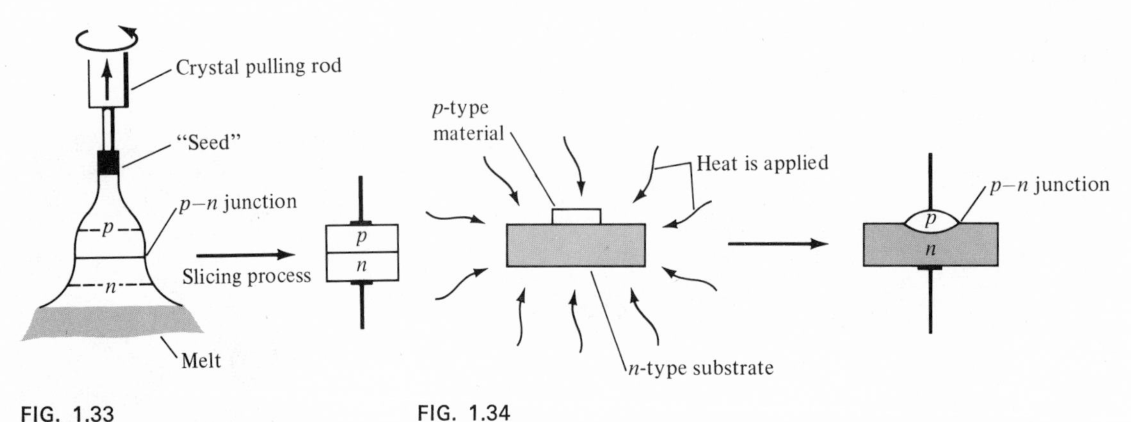

FIG. 1.33
Grown junction diode.

FIG. 1.34
Alloy process diode.

cooled, will produce a *p-n* junction at the boundary of the alloy and substrate. The roles played by the *n*- and *p*-type materials can be interchanged.

DIFFUSION

The diffusion process of forming semiconductor junction diodes can employ either solid or gaseous diffusion. This process requires more time than alloy process but it is relatively inexpensive and can be quite accurately controlled. Diffusion is a process by which a heavy concentration of particles will "diffuse" into a surrounding region of lesser concentration. The primary difference between the diffusion and alloy process is the fact that liquefaction is not reached in the diffusion process. Heat is applied in the diffusion process only to increase the activity of the elements involved.

The process of solid diffusion commences with the "painting" of an acceptor impurity on an *n*-type substrate and heating the two until the impurity diffuses into the substrate to form the *p*-type layer (Fig. 1.35a).

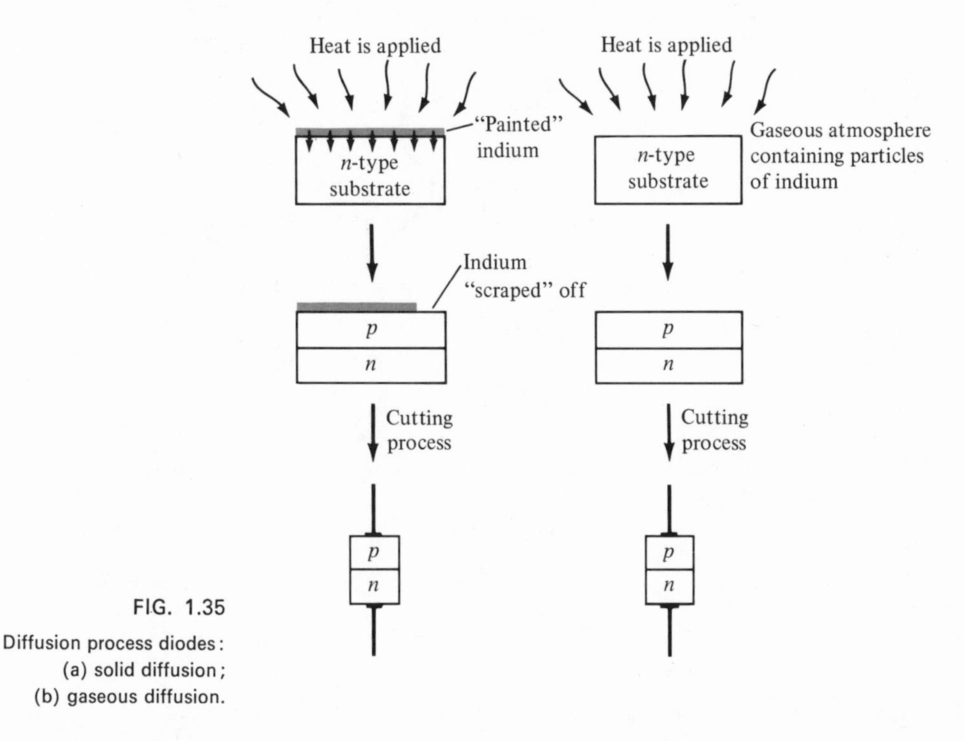

FIG. 1.35
Diffusion process diodes: (a) solid diffusion; (b) gaseous diffusion.

In the process of gaseous diffusion, an *n*-type material is submerged in a gaseous atmosphere of acceptor impurities and then heated (Fig. 1.35b). The impurity diffuses into the substrate to form the *p*-type layer of the semiconductor diode. The roles of the *p*- and *n*-type materials can also be interchanged in each case. The diffusion process is the most frequently used today in the manufacture of semiconductor diodes.

POINT CONTACT

The point-contact semiconductor diode is constructed by pressing a phosphor bronze spring (called a cat whisker) against an *n*-type substrate (Fig. 1.36). A high current is then passed through the whisker and substrate for a short period of time, resulting in a number of atoms passing from the wire into the *n*-type material to create a *p*-region in the wafer. The small area of the *p-n* junction results in a very small junction capacitance (typically 1 pF or less). For this reason, the point-contact diode is frequently used in applications where very high frequencies are encountered, such as in microwave mixers and detectors. The disadvantage of the small contact area is the resulting low current ratings and characteristics less ideal than those obtained from junction-type semiconductor diodes. The basic construction and various types of point-contact diodes are indicated in Fig. 1.37. Various types of junction diodes (grown junction, alloy, and diffused) appear in Fig. 1.38.

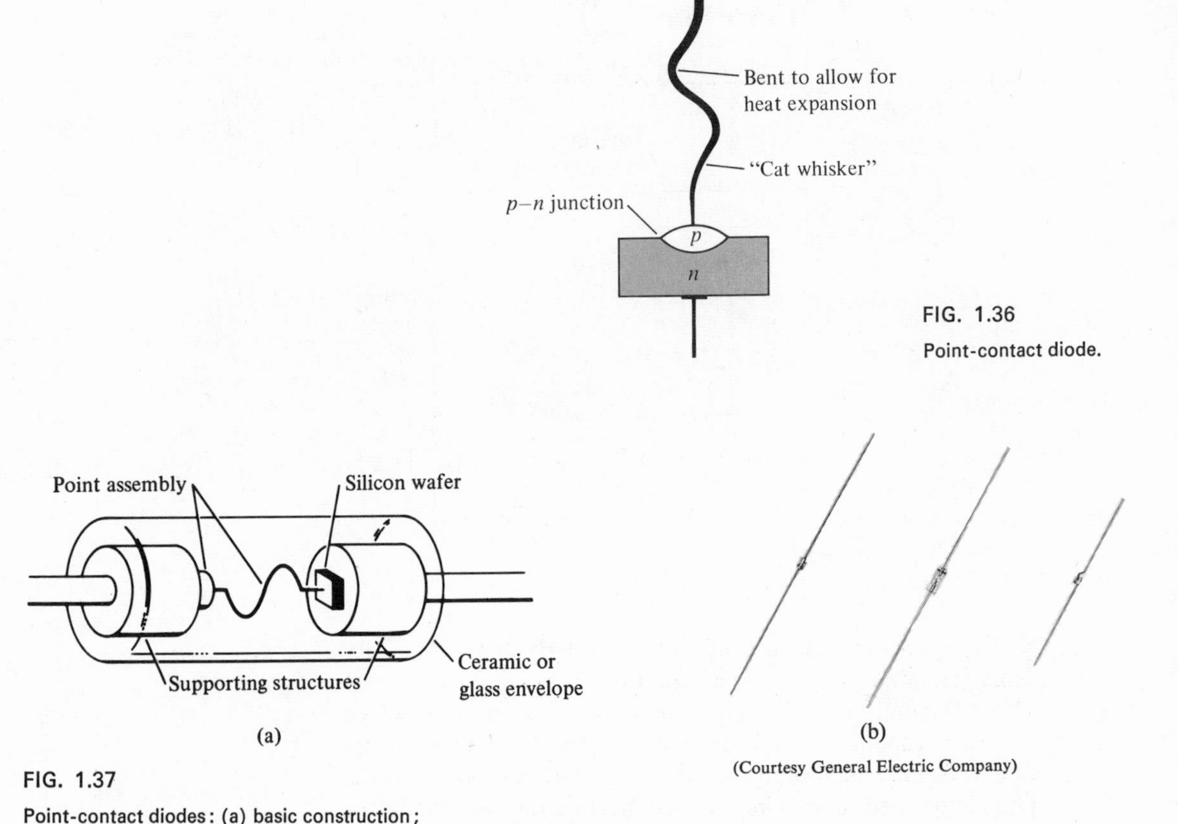

FIG. 1.36
Point-contact diode.

FIG. 1.37
Point-contact diodes: (a) basic construction; (b) various types.

Semiconductor Diode Characteristics

The characteristics of a point-contact and junction diode (silicon and germanium) appear in Fig. 1.39. Note the differences in the PIV

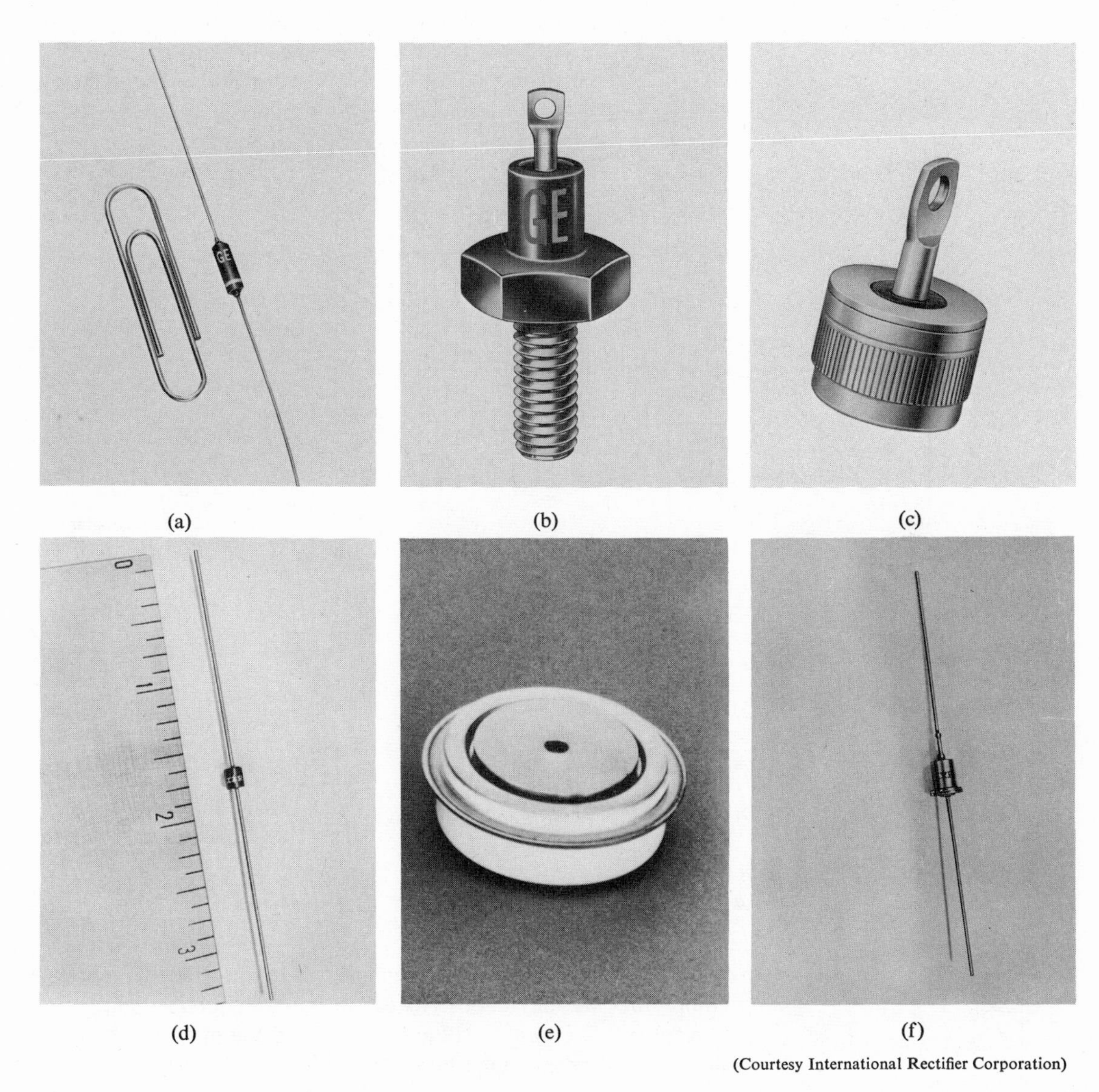

(Courtesy International Rectifier Corporation)

FIG. 1.38
Various types of junction diodes.

values and the forward bias potential required to reach the region of upward swing. Also consider which type of diode best fits the ideal characteristics.

SEMICONDUCTOR DIODE NOTATION

The notation most frequently used for semiconductor diodes is provided in Fig. 1.40. Note the carryover of the anode-cathode terminology from the vacuum tube.

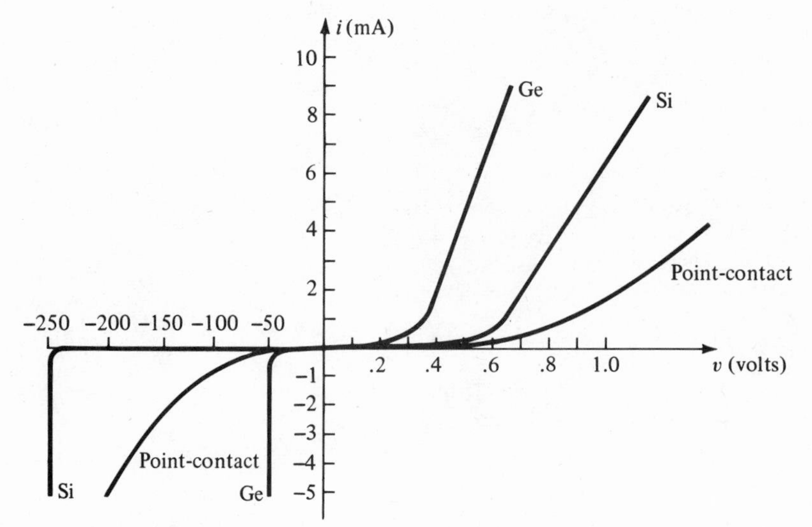

FIG. 1.39
Diode characteristics.

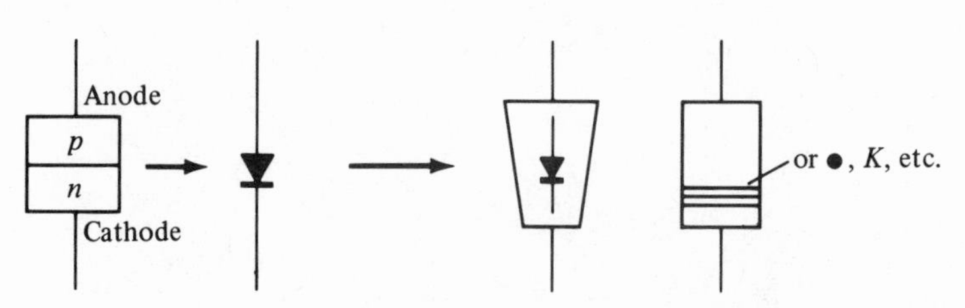

FIG. 1.40
Semiconductor diode notation.

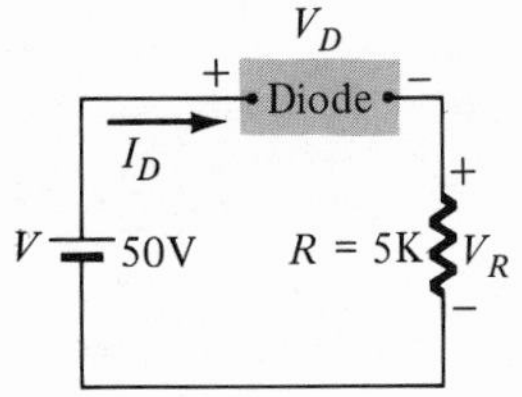

FIG. 1.41
Fundamental diode circuit.

1.6 LOAD LINE AND QUIESCENT CONDITIONS

In Fig. 1.41 a block symbol has been inserted in the circuit to represent *any* diode that we may choose to use in this particular circuit.

Applying Kirchhoff's voltage law around the indicated loop will result in the following equation:

$$\boxed{V = V_D + V_R} \tag{1.2}$$

Solving for V_D and substituting $V_R = I_D R$ we have

$$V_D = V - V_R$$

and

$$\boxed{V_D = V - I_D R} \tag{1.3a}$$

Eq. (1.3a) has two dependent variables (V_D and I_D) and two fixed values (V and R). Since a minimum of two equations are required to solve for two unknown dependent variables, Eq. (1.3a) is not sufficient for a complete solution. The second equation necessary to determine the value of V_D and I_D determined by V and R is provided by the characteristics of the diode element in the enclosed container; that is, for the diode employed, we know that the current is a function of the voltage across the diode, or mathematically,

$$\boxed{I_D = f(V_D)} \tag{1.3b}$$

It is necessary, therefore, to find the common solution of the equation determined by the load circuit [Eq. (1.3a)] and the characteristics of the diode. One method of finding this solution is the graphical method, which will now be outlined. It is extremely important that the procedure described in the next few paragraphs be fully understood since similar operations will appear when we consider other devices such as the transistor, vacuum triode, and FET.

Rewriting Eq. (1.3) in a slightly different form

$$\boxed{I_D = -\frac{1}{R}V_D + \frac{V}{R}} \tag{1.4}$$

$$\begin{array}{ccccccc} \downarrow & & \downarrow & \downarrow & & \downarrow & \\ y & = & m & x & + & b & \text{(straight line equation)} \end{array}$$

Below this newly formed equation, the general equation for a straight line has been included. Note that the slope of the line is negative (I_D decreases in magnitude with increase in V_D) with a magnitude $1/R$, while the y-intercept is V/R and I_D and V_D are the y- and x-variables, respectively. The intercepts of this straight line with the axes of the graph of Fig. 1.42 can be found rather quickly by applying the following conditions. If we consider first that if $I_D = 0$ mA we must be somewhere along the horizontal axis of Fig. 1.42 and if we apply this condition to Eq. (1.4), then

$$I_D = 0 = -\frac{V_D}{R} + \frac{V}{R}$$

and solving for V_D

$$\boxed{V_D = V|_{I_D=0}} \tag{1.5a}$$

The intersection of the straight line with the horizontal axis is the applied voltage V. If we then consider that if $V_D = 0$ we must be

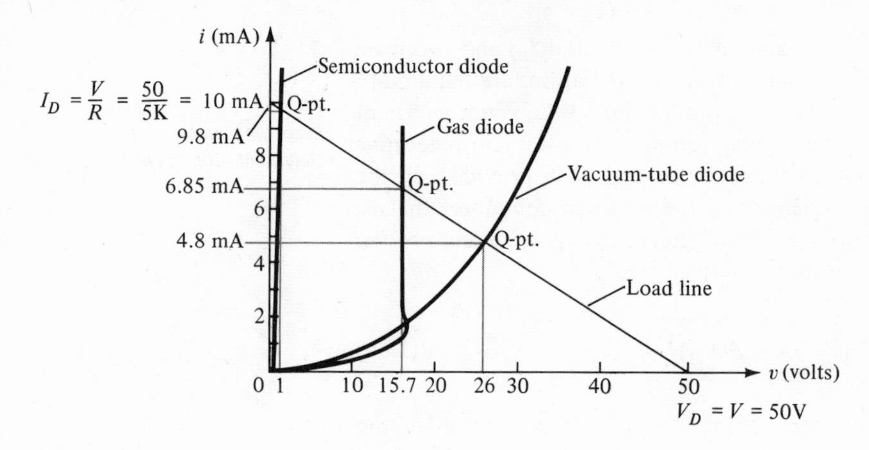

FIG. 1.42

Load line and quiescent value determination.

somewhere along the vertical axis, and apply this condition to Eq. (1.4), then

$$I_D = -\frac{V_D}{R} + \frac{V}{R} = 0 + \frac{V}{R} \tag{1.5b}$$

and

$$\boxed{I_D = \frac{V}{R}\bigg|_{V_D=0}} \tag{1.5b}$$

The intersection, therefore, of the straight line with the vertical axis is determined by the ratio of the applied voltage and load. Both intersections have been indicated in Fig. 1.42. All that remains is to connect these two points by a straight line to obtain a graphical representation of Eq. (1.3). This resulting line is called the *load line* since it represents the properties of the applied voltage and load and tells us nothing about the diode's characteristics.

In Fig. 1.42 the characteristics of a semiconductor, gas, and vacuum-tube diode have also been included. For each, the intersection of the load line and the diode's characteristic curve will determine the point of operation for that diode. This point, due only to the dc input, is called the *quiescent* point—quiescent meaning still, quiet, or inactive. The voltage across and current through the diode can now be found by simply drawing a vertical and horizontal line, respectively, to the voltage and current axis as indicated in Fig. 1.42. Note that each diode has a different potential drop across it, resulting in a different voltage across the load and consequently different current through the load.

The subscript Q is used to denote quiescent values of current and voltage as shown in Fig. 1.42. For the semiconductor diode:

$$V_Q = 1V \qquad \text{and} \qquad I_Q = 9.8 \text{ mA}$$

Substituting into Eq. (1.2)

$$V_R = V - V_D = V - V_Q = 50 - 1 = 49 \text{ V}$$

or

$$V_R = I_Q R = (9.8 \times 10^{-3})(5 \times 10^3) = 49 \text{ V}$$

The power delivered to the load is

$$P_L = I_Q^2 R = (9.8 \times 10^{-3})^2(5 \times 10^3) = 480.2 \text{ mW}$$

or

$$P_L = P_S - P_D$$

where P_S is the power supplied by the source and P_D is the power dissipation of the diode, so that

$$\begin{aligned} P_L &= VI_Q - V_Q I_Q = I_Q(V - V_Q) \\ &= (4.8 \times 10^{-3})(50 - 1) = 480.2 \text{ mW} \end{aligned}$$

The calculations for the vacuum-tube and gas diodes are left to the reader as an exercise. Take special note of how closely the semiconductor diode approaches that of the ideal diode for the magnitudes of current and voltage indicated.

1.7 STATIC RESISTANCE

A second glance at Fig. 1.42 will reveal that each diode has a fixed voltage and current associated with the diode at the point of operation. Applying Ohm's law to these values for each diode will result in the *static* or *dc* resistance of each diode at the quiescent point.

$$\boxed{R_{dc} = \frac{V_D}{I_D}} \qquad (1.6)$$

For the semiconductor diode:

$$R_{dc} = \frac{V_D}{I_D} = \frac{1}{9.8 \times 10^{-3}} = 102\ \Omega$$

and the vacuum tube diode:

$$R_{dc} = \frac{V_D}{I_D} = \frac{26}{4.8 \times 10^{-3}} = 5.41 \text{ K}$$

and the gas tube:

$$R_{dc} = \frac{V_D}{I_D} = \frac{15.7}{6.85 \times 10^{-3}} = 2.29 \text{ K}$$

Once the dc resistance has been determined the diode can be replaced by a resistor of this value. Any change in the applied voltage or load resistor, however, will result in a different Q-point and therefore different dc resistance.

1.8 DYNAMIC RESISTANCE

It is quite obvious from Fig. 1.42 that the dc resistance of a diode is independent of the shape of the characteristic in the region surrounding the point of interest. If a sinusoidal rather than dc input is applied to the circuit of Fig. 1.42 the situation will change completely. Consider the circuit of Fig. 1.43a, which has as its input a sinusoidal signal on a dc level. Since the magnitude of the dc level is much greater than that of the sinusoidal signal at any instant of time, the diode will always be forward biased and current will exist continuously in the circuit in the direction shown.

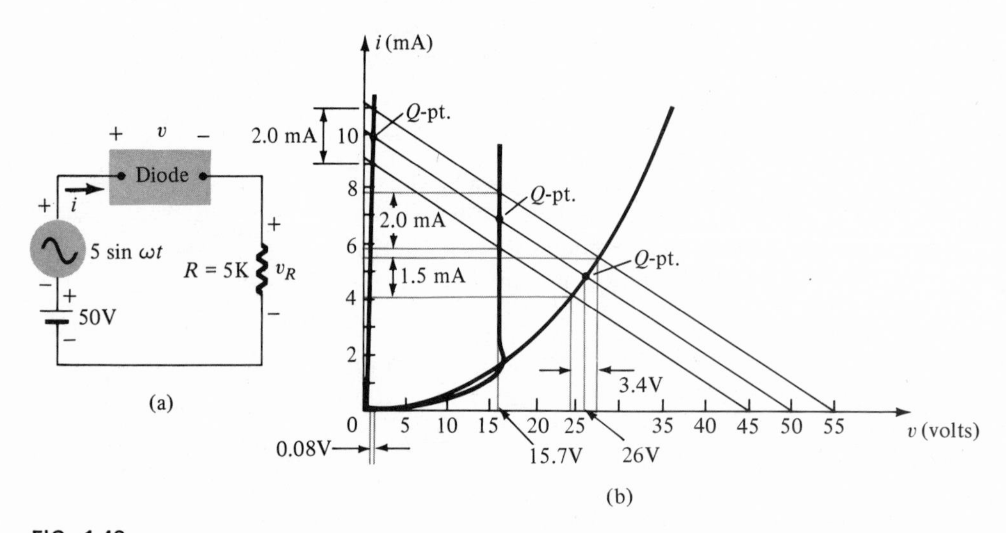

FIG. 1.43
ac resistance: (a) circuit;
(b) resulting region of operation.

The load line resulting from the dc input is shown in Fig. 1.43b. The effect of the ac signal is also demonstrated pictorially in the same figure. Note that two additional load lines have been drawn at the positive and negative peaks of the input signal. At the instant the sinusoidal signal is at its positive peak value the input could be replaced by a dc battery with a magnitude of 55 V and the resultant load line drawn as shown. For the negative peak, $V_{dc} = 45$ V. A moment of thought, however, should reveal the relative simplicity of superimposing the sinusoidal signal on the dc load line and drawing the load lines coinciding with the positive and negative peaks of the sinusoidal signal.

Note that we are now interested in a region of the diode characteristics as determined by the sinusoidal signal rather than a single point, as was the case for purely dc inputs. Since the resistance will vary from point to point along this region of interest, which value should we use to represent this portion of the characteristic curve? The value chosen is determined by drawing a straight line tangent to the curve at the quiescent point as shown in Fig. 1.44 for the vacuum-

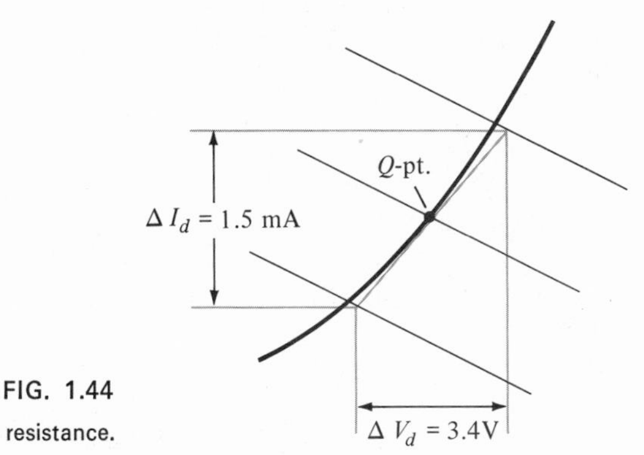

FIG. 1.44
Vacuum-tube ac resistance.

tube diode. The tangent line should "best fit" the characteristics in the region of interest as shown. The resultant resistance, called the *dynamic* or *ac* resistance, is then calculated on an approximate basis, in the following manner:

$$\boxed{r_{ac} = \frac{\Delta V_d}{\Delta I_d}} \tag{1.7}$$

For the vacuum-tube diode,

$$r_{ac} = \frac{\Delta V_d}{\Delta I_d} \cong \frac{3.4}{1.5 \times 10^{-3}} = 2.27 \text{ K}$$

for the semiconductor diode,

$$r_{ac} = \frac{\Delta V_d}{\Delta I_d} \cong \frac{0.08}{2 \times 10^{-3}} = 40\ \Omega$$

and for the gas diode,

$$r_{ac} = \frac{\Delta V_d}{\Delta I_d} \cong \frac{0}{2.0 \times 10^{-3}} = 0\ \Omega$$

In summary, both the dc and ac resistances of a diode will vary along the characteristic curve. The ac resistance, however, will depend on the shape of the characteristics in the region of operation.

1.9 AVERAGE AC RESISTANCE

If the input signal is sufficiently large to produce the type of swing indicated in Fig. 1.45 the resistance associated with the device for this region is called the *average ac resistance*. Three values of ac resistance have been calculated and indicated in Fig. 1.45.

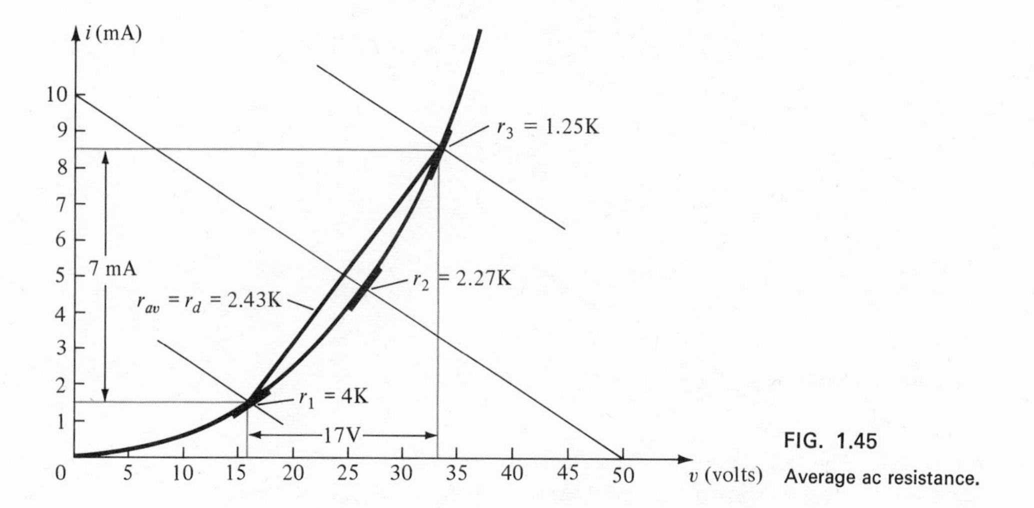

FIG. 1.45 Average ac resistance.

The average ac resistance is, by definition, the resistance determined by a straight line drawn between the two intersections determined by the maximum and minimum values of input voltage. In equation form (note Fig. 1.45)

$$r_{av} = r_d = \left.\frac{\Delta V_d}{\Delta I_d}\right|_{\text{pt. to pt.}} \tag{1.8}$$

For the situation indicated by Fig. 1.45

$$r_d = \left.\frac{\Delta V_d}{\Delta I_d}\right|_{\text{pt. to pt.}} = \frac{17}{7 \times 10^{-3}} \cong 2.43\ \text{K}$$

The average value of the three ac resistances is

$$\text{Average} = \frac{4.0\text{ K} + 2.27\text{ K} + 1.25\text{ K}}{3} \cong 2.5\text{ K}$$

Note how closely this average value compares to that determined by Eq. (1.8).

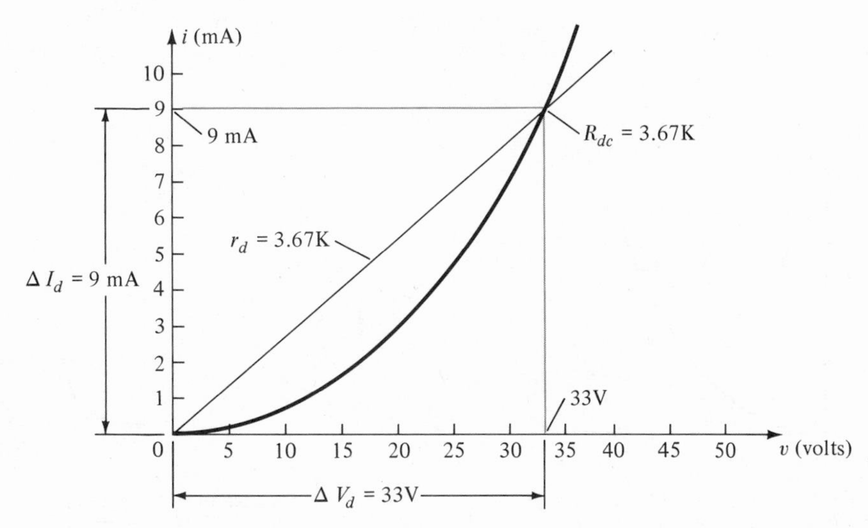

FIG. 1.46
Average ac resistance of a region from zero to some positive diode voltage or current.

If the type of input were such that the region indicated in Fig. 1.46 were employed the average ac resistance would equal the previously defined dc resistance; that is,

$$r_d = \left.\frac{\Delta V_d}{\Delta I_d}\right|_{\text{pt. to pt.}} = \frac{33}{9 \times 10^{-3}} = 3.67\text{ K}$$

and

$$R_{dc} = \frac{V_D}{I_D} = \frac{33}{9 \times 10^{-3}} = 3.67\text{ K}$$

It is important to note in this discussion of average ac resistance that the resistance to be associated with the element is determined *only* by the region of interest, *not* by the entire characteristic.

1.10 EQUIVALENT CIRCUITS

It is sometimes a long and tedious process to find the response of a network having one or more diodes using only the graphical method.

A method frequently employed to obtain a *first approximation* to the actual response, in a much shorter period of time, includes the use of equivalent circuits. An equivalent circuit is a circuit composed of linear elements that has terminal characteristics very similar to the actual nonlinear terminal characteristics of the device. Once the equivalent circuit has been obtained, the characteristics of the device can be ignored in the ensuing analysis.

The equivalent circuit is obtained by first replacing the nonlinear characteristics of a diode by straight line segments as shown in Fig. 1.47, for the vacuum, gas, and semiconductor diodes. In each case,

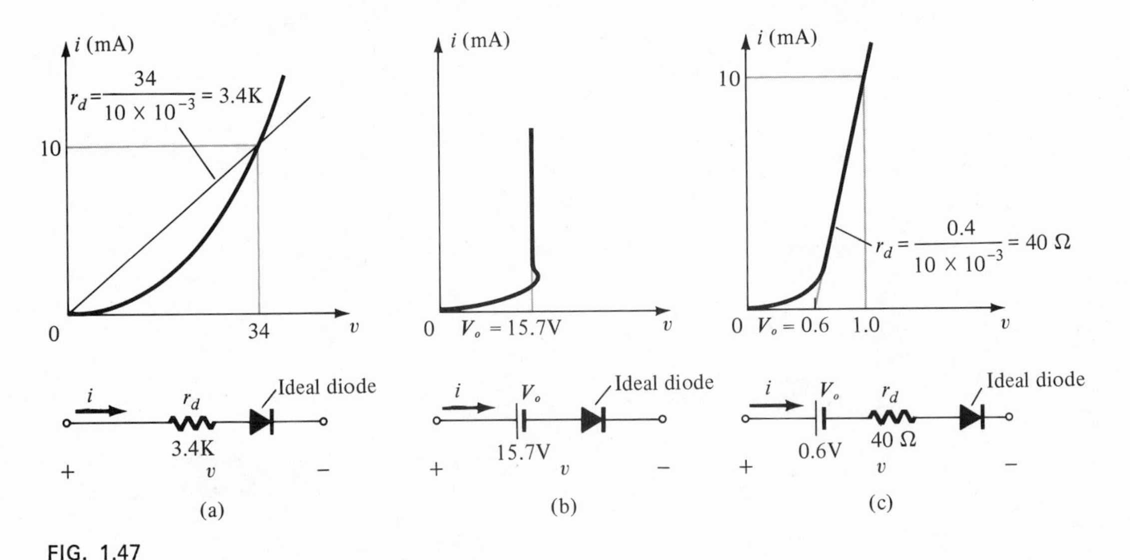

FIG. 1.47
Equivalent circuits: (a) vacuum-tube diode; (b) gas-tube diode; (c) semiconductor diode.

note that only ideal diodes are employed and the resistance is the average ac resistance for each nonlinear segment. As demonstrated in Fig. 1.47 the vacuum diode is a resistor in the region of conduction (the series ideal diode is a short) and an open circuit in the region of nonconduction (the series ideal diode is an open circuit). The semiconductor and gas diodes intersect the horizontal axis at 0.6 and 15.7 V, respectively.

The reader must always keep in mind that the dc battery V_o is not an isolated source of *emf* but simply an element that is necessary to best represent the actual diode characteristics.

The following example will demonstrate the equivalent circuit technique.

EXAMPLE 1.2 Find v_o for the input shown using the vacuum-tube, gas-tube, and semiconductor diodes of Fig. 1.47. Use the equivalent circuit technique. See Fig. 1.48.

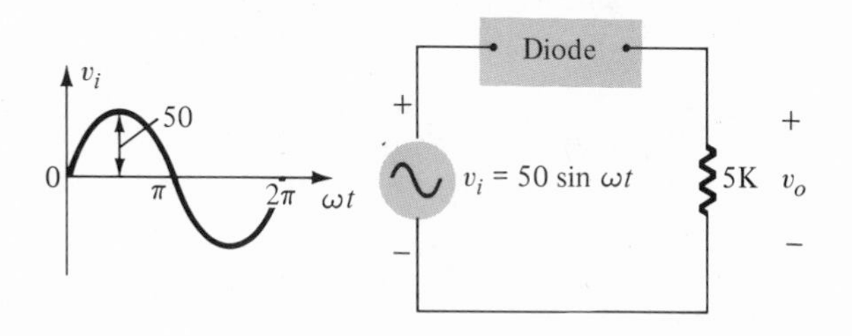

FIG. 1.48
Equivalent circuit example.

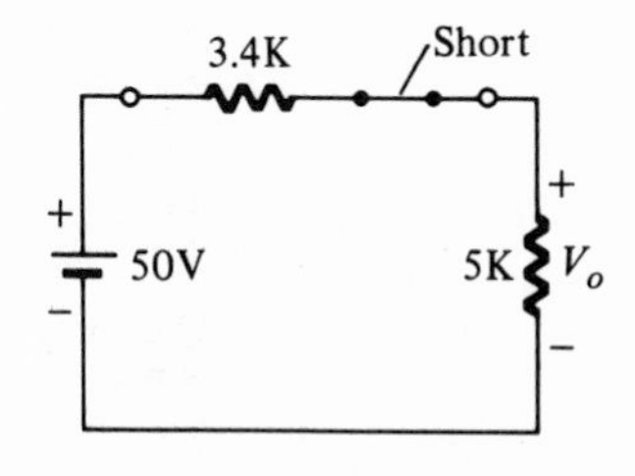

FIG. 1.49
Circuit of Fig. 1.48 at instant $v_i = 50$ V with the vacuum-tube diode.

Solution: *Vacuum-Tube Diode:* For $0 \rightarrow \pi$, the ideal diode is forward biased. At the peak value of input voltage, the circuit will appear as in Fig. 1.49.

$$V_o = \frac{5\text{ K}(50)}{5\text{ K} + 3.4\text{ K}} = \frac{250}{8.4} = 29.8\text{ V}$$

For $\pi \rightarrow 2\pi$, the ideal diode is reverse biased and the circuit will appear as in Fig. 1.50 at the negative peak.

Gas Diode: For any value of input voltage less than 15.7 V the ideal diode will be reverse biased. This is demonstrated in Fig. 1.51 for $v_i = 15$ V.

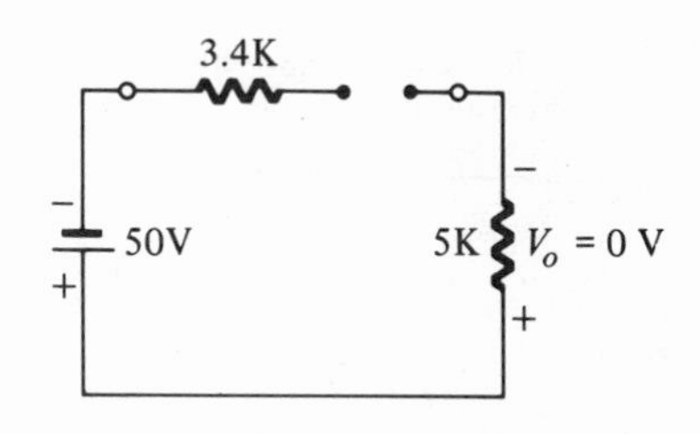

FIG. 1.50
Circuit of Fig. 1.48 at instant $v_i = -50$ V with the vacuum-tube diode.

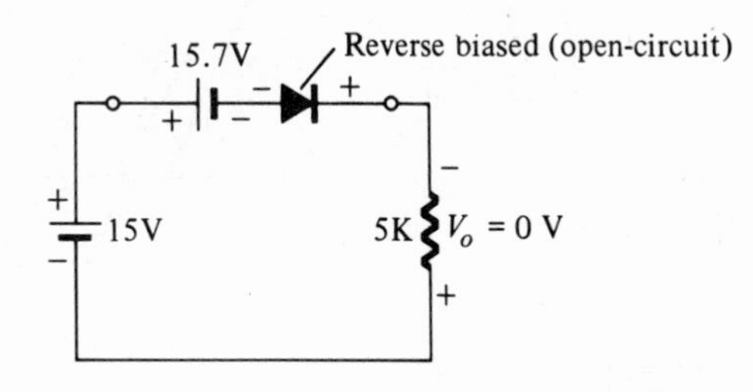

FIG. 1.51
Circuit of Fig. 1.48 at instant $v_i = 15$ V with the gas diode.

For $v_i > 15.7$ V, the ideal diode is forward biased, resulting in the configuration of Fig. 1.52 at $v_i = 50$ V.

Semiconductor Diode: For any value of input voltage less than 0.6 V the ideal diode will be reverse biased and $v_o = 0$. This is the same type of situation demonstrated above for the gas diode.

For $v_i > 0.6$ V the following circuit of Fig. 1.53 results if we consider the instant $v_i = 50$ V.

$$V_o = \frac{5\,\text{K}(50 - 0.6)}{5\,\text{K} + 40} = \frac{5\,\text{K}(49.4)}{5040} \cong 49.1\ \text{V}$$

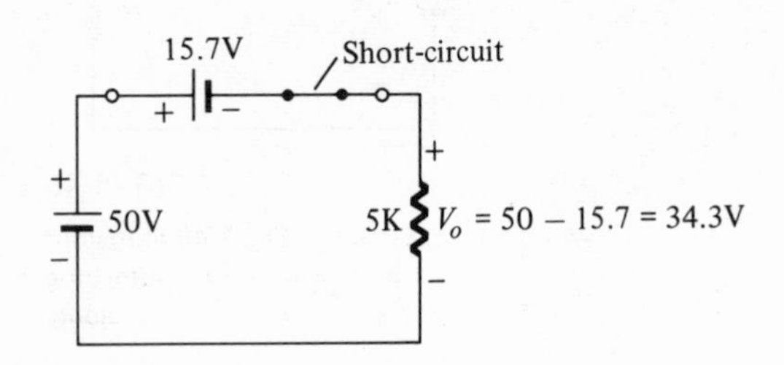

FIG. 1.52

Circuit of Fig. 1.48 at instant $v_i = 50$ V with the gas diode.

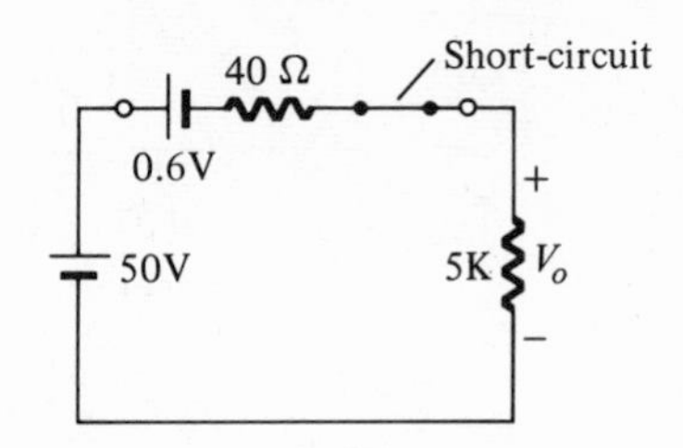

FIG. 1.53

Circuit of Fig. 1.48 at instant $v_i = 50$ V with the semiconductor diode.

The output curves for each appear in Fig. 1.54.

Take careful note of the differences between the various curves.

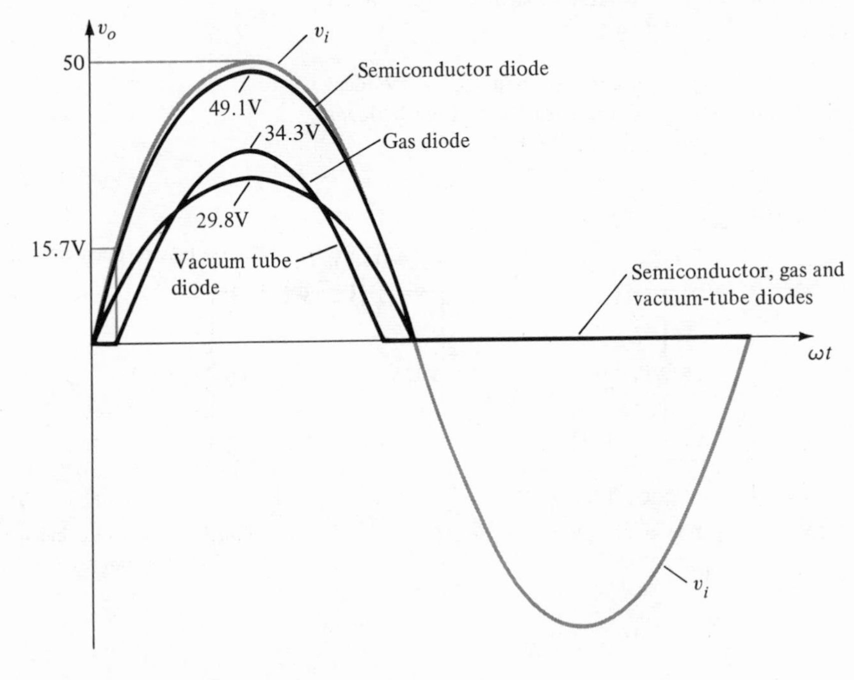

FIG. 1.54

The output wave form, v_o, for the circuit of Fig. 1.48 using semiconductor, gas, and vacuum-tube diode.

Clippers and clampers are diode waveshaping circuits. Each performs the function indicated by its name. The output of clipping circuits appears as if a portion of the input signal were clipped off. Clamping circuits simply clamp the waveform to a different dc level.

Clippers

A clipping circuit requires at least two fundamental components, a diode and a resistor. A dc battery, however, is also frequently employed. The output waveform can be clipped at different levels simply by interchanging the position of the various elements and changing the magnitude of the dc battery.

EXAMPLE 1.3

Clipper: Find the output voltage waveshape (v_o) for the inputs shown in Fig. 1.55.

Solution: *Input 1:* For any value of $v_i > 10$ V the ideal diode is forward biased and $v_o = v_i - 10$. For example, at $v_i = 15$ V (Fig. 1.56), the result is $v_o = 15 - 10 = 5$ V.

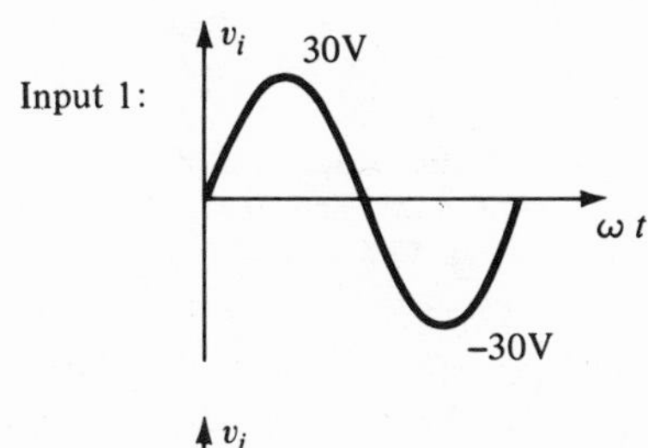

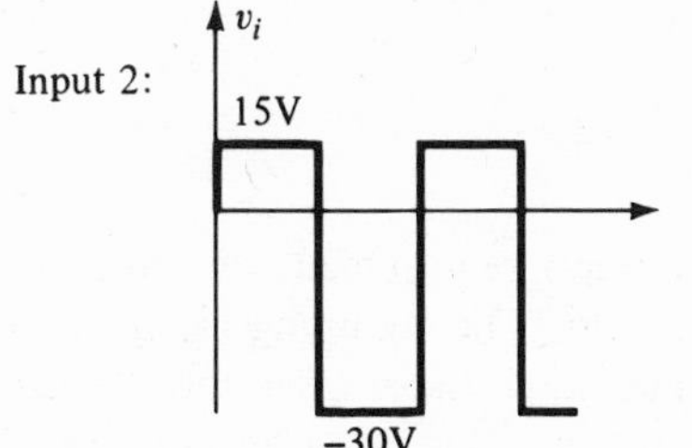

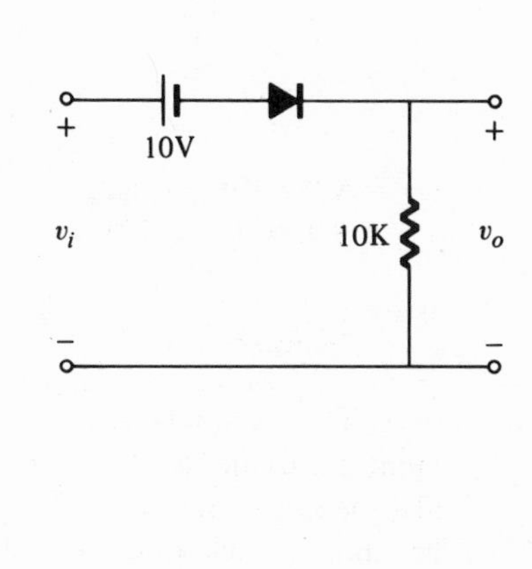

FIG. 1.55
Clipping circuit and inputs for Example 1.3.

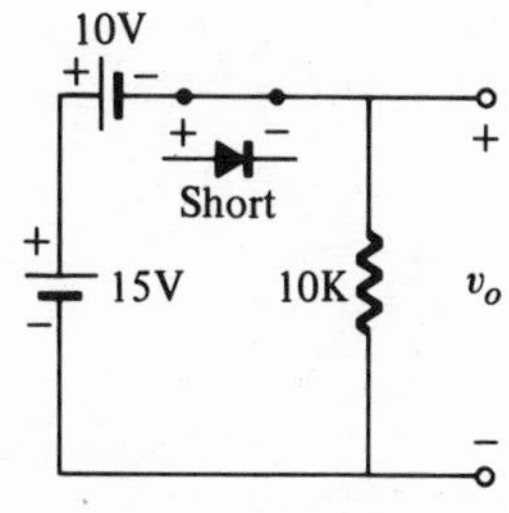

FIG. 1.56
Clipping circuit of Fig. 1.55 at instant $V_i = 15$ V of input 1.

For any value of $v_i < 10$ V the ideal diode is reverse biased and $v_o = 0$ since the current in the circuit is zero. For example, at $v_i =$ 5 V (Fig. 1.57).

The output waveform v_o appears as if the entire input were clipped off except the positive peak (Fig. 1.58).

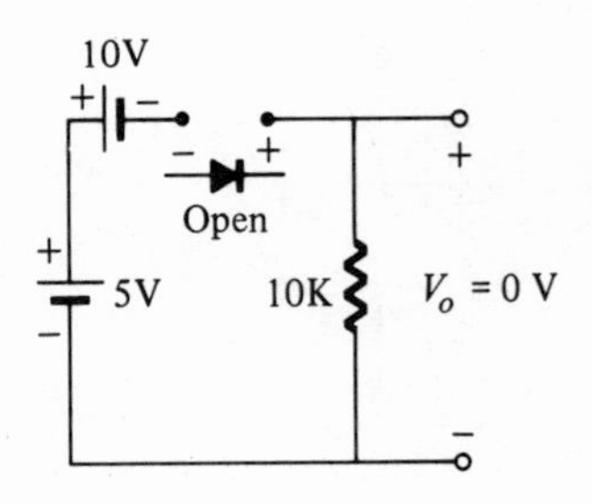

FIG. 1.57

Clipping circuit of Fig. 1.55 at instant V_i = 5 V of input 1.

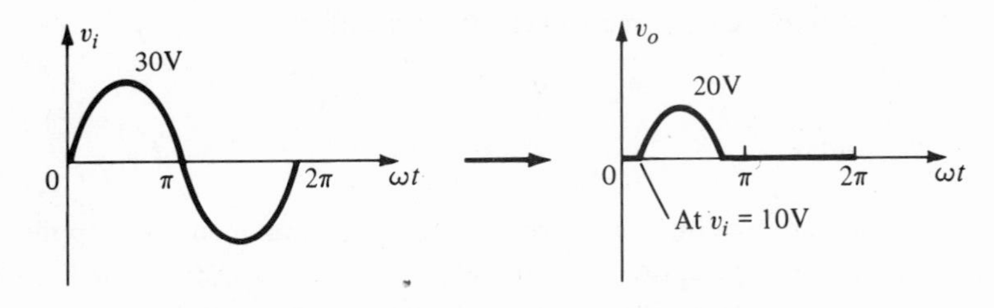

FIG. 1.58

Output wave form (v_o) for input 1 to the clipping circuit of Fig. 1.55.

Input 2: The diode will change state at the same levels indicated for the first input. The output waveform appears as shown in Fig. 1.59.

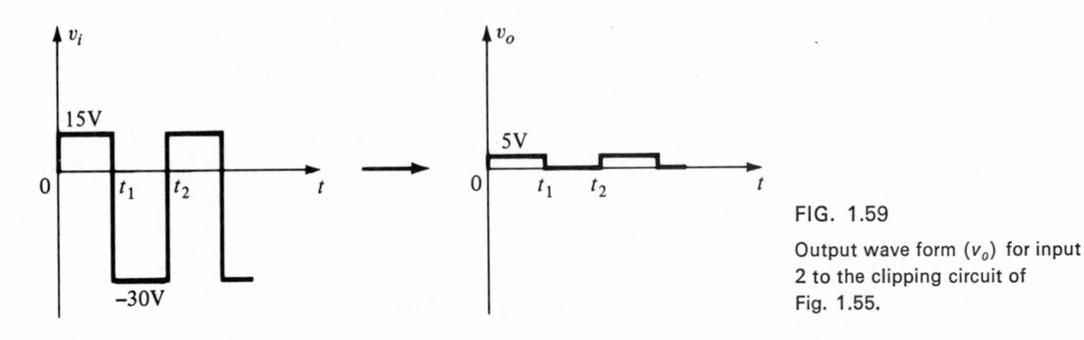

FIG. 1.59

Output wave form (v_o) for input 2 to the clipping circuit of Fig. 1.55.

A number of clipping circuits and their effect on the input signal appear in Fig. 1.60.

Clampers

The clamping circuit has a minimum requirement of three elements: a diode, a capacitor, and a resistor. The clamping circuit may also be augmented by a dc battery. The magnitudes of R and C must be chosen such that the time constant $\tau = RC$ is large enough to ensure that the voltage across the capacitor does not change significantly during the interval of time, determined by the input, that both R and C affect the output waveform. The need for this condition will be demonstrated in Example 1.4. Throughout the discussion, we shall

SIMPLE SERIES CLIPPERS

POSITIVE

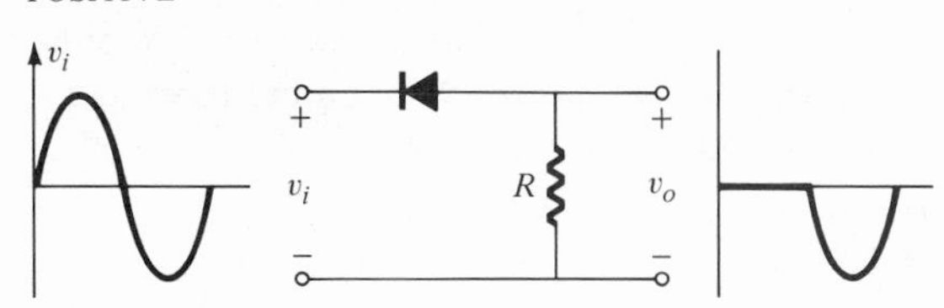

NEGATIVE

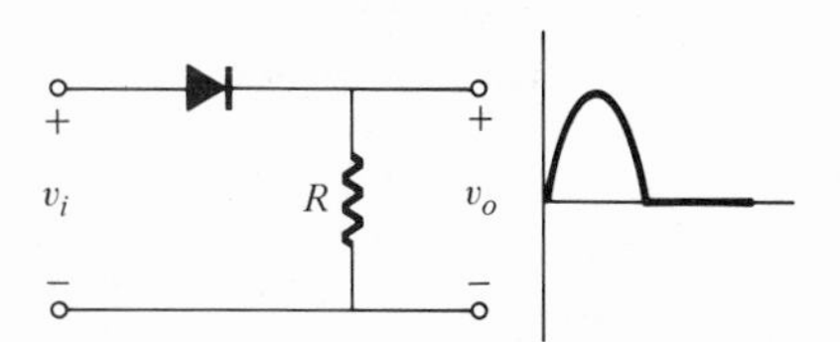

BIASED SERIES CLIPPERS

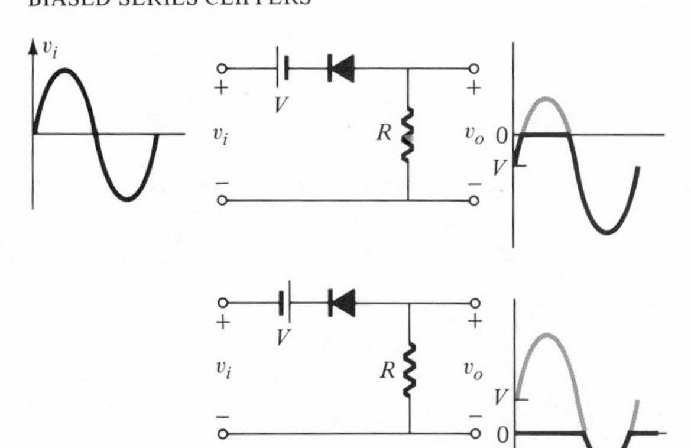

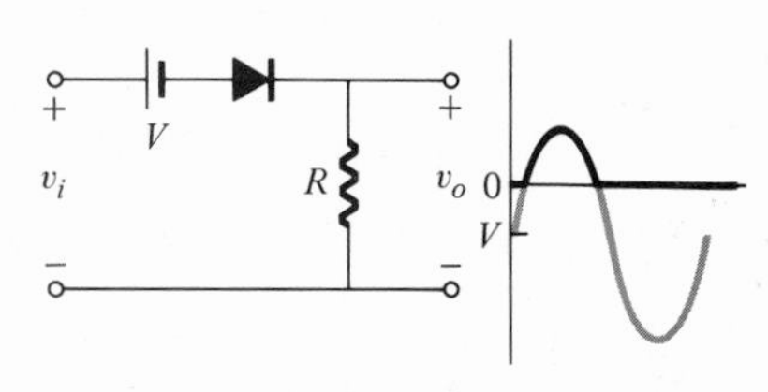

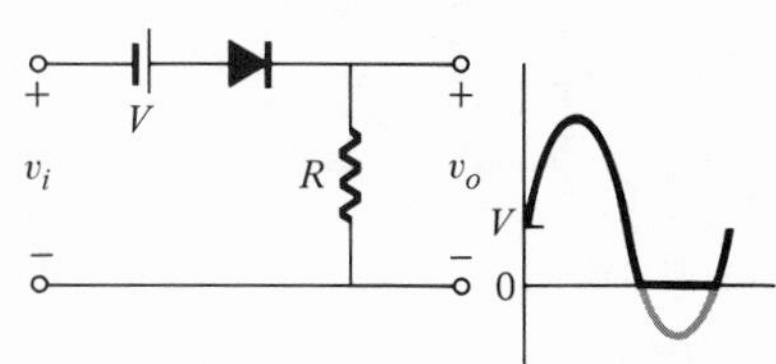

SIMPLE PARALLEL CLIPPERS

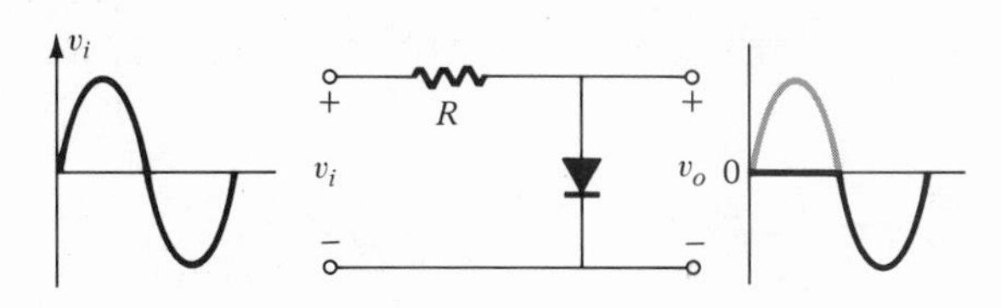

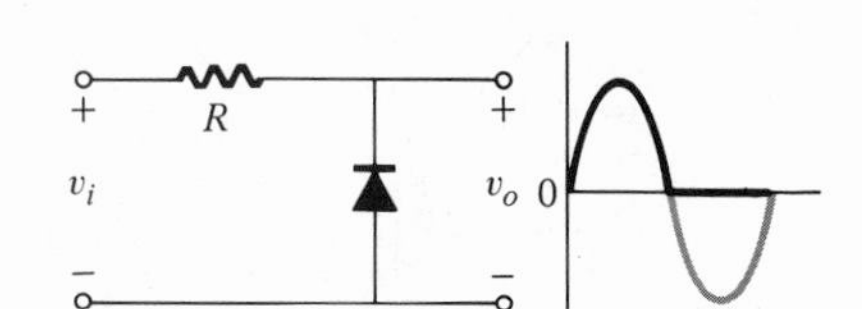

BIASED PARALLEL CLIPPERS

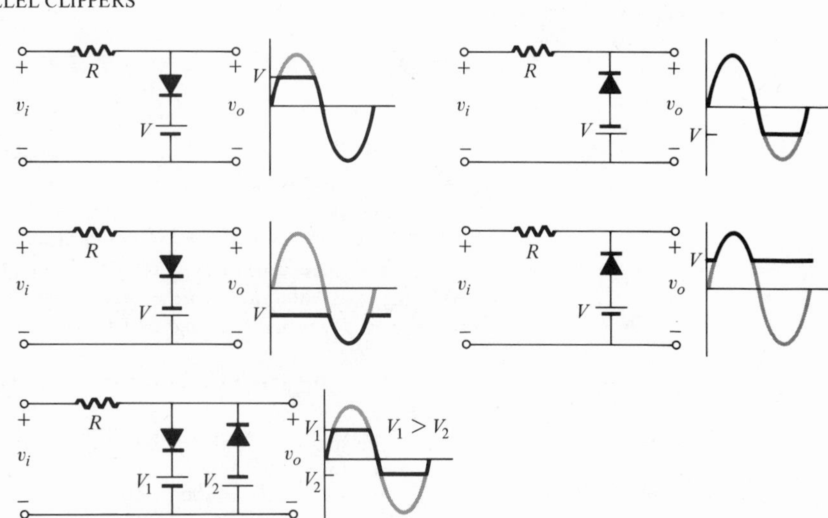

FIG. 1.60

Clipping circuits ($RC \gg \tau$).

assume that for all practical purposes a capacitor will charge to its final value in five time constants.

It is usually advantageous when examining clamping circuits to first consider the conditions that exist when the input is such that the diode is forward biased.

EXAMPLE 1.4

Clamper: Draw the output voltage waveform (v_o) for the input shown (Fig. 1.61).

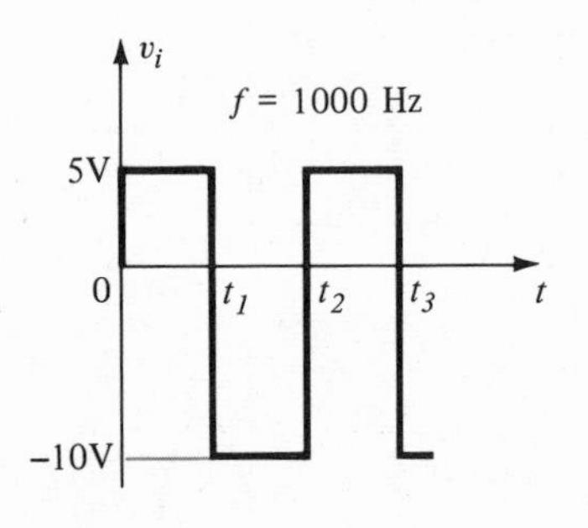

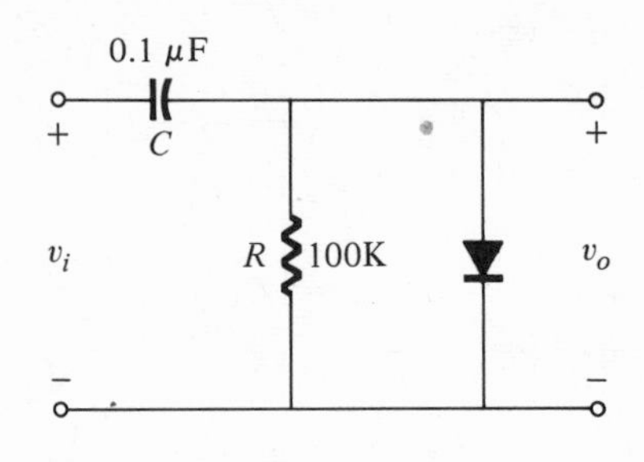

FIG. 1.61
Clamping circuit and input for Example 1.4.

Solution: At the instant the input switches to the +5-Vt state the circuit will appear as shown in Fig. 1.62. The input will remain in the

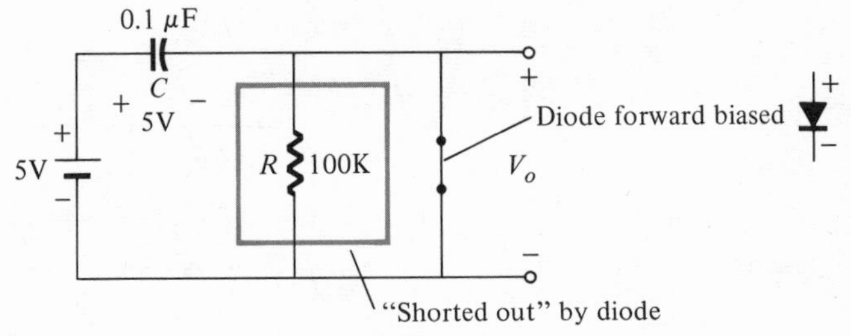

FIG. 1.62
Clamping circuit of Fig. 1.61, when v_i = 5 V (0→t_1).

+5-V state for an interval of time equal to one-half the period of the waveform since the time interval $0 \rightarrow t_1$ is equal to the interval $t_1 \rightarrow t_2$.

The period of v_i is $T = 1/f = 1/1000 = 1$ ms and the time interval of the +5-V state is $T/2 = 0.5$ ms.

Since the output is taken from directly across the diode it is 0 V for this interval of time. The capacitor, however, will rapidly charge to 5 V since the time constant of the network is now $\tau = RC \cong 0C = 0$.

When the input switches to −10 V the circuit of Fig. 1.63 will result.

The time constant for the circuit of Fig. 1.63 is

$$\tau = RC = 100 \times 10^3 \times 0.1 \times 10^{-6} = 10 \text{ ms}$$

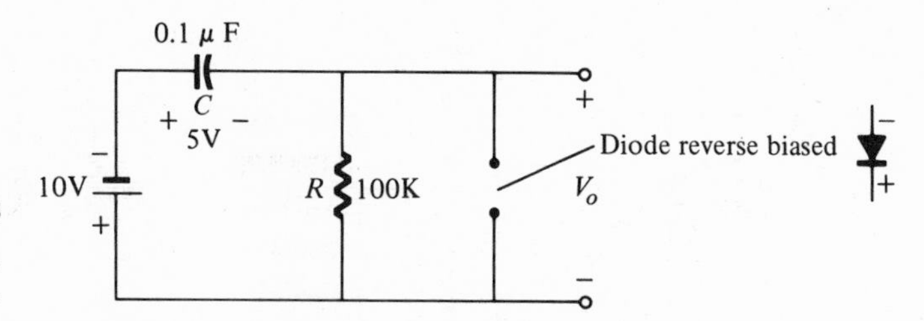

FIG. 1.63
Clamping circuit of Fig. 1.61, when $v_i = -10$ V ($t_1 \to t_2$).

Since it takes approximately five time constants or 50 ms for a capacitor to discharge, and the input is only in this state for 0.5 ms, to assume the voltage across the capacitor does not change appreciably during this interval of time is certainly a reasonable approximation. The output is therefore

$$V_o = \underbrace{-10}_{\text{supply}} - \overbrace{5}^{\text{capacitor}} = -15\text{ V}$$

The resulting output waveform (v_o) is provided in Fig. 1.64. As indicated, the output is clamped to the negative region and will

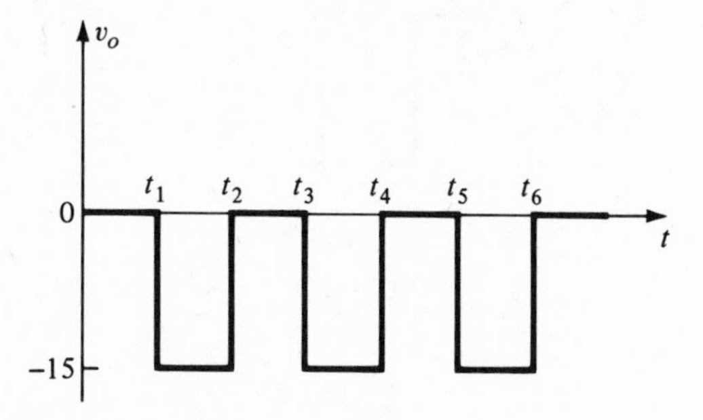

FIG. 1.64
Output (v_o) for clamping circuit and input of Fig. 1.61.

repeat itself at the same frequency as the input signal. Note that the swing of the input and output voltages is the same: 15 V. For all clamping circuits the voltage swing of the input and output waveforms will be the same. This was certainly not the case for clipping circuits.

A number of clamping circuits and their effect on the input signal appear in Fig. 1.65.

1.12 ZENER DIODES

The Zener or avalanche region of the semiconductor diode was discussed in detail in Section 1.5. It occurs at a reverse-bias potential of 10 V for the diode of Fig. 1.66a. For the purposes of introducing notation for the Zener diode and comparing its characteristics to those of the gas diode, a semiconductor Zener diode characteristic has been drawn as shown in Fig. 1.66b.

The location of the Zener region can be controlled by varying the doping levels. An increase in doping, producing an increase in the

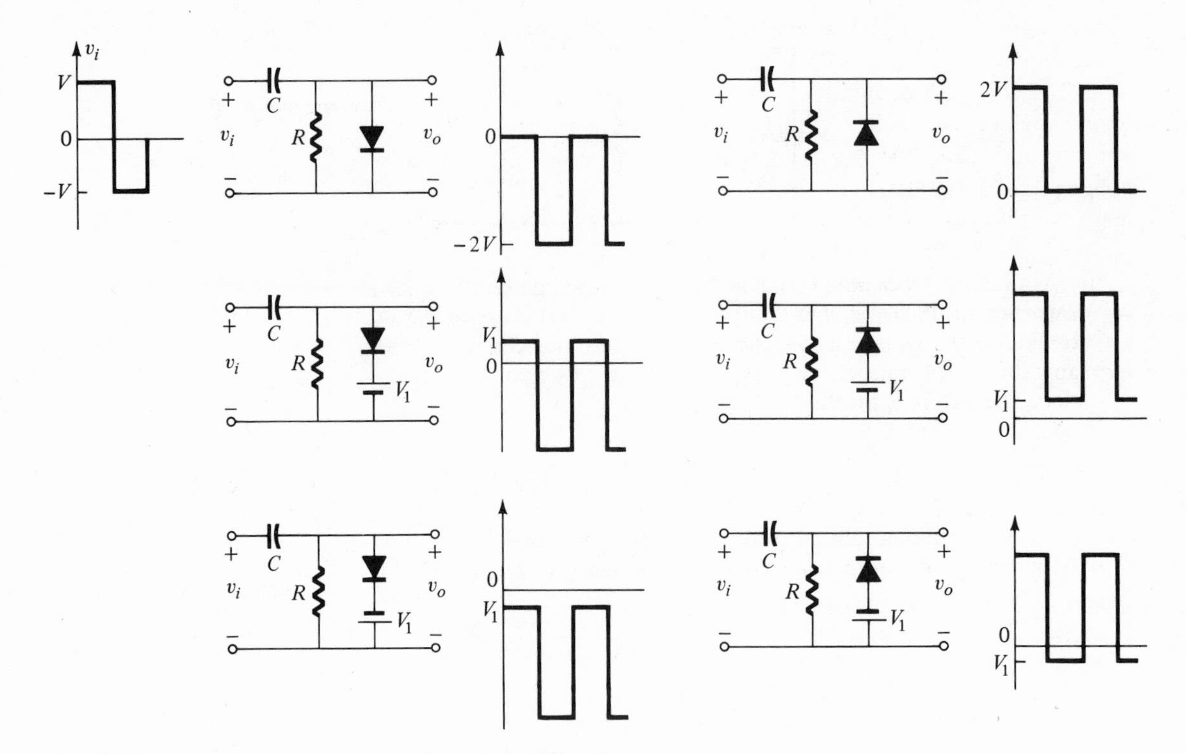

FIG. 1.65

Clamping circuits ($RC \gg \tau$).

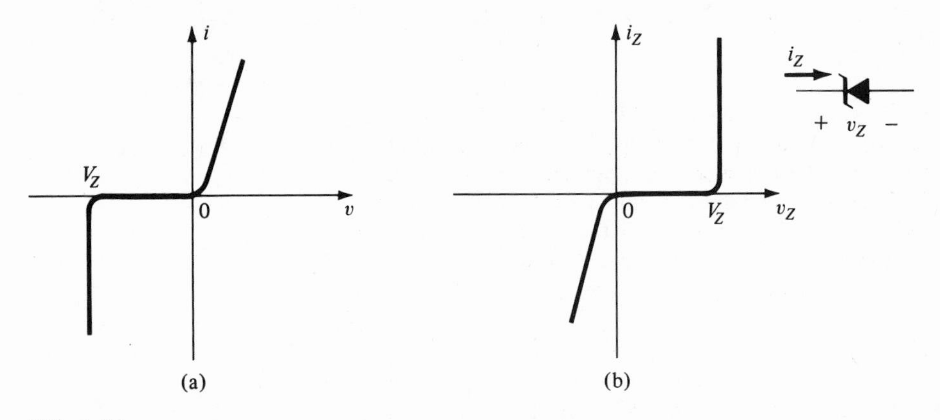

FIG. 1.66

Zener diodes: (a) Zener potential; (b) characteristics and notation.

number of added impurities, will decrease the Zener potential. Zener diodes are available having Zener potentials of 2.4 to 200 V with power ratings from $\frac{1}{4}$ to 50 W. Because of its higher temperature and current capability, silicon is usually preferred in the manufacture of Zener diodes.

The complete equivalent circuit of the Zener diode in the Zener region includes a small resistor and dc battery equal to the Zener potential as shown in Fig. 1.67a. For all applications to follow, however, we shall assume that the external resistors are much larger in magnitude than the Zener-equivalent resistor and the equivalent circuit is simply that indicated in Fig. 1.67b.

Three areas of typical operation for the Zener diodes are in dc and ac voltage regulators, voltage references, and clipping networks. In Fig. 1.68 the basic configuration of a dc and ac voltage regulator is provided.

Note the similarities between the dc Zener diode voltage regulator and the gas-tube regulator. Since the characteristics of each are also very similar the analysis of this circuit is exactly the same as that for the gas tube if the left half-plane of Fig. 1.66b is avoided. The ac voltage regulator, however, will require some explaining. For the sinusoidal signal shown in Fig. 1.69a the circuit of Fig. 1.68b will appear as shown in Fig. 1.69b at the instant $v_i = 10$ V. The region of

For 10-V Zener, typically less than 10 Ω

V_Z R_Z ≅ V_Z

(a) (b)

FIG. 1.67
Zener equivalent circuit: (a) complete; (b) approximate.

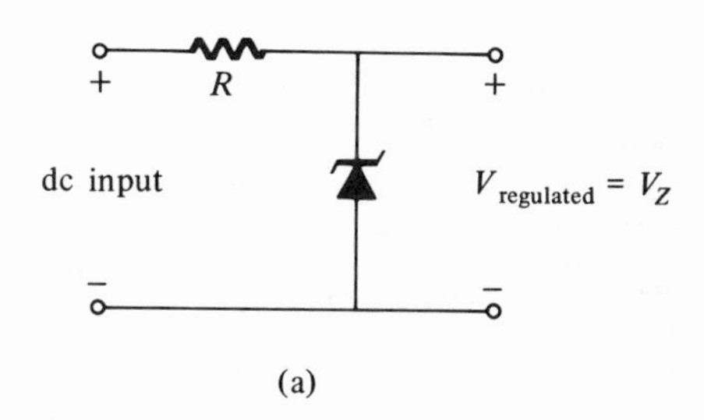

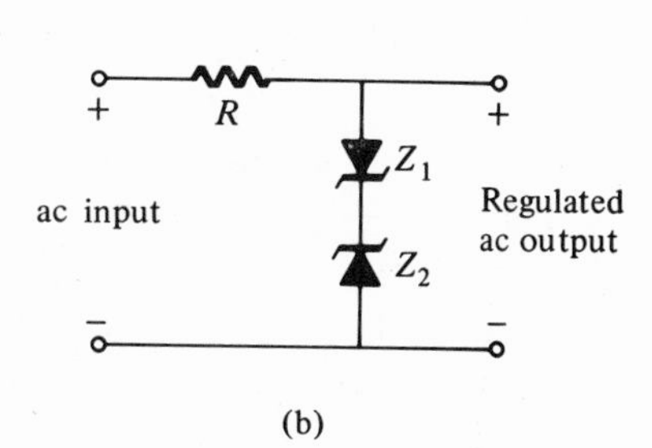

FIG. 1.68
Regulators: (a) dc; (b) sinusoidal ac.

operation for each diode is indicated in the adjoining figure. Note that the impedance associated with Z_1 is very small, or essentially a short, since it is in series with 5 K, while the impedance of Z_2 is very large corresponding to the open-circuit representation. Since Z_2 is an open circuit, $v_o = v_i = 10$ V. This will continue to be the case until v_i is slightly greater than 20 V. Then Z_2 will enter the low-resistance region (Zener region) and Z_1 will for all practical purposes be a short circuit and Z_2 will be replaced by $V_Z = 20$ V. The resultant output waveform is indicated in the same figure. Note that the waveform is not purely sinusoidal, but its rms (effective) value is closer to the desired 20-V peak sinusoidal waveform than the sinusoidal input having a peak value of 22 V.

The circuit of Fig. 1.68b can be extended to that of a simple square-wave generator (due to its clipping action) if the signal v_i is

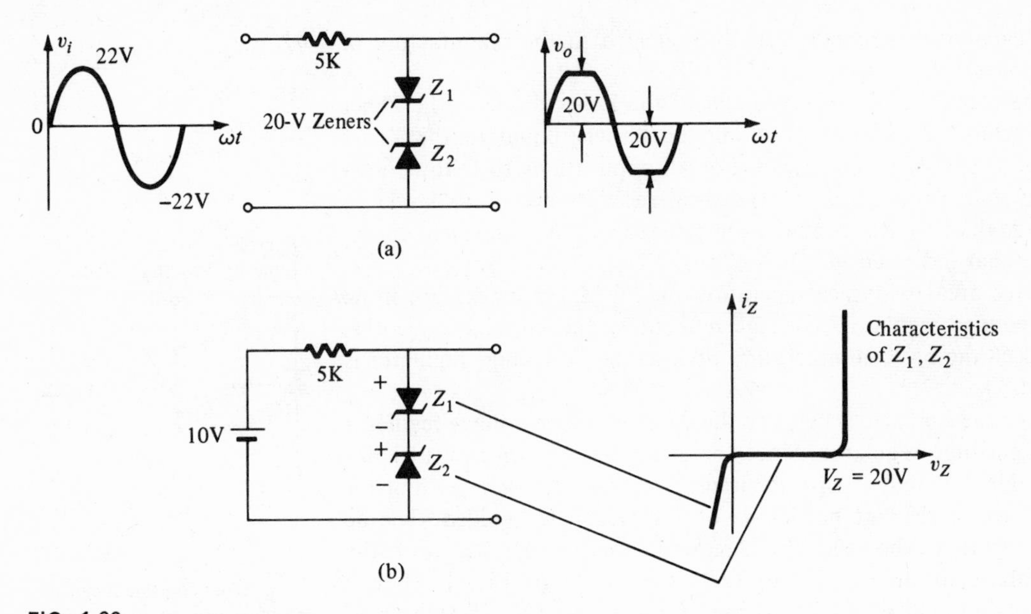

FIG. 1.69

Sinusoidal ac regulation: (a) 40-V peak-to-peak sinusoidal ac regulator; (b) circuit operation at $V_i = 10$ V.

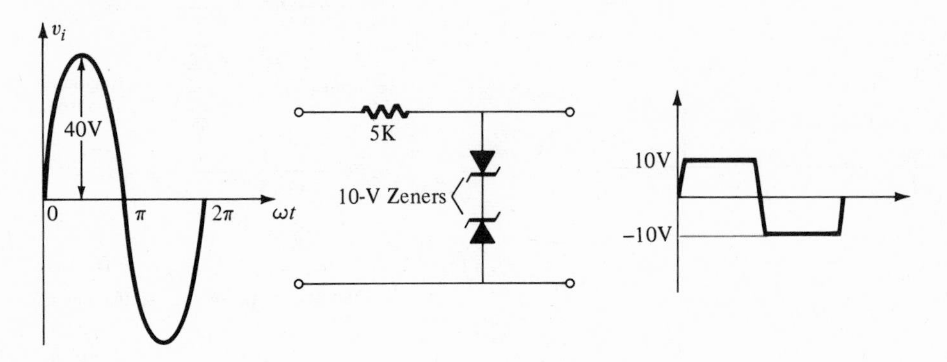

FIG. 1.70

Simple square-wave generator.

increased to perhaps 40-V peak with 10-V Zeners. The resultant waveform is indicated in Fig. 1.70.

It is often necessary to have a fixed reference voltage in a network for biasing and comparison purposes. This can be accomplished using a Zener diode as shown in Fig. 1.71. The variation in dc supply voltage due to any number of reasons has been included as a small sinusoidal signal.

Since v_i is always greater than 10 V, the Zener diode will always be in the "on" state. The output voltage v_o, therefore, will remain fixed at the Zener potential of 10 V, our reference potential.

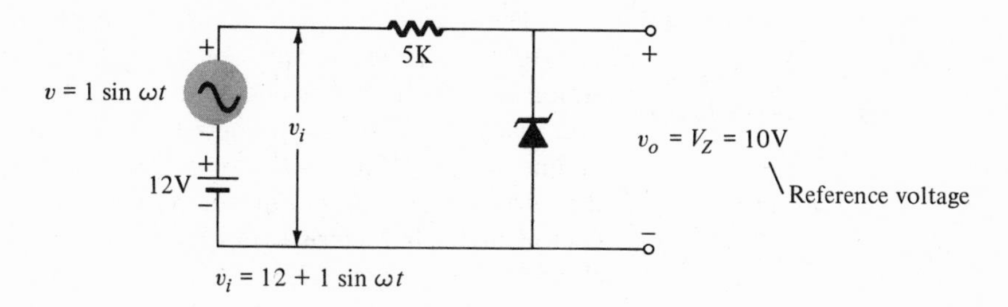

FIG. 1.71
Reference voltage.

1.13 TUNNEL DIODES

The tunnel diode was first introduced by Dr. Leo Esaki in 1958. Its characteristics, shown in Fig. 1.72, are quite different from any diode discussed thus far. A portion of the characteristics has a negative resistance region as shown in Fig. 1.72. In this region, an increase in terminal voltage results in a reduction in diode current.

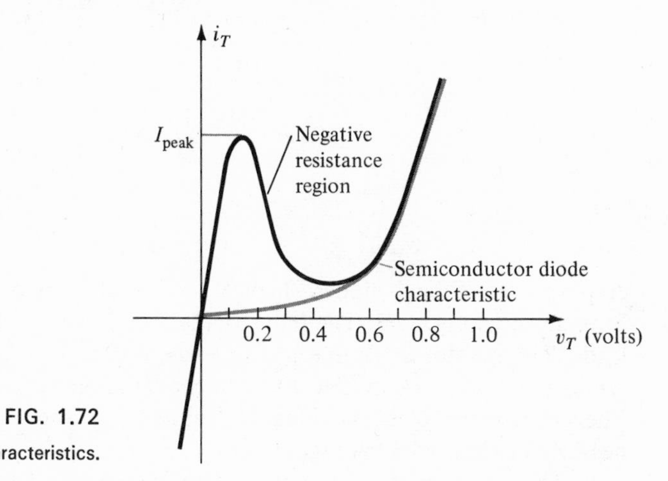

FIG. 1.72
Tunnel diode characteristics.

The tunnel diode is fabricated by doping the semiconductor materials that will form the *p-n* junction at a level one hundred to several thousand times that of a typical semiconductor diode. You will recall in the previous section on Zener diodes that an increase in the doping level will drop the Zener potential. Note the effect of a very high doping level on this region in Fig. 1.72. The semiconductor materials most frequently used in the manufacture of tunnel diodes are germanium and gallium arsenide.

The depletion region of the tunnel diode is in the neighborhood of $1/10^6$ of an inch in width or typically about $\frac{1}{100}$ the width of this region for a typical semiconductor diode. It is this thin depletion region that many carriers can "tunnel" through, rather than attempt to surmount, at low forward-bias potentials, that accounts for the peak in the curve of Fig. 1.72. For comparison purposes, a typical semiconductor diode characteristic has been superimposed on the tunnel diode characteristic of Fig. 1.72.

The peak current of a tunnel diode can vary from a few microamperes to several hundred amperes. The peak voltage, however, is limited to about 600 mV. For this reason, a simple VOM with an internal dc battery potential of 1.5 V can severely damage a tunnel diode if used improperly.

The tunnel diode equivalent circuit in the negative-resistance region is provided in Fig. 1.73, with the symbols most frequently

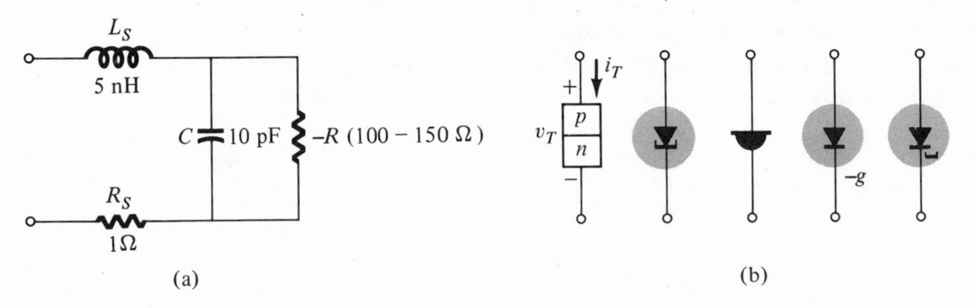

FIG. 1.73
Tunnel diode: (a) equivalent circuit; (b) symbols.

employed for tunnel diodes. Typical values of each parameter have been given in Fig. 1.73. The inductor L_s is due mainly to the terminal leads. The resistor R_s is due to the leads, ohmic contact at the lead-semiconductor junction, and the semiconductor materials themselves. The capacitance C is the junction capacitance and the $-R$ is the negative resistance of the region.

The tunnel diode is employed in a wide variety of areas, including oscillators, amplifiers, high-speed switching, logic circuitry, mixers, and detectors.

1.14 POWER DIODES

There are a number of diodes designed specifically to handle the high-power and high-temperature demands of some applications. The most frequent use of power diodes occurs in the rectification process, in which ac signals (having zero average value) are converted to one having an average or dc level. When used in this capacity, diodes are normally referred to as rectifiers.

The majority of the power diodes are constructed using silicon because of its higher current, temperature, and PIV ratings. The higher-current demands require that the junction area be larger to ensure that there is a low forward diode resistance. If the forward resistance were too large, the I^2R losses would be excessive. The current capability of power diodes can be increased by placing two or more in parallel and the PIV rating can be increased by stacking the diodes in series.

Various types of power diodes and their current rating have been provided in Fig. 1.74a. The high temperatures resulting from the heavy current flow require, in many cases, that heat sinks be used to draw the heat away from the element. A few of the various types of heat

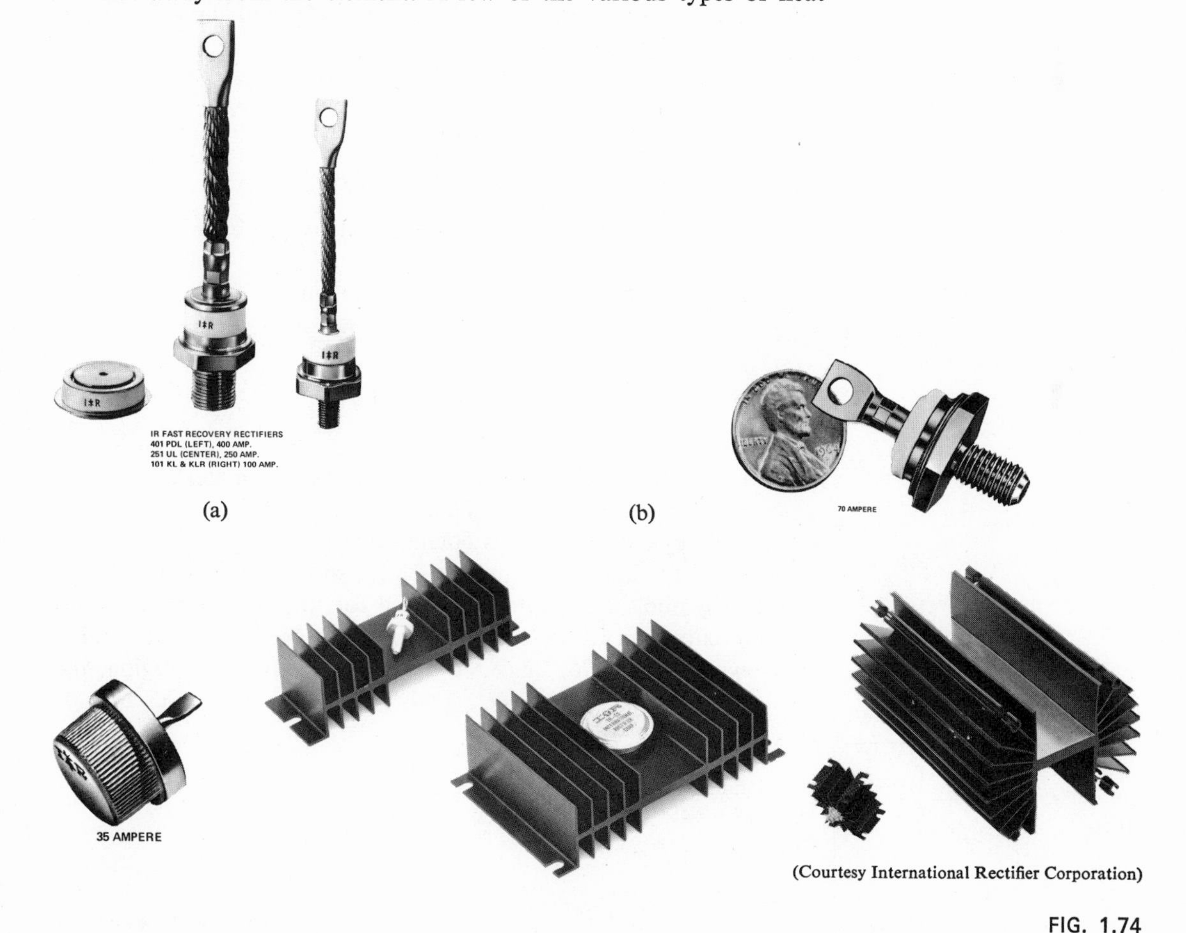

(Courtesy International Rectifier Corporation)

FIG. 1.74
Power diodes and heat sinks.

sinks available are shown in Fig. 1.74b. If heat sinks are not employed, stud diodes are designed to be attached directly to the chassis, which in turn will act as the heat sink.

1.15 VARICAP (VARACTOR) DIODES

Varicap diodes are semiconductor, voltage-dependent, variable capacitors. Their mode of operation depends on the small capacitance that exists at the *p-n* junction when the element is reverse biased. An increase in reverse-bias potential will increase the width of the depletion region resulting in a decrease in diode capacitance. In equation form, the junction capacitance (C_j) is related to the permittivity of the semiconductor materials (ϵ), *p-n* junction area (A), and depletion width (W_d) in the following manner:

$$C_j = \frac{\epsilon A}{W_d} \tag{1.9a}$$

Note the similarities between Eq. (1.9a) and the basic equation for capacitance: $C = \epsilon A/d$.

In terms of the applied reverse bias potential

$$C_j = \frac{1}{K(V_o + V)^n} \tag{1.9b}$$

where K = constant determined by semiconductor material;
V_o = knee potential as defined in Section 1.5;
V = applied reverse-bias potential; and
$n = \frac{1}{2}$ for alloy junctions and $\frac{1}{3}$ for diffused junctions.

The symbols most commonly used for the varicap diode and a first approximation for its equivalent circuit in the reverse-bias region are shown in Fig. 1.75. Since we are in the reverse-bias region, the

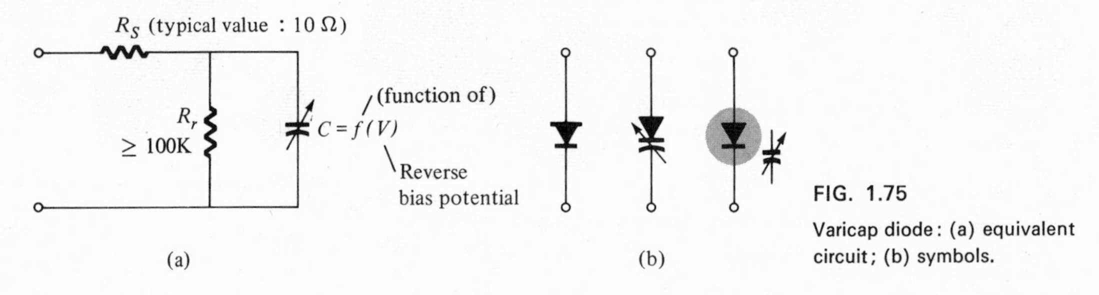

FIG. 1.75

Varicap diode: (a) equivalent circuit; (b) symbols.

resistor in the equivalent circuit is quite large in magnitude—typically 100 K or larger—while R_s, the geometric resistance of the diode, is, as indicated in Fig. 1.75, quite small. The magnitude of C will vary from about 2 to 100 pF depending on the varicap considered. To ensure that R_r is as large (for minimum leakage current) as possible, silicon is normally used in varicap diodes.

Some of the areas of application for the varicap diode include FM modulators, adjustable band-pass filters, automatic frequency control devices, and parametric amplifiers.

1.16 PHOTOTUBES

The phototube is a two-terminal, light-sensitive device having the basic construction and symbol appearing in Fig. 1.76. The cathode is design-

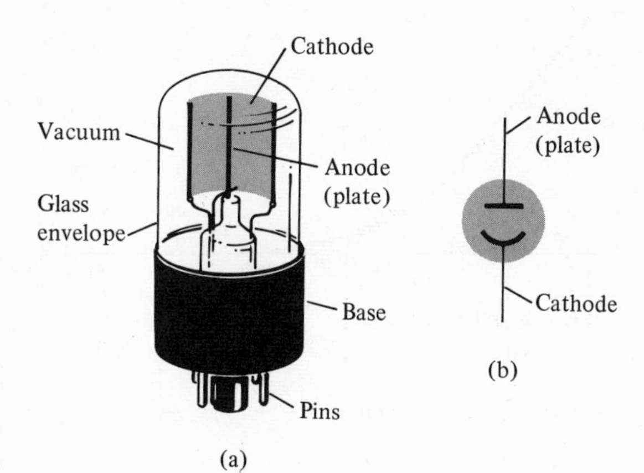

FIG. 1.76
Vacuum phototube: (a) basic construction; (b) symbol.

ed to pick up the maximum incident light possible. It is either constructed of, or coated with, a metallic material having photoemissive properties, such as sodium, potassium, and cesium. The material employed will be determined by the wavelength of the incident light waves. The wavelength, being the distance between successive peaks of the traveling light wave, usually is measured in angstrom units (Å) or microns (μ), where

$$1\text{Å} = 10^{-4}\mu$$

and

$$1\mu = 10^{-6}\ \text{m}$$

It is related to the frequency of the traveling wave by

$$\boxed{\lambda = \frac{v}{f}} \tag{1.10}$$

where λ = wavelength in meters;
v = velocity of light, 3×10^8 m/s; and
f = frequency in hertz of the traveling wave.

The spectral response of a frequently employed phototube appears in Fig. 1.77. Note the effect of wavelength on the color of the

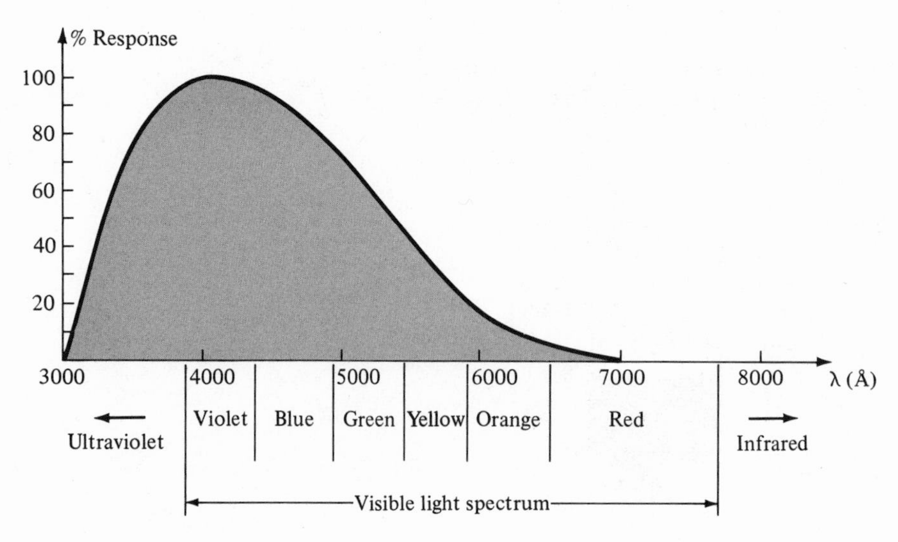

FIG. 1.77
Phototube spectral response.

visible light. The energy associated with the incident light is directly related to the *frequency* of the traveling wave by

$$\boxed{W_{(\text{joules})} = \hbar f} \tag{1.11}$$

where $\hbar$ is called Planck's constant and is equal to 6.624×10^{-3} Joule-seconds

It states quite clearly that since $\hbar$ is a constant the increase in energy associated with incident light waves is directly related to the frequency of the traveling wave. It has been further theorized that this light energy exists in the form of discrete packages of energy called *photons* rather than in a continuous distribution.

The process of photoemission is clearly depicted in Fig. 1.78. The photons of energy in the incident light are absorbed by the relatively "free" electrons near and on the surface of the photoemissive material. A number of these electrons will then possess sufficient energy to leave the surface of the material as indicated in the figure. The energy associated with these free electrons will be directly proportional to the *frequency* of the incident light as determined by Eq. (1.11). The number of free electrons, however, is proportional to the *intensity* of the incident light. Light intensity is a measure of the amount of luminous flux falling on a particular surface area. Luminous flux is normally measured in lumens (lm) or watts. The two units are related by

$$1 \text{ lm} = 1.496 \times 10^{-3} \text{ W}$$

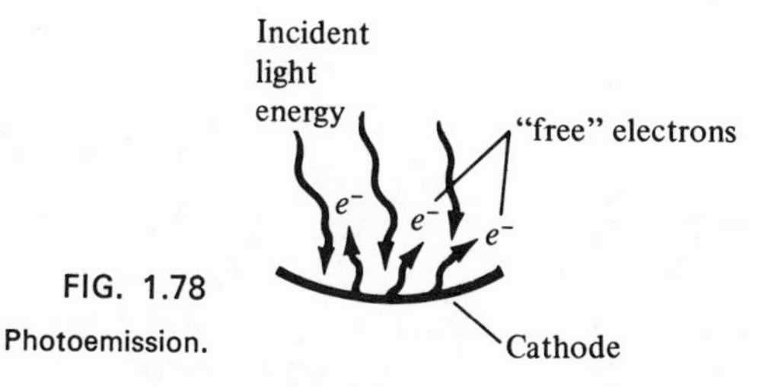

FIG. 1.78
Photoemission.

The light intensity is normally measured in lm/ft², footcandles (fc), or W/m², where

$$1 \text{ lm/ft}^2 = 1 \text{ fc} = 1.609 \times 10^{-12} \text{ W/m}^2$$

Let us now examine how this photoemissive effect will affect the behavior of the relatively simple photoelectric circuit of Fig. 1.79. Note

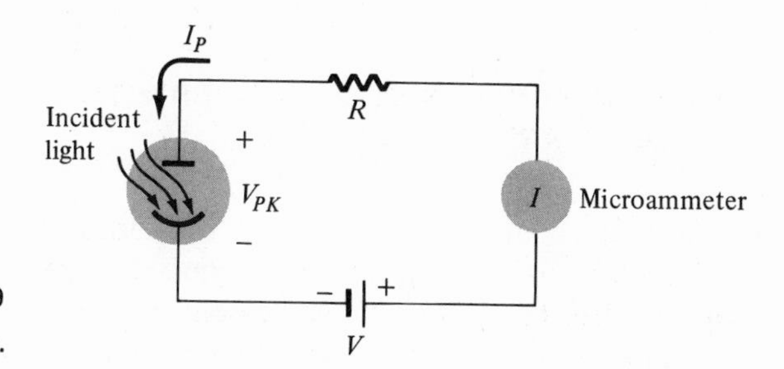

FIG. 1.79
Basic phototube circuit.

the necessity for a separate dc supply. Consider also that the plate of the tube is, before emission takes place, positive with respect to the cathode by V volts (the battery potential). When a light source of the proper wavelength is applied, photoemission will result and the negatively charged free electrons will travel directly to the positive plate of the tube. The microammeter (I) will indicate the strength of the resulting current flow. The effect of a change in light intensity is clearly indicated by the typical set of characteristics of Fig 1.80. A luminous flux of 0.1 lm at 200 V (V_{PK}) will result in approximately 4 times the anode current at a luminous flux of 0.02 lm (maintaining $V_{PK} = 200$ V).

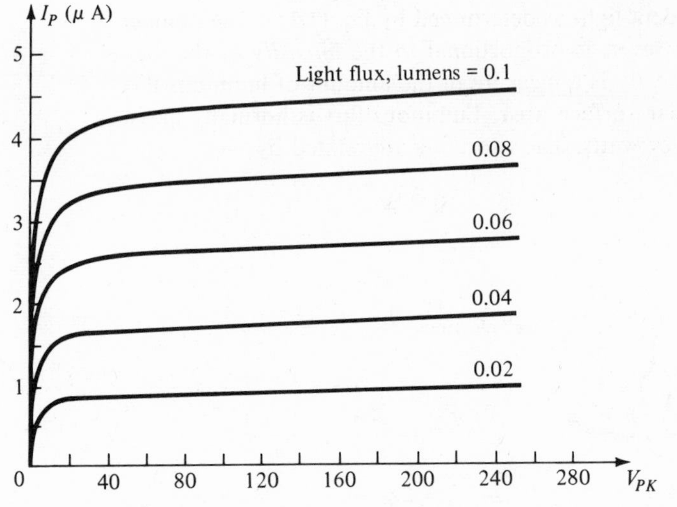

FIG. 1.80

Typical set of phototube characteristics.

A direct application of the circuit of Fig. 1.79 is in a photographic exposure meter. A variable resistor in parallel with the movement would permit the measurement of higher light intensities. A photograph of the phototube having the characteristics of Fig. 1.80 appears in Fig. 1.81.

(Courtesy Radio Corporation of America)

FIG. 1.81

Photograph of a 929 high-vacuum phototube.

1.17 SEMICONDUCTOR PHOTOCONDUCTIVE CELL AND PHOTODIODE

The photoconductive cell is simply a two-terminal semiconductor device whose terminal resistance will vary (quite linearly) with the intensity of the incident light. For obvious reasons, it is frequently called a photoresistive device. A few typical photoconductive cells and the most widely used graphic symbol for each appears in Fig. 1.82.

As the illumination on the device increases in intensity, the energy state of a larger number of electrons in the structure will also increase due to the increased availability of the photon packages of energy. The result is an increasing number of relatively "free" electrons in the structure and a decrease in the terminal resistance. The sensitivity curve for a typical photoconductive device appears in Fig. 1.83.

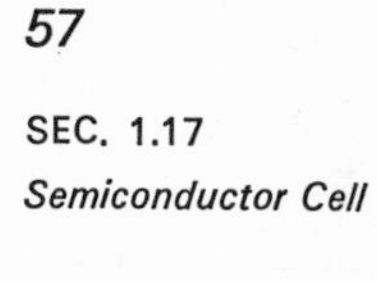

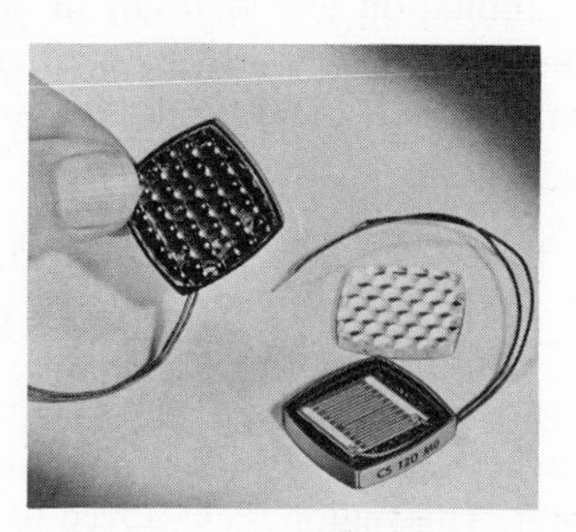

(Courtesy International Rectifier Corporation)

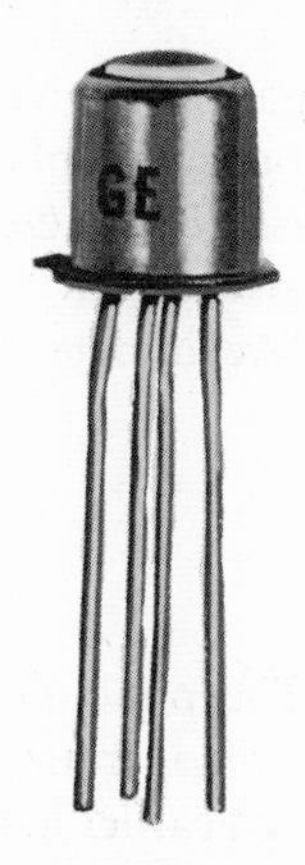

(Courtesy General Electric Company)

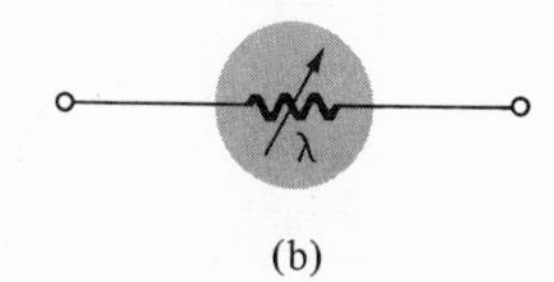

(b)

FIG. 1.82
Photoconductive cell: (a) types; (b) symbol.

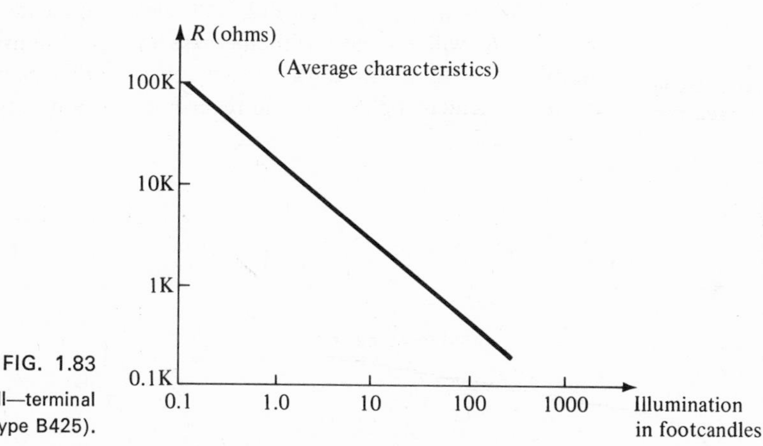

FIG. 1.83
Photoconductive cell—terminal characteristics (GE type B425).

Note the linearity of the resulting curve and the large change in resistance (100 K → 100 Ω) for the indicated change in illumination.

One rather simple, but interesting, application of the device appears in Fig. 1.84. The purpose of the system is to maintain V_{out} at a fixed level even though V_i may fluctuate from its rated value. As indicated in the figure, the photoconductive cell, bulb, and resistor, all

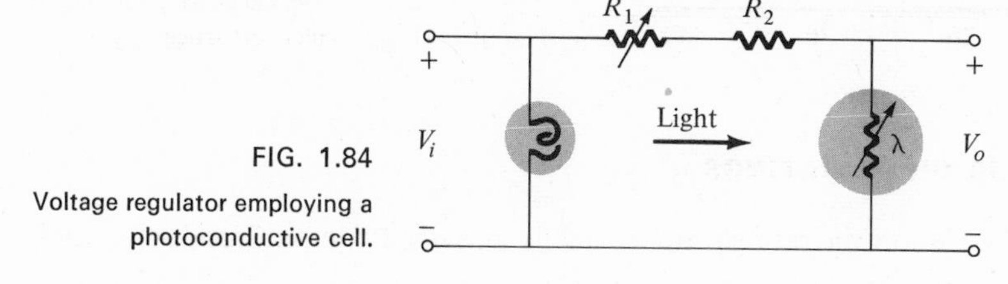

FIG. 1.84
Voltage regulator employing a photoconductive cell.

form part of this voltage regulator system. If V_i should drop in magnitude for any number of reasons, the brightness of the bulb would also decrease. The decrease in illumination would result in an increase in the resistance (R_λ) of the photoconductive cell to maintain V_o at its rated level as determined by the voltage divider rule; that is,

$$\boxed{V_o = \frac{R_\lambda V_i}{R_\lambda + R}} \tag{1.12}$$

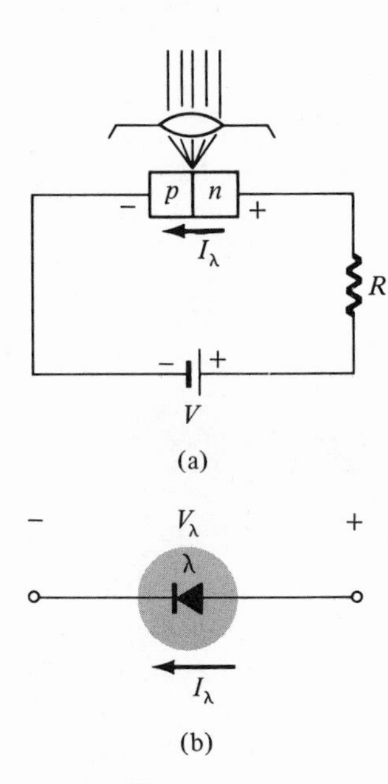

FIG. 1.85

Photodiode: (a) basic biasing arrangement and construction; (b) symbol.

The photodiode is a semiconductor *p-n* junction device whose region of operation is the reverse-bias region of the junction diode discussed earlier in this chapter. This region is employed to take advantage of the fact that the reverse current increases almost linearly with the increase in the incident light. The basic biasing arrangement, construction, and symbol for the device appear in Fig. 1.85. Compare the defined direction of I_λ to that employed for the junction diode of Section 1.5.

The characteristics of Fig. 1.86 clearly indicate that the reverse current I_λ will increase with increase in light intensity for the same applied potential. The dark current refers to that current which flows with no incident light. It is the reverse saturation current discussed in Section 1.5.

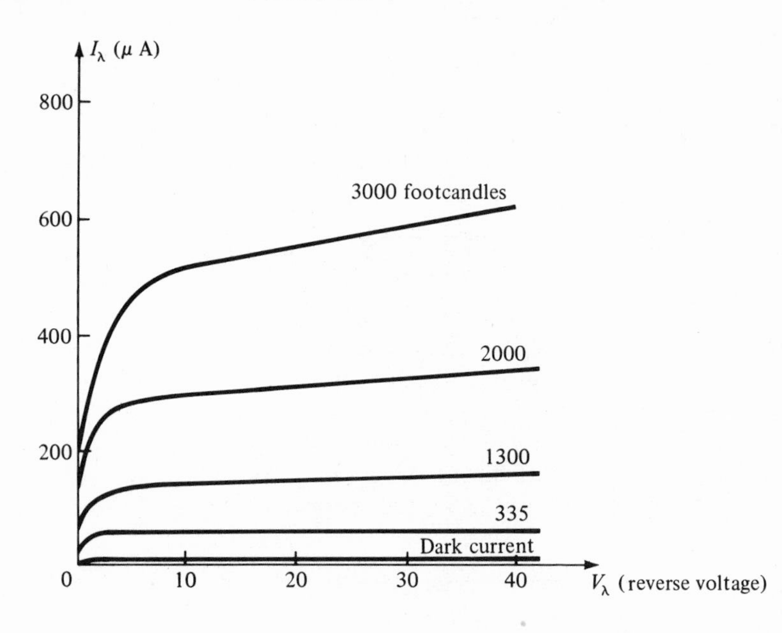

FIG. 1.86

Typical set of photodiode characteristics.

1.18 DIODE RATINGS

The following ratings are normally provided for semiconductor and tube diodes:

1. the maximum forward voltage (at a specified current and temperature);
2. the maximum forward current (at a specified temperature);
3. the maximum reverse current (at a specified temperature);
4. the reverse voltage rating (PIV) (at a specified temperature); and
5. the maximum operating temperature.

Depending on the type of diode being considered, additional data will also be provided, such as frequency range, noise level, capacitance, switching time, etc. If the maximum power rating is also provided, it is understood to be the equal to the following product:

$$P_{D_{\max}} = V_D I_D \tag{1.13}$$

where I_D and V_D are the diode current and voltage at a particular point of operation.

1.19 THERMISTORS

The thermistor is, as the name implies, a temperature-sensitive resistor; that is, its terminal resistance is somehow related to its body temperature. It has a negative temperature coefficient, indicating that its resistance will decrease with an increase in its body temperature. Some of the materials employed in the manufacture of thermistors include oxides of cobalt, nickel, strontium, and manganese.

The characteristics of a fairly representative thermistor are provided in Fig. 1.87 with a commonly used symbol for the diode. Note, in particular, that at room temperature (20°C) the resistance of the thermistor is approximately 5000 Ω, while at 100°C (212°F) the resistance has decreased to 100 Ω. A temperature span of 80°C has therefore resulted in a 50:1 change in resistance. There are, fundamentally, two ways to change the temperature of the device: internally and externally. A simple change in current through the device will result in an internal change in temperature. Externally would require changing the temperature of the surrounding medium or immersing the device in a hot or cold solution.

A photograph of a number of commercially available types of thermistors is provided in Fig. 1.88.

A simple temperature-indicating circuit appears in Fig. 1.89. Any increase in the temperature of the surrounding medium will result in a decrease in the resistance of the thermistor and an increase in the current I_T. An increase in I_T will produce an increased movement

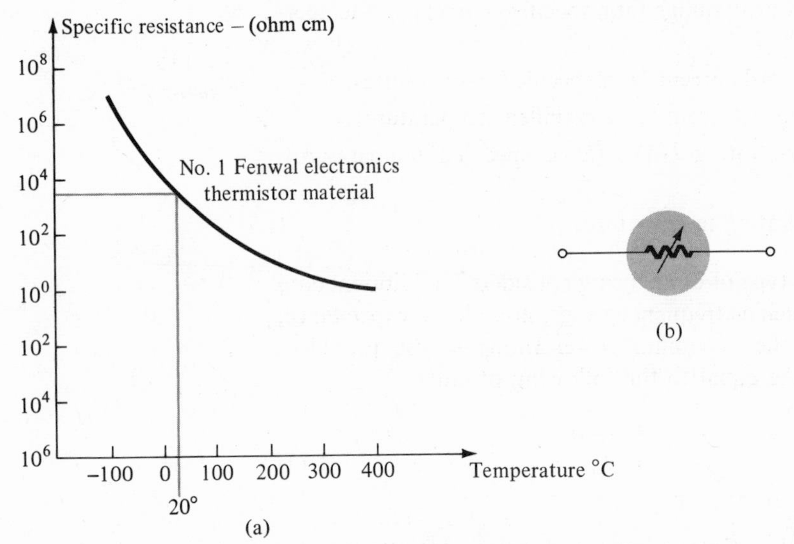

FIG. 1.87

Thermistor: (a) typical set of characteristics; (b) symbol.

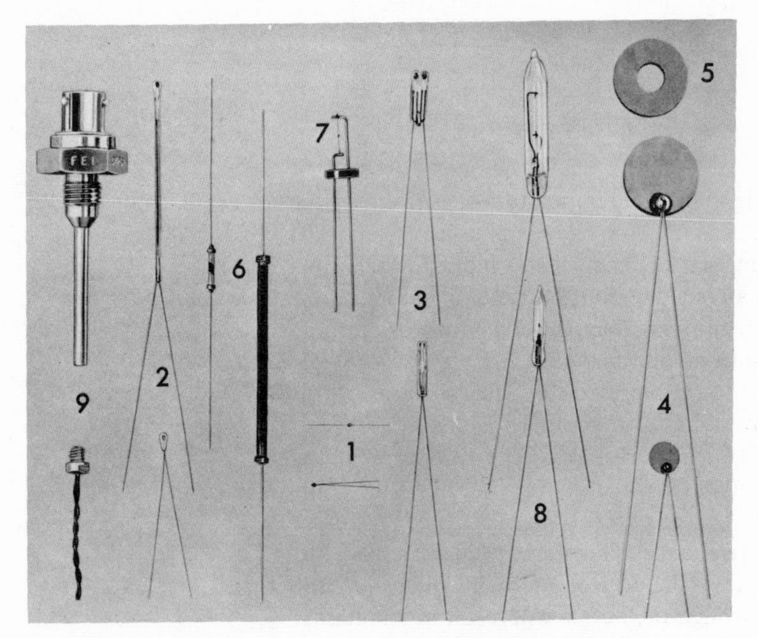

(Courtesy Fenwal Electronics, Inc.)

FIG. 1.88

Various types of thermistors: (1) beads; (2) glass probes; (3) Iso-curve interchangeable probes and beads; (4) discs; (5) washers; (6) rods; (7) specially-mounted beads; (8) vacuum and gas filled probes; (9) special probe assemblies.

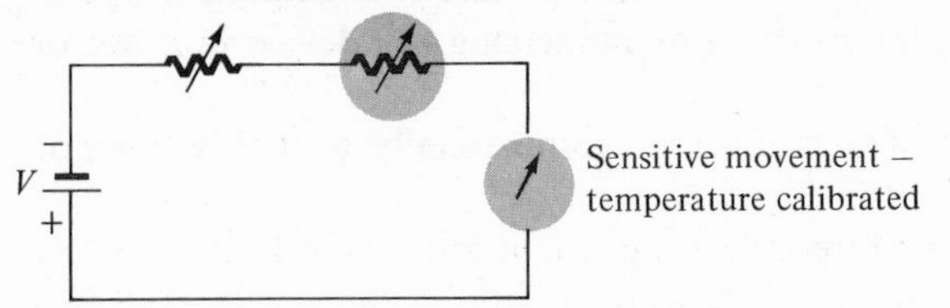

FIG. 1.89

Temperature-indicating circuit.

deflection, which when properly calibrated will accurately indicate the higher temperature. The variable resistance was added to compensate for battery voltage and circuit element fluctuations.

PROBLEMS

§ 1.2

1. Describe the characteristics of an ideal diode.
2. (a) Draw the waveform across the resistor of Fig. 1.2 for an input sinusoidal signal of 120 V rms. Indicate peak voltage magnitude on the waveform.
   (b) Repeat for the case where the diode is connected opposite to that shown in Fig. 1.2.
3. Draw the output waveform for the input signal and rectifier circuit of Fig. 1.90.

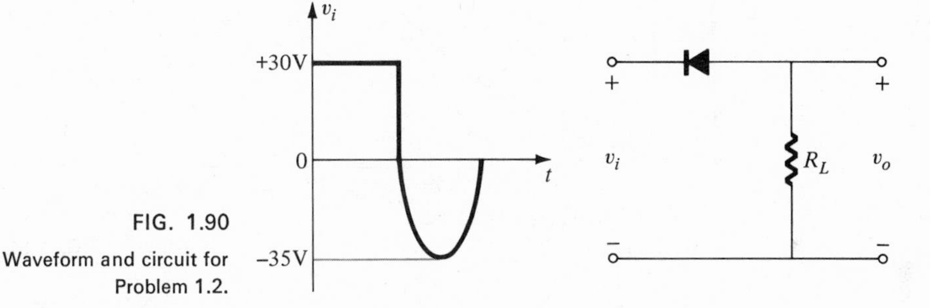

FIG. 1.90
Waveform and circuit for Problem 1.2.

§ 1.3

4. How does a vacuum-tube diode compare to an ideal diode?
5. Is it necessary to apply a heater (filament) voltage to operate a vacuum-tube diode?
6. What would be the effect of reducing the filament voltage on a vacuum tube's operation?

§ 1.4

7. What are the differences between a cold cathode gas diode, a hot cathode gas diode and a vacuum-tube diode?
8. How does a gas diode compare to the ideal diode?
9. For the circuit of Fig. 1.91, calculate the voltage across the resistor load for input voltages of magnitude 10, 15, and 20 V.

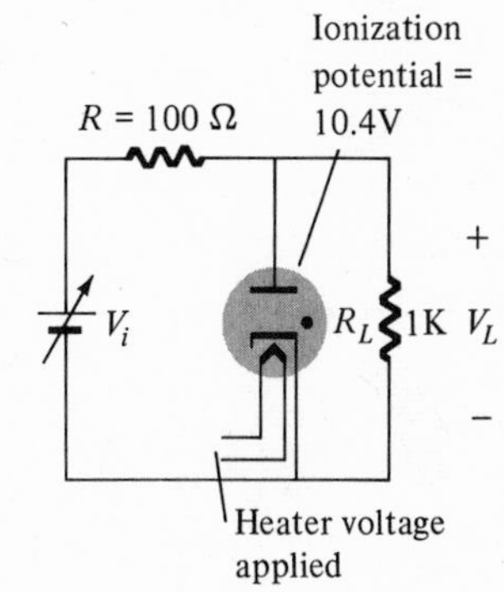

FIG. 1.91
Circuit for Problem 1.9.

§ 1.5

10. Describe the difference between *n*-type and *p*-type semiconductor material.

11. Describe the difference between donor and acceptor impurities.

12. Describe the difference between majority and minority carriers.

13. (a) What is meant by forward bias in a diode?
(b) What is meant by reverse bias in a diode?

§ 1.6

14. Draw a load line on the characteristics of Fig. 1.92 for a circuit as in Fig. 1.41 having values of $V = 25$ V and $R = 10$ K. What is the quiescent operating point for this load line for each diode of Fig. 1.92?

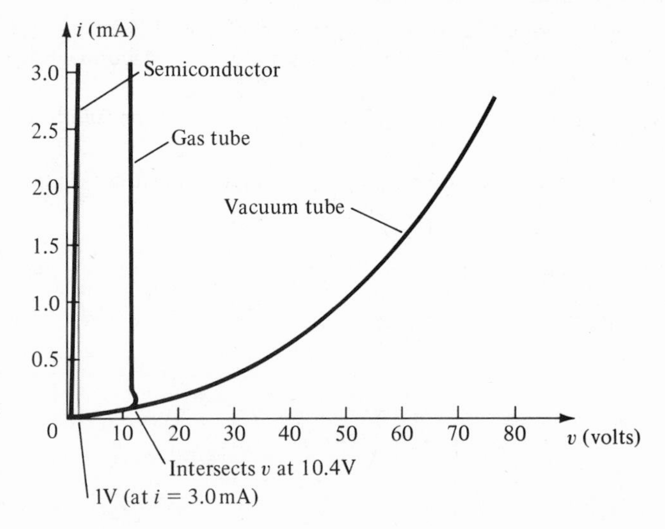

FIG. 1.92

§ 1.7

15. For the diode characteristics of Fig. 1.92 determine the values of dc resistance (R_{dc}) at the quiescent points determined in Problem 1.14.

§ 1.8

16. At the quiescent points of Prob. 1.14, calculate the ac resistance (r_{ac}), (selecting a reasonable interval around the quiescent point).

§ 1.9

17. Calculate the average ac resistance (r_d) for each curve of Fig. 1.92.

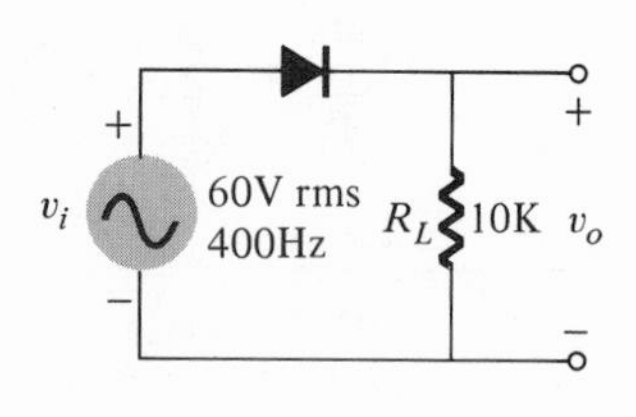

FIG. 1.93
Circuit for Problem 1.18.

§ 1.10

18. A vacuum-tube diode is used in the circuit of Fig. 1.93. If the diode equivalent circuit resistance (r_d) is 4.6 K determine the output voltage signal for one full cycle. Draw the output waveform obtained.

19. If the diode used in Problem 1.18 is a gas diode ($V_f = 10.4$ V) draw the output waveform obtained.

20. If the diode in Problem 1.18 is a semiconductor diode having an average forward resistance of 120 Ω draw the output voltage waveform ($V_o = 0.3$ V).

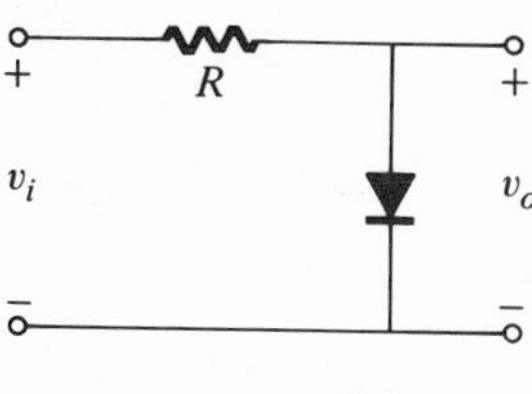

FIG. 1.94

§ 1.11

21. Assuming an ideal diode in the circuit of Fig. 1.94 determine the output waveform for each of the input signals of Fig. 1.95.
22. Repeat Problem 1.21 with the diode of Fig. 1.94 reversed.

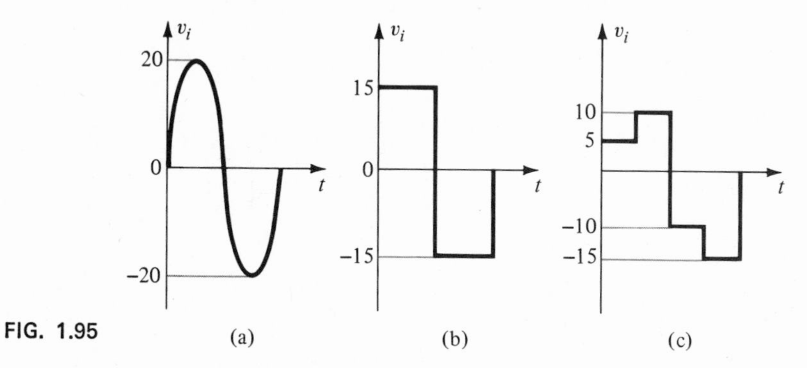

FIG. 1.95

23. Draw the output waveform for the circuit of Fig. 1.96a for each of the input signals of Fig. 1.95.
24. Draw the output waveform for the circuit of Fig. 1.96b for each of the input signals of Fig. 1.95. Use $V = 5$ V.
25. Repeat Problem 1.24 for $V = -10$ V.
26. Draw the output waveforms for the circuit of Fig. 1.97a for the input of Fig. 1.95b.
27. Draw the output waveforms for the circuit of Fig. 1.97b for the input Fig. 1.95b. Use $V = 5$ V.

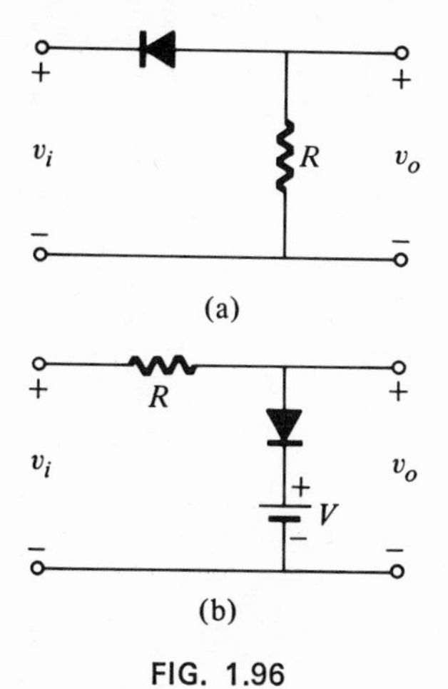

FIG. 1.96

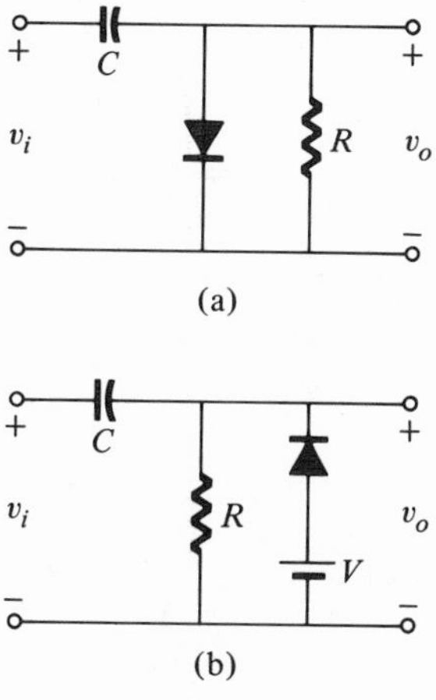

(a)

(b)

FIG. 1.97

28. Repeat Problem 1.27 for $V = -10$ V.

§ 1.12

29. Calculate the voltage across the load resistor (R_L) and the circuit currents for the conditions of input voltage at -10 V and at -14 V (Fig. 1.98).

§ 1.13

30. How does a tunnel diode differ from a semiconductor junction diode?

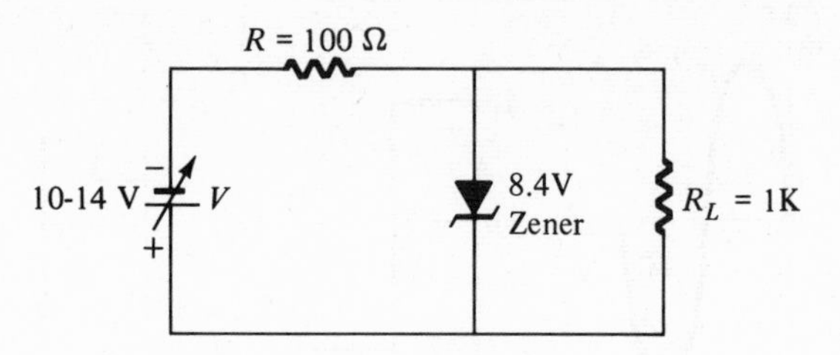

FIG. 1.98

2

Diode Rectifiers and Filters

2.1 DIODE RECTIFICATION

A rectifying circuit converts ac voltage into pulsating dc voltage. For example, a rectifying circuit will convert the 60-Hz ac voltage obtained from the power line from one having an average voltage of zero to one which has an average value.

Figure 2.1 shows a half-wave rectifier circuit using an ideal diode and the resulting half-wave rectified output voltage developed. When the input ac voltage is positive (positive to negative is measured from top to bottom of the voltage generator) the polarity of voltage across the diode will cause the diode to conduct; that is, the voltage across the diode is positive to negative from anode to cathode, and, in the case of the ideal diode, the forward resistance is zero. The positive half-cycle of the input signal then appears across the resistor as shown in Fig. 2.1.

When the input voltage is negative (measured from top to bottom of the generator) the diode is reverse biased, having then infinite resistance and appearing as an open circuit. Since there can be no current flow during the complete time that the voltage at the input causes the diode to be reverse biased, the voltage across the resistor is zero.

The resulting output signal across the resistor due to the half-cycle of diode conduction and the lack of signal during the half-cycle of diode nonconduction is shown in Fig. 2.1. Notice that although this signal is not steady dc (it is pulsating dc) it nevertheless has an average positive value. If the sinusoidal voltage from the power line were applied to a dc voltmeter the reading obtained would be zero. With the pulsating

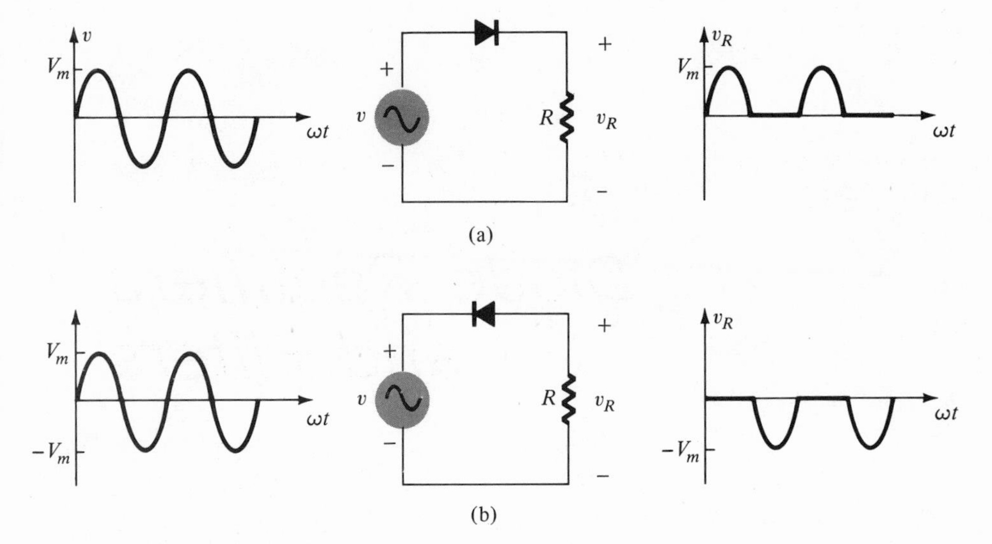

FIG. 2.1
Half-wave rectifier circuits: (a) positive cycle; (b) negative cycle.

dc applied to a dc meter there will be a reading representing the average of the applied signal.

To determine the average value of the rectified signal we can calculate the area under the curve of Fig. 2.2 and divide this value by the

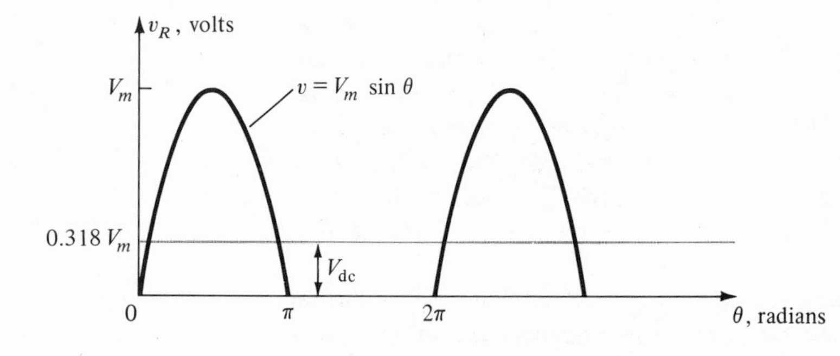

FIG. 2.2
Half-wave rectified voltage showing dc value.

period of the rectified waveform. To calculate the area under the half-cycle curve of the rectified signal we must integrate the rectified signal.[1] Doing this integration procedure (and dividing by the period)

[1] The dc signal can be expressed as $v = V_m \sin\theta$ (see Fig. 2.2). For θ from 0 to 2π rad the average value is calculated to be

$$V_{dc} = V_{av} = \frac{1}{T}\int v\,dt = \frac{1}{2\pi}\int_0^{\pi}(V_m \sin\theta)\,d\theta$$

$$V_{dc} = \frac{V_m}{2\pi}[-\cos\theta]_0^{\pi} = \frac{V_m}{2\pi}[-1(-1)-(-1)] = \frac{V_m}{\pi} = 0.318\,V_m$$

results in

$$\boxed{V_{dc} = 0.318V_m} \tag{2.1}$$

where V_m = maximum (peak) value of ac voltage, and
V_{dc} = average value of rectified voltage.

EXAMPLE 2.1 Calculate the dc (average) value of the rectified half-wave signal obtained by rectifying a signal of 240 V, peak value.

Solution

$$V_{dc} = 0.318V_m = 0.318(240) = \mathbf{76.3\ V}$$

EXAMPLE 2.2 Calculate the average voltage of the rectified signal obtained from the circuit of Fig. 2.3.

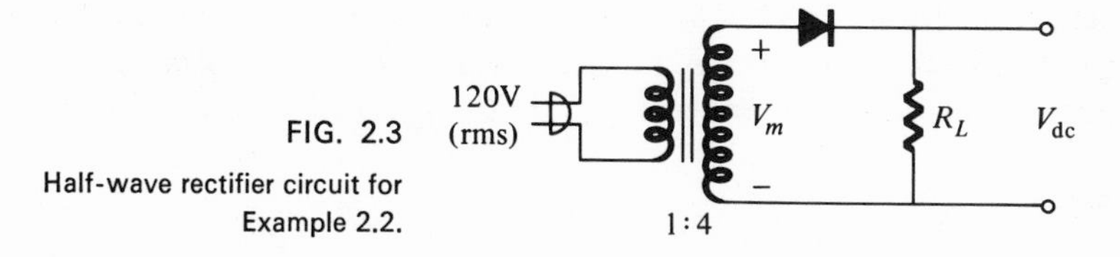

FIG. 2.3
Half-wave rectifier circuit for Example 2.2.

Solution

$$V_m = \frac{4}{1}(1.414 \times 120) = 678.8\ \text{V}$$

$$V_{dc} = 0.318V_m = (0.318)(678.8) = \mathbf{215.9\ V}$$

The important diode ratings for a half-wave rectifier circuit (as in Fig. 2.1) include the maximum forward current, I_{max} (current flowing in direction of diode arrow), and the peak inverse voltage, PIV (the maximum voltage across the diode in the direction to block current flow). For the half-wave rectifier circuit of Fig. 2.4 the peak voltage across the diode when the diode is reverse biased is shown to be V_m in value.

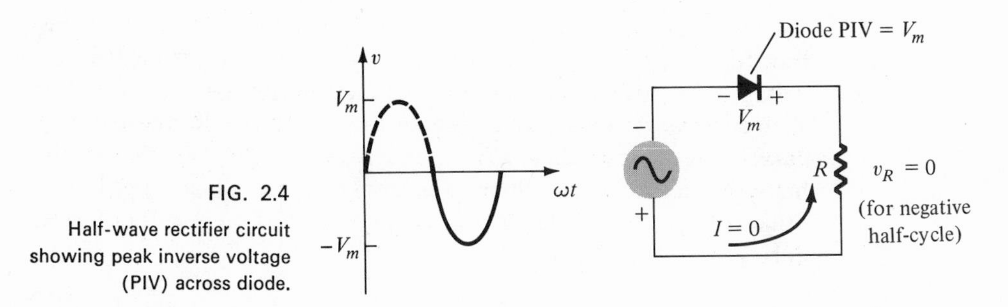

FIG. 2.4
Half-wave rectifier circuit showing peak inverse voltage (PIV) across diode.

In Example 2.2 the peak inverse voltage was $V_m = 678.8$ V. The dc voltage obtained, however, was only $V_{dc} = 215.9$ V. This clearly points out one poor feature of the half-wave circuit, namely, the diode PIV rating must be considerably larger than the dc voltage obtained using the circuit.

The forward current rating of the diode must equal, at least, the average current through it — V_{dc}/R. The peak current through the diode is V_m/R and must be less than the peak current rating for the diode.

2.2 FULL-WAVE RECTIFICATION

Center-Tapped Full-Wave Rectifier

It would be preferable to obtain a larger dc voltage compared to the maximum voltage than that of 0.318 V_m for a half-wave rectified signal. In addition, we note that although an average voltage is obtained using a half-wave rectifier no voltage is developed for half of the cycle. Using two diodes, as shown in Fig. 2.5, it is possible to rectify a sinu-

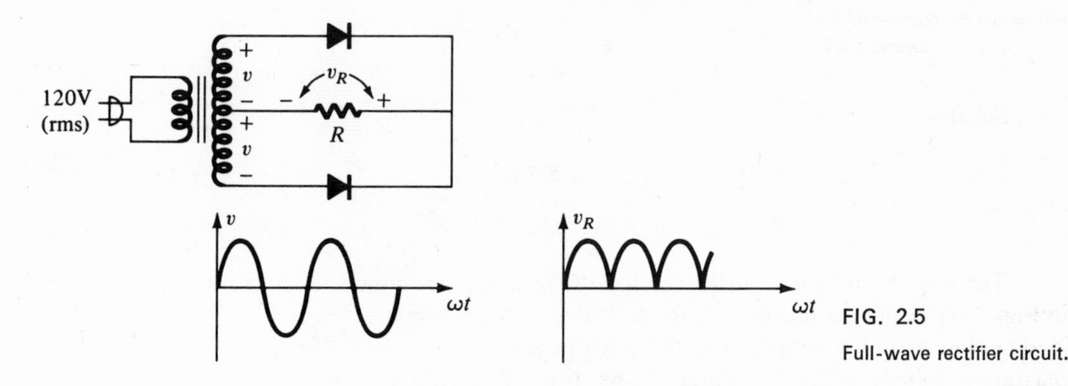

FIG. 2.5
Full-wave rectifier circuit.

soidal signal to obtain one having the same polarity half-cycle for *each* of the half-cycles of input signal. This *full-wave* rectified signal provides a "nicer"-looking dc signal and is twice the dc value of the comparable half-wave rectified signal.

The full-wave rectifier circuit of Fig. 2.5 requires a center-tapped transformer and two diodes to develop a full-wave rectified output voltage. To understand how the output waveform is developed we shall consider the detailed circuit operation for each half-cycle of secondary voltage. Figure 2.6a, shows the circuit operation for the positive half-cycle of secondary voltage. The transformer is center tapped and a peak voltage, V_m, is developed across each half of the transformer during the positive cycle.

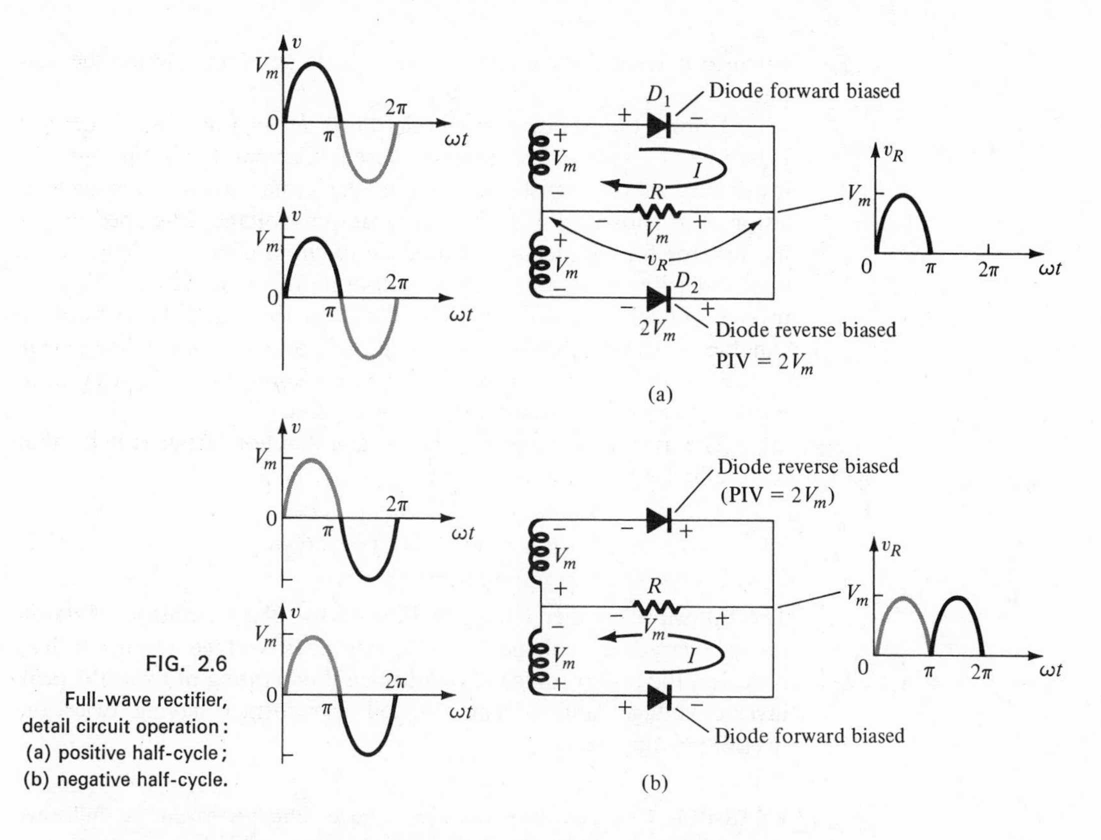

FIG. 2.6
Full-wave rectifier, detail circuit operation: (a) positive half-cycle; (b) negative half-cycle.

During the entire positive half-cycle the polarity of the signal across the upper half of the transformer is in a direction to forward bias diode D_1 causing it to conduct. With diode D_1 conducting, a positive half-cycle of voltage is developed across resistor R as shown in Fig. 2.6a. The figure shows the voltages in the circuit at the time of the peak positive voltage and as shown, there is a voltage V_m across the resistor at this time.

The current in the upper transformer half flows through the transformer, diode D_1, and the load resistor. For a perfect diode ($V_D =$ 0, when conducting) the voltage across the resistor will equal that of the transformer. At the time the transformer voltage is V_m the voltage across the resistor is also V_m in magnitude as shown in Fig. 2.6a. The voltage developed across the resistor is thus a half-cycle of signal.

The polarity of the voltage developed across the lower half of the transformer results in diode D_2 being back biased. In addition the reverse-bias voltage across the diode, which is maximum at the time the maximum voltage V_m is present, is $2V_m$. We can see that this is so by noting that the voltage across reverse-biased diode D_2 is equal to the sum of the voltages across the lower half of the transformer and the load resistor, these voltages being of the same polarity. A diode in this circuit must therefore be capable of handling a reverse-bias voltage equal to twice the value of the peak voltage developed across the output.

During the negative half-cycle diode D_2 in Fig. 2.6b is forward biased, and diode D_1 is reverse biased. Current flows through the lower half of the transformer but in the same direction through resistor R as shown in Fig. 2.6b. The output voltage developed across the resistor for the negative half-cycle of input signal is, then, of the same polarity as for the positive half-cycle of input signal. The peak inverse voltage across diode D_1 is $2V_m$, so that each diode must be capable of withstanding a reverse bias voltage of $2V_m$ sometime during a cycle of operation. The resulting output voltage for a full cycle of input voltage is two positive-going half-cycles.

The average voltage for a full-wave rectified signal is twice that for the half-wave rectified, so that

$$\boxed{V_{dc} = 2(0.318V_m) = 0.636V_m} \tag{2.2}$$

The full-wave rectifier circuit of Fig. 2.5 has the advantage of developing a larger dc voltage for the same peak voltage rating. It has, however, the disadvantage of requiring a diode rating of twice the peak inverse voltage, and a center-tapped transformer having twice the overall voltage rating.

EXAMPLE 2.3 Calculate the dc voltage obtained from a full-wave rectifier for which the peak rectified voltage is 100 V.

Solution

$$V_{dc} = 0.636V_m = 0.636(100) = \mathbf{63.6\ V}$$

EXAMPLE 2.4 Calculate the value of the diode PIV rating necessary for a center-tapped full-wave rectifier developing a dc voltage of 75 V.

Solution

$$V_{dc} = 0.636V_m$$

$$V_m = \frac{V_{dc}}{0.636} = \frac{75}{0.636} = 117.9\ \text{V}$$

$$\text{diode PIV} = 2V_m = 2(117.9) = \mathbf{235.8\ V}$$

Bridge Rectifier Circuit

Another circuit variation of a full-wave rectifier is the bridge circuit of Fig. 2.7. The circuit requires four diodes instead of two for

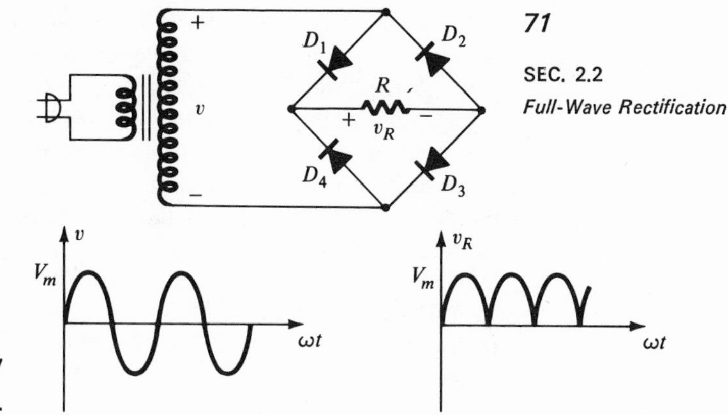

FIG. 2.7
Full-wave bridge rectifier circuit.

full-wave rectification but the transformer required is not center tapped and must develop only V_m as the maximum voltage across it. In addition the diode rating for peak inverse voltage will be shown to be only V_m, rather than $2V_m$.

In considering how the circuit operates we must understand how the conduction and nonconduction paths are formed during each half of the ac cycle. During the positive half-cycle the voltage across the transformer (measured from top to bottom) is positive and the conduction path is shown in Fig. 2.8. Figure 2.8a shows the voltages at

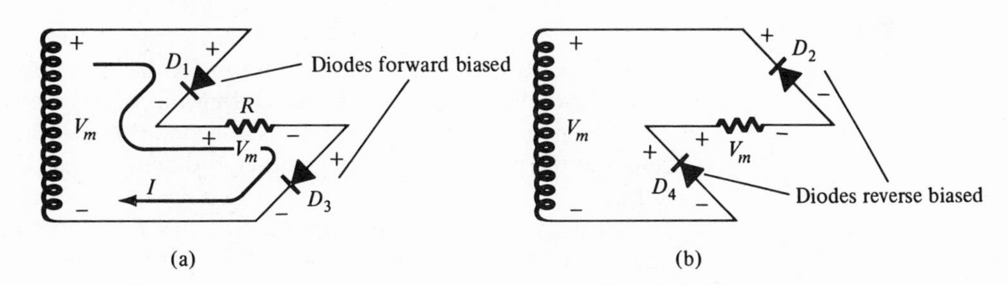

FIG. 2.8
Bridge circuit, positive half-cycle operation: (a) conduction path; (b) nonconduction path.

the time of the peak positive voltage, V_m. Since the diodes shown are forward biased the voltage drop across each is 0 V and the peak voltage from the transformer appears across resistor, R, at this time.

At the same time the voltage polarity is such as to reverse bias diodes D_2 and D_4, as shown in Fig. 2.8b. This represents the nonconduction path during the positive half-cycle of the input ac signal. Resistor R has a voltage developed across it by the current flowing through the conducting path of diodes D_1 and D_3. Summing the voltage drops around the nonconducting loop, the transformer voltage and resistor voltage at the time of the peak voltage add up to $2V_m$. Since there are two diodes in the path, the voltage across each reverse-biased diode is V_m. This is half the developed peak inverse voltage across the diodes in the previous full-wave rectifier circuit (Fig. 2.5).

During the negative half-cycle the conduction and nonconduction paths are shown in Fig. 2.9. Figure 2.9a shows that diodes D_4 and D_2

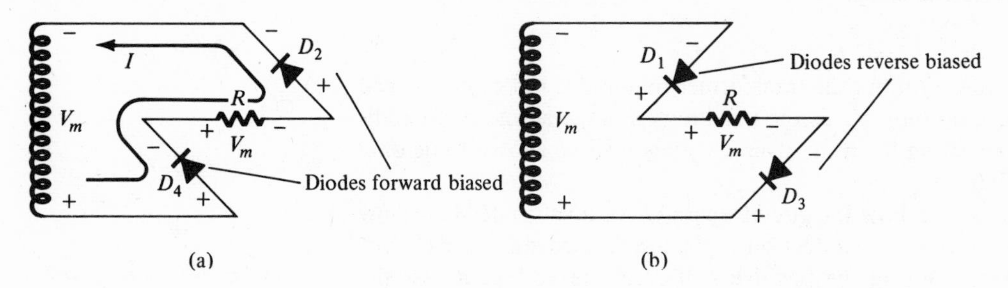

FIG. 2.9
Bridge circuit, negative half-cycle operation: (a) conduction path; (b) nonconduction path.

are forward biased. Note carefully that the current, I, goes through resistor R in the same direction as did the current on the previous half-cycle. The voltage across resistor R is thus of the same polarity during each half-cycle of the input signal. During the negative-polarity half-cycle the path of diodes D_1 and D_3 is nonconducting as shown in Fig. 2.9b and the peak inverse voltage developed across each of the diodes is V_m.

To summarize the operation of the bridge rectifier circuit, the addition of two extra diodes above the number in the center-tapped full-wave circuit provides improvement in two main factors. One, the transformer used need not be center tapped and requiring in this case, a maximum voltage across the transformer of V_m. Two, the peak inverse voltage (PIV) required of each diode is half that for the center-tapped full-wave circuit only; V_m, in magnitude, in this case. For low values of secondary maximum voltage the center-tapped full-wave circuit will be acceptable, whereas for high values of maximum secondary voltage the use of the bridge to reduce the maximum transformer rating and diode PIV rating is usually necessary.

EXAMPLE 2.5 The dc voltage developed by a full-wave bridge rectifier circuit is 325 V. Calculate the diode peak inverse voltage rating required for the diodes selected for this circuit.

Solution

$$V_m = \frac{V_{dc}}{0.636} = \frac{325}{0.636} = 511 \text{ V}$$

For the bridge rectifier the value of diode PIV is V_m so that diode PIV $= V_m =$ **511 V.**

2.3 GENERAL FILTER CONSIDERATIONS

A rectifier circuit is necessary to convert a signal having no average value to one that has an average. However, the resulting signal is pulsating dc and is not true dc or even a good representation of it. Of course, for a circuit such as a battery charger the pulsating nature of the signal is no great detriment as long as the dc level provided will result in charging of the battery. On the other hand, for voltage supply circuits for a tape recorder or radio the pulsating dc will result in a 60- (or 120-) Hz signal appearing in the output, thereby making the operation of the overall circuit quite poor. For these applications, as well as for many more, the output dc developed will have to be much "smoother" than that of the pulsating dc obtained directly from half-wave or full-wave rectifier circuits.

A number of different types of filter or smoothing circuits will be considered in this chapter. These will include the popular simple-capacitor filter, the RC filter, the choke filter, the LC filter, and the π-type filter circuits. Voltage regulator circuits using Zener diodes and transistors will be covered in Chapter 14.

Filter Voltage Regulation and Ripple Voltage

Before going into any of the filter circuits it would be appropriate to consider the usual method of rating the circuits so that we are able to compare a circuit's effectiveness as a filter. Figure 2.10 shows a typical filter output voltage plot, which will be used to define some of the signal factors. The filtered output voltage of Fig. 2.10 has a dc value and some ac variation (*ripple*). Although a battery has essentially a constant or dc output voltage, the dc voltage derived from an ac source signal by rectifying and filtering will have some instantaneous variation (ripple). The smaller the ac variation *with respect to* the dc level the better the filter circuit operation.

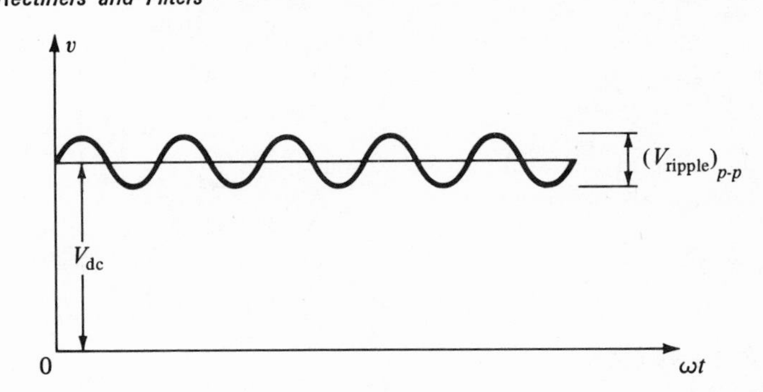

FIG. 2.10
Filter voltage waveform showing dc and ripple voltages.

Consider measuring the output voltage of the filter circuit using a dc voltmeter and an ac (rms) voltmeter. The dc voltmeter will read only the average or dc level of the output voltage. The ac (rms) meter will read only the rms value of the ac component of the output voltage (assuming the signal is coupled to the meter through a capacitor to block out the dc level).

DEFINITION: Ripple factor:

$$r = \text{ripple factor} \equiv \frac{\text{ripple voltage (rms)}}{\text{dc voltage}} = \frac{V_r\,(\text{rms})}{V_{dc}} \tag{2.3a}$$

DEFINITION: Percent of Ripple:

$$\%r = \%\text{ripple} \equiv \frac{V_r\,(\text{rms})}{V_{dc}} \times 100 \tag{2.3b}$$

EXAMPLE 2.6 Using a dc and ac voltmeter to measure the output signal from a filter circuit, a dc voltage of 25 V and an ac ripple voltage of 2.5 V (rms) are obtained. Calculate the ripple of the filter output.

Solution

$$r = \text{ripple factor} = \frac{V_r\,(\text{rms})}{V_{dc}} = \frac{2.5}{25} = 0.1$$

$$\%r = \%\ \text{Ripple} = \frac{V_r\,(\text{rms})}{V_{dc}}(100) = 0.1(100) = \mathbf{10\%}$$

Another factor of importance in a voltage supply is the amount of change in the output dc voltage over the range of the circuit operation. The voltage provided at the output at no-load (no current drawn from the supply) is reduced when load current is drawn from the supply. How much this voltage changes with respect to either the loaded or unloaded voltage value is of considerable interest to anyone using the supply. This voltage change is described by a factor called *voltage regulation*.

DEFINITION: Voltage Regulation:

$$\text{voltage regulation} \equiv \frac{\text{voltage at no-load} - \text{voltage at full-load}}{\text{voltage at full-load}}$$

$$V.R. = \frac{V_{NL} - V_{FL}}{V_{FL}} \tag{2.4a}$$

DEFINITION: Percent of Voltage Regulation:

$$\%\,V.R. = \%\text{ voltage regulation} \equiv \frac{V_{NL} - V_{FL}}{V_{FL}} \times 100 \tag{2.4b}$$

EXAMPLE 2.7 A dc voltage supply provides 60 V when the output is unloaded. When full-load current is drawn from the supply the output voltage drops to 50 V. Calculate the values of voltage regulation and percent of voltage regulation.

Solution

$$V.R. = \frac{V_{NL} - V_{FL}}{V_{FL}} = \frac{60 - 50}{50} = 0.20$$

$$\%\,V.R. = V.R. \times 100 = 0.20 \times 100 = \mathbf{20\%}$$

If the value of full-load voltage is the same as the no-load voltage, the *V.R.* calculated is 0%, which is the best to expect. This value means that the supply is a true-voltage source for which the output voltage is independent of current drawn from the supply. Actually, all voltage supplies drop in voltage as the amount of current drawn from the voltage supply is increased. The smaller this voltage drop the smaller the percent of *V.R.* and the better the operation of the voltage supply circuit.

Although the rectified voltage is not a filtered voltage it nevertheless contains a dc component and a ripple component. We can calculate these values of dc voltage and ripple voltage (rms) and from them obtain the ripple factor for the half-wave and full-wave rectified voltages. The calculations will show that the full-wave rectified signal has less percent of ripple and is therefore a better rectified signal than the half-wave rectified signal, if lowest percent of ripple is desired. The percent of ripple is not always the most important concern. If circuit complexity or cost considerations are important (and the percent of ripple is secondary) then a half-wave rectifier may be satisfactory. Also, if the filtered output supplies only a small amount of current to the load and the filtering circuit is not critical then a half-wave rectified signal may be acceptable. On the other hand, when the supply must have as low a ripple as possible, it is best to start with a full-wave rectified signal since it has a smaller ripple factor, as will now be shown.

For the half-wave rectified signal the output dc voltage is $V_{dc} = 0.318V_m$. The rms value of the ac component of output signal can be calculated (see Appendix B), and is $V_r\,(\text{rms}) = 0.385V_m$. Calculating the per cent of ripple,

$$\%\ \text{ripple} = \frac{V_r\,(\text{rms})}{V_{dc}}(100) = \frac{0.385V_m}{0.318V_m}(100)$$
$$= 1.21(100) = 121\%\ \text{(half-wave)}$$

For the full-wave rectifier the value of V_{dc} is $V_{dc} = 0.636V_m$. From the results obtained in Appendix B the ripple voltage of a full-wave rectified signal is $V_r(\text{rms}) = 0.305V_m$. Calculating the percent of ripple,

$$\%\ \text{ripple} = \frac{V_r\,(\text{rms})}{V_{\text{dc}}}(100) = \frac{0.305V_m}{0.636V_m}(100)$$
$$= 48\%\ \text{(full-wave)}$$

The amount of ripple factor of the full-wave rectified signal is about 2.5 times smaller than that of the half-wave rectified signal and provides a better filtered signal. Note that these values of ripple factors are absolute values and do not depend at all on the peak voltage. If the peak voltage is made larger the dc value of the output increases but then so does the ripple voltage. The two increase in the same proportion so that the ripple factor stays the same.

A popular filter circuit is the simple-capacitor filter circuit shown in Fig. 2.11. The capacitor is connected across the rectifier output and the

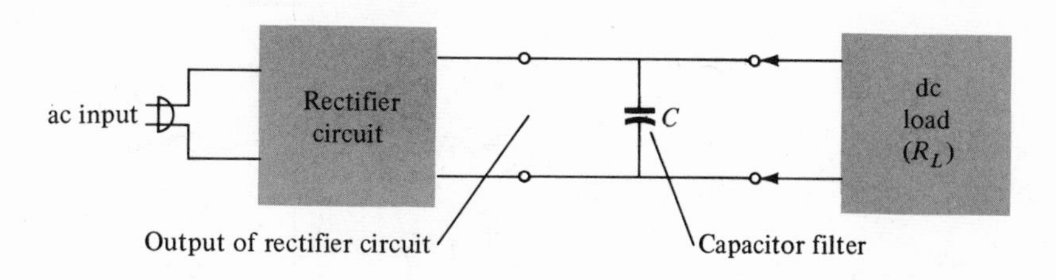

FIG. 2.11
Simple-capacitor filter.

dc output voltage is available across the capacitor. Figure 2.12a shows the rectifier output voltage of a full-wave rectifier circuit before the signal is filtered. Figure 2.12b shows the resulting waveform after the capacitor is connected across the rectifier output. As shown this filtered voltage has a dc level with some ripple voltage riding on it.

Figure 2.13a shows a full-wave rectifier and the output waveform obtained from the circuit when connected to an output load. If no

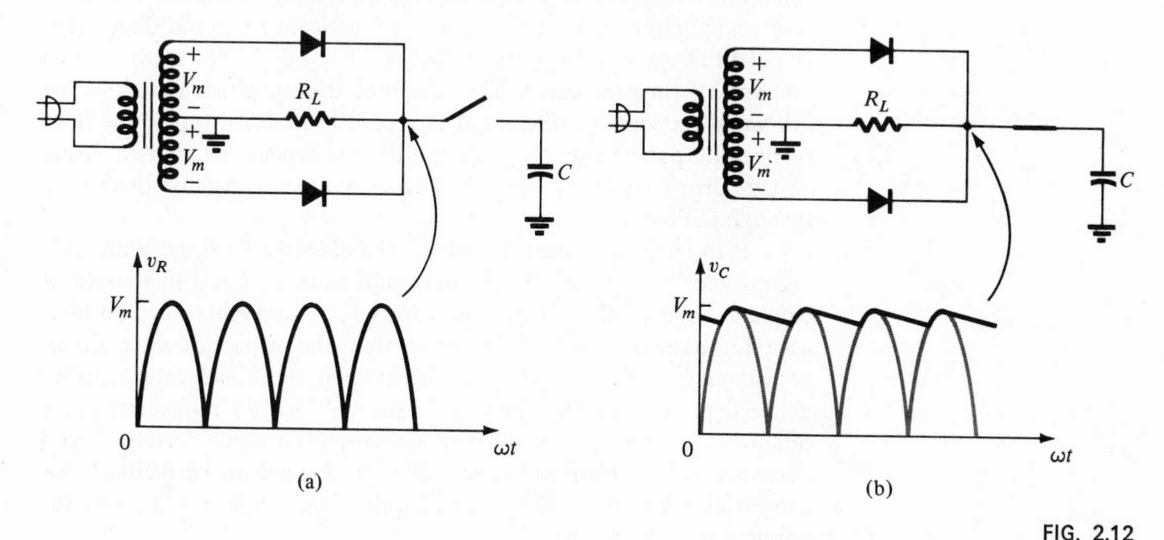

FIG. 2.12
Capacitor filter operation: (a) full-wave rectifier voltage; (b) filtered output voltage.

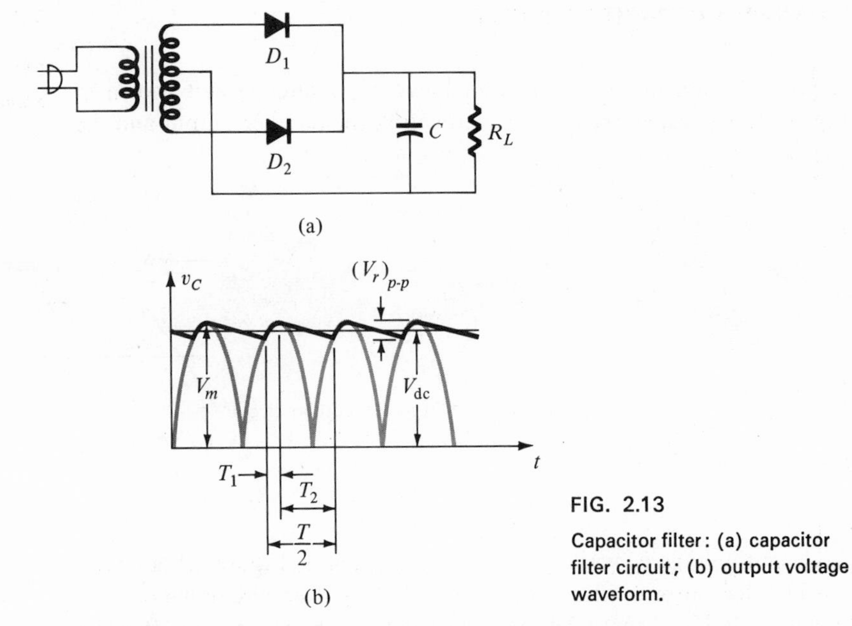

FIG. 2.13

Capacitor filter: (a) capacitor filter circuit; (b) output voltage waveform.

load were connected to the filter the output waveform would ideally be a constant dc level equal in value to the peak voltage (V_m) from the rectifier circuit. However, the purpose of obtaining a dc voltage is to provide this voltage for use by other electronic circuits, which then constitute a load on the voltage supply. Since there will always be some load on the filter we must consider this practical case in our discussion. For the full-wave rectified signal indicated in Fig. 2.13b, there are two intervals of time indicated. T_1 is the time during which a diode of the full-wave rectifier conducts and charges the capacitor up to the peak rectifier output voltage (V_m). T_2 is the time during which the rectifier voltage drops below the peak voltage, and the capacitor discharges through the load.

If the capacitor were to discharge only slightly, the average voltage would be very close to the optimum value of V_m. The amount of ripple voltage would be very slight for a light load and quite considerable for a heavy load. This shows that the capacitor filter circuit provides a large dc voltage with little ripple for light loads and a smaller dc voltage with larger ripple for heavy loads. To appreciate these quantities better we must further examine the output waveform and determine some relations between the input signal to be rectified, the capacitor value, the resistor (load) value, the ripple factor, and the regulation of the circuit.

Figure 2.14 shows the output waveform approximated by straight line charge and discharge. This is reasonable since the analysis with the nonlinear charge and discharge that actually takes place is complex to analyze and because the results obtained will yield values that agree

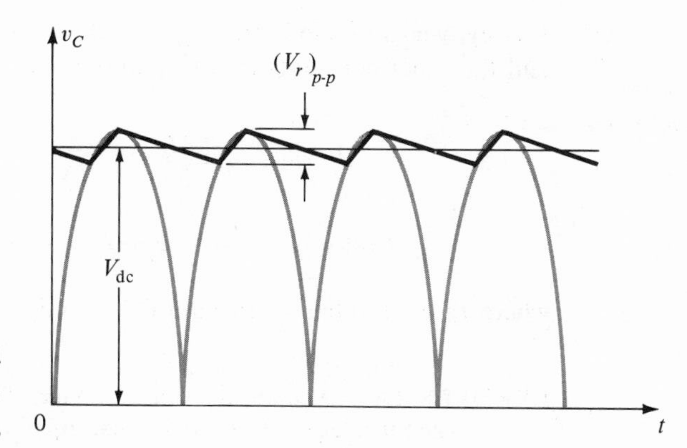

FIG. 2.14
Approximate output voltage of capacitor filter circuit.

well with actual measurements made on circuits. The waveform of Fig. 2.14 shows the approximate output voltage waveform for a full-wave rectified signal. From an analysis of this voltage waveform the following relations are obtained:

$$V_{dc} = V_m - \frac{V_r(p-p)}{2} \tag{2.5}$$

$$V_r\,(\text{rms}) = \frac{V_r(p-p)}{2\sqrt{3}} \tag{2.6}$$

These relations, however, are only in terms of the waveform voltages and we must further relate them to the different components in the circuit.

Ripple Voltage, V_r (rms)

Appendix B provides the details for determining the value of the ripple voltage in terms of the other circuit parameters. The result obtained for V_r (rms) is the following:

$$V_r\,(\text{rms}) \cong \frac{I_{dc}}{4\sqrt{3}\,fC} \times \frac{V_{dc}}{V_m} \qquad \text{(full-wave)} \tag{2.7a}$$

where f is the frequency of the sinusoidal ac power supply voltage (usually 60 Hz), I_{dc} is the average current drawn from the filter by the load, and C is the filter capacitor value.

Another simplifying approximation that can be made is to assume that when used typically for light loads the value of V_{dc} is only slightly less than V_m so that $V_{dc} \cong V_m$, and the equation can be written as

$$V_r\,(\text{rms}) \cong \frac{I_{dc}}{4\sqrt{3}\,fC} \qquad \text{(full-wave, light load)} \tag{2.7b}$$

Finally, we can include the typical value of line frequency ($f = 60$ Hz) and the other constants into the simpler equation

$$V_r\,(\text{rms}) = \frac{2.4 I_{dc}}{C}$$

(full-wave capacitor filter circuit, light load) (2.7c)

where I_{dc} is in milliamperes and C is in microfarads.

EXAMPLE 2.8 Calculate the ripple voltage of a full-wave rectifier with a capacitor filter of 20 μF and a load of 40 mA.

Solution: Using Eq. (2.7c),

$$V_r\,(\text{rms}) = \frac{2.4(40)}{20} = \mathbf{4.8\ V}$$

dc Voltage, V_{dc}

Using Eqs. (2.5), (2.6) and (2.7a) the dc voltage of the filter is

$$V_{dc} = V_m - \frac{V_r(p-p)}{2} = V_m - \frac{I_{dc}}{4fC} \times \frac{V_{dc}}{V_m}$$

and

$$V_{dc} = \frac{V_m}{1 + I_{dc}/4fCV_m} \quad \text{(full-wave)} \qquad (2.8a)$$

Again, using the simplifying assumption that V_{dc} is about the same as V_m for light loads we get an approximate value of V_{dc} (which is less than V_m), of

$$V_{dc} = V_m - \frac{I_{dc}}{4fC} \quad \text{(full-wave, light load)} \qquad (2.8b)$$

which can be written (using $f = 60$ Hz):

$$V_{dc} = V_m - \frac{4.16 I_{dc}}{C}$$

(full-wave capacitor filter circuit, light load) (2.8c)

where V_m is the peak rectified voltage, in volts; I_{dc} is the load current in milliamperes, and C is the filter capacitor in microfarads.

EXAMPLE 2.9 If the peak rectified voltage for the filter circuit of Example 2.8 is 100 V, calculate the filter dc voltage.

Solution: Using Eq. (2.8c):

$$V_{dc} = V_m - \frac{4.16 I_{dc}}{C} = 100 - \frac{4.16(40)}{20} = 100 - 8.32$$

$$V_{dc} = \mathbf{91.68\ V}$$

The value of dc voltage is less than the peak rectified voltage. Note, also, from Eq. (2.8c), that the larger the value of average current drawn from the filter the less the value of output dc voltage, and the larger the value of the filter capacitor, the closer the output dc voltage approaches the peak value of V_m.

Filter Capacitor Ripple

Using the definition of ripple [Eq. (2.3)] and the equation for ripple voltage [Eq. (2.7c)] we obtain the expression for the ripple factor of a full-wave capacitor filter

$$\boxed{r = \frac{V_r\,(\text{rms})}{V_{dc}} \cong \frac{2.4 I_{dc}}{C V_{dc}}} \quad \text{(full-wave, light load)} \quad (2.9a)$$

Since V_{dc} and I_{dc} relate to the filter load R_L we can also express the ripple as

$$\boxed{r = \frac{2.4}{R_L C}} \quad \text{(full-wave, light load)} \quad (2.9b)$$

where I_{dc} is in milliamperes, C is in microfarads, V_{dc} is in volts, and R_L is in kilohms.

This ripple factor is seen to vary directly with the load current (larger load current, larger ripple factor), and inversely with the capacitor size. This agrees with the previous discussion of the filter circuit operation.

EXAMPLE 2.10 A capacitor filter circuit ($C = 2\ \mu F$) provides 15-mA load current. If the peak rectified voltage is 185 V, calculate:

(a) V_r (rms)
(b) V_{dc}
(c) $\% r$

Solution

(a) $$V_r\,(\text{rms}) = \frac{2.4I_{\text{dc}}}{C} = \frac{2.4(15)}{2} = \mathbf{18\ V}$$

(b) $$V_{\text{dc}} = V_m - \frac{4.16I_{\text{dc}}}{C} = 185 - \frac{4.16(15)}{2} = 185 - 31.2 = \mathbf{153.8\ V}$$

(c) $$\%\,r = \frac{2.4I_{\text{dc}}}{CV_{\text{dc}}} \times 100 = \frac{2.4(15)}{(2)(153.8)} \times 100 = \mathbf{11.7\%}$$

Using the basic definition of $\%\,r$,

$$\%\,r = \frac{V_r\,(\text{rms})}{V_{\text{dc}}}(100) = \frac{18}{153.8}(100) = 11.7\%$$

Since the problem is often that of selecting a capacitor for a filter circuit an example will now follow showing how the value of C can be determined.

EXAMPLE 2.11 Calculate the required capacitor filter for a full-wave rectifier of 120 V, peak voltage derived from the 60-Hz supply line through a transformer. The filter is to provide a load current of 10 mA with a ripple of 8%.

Solution: Assuming that $V_m \cong V_{dc} = 120$ V the ripple voltage can be calculated:

$$V_r\,(\text{rms}) = rV_{\text{dc}} = (0.08)(120) = 9.6\ \text{V}$$

Using Eq. (2.7c) to solve for the value of C:

$$C = \frac{2.4I_{\text{dc}}}{V_r\,(\text{rms})} = \frac{2.4(10)}{9.6} = \mathbf{2.5\ \mu F}$$

EXAMPLE 2.12 A capacitor filter is to be used to provide 5-mA dc to a 10-KΩ load with ripple of less than 10%. Calculate the value of the capacitor needed.

Solution

$$V_{\text{dc}} = I_{\text{dc}}R_L = (5\ \text{mA})(10\ \text{K}) = 50\ \text{V}$$

$$V_r\,(\text{rms}) = (r)(V_{\text{dc}}) = (0.10)(50) = 5\ \text{V}$$

$$C = \frac{2.4I_{\text{dc}}}{V_r\,(\text{rms})} = \frac{(2.4)(5)}{5} = \mathbf{2.4\ \mu F}$$

Diode Conduction Period and Peak Diode Current

From the previous discussion it should be clear that larger values of capacitance provide less ripple and higher average voltages, thereby

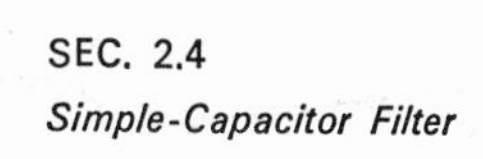

providing better filter action. From this one may conclude that to improve the performance of a capacitor filter it is only necessary to increase the size of the filter capacitor. However, the capacitor also affects the peak current through the rectifying diode and, as will now be shown, the larger the value of capacitance used the larger the peak current through the rectifying diode.

Referring back to the operation of the rectifier and capacitor filter circuit there are two periods of operation to consider. After the capacitor is charged to the peak rectified voltage (see Fig. 2.13b) a period of diode nonconduction elapses (time T_2) while the output voltage discharges through the load. After T_2 the input rectified voltage becomes greater than the capacitor voltage and for a time, T_1, the capacitor will charge back up to the peak rectified voltage. The average current supplied to the capacitor during this charge period must equal the average current drawn from the capacitor during the discharge period. Figure 2.15 shows the diode current waveform for half-wave rectifier operation. Notice that the diode conducts for only a short period of the cycle. In fact, it should be seen that the larger the capacitor the less that amount of voltage decay and the shorter the interval during which charging takes place. In this shorter charging interval the diode will have to pass the same amount of *average current*, and can do so only by passing larger peak current. Figure 2.16 shows the output current and voltage waveforms for small and large capacitor values. The important factor to note is the increase in peak current through the

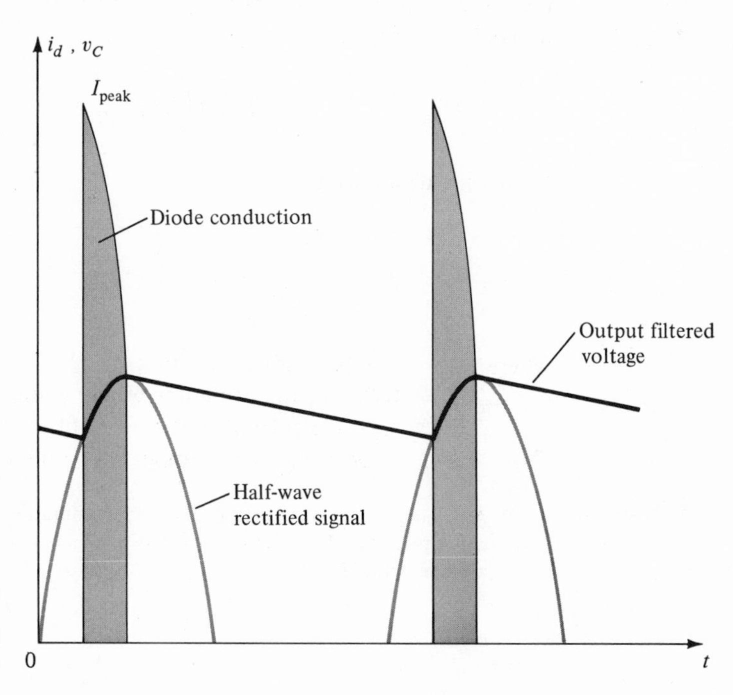

FIG. 2.15
Diode conduction during charging part of cycle.

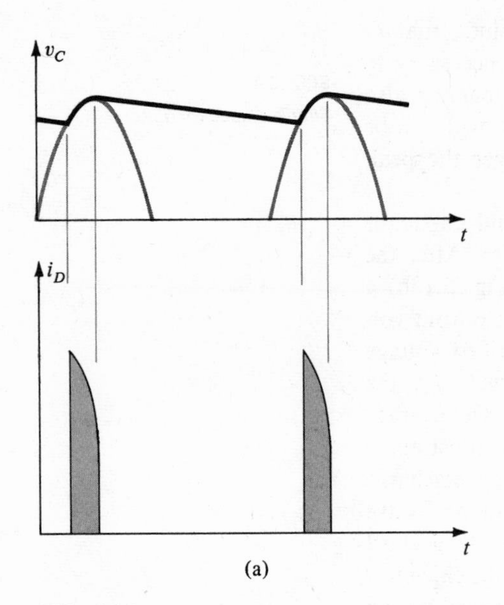

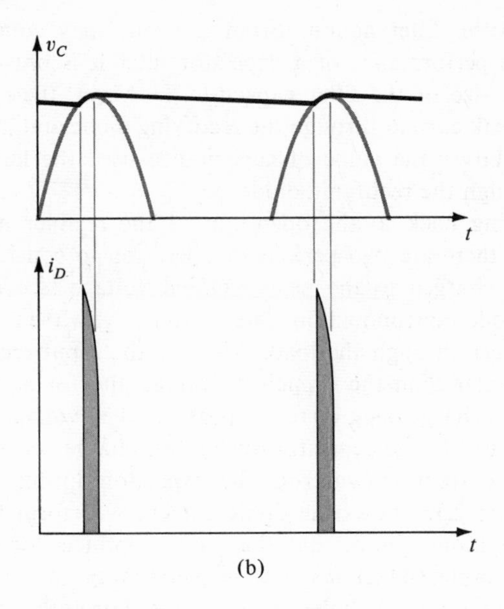

FIG. 2.16

Output voltage and diode current waveforms: (a) small C; (b) large C.

diode for the larger values of capacitance. Since the average current drawn from the supply must equal the average of the current through the diode during the charging period the following relation can be derived from Fig. 2.16*:

$$I_{dc} = \left(\frac{T_1}{T}\right)(I_{peak}) \tag{2.10a}$$

from which we obtain

$$\boxed{I_{peak} = \left(\frac{T}{T_1}\right) I_{dc}} \tag{2.10b}$$

where T_1 = diode conduction time;
$T = 1/f = \frac{1}{60}$ for usual line 60-Hz voltage;
I_{dc} = average current drawn from filter circuit; and
I_{peak} = peak current through the conducting diode.

* Assuming a rectangular-shaped pulse of duration T_1, peak value, I_{peak}, and a period of T, the area under the pulse divided by the period gives the average value, I_{dc}:

$$\frac{1}{T}(I_{peak}T_1) = I_{dc}$$

EXAMPLE 2.13 What peak current will flow through a diode used in the rectifier part of a capacitor filter circuit if the average current drawn from the filter is 50 mA, and the diode conducts for one-quarter of the cycle?

Solution: For conduction of one-quarter of the cycle $T_1/T = \frac{1}{4}$, and solving for the peak current we obtain

$$I_{\text{peak}} = \left(\frac{T}{T_1}\right)(I_{\text{dc}}) = \left(\frac{4}{1}\right) 50 \text{ mA} = \mathbf{200\ mA}$$

SUMMARY OF CAPACITOR FILTER OPERATION

The simple-capacitor filter circuit provides high dc voltage (near V_m) and low ripple for *light* loads. The primary disadvantages of the circuit are the relatively poor voltage regulation, high ripple at heavy loads, and the high peak currents drawn through the rectifier diode(s).

2.5 RC FILTER

An improvement, in terms of lower ripple, is provided by using additional sections of resistance and capacitance as shown in Fig. 2.17.

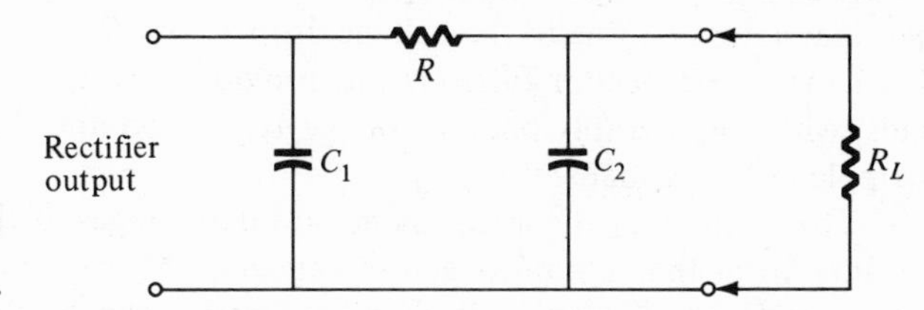

FIG. 2.17
RC filter stage.

The purpose of the added network is to pass as much of the dc component of the voltage developed across the first filter capacitor C_1 and to attenuate as much of the ac component of the ripple voltage developed across C_1 as possible. This action would reduce the amount of ripple in relation to the dc level, providing better filter operation than for the simple capacitor filter. There is a price to pay for this improvement, as will be shown; this includes a lower dc output voltage due to the dc voltage drop across the resistor, and the cost of the two additional components in the circuit.

Figure 2.18 shows the rectifier filter circuit for full-wave operation. Since the rectifier feeds directly into a capacitor the peak currents through the diodes are many times the average current drawn from the

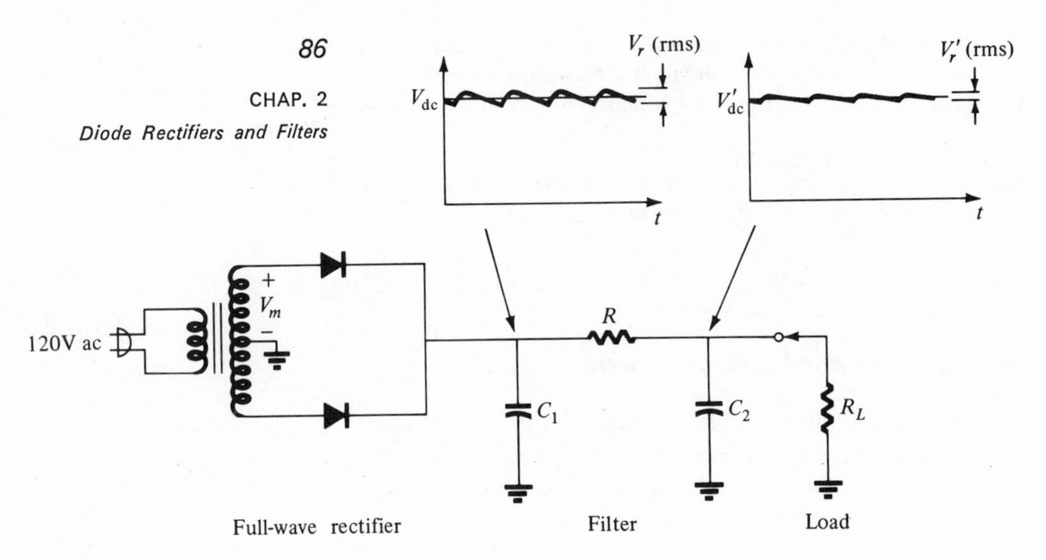

FIG. 2.18
Full-wave rectifier and RC filter circuit.

supply [as specified by Eq. (2.10)]. The voltage developed across capacitor C_1 is further filtered by the resistor-capacitor section (R, C_2) providing an output voltage having less percent of ripple than that across C_1. The load, represented by resistor R_L, draws dc current through resistor R with an output dc voltage across the load being somewhat less than that across C_1 due to the voltage drop across R. This filter circuit, like the simple-capacitor filter circuit, provides best operation at light loads, with considerably poorer voltage regulation and higher percent of ripple at heavy loads.

The analysis of the resulting ac and dc voltages at the output of the filter from that obtained across capacitor C_1 can be carried out simply, using superposition. We can separately consider the RC circuit acting on the dc level of the voltage across C_1 and then the RC circuit action on the ac (ripple) portion of the signal developed across C_1. The resulting values can then be used to calculate the overall circuit voltage regulation and percent of ripple.

dc Operation of RC Filter Section

Figure 2.19a shows the equivalent circuit for consideration of the dc voltage and currents in the filter and load. The two filter capacitors are open circuit for dc and are thus removed from consideration at this time. Calculation of the dc voltage across filter capacitor C_1 was essentially covered in Section 2.4 and the treatment of the additional RC filter stage will proceed from there. Knowing the dc voltage across

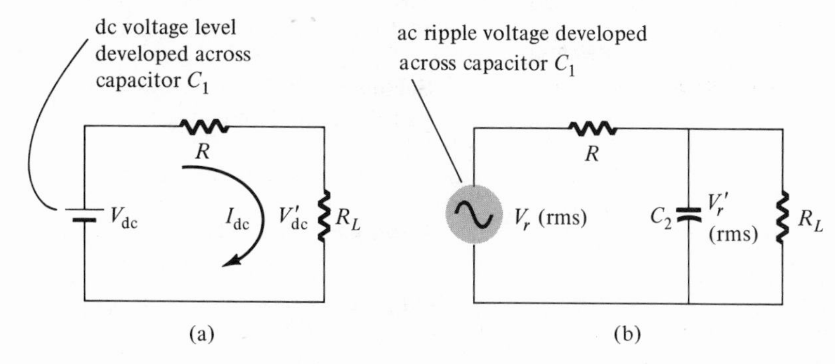

FIG. 2.19
dc and ac equivalent circuits of RC filter: (a) dc equivalent circuit; (b) ac equivalent circuit.

the first filter capacitor (C_1) we can calculate the dc voltage at the output of the additional RC filter section. From Fig. 2.19a we see that the voltage, V_{dc}, across capacitor C_1 is attenuated by a resistor-divider network of R and R_L (the equivalent load resistance); the resulting dc voltage across the load being V'_{dc}. Using the voltage divider rule we can obtain the value of V'_{dc}:

$$\boxed{V'_{dc} = \left(\frac{R_L}{R + R_L}\right)(V_{dc})} \tag{2.11}$$

EXAMPLE 2.14 The addition of an RC filter section with $R = 400\ \Omega$, reduces the dc voltage across the initial filter capacitor from 180 V (V_{dc}). If the load resistance is 4 K, calculate the value of the output dc voltage (V'_{dc}) from the filter circuit.

Solution: Using the result of Eq. (2.11):

$$V'_{dc} = \frac{R_L}{R + R_L} \times V_{dc} = \frac{4000}{400 + 4000} \times 180 = \frac{4000}{4400}(180)$$
$$= \mathbf{163.6\ V}$$

In addition we may calculate the drop across the filter resistor and the load current drawn:

$$V_R = V_{dc} - V'_{dc} = 180 - 163.6 = 16.4\ \text{V}$$
$$I_{dc} = \left(\frac{V'_{dc}}{R_L}\right) = \left(\frac{163.6}{4 \times 10^3}\right) = 40.9\ \text{mA}$$

EXAMPLE 2.15 An RC filter stage connected to a simple-capacitor filter supplies a 10-K load with 20 mA. The RC stage is composed of a

1-K resistor and a 10-μF capacitor. Calculate the value of dc voltage that must be developed by filter capacitor C_1.

Solution: From the values of load resistance and load current the value of output dc voltage is

$$V'_{dc} = I_{dc}R_L = 20 \times 10^{-3}(10 \times 10^3) = 200 \text{ V}$$

Reworking Eq. (2.11) to solve for V_{dc} we get

$$V_{dc} = \frac{R + R_L}{R_L} \times V'_{dc} = \frac{(1 + 10) \times 10^3}{10 \times 10^3} \times 200 = \frac{11}{10} 200 = \mathbf{220 \text{ V}}$$

ac Operation of RC Filter Section

Figure 2.19b shows the equivalent circuit for analyzing the ac operation of the filter circuit. The input to the filter stage from the first filter capacitor (C_1) is the ripple or ac signal part of the voltage across C_1, V_r (rms). Both the RC filter stage components and the load resistance affect the ac signal at the output of the filter.

For a filter capacitor (C_2) value of 10 μF at a ripple voltage frequency (f) of 60 Hz, the ac impedance of the capacitor is

$$X_C = \frac{1}{\omega C} = \frac{1}{2\pi f C} = \frac{1}{6.28(60)(10 \times 10^{-6})} = 0.265 \text{ K}$$

Referring to Fig. 2.19b, this capacitive impedance is in parallel with the load resistance. For a load resistance of 2 K, for example, the parallel combination of the two components would yield an impedance of magnitude:

$$Z = \frac{RX_C}{\sqrt{R^2 + X_C^2}} = \frac{2(0.265)}{\sqrt{2^2 + (0.265)^2}} = \frac{2}{2.02}(0.265) = 0.263 \text{ K}$$

This is quite close to the value of the capacitive impedance alone, as expected, since the capacitive impedance is much less than the load resistance and the parallel combination of the two would be less than the value of either. As a rule of thumb we can consider neglecting the loading by the load resistor on the capacitive impedance as long as the load resistance is at least 5 times as large as the capacitive impedance. Because of the limitation of light loads on the filter circuit the effective value of load resistance is usually large compared to the impedance of capacitors in the range of microfarads.

In the above discussion it was stated that the frequency of the ripple voltage was 60 Hz. Assuming that the line frequency was 60 Hz the ripple frequency will also be 60 Hz for the ripple voltage from a half-wave rectifier. The ripple voltage from a full-wave rectifier, however, will be double since there are twice the number of half-cycles and the ripple frequency will then be 120 Hz. Referring to the relation for capacitive impedance $X_C = 1/\omega C$, we have values of $\omega = 377$ for 60 Hz, and $\omega = 754$ for 120 Hz. Using values of capacitance in μF we can express the relation for capacitive impedance as

$$X_C = \frac{2.65}{C} \quad \text{(half-wave)} \tag{2.12a}$$

$$X_C = \frac{1.33}{C} \quad \text{(full-wave)} \tag{2.12b}$$

where C is in microfarads and X_C is in kilohms.

EXAMPLE 2.16 Calculate the impedance of a 15-μF capacitor used in the filter section of a circuit using full-wave rectification.

Solution

$$X_C = \frac{1.33}{C} = \frac{1.33}{15} = 0.0886\text{K} = \mathbf{88.6\ \Omega}$$

Using the simplified relation that the parallel combination of the load resistor and the capacitive impedance equals, approximately, the capacitive impedance, we can calculate the ac attenuation in the filter stage:

$$V_r'\text{(rms)} = \left(\frac{X_C}{\sqrt{R^2 + X_C^2}}\right)(V_r\text{(rms)}) \tag{2.13a}$$

The use of the square root of the sum of the squares in the denominator was necessary since the resistance and capacitive impedance must be added vectorially, not algebraically. If the value of the resistance is larger by a factor of 5 than that of the capacitive impedance, then a simplification of the denominator may be made yielding the following result:

$$V_r'\text{(rms)} \cong \left(\frac{X_C}{R}\right)(V_r\text{(rms)}) \tag{2.13b}$$

EXAMPLE 2.17 The output of a full-wave rectifier and capacitor filter is further filtered by an RC filter section (see Fig. 2.20). The component values of the RC section are $R = 500\ \Omega$ and $C = 10\ \mu$F. If the initial capacitor filter develops 150 V dc with a 15 V ac ripple voltage, calculate the resulting dc and ripple voltage across a 5-K load.

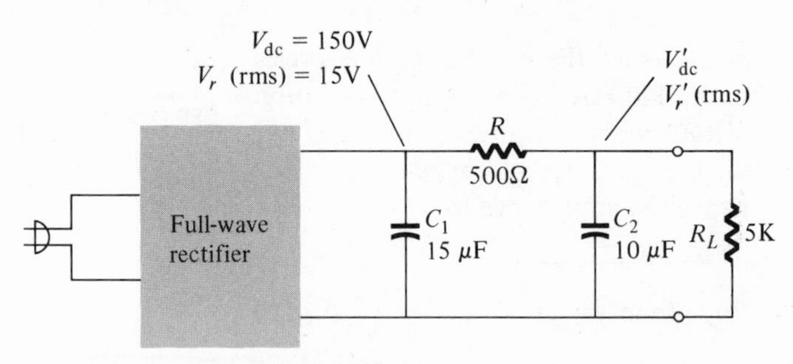

FIG. 2.20
RC filter circuit for Example 2.17.

Solution

dc calculations: Calculating the value of V'_{dc} from Eq. (2.11):

$$V'_{dc} = \frac{R_L}{R + R_L} \times V_{dc} = \frac{5000}{500 + 5000}(150) = \frac{5000}{5500}(150) = \mathbf{136.4\ V}$$

ac calculations: Calculating the value of the capacitive impedance first (for full-wave operation):

$$X_C = \frac{1.33}{C} = \frac{1.33}{10} = 0.133\ \text{K} = 133\ \Omega$$

Since this impedance is not quite 5 times smaller than that of the filter resistor ($R = 500\ \Omega$) we shall use Eq. (2.13a) for the calculation, and then repeat the calculation to show what the difference would have been using Eq. (2.13b) (since the components are almost 5 times different in size). Using Eq. (2.13a),

$$V'_r\ (\text{rms}) = \frac{X_C}{\sqrt{R^2 + X_C^2}}(V_r\ (\text{rms})) = \frac{0.133}{\sqrt{(0.5)^2 + (0.133)^2}}(15)$$

$$= \frac{0.133}{0.518}(15) = 3.86\ \text{V}$$

Now using Eq. (2.13b),

$$V'_r\ (\text{rms}) = \left(\frac{X_C}{R}\right)(V_r\ (\text{rms})) = \frac{0.133}{0.500}(15) = \mathbf{3.99\ V}$$

Comparing the results of 3.86 V and 3.99 V, using Eq. (2.14b), would have yielded an answer within 3.5% of the more exact solution. For values of R much larger than 5 tims X_C it should be clear that the answer obtained using the simplified relation of Eq. (2.13b) would be quite satisfactory.

Ripple Factor and Percent of Ripple with RC Filter Section

One of the more important gains of the added RC filter section is the improvement in ripple factor, which we consider next. The ripple factor at the output of the complete filter circuit (including the RC filter stage) can be expressed as

$$r' = \frac{V'_r\,(\text{rms})}{V'_{\text{dc}}} \tag{2.14}$$

Using Eqs. (2.11) and (2.13) we can express the output ripple factor in terms of the ripple factor at the first filter capacitor and the values of the RC filter stage and load resistance. Doing so we obtain

$$r' = \frac{V'_r\,(\text{rms})}{V'_{\text{dc}}} = \frac{\left(\frac{X_C}{R}\right)(V_r\,(\text{rms}))}{\left(\frac{R_L}{R + R_L}\right)(V_{\text{dc}})}$$

$$= \left(\frac{X_C}{(R \times R_L)/(R + R_L)}\right)\left(\frac{V_r\,(\text{rms})}{V_{\text{dc}}}\right) \tag{2.15a}$$

$$\boxed{r' = \frac{X_C}{R'} \times r} \tag{2.15b}$$

where

$$R' = \frac{RR_L}{R + R_L}$$

EXAMPLE 2.18 Calculate the ripple at the first filter capacitor and the output of the complete filter circuit for the circuit and component values of Example 2.17.

Solution

$$r = \frac{V_r\,(\text{rms})}{V_{\text{dc}}} = \frac{15}{150} = 0.10; \qquad \%r = r \times 100 = 10\%$$

$$r' = \frac{V'_r\,(\text{rms})}{V'_{\text{dc}}} = \frac{3.86}{136.4} = 0.0283; \qquad \%r' = r' \times 100 = 2.83\%$$

Using the approximate relation of Eq. (2.15b):

$$r' = \frac{X_C}{R'}r = \frac{0.133}{0.455}(0.10) = 0.0292; \qquad \%r' = r' \times 100 = 2.92\%$$

where

$$R' = \frac{RR_L}{R + R_L} = \frac{0.5(5)}{0.5 + 5} = \mathbf{0.455\,K}$$

From the above results we see that the ripple was reduced from 10% to either 2.83 or 2.92% (depending on the simplification used in the calculation). This is improvement by a factor of about 3, as the ratio of X_C to R' indicates.

Voltage Regulation for RC Filter Circuit

The calculation of voltage regulation requires calculating the output voltage at no-load and at full-load. If no-load were connected to the output of the filter circuit the capacitors would charge up to the peak voltage of the rectifier output (V_m). The relation for voltage regulation is

$$\% \text{ voltage regulation} = \frac{V_{NL} - V_{FL}}{V_{FL}} \times 100$$

For the general RC filter circuit this can be written

$$\% V.R. = \frac{(V'_{dc})_{NL} - (V'_{dc})_{FL}}{(V'_{dc})_{FL}} \times 100 \qquad (2.16a)$$

For the case where the no-load voltage is the peak rectified voltage Eq. (2.16a) can be written:

$$\boxed{\% V.R. = \frac{V_m - V'_{dc}}{V'_{dc}} \times 100} \qquad (2.16b)$$

where

$$V_{NL} = V_m \qquad \text{and} \qquad V_{FL} = V'_{dc}$$

EXAMPLE 2.19 The following filter circuit (Fig. 2.21) is used to supply voltage to a load resistance of 10 K. Calculate the voltage regulation of the circuit at full-load.

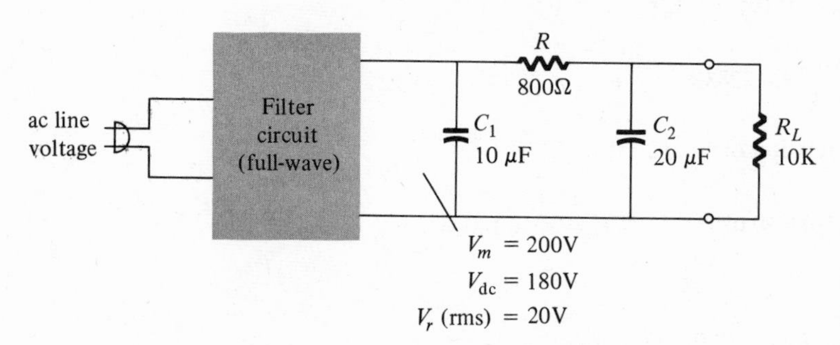

FIG. 2.21
RC filter circuit for Example 2.19.

Solution

No-load:

$$(V'_{dc})_{NL} \cong V_m = 200 \text{ V}$$

Full-load:

$$(V'_{dc})_{FL} = \frac{R_L}{R + R_L}(V_{dc}) = \frac{10}{10.8}(180) = 166.7 \text{ V}$$

$$\% \ V.R. = \frac{V_m - V'_{dc}}{V'_{dc}}(100) = \frac{200 - 166.7}{166.7}(100) = \mathbf{20\%}$$

2.6 π-Type Filter

The addition of an RC filter section improved the ripple factor by decreasing the ac ripple voltage by a greater amount than it decreased the dc voltage. To decrease the dc output as little as possible the series resistor should be as small as possible. On the other hand, the attenuation of the ripple voltage requires that the RC section capacitor have an impedance that is much smaller than that of the series resistor. Making the series resistor R smaller to pass most of the dc voltage will not provide a large reduction of the ripple voltage. Keeping R large compared to the RC section capacitor will provide attenuation of the ripple voltage but will then result in decrease of the dc voltage.

The ripple factor of the voltage across the first filter capacitor will be reduced most if the filter (attenuation) section following it provides little series dc resistance, while at the same time providing large series ac impedance. A resistor is the same to both ac and dc signals. An inductor, however, can have very low dc resistance, while at the same time having a large ac impedance. Fig. 2.22(a) shows just such an

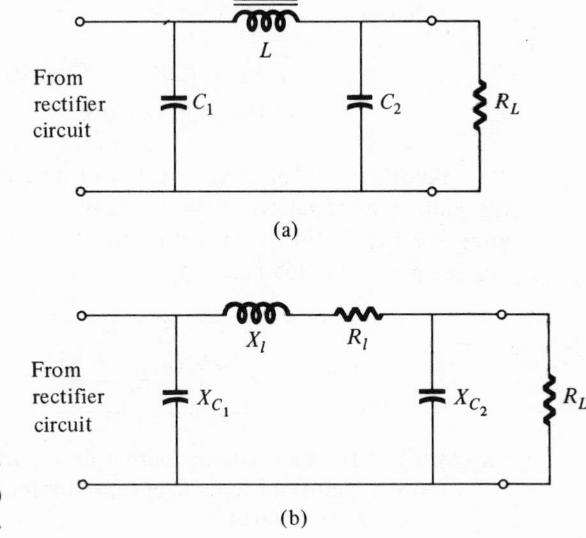

FIG. 2.22

π-type filter circuit: (a) components; (b) impedances.

arrangement, the entire filter circuit of capacitor C_1, inductor L, and capacitor C_2, comprising a π-*type filter* circuit.

The inductor will be considered as a practical component having a dc resistance, R_l (due to wire resistance and core losses), and an inductance, L (see Fig. 2.22b). In analyzing the operation of inductor L, and capacitor C_2, in reducing the ripple factor of the initial filtered signal (across C_1) we shall separately consider the dc and ac operation of the circuit.

dc Calculations

The output dc voltage (V'_{dc}) is less than the dc voltage developed across the first filter capacitor (C_1) by the drop in voltage across the resistance of the inductor. This voltage drop depends on the amount of load current being drawn from the filter circuit. If no current is drawn from the filter the output dc voltage is the same as that across filter capacitor C_1, V_m. In this case, when load current is drawn the output dc voltage can be calculated from

$$V'_{dc} = V_{dc} - I_{dc}R_l \tag{2.17a}$$

where R_l is the dc resistance of the inductor.

EXAMPLE 2.20 A π-type filter has an input voltage of 150 V dc across the input capacitor. If the filter inductor is a 5-H choke having 300-Ω resistance calculate the output dc voltage at a load current of 100 mA.

Solution

$$V'_{dc} = V_{dc} - I_{dc}R_l = 150 - 100(10^{-3}) \times 300$$
$$= 150 - 30 = \mathbf{120\ V}$$

Another way of obtaining the output dc voltage is to calculate the voltage attenuation of the inductor resistance and the load resistance (see Fig. 2.23a). Considering the dc voltage divider network, the voltage across the load resistor is

$$V'_{dc} = \frac{R_L}{R_L + R_l} V_{dc} \tag{2.17b}$$

EXAMPLE 2.21 Calculate the output dc voltage of a π-type filter having 150 V dc across the input filter capacitor for a 1.2 K load. The inductor dc resistance is 300 Ω.

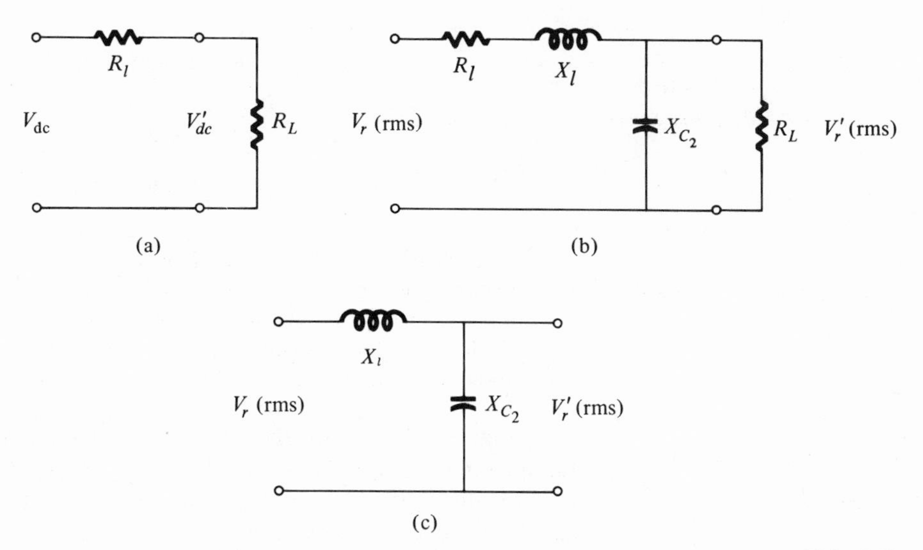

FIG. 2.23

Equivalent circuits for π-type filter: (a) dc equivalent circuit; (b) ac equivalent circuit; (c) approximate ac voltage divider circuit.

Solution

$$V'_{dc} = \frac{R_L}{R_L + R_l} V_{dc} = \frac{1200}{1200 + 300}(150) = \mathbf{120\ V}$$

ac Calculations

The output ripple voltage can be calculated by considering the action of the ac divider network of Fig. 2.23b. Carrying out the calculation in detail would require using the divider impedance of X_{C_2} in parallel with R_L, and the impedance of R_l in series with X_l. This detail is usually not necessary since proper selection of filter components should result in X_l being much larger than R_l and in X_{C_2} being much smaller than R_L. An example will help reinforce this consideration.

EXAMPLE 2.22 Calculate the impedance of the series combination of R_l and X_l and the parallel combination of X_{C_2} and R_L, where, $R_l = 300\ \Omega$, $L = 10$ H, $C_2 = 10\ \mu$F, and $R_L = 5$ K. The ripple frequency is 60 Hz.

Solution

$$X_l = 2\pi f L = 6.28(60)(10) = 3768\ \Omega$$
$$Z_l = \sqrt{X_l^2 + R_l^2} = \sqrt{(3768)^2 + (300)^2} = \mathbf{3780\ \Omega}$$

$$X_{C_2} = \frac{1}{2\pi f C_2} = \frac{1}{6.28(60)(10 \times 10^{-6})} = 0.265\text{ K}$$

$$X_{C_2} \parallel R_L = \frac{X_{C_2} R_L}{\sqrt{X_{C_2}^2 + R_L^2}} = \frac{(0.265)(5)}{\sqrt{(0.265)^2 + (5)^2}}$$

$$= \frac{5}{\sqrt{25.09}}(0.265) \cong \mathbf{0.264\text{ K}}$$

In comparison, the series impedance of the coil inductance and resistance can be approximated by the inductive impedance of 3768 Ω with less than 1% error. The parallel impedance of filter capacitor C_2 and the load resistance is equal to the impedance of the capacitor alone to within 0.1% error.

Figure 2.23c shows the simplified ac circuit representation with the inductor replaced by an impedance X_l and the filter capacitor and load resistor replaced by a capacitive impedance of X_{C_2}. Using the voltage divider rule to calculate the ac voltage across the output due to an input ac voltage V_r (rms):

$$V_r' \text{ (rms)} = \frac{X_{C_2}}{|X_l - X_{C_2}|} V_r \text{ (rms)} \tag{2.18a}$$

Again we are able to resort to a simplification replacing the denominator term above by the X_l value alone:

$$\boxed{V_r' \text{ (rms)} \cong \frac{X_{C_2}}{X_l} V_r \text{ (rms)}} \tag{2.18b}$$

The following example will show this to be a reasonable simplification.

EXAMPLE 2.23 Compare the impedance ratio obtained from Eq. (2.18a) and that from Eq. (2.18b) using the impedance values of Example 2.22.

Solution: Using $X_{C_2} = 0.265$ K and $X_l = 3.77$ K,

$$\frac{X_{C_2}}{|X_l - X_{C_2}|} = \frac{0.265}{3.77 - 0.265} \cong 0.0756$$

$$\frac{X_{C_2}}{X_l} = \frac{0.265}{3.77} = 0.0703$$

The difference, less than 10% error, within slide-rule accuracy, and the simpler expression of Eq. (2.18b) may be used.

EXAMPLE 2.24 A 5-V ripple voltage is developed across the initial capacitor of a π-type filter. If the filter components are $L = 20$ H and

$C_2 = 20\ \mu F$, calculate the ripple voltage across the second capacitor filter (C_2). The ripple frequency is 60 Hz.

Solution: Using Eq. (2.18b),

$$V'_r\,(\text{rms}) = \frac{X_{C_2}}{X_l} V_r\,(\text{rms}) = \frac{1/[6.28(60)(20 \times 10^{-6})]}{6.28(60)(20)}(5)$$

$$= 0.088\ \text{V} = \mathbf{88\ mV}$$

The frequency of the ripple voltage developed across the first filter capacitor depends on the original sinusoidal signal frequency and on whether the rectifier circuit was half- or full-wave. Using the line ac frequency of 60 Hz, the ripple frequency for half-wave rectification will also be 60 Hz but that with full-wave rectification will be 120 Hz. Referring back to Eq. (2.18b) we can include the frequency and component dimensional units in the equation as a constant to obtain the following equation for the action with a full-wave rectified signal ($f = 120$ Hz):

$$V'_r\,(\text{rms}) = \frac{X_{C_2}}{X_l} V_r\,(\text{rms}) = \frac{1/2\pi f C_2}{2\pi f L} = V_r\,(\text{rms}) = \frac{1.76}{LC_2} V_r\,(\text{rms})$$

$$\boxed{V'_r\,(\text{rms}) = \frac{1.76}{LC_2} V_r\,(\text{rms}), \qquad \text{full-wave}} \tag{2.19a}$$

and for a half-wave rectified signal ($f = 60$ Hz):

$$\boxed{V'_r\,(\text{rms}) = \frac{7.04}{LC_2} V_r\,(\text{rms}), \qquad \text{half-wave}} \tag{2.19b}$$

In Eqs. (2.19a) and (2.19b) the calculation holds for a 60-Hz sinusoidal supply signal, the inductance being in units of henrys, and the capacitance in units of microfarads.

EXAMPLE 2.25 Calculate the ripple voltage out of a π-type filter ($L = 8$ H and $C_2 = 2\ \mu F$) for an input ripple voltage of 4 V (rms) across the initial filter capacitor. The ac signal (supply frequency of 60 Hz) is fed to the filter from a full-wave rectifier.

Solution: Using Eq. (2.19a),

$$V'_r\,(\text{rms}) = \frac{1.76}{LC_2} V_r\,(\text{rms}) = \frac{1.76}{(8)(2)}(4) = \mathbf{0.44\ V}$$

Combining the separate calculations for dc and ripple (ac) voltage at the filter output, the output ripple factor (r) is then obtained. The equations for these calculations using separate calculations of dc and ac voltage division by the filter are summarized below:

$$V'_{dc} = \frac{R_L}{R_l + R_L} V_{dc} \tag{2.20}$$

$$V'_r\,(\text{rms}) = \frac{X_{C_2}}{X_l} V_r\,(\text{rms}) \tag{2.21a}$$

$$= \frac{1.76}{LC_2} V_r\,(\text{rms}), \qquad \text{full-wave} \tag{2.21b}$$

$$= \frac{7.04}{LC_2} V_r\,(\text{rms}), \qquad \text{half-wave} \tag{2.21c}$$

EXAMPLE 2.26 Calculate the ripple factor of the output voltage of a π-type filter ($L = 5$ H, $C_2 = 4\ \mu$F, $R_l = 250\ \Omega$) connected to a 4-K load. The voltage across the first filter capacitor (C_1) is 80 V dc with 10 V (rms) ripple at 120 Hz. Compare this to the input voltage ripple factor.

Solution

$$V'_{dc} = \frac{R_L}{R_l + R_L} V_{dc} = \frac{4000}{250 + 4000}(80) = \frac{4}{4.25}(80) = 75.3\text{ V}$$

$$V'_r\,(\text{rms}) = \frac{1.76}{LC_2} V_r\,(\text{rms}) = \frac{1.76}{(5)(4)}(10) = 0.88\text{ V}$$

$$r' = \frac{V'_r\,(\text{rms})}{V'_{dc}} = \frac{0.88}{75.3} = \mathbf{0.0117} \qquad (\%r' = 1.17\%)$$

For the input voltage:

$$r = \frac{V_r\,(\text{rms})}{V_{dc}} = \frac{10}{80} = 0.125 \qquad (\%r = 12.5\%)$$

EXAMPLE 2.27 Design part of a π-type filter to provide 200 V dc at 50 mA. The voltage across the initial filter capacitor (C_1) is 220 V dc with a ripple voltage of 12 V rms ($f = 120$ Hz). The desired ripple factor at 50-mA load current is to be less than 0.02.

Solution

$$R_L = \frac{V'_{dc}}{I_{dc}} = \frac{200\text{ V}}{50\text{ mA}} = 4\text{ K}$$

Using Eq. (2.20)

$$V'_{dc} = \frac{R_L}{R_l + R_L} V_{dc}$$

$$200 = \frac{4}{R_l + 4}(220)$$

Solving for R_l:

$$200(R_l + 4) = 880$$

$$R_l = 0.4\text{ K} = \mathbf{400\ \Omega}$$

Since

$$r' = \frac{V_r'(\text{rms})}{V_{\text{dc}}'} \leq 0.02, \qquad V_r'(\text{rms}) \leq r(V_{\text{dc}}') = 0.02(200) = 4\text{ V}$$

Using Eq. (2.21b):

$$V_r'(\text{rms}) = \frac{1.76}{LC_2} V_r(\text{rms})$$

$$4 = \frac{1.76}{LC_2}(12)$$

$$LC_2 = 5.28$$

Selecting X_l to be 5 times R_l

$$X_l = 5R_l = 5(400\ \Omega) = 2K$$

$$L = \frac{X_l}{2\pi f} = \frac{2000}{754} = \mathbf{2.66\ H}$$

Then C_2 is

$$C_2 = \frac{5.28}{L} = \frac{5.28}{2.66} = \mathbf{1.96\ \mu F} \quad (\text{use } C_2 = 2\ \mu\text{F})$$

2.7 *L*-TYPE FILTER (CHOKE FILTER)

Another filter circuit that provides low ripple voltage at heavy load currents is the *L*-type filter (see Fig. 2.24). An important advantage of this circuit over the previous filter circuits is the absence of a filter capacitor immediately following the rectifier circuit. Recall that the capacitor placed directly after the rectifier circuit resulted in the rectifier diode(s) conducting for only a part of the cycle, therefore carrying peak currents many times the average load current. As will soon be shown the presence of the inductor in series with a rectifier diode can result in the diode conducting continuously, the peak diode current being about the same as the average current (for sufficiently large values of *L*).

Whereas the dc voltage developed with a filter capacitor following the rectifier circuit is equal to V_m under no-load, the dc voltage

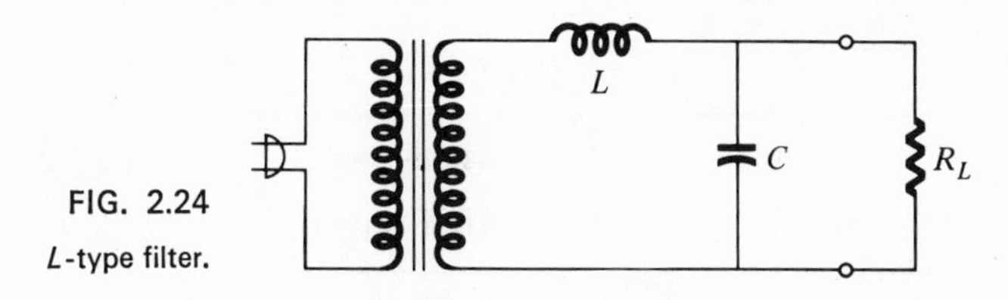

FIG. 2.24
L-type filter.

developed by the L-type filter is the average of the input rectified signal ($0.318V_m$ for half-wave and $0.636V_m$ for full-wave rectified signals). Thus, the output voltage from the L-type filter is initially lower than from those using a capacitor following the rectifier circuit.

Selecting a choke (inductor) having a low value of resistance allows the circuit to pass heavy currents to the load (R_L) with comparatively small dc voltage drop across the choke. The voltage regulation of this circuit is quite good since the dc voltage changes slightly from no-load to full-load operation.

The action of the inductor and capacitor on the ac or ripple voltage is to provide a large amount of attenuation since the ac impedance of the capacitor, X_C, is designed to be much less than the impedance of the inductor, X_l. For dc operation the choke resistance is small compared to the load resistance and the output dc voltage is nearly the average rectified value.

To analyze the operation of the filter circuit we can again separately calculate the ac and dc voltages through the filter circuit. Consider the dc operation first.

dc Operation

Figure 2.25a shows the dc equivalent of the L-type filter circuit connected to an output load resistor (R_L). Notice that the input dc voltage is either $0.318V_m$ or $0.636V_m$ depending on whether half-wave or full-wave rectification is used.

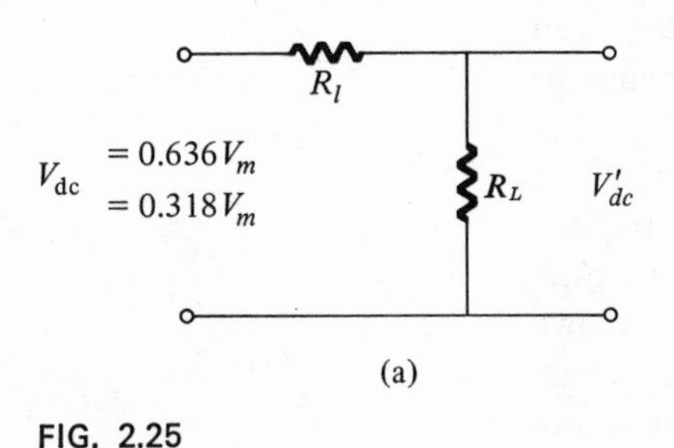

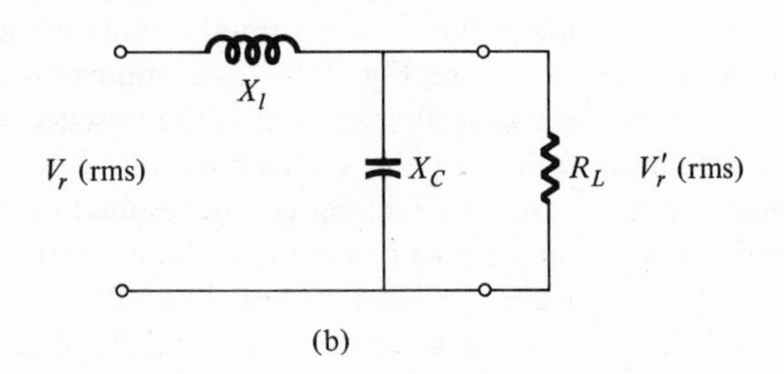

FIG. 2.25
Equivalent circuits of L-type filter: (a) dc equivalent circuit; (b) ac equivalent circuit.

Calculating the output dc voltage in terms of the drop in voltage across the coil resistance,

$$V'_{dc} = 0.318V_m - I_{dc}R_l, \quad \text{(half-wave)} \tag{2.22a}$$

$$V'_{dc} = 0.636V_m - I_{dc}R_l, \quad \text{(full-wave)} \tag{2.22b}$$

Using the voltage divider rule the same result can be expressed as

$$V'_{dc} = \frac{R_L}{R_l + R_L} V_{dc} \tag{2.23}$$

EXAMPLE 2.28 Calculate the output dc voltage from an L-type filter fed by a full-wave rectifier with peak rectified voltage of 150 V. The choke resistance is 400 Ω and the full-load current is 60 mA.

Solution

$$V'_{dc} = 0.636V_m - I_{dc}R_l = 0.636(150) - (60 \times 10^{-3})(400)$$
$$= 95.4 - 24 = \mathbf{71.4\ V}$$

EXAMPLE 2.29 Calculate the output dc voltage across a 1.5-K load fed by an L-type filter. The filter ($R_l = 250\ \Omega$, $L = 5$ H, $C = 6\ \mu$F) is fed a half-wave rectified signal of 200 V peak value. Calculate I_{dc}.

Solution

$$V_{dc} = 0.318V_m = 0.318\ (200) = 63.3\ V$$

$$V'_{dc} = \frac{R_L}{R_l + R_L} V_{dc} = \frac{1500}{250 + 1500}(63.3)$$

$$= \frac{1500}{1750}(63.3) = \mathbf{54.3\ V}$$

The output load current is then

$$I_{dc} = \frac{V'_{dc}}{R_L} = \frac{54.3V}{1.5\ K} = \mathbf{36.1\ mA}$$

ac Operation

The ac operation of the circuit can be obtained using the ac equivalent circuit of Fig. 2.25b. Recall that a good approximation is obtained by replacing the total impedance of the choke inductive impedance and resistance by the inductive impedance alone and also replacing the parallel combination of the capacitive impedance and load resistance by the capacitive impedance alone. Using the voltage divider rule for the equivalent ac circuit we get

$$V'_r\,(\text{rms}) \cong \frac{X_C}{X_l} V_r\,(\text{rms}) \tag{2.24}$$

We are not completely finished with our considerations. The ripple voltage is not a pure sinusoidal voltage and this factor will have to be included in our final calculations. Considering the waveform developed by a full-wave rectifier we note that the voltage is not purely sinusoidal. Using the mathematical technique of Fourier analysis, a Fourier series representation of the waveform can be obtained. The significance of this series is that the actual nonsinusoidal periodic waveform can be represented by a series of sinusoidal waveforms whose frequencies are multiples of the basic signal frequency. For the sake of simplicity (and since reasonable results can be obtained) we shall consider only the first two terms of the series representing the full-wave rectified voltage waveform.

$$v = \frac{2V_m}{\pi} - \frac{4V_m}{3\pi}\cos 2\omega t + \cdots \tag{2.25}$$

The significance of this relation is that the dc level of the output is $2V_m/\pi$, or $0.636V_m$, as expected and that the basic frequency component of the ripple voltage is at twice the power supply frequency (or twice 60 Hz). The magnitude of this second harmonic (twice the basic frequency) signal is $4V_m/3\pi$, or $0.424V_m$. Thus, we can consider the input ripple voltage from the full-wave rectifier as a sinusoidal signal at frequency, $2f$, and at peak amplitude $0.424V_m$. Now having a sinusoidal signal to deal with we can use the simple voltage divider method stated above to obtain the output ripple voltage. Doing this we get

$$V'_r\,(\text{rms}) = \frac{X_C}{X_l}V_r\,(\text{rms}) = \frac{1/2\omega C}{2\omega L}[0.707(0.424V_m)]$$

$$\boxed{V'_r\,(\text{rms}) = \frac{0.529}{LC}V_m, \qquad \text{full-wave}} \tag{2.26}$$

where L is in henrys and C is in microfarads.

The above relations are correct for full-wave rectification only. For half-wave rectified input the analysis of the rectified signal would yield the following Fourier series.

$$v = \frac{V_m}{\pi} + 0.5V_m \sin \omega t + \cdots \tag{2.27}$$

The dc value of the input is then $0.318V_m$. The output ripple voltage is calculated to be

$$V'_r\,(\text{rms}) = \frac{X_C}{X_l}V_r\,(\text{rms}) = \frac{1/\omega C}{\omega L}[(0.707)(0.5V_m)]$$

$$\boxed{V'_r\,(\text{rms}) = \frac{2.5V_m}{LC}, \qquad \text{half-wave}} \tag{2.28}$$

where L is in henrys and C is in microfarads.

Using the above relations we can calculate the ripple factor for the choke filter for both half- and full-wave rectified input signals. Using Eqs. (2.25), (2.26), and the equations for the dc voltage of the rectified signal we get

$$r' = \frac{V'_r\,(\text{rms})}{V'_{dc}} = \frac{0.529V_m/LC}{0.636V_m} = \frac{0.83}{LC}, \quad (\text{full-wave}) \tag{2.29a}$$

where L is in henrys and C is in microfarads.

$$r' = \frac{V'_r\,(\text{rms})}{V'_{dc}} = \frac{2.5V_m/LC}{0.318V_m} = \frac{7.85}{LC}, \quad (\text{half-wave}) \tag{2.29b}$$

where L is in henrys and C is in microfarads.

EXAMPLE 2.30 Calculate the dc output voltage, ripple voltage, and ripple factor of a choke filter with $L = 10$ H, $R_l = 350\ \Omega$, $C = 20\ \mu$F, and a load of 4 K. The input is a full-wave rectified signal ($V_m = 180$ V) derived from the 60-Hz ac line.

Solution

$$V'_{dc} = \frac{R_L}{R_l + R_L} V_{dc} = \frac{4000}{350 + 4000}(0.636 \times 180) = \mathbf{105\ V}$$

$$V'_r\,(\text{rms}) = \frac{0.529}{LC} V_m = \frac{0.529(180)}{(10)(20)} = \mathbf{0.476\ V}$$

$$r' = \frac{V'_r\,(\text{rms})}{V'_{dc}} = \frac{0.476}{105} = \mathbf{0.0045}\ (\%\,r' = 0.45\%)$$

Using Eq. (2.29a):

$$r' = \frac{0.83}{LC} = \frac{0.83}{(10)(20)} = 0.00415\ (\%\,r' = 0.415\%)$$

In terms of percent of ripple the answers obtained are 0.45 and 0.415% respectively. The slight difference in these values is due to the dc voltage drop across the coil resistance, which is included in Eq. (2.22b) or (2.23) but not in Eq. (2.29a). Since the dc drop should usually be small the simplification by Eq. (2.29a) is reasonable.

EXAMPLE 2.31 A filter circuit (L-type) is to be built to provide a dc output voltage of 140 V with a ripple of 0.2%. If the filter is fed a full-wave rectified signal, calculate

(a) value of capacitor needed if a 5-H inductor is used, and
(b) peak rectified voltage (V_m) if coil resistance is negligible.

Solution: (a) Using Eq. (2.29a),

$$r' = \frac{0.83}{LC}, \qquad C = \frac{0.83}{Lr'} = \frac{0.83}{(5)(0.002)} = \frac{0.83}{0.01} = \mathbf{83\ \mu F}$$

(b) Using Eq. (2.22a), with $R_l = 0$,

$$V'_{dc} = 0.636 V_m$$

from which

$$V_m = \frac{V'_{dc}}{0.636} = \frac{140}{0.636} = \mathbf{220\ V}$$

EXAMPLE 2.32 An L-type filter ($L = 2$ H, $R_l = 40\ \Omega$, $C = 40\ \mu$F) is used to provide 18 V dc at a load current of 100 mA. If full-wave rectification was used calculate (a) the ripple factor of the voltage developed, (b) the output voltage at no-load, and (c) the circuit voltage regulation.

Solution: (a) Using Eq. (2.29a),

$$r' = \frac{0.83}{LC} = \frac{0.83}{(2)(40)} = \mathbf{0.0104} \qquad (r' = 1.04\%)$$

(b) Using Eq. (2.22b),

$$V_{dc} = 0.636 V_m - I_{dc} R_l$$
$$18 = 0.636 V_m - 40(100 \times 10^{-3})$$
$$V_m = \frac{22}{0.636} = 34.6\ \text{V}$$

At no-load,

$$V'_{dc} = 0.636 V_m = 0.636(34.6) = \mathbf{22\ V}$$

(c) $$\%\ V.R. = \frac{V_{NL} - V_{FL}}{V_{FL}}(100) = \frac{22 - 18}{18}(100) = \mathbf{22.2\%}$$

Critical Inductance in *L*-Type Filter

In all the previous analysis of the L-type filter circuit it was assumed that the diode always conducts. If at any time during the ac cycle the diode does not conduct, the previous material on the L-type filter is no longer correct. This implies that there is a minimum value of inductance that must be used for the desired operation as an L-type filter. Consider the extreme case of the inductor value being zero. This would leave a capacitor filter circuit, which we know causes the rectifier diode to conduct only part of the time. If a small inductor is used in the filter we may still expect that it cannot override the capacitor charging up to the peak voltage and then causing the diode to be cut off as the input ac voltage becomes less positive. If, however, a large enough inductor is used, it can prevent the current from changing too quickly. This will then delay the capacitor charging up to the peak voltage, and it should be possible to keep the diode conducting for the full cycle for a more limited range of current swing than occurs with the capacitor alone.

For inductor value less than the critical value the diodes in the rectifier section conduct for part of the cycle, not the complete cycle, and the previous material on the L-filter does not apply. If the inductance is sufficiently large (greater than a critical value to be defined shortly) the previous material is correct.

For the diode to conduct over the entire cycle the ac component of the current must not exceed the dc current. Figure 2.26 shows current waveforms for values of inductance below critical and above critical.

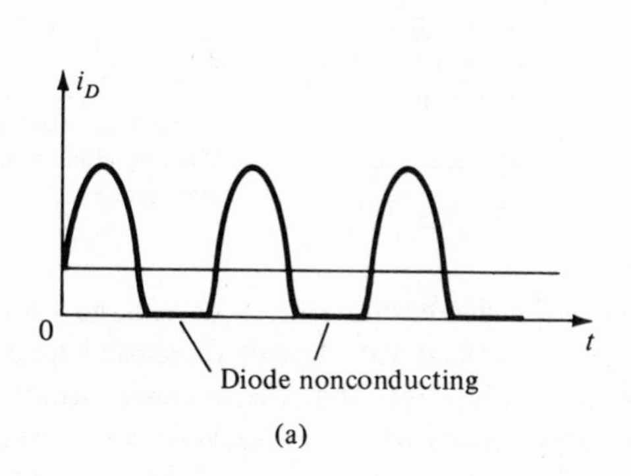

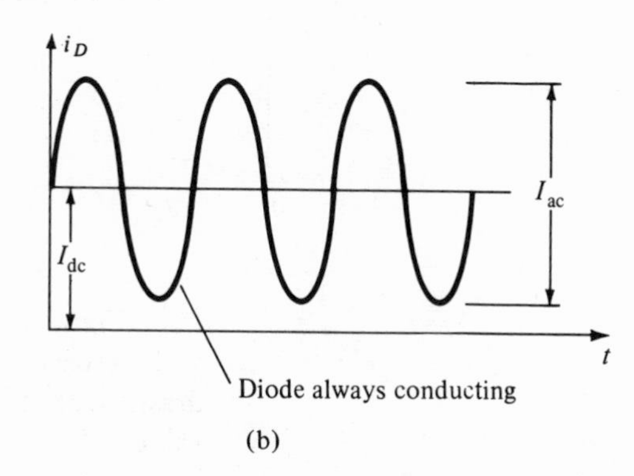

FIG. 2.26
Diode conduction and L-filter critical inductance: (a) $L < L_c$; (b) $L > L_c$.

Setting the dc current (V'_{dc}/R_L) larger than the peak of the ac component ($2V_{dc}/3X_l$) for a full-wave rectifier input the result obtained is

$$L_C \geq \frac{R_L}{1130} \tag{2.30a}$$

This result is based on an approximation of the input signal so that a slightly larger value of inductance would be safer to choose. Doing so, the determination of an approximate critical inductance value (the minimum value of inductance allowed) is obtained from

$$\boxed{L_C \geq \frac{R_L}{1000}} \tag{2.30b}$$

where R_L is in ohms and L_C is in henrys.

The result shows that the value of critical inductance depends solely on the load connected to the power supply. The larger the load current (the smaller the load resistance), the smaller the value of critical inductance. This should make sense since it is necessary to use a larger inductor to limit the ac current to a smaller swing for smaller values of dc load current, thereby ensuring that the diode is always conducting.

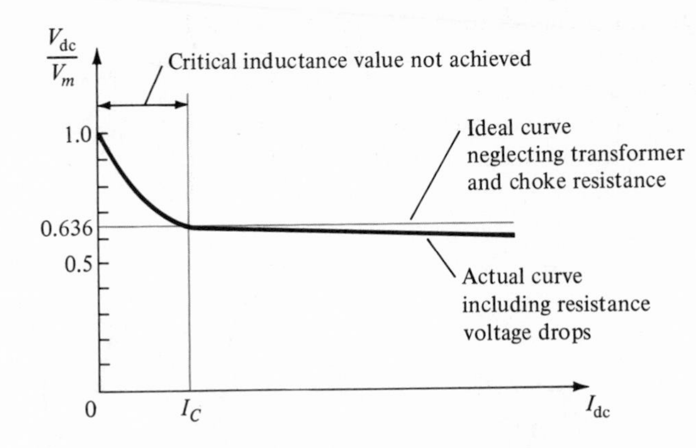

FIG. 2.27

L-type filter regulation curve showing effect of critical inductance.

Figure 2.27 shows the decrease in output dc voltage (normalized with respect to the peak rectified value) with increasing load current drawn from the power supply. Note that the decrease occurs for low values of current or large values of load resistance. In operation, the load current must exceed a critical current value for the circuit to operate properly and maintain the output dc voltage at the average value of the input rectified signal—$0.636V_m$ for the full-wave rectifier. As shown in Fig. 2.27, the output voltage will actually drop slightly with increasing load due to the voltage drop in the transformer and choke.

To ensure that there is a minimum amount of current flowing even for no external load on the filter, thereby ensuring a minimum current drawn through the choke, a bleeder resistor is connected across the filter output. This value of resistance sets the largest value of R_L, and thereby determines the largest value of inductance that must be used.

Another approach to obtaining the critical inductance value for a wide range of load currents, is the use of a *swinging choke* inductor. The swinging choke inductor is designed in terms of wire size, number of turns, core size, and core material, among other factors, so that the effective inductance decreases when increasing current flows through the component. The analysis of the *L*-type filter showed that the coil inductance should increase for smaller values of load current, which is exactly what the swinging choke provides.

EXAMPLE 2.33 An L-type filter circuit providing from 20 to 350-mA load current operates from a full-wave rectified signal of 100 V peak. Calculate the minimum inductance to be used in the filter.

Solution: If the circuit operates properly (diodes always conduct) the output dc voltage is the average of the rectified voltage

$$0.636V_m = 0.636(100) = 63.6 \text{ V dc}$$

Since the critical inductance must be calculated at the smallest load current, we can obtain the largest load resistance from $R_L = V'_{dc}/I_{dc} = 63.6/20 \times 10^{-3} = 3180\ \Omega$. Using Eq. (2.30b).

$$L_C \geq \frac{R_L}{1000} = \frac{3180}{1000} = \mathbf{3.18\ H}$$

A swinging choke can be selected that has an inductance of 3.2 H at 20 mA (with inductance decreasing to as low as 0.2 H at 350 mA).

EXAMPLE 2.34 An L-type filter is to be built to provide 120 mA at 40 V with a ripple of less than 2%. Design the circuit and then check the resulting circuit.

Solution: From the description of the problem we may assume that the supply current will be a fixed value of 120 mA. For this current at 40 V, the value of the load resistance is calculated as follows.

$$R_L = \frac{V'_{dc}}{I_{dc}} = \frac{40}{120 \text{ mA}} = 0.333 \text{ K} = 333\ \Omega$$

Calculating the critical inductance using Eq. (2.30b):

$$L_C \geq \frac{R_L}{1000} = \frac{333}{1000} = \mathbf{0.333\ H}, \qquad \text{or} \qquad 333 \text{ mH}$$

Using Eq.(2.29a)we can calculate the required capacitance:

$$C = \frac{0.83}{rL} = \frac{0.83}{(0.02)(0.333)} = \mathbf{125\ \mu F}$$

Selecting, instead, a 50-μF capacitor and a 1-H choke, we are assured that the choke is larger than the critical inductance value and need only check that the ripple is less than 2%.

$$r = \frac{0.83}{LC} = \frac{0.83}{(1)(50)} = 0.0166$$

$$\%\, r = r(100) = 0.0166(100) = 1.66\%$$

(less than 2%, therefore OK)

A summary of the important equations for the rectifier and filter circuits covered in this chapter is provided in Table 2-1.

TABLE 2.1

Summary of Filter Circuit Operation

FILTER TYPE	NO-LOAD DC VOLTAGE $(V_{dc})_{NL}$	DC VOLTAGE V_{dc}	RMS VALUE OF AC COMPONENT OF RIPPLE VOLTAGE V_r (rms)	RIPPLE FACTOR r	IMPORTANT FACTORS
Capacitor (C)	V_m	$V_{dc} = \frac{V_m}{1 + (I_{dc}/4fCV_m)}$ $= V_m - \frac{4.16 I_{dc}}{C}$	$V_r\text{(rms)} = \frac{I_{dc}}{4\sqrt{3}fC} \times \frac{V_{dc}}{V_m}$ $\cong \frac{2.4 I_{dc}}{C}$	$r = \frac{I_{dc}}{4\sqrt{3}fCV_m}$ $\cong \frac{2.4 I_{dc}}{CV_{dc}} = \frac{2.4}{R_L C}$	full-wave
	V_m	$V_{dc} = \frac{V_m - (I_{dc}/4fC)}{1 + (I_{dc}/4fCV_m)}$	$V_r\text{(rms)} = \frac{I_{dc}}{4\sqrt{3}fC}\left[1 + \frac{V_{dc}}{V_m}\right]$	$r = \frac{I_{dc}}{4\sqrt{3}fC}\left(\frac{1}{V_m} + \frac{1}{V_{dc}}\right)$	half-wave
RC (following C-filter)	V_m	$V_{dc}' = \frac{R_L}{R + R_L} V_{dc}$	$V_r'\text{(rms)} \cong \frac{X_c}{R} V_r\text{(rms)}$	$r' = \frac{X_c}{R'} r$ $\left(R' = \frac{RR_L}{R + R_L}\right)$	$f_r = 120\text{Hz}$ full-wave $f_r = 60$ Hz, half-wave
π-type $(C_1 - LC_2)$	V_m	$V_{dc}' = \frac{R_L}{R_l + R_L} V_{dc}$	$V_r'\text{(rms)} = \frac{1.77}{LC_2} V_r\text{(rms)}$, half-wave $= \frac{7.09}{LC_2} V_r\text{(rms)}$, full-wave	$r' = \frac{3300}{C_1 C_2 L R_L}$,	full-wave
L-type (choke)	$0.636V_m$	$V_{dc}' = 0.636V_m - I_{dc} R_l$	$V_r'\text{(rms)} = \frac{0.529}{LC} V_r\text{(rms)}$	$r' = \frac{0.83}{LC}$	full-wave $L_C > \frac{R_L}{1000}$
	$0.318V_m$	$V_{dc}' = 0.318V_m - I_{dc} R_l$	$V_r'\text{(rms)} = \frac{2.49\ V_m}{LC}$	$r' \cong \frac{7.83}{LC}$	half-wave
Half-wave rectifier	$0.318V_m$	$0.318V_m$	$0.385V_m$	1.21	
Full-wave rectifier	$0.636V_m$	$0.636V_m$	$0.305V_m$	0.48	

2.8 VOLTAGE MULTIPLIER CIRCUITS

Voltage Doubler

A modification of the capacitor filter circuit allows building up a larger voltage than the peak rectified voltage (V_m). The use of this type of circuit allows keeping the transformer peak voltage rating low while stepping up the peak output voltage to 2, 3, 4, or more times the peak rectified voltage.

Figure 2.28 shows a half-wave voltage doubler. During the positive-voltage half-cycle across the transformer, secondary diode D_1 conducts (and diode D_2 is cut off), charging capacitor C_1 up to the peak rectified voltage (V_m). Diode D_1 is ideally a short during this half-cycle and the input voltage charges capacitor C_1 to V_m with the polarity shown in Fig. 2.29a. During the negative half-cycle of the secondary voltage, diode D_1 is cutoff and diode D_2 conducts charging capacitor

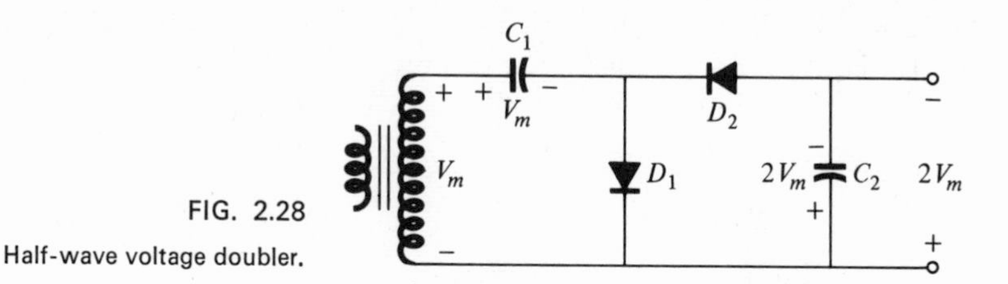

FIG. 2.28
Half-wave voltage doubler.

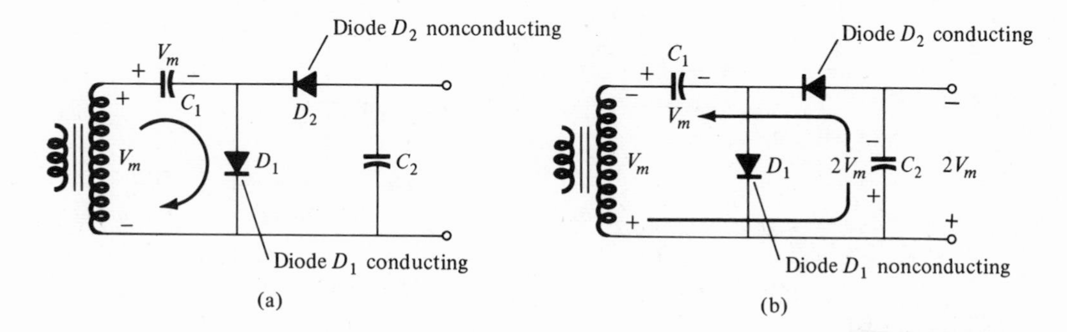

FIG. 2.29
Double operation, showing each half-cycle of operation: (a) positive half-cycle; (b) negative half-cycle.

C_2. Since diode D_2 acts as a short during the negative half-cycle (and diode D_1 is open) we can sum the voltages around the outside loop (see Fig. 2.29b):

$$-V_{C_2} + V_{C_1} + V_m = 0$$
$$-V_{C_2} + V_m + V_m = 0$$

from which

$$V_{C_2} = 2V_m$$

On the next positive half-cycle, diode D_2 is nonconducting and capacitor C_2 will discharge through the load. If no load is connected across capacitor C_2 both capacitors stay charged—C_1 to V_m and C_2 to $2V_m$. If, as would be expected, there is a load connected to the output of the voltage doubler, the voltage across capacitor C_2 drops during the positive half-cycle (at the input) and the capacitor is recharged up to $2V_m$ during the negative half-cycle.

The output waveform across capacitor C_2 is that of a half-wave signal filtered by a capacitor filter. The peak inverse voltage across each diode is $2V_m$.

Another doubler circuit is the full-wave doubler of Fig. 2.30. During the positive half-cycle of transformer secondary voltage (see Fig. 2.31a) diode D_1 conducts charging capacitor C_1 to a peak voltage V_m. Diode D_2 is nonconducting at this time.

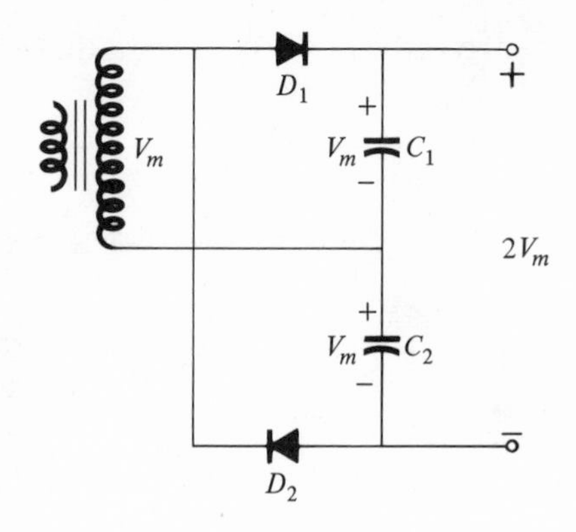

FIG. 2.30
Full-wave voltage doubler.

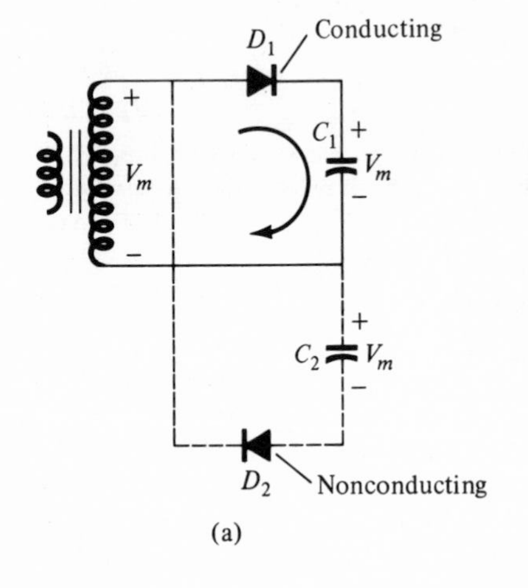

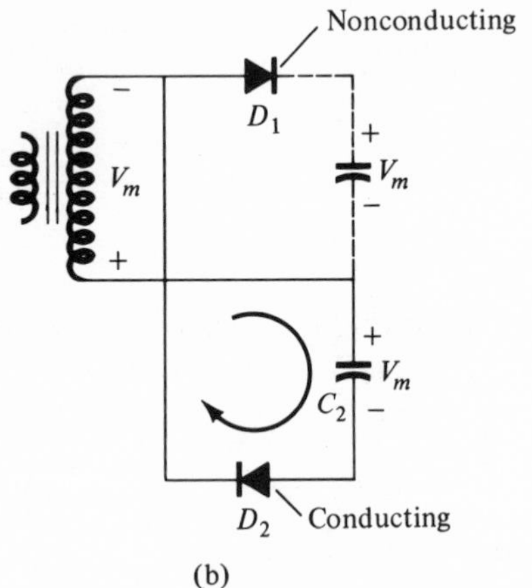

FIG. 2.31
Alternate half-cycles of operation for full-wave voltage doubler.

During the negative half-cycle (see Fig. 2.31b) diode D_2 conducts charging capacitor C_2 while diode D_1 is nonconducting. If no load current is drawn from the circuit the voltage across capacitors C_1 and C_2 is $2V_m$. If load current is drawn from the circuit the voltage across capacitors C_1 and C_2 is the same as that across a capacitor fed by a full-wave rectifier circuit. One difference is that the effective capacitance is that of C_1 and C_2 in series, which is less than the capacitance of either

C_1 or C_2 alone. The lower capacitor value will provide poorer filtering action than the single-capacitor filter circuit.

The peak inverse voltage across each diode is $2V_m$ as it is for the filter capacitor circuit. In summary, the full-wave voltage doubler circuit provides twice the peak voltage of the transformer secondary while requiring no center-tapped transformer and only $2V_m$ PIV rating for the diodes.

Voltage Tripler and Quadrupler

Figure 2.32 shows an extension of the half-wave voltage doubler, which develops 3 and 4 times the peak input voltage. It should be obvious from the pattern of the circuit connection how additional diodes and capacitors may be connected so that the output voltage may also be 5, 6, 7, etc., times the basic peak voltage (V_m).

In operation capacitor C_1 charges through diode D_1 to a peak voltage, V_m, during the positive half-cycle of the transformer secondary voltage. Capacitor C_2 charges to twice the peak voltage $2V_m$ developed by the sum of the voltages across capacitor C_1 and the transformer, during the negative half-cycle of the transformer secondary voltage.

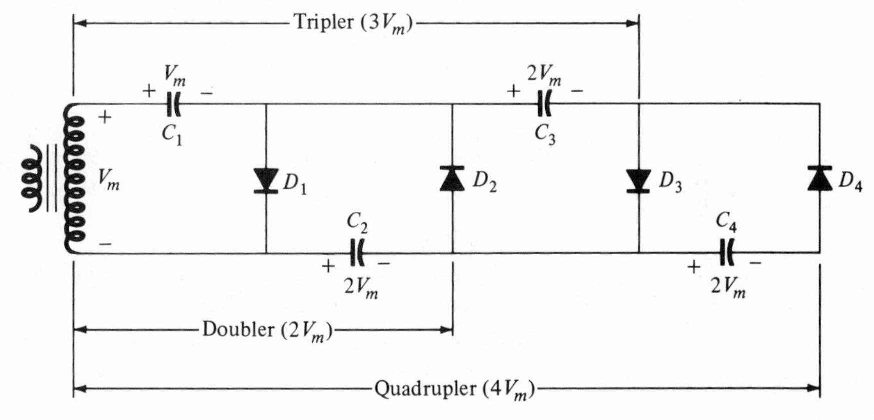

FIG. 2.32

Voltage tripler and quadrupler.

During the positive half-cycle, diode D_3 conducts and the voltage across capacitor C_2 charges capacitor C_3 to the same $2V_m$ peak voltage. On the negative half-cycle, diodes D_2 and D_4 conduct with capacitor C_3, charging C_4 to $2V_m$.

The voltage across capacitor C_2 is $2V_m$, across C_1 and C_3 it is $3V_m$, and across C_2 and C_4, it is $4V_m$. If additional sections of diode and capacitor are used each capacitor will be charged to $2V_m$. Measuring from the top of the transformer winding (Fig. 2.32) will provide odd multiples of V_m at the output, whereas measuring from the bottom of the transformer the output voltage will provide even multiples of the peak voltage, V_m.

The transformer rating is only V_m maximum and each diode in the circuit must be rated at $2V_m$ PIV. If the load is quite small and the capacitors have little leakage, extremely high dc voltages may be developed by this type of circuit, using many sections to step up the dc voltage.

PROBLEMS

§ 2.1

1. Calculate the average (dc) voltage of a 150-V, peak rectified half-wave signal.
2. A half-wave rectifier operates off the 120-V (rms) line voltage through a 3:1 step-down transformer. Calculate the dc voltage of the rectified signal.
3. A half-wave rectifier circuit developes a dc voltage of 180 V. What is the minimum diode peak inverse voltage rating for this circuit?
4. Draw the circuit diagram of a transformer-fed half-wave rectifier providing an output having negative half-cycles.
5. A half-wave rectifier circuit provides 10 mA (dc) to a 5-K load resistance. Calculate the output dc voltage and diode PIV rating for this circuit.
6. A half-wave diode rectifier circuit is transformer fed from the 120-V line. Calculate the transformer peak voltage rating, turns ratio, and diode PIV rating, if the circuit provides an output of 12 V dc.

§ 2.2

7. Calculate the dc voltage obtained from a full-wave rectifier having a peak rectified voltage of 150 V.
8. Calculate the peak voltage rating of each half of a center-tapped transformer used in a full-wave rectifier circuit whose output dc voltage is 120 V.
9. Calculate the PIV rating for the diodes of a center-tapped full-wave rectifier circuit having an output dc voltage of 180 V.
10. Draw the circuit diagram of a center-tapped full-wave rectifier circuit developing an output rectified voltage of negative-going voltage cycles.
11. Draw the circuit diagram of a bridge full-wave rectifier circuit developing an output rectified voltage of negative-going voltage cycles. Show the conduction path through the circuit for each half-cycle of the input sinusoidal ac voltage.
12. Calculate the diode PIV rating for a bridge rectifier developing 50 V dc.
13. Design rectifier circuits to provide an output of 100 V dc using (a) half-wave; (b) center-tapped full-wave; and (c) bridge full-wave cir-

cuits. For each circuit calculate the transformer peak voltage rating, the diode PIV ratings, and the transformer turns ratio, if power is taken from the 120-V line ac supply.

§ 2.3

14. What is the ripple factor of a filter signal having a peak ripple of 10 V on an average of 150 V?
15. A filter circuit provides an output of 80 V unloaded and 70 V under full-load operation. Calculate the voltage regulation.
16. A half-wave rectifier develops 100 V dc. What is the rms value of the ripple voltage?
17. What is the ripple voltage (rms) of a full-wave rectifier whose output voltage is 80 V dc?

§ 2.4

18. A simple-capacitor filter fed by a full-wave rectifier develops 200 V dc at a ripple factor of 10%. What is the output ripple voltage (rms)?
19. A full-wave rectified signal of 180 V peak is fed into a capacitor filter. What is the voltage regulation of the filter circuit, if the output dc voltage is 150 V at full-load?
20. A full-wave rectified voltage of 80 V peak is connected to a 10-μF filter capacitor. What is the dc voltage at 20-mA load?
21. A full-wave rectifier operating from the 60-Hz ac supply line produces a 70-V peak rectified voltage. If a 20-μF filter capacitor is used, calculate the ripple factor and percent of ripple at 40-mA load.
22. A capacitor filter circuit ($C = 4\ \mu$F) develops a dc voltage of 100 V when connected to a load of 5 K. The full-wave rectifier operating from a 60-Hz supply produces a rectified signal of 140 V peak. Calculate the percent of ripple of the output voltage.
23. Calculate the size of the filter capacitor to obtain a filtered voltage with 12% ripple at a load of 15 mA. The full-wave rectified voltage of 80 V peak is obtained from a 60-Hz ac supply.
24. A 2-μF filter capacitor provides 12 mA load current at 15% ripple. Calculate the peak rectified voltage obtained from the 60-Hz line and the dc voltage across the filter capacitor.
25. Calculate the percent of ripple for the voltage developed across an 8-μF filter capacitor providing 35-mA load current. The full-wave rectifier operating from the 60-Hz line develops a peak rectified voltage of 65 V.
26. Calculate the amount of peak diode current through the rectifier diode of a half-wave rectifier feeding a capacitor filter, if the average current drawn from the filter is 20 mA and the diode conducts for only one quarter of the cycle.

§ 2.5

27. An RC filter stage is added after a capacitor filter to reduce the percent of ripple to 2%. Calculate the ripple voltage at the output of the RC filter stage providing 80 V dc.

28. An RC filter stage ($R = 320\ \Omega$, $C = 20\ \mu F$) is used to filter a signal of 120 V dc with 8 V ripple voltage (rms). Calculate the percent of ripple at the output of the RC section for a load of 10 mA. Calculate, also, the ripple of the filtered signal applied to the RC stage.
29. A simple-capacitor filter has an input voltage of 40 V. If this voltage is fed through an RC filter section ($R = 500\ \Omega$, $C = 40\ \mu F$), what is the load current for a load resistance of 2 K?
30. Calculate the ripple voltage (rms) at the output of an RC filter section (full-wave) feeding a 4-K load, if the filter input is 250 V dc with 25 V (rms) ripple. The filter components are $R = 750\ \Omega$ and $C = 6\ \mu F$.
31. If the output no-load voltage for the circuit of Problem 2.30 is 300 V, calculate the percent of voltage regulation with the 4-K load.

§ 2.6

32. A π-type filter of $L = 5$ H ($R_l = 200\ \Omega$) has an input voltage of 190 V. Calculate the filter output dc voltage at a load current of 30 mA.
33. At a ripple voltage frequency of 120 Hz calculate the ac impedance of inductor ($R_l + X_l$) and the effective load impedance ($R_L \| X_{C_2}$) for $L = 8$ H, $R_l = 350\ \Omega$, $C_2 = 25\ \mu F$, and $R_L = 4$ K.
34. A π-type filter operating from a full-wave rectifier has components $L = 8$ H, $R_l = 350\ \Omega$, and $C_2 = 25\ \mu F$. At a load of $R_L = 4$ K calculate the output ripple voltage, if the ripple voltage input (across first filter capacitor) is 4 V.
35. Calculate the ripple voltage from a π-type filter operating from a half-wave rectifier with 10 V rms across the initial filter capacitor. Filter values are $C_1 = 20\ \mu F$, $C_2 = 40\ \mu F$, $L = 2$ H, and $R_L = 2$ K.
36. Design a π-type filter to provide 100-V dc to a 5-K load. Capacitor C_1 (10 μF) has 120 V dc with a ripple voltage of 8 V. A full-wave rectifier is used and the filter output is to have a ripple of less than 1%.
37. A capacitor filter fed by a half-wave rectifier develops 40 V dc with a ripple of 8%. Determine the values of an additional LC section (making a π-type filter) so that with a 2-K load, the dc voltage developed is at least 36 V with a ripple of less than 1%.

§ 2.7

38. Calculate the output dc voltage of an L-type filter operating from a half-wave rectifier, if the peak rectified voltage is 250 V, choke resistance is 200 Ω, and load current is 80 mA.
39. What is the ripple factor of an L-type filter ($L = 0.8$ H, $C = 40\ \mu F$) when operated from a half-wave rectifier?
40. Repeat Problem 2.39 for full-wave rectification.
41. A full-wave rectifier ($V_m = 60$ V) and choke filter ($L = 8$ H, $R_l = 175\ \Omega$, $C = 20\ \mu F$) provide dc voltage to a 1-K load. Calculate the output dc voltage, load current, ripple voltage, and ripple factor.
42. Calculate the value of capacitor needed in an L-type filter having L = 0.5 H to provide a per cent of ripple of less then 0.5% when operating with a full-wave rectifier.

43. An L-type filter (and full-wave rectifier) is to be built to provide 24 V, dc with less than 1% ripple at a load of 100 mA. If the choke selected is 2.5 H with 150-Ω resistance, determine
    (a) The value of C
    (b) Transformer and diode peak voltage rating, if a bridge rectifier is used
44. An L-type filter providing 15 mA at 30 V is made of a 2.5-H inductor and a 10-μF capacitor. Is the inductance value sufficiently large to operate properly in the circuit?
45. Design an L-type filter to provide 60 V dc at 75 mA with ripple of less than 1%, when operating from a full-wave rectifier.

§ 2.8

46. Draw the circuit diagram of a voltage doubler. Indicate the value of the diode PIV rating in terms of the transformer peak voltage, V_m.
47. Draw a voltage tripler circuit indicating diode PIV ratings and voltages across each circuit capacitor, including polarity.
48. Repeat Problem 2.47 for a voltage quadrupler.

Transistors and Vacuum Tubes

3.1 INTRODUCTION

During the period 1904–1947, the tube was undoubtedly the electronic device of interest and development. In 1904, as discussed in Chapter 1, the vacuum-tube diode was introduced by J. A. Fleming. Shortly thereafter, in 1906, Lee De Forest added a third element, called the control grid, to the vacuum diode, resulting in the first amplifier, the triode. In the following years, radio and television provided great stimulation to the tube industry. Production rose from about 1 million tubes in 1922 to about 100 million in 1937. In the early 1930s the four-element tetrode and five-element pentode gained prominence in the electron-tube industry. In the years to follow, the electronic industry became one of primary importance and rapid advances were made in design, manufacturing techniques, high-power and high-frequency applications, and miniaturization.

On December 23, 1947, however, the electronic industry was to experience the advent of a completely new direction of interest and development. It was on the afternoon of this day that Walter H. Brattain and John Bardeen demonstrated the amplifying action of the first transistor at the Bell Telephone Laboratories. The original transistor (a point-contact transistor) is shown in Fig. 3.1. The advantages of this three-terminal solid-state device over the tube were immediately obvious: it was smaller and lightweight; had no heater requirement or heater loss; it had rugged construction; and was more efficient since less power was absorbed by the device itself; it was instantly

FIG. 3.1
The first transistor.
(Courtesy Bell Telephone Laboratories)

available for use, requiring no warm up period; and lower operating voltages were possible. In the early stages of development, transistors were limited to the low-power and low-frequency type. If recent developments continue, however, the semiconductor element will in all probability equal and eventually surpass the power and frequency ratings of today's tubes. Curves indicating which is used more frequently today at various power and frequency levels are indicated in Fig. 3.2.

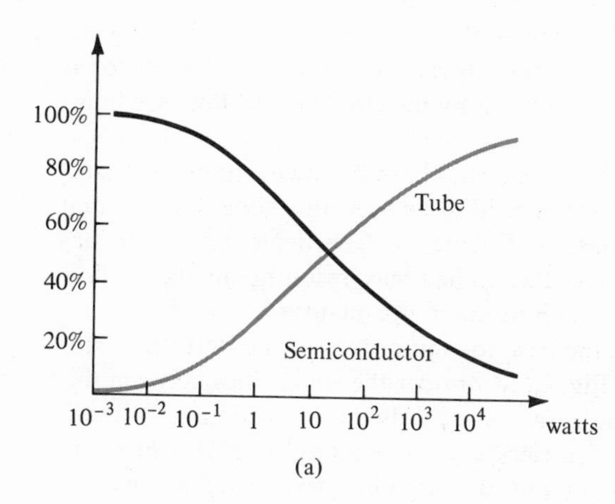

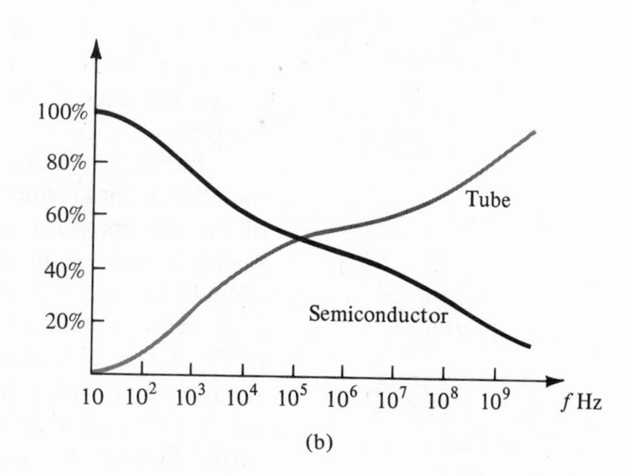

FIG. 3.2
Transistors vs. tubes: (a) power; (b) frequency.

Just as the triode led the way to four-, five-, and more element tubes, the transistor has paved the way toward four-, five-, and more terminal semiconductor devices, some of which will be discussed in Chapter 9.

3.2 TRANSISTOR CONSTRUCTION

The transistor is a three-layer semiconductor device consisting of either two *n*- and one *p*-type layers of material or two *p*- and one *n*-type layers of material. The former is called an NPN transistor, while the latter is called a PNP transistor. Both are shown in Fig. 3.3 with the proper dc biasing. The outer layers of the transistor are heavily doped semiconductor materials having widths much greater than that of the sandwiched *p*- or *n*-type material. For the transistors shown in Fig. 3.3 the ratio of the total width to that of the center layer is 0.150/0.001 = 150:1. The doping of the sandwiched layer is also considerably less than that of the outer layers (typically 10:1 or less). This lower doping level decreases the conductivity or increases the resistance of this material by limiting the number of "free" carriers.

For the biasing shown in Fig. 3.3 the terminals have been indicated by the capital letters *E* for emitter, *C* for collector, and *B* for base. An appreciation for this choice of notation will develop when we discuss the basic operation of the transistor.

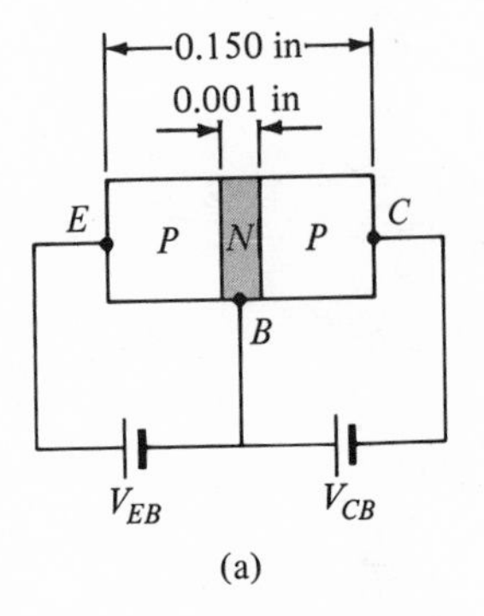

(a)

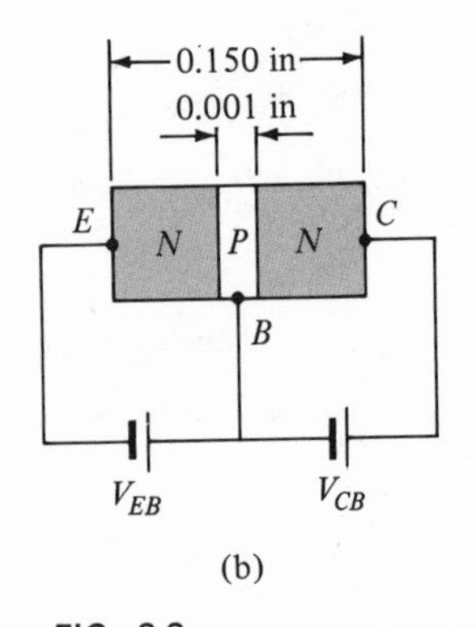

(b)

FIG. 3.3
Types of transistors: (a) PNP; (b) NPN.

3.3 TRANSISTOR OPERATION

The basic operation of the transistor will now be described using the PNP transistor of Fig. 3.3a. The operation of the NPN transistor is exactly the same if the roles played by the electron and hole are interchanged.

In Fig. 3.4, the PNP transistor has been redrawn without the base-to-collector bias. Note the similarities between this situation and that of the *forward-biased* diode in Chapter 1. The depletion region has been reduced in width due to the applied bias, resulting in a heavy flow of majority carriers from the *p*-to the *n*-type material.

Let us now remove the base-to-emitter bias of the PNP transistor of Fig. 3.3a as shown in Fig. 3.5. Consider the similarities between this situation and that of the *reverse-biased* diode of Section 1.5. Recall that the flow of majority carriers is zero resulting in only a minority carrier flow as indicated in Fig. 3.5. *In summary, therefore, one p-n junction of a transistor is reverse biased, while the other is forward biased.*

In Fig. 3.6 both biasing potentials have been applied to a PNP transistor with the resulting majority and minority carrier flow indicated. Note in Fig. 3.6 the widths of the depletion regions, indicating quite clearly which junction is forward biased and which is reverse biased. As indicated in Fig. 3.6, a large number of majority carriers will diffuse across the forward-biased *p-n* junction into the *n*-type material. The

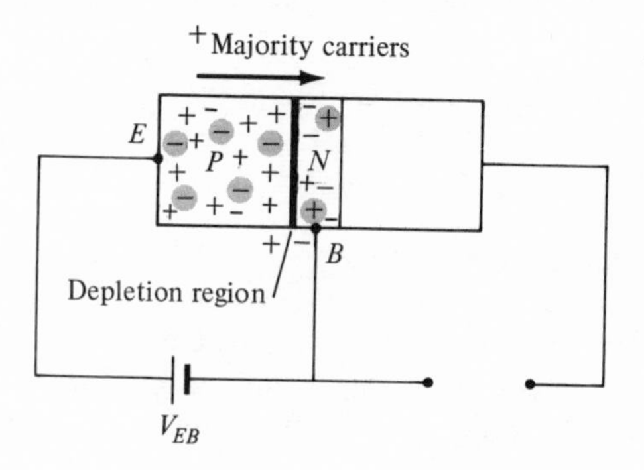

FIG. 3.4
Forward-biased junction of a PNP transistor.

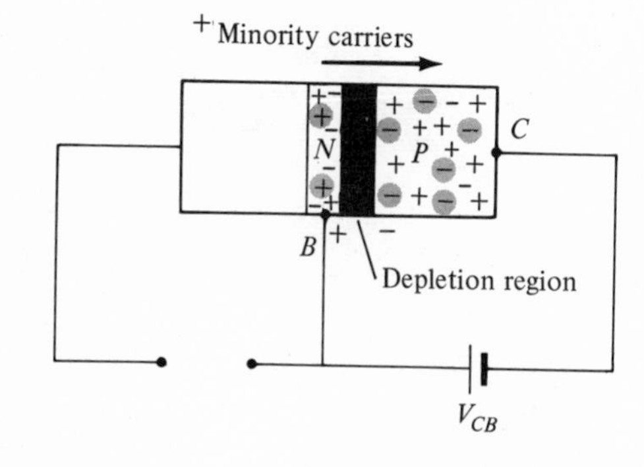

FIG. 3.5
Reverse-biased junction of a PNP transistor.

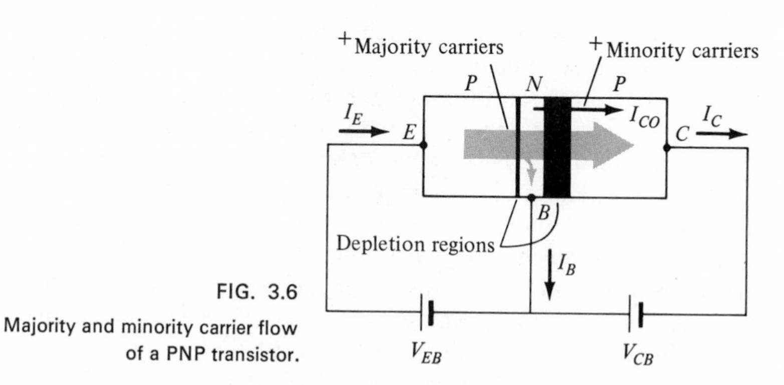

FIG. 3.6
Majority and minority carrier flow of a PNP transistor.

question then is whether these carriers will contribute directly to the base current I_B or pass directly into the *p*-type material. Since the sandwiched *n*-type material is very thin and has a low conductivity a very small number of these carriers will take this path of high resistance to the base terminal. The magnitude of the base current is typically on the order of microamperes as compared to milliamperes for the emitter and collector currents. The larger number of these majority carriers will diffuse across the reverse-biased junction into the *p*-type material connected to the collector terminal as indicated in Fig. 3.6. The reason for the relative ease with which the majority carriers can cross the reverse-biased junction is quite easily understood if we consider that for the reverse-biased diode the injected majority carriers will appear as minority carriers in the *n*-type material. In other words, there has been an *injection* of minority carriers into the *n*-type base region material. Combining this with the fact stated in Section 1.5, that all the minority carriers in the depletion region will cross the reverse-biased junction of a diode, accounts for the flow indicated in Fig. 3.6.

Applying Kirchhoff's voltage law to the transistor of Fig. 3.6 as if it were a single node

$$I_E = I_C + I_B \tag{3.1}$$

and we find that the emitter current is the sum of the collector and base currents. The collector current, however, is composed of two components, due to majority and minority carriers as indicated in Fig. 3.6. The minority current component is called the *leakage current* and is given the symbol I_{CO}. The collector current, therefore, is determined in total by Eq. (3.2).

$$I_C = I_{C_{\text{majority}}} + I_{CO_{\text{minority}}} \tag{3.2}$$

In general, I_C is measured in milliamperes, while I_{CO} is measured in microamperes or nanoamperes. I_{CO}, like I_S for a reverse-biased diode, is temperature sensitive, and must be examined carefully when applications of wide temperature ranges are considered. The effects of I_{CO} will be considered in detail when stability is discussed in Chapter 4.

The configuration shown in Fig. 3.3 for the PNP and NPN transistors is called the *common-base* configuration since the base is common to both the emitter and collector terminals. For fixed values of V_{CB} in the common-base configuration the ratio of a small change in I_C to a small change in I_E is commonly called the *common-base, short-circuit amplification factor* and is given the symbol α (alpha).

In equation form, the magnitude of α is given by

$$\alpha = \left.\frac{\Delta I_C}{\Delta I_E}\right|_{V_{CB}=\text{constant}} \tag{3.3}$$

The term short circuit indicates that the load is short circuited when α is determined. More will be said about the necessity for shorting the load and the operations involved with using equations of the type indicated by Eq. (3.3) when we consider equivalent circuits in Chapter 5. Typical values of α vary from 0.90 to 0.998. For most practical applications, a first approximation for the magnitude of α, usually correct to within a few per cent, can be obtained using the following equation:

$$\alpha \cong \frac{I_C}{I_E} \tag{3.4}$$

where I_C and I_E are the magnitude of the collector and emitter currents, respectively, at a particular point on the transistor characteristics.

3.4 TRANSISTOR AMPLIFYING ACTION

The basic voltage-amplifying action of the common-base configuration can now be described using the circuit of Fig. 3.7. For the common-base configuration, the input resistance between emitter and base of a

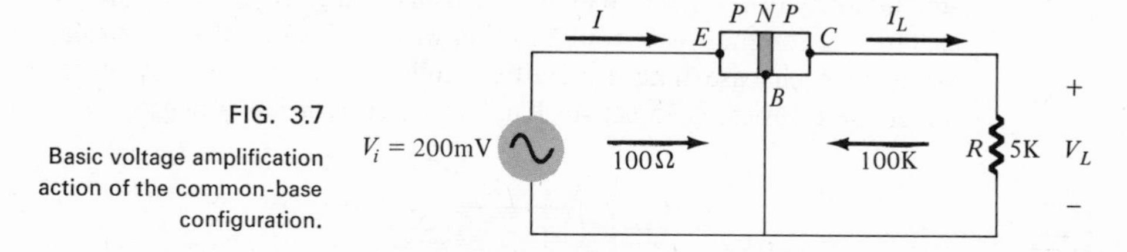

FIG. 3.7

Basic voltage amplification action of the common-base configuration.

transistor will typically vary from 20 to 200 Ω, while the output resistance may vary from 100 K to 1 M. The difference in resistance is due to the forward-biased junction at the input (base to emitter) and the reverse-biased junction at the output (base to collector). Using effective values and an average value of 100 Ω for the input resistance

$$I = \frac{200 \times 10^{-3}}{100} = 2 \text{ mA}$$

If we assume for the moment that $\alpha = 1$ ($I_C = I_E$),

$$I_L = I = 2 \text{ mA}$$

and

$$V_L = I_L R$$
$$= 2 \times 10^{-3} \times 5 \times 10^{+3}$$
$$V_L = 10 \text{ V}$$

The voltage amplification is

$$A_v = \frac{V_L}{V_i} = \frac{10}{200 \times 10^{-3}} = \mathbf{50}$$

Typical values of voltage amplification for the common-base configuration vary from 20 to 100. The current amplification (I_C/I_E) is always less than one for the common-base configuration. This latter characteristic should be obvious since $I_C = \alpha I_E$ and α is always less than one.

The basic amplifying action was produced by *transferring* a current I from a low- to a high-*resistance* circuit. The combination of the two terms in italics results in the name transistor; that is,

$$\textit{trans}\text{fer} + \text{re}\textit{sistor} \longrightarrow \textit{transistor}$$

3.5 COMMON-BASE CONFIGURATION

The notation and symbols used in conjunction with the transistor in the majority of texts and manuals published today are indicated in Fig. 3.8 for the common-base configuration with PNP and NPN transistors. Note in each case that the emitter, collector, and base currents all enter the transistor. To satisfy Kirchhoff's current law a negative sign

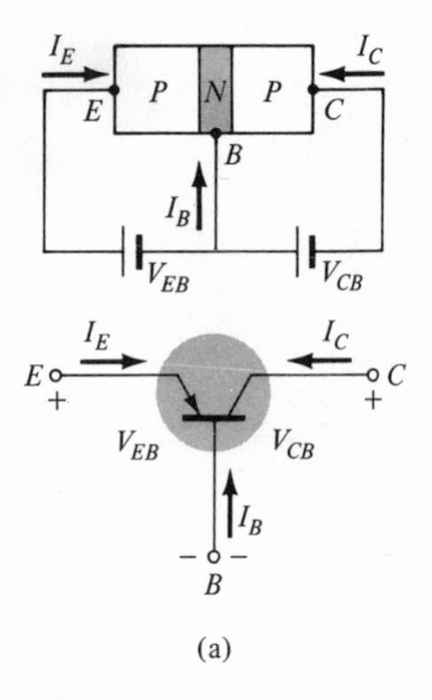

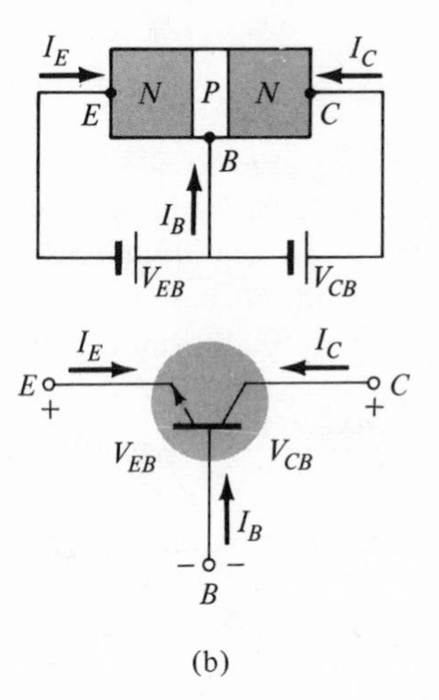

FIG. 3.8

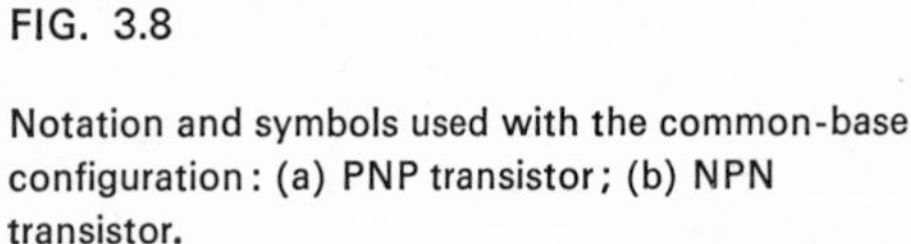

Notation and symbols used with the common-base configuration: (a) PNP transistor; (b) NPN transistor.

must be associated with the magnitude of at least one of these quantities. Throughout this text all current directions will refer to the conventional (hole flow) rather than the electron flow. This choice was based partly on the fact that the vast majority of past and present publications in electrical engineering use conventional current flow.

For the PNP transistor of Fig. 3.8 I_B and I_C will be negative in sign since the conventional flow for each as indicated in Fig. 3.6 is in the opposite direction. For the common-base PNP characteristics of Fig. 3.9 the negative sign has been associated with each quantity as

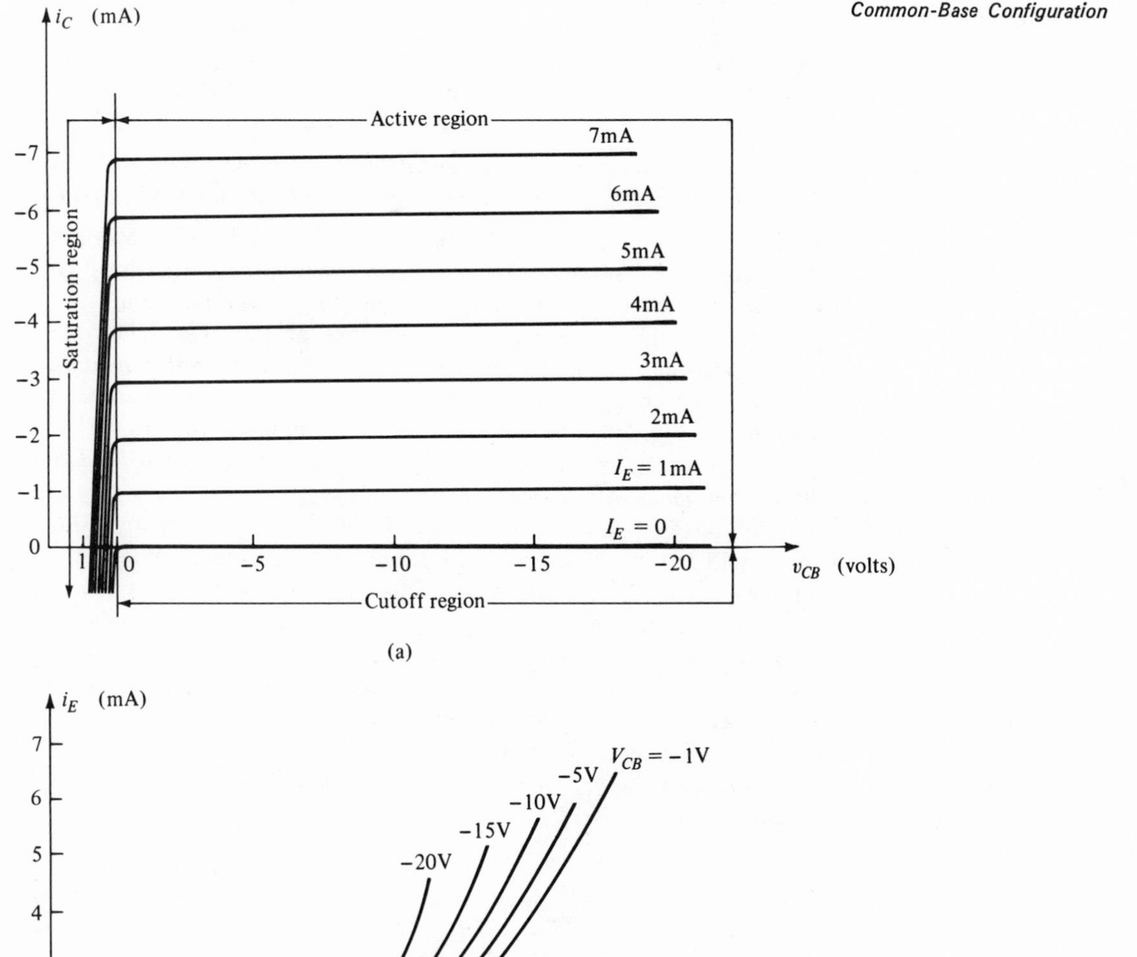

FIG. 3.9
Characteristics of a PNP transistor in the common-base configuration: (a) collector characteristics; (b) emitter characteristics.

dictated by the notation of Fig. 3.8. I_E appears as a positive quantity in the characteristics of Fig. 3.9 since the direction of I_E in Fig. 3.8 is the same as that indicated in Fig. 3.6. For the NPN transistor, I_E will be negative and I_B and I_C will be positive on the device characteristics.

For the common-base configuration, the applied potentials are written with respect to the base potential resulting in V_{EB} and V_{CB}. In other words, the second subscript will always indicate the transistor configuration. In all cases the first subscript refers to the point of higher potential. For the PNP transistor, therefore, V_{EB} is positive and V_{CB} is negative as indicated on the characteristics of Fig. 3.9. For the NPN transistor V_{EB} is negative and V_{CB} is positive.

Note that the arrow drawn in the symbol for the PNP and NPN transistor is in the direction of conventional flow for the emitter current.

In addition, note that two sets of characteristics are necessary to represent the behavior of the PNP common-base transistor of Fig. 3.8, one for the input circuit and one for the output circuit.

The output or collector characteristics of Fig. 3.9a relate the collector current to the collector-to-base voltage and emitter current. The collector characteristics have three basic regions of interest as indicated in Fig. 3.9a: the *active*, *cutoff*, and *saturation* regions.

In the active region the collector junction is reverse biased, while the emitter junction is forward biased. These conditions refer to the situation of Fig. 3.6. The active region is the only region employed for the amplification of signals with minimum distortion. When the emitter current (I_E) is zero, the collector current is simply that due to the reverse saturation current I_{CO} as indicated in Fig. 3.9a. The current I_{CO} is so small (microamperes) in magnitude compared to the vertical scale of I_C (milliamperes) that it appears on virtually the same horizontal line as $I_C = 0$. The circuit conditions that exist when $I_E = 0$ for the common-base configuration are shown in Fig. 3.10. The notation most frequently used for I_{CO} on data and specification sheets is, as indicated in Fig. 3.10, I_{CBO}.

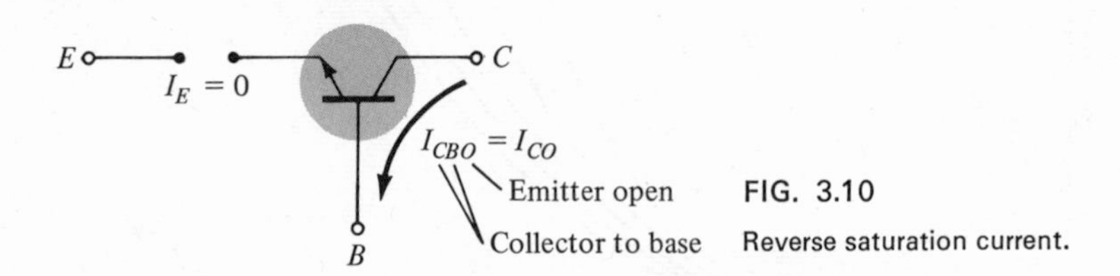

FIG. 3.10
Reverse saturation current.

Note in Fig. 3.9a that as the emitter current increases above zero, the collector current increases to a magnitude slightly less ($\alpha < 1$) than that of the emitter current as determined by the basic transistor-current relations. Note also the almost negligible effect of V_{CB} on the collector current for the active region.

In the cutoff region, the collector and emitter junctions are both reverse biased resulting in negligible collector current as demonstrated in Fig. 3.9a.

The horizontal scale for V_{CB} has been expanded to the left of 0 V to represent clearly the characteristics in this region. *In this region, called the saturation region, the collector and emitter junctions are forward biased* resulting in the exponential change in collector current with small changes in collector-to-base potential.

The input or emitter characteristics have only one region of interest as illustrated by Fig. 3.9b. For fixed values of collector voltage (V_{CB}), as the emitter-to-base potential increases the emitter current increases as shown. In other words, for a fixed collector potential, as the forward-bias potential (V_{EB}) increases the flow of majority carriers (I_E) across the forward-biased junction will increase.

EXAMPLE 3.1 Using the characteristics of Fig. 3.9,

(a) Find the resulting collector current if $I_E = 3$ mA and $V_{CB} = -10$ V.

(b) Find the resulting collector current if $V_{EB} = 200$ mV and $V_{CB} = -10$ V.

(c) Find V_{EB} for the conditions $I_C = 2.5$ mA and $V_{CB} = -5$ V.

Solution

(a)
$$I_C \cong I_E = \mathbf{3\ mA}$$

(b) On the input characteristics $I_E = 1.6$ mA at the intersection of $V_{EB} = 200$ mV, $V_{CB} = -10$ V, and $I_C \cong I_E = \mathbf{1.6\ mA}$.

(c)
$$I_E \cong I_C = 2.5 \text{ mA}$$

On the input characteristics the intersection of $I_E = 2.5$ mA and $V_{CB} = -5$ V results in $V_{EB} = \mathbf{240\ mV}$.

3.6 COMMON-EMITTER CONFIGURATION

The most frequently encountered transistor configuration is shown in Fig. 3.11 for the PNP and NPN transistors. It is called the *common-emitter configuration* since the emitter is common to both the base and collector terminals. Two sets of characteristics are again necessary to describe fully the behavior of the common-emitter configuration: one for the input or base circuit and one for the output or collector circuit. Both are shown in Fig. 3.12

By convention, the emitter, collector, and base currents enter the transistor, while the potentials have the capital letter E as the second subscript to indicate the configuration. Even though the transistor

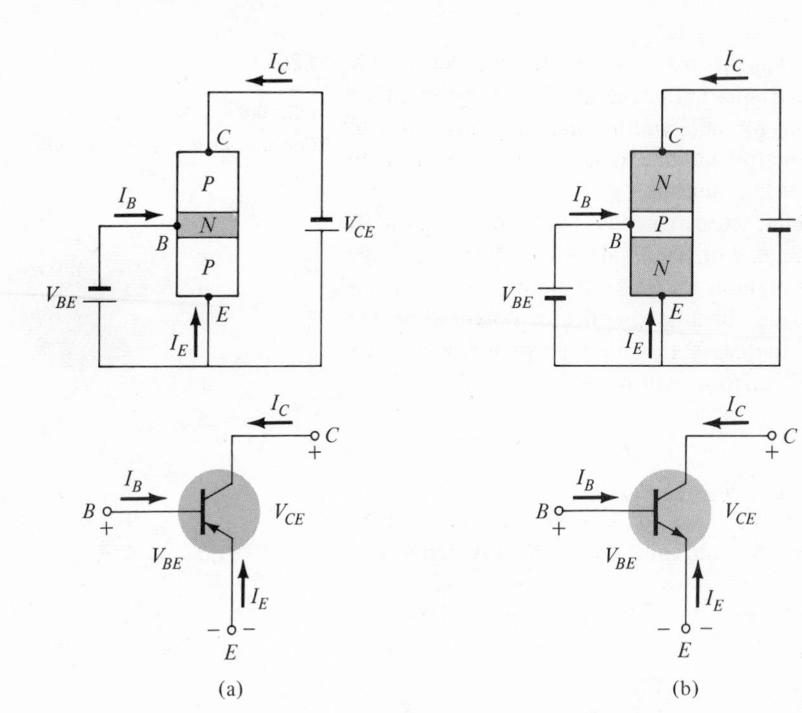

FIG. 3.11
Notation and symbols used with the common-emitter configuration: (a) PNP transistor; (b) NPN transistor.

configuration has changed, the current relations developed earlier for the common-base configuration are still applicable. For the PNP transistor, in the common-emitter configuration, the collector and base currents still have a negative sign associated with each in the characteristics of Fig. 3.12, while the emitter current will have a positive sign.

For the common-emitter configuration, the output characteristics will be a plot of the input current (I_B) versus the output voltage (V_{CE}) and output current (I_C). The input characteristics are a plot of the output voltage (V_{CE}) vs. the input voltage (V_{BE}) and input current (I_B).

Note that on the characteristics of Fig. 3.12 the magnitude of I_B is in microamperes as compared to milliamperes for I_C. Consider also that the curves of I_B are not as horizontal as those obtained for I_E in the common-base configuration indicating that the collector-to-emitter voltage will influence the magnitude of the collector current.

The active region for the common-emitter configuration is that portion of the upper right quadrant that has the greatest linearity, that is, that region in which the curves for I_B are nearly straight and equally spaced. In Fig. 3.12a this region exists to the right of the vertical dashed line at $V_{CE_{sat}}$ and above the curve for I_B equal to zero. The region to the left of $V_{CE_{sat}}$ is called the saturation region. *In the active region the collector junction is reverse biased, while the base junction is*

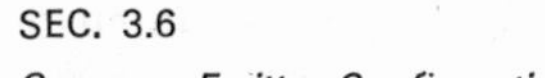

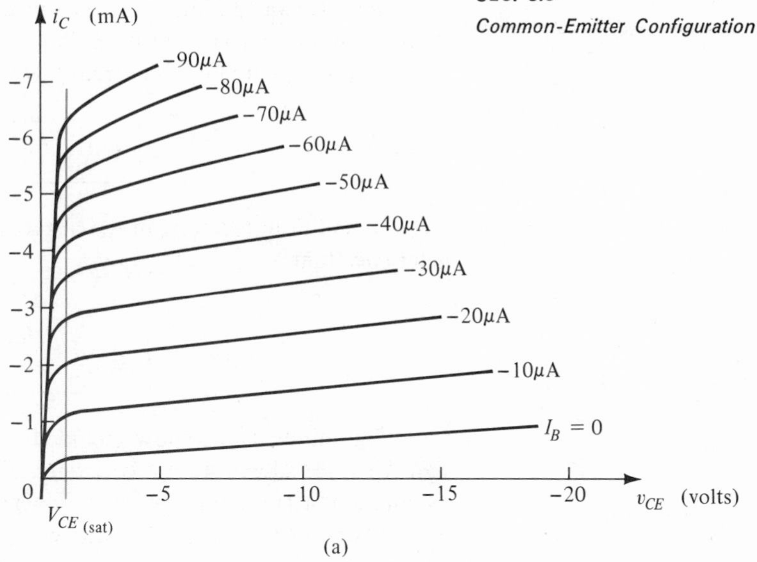

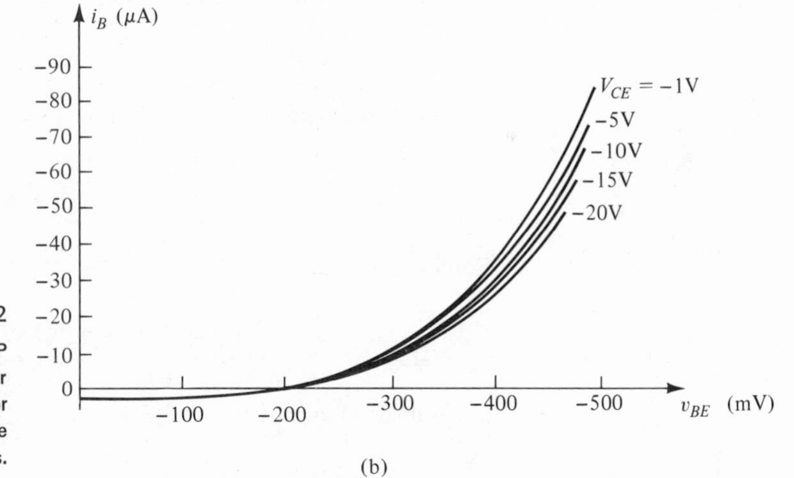

FIG. 3.12

Characteristics of a PNP transistor in the common-emitter configuration: (a) collector characteristics; (b) base characteristics.

forward biased. You will recall that these were the same conditions that existed in the active region of the common-base configuration. The active region of the common-emitter configuration can be employed for voltage, current, or power amplification.

The cutoff region for the common-emitter configuration is not as well defined as for the common-base configuration. Note on the collector characteristics of Fig. 3.12 that I_C is not equal to zero when I_B is zero. For the common-base configuration, when the input current I_E was equal to zero the collector current was equal only to the reverse saturation current I_{CO} so that the curve $I_E = 0$ and the voltage axis were, for all practical purposes, one.

The reason for this difference in collector characteristics can be derived on an approximate basis in the following manner.

Applying Kirchhoff's current law to either transistor of Fig. 3.11,

$$\boxed{I_E + I_C + I_B = 0} \tag{3.5}$$

In addition, recall from the discussion of the common-base characteristics that

$$\boxed{\alpha \cong -\frac{I_C}{I_E}} \tag{3.6}$$

The minus sign is now included to ensure that α has a positive sign when we substitute for I_C and I_E. For the PNP common-base, or common-emitter, configuration, I_C is negative, while I_E is positive using the convention of Fig. 3.11, which will result in a positive value for α in each case.

If we rewrite Eq. (3.6) in the following manner:

$$I_C = -\alpha I_E$$

and include the effects of I_{CO}, then

$$\boxed{I_C = -\alpha I_E + I_{CO}} \tag{3.7}$$

Solving for I_E in Eq. (3-7),

$$I_E = -\frac{(I_C - I_{CO})}{\alpha}$$

and substituting into Eq. (3.5),

$$I_B = -(I_C + I_E) = -\left[I_C - \frac{(I_C - I_{CO})}{\alpha}\right]$$

$$= -I_C + \frac{I_C - I_{CO}}{\alpha}$$

and

$$\alpha(I_B + I_C) = I_C - I_{CO}$$

so that

$$I_C - \alpha I_C = I_{CO} + \alpha I_B$$

resulting in

$$I_C = \frac{I_{CO}}{1-\alpha} + \frac{\alpha I_B}{1-\alpha} \tag{3.8}$$

If we consider the case discussed earlier, where $I_B = 0$, and substitute this value into Eq. (3.8) then

$$I_C = \frac{I_{CO}}{1-\alpha}\bigg|_{I_B=0} \tag{3.9}$$

For $\alpha = 0.98$,

$$I_C = \frac{I_{CO}}{1-0.98} = \frac{I_{CO}}{0.02}$$

and

$$I_C = 50\ I_{CO}$$

which accounts for the vertical shift in the $I_B = 0$ curve from the horizontal voltage axis.

For future reference, the collector current defined by Eq. (3.9) will be assigned the notation indicated by Eq. (3.10).

$$I_{CEO} = \frac{I_{CO}}{1-\alpha} \tag{3.10}$$

In Fig. 3.13, the conditions surrounding this newly defined current are demonstrated with its assigned reference direction.

The magnitude of I_{CEO} is typically much smaller for silicon materials than germanium materials. For transistors with similar ratings I_{CEO} would typically be 2 μA for silicon but 200 μA for germanium.

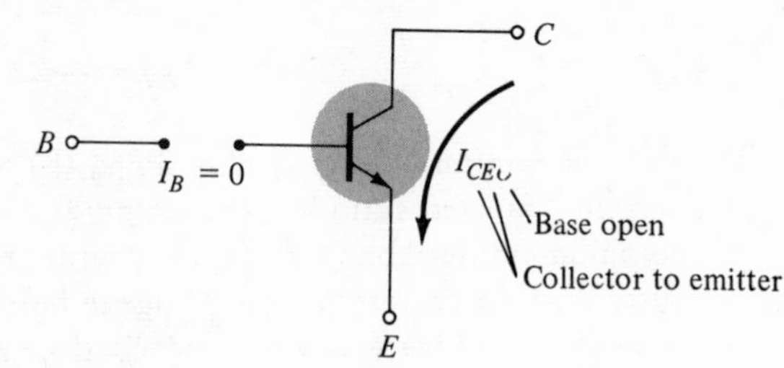

FIG. 3.13
Circuit conditions related to I_{CEO}.

For linear (least distortion) amplification purposes, cutoff for the common-emitter configuration will be (for this text) determined by $I_C = I_{CEO}$. In other words, the region below $I_B = 0$ is to be avoided if an undistorted output signal is required.

When employed as a switch in the logic circuitry of a computer a transistor will have two points of operation of interest: one in the cutoff and one in the saturation region. The cutoff condition should ideally be $I_C = 0$ for the chosen V_{CE} voltage. Since I_{CEO} is typically quite low in magnitude for silicon materials, *cutoff will exist for switching purposes when* $I_B = 0$ *or* $I_C = I_{CEO}$ *for silicon transistors only. For germanium transistors, however, cutoff for switching purposes will be defined as those conditions that exist when* $I_C = I_{CBO} = I_{CO}$. This condition can normally be obtained for germanium transistors by reverse biasing the normally forward-biased base-to-emitter junction a few tenths of a volt.

EXAMPLE 3.2 Using the characteristics of Fig. 3.12,

(a) Find the value of I_C corresponding to $V_{BE} = -400$ mV and $V_{CE} = -5$ V.

(b) Find the value of V_{CE} and V_{BE} corresponding to $I_C = -4$ mA and $I_B = -40\ \mu$A.

Solution: (a) On the input characteristics, the intersection of $V_{BE} = -400$ mV and $V_{CE} = -5$ V results in

$$I_B = -34\ \mu\text{A}$$

On the output characteristics, the intersection of $I_B = -34\ \mu$A and $V_{CE} = -5$ V results in

$$I_C = \mathbf{-3.4\ mA}$$

(b) On the output characteristics, the intersection of $I_C = -4$ mA and $I_B = -40\ \mu$A results in

$$V_{CE} = \mathbf{-6.2\ V}$$

On the input characteristics, the intersection of $I_B = -40\ \mu$A and $V_{CE} = -6.2$ V results in

$$V_{BE} = \mathbf{-425\ mV}$$

In Section 3.3 the symbol alpha (α) was assigned to the forward current transfer ratio of the common-base configuration. For the common-emitter configuration, the ratio of a small change in collector current to the corresponding change in base current at a fixed collector-to-emitter voltage (V_{CE}) is assigned the Greek letter beta (β) and is commonly called the *common-emitter forward current amplification factor*. In equation form, the magnitude of β is given by

$$\beta = \frac{\Delta I_C}{\Delta I_B}\bigg|_{V_{CE}=\text{constant}} \tag{3.11}$$

As a first, but close, approximation, the magnitude of beta (β) can be determined by the following equation:

$$\beta \cong \frac{I_C}{I_B} \tag{3.12}$$

where I_C and I_B are collector and base currents of a particular operating point in the linear region (i.e., where the horizontal base current lines of the common-emitter characteristics are closest to being parallel and equally spaced). Since I_C and I_B in Eq. (3.12) are fixed or dc values, the value obtained for β from Eq. (3.12) is frequently called the *dc beta*, while that obtained by Eq. (3.11) is called the *ac* or *dynamic* value. Typical values of β vary from 20 to 200.

Since

$$I_E + I_C + I_B = 0$$

and

$$I_E = -\frac{I_C}{\alpha}$$

then

$$\frac{I_C}{\alpha} = I_C + I_B$$

so that

$$I_C = \alpha I_C + \alpha I_B$$

or

$$I_C(1 - \alpha) = \alpha I_B$$

and

$$\frac{I_C}{I_B} = \frac{\alpha}{1 - \alpha}$$

resulting in

$$\boxed{\beta = \frac{\alpha}{1-\alpha}} \tag{3.13a}$$

or

$$\boxed{\alpha = \frac{\beta}{\beta+1}} \tag{3.13b}$$

In addition, since

$$I_{CEO} = \frac{I_{CO}}{1-\alpha} = \frac{I_{CBO}}{1-\alpha}$$

then

$$\boxed{I_{CEO} = (\beta+1)I_{CBO} \cong \beta I_{CBO}} \tag{3.14}$$

EXAMPLE 3.3 (a) Find the dc beta at an operating point of $V_{CE} = -10$ V and $I_C = -3$ mA on the characteristics of Fig. 3.12.
(b) Find the value of α corresponding with this operating point.
(c) At $V_{CE} = -10$ V find the corresponding value of I_{CEO}.
(d) Calculate the approximate value of I_{CBO} using the β_{dc} obtained in part (a).

Solution: (a) At the intersection of $V_{CE} = -10$ V and $I_C = -3$ mA, $I_B = -25\ \mu$A,

so that
$$\beta_{dc} = \frac{I_C}{I_B} = \frac{3\times10^{-3}}{25\times10^{-6}} = \mathbf{120}$$

(b)
$$\alpha = \frac{\beta}{\beta+1} = \frac{120}{121} \cong \mathbf{0.992}$$

(c)
$$I_{CEO} = \mathbf{600\ \mu A}$$

(d)
$$I_{CBO} \cong \frac{I_{CEO}}{\beta} = \frac{600\ \mu\text{A}}{120} = \mathbf{5\ \mu A}$$

The input characteristics for the common-emitter configuration are very similar to those obtained for the common-base configuration (Fig. 3.12). In both cases, the increase in input current is due to an increase in majority carriers crossing the base-to-emitter junction with increasing forward-bias potential.

In manuals, data sheets, and other transistor publications, the common-emitter characteristics are the most frequently presented. The common-base characteristics can be obtained directly from the common-emitter characteristics using the basic current relations derived in the past few sections. In other words, for each point on the characteristics of the common-emitter configuration a sufficient number of variables can be obtained to substitute into the equations derived to come up with a point on the common-base characteristics. This process is, of course, time consuming, but it will result in the desired characteristics.

3.7 COMMON-COLLECTOR CONFIGURATION

The third and final transistor configuration is the *common-collector configuration*, shown in Fig. 3.14 with the reference current directions and voltage notation.

The common-collector configuration is used primarily for impedance matching purposes since it has a high input impedance, and low output impedance, opposite to that which is true of the common-base and common-emitter configurations.

The common-collector circuit configuration is generally as shown in Fig. 3.15 with the load resistor from emitter to ground. Note that the collector is tied to ground even though the transistor is connected in a manner similar to the common-emitter configuration. From a

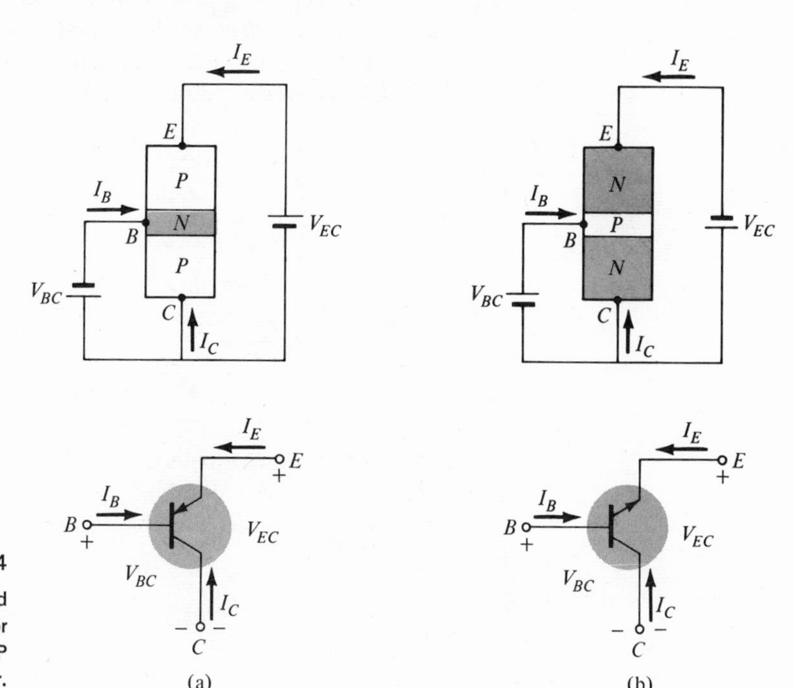

FIG. 3.14
Notation and symbols used with the common-collector configuration: (a) PNP transistor; (b) NPN transistor.

design viewpoint, there is no need for a set of common-collector characteristics to choose the parameters of the circuit of Fig. 3.15. It can be designed using the common-emitter characteristics of Section 3.6. For all practical purposes, the output characteristics of the common-collector configuration are the same as for the common-emitter configuration. For the common-collector configuration the output characteristics are a plot of the input current I_B vs. V_{EC} and I_E. The input current, therefore, is the same for both the common-emitter and common-collector characteristics. The horizontal voltage axis for the common-collector configuration is obtained by simply changing the sign of the collector-to-emitter voltage of the common-emitter characteristics since $V_{EC} = -V_{CE}$. Finally, there is an almost unnoticeable change in the vertical scale of I_C of the common-emitter characteristics if I_C is replaced by I_E for the common-collector characteristics (since $\alpha \cong 1$). For the input circuit of the common-collector configuration the common-emitter base characteristics are sufficient for obtaining any required information by simply writing Kirchhoff's voltage law around the loop indicated in Fig. 3.15 and performing the proper mathematical manipulations.

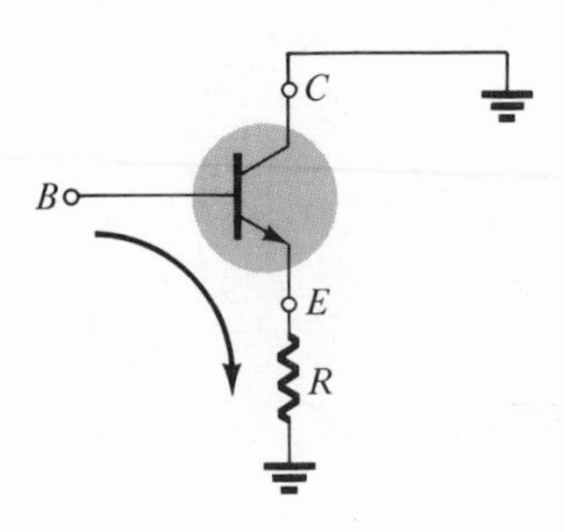

FIG. 3.15
Common-collector configuration used for impedance matching purposes.

3.8 TRANSISTOR MAXIMUM RATINGS

The standard transistor data sheet will include three maximum ratings: *collector dissipation, collector voltage, and collector current.*

For the transistor whose characteristics were presented in Fig. 3.12 the following maximum ratings were indicated.

$$P_{C_{\max}} = 30 \text{ mW}$$
$$I_{C_{\max}} = 6 \text{ mA}$$
$$V_{CE_{\max}} = 20 \text{ V}$$

The power or dissipation rating is the product of the collector voltage and current. For the common-emitter configuration:

$$\boxed{P_{C_{\max}} = V_{CE} I_C} \tag{3.15}$$

The nonlinear curve determined by this equation is indicated in Fig. 3.16. The curve was obtained by choosing various values of V_{CE} or I_C and finding the other variable using Eq. (3.15). For example, at $V_{CE} = 10$ V

$$I_C = \frac{P_{C_{\max}}}{V_{CE}} = \frac{30 \times 10^{-3}}{10} = 3 \text{ mA}$$

as indicated in Fig. 3.16. The region above this curve must be avoided in the design of systems using this particular transistor if the maximum power rating is not to be exceeded. The maximum collector voltage, in this case V_{CE}, is indicated as a vertical line in Fig. 3.16. The maximum collector current is also indicated in Fig. 3.16 as a horizontal line.

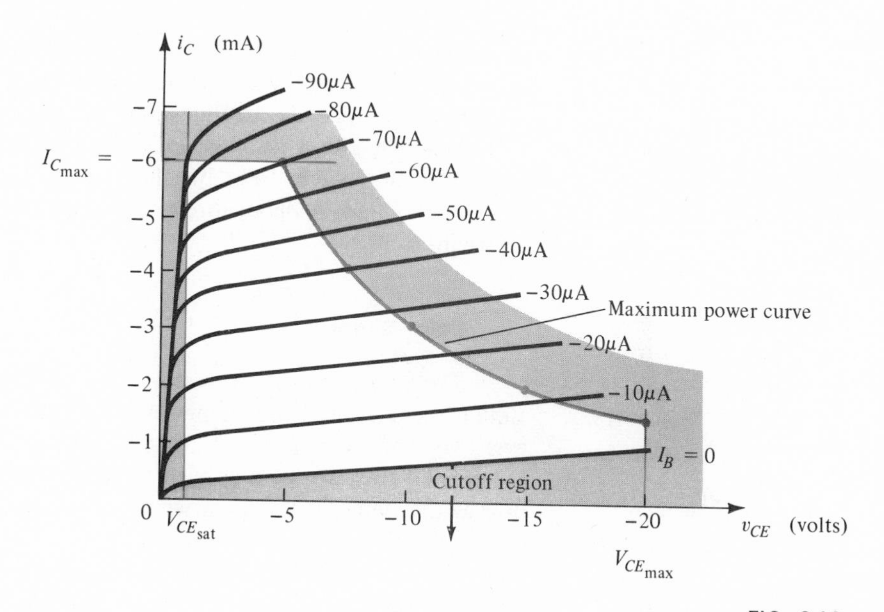

FIG. 3.16
Region of operation for amplification purposes.

For the common-base configuration, the collector dissipation is determined by the following equation. The maximum collector voltage would refer to V_{CB}.

$$P_{C_{max}} = V_{CB} I_C \tag{3.16}$$

For amplification purposes, the nonlinear characteristics of the saturation and the cutoff regions are also avoided. The saturation region has been indicated by the vertical line at $V_{CE_{sat}}$ and the cutoff region by $I_B = 0$ in Fig. 3.16. The unshaded region remaining is the region employed for amplification purposes. Although it appears as though the area of operation has been drastically reduced we must keep in mind that many signals are in the microvolt or millivolt range, while the horizontal axis of the characteristics is measured in volts. In addition to maximum ratings, data and specification sheets on transistors also include other important information about its operation. The discussion of this additional data will not be considered until each parameter is fully defined.

3.9 TRANSISTOR FABRICATION

The majority of the methods used to fabricate transistors are simply extensions of the methods used to manufacture semiconductor diodes. The methods most frequently employed today include *point-contact, alloy junction, grown junction, and diffusion*. The following discussion of each method will be brief, but the fundamental steps included in each will be presented. A detailed discussion of each method would require a text in itself.

Point-Contact

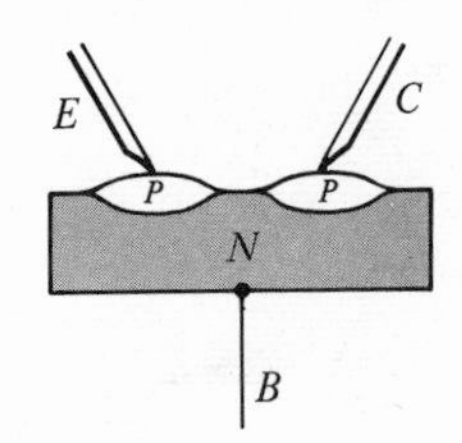

FIG. 3.17
Point-contact transistor.

The point-contact transistor is manufactured in a manner very similar to that used for point-contact semiconductor diodes. In this case two wires are placed next to an *n*-type wafer as shown in Fig. 3.17. Electrical pulses are then applied to each wire resulting in a *p-n* junction at the boundary of each wire and the semiconductor wafer. The result is a PNP transistor as shown in Fig. 3.17. This method of fabrication is today limited to high-frequency, low-power devices. It was the method used in the fabrication of the first transistor shown in Fig. 3.1.

Alloy Junction

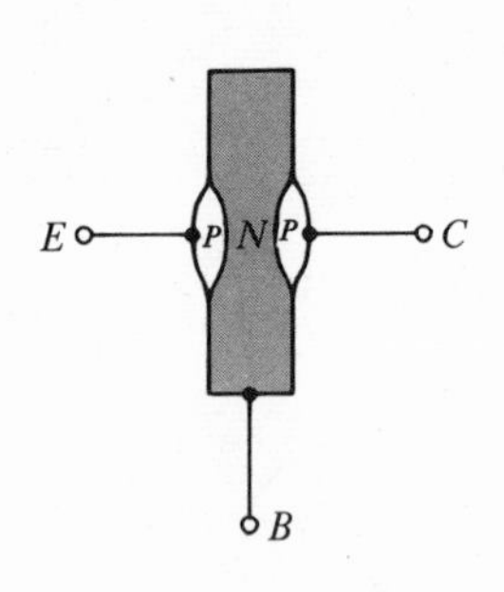

FIG. 3.18
Alloy junction transistor.

The alloy junction technique is also an extension of the alloy method of manufacturing semiconductor diodes. For a transistor, however, two dots of the same impurity are deposited on each side of a semiconductor wafer having the opposite impurity as shown in Fig. 3.18. The entire structure is then heated until melting occurs and each dot is alloyed to the base wafer resulting in the *p-n* junctions indicated in Fig. 3.18 as described for semiconductor diodes.

The collector dot and resulting junction are larger to withstand the heavy current and power dissipation at the collector-base junction. This method is not employed as much as the diffusion technique to be described shortly, but it is still used quite extensively in the manufacture of high-power diodes.

Grown Junction

The Czochralski technique (Section 1.5) is used to form the two *p-n* junctions of a grown junction transistor. The process, as depicted in Fig. 3.19, requires that the impurity control and withdrawal rate be such as to ensure the proper base width and doping levels of the *n*- and *p*-type materials. Transistors of this type are, in general, limited to less than $\frac{1}{4}$-W rating.

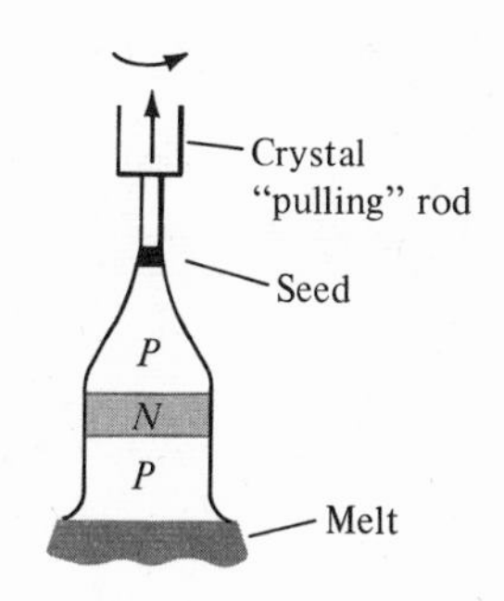

FIG. 3.19
Grown junction transistor.

Diffusion

The most frequently employed method of manufacturing transistors today is the diffusion technique. The basic process was introduced in the discussion of semiconductor diode fabrication. The diffusion technique is employed in the production of *mesa* and *planar* transistors, each of which can be of the *diffused* or *epitaxial* type.

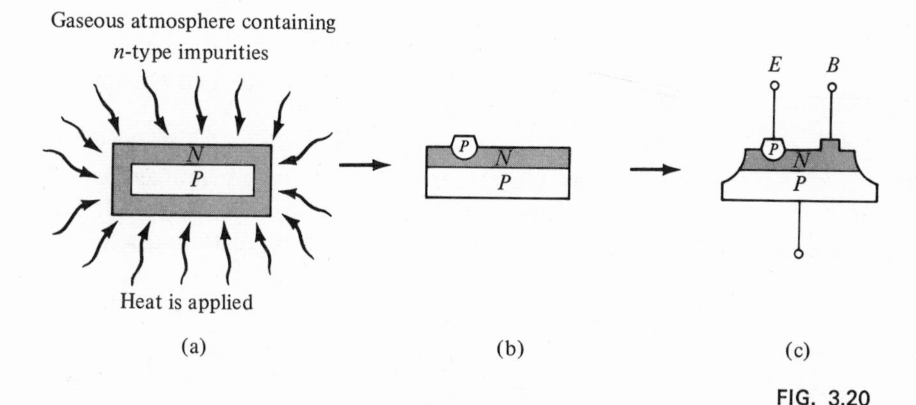

FIG. 3.20
Mesa transistor: (a) diffusion process; (b) alloy process; (c) etching process.

In the PNP, diffusion-type mesa transistor the first process is an *n*-type diffusion into a *p*-type wafer, as shown in Fig. 3.20, to form the base region. Next, the *p*-type emitter is diffused or alloyed to the *n*-type base as shown in the figure. Etching is done to reduce the capacitance of the collector junction. The term mesa is derived from its similarities with the geographical formation. As mentioned earlier in the discussion of diode fabrication the diffusion technique permits very tight control of the doping levels and thicknesses of the various regions.

The major difference between the epitaxial mesa transistor and the mesa transistor is the addition of an epitaxial layer on the original collector substrate. The term epitaxial is derived from the Greek words *epi*—upon, and *taxi*—arrange, which describe the process involved in forming this additional layer. The original *p*-type substrate (collector of Fig. 3.21) is placed in a closed container having a vapor of the same impurity. Through proper temperature control, the atoms of the vapor

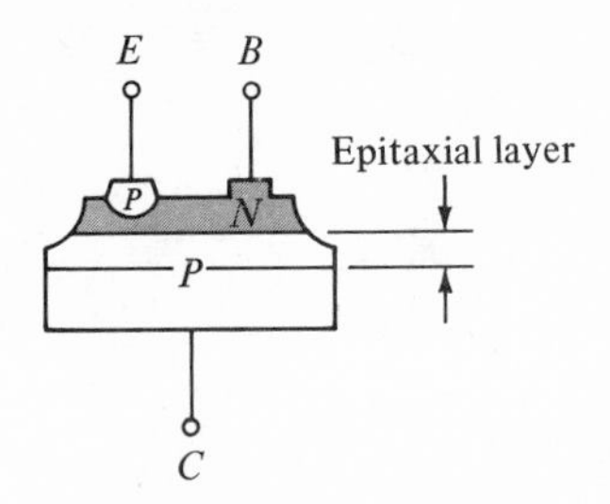

FIG. 3.21
Epitaxial mesa transistor.

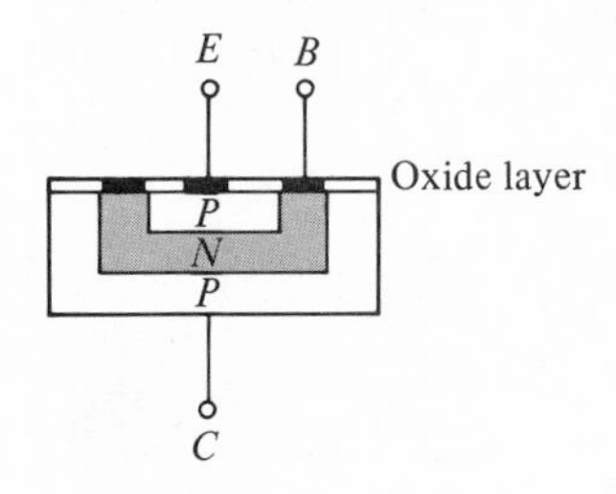

FIG. 3.22
Planar transistor.

will *fall upon* and *arrange* themselves on the original *p*-type substrate resulting in the epitaxial layer indicated in Fig. 3.21. Once this layer is established, the process continues, as above for the mesa transistor, to form the base and emitter regions. The original *p*-type substrate will have a higher doping level and correspondingly less resistance than the epitaxial layer. The result is a low-resistance connection to the collector lead that will reduce the dissipation losses of the transistor.

The planar and epitaxial planar transistor are fabricated using two diffusion processes to form the base and emitter regions. The planar transistor, as shown in Fig. 3.22, has a flat surface, which accounts for the term planar. An oxide layer is added as shown in Fig. 3.22 to eliminate exposed junctions, which will reduce substantially the surface leakage loss (leakage currents that flow on the surface rather than through the junction).

3.10 TRANSISTOR CASING AND TERMINAL IDENTIFICATION

After the transistor has been manufactured using one of the techniques indicated in Section 3.9 leads of, typically, gold, aluminum, or nickel are then attached and the entire structure is encapsulated in a con-

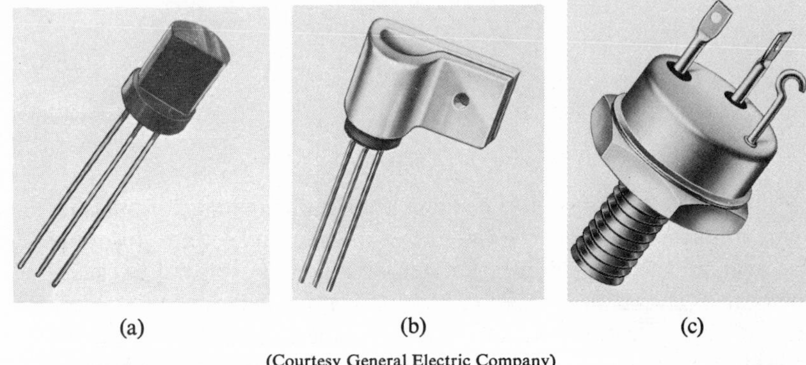

(a) (b) (c)

(Courtesy General Electric Company)

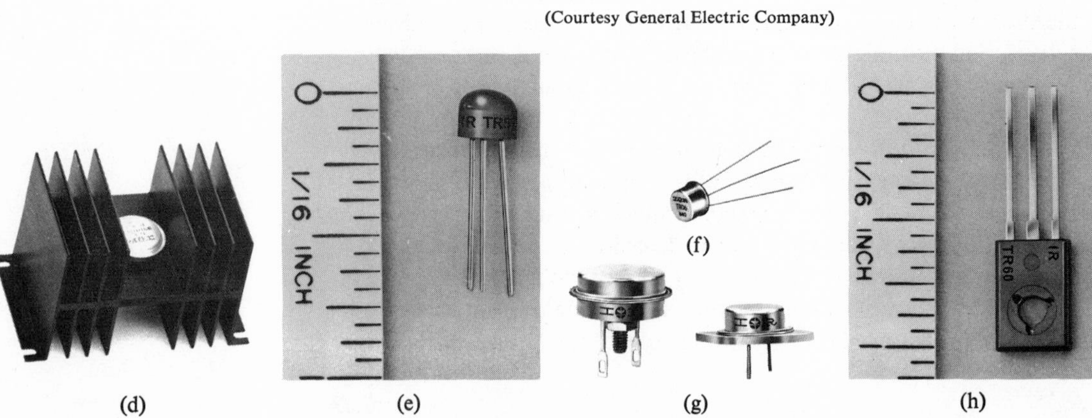

(d) (e) (f) (g) (h)

(Courtesy International Rectifier Corporation)

FIG. 3.23
Various types of transistors.

tainer such as that shown in Fig. 3.23. Those with the studs and heat sinks are high-power devices, while those with the small can (top hat) or plastic body are low- to medium-power devices.

Whenever possible, the transistor casing will have some marking to indicate which leads are connected to the emitter collector or base of a transistor. A few of the methods commonly used are indicated in Fig. 3.24.

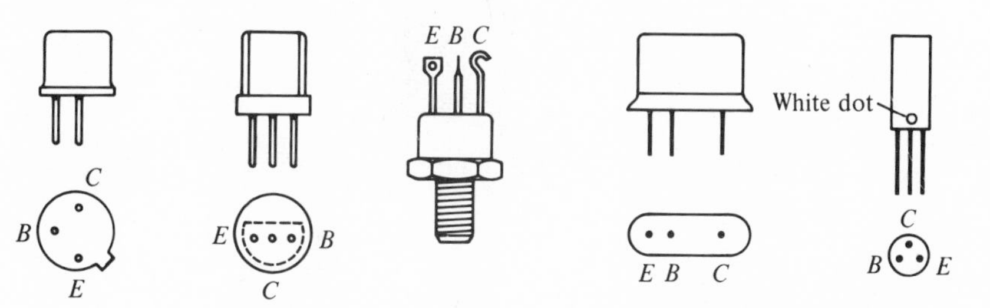

FIG. 3.24
Transistor terminal identification.

3.11 TRIODE

The *triode* is a three-terminal, high-vacuum-tube device, which, like the transistor, has an amplifying capability. However, there is one distinct difference between the two, that must be made clear. The transistor, having the input base current controlling the resulting point of operation in the collector characteristics, is called a *current-controlled device*. The triode, which has a grid voltage controlling the point of operation in the output or plate characteristics, is called a *voltage-controlled device*. The ac analysis of each, to follow in Chapter 5, will be quite different due to the variation in the mode of operation. The triode construction consists of one more element, called the *control grid*, than was present in the vacuum-tube diode. The relative position of each structure in the vacuum-tube triode is shown in Fig. 3.25 along with a typical triode glass envelope tube. Note that the control grid is placed considerably closer to the cathode than the plate to increase its effectiveness. The control grid structure is very similar to that of a wire mesh fence; that is, the open area is many times greater than that of the wire itself. In this way, the *negative* potential to be applied to the grid can control the flow of charge from cathode to plate without adversely affecting the tube response due to a large number of electrons hitting the grid structure.

The triode symbol and biasing are provided in Fig. 3.26a. Note that a negative potential is applied to the grid and a *positive* potential

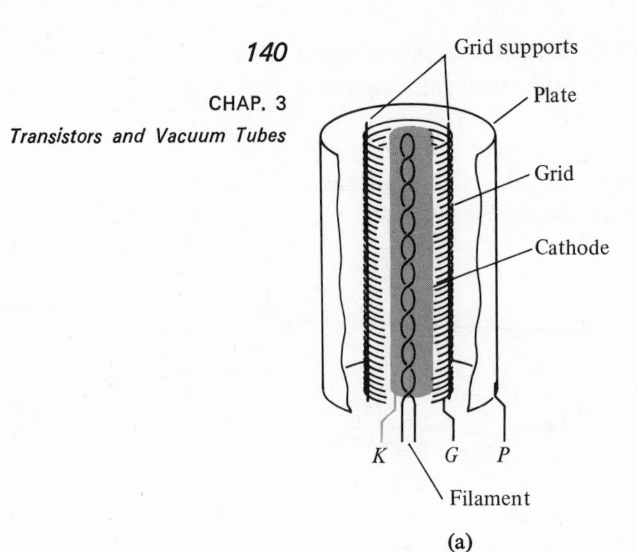

(b)

(Courtesy Radio Corporation of America)

FIG. 3.25

Triode: (a) basic construction for octal base type; (b) photograph of high-mn miniature vacuum tube triode.

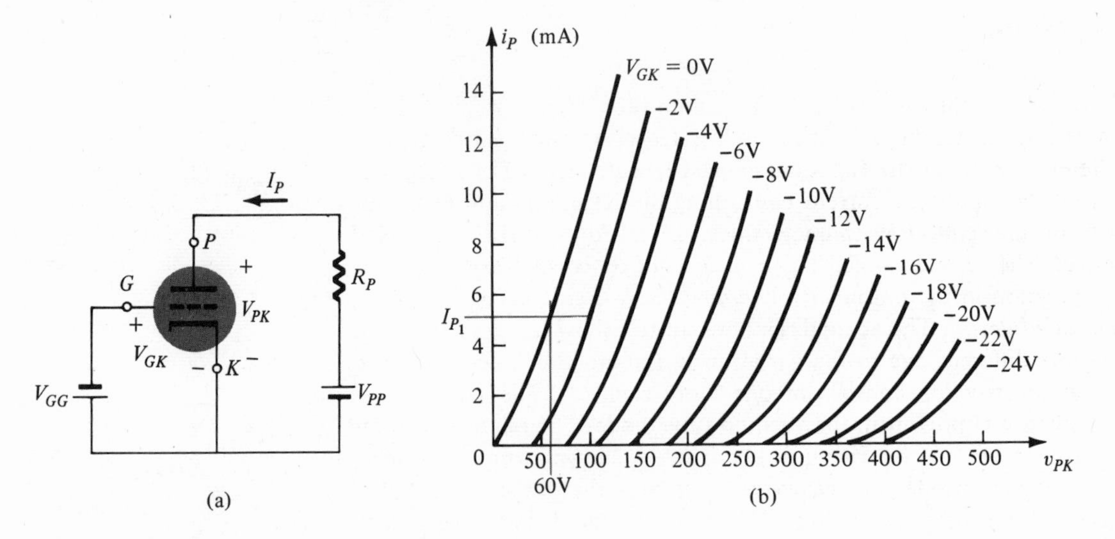

FIG. 3.26

(a) Basic triode biasing circuit; (b) triode plate characteristics.

to the plate. The choice of load resistor and battery potentials will be discussed in detail in Chapter 4. In Fig. 3.26b, the *plate* characteristics of the triode indicate quite clearly the effect that the *grid* potential has on the characteristics. For $V_{GK} = 0$ V, the resulting characteristics are essentially the same as those obtained for the vacuum-tube diode. If a negative potential were applied to the grid, the grid would repel a number of the negatively charged electrons emitted by the cathode back in that direction. This would in turn reduce the plate current. For $V_{GK} = -2$ V the curve has shifted somewhat to the right. To demonstrate the effect just discussed consider the vertical dashed line at $V_{PK} = 60$ V. The plate current corresponding with the intersection of this line with $V_{GK} = 0$ is certainly much higher than that for $V_{GK} = -2$ V. To establish the same plate current (I_{P_1} in Fig. 3.26b) for different values of V_{GK} (0, −2) a higher plate-to-cathode potential would be required. The effect of increasing the plate potential is to increase the attraction for the negatively charged electrons being emitted by the cathode. In other words, the higher positive potential is overriding some of the effect of the increased negative bias on the grid resulting in more electrons being able to pass through to the grid and reach the plate.

EXAMPLE 3.4 Using the characteristics of Fig. 3.26:

(a) Find the change in plate current if the grid-to-cathode (V_{GK}) potential is reduced from $V_{GK} = -12$ V to $V_{GK} = -8$ V at $V_{PK} = 240$ V.

(b) Find the increase in plate-to-cathode potential required to maintain the plate current at 8 mA if V_{GK} is changed from −2 to −8 V.

Solution: (a) At $V_{GK} = -12$ V, $I_{P_1} = 0.6$ mA and at $V_{GK} = -8$ V, $I_{P_2} = 6.6$ mA.
The resulting change in plate current is

$$\Delta I_P = I_{P_2} - I_{P_1} = 6.6 - 0.6 = \mathbf{6\ mA}$$

(b) At $V_{GK} = -2$ V, $V_{PK_1} = 123$ V, and at $V_{GK} = -8$ V, $V_{PK_2} = 248$ V.
The change in plate potential is

$$\Delta V_{PK} = V_{PK_2} - V_{PK_1} = 248 - 123 = \mathbf{125\ V}$$

A second glance at the characteristics of Fig. 3.26 will reveal that the *family* of curves for $V_{GK} = 0, -2, -4, -6$, etc., are fairly linear, parallel, and equidistant from each other in a major portion of the characteristics. It is characteristics of this type that result in the least distorted amplification of signals. For the transistor, the input characteristics were of considerable importance. For the triode, however, we assume that there exists an open circuit between the grid and plate eliminating the necessity for input characteristics. In actual practice, there is a small grid current, even with the negative bias, due to those electrons that bombard the grid structure. In addition, the interelectrode capacitances tie the grid to the plate and cathode with low-impedance values at very high frequencies. The grid of a triode can be made positive with respect to the cathode, but this will result in a heavy grid current and higher internal losses than would result with negative grid voltages for the same amplification.

The extent to which the region described in the previous paragraph can be employed is limited by the ratings of the tube. The *maximum ratings* normally provided on a data sheet include the maximum power or plate dissipation, plate current, and plate-to-cathode potential. For the family of plate characteristics provided in Fig. 3.26b, the following maximum values were provided.

$$P_{D_{max}} = 2.1 \text{ W}$$
$$I_{P_{max}} = 12 \text{ mA}$$
$$V_{PK_{max}} = 350 \text{ V}$$

The plate dissipation is equal to the product of the plate potential and corresponding plate current; that is,

$$\boxed{P_{D_{max}} = V_{PK}I_P} \qquad (3.17)$$

The calculations required with Eq. (3.17) to obtain the maximum power curve are exactly the same as those employed with the maximum transistor ratings. The resulting curve is shown in Fig. 3.27, along with the maximum voltage and current ratings. In addition, a horizontal line has been drawn above that nonlinear region to be avoided if distortion in the output waveform is to be kept to a minimum. The resulting unshaded region is that region remaining for the linear (least distorted) amplification of signals within the device limitations.

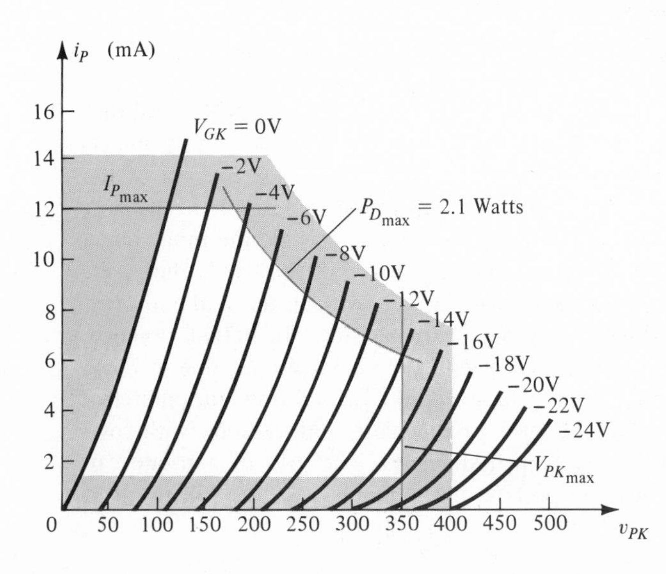

FIG. 3.27
Region of operation.

A second family of curves of interest for the triode amplifier is called the *mutual* or *transfer* characteristics (Fig. 3.28). They show, at a glance, the effect of an input quantity (V_{GK}) on an output quantity (I_P). For *fixed* values of plate voltage (rather than grid voltages as in the plate characteristics) the curve is a plot of I_P vs. V_{GK}. Note, for example, that for $V_{PK} = 400$ V, the less negative the grid potential, the higher the plate current.

The *pin connections* for the triode amplifier will require, as was true for the vacuum diode, the use of the tube manual. The pin connections and some of the associated data, as they would appear in a tube manual for the 6J5 tube, are provided in Fig. 3.29 along with some of their typical areas of application.

The *interelectrode capacitance* values in the specifications of Fig. 3.29 are the capacitance values that exist between the terminals indicated. At high frequencies, these interelectrode capacitances can have a pronounced affect on the triode performance. Consider, for example, the resulting plate-to-cathode capacitive reactance at 1 MHz for the 6J5 triode:

$$X_C = \frac{1}{2\pi fC} = \frac{1}{(6.28)(10^6)(3.6 \times 10^{-12})} \cong 45 \text{ K}$$

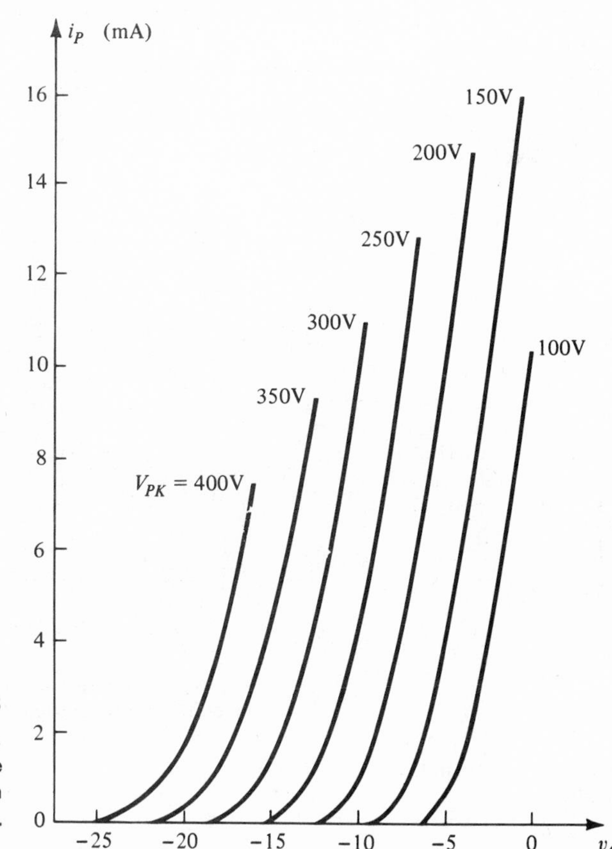

FIG. 3.28
Mutual or transfer characteristics, the triode having the plate characteristics indicated in Fig. 3.26b.

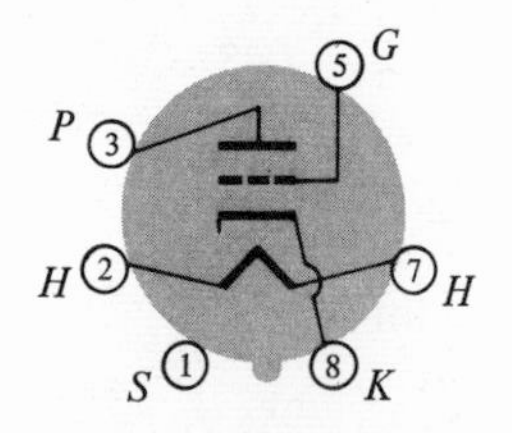

6J5 Triode

Metal and glass octal-type 6J5 used as detectors, amplifiers, or oscillators in radio equipment

Heater voltage (ac/dc) 6.3V
Heater current 0.3A
Direct interelectrode capacitances (approx.)
Grid to plate 3.4pF
Grid to cathode and heater 3.4pF
Plate to cathode and heater 3.6pF

FIG. 3.29

Pin connections and associated data for the 6J5 triode.

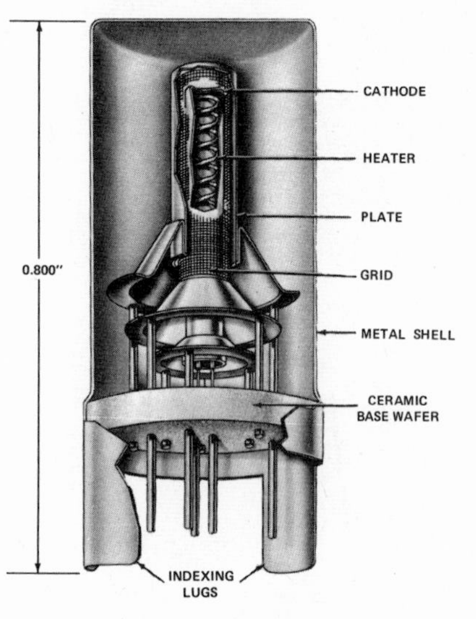

(a) Nuvistor

(Courtesy Radio Corporation of America)

(power amplifier and frequency multiplier):
frequency range—up to 250 MHz
direct interelectrode capacitances:

$$C_{gp} = 2.2 \text{ pf}$$
$$C_{gk} = 4.2 \text{ pf}$$
$$C_{pk} = 0.26 \text{ pf}$$

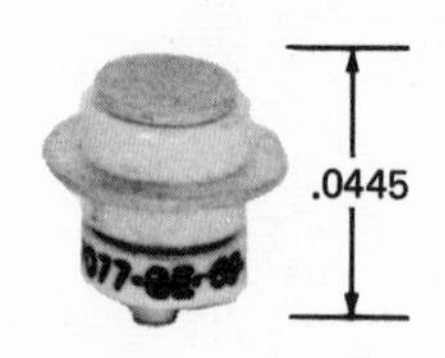

(b) Ceramic

(Courtesy General Electric Company)

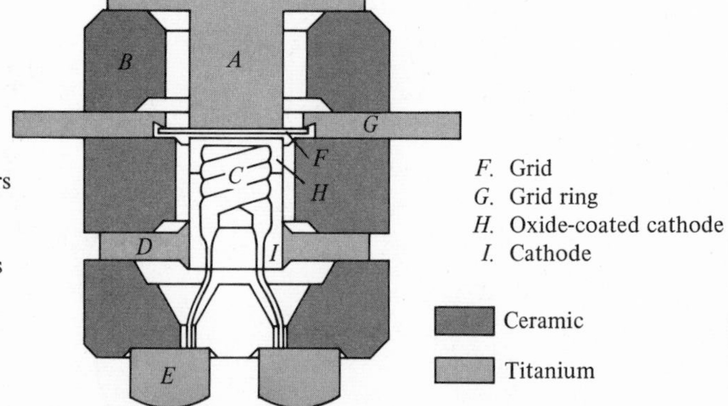

A. Anode
B. Ceramic spacers
C. Heater
D. Cathode ring
E. Heater buttons
F. Grid
G. Grid ring
H. Oxide-coated cathode
I. Cathode

frequency range (grounded grid amplifier)—to 450 MHz
direct interelectrode capacitances:

$$C_{gp} = 1.0 \text{ pf}$$
$$C_{gk} = 1.7 \text{ pf}$$
$$C_{pk} = 0.01 \text{ pf}$$

FIG. 3.30

General construction and pertinent data of the nuvistor and ceramic triode.

A capacitive reactance of 45 K between plate and cathode will in most applications disallow the use of the open-circuit approximation. In recent years, the frequency range has been substantially increased through improved design techniques. Two such triodes with very-high-frequency ranges are the *nuvistor* and *ceramic* triode shown in Fig. 3.30. Note the small size of each triode. Each, by virtue of its construction, can withstand a higher level of shock and vibration. The Nuvistor and 6J5 triode have a *cylindrical*-type assembly, while the Ceramic triode has a *planar* construction. Note in Fig. 3.30 that the grid, plate, and cathode have a single planar surface and do not completely encompass the heater as shown in the cylindrical structure of Fig. 3.25. The data provided in the same figure indicate quite clearly the wide frequency range of each device. Note also the low interelectrode capacitance values of each device. The small-size, high-frequency capabilities of each make it suitable for instrumentation equipment, audio and video equipment, and communication systems.

In many cases, the same heater is used to operate two or more devices in the same package. This is shown in Fig. 3.31 for a 6GU7 double triode. Another variation is the 6SQ7 twin diode-triode, shown in the same figure.

The various biasing techniques, paramaters, and equivalent circuits employed with triodes will be discussed in detail in Chapters 4 and 5.

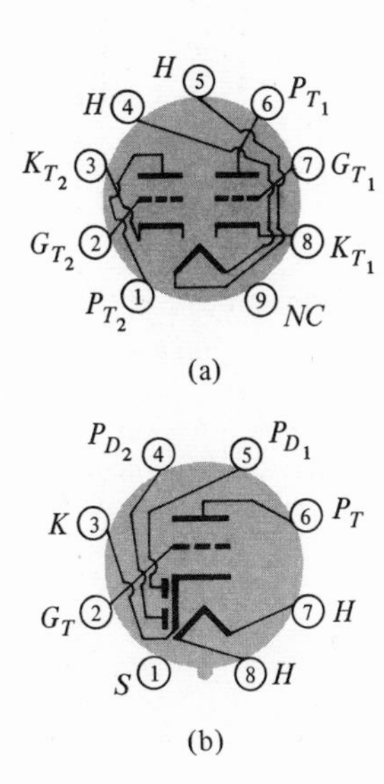

FIG. 3.31
(a) 6GU7 twin triode; (b) 6SQ7 twin diode-triode.

3.12 TETRODE

The *tetrode* is a four-element vacuum-tube device having smaller interelectrode capacitance values and better amplifying characteristics at higher voltages than the triode. The fourth grid, called the *screen* grid, is placed between the control grid and the plate of the tube, closer to the latter as shown in Fig. 3.32a. The screen grid of a tetrode is always typically about 100 V positive with respect to the cathode as shown in the basic biasing circuit of Fig. 3.32b.

The function of the screen grid is to electrostatically *screen* the plate from the control grid and cathode. To understand fully the effects of this positive grid structure in the tetrode tube would require a knowledge of electrostatic theory beyond the scope of this text. However, some appreciation for its isolating effect can be developed by considering an isolated negatively charged electron that has just passed through the control grid structure. As far as the electron is concerned the positive potential to which it would now be attracted is that of the screen grid, even if the plate were at the same or a higher potential. For the 6J5 triode the plate-to-grid capacitance is 3.4 pF, while a typical value for the same capacitance of a tetrode (in this case, the

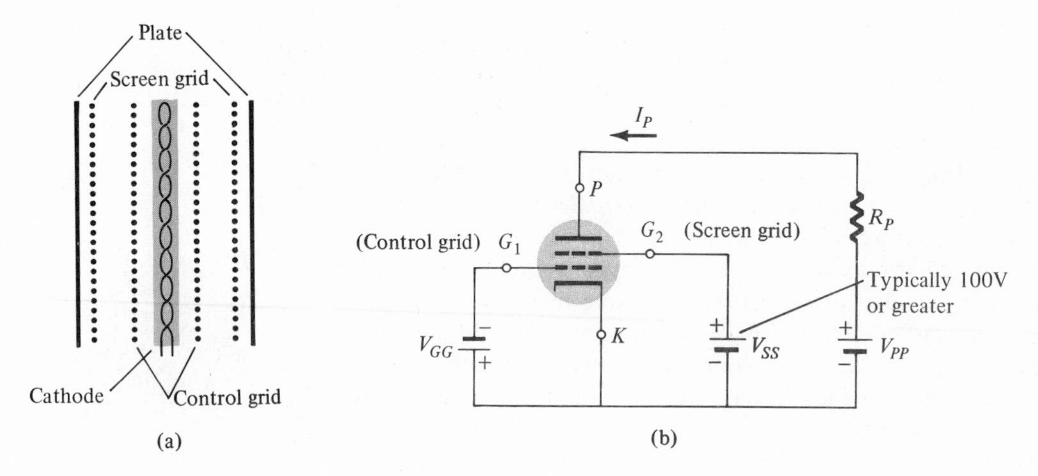

FIG. 3.32

Tetrode: (a) basic construction; (b) basic biasing circuit.

6FV6) is 0.03 pF (less than $\frac{1}{100} = 0.01$ of the triode value). This reduction in interelectrode capacitance will substantially increase the frequency range of the device. The graphical symbol for the tetrode, as shown in Fig. 3.32b, is simply an extension of that used for the triode. In general, the control grid is referred to as G_1 and the screen grid as G_2.

The effect of the screen grid on the characteristics of the device is shown in Fig. 3.33. Note that the most linear region has shifted to a horizontal configuration as compared to the more rising characteristics of the triode. This indicates that the plate potential has less effect on the plate current in the tetrode than it had in the triode. This result certainly agrees with our earlier discussion on the effect of the screen

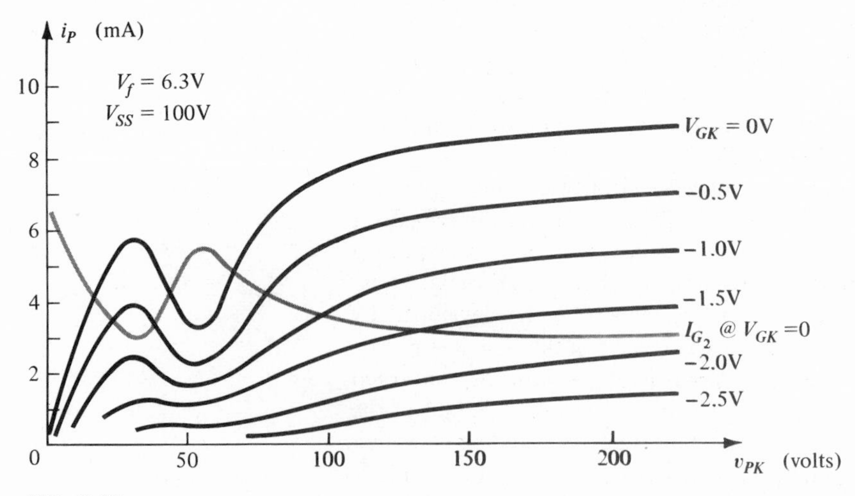

FIG. 3.33

Tetrode plate characteristics.

grid. Note also that the characteristics have a negative-resistance region similar to that of the tunnel diode. This and the neighboring region must be avoided if the output waveform is not to be completely distorted. The negative-resistance region is due to the secondary emission that occurs at the plate of the tube at plate potentials less than the screen potential. If we consider the fictitious case of two electrons being liberated through secondary emission at the plate due to the impact of a single electron, the resulting two electrons will naturally proceed to the structure of highest potential. If the screen grid is at a higher potential than the plate the resulting flow of charge for the situation just described would be increasing for the screen circuit and decreasing for the plate circuit (one electron contributes to the plate current, while two have the opposite effect). The curve continues its downward trend with increasing plate potential because more electrons are liberated through secondary emission due to the increased velocity of the bombarding electron. The velocity increases due to the greater attraction for the plate. Eventually, however, the plate potential will be sufficiently high to attract a number of the secondary emitted electrons *back* to the plate resulting in an increase in plate current as shown in Fig. 3.33. For a fixed grid potential, the plate current will gradually level off resulting in the linear region of Fig. 3.33.

In almost all instances, the tetrode has been replaced by the next device to be considered, the pentode. The tetrode, however, is still used in a limited number of areas, such as RF transmitters and receivers, and it does form a bridge between the triode and pentode, so it has been discussed in detail here.

3.13 PENTODE

The *pentode* vacuum-tube amplifier has improved characteristics over both the triode and tetrode. Its interelectrode capacitance values are in general less than those of the tetrode and its amplifying capabilities surpass either the triode or tetrode by as much as 1000:1 and 5:1, respectively. In addition, its characteristics, shown in Fig. 3.34 with the basic biasing circuit, do not have the nonlinear, negative-resistance region of the tetrode, thereby permitting lower operating potentials. The pentode, as the name implies, has five elements as shown in Fig. 3.35. The fifth structure to be added, called the *suppressor* grid, is placed between the plate and screen grid and is normally connected directly to the cathode (normally grounded) as shown in Fig. 3.34.

The purpose of this additional grid is to "suppress" the secondary emission current from the plate to the screen grid. Any electron that now finds itself between the plate and suppressor grid due to secondary emission will have to choose between the normally grounded suppressor grid (zero potential) and the positive plate. The natural result will be a

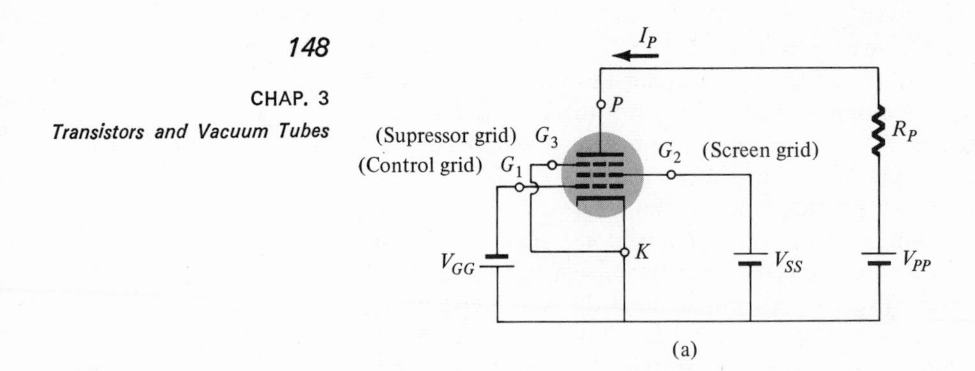

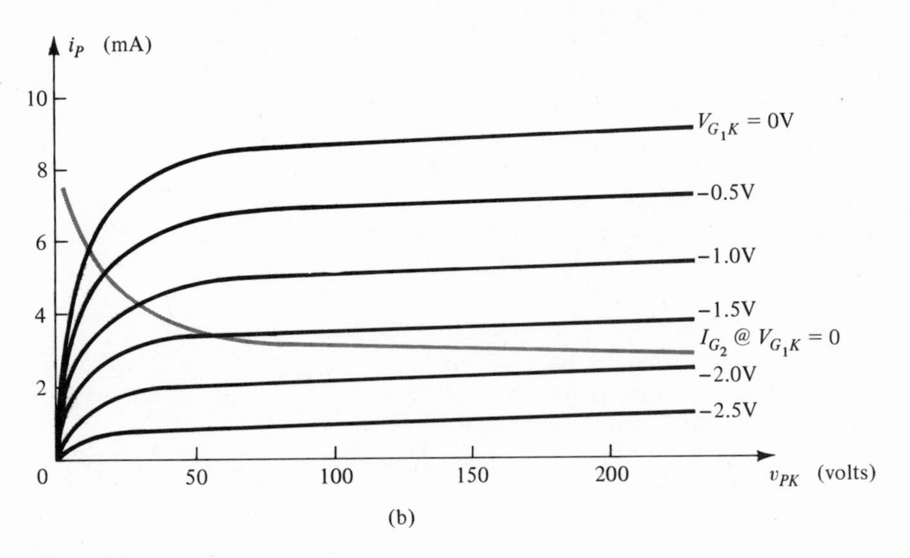

FIG. 3.34
Pentode: (a) basic biasing circuit; (b) plate characteristics.

return to the plate as indicated in Fig. 3.36. The net result is the elimination of the negative-resistance region as shown in Fig. 3.34.

The pin connections, pertinent data, and photographs of a miniature pentode and a pentode-triode tube are shown in Fig. 3.37 with an area of application. Additional data are also normally available, but, as before, data of the triode and pentode are only slightly different due to the addition of the screen and suppressor grid, so that the ac analysis of each will be quite similar. Due to the present-day lack of interest in the development of the tetrode it will not be considered in the discussions to follow. However, other four-element tube devices, such as the *beam power tube* which has a basic structure somewhat different from the tetrode, will be considered in Chapter 9.

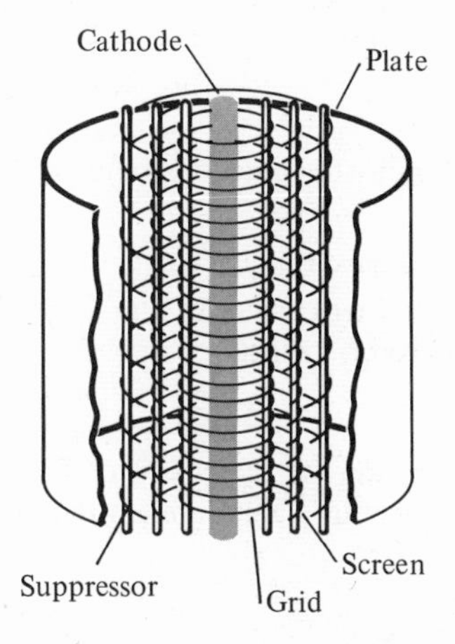

FIG. 3.35

Basic pentode construction.

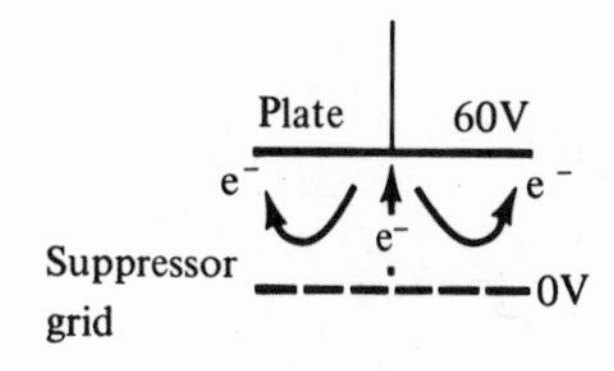

FIG. 3.36

Effect of the suppressor grid in the vacuum-tube pentode.

Heater voltage = 6.3 volts
C_{pk} = 5.0 pf
Plate voltage = 330 max.
Plate dissipation = 3.5 watts

(a)

(Courtesy Radio Corporation of America)

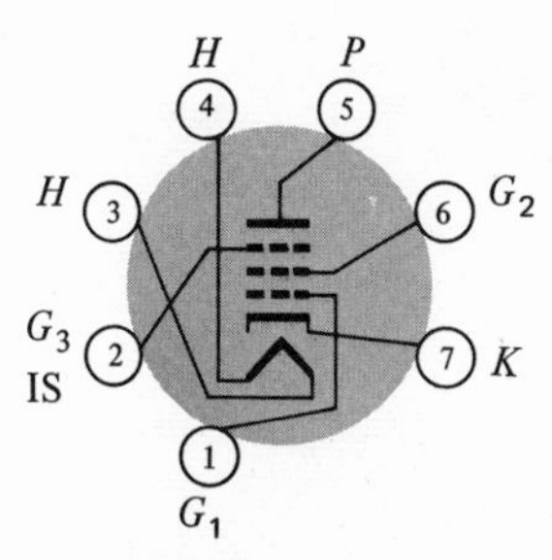

Application:
In limiter stages of FM receivers

Heater voltage = 6.3 volts
C_{gp} = 2 pf (triode)
= 0.06 pf (pentode)
Plate voltage = 330 V max (triode and pentode)
Plate dissipation = 2.4 watts (triode)
3 watts (pentode)

(b)

(Courtesy Radio Corporation of America)

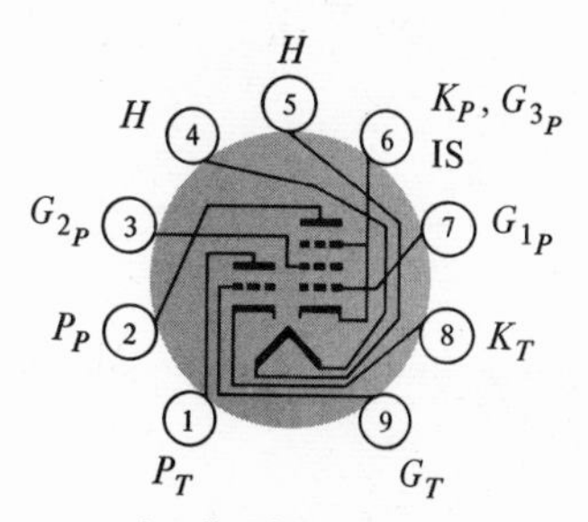

Application:
High quality, high-fidelity audio applications

FIG. 3.37

(a) Miniature pentode vacuum tube; (b) pentode-triode vacuum tube.

§ 3.3

1. How must the two transistor junctions be biased for transistor operation?

2. Why does a transistor have leakage current?

3. A transistor has an emitter current of 8 mA and an alpha (α) of 0.99. How large is the collector current?

§ 3.4

4. Calculate the voltage gain for the circuit of Fig. 3.7, if $V_i = 500$ mV and $R = 1$ K. (Other circuit values remain the same.)

§ 3.5

5. Using the characteristics of Fig. 3.9:

(a) Find the resulting collector current if $I_E = 5$ mA and $V_{CB} = -10$ V.

(b) Find the resulting collector current if $V_{EB} = 250$ mV and $V_{CB} = -10$ V.

(c) Find V_{EB} for the conditions $I_C = 4$ mA and $V_{CB} = -15$ V.

§ 3.6

6. Is there a difference between I_{CO} and I_{CEO}? If so, what is it?

7. Using the characteristics of Fig. 3.12:

(a) Find the value of I_C corresponding to $V_{BE} = -350$ mV and $V_{CE} = -5$ V.

(b) Find the value of V_{CE} and V_{BE} corresponding to $I_C = -3$ mA and $I_B = -30\ \mu$A.

8. (a) For the common-emitter characteristics of Fig. 3.12 find the dc beta at an operating point of $V_{CE} = -8$ V and $I_C = -2$ mA.

(b) Find the value of α corresponding to this operating point.

(c) At $V_{CE} = -8$ V find the corresponding value of I_{CEO}.

(d) Calculate the approximate value of I_{CBO} using the dc beta value obtained in part (a).

§ 3.7

9. An input voltage of 2 V rms (measured from base to ground) is applied to the circuit of Fig. 3.15. Assuming that the emitter voltage follows the base voltage exactly, calculate the circuit voltage amplification and emitter current for $R_E = 1$ K.

§ 3.11

10. Using the characteristics of Fig. 3.26:

(a) Find the change in plate current if the grid-to-cathode (V_{GK}) potential is reduced from $V_{GK} = -6$ V to $V_{GK} = -10$ V at $V_{PK} = 200$ V.

(b) Find the increase in plate-to-cathode potential required to maintain the plate current at 6 mA, if V_{GK} is changed from -2 to -12 V.

11. Determine the ac resistance of a triode using the characteristic of Fig. 3.27 for operation between plate currents of 2–4 mA for the following two grid-to-cathode voltages. $r_p = \left.\frac{\Delta V_{PK}}{\Delta I_P}\right|_{V_{GK} = \text{constant.}}$
(a) $V_{GK} = -2$ V.
(b) $V_{GK} = -20$ V.
12. Using the mutual characteristic of Fig. 3.28 determine the change in plate current at a plate-cathode voltage of 200 V for a grid-cathode voltage change from −5 to −10 V.
13. Using the mutual characteristic of Fig. 3.28 determine the necessary change in grid-cathode voltage to cause a plate current change from 2 to 10 mA at a plate-cathode voltage of 250 V.

§ 3.12

14. Describe two differences between a triode and a tetrode.

§ 3.13

15. How does the pentode differ from the triode?
16. Determine the change in plate current for a grid-cathode voltage change from −1.5 to −2.0 V at a plate-cathode voltage of 200 V using the pentode characteristic of Fig. 3.34.
17. Determine the approximate change in plate current using the characteristic of Fig. 3.34 for a plate-cathode voltage change from 100 to 200V at a grid-cathode voltage of −1 V.

4

DC Biasing

4.1 GENERAL

Transistors and vacuum tubes are used in a large variety of applications and in many different ways. It would be difficult if not impossible to learn each area and application of these devices. Instead, one studies the more fundamental properties and aspects of these devices so that enough is known to carry over this knowledge to slightly different or even completely different applications. The present chapter covers the basic concepts in the dc biasing of transistor and tube devices.

To use these devices for amplification of voltage or current signals, or as control (ON or OFF) elements, or in any other application, it is necessary first to *bias* the device. The usual reason for this biasing is to turn the device on, and in particular, to place it in operation in the region of its characteristic where the device operates most linearly, providing a constant amount of gain.

Although the purpose of the bias network or biasing circuit is to cause the device to operate in this desired *linear* region of operation (which is defined by the manufacturer for each device type) the bias components are still part of the overall application circuit—amplifier, waveform shaper, logic circuit, etc. We could treat the overall circuit and consider all aspects of the operation at once, but this is more complex and more confusing. Each type of circuit application would have to be studied for all aspects of operation without a more basic understanding of those common features of operation of most other application circuits. This chapter therefore provides basic concepts of dc biasing of the transistor and tube devices, with the understood aim of getting the device operating in a desired region of the device characteristic. If these concepts are well understood, many different circuits, even new circuit applications, can be studied and analyzed more easily because a basic understanding of the circuit has been established. Amplifier gain and other factors affecting ac operation will be considered in Chapter 5. With this breakdown the basic concepts can be presented and consideration can then be given in Chapters 7, 8, 11, 12, and 14 to the use of these devices in specific areas of application, with the main concern being the problems and operations peculiar to that area of interest.

The dc biasing is a *static* operation since it deals with setting a fixed (steady) level of current flow (through the device) with a desired fixed voltage drop across the device. The necessary information about the device can be obtained from the device's static characteristics, both input and output. As will be shown, the transistor biasing considerations can be obtained from the manufacturer's listing for a particular device, whereas the tube is best described by a set or family of curves of output plate-to-cathode voltage and plate current. Thus, we shall be able, with these two devices, to cover the essential features of the two basic techniques (analytical or mathematical, and graphical) used to determine the operation of the circuit, and to set the desired operating voltages and currents in the circuit.

4.2 OPERATING POINT

Since the aim of biasing is to achieve a certain condition of current and voltage called the *operating point* (or *quiescent* point) some attention is given to the selection of this point in the device characteristic. Figure 4.1 shows a general device characteristic with four indicated operating points. The biasing circuit may be designed to set the device operation at any of these points or others within the *operating region*. The operating region is the area of current or voltage within the maximum limits for the particular device. These maximum ratings are indicated on the characteristic of Fig. 4.1 by a horizontal line for the maximum current, I_{max}, and a vertical line for the maximum voltage, V_{max}. An additional consideration of maximum power (product of voltage and current) must also be taken into consideration in defining the operating region of a particular device, as shown by the dotted line marked P_{max} on Fig. 4.1.

It should be realized that the device could be biased to operate outside these maximum limit points but that the result of such operation would be either a considerable shortening of the lifetime of the device, or destruction of the device. Confining ourselves to the safe operating region we may select many different operating areas or points. The exact point or area often depends on the intended use of the circuit.

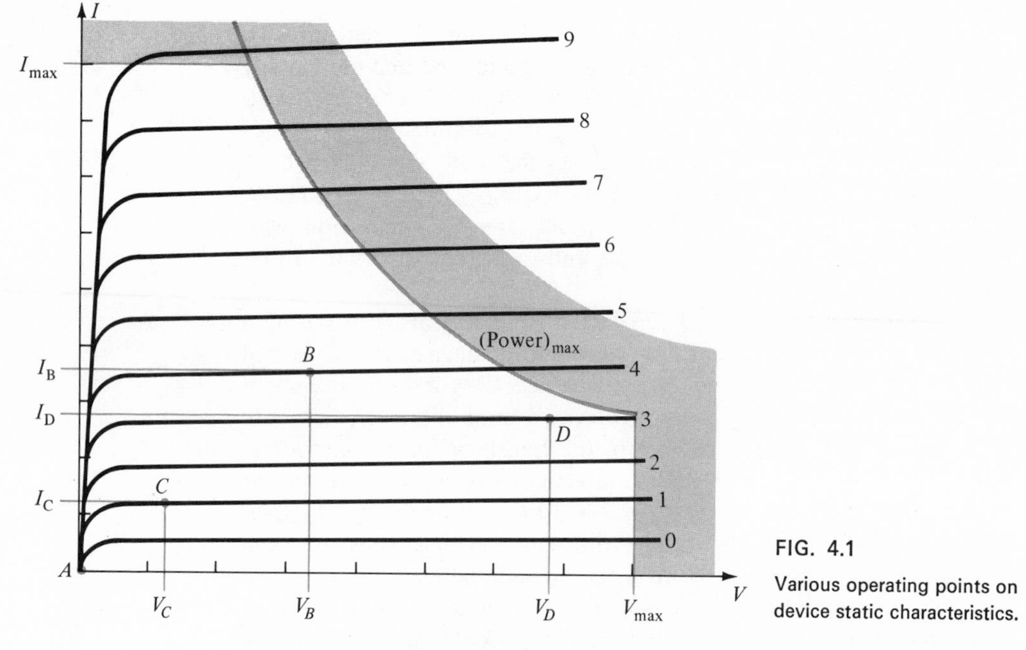

FIG. 4.1 Various operating points on device static characteristics.

Still, we can consider some differences between operation at the different points shown in Fig. 4.1 to present some basic ideas about the operating point and, thereby, the bias circuit.

If no bias were used the device would initially be completely off, which would result in the current of point *A*—namely, zero current through the device (and zero voltage across it). The undesirable aspect of operating point *A* is that it will react only to the part of an input signal that causes the device to conduct. Thus, a sinusoidal voltage going both positive and negative around 0 V would be amplified only during the signal swing time when the active device (transistor or tube) is caused to operate and conduct. When the input signal goes negative the device remains off and no change occurs in the current or voltage of the device. It is necessary to bias the device so that it can respond or change in current and voltage for the entire range of the input signal. While point *A* would not be suitable, point *B* provides this desired operation. If a signal is applied to the circuit, *in addition to the bias level*, the device will vary in current and voltage from operating point *B*, allowing the device to react to (and possibly amplify) both the positive and negative part of the input signal. If, as could be the case, the input signal is small, the voltage and current of the device will vary but not enough to drive the device into *cutoff* or *saturation*. Cutoff is the condition in which the device no longer conducts. Saturation is the condition wherein voltage across the device is as small as possible with the current flow in the device path reaching a limiting value depending on the external circuit. The usual amplifier action desired occurs within the operating region of the device—that is, between saturation and cutoff.

Point C would also allow some positive and negative variation with the device still operating, but the output could not vary too negatively (left of V_C) because bias point C is lower in voltage than point B. Point C is also in a region of operation in which the current level in the device is smaller and the device gain is *not* linear—that is, the spacing in going from one curve to the next is unequal. This nonlinearity shows that the amount of gain of the device is smaller lower on the characteristic and larger higher up. It is preferable to operate where the gain of the device is most constant (or linear) so that the amount of amplification over the entire swing of input signal is the same. Point B is in a region of more linear spacing and therefore more linear operation, as shown in Fig. 4.1. Point D sets the device operating point near the maximum voltage level. The output voltage swing in the positive direction is thus limited if the maximum voltage is not to be exceeded. Point B, therefore, seems the best operating point in terms of linear gain or largest possible voltage and current swing. This is usually the desired condition for small-signal amplifiers (Chapter 5) but not necessarily for power amplifiers and logic circuits, which will be considered in Chapters 8 and 13, respectively. In this discussion, let us concentrate mainly on biasing the device for *small-signal* amplification operation.

One other very important biasing factor must be considered. Having selected and biased for a desired operating point, the effect of temperature must also be taken into account. Temperature causes the device characteristic to change. Higher temperature results in more current flow in the device than at room temperature, thereby upsetting the operating condition set by the bias circuit. Because of this, the bias circuit must also provide a degree of *temperature stability* to the circuit so that temperature changes at the device produce minimum change in its operating point. This maintenance of operating point may be specified by a *stability factor*, S, indicating the amount of change in operating-point current due to temperature. A highly stable circuit is desirable and the stability factors of a few basic bias circuits will be compared.

Tube operation is described by a graphical static characteristic and the selection of a biasing point is obtained from this characteristic. Transistor operation, however, may be specified sufficiently well by device parameters and more mathematical (rather than graphical) techniques can be used to determine its biasing. The transistor characteristic, nevertheless, still provides a convenient picture for understanding device operation and will be used on occasion.

4.3 COMMON-BASE (CB) BIAS CIRCUIT

The common-base (CB) configuration provides a relatively straightforward and simple starting point in our dc bias considerations. Figure

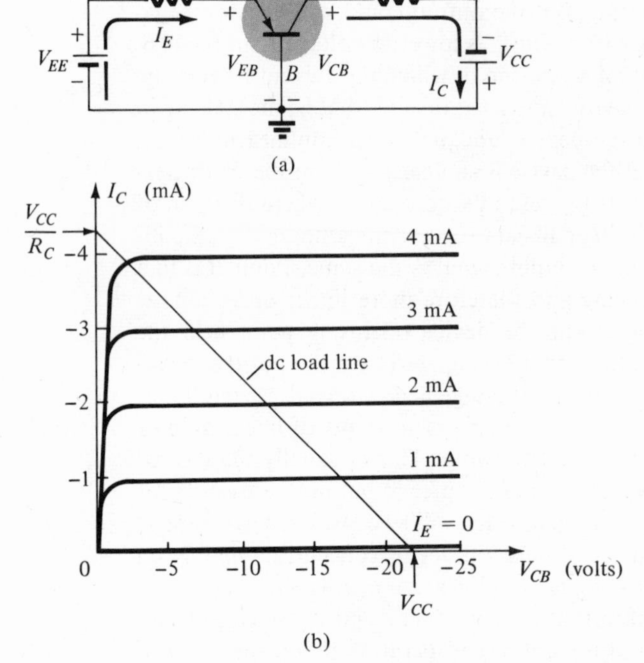

FIG. 4.2

Common-base circuit and transistor characteristic: (a) common-base circuit; (b) dc load line on common-base characteristic.

4.2a shows a common-base circuit configuration. By CB we mean that the base is the reference point for measurement for both input (emitter, in this case) and output (collector).

The dc supply (battery) terminals are marked with a double-letter designation. V_{EE} is the dc voltage supply associated with the emitter section, and V_{CC} is the dc supply for the collector section of the circuit. Two separate voltage supplies are required for the CB configuration, one of the detrimental factors of this connection. Resistor R_E is basically a current-limiting resistor for setting the emitter current I_E. Resistor R_C is the collector (output or load) resistor and the output ac signal is developed across it. It is also one of the components that are used to set a desired operating point.

Figure 4.2b shows the CB collector characteristic, which is the output characteristic for this circuit connection. The abscissa is the collector-to-base voltage V_{CB}, which is a negative voltage for the PNP circuit of Fig. 4.2a. The ordinate is the collector current, I_C. Negative-current values are given, using the standard definition that all currents flowing into the transistor are taken as the *positive*-current direction. Since collector current actually flows out of the PNP transistor, the current polarity marked on the characteristic is negative. The family of curves for the characteristic are for various emitter currents and are positive values since the emitter current flows into the PNP transistor.

The theory developed for either the PNP or NPN transistor

device can be equally applied to the other device by merely changing all current directions and all voltage polarities. We shall continually change back and forth between NPN and PNP units in developing concepts, so that circuit design or analysis can be made equally well with either.

It is possible to consider biasing the CB circuit by analyzing separately the input (base-emitter loop) and the output (base-collector loop) sections of the circuit. Although in reality there is some interaction between the operation of the base-collector section and that of the base-emitter section this can be neglected with excellent practical results obtained.

Input Section

The input loop (see Fig. 4.3a) is composed of the V_{EE} battery, the resistor, R_E, and the emitter-base of the transistor. Writing the voltage loop equation (using Kirchhoff's voltage law) for the input loop

$$+V_{EE} - I_E R_E - V_{EB} = 0$$

from which we get

$$\boxed{I_E = \frac{V_{EE} - V_{EB}}{R_E}} \tag{4.1}$$

where all terms have been defined above.

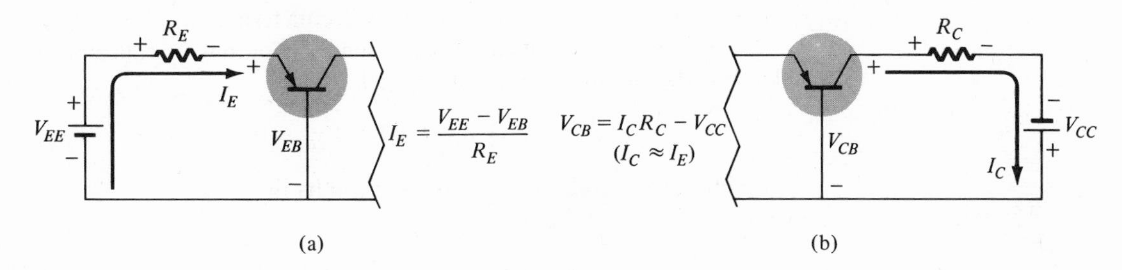

FIG. 4.3
Input and output sections for common-base circuit: (a) input (emitter-base) section only; (b) output (collector-base) section only.

When forward biased, the emitter-base voltage V_{EB} is quite small —on the order of 0.3 V for germanium transistors and 0.7 V for silicon. Although the actual emitter-base voltage is slightly affected by the collector-base voltage, this effect can be neglected for practical considerations. In fact, if the supply voltage V_{EE} is, say, 10 V or more, the emitter-base voltage could be neglected, giving

$$I_E \cong \frac{V_{EE}}{R_E} \tag{4.2}$$

as a good approximation. Observe that the emitter current is set essentially by the emitter supply voltage and the emitter resistor. But the supply voltage is usually fixed since it is required to provide voltage to other parts of the electronic circuit. The emitter current, therefore, is specifically determined by the emitter resistor R_E, whose value is selected to give the desired emitter current.

Output Section

The output loop (see Fig. 4.3b) consists of the V_{CC} battery, the resistor R_C, and the collector-base junction of the transistor. For operation as an amplifier the *collector-base* junction must be *reverse biased* in addition to the *emitter-base* being *forward biased.* This *reverse bias* is provided by the V_{CC} battery voltage connected in polarity so that the ***p**ositive* battery terminal connects to the ***n**-material* and the ***n**egative* battery terminal to the ***p**-material*. (Note the letter opposites for *p* and *n* for battery and transistor type.) The result of this consideration is that for PNP transistors the battery polarity should be positive terminal to the common-base point and negative terminal to the resistor connected to the collector terminal. (For NPN transistors we have the opposite—the negative battery terminal connected to the CB terminal and the positive terminal of the battery connected to the resistor, which then connects to the collector terminal.)

Summing the voltage drops around the output or collector-base loop of the circuit of Fig. 4.3b, we get

$$+V_{CC} - I_C R_C + V_{CB} = 0$$

Solving for the collector-base voltage results in

$$V_{CB} = -V_{CC} + I_C R_C \tag{4.3}$$

The collector current I_C is approximately the same magnitude as the emitter current I_E [obtained from Eqs. (4.1) or (4.2)]. This is a very good approximation that will be true for any type of transistor connection used. For the purposes of calculating circuit bias values we may write the relation as

$$I_C \cong I_E \tag{4.4}$$

actually, $I_C = \alpha I_E$, where α (alpha) is typically 0.9–0.998 in value.

Having presented the essential circuit operation of the CB connection we can consider the complete solution of bias currents and voltages. The results obtained will apply to both NPN and PNP CB transistor circuits. To help in the solution of bias conditions for a CB circuit as in Fig. 4.2a and to provide a structured procedure to be followed in later types of circuit connections we shall formalize the solution into a step-by-step calculation. To solve for the bias voltages and currents of a CB bias circuit as in Fig. 4.2a proceed as follows:

1. With emitter-base voltage polarity providing forward bias, assume approximate voltages of

$$V_{EB} \cong 0.3 \text{ V (germanium), or}$$
$$V_{EB} \cong 0.7 \text{ V (silicon)}$$

   (Forward bias is provided if the battery voltage connection results in ***p***ositive voltage at *p*-type material and ***n***egative voltage at *n*-type material of the transistor.)

2. Calculate the emitter current I_E using

$$I_E = \frac{V_{EE} - V_{EB}}{R_E} \cong \frac{V_{EE}}{R_E} \quad \text{(PNP circuit)}$$
$$I_E = \frac{V_{EE} + V_{EB}}{R_E} \cong \frac{V_{EE}}{R_E} \quad \text{(NPN circuit)} \tag{4.5}$$

3. Collector current is approximately the emitter current calculated in step (2) above:

$$I_C \cong I_E \tag{4.6}$$

4. Collector-base voltage, is calculated from

$$V_{CB} = -V_{CC} + I_C R_C \quad \text{(PNP circuit)}$$
$$V_{CB} = V_{CC} - I_C R_C \quad \text{(NPN circuit)} \tag{4.7}$$

EXAMPLE 4.1 Calculate bias voltages V_{EB} and V_{CB} and currents I_E and I_C for the circuit of Fig. 4.4. The circuit contains a PNP Germanium transitor with alpha (α) of 0.99.

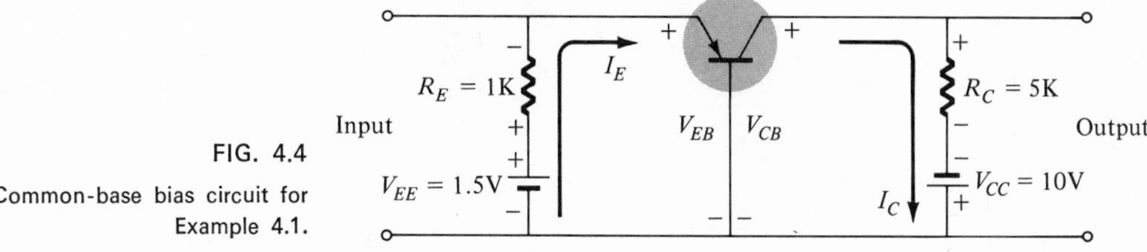

FIG. 4.4 Common-base bias circuit for Example 4.1.

Solution: Using the step-by-step procedure outlined previously:

(a) $V_{EB} \cong 0.3$ (germanium).

(b) $I_E = (V_{EE} - V_{EB})/R_E = (1.5 - 0.3\text{ V})/1\text{ K} = 1.2\text{ mA}$.

(c) $I_C \cong I_E = 1.2\text{ mA}$.

(d) $V_{CB} = -V_{CC} + I_C R_C = -10 + 1.2\text{ mA} \times 5\text{ K} = -10 + 6 = -4\text{ V}$.

EXAMPLE 4.2 Repeat Example 4.1 for the NPN silicon transistor circuit of Fig. 4.5.

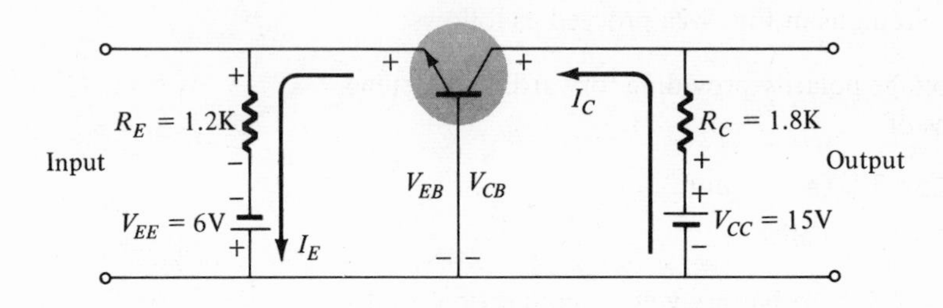

FIG. 4.5 Common-base bias circuit for Example 4.2.

Solution

(a) $V_{EB} \cong -0.7$ (silicon).

(b) $I_E = (V_{EE} + V_{EB})/R_E = (6 - 0.7\text{ V})/1.2\text{ K} = 4.4\text{ mA}$.

(c) $I_C \cong I_E = 4.4\text{ mA}$.

(d) $V_{CB} = V_{CC} - I_C R_C = 15 - (4.4\text{ mA})(1.8\text{ K}) = 15 - 7.9 = 7.1\text{ V}$.

4.4 COMMON-EMITTER (CE) CIRCUIT CONNECTION—GENERAL BIAS CONSIDERATIONS

A more popular amplifier connection applies the input signal to the base of the transistor with the emitter as common terminal. The CE circuit of Fig. 4.6 shows one advantage of this connection—only one supply voltage. Recall that the CB circuit requires two supply voltages,

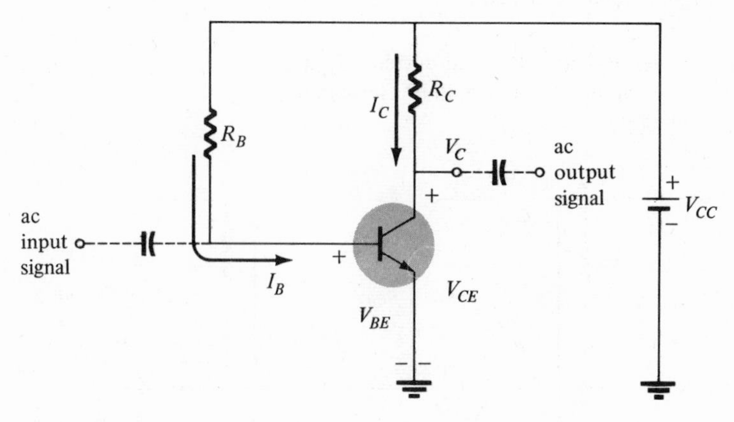

FIG. 4.6 Common-emitter fixed-bias circuit.

one to forward bias the base-emitter and the second to provide reverse bias for the base-collector. Both forward- and reverse-bias conditions are achieved in the CE connection using one voltage supply. In addition, we shall show a number of other important advantages of the CE circuit relating to input and output impedances, current and voltage gain, etc., which apply to the ac operation of the circuit. The present chapter deals with the dc biasing of the circuit and these other important circuit factors will be considered in Chapter 5.

Although a mathematical (rather than graphical) technique is used to obtain the bias voltages and currents in the circuit, we shall refer to the CE collector characteristic of the transistor to provide a reason for choosing a particular operating point. Figure 4.7 shows the

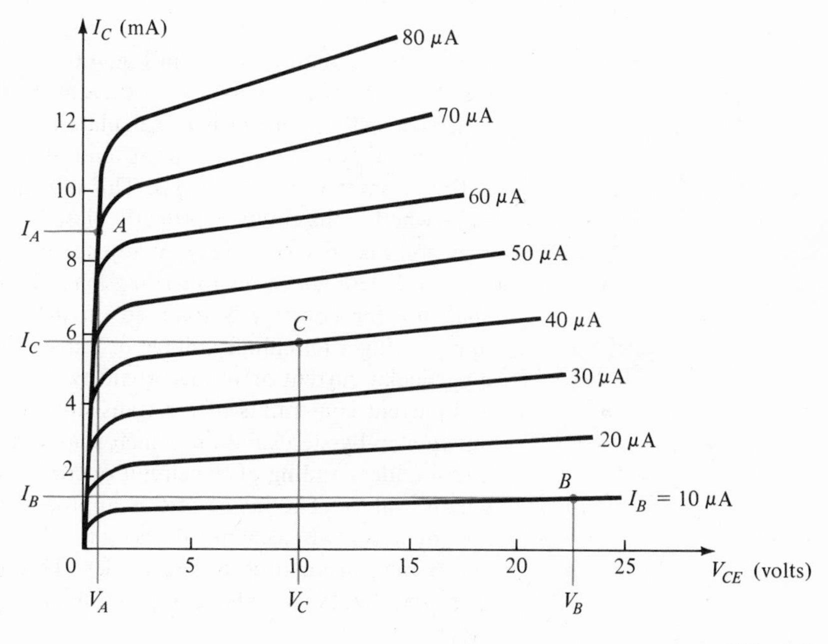

FIG. 4.7
Common-emitter collector characteristics showing typical bias points.

CE collector characteristic of a transistor indicating a few bias points. The result of connecting a supply voltage V_{CC} and resistors R_B and R_C, as in Fig. 4.6, is to cause both base and collector current flow with a resulting bias voltage V_{CE} measured from collector to emitter (common lead).

It would seem from the discussion in Section 4.2 about selecting a suitable bias point that point C in Fig. 4.7 would be satisfactory. It would allow a range of voltage for the collector-emitter voltage to be either increased or decreased by the input ac signal.

The remainder of this chapter will concentrate on (a) determining what bias point actually will result for a given circuit containing given circuit elements (*analysis*) and (b) how to obtain the circuit elements to provide a desired bias point of collector current and collector-emitter voltage (*synthesis*). To present the basic circuit theory and provide some appreciation of the circuit operation, analysis of a given circuit will be considered first. After that, the design (synthesis) of a circuit to obtain a desired bias condition will be covered.

4.5 BIAS CONSIDERATIONS FOR A FIXED-BIAS CIRCUIT

A fixed-bias circuit is shown in Fig. 4.6 with typical component values. An NPN transistor is used with current gain (beta or h_{FE}) of 50. Our objective at the moment is to calculate the base current I_B, the collector current I_C in the given circuit, and also the base-emitter (V_{BE}) and collector-emitter (V_{CE}) voltages. This information will be used to determine whether the circuit is properly biased and also to consider how the bias point is affected in case an adjustment is desired.

We shall *not* resort to writing many loop equations but rather we shall use the constraints imposed on the circuit by the transistor to apply a single Kirchhoff voltage or current loop equation to solve for a particular current or voltage quantity. Using, in addition, the voltage and current constraints of the transistor (which are quite simple) permits a step-by-step solution, which should also provide a more meaningful understanding of the circuit operation and expected current and voltage values. The intent of this approach is that numerical answers are not arrived at mechanically using involved mathematical techniques, which then mean little to the reader. Throughout the procedure the question, "Does the value computed make sense?," will and should be asked.

The transistor in Fig. 4.6 can be used as an amplifying device if the base-emitter is forward biased and the base-collector reverse biased. Let us see how this is accomplished in Fig. 4.6 using a single voltage supply. The positive supply voltage V_{CC} connected through resistor R_B provides a positive voltage at the base of the transistor with respect to the common-emitter point (ground). With the base-emitter junction of the transistor forward biased, the voltage drop across the junction is determined mainly by the transistor type and only very slightly by the supply voltage used or even by the base resistor value. What is essential is that the supply voltage polarity be such that the base-emitter is forward biased (and not reverse biased). As a means of remembering the polarity needed for forward bias consider the following: *p*-material is connected toward the ***p***ositive-voltage side and *n*-

material is connected toward the ***n***egative-voltage side of the supply battery. For the NPN transistor in Fig. 4.6 this would require the base (p-material) to be connected toward the positive of the supply voltage and the emitter (n-material) to be connected toward the negative terminal of the battery. Since this is precisely the connection in Fig. 4.6 the base-emitter will be forward biased. For a silicon transistor the forward-bias voltage developed across the transistor base-emitter will be around 0.7 V, typically, with possible values from 0.5 up to 1.0 V, depending on base current, temperature, etc. For present purposes typical values of 0.6–0.7 V will be quite suitable. For germanium transistors typical emitter-base forward-bias voltage will be 0.2–0.3 V.

A reverse-bias voltage must also be provided by the single voltage supply. For the NPN transistor this would require a collector voltage more positive than the base voltage; that is, the collector-base voltage must be ***p***ositive toward ***n***-material and ***n***egative toward p-material for reverse bias. Since the supply voltage is connected in the collector-emitter circuit, how does the collector-base become reverse biased? Referring to Fig. 4.6, the base-emitter will be forward biased with a few tenths volt from base to emitter (or from base to ground). As long as the collector voltage is more positive than V_{BE} (a few tenths volt), the voltage measured from collector to base V_{CB} will be positive to negative (more positive to less positive) and the collector-base junction will be reverse biased. If the collector voltage drops below the few tenths base-emitter voltage, the collector-base is no longer reverse biased and the device no longer acts as an amplifier. If V_{CB} becomes forward biased, the transistor is in saturation—one of the voltage states desired in computer switching circuits, but not in amplifier circuits. This bias condition will be considered in Chapter 13 when switching circuits are covered.

Thus, the single supply V_{CC} provides forward bias for the base-emitter junction and reverse bias for the collector-base junction to bias the transistor into operation as an amplifier device. A PNP transistor would be biased in much the same way as that discussed above. The only difference would be that a single negative battery voltage polarity would be used, all voltages would be negative, and all currents would be opposite those shown in Fig. 4.6. The magnitudes of the calculated values would, however, be the same for either transistor type.

4.6 CALCULATION OF BIAS POINT FOR FIXED-BIAS CIRCUIT

Given the fixed-bias circuit of Fig. 4.6, how can we determine the dc bias currents and voltages in the base and collector of the transistor? The present section will develop a step-by-step procedure, which will determine the answers to the above question, thereby providing analysis of a given fixed-bias circuit.

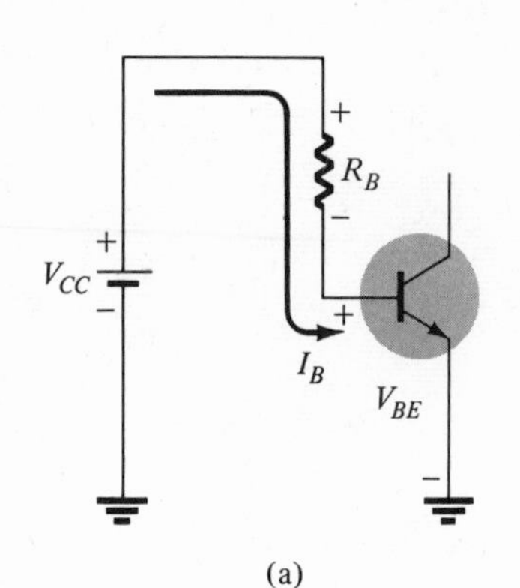

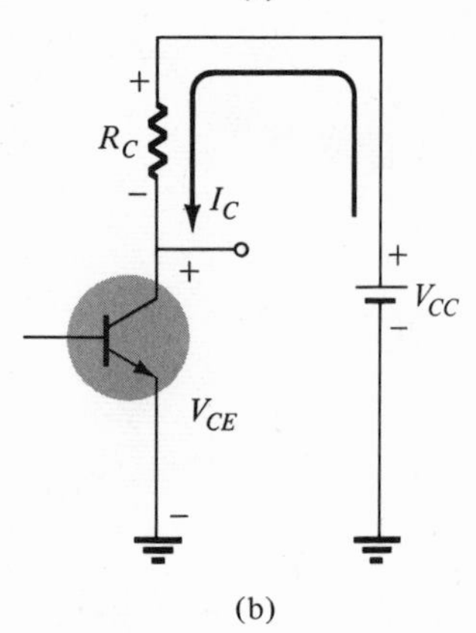

FIG. 4.8
Separate input and output loops for fixed-bias circuit: (a) input base-emitter loop; (b) output collector-emitter loop.

Input Section

To provide for simple step-by-step analysis consider only the base-emitter circuit loop shown in the partial circuit diagram of Fig. 4.8a. Writing the Kirchhoff voltage equation for the given loop we get

$$+V_{CC} - I_B R_B - V_{BE} = 0$$

We can solve the above equation for the base current I_B

$$\boxed{I_B = \frac{V_{CC} - V_{BE}}{R_B}} \tag{4.8}$$

Since the supply voltage V_{CC} and the base-emitter voltage V_{BE} are fixed values of voltage the selection of a base bias resistor fixes the value of the base current. As a good approximation we may even neglect the few tenths volt drop across the forward-biased base-emitter V_{BE}, obtaining the simplified form for calculating base current,

$$\boxed{I_B \cong \frac{V_{CC}}{R_B}} \tag{4.9}$$

Output Section

The output section of the circuit (Fig. 4.8b) consists of the supply battery, the collector (load) resistor, and the transistor collector-emitter junctions. The currents in the collector and emitter are about the same since I_B is quite small in comparison to either. For linear amplifier operation the collector current is related to the base current by the transistor current gain, beta (β) or h_{FE}. Expressed mathematically

$$\boxed{I_C = \beta I_B} \tag{4.10}$$

The base current is determined from the operation of the base-emitter section of the circuit as provided by Eq. (4.8) or (4.9). The collector current as shown by Eq. (4.10) is β times greater than the base current *and* not at all dependent on the resistance in the collector circuit. From the previous consideration of the common-base circuit, we know that the collector current is controlled in the base-emitter section of the circuit and not in the collector-base (or collector-emitter, in this case) section of the circuit.

Calculating voltage drops in the output loop

$$V_{CC} - I_C R_C - V_{CE} = 0$$

$$\boxed{V_{CE} = V_{CC} - I_C R_C} \tag{4.11}$$

Equation (4.11) shows that the sum of voltages across the collector-emitter and across the collector resistor is the supply voltage value. This can also be stated as: The supply voltage provides the voltages across the collector resistor and across the collector-emitter—or the voltage across the collector-emitter V_{CE} is the remaining voltage from that of the voltage supply minus that voltage dropped across the collector resistor.

Thus, the analysis of the dc bias point for a given fixed-bias circuit can be provided as a step-by-step procedure with separate consideration of the input and output loops for the circuit of Fig. 4.6. To summarize the method of solution the equations used in the step-by-step calculation are rewritten below.

1. Calculate the base current using

$$I_B = \frac{V_{CC} \pm V_{BE}}{R_B} \cong \frac{V_{CC}}{R_B} \tag{4.12}$$

Use the absolute value of the supply voltage with the understanding that the base current flows in the direction specified by the arrow of the transistor emitter symbol (into the base for NPN and out of the base for PNP transistors). Use a plus sign for the PNP circuit and a minus sign for the NPN circuit in Eq. (4.12).

2. Calculate the collector current from

$$I_C = \beta I_B \tag{4.13}$$

(Again the collector current direction is that specified by the transistor emitter symbol—into the collector for NPN and out of the collector for PNP transistors.)

3. Calculate the voltage across the collector-emitter using

$$\begin{aligned} V_{CE} &= V_{CC} - I_C R_C \quad \text{(NPN transistor)} \\ V_{CE} &= -V_{CC} + I_C R_C \quad \text{(PNP transistor)} \end{aligned} \tag{4.14}$$

(Recalling that the collector-emitter voltage and the voltage across the collector resistor add up to the supply voltage it would be unreasonable to obtain—and accept—a value of collector-emitter dc bias voltage *larger* than the supply voltage. Probably a sign error has been made and in no case would it be reasonable merely to accept the number as correct.)

One additional consideration must be included in the above solution steps. The relation between collector and base current, namely, that $I_C = \beta I_B$, is true *only* if the transistor is properly biased in the linear region of the transistor's operation. If the transistor, for example, is biased in the *saturation* region (too large an amount of base bias current), the use of Eqs. (4.13) and (4.14) leads to incorrect results.

For the transistor to be biased in a region of linear amplifier operation (as opposed to regions of cutoff or saturation) the base-emitter junction must be forward biased and the base-collector junction reverse biased. Our concern here is with the second bias condition—that the collector-base be properly reverse biased. This is true only as long as the collector-emitter voltage V_{CE} is larger in value than the base-emitter forward-bias voltage V_{BE}. Since the collector-emitter voltage V_{CE} given by Eq. (4.14) is the difference between the supply voltage V_{CC} and the voltage drop across the collector resistor ($I_C R_C$), the latter must be less than V_{CC} or in terms of the collector current, I_C must be less than V_{CC}/R_C. Stated mathematically,

$$I_C \leqq \frac{V_{CC}}{R_C} \tag{4.15}$$

for the transistor to be biased in the active (linear) region of operation. A quick check of Eq. (4.13) would therefore be in order using Eq. (4.15) when performing the calculations of collector-emitter voltage, to make sure that the condition just stated is correct in the circuit under consideration. If so, the three solution steps outlined above can be carried out, as representing the operation of the circuit. If, however, the above relation of maximum I_C allowable for operation in the transistor linear region is exceeded, the transistor is operating in the saturation region. In this case the collector current will be the maximum value set by the circuit:

$$I_{C_{sat}} = \frac{V_{CC}}{R_C}$$

and

$$V_{CE_{sat}} \cong 0\text{ V} \qquad \text{(actually a few tenths volt)} \tag{4.16}$$

The base current calculated from Eq. (4.12) is correct in any case.

If the circuit to be analyzed is used as an amplifier we shall not expect it to be biased in the saturation region. If, however, some value used is incorrect, or some wiring error occurs, the resulting operation might possibly bias the transistor into saturation and we must be aware of this condition. (Keep in mind that the saturation condition is undesirable only for amplifier operation. For operation in computer switching circuits, the saturation region of operation is quite important and we shall consider it fully under that topic in Chapter 13.)

EXAMPLE 4.3 Compute the dc bias voltages and currents for the NPN CE circuit of Fig. 4.9.

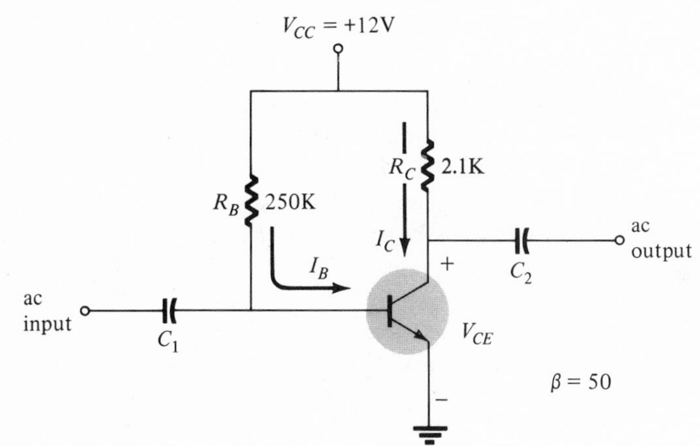

FIG. 4.9
dc fixed-bias circuit for Example 4.3.

Solution

(a) $I_B = (V_{CC} - V_{BE})/R_B \cong V_{CC}/R_B = 12\text{ V}/250\text{ K} = 48\ \mu\text{A}$

(b) $I_C = \beta I_B = 50(48\ \mu\text{A}) = 2.4\text{ mA}.$

(c) $V_{CE} = V_{CC} - I_C R_C = 12 - (2.4\text{ mA})(2.1\text{ K}) = 12 - 5 = 7\text{ V}.$

EXAMPLE 4.4 Calculate the collector-emitter voltage, V_{CE} for the PNP CE circuit of Fig. 4.10.

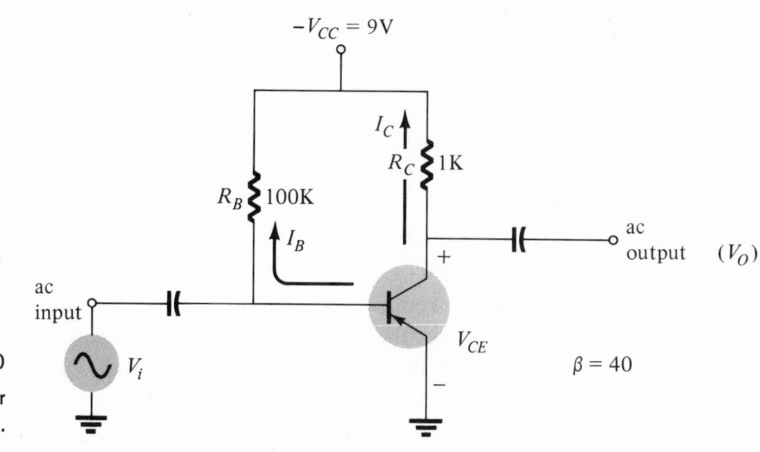

FIG. 4.10
dc fixed-bias circuit for Example 4.4.

Solution

(a) $I_B \cong V_{CC}/R_B = 9\text{ V}/100\text{ K} = 90\ \mu\text{A}$.

(b) $I_C = \beta I_B = 40(90\ \mu\text{A}) = 3.6\text{ mA}$.

(c) $V_{CE} = -V_{CC} + I_C R_C = -9 + (3.6\text{ mA})(1\text{ K}) = -9 + 3.6$
$= \mathbf{-5.4\ V}$.

4.7 NEED FOR BIAS STABILIZATION

Analysis of the fixed-bias circuit previously considered is quite simple. But the circuit has some very poor operating features in its failure to retain a fixed-bias point over a wide range of temperatures and with substitution of different transistors (even of the same type). Two transistor parameters that vary with temperature (or are not well defined in value) are leakage current (I_{CO}) and current gain (β). Germanium transistors, for example, have considerable leakage current at room temperature (typically, a few microamperes at 25°C) and considerably more leakage current at high temperatures (hundreds of microamperes to milliamperes at up to 110°C). Additional leakage current produces a shift of the dc bias point, as will be shown later in this section. The fixed-bias circuit of Fig. 4.8 cannot maintain a constant dc bias point with increased leakage current and, therefore, is a poor circuit over a wide range of operating temperatures. Silicon transistors, on the other hand, have little leakage current at room temperature, typically a few nanoamperes, and still small leakage at high temperatures—say, a few microamperes at about 100°C. However, where the germanium transistor current gain remains fairly constant with temperature, the silicon transistor current gain does not. Since the value of beta is important to the bias point for the fixed-bias circuit some means of maintaining a dc bias condition will be necessary over a range of temperature operation.

The present discussion will concern itself with demonstrating the effect of leakage current and current gain change on the dc bias point initially set by the circuit, and some general means for adjusting the bias point. Sections 4.8, 4.9, and 4.10 will provide the detail analysis of better stabilized dc bias circuits. Consider the graphs of Fig. 4.11a and b, which show a transistor collector characteristic at room temperature (25°C) and the same transistor at some elevated temperature (100°C). Notice that the significant increase of leakage current not only causes the curves to rise but also that an increase in beta occurs as shown by the larger spacing between the curves at the higher temperature.

The operating point may be specified by drawing the circuit dc load line on the graph of the collector characteristic and noting the intersection of the load line and the dc base current set by the input circuit. An arbitrary point is marked as an example in Fig. 4.12a.

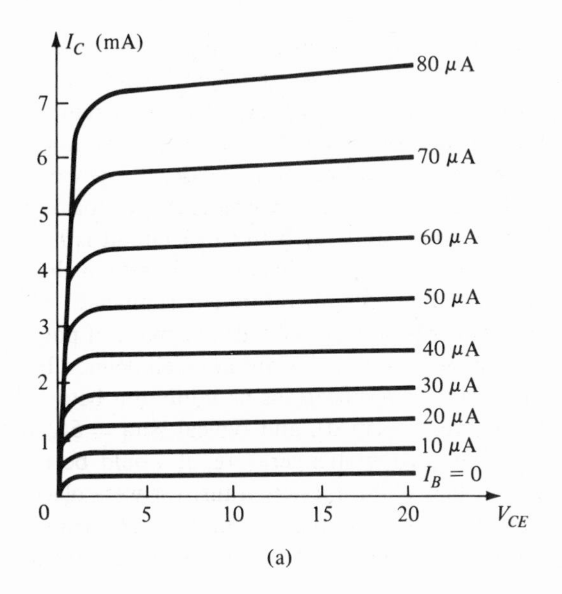

(a)

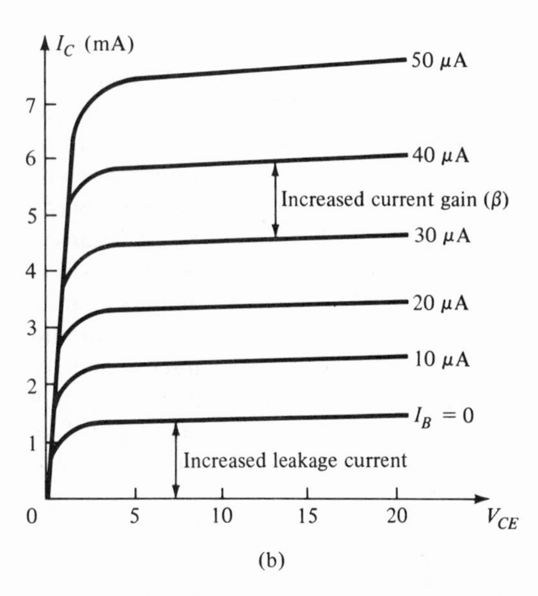

(b)

FIG. 4.11

Transistor common-emitter collector characteristics at: (a) 25°C; (b) 100°C.

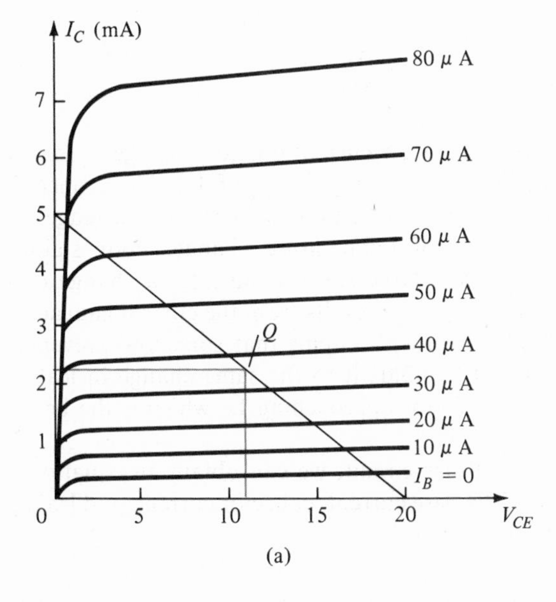

(a)

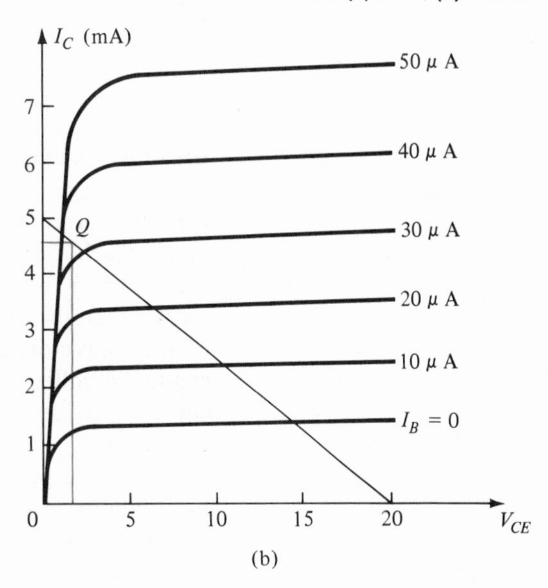

(b)

FIG. 4.12

Shift in dc bias point (*Q*-point) due to change in temperature: (a) 25°C; (b) 100°C.

Since the fixed-bias circuit provides a base current whose value depends approximately on the supply voltage and base resistor, neither of which are affected by temperature or the change in leakage current or beta,

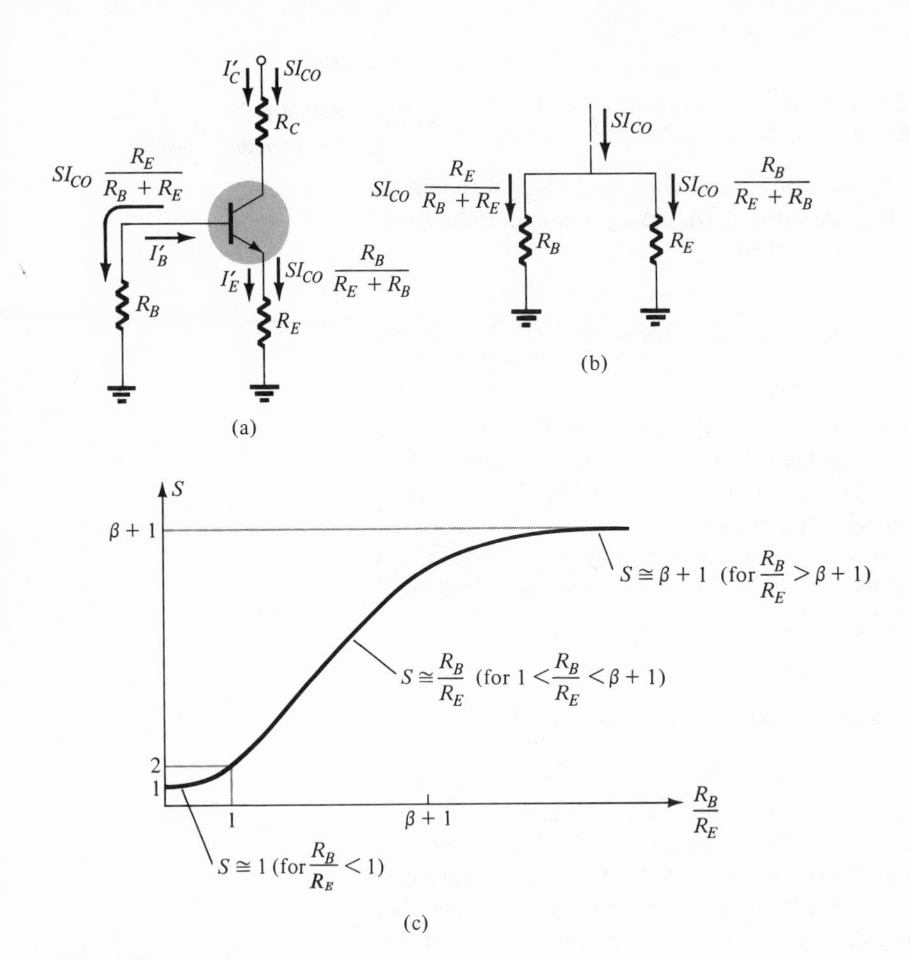

FIG. 4.13
Effect of I_{CO} on bias point.

To show how the emitter and base resistance values affect the amount of stability, a plot of stability vs. the ratio of R_B/R_E is provided in Fig. 4.13c. The figure shows that at best the stability (S) is 1 for values of $R_B \ll R_E$. The stability increases as R_E is made larger (or R_B smaller) with the largest stability being limited to the current gain of the transistor (actually to $\beta + 1$). Usually, the values of R_B and R_E are such that the stability falls somewhere within the limits of 1 and $\beta + 1$, but hopefully closer to the lower limit of 1.

EXAMPLE 4.5 For the CE circuit shown in Fig. 4.14, calculate

(a) the stability factor.
(b) the ratio of R_B/R_E for a stability of 10.
(c) the value of R_E required for R_B as given in Fig. 4.14.

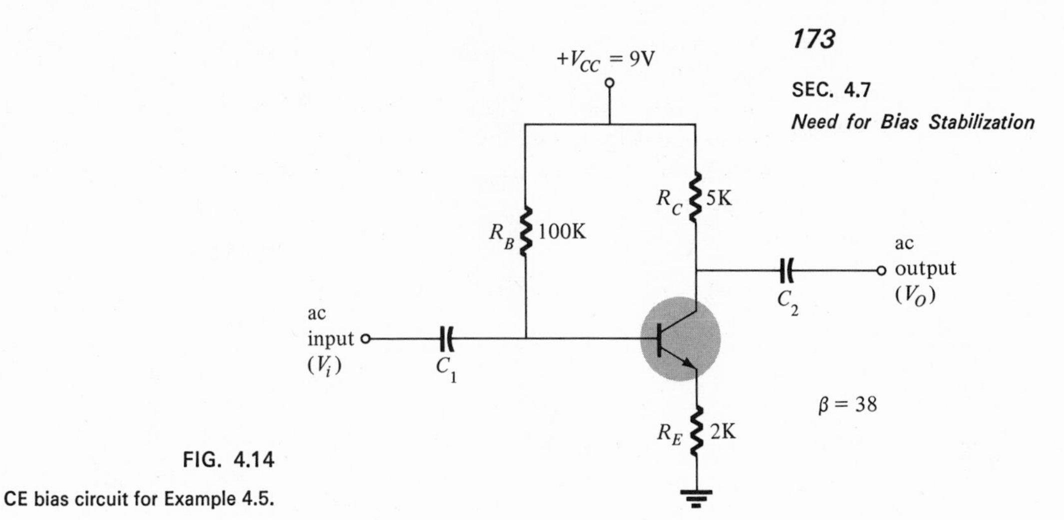

FIG. 4.14
CE bias circuit for Example 4.5.

Solution

(a) $S' \cong \beta + 1$ for $R_B/R_E = 100/2 = 50 > \beta + 1 = 39$.
(b) For $S = 10$, $R_B/R_E \cong 10 < \beta + 1 = 39$.
(c) With $R_B = 100$ K as given, $R_E = 100$ K$/10 = 10$ K.

Voltage Feedback Stabilization

Another method to improve the dc bias stability is to feed back some of the collector voltage to the base circuit. Thus an increase in collector voltage would provide an increased amount of voltage feedback to the base. Since there is a voltage phase reversal between base and collector an increase in base voltage would oppose the increased collector voltage to provide some stabilization. Figure 4.15 shows a dc bias circuit with voltage feedback. Solving for the stability provides the approximate relation

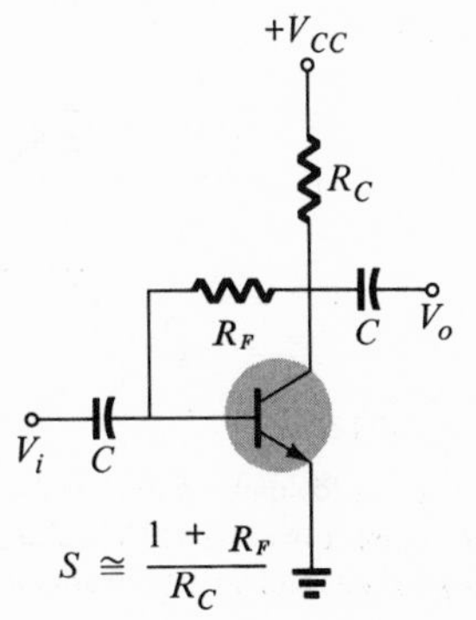

$$S \cong 1 + \frac{R_F}{R_C} \tag{4.22}$$

which shows that the stability is smallest for values of R_F much smaller than R_C. If $R_F = 0$, for example, the stability is 1, which is excellent. However, if R_F is very small (compared to R_C) the voltage gain of the resulting circuit is quite small, making the circuit operation quite poor. A compromise must be made between the voltage gain for the amplifier circuit and the amount of dc bias stability. Practical examples of voltage feedback stabilized circuits will be considered in detail in Section 4-10.

$+V_{CC}$, R_C, R_F, C, V_o, V_i, C

$S \cong \frac{1 + R_F}{R_C}$

FIG. 4.15
Voltage feedback stabilized dc bias circuit.

As a summary of the methods of providing better dc bias stability, Fig. 4.16 shows circuits using the methods previously considered in this section. The first (Fig. 4.16a) shows the use of current feedback; the second (Fig. 4.16b), the use of voltage feedback; and the third (Fig. 4.16c), the combined use of both current and voltage feedback in a single circuit.

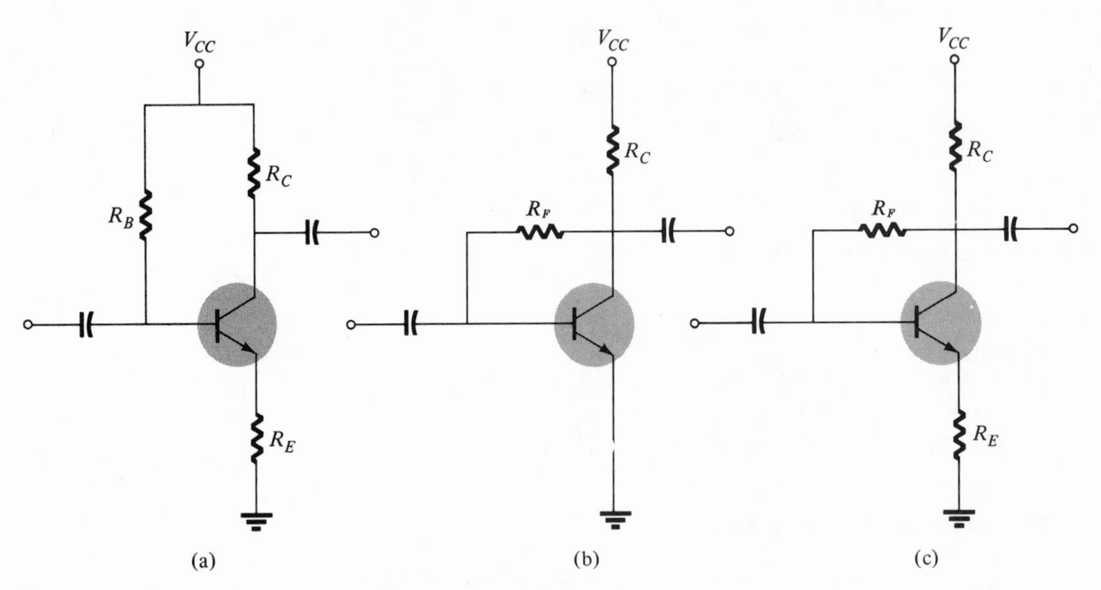

FIG. 4.16
Circuits providing voltage and current feedback: (a) current feedback; (b) voltage feedback; (c) current and voltage feedback.

4.8 DC BIAS CIRCUIT WITH EMITTER RESISTOR

The dc bias circuit of Fig. 4.17 contains an emitter resistor to provide better bias stability than the fixed-bias circuit considered in Section 4.6. For the analysis of the circuit operation we shall deal separately with the base-emitter loop of the circuit and the collector-emitter loop of Fig. 4.17.

Input Section (Base-Emitter Loop)

A partial circuit diagram is shown in Fig. 4.18a of the base-emitter loop. Writing Kirchhoff's voltage equation for the loop

$$V_{CC} - I_B R_B - V_{BE} - I_E R_E = 0$$

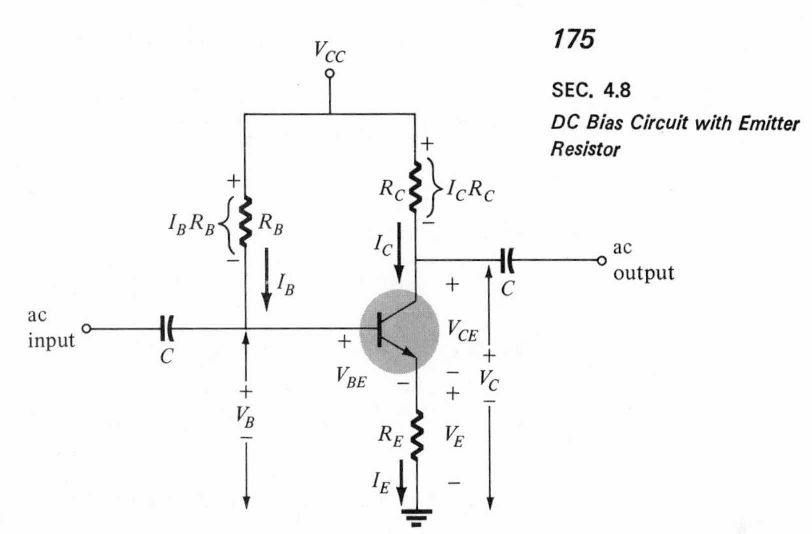

FIG. 4.17
dc bias circuit with emitter stabilization resistor.

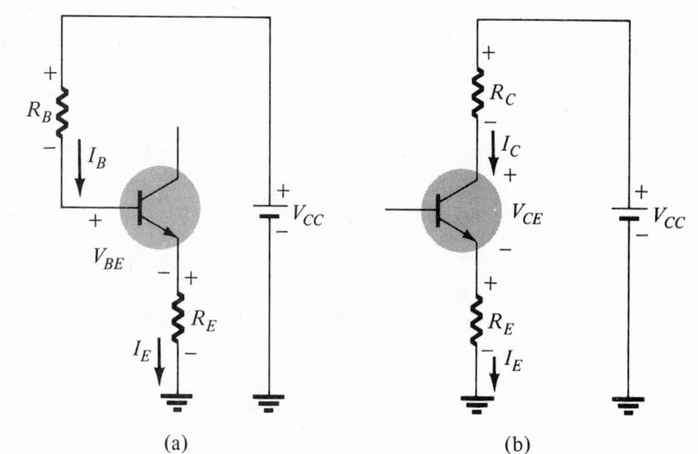

FIG. 4.18
Input and output loops of the circuit of Fig. 4.17: (a) input loop; (b) output loop.

We can replace I_E with $(\beta + 1)I_B$ so that the above equation can be written as

$$V_{CC} - I_BR_B - V_{BE} - (\beta + 1)I_BR_E = 0$$

Solving for the base current we get

$$\boxed{I_B = \frac{V_{CC} - V_{BE}}{R_B + (\beta + 1)R_E}} \qquad (4.23)$$

Note that the difference between the fixed-bias current calculation [Eq. (4.8)] and Eq. (4.23) is the additional term of $(\beta + 1)R_E$ in the denominator.

The collector-emitter loop is shown in Fig. 4.18b. Writing the Kirchhoff voltage equation for this loop

$$V_{CC} - I_C R_C - V_{CE} - I_E R_E = 0 \tag{4.24}$$

Using the relation

$$I_C \cong I_E \tag{4.25}$$

we can solve for the voltage across the collector-emitter:

$$\boxed{V_{CE} \cong V_{CC} - I_C(R_C + R_E)} \tag{4.26}$$

The voltage measured from emitter to ground is

$$V_E = I_E R_E \cong I_C R_E \tag{4.27}$$

and the voltage measured from collector to ground is

$$V_C = V_{CC} - I_C R_C \tag{4.28}$$

The voltage at which the transistor is biased is that measured from collector to emitter, V_{CE}, which is given by Eq. (4.26) and may also be calculated as

$$V_{CE} = V_C - V_E \tag{4.29}$$

The dc bias analysis of the currents and voltages for the circuit of Fig. 4.17 can be summarized as follows:

1. Calculate the input base current using

$$I_B = \frac{V_{CC} - V_{CE}}{R_B + (\beta + 1)R_E} \cong \frac{V_{CC}}{R_B + \beta R_E} \tag{4.30}$$

2. Calculate the collector and emitter currents using

$$I_C = \beta I_B \cong I_E \tag{4.31}$$

3. Calculate the collector-emitter voltage using

$$\begin{aligned} V_{CE} &= V_{CC} - I_C R_C - I_E R_E \quad \text{(NPN)} \\ V_{CE} &= -V_{CC} + I_C R_C + I_E R_E \quad \text{(PNP)} \end{aligned} \tag{4.32}$$

EXAMPLE 4.6 Calculate all dc bias voltages and currents in the circuit of Fig. 4.19.

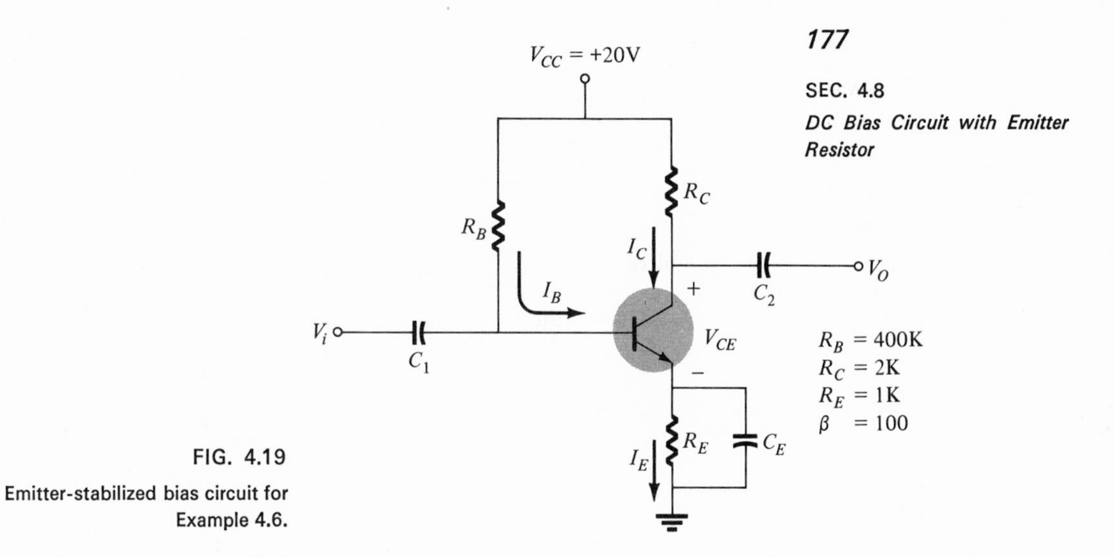

FIG. 4.19

Emitter-stabilized bias circuit for Example 4.6.

Solution

(a) $I_B \cong V_{CC}/(R_B + \beta R_E) = 20\text{ V}/[400\text{ K} + 100(1\text{ K})] = 20\text{ V}/500\text{ K}$
$= \mathbf{40\ \mu A}$.

(b) $I_C = \beta I_B = 100(40\ \mu\text{A}) = \mathbf{4\ mA} \cong I_E$.

(c) $V_{CE} = V_{CC} - I_C R_C - I_E R_E = 20 - (4\text{ mA})\,2\text{ K} - (4\text{ mA})\,1\text{ K}$
$= 20 - 8 - 4 = \mathbf{8\ V}$.

EXAMPLE 4.7 Calculate all dc bias voltages and currents in the circuit of Fig. 4.20.

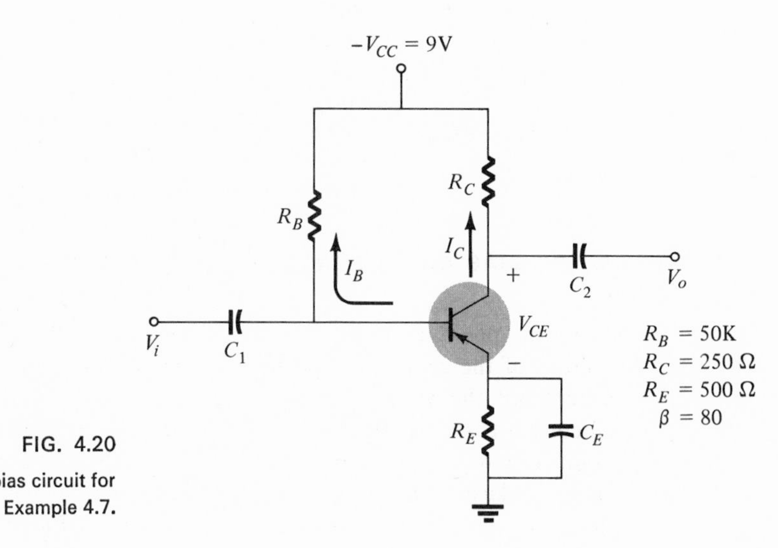

FIG. 4.20

Emitter-stabilized bias circuit for Example 4.7.

$$V_B = \frac{R_2}{R_1 + R_2} V_{CC} \tag{4.34}$$

where V_B is the voltage measured from base to ground.

Since the voltage drop across the base-emitter, when forward biased, is quite small (and slightly dependent on the base current) we can calculate the voltage at the emitter (Fig. 4.21) from

$$V_E = V_B - V_{BE} \cong V_B \tag{4.35}$$

The current in the emitter may then be calculated (Fig. 4.21) from

$$I_E = \frac{V_E}{R_E} \tag{4.36}$$

and the collector current is then

$$I_C \cong I_E \tag{4.37}$$

The voltage drop across the collector resistor is

$$V_{R_C} = I_C R_C \tag{4.38}$$

The voltage at the collector (measured with respect to ground) can be obtained

$$V_C = V_{CC} - V_{R_C} = V_{CC} - I_C R_C \tag{4.39}$$

and finally, the voltage from collector to emitter is calculated from

$$V_{CE} = V_C - V_E$$

$$V_{CE} = V_{CC} - I_C R_C - I_E R_E \tag{4.40}$$

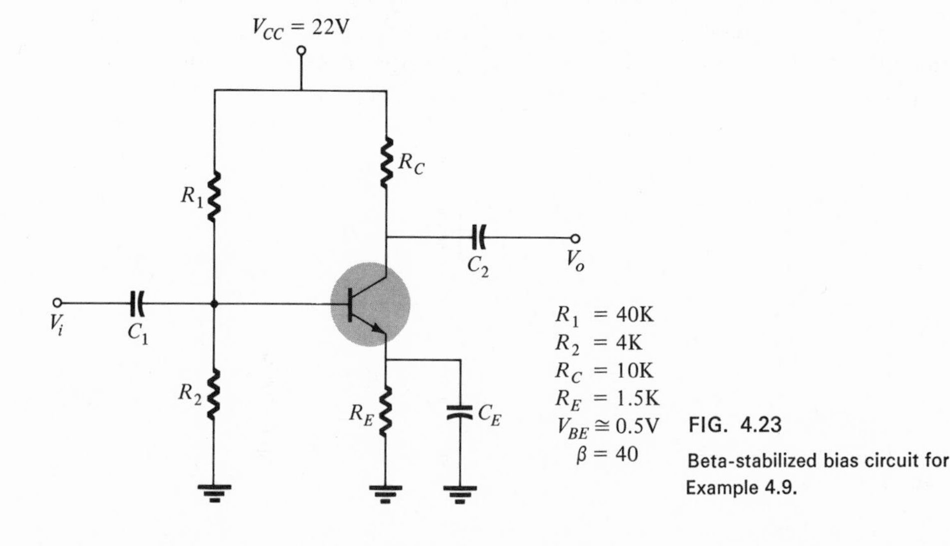

FIG. 4.23
Beta-stabilized bias circuit for Example 4.9.

Look back at the procedure just outlined and notice that the value of beta was never used in Eqs. 4.34–4.40. The base voltage is set by resistors R_1 and R_2 and the supply voltage. The emitter voltage is fixed at approximately the same voltage value as the base. Resistor R_E then determines emitter and collector currents. Finally, R_C determines the collector voltage and, thereby, the collector-emitter bias voltage.

The base voltage V_B is best adjusted using resistor R_2, the collector current by resistor R_E, and the collector-emitter voltage by resistor R_C. Varying other components will have much less effect on the dc bias adjustments mentioned. The capacitor components are part of the ac amplifier operation but have no effect on and actually help to maintain the dc bias and will not be discussed at this time.

A step-by-step summary of the dc bias calculations is helpful in solving problems requiring analysis of the dc bias voltages and currents for a given circuit of the same type as Fig. 4.23.

1. Calculate the base voltage (with respect to ground):

$$V_B = \frac{R_2}{R_1 + R_2}(V_{CC}) \quad \text{(NPN)}$$
$$V_B = \frac{R_2}{R_1 + R_2}(-V_{CC}) \quad \text{(PNP)} \tag{4.41}$$

2. Calculate the emitter voltage:

$$V_E = V_B - V_{BE} \cong V_B \tag{4.42}$$

3. Calculate the emitter and collector currents:

$$I_E = \frac{V_E}{R_E} \cong I_C \tag{4.43}$$

4. Calculate the collector voltage:

$$V_C = V_{CC} - I_C R_C \quad \text{(NPN)}$$
$$V_C = -V_{CC} + I_C R_C \quad \text{(PNP)} \tag{4.44}$$

5. Calculate the collector-emitter voltage:

$$V_{CE} = V_C - V_E \tag{4.45}$$

EXAMPLE 4.9 Calculate the bias voltages and currents for the NPN circuit of Fig. 4.23 using steps (1)–(5) outlined above.

Solution

(a) $V_B = [R_2/(R_1 + R_2)](V_{CC}) = [4/(40 + 4)](22) = 2$ V.
(b) $V_E = V_B - V_{BE} = 2 - 0.5 = 1.5$ V.

(c) $I_E = V_E/R_E \cong I_C = 1.5\ \text{V}/1.5\ \text{K} = 1\ \text{mA}$.
(d) $V_C = V_{CC} - I_C R_C = 22 - (1\ \text{mA})(10\ \text{K}) = 22 - 10 = 12\ \text{V}$.
(e) $V_{CE} = V_C - V_E = 12 - 1.5 = 10.5\ \text{V}$.

EXAMPLE 4.10. Repeat Example 4.9 for the PNP circuit of Fig. 4.24.

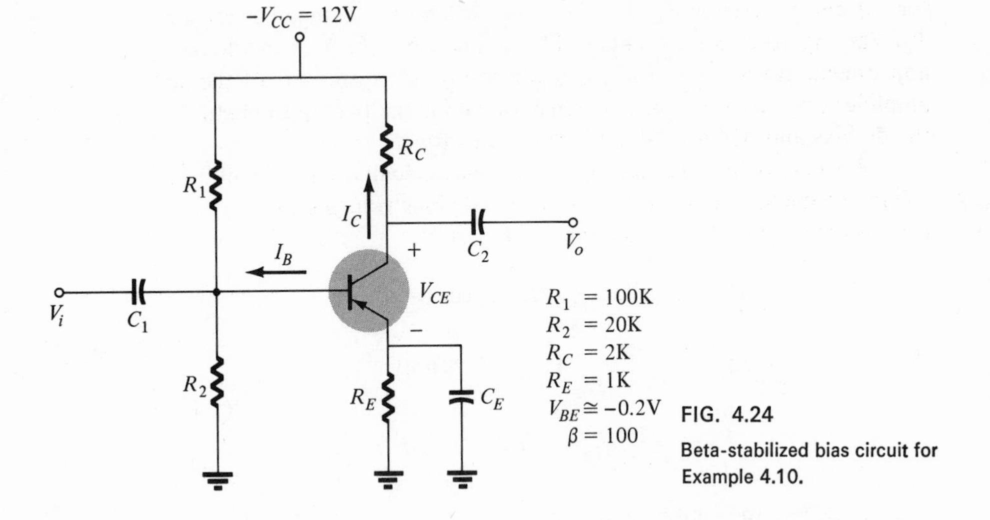

FIG. 4.24
Beta-stabilized bias circuit for Example 4.10.

Solution
(a) $V_B = [R_2/(R_1 + R_2)](-V_{CC}) = [20/(100 + 20)](-12) = -2\ \text{V}$
(b) $V_E = V_B - V_{BE} = -2 - (-0.2) = -1.8\ \text{V}$.
(c) $I_E = V_E/R_E = 1.8\ \text{V}/1\ \text{K} = 1.8\ \text{mA} \cong I_C$.
(d) $V_C = -V_{CC} + I_C R_C = -12 + (1.8\ \text{mA})(2\ \text{K}) = -12 + 3.6$
$= -8.4\ \text{V}$.
(e) $V_{CE} = V_C - V_E = -8.4 - (-1.8) = -6.6\ \text{V}$.

Stability

The stability of the bias circuit of Figs. 4.23 and 4.24 can be equated with the stability of the simpler emitter-stabilized circuit of Fig. 4.17. Calculating a Thévenin equivalent of the input section (resistors R_1 and R_2) provides a value of base resistance, R_B, that can be used in the stability calculation of Eq. (4.33).

Calculate R_B from the parallel combination of resistors R_1 and R_2

$$R_B = \frac{R_1 R_2}{R_1 + R_2} \tag{4.46}$$

The following example will show how R_B is used to calculate the circuit bias stability as done previously.

EXAMPLE 4.11 Calculate the stability of the circuits of (a) Fig. 4.23 and (b) Fig. 4.24.

Solution
(a) From Fig. 4.23 we have the values $\beta = 40$, $R_1 = 40$ K, $R_2 = 4$ K, and $R_E = 1.5$ K.
First calculate R_B

$$R_B = \frac{R_1 R_2}{R_1 + R_2} = \frac{40 \times 4}{40 + 4} = 3.64 \text{ K}$$

Then calculate the stability

$$S = \frac{\beta + 1}{1 + \beta R_E/(R_B + R_E)} = \frac{41}{1 + 40(1.5)/[3.64 + 1.5]} = \mathbf{3.24}$$

(b) From Fig. 4.24 we have $\beta = 100$, $R_1 = 100$ K. $R_2 = 20$ K, and $R_E = 1$ K.

$$R_B = \frac{R_1 R_2}{R_1 + R_2} = \frac{100 \times 20}{100 + 20} = 16.7 \text{ K}$$

$$S = \frac{\beta + 1}{1 + \beta R_E/(R_E + R_B)} = \frac{101}{1 + 100(1)/17.7} = \frac{101}{6.65} = \mathbf{15.2}$$

4.10 DC BIAS CALCULATIONS FOR VOLTAGE FEEDBACK CIRCUITS

Apart from the use of an emitter resistor to provide improved bias stability, voltage feedback also provides improved dc bias stability. Figure 4.25 shows a dc bias circuit with voltage feedback. This section shows how to calculate the dc currents and voltages of this circuit.

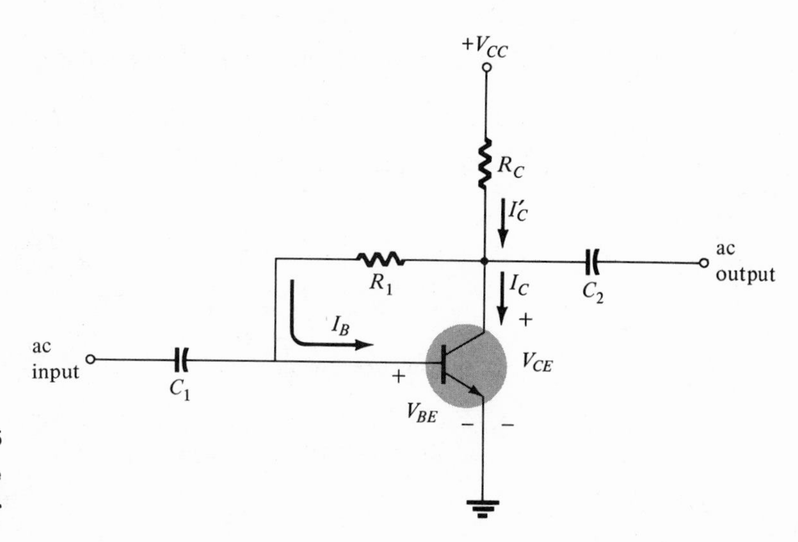

FIG. 4.25
dc bias circuit with voltage feedback.

Figure 4.26a shows the input section (base-emitter) loop of the voltage feedback circuit. Writing the Kirchhoff voltage equation around the loop gives

$$+V_{CC} - I'_C R_C - I_B R_1 - V_{BE} = 0$$

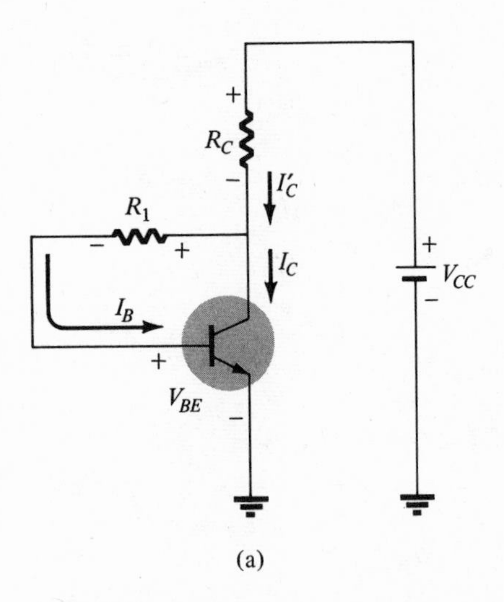

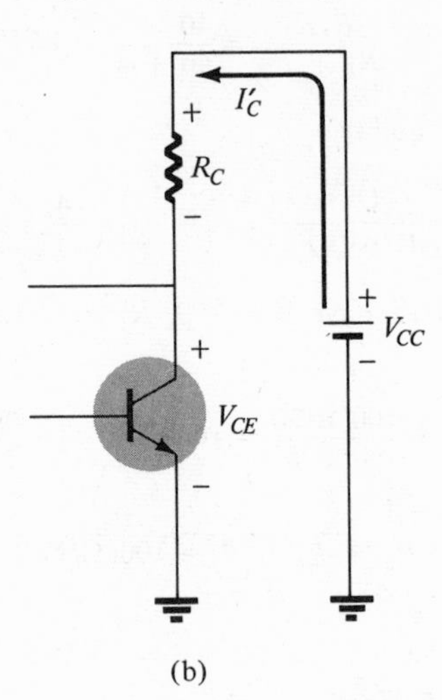

FIG. 4.26
Input and output sections of a voltage feedback dc bias circuit: (a) input section; (b) output section.

The current I'_C is the sum of I_C and I_B, but I_B is so much smaller than I_C that we can write the approximation:

$$I'_C \cong I_C = \beta I_B \tag{4.47}$$

which, substituted into the Kirchhoff voltage equation, yields

$$V_{CC} - \beta I_B R_C - I_B R_1 - V_{BE} = 0$$

Solving for the base current I_B:

$$\boxed{I_B = \frac{V_{CC} - V_{BE}}{R_1 + \beta R_C} \cong \frac{V_{CC}}{R_1 + \beta R_C}} \tag{4.48}$$

Output Section

From the partial circuit diagram of the output section shown in Fig. 3.25b the Kirchhoff voltage equation is

$$+V_{CC} - I'_C R_C - V_{CE} = 0$$

and using $I'_C = I_C$ we can solve for V_{CE}:

$$\boxed{V_{CE} \cong V_{CC} - I_C R_C} \qquad (4.49)$$

The procedure is summarized in the following step-by-step method.

1. Calculate the base current using

$$I_B = \frac{V_{CC} - V_{BE}}{R_1 + \beta R_C} \qquad (4.50)$$

2. The collector and collector resistor currents are next calculated:

$$I'_C \cong I_C = \beta I_B \qquad (4.51)$$

3. Finally, the collector-emitter voltage V_{CE} is calculated using

$$\begin{aligned} V_{CE} &\cong V_{CC} - I_C R_C \quad \text{(NPN)} \\ V_{CE} &\cong -V_{CC} + I_C R_C \quad \text{(PNP)} \end{aligned} \qquad (4.52)$$

EXAMPLE 4.12 Calculate the dc bias currents and voltages for the practical dc bias circuit of Fig. 4.27 using voltage feedback.

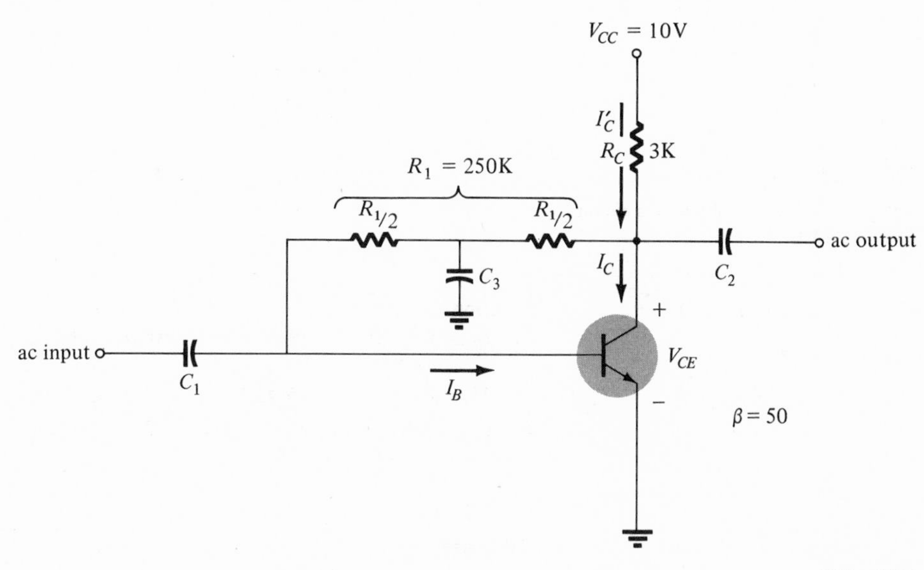

FIG. 4.27

Voltage feedback stabilized bias circuit for Example 4.12.

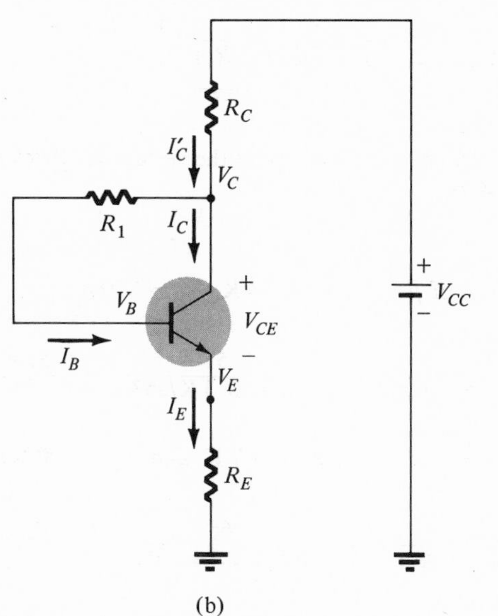

FIG. 4.29
(b) dc bias part of circuit.

the dc bias currents and voltages. Again, the circuit may be analyzed by separately considering the operation of the base-emitter loop and the collector-emitter loop of the circuit.

Input Section

Writing the Kirchhoff voltage equation around the input loop of the circuit results in

$$+V_{CC} - I'_C R_C - I_B R_1 - V_{BE} - I_E R_E = 0$$

Using the current relations

$$I'_C \cong I_C = \beta I_B \cong I_E \tag{4.54}$$

we can solve for the base current, obtaining

$$\boxed{I_B = \frac{V_{CC} - V_{BE}}{R_1 + \beta R_C + (\beta + 1)R_E} \cong \frac{V_{CC}}{R_1 + \beta(R_C + R_E)}} \tag{4.55}$$

Output Section

Solving for the base-emitter voltage from the voltage equation of the output loop results in

$$V_{CE} = V_{CC} - I'_C R_C - I_E R_E \cong V_{CC} - I_C(R_C + R_E) \quad (4.56)$$

A summary of the step-by-step method for obtaining the dc bias currents and voltages for the circuit of Fig. 4.29 follows:

1. Calculate the base current using

$$I_B \cong \frac{V_{CC}}{R_1 + \beta(R_C + R_E)} \quad (4.57)$$

2. Calculate the collector and emitter currents:

$$I'_C \cong I_C = \beta I_B \cong I_E \quad (4.58)$$

3. Calculate the collector-emitter voltage using

$$\begin{aligned} V_{CE} &\cong V_{CC} - I_C(R_C + R_E) \quad \text{(NPN)} \\ V_{CE} &\cong -V_{CC} + I_C(R_C + R_E) \quad \text{(PNP)} \end{aligned} \quad (4.59)$$

EXAMPLE 4.15 Calculate dc bias currents and voltages for the CE stabilized circuit of Fig. 4.30.

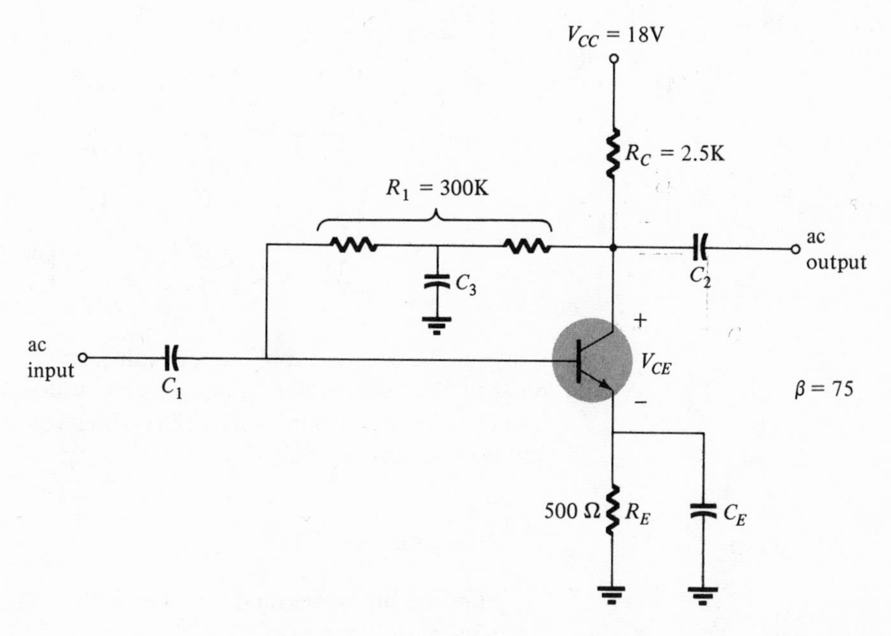

FIG. 4.30
Voltage feedback and emitter resistor stabilized bias circuit for Example 4.15.

Solution

(a) $I_B \cong V_{CC}/[R_1 + \beta(R_C + R_E)] = 18\text{ V}/[300 + 75(0.5 + 2.5)]$
$= 18\text{ V}/525\text{ K} = 34.3\ \mu\text{A}.$

(b) $I_C' \cong I_C = \beta I_B = 75(34.3\ \mu\text{A}) = 2.57\text{ mA}.$

(c) $V_{CE} \cong V_{CC} - I_C(R_C + R_E) \cong 18 - (2.6\text{ mA})(3\text{ K}) = 18 - 7.8$
$= 10.\ 2\text{V}.$

4.11 COMMON-COLLECTOR (EMITTER-FOLLOWER) DC BIAS CIRCUIT

A third connection for the transistor provides input to the base circuit and output from the emitter circuit, with the collector common (CC) to the ac input and output signals. A simple CC circuit (usually referred to as emitter-follower) is shown in Fig. 4.31. The collector voltage

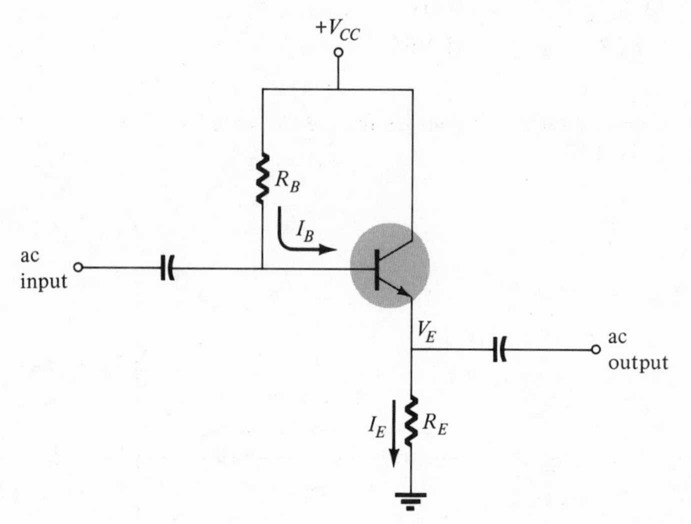

FIG. 4.31
Emitter-follower dc bias circuit.

is fixed at the positive supply voltage value. For V_{CE} to be approximately one-half the voltage of V_{CC}, allowing the widest voltage swing in the output before distortion occurs, the emitter voltage should be set at a voltage of about one-half V_{CC}.

Input Section

For the input section of the circuit the voltages summed around the base-emitter loop give

$$+V_{CC} - I_B R_B - V_{BE} - I_E R_E = 0$$

Using the current relation:

$$I_E = (\beta + 1)I_B \cong \beta I_B \tag{4.60}$$

we can solve for the base current, getting

$$I_B = \frac{V_{CC} - V_{BE}}{R_B + (\beta + 1)R_E} \cong \frac{V_{CC}}{R_B + \beta R_E} \tag{4.61}$$

Output Section

The voltage from emitter to ground is

$$V_E = I_E R_E \tag{4.62}$$

and the collector-emitter voltage is

$$V_{CE} = V_{CC} - V_E = V_{CC} - I_E R_E \tag{4.63}$$

A summary of the step-by-step method of analyzing the emitter-follower circuit of Fig. 4.31 follows

1. Calculate the base current using:

$$I_B = \frac{V_{CC} - V_{BE}}{R_B + (\beta + 1)R_E} \cong \frac{V_{CC}}{R_B + \beta R_E} \tag{4.64}$$

2. Calculate the emitter current using:

$$I_E = (\beta + 1)I_B \tag{4.65}$$

3. Calculate the collector-emitter voltage using:

$$\begin{aligned} V_{CE} &= V_{CC} - I_E R_E \quad \text{(NPN)} \\ V_{CE} &= -V_{CC} + I_E R_E \quad \text{(PNP)} \end{aligned} \tag{4.66}$$

4. If desired, the emitter voltage can be calculated:

$$V_E = I_E R_E \quad \text{(positive voltage for NPN, negative for PNP transistors)} \tag{4.67}$$

EXAMPLE 4.16 Calculate all dc bias currents and voltages for the circuit of Fig. 4.32.

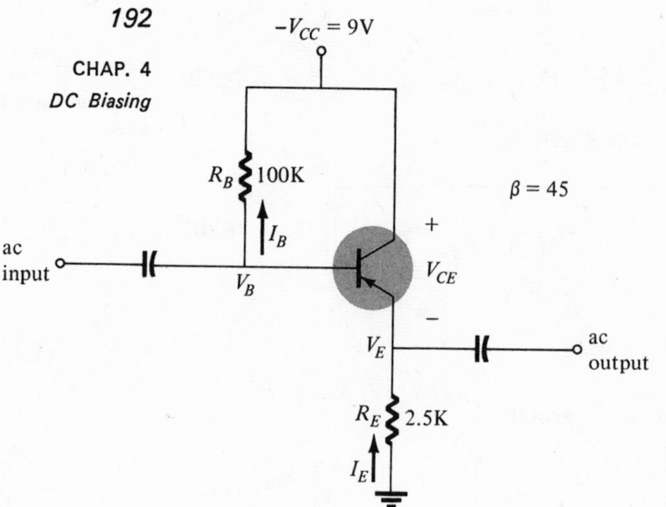

FIG. 4.32
Emitter follower bias circuit for Example 4.16.

Solution

(a) $I_B \cong V_{CC}/(R_B + \beta R_E) = 9\text{ V}/(100 + 45 \times 2.5) = 9\text{ V}/212.5\text{ K}$
$= 42.5\ \mu\text{A}$.

(b) $I_E = (\beta + 1)I_B = 46(42.5\ \mu\text{A}) = 1.94\text{ mA} \cong 2\text{ mA}$.

(c) $V_{CE} = -V_{CC} + I_C R_C = -9 + (2\text{ mA})(2.5\text{ K}) = -4\text{ V}$.

(d) $V_E = -I_E R_E = -(2\text{ mA})(2.5\text{ K}) = -5\text{ V}$.

A second emitter-follower dc bias circuit is shown in Fig. 4.33. Like the similar CE dc bias circuit of Fig. 4.21, this circuit provides a bias operating condition that depends not on the current gain (β) of the transistor, but only on the resistor components and the supply voltage.

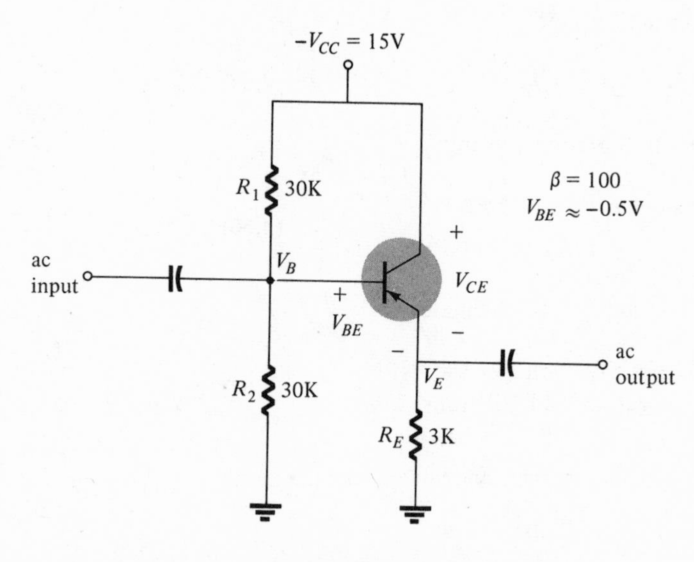

FIG. 4.33
Beta-stabilized emitter follower bias circuit for Example 4.17.

The voltage measured from base to ground can be calculated from the voltage divider made of resistors R_1 and R_2 and the supply voltage. As long as the input resistance looking into the transistor base, $(\beta + 1)R_E$ is sufficiently larger than R_2, the voltage divider relation is correct.

$$\boxed{V_B = \frac{R_2}{R_1 + R_2} V_{CC}} \tag{4.68}$$

The voltage at the emitter is less than that of the base by the small voltage drop across the base-emitter:

$$\boxed{V_E = V_B - V_{BE}} \tag{4.69}$$

Using Ohm's law the current in the emitter can be calculated:

$$\boxed{I_E = \frac{V_E}{R_E}} \tag{4.70}$$

and

$$\boxed{V_{CE} = V_{CC} - I_E R_E} \tag{4.71}$$

A summary of the steps in solving for all dc bias currents and voltages in this CC circuit follows.

1. Calculate the base voltage:

$$\begin{aligned} V_B &= \frac{R_2}{R_1 + R_2}(V_{CC}) \qquad \text{(NPN)} \\ V_B &= \frac{R_2}{R_1 + R_2}(-V_{CC}) \qquad \text{(PNP)} \end{aligned} \tag{4.72}$$

2. Calculate the emitter voltage:

$$V_E = V_B - V_{BE} \tag{4.73}$$

3. Calculate the emitter current:

$$I_E = \frac{V_E}{R_E} \tag{4.74}$$

4. Calculate the collector-emitter voltage:

$$\begin{aligned} V_{CE} &= V_{CC} - V_E = V_{CC} - I_E R_E \qquad \text{(NPN)} \\ V_{CE} &= -V_{CC} + I_E R_E \qquad \text{(PNP)} \end{aligned} \tag{4.75}$$

EXAMPLE 4.17 Calculate all dc bias currents and voltages for the CC circuit of Fig. 4.33.

Solution

(a) $V_B = (R_2/R_1 + R_2)(-V_{CC}) = (30/30 + 30)(-15) = -7.5$ V.
(b) $V_E = V_B - V_{BE} = -7.5 - (-0.5) = -7$ V.
(c) $I_E = (V_E/R_E) = 7\text{V}/3\text{ K} = 2.33$ mA.
(d) $V_{CE} = -V_{CC} + I_E R_E = -15 + (7/3)(3) = -8$ V.

4.12 GRAPHICAL DC-BIAS ANALYSIS

The previous analysis of the dc bias currents and voltages was carried out mathematically for a number of transistor circuits. The only factors of interest used, were the current gain (β) and base-emitter voltage (V_{BE}) when forward biased. This section shows a graphical technique for finding the operating point of a biased transistor circuit. The graphical method demonstrated provides additional insight into the choice of operating point and leads into Section 4.14 on the design (or synthesis) of a dc bias circuit.

The typical CE collector characteristic, shown in Fig. 4.34, only defines the overall operation of the transistor device. The circuit constraints must also be taken into account in obtaining the actual operating point (called the *quiescent operating point* or Q-point). Eq. (4.76) for the output loop of the circuit is a straight line in a voltage-current plot (as in Fig. 4.35).

$$\begin{aligned} I_C &= -\frac{1}{R_C} V_{CE} + \frac{V_{CC}}{R_C} \\ \downarrow \quad & \qquad \downarrow \quad \downarrow \qquad \downarrow \\ y &= \quad m \quad x \; + \; b \end{aligned} \tag{4.76}$$

The straight line representing Eq. (4.76) can be drawn on the graph of Fig. 4.34 by obtaining the two extreme points of the straight line as follows:

1. For $I_C = 0$, $V_{CE} = V_{CC}$ in Eq. (4.76).
2. For $V_{CE} = 0$, $I_C = V_{CC}/R_C$ in Eq. (4.76).

These points are marked in Fig. 4.35 as (1) and (2), respectively, and the straight line connecting them is called the *dc load line*. Although the same voltage-current axis as that of the transistor collector characteris-

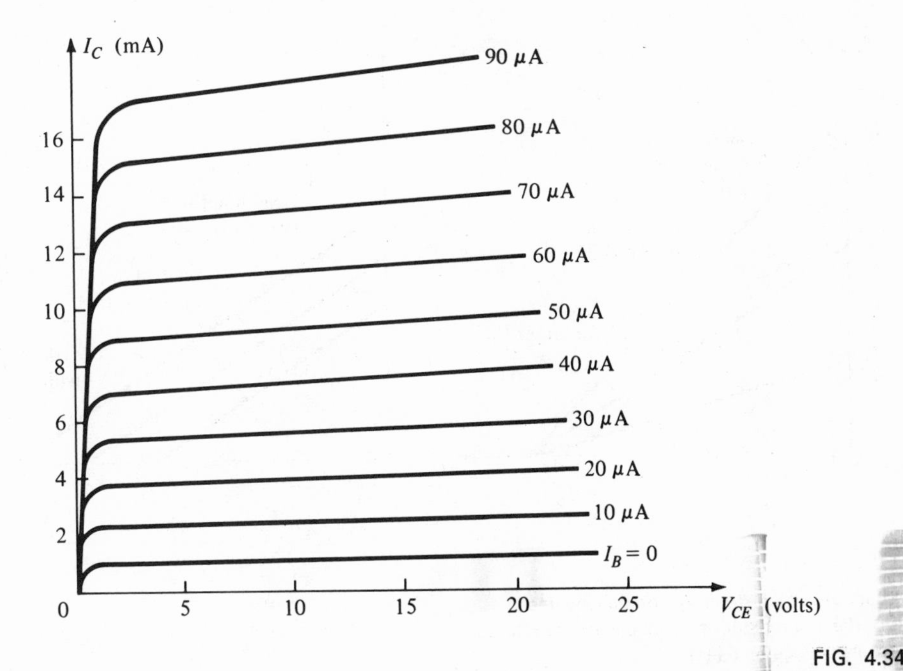

FIG. 4.34
Transistor collector characteristics.

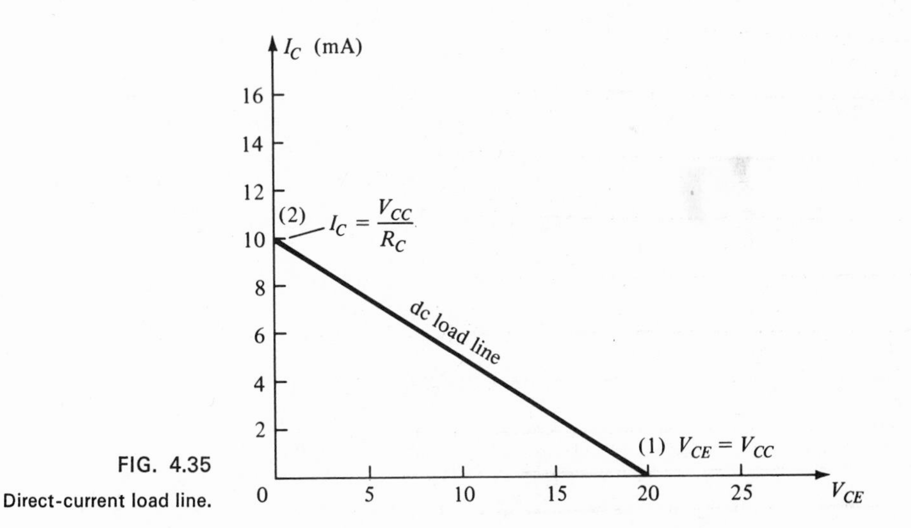

FIG. 4.35
Direct-current load line.

tic is used, no characteristic is shown to reinforce the fact that the dc load line has nothing to do with the device itself. The load line drawn depends only on the supply voltages, V_{CC}, and the value of R_C, the collector resistor.

The slope of the load line depends only on the value of R_C. Figure 4.36a shows the load line slopes for R_C values smaller and larger than that of Fig. 4.35. Figure 4.36b shows that changing only the

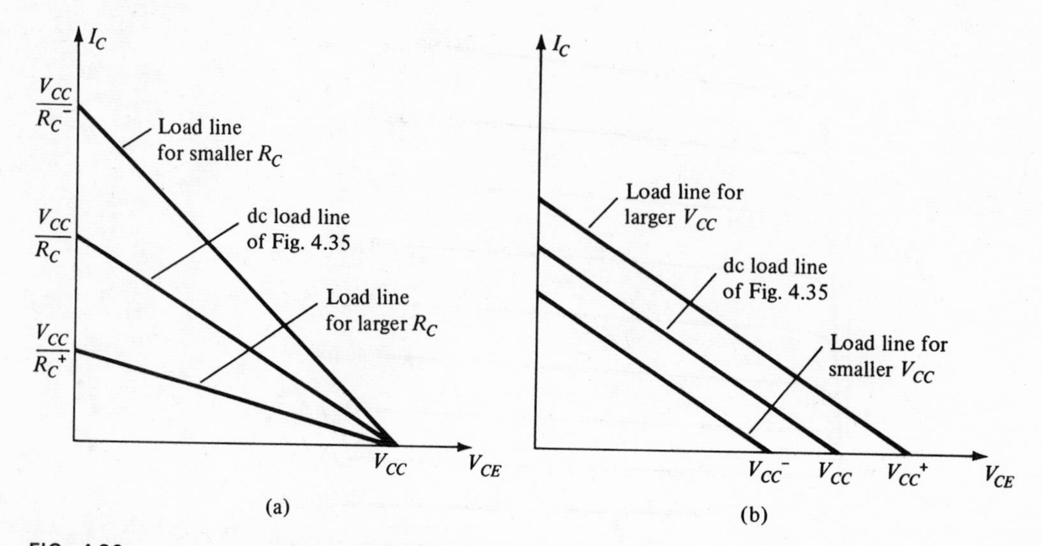

FIG. 4.36

Effect of varying R_C or V_{CC} on dc load line: (a) effect of resistor on dc load line; (b) effect of supply voltage on dc load line.

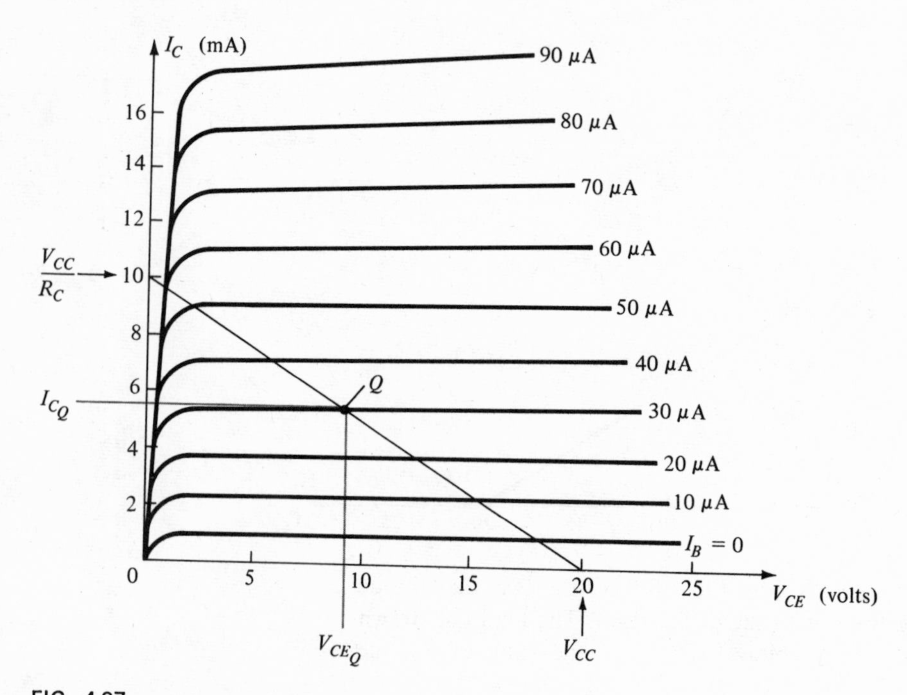

FIG. 4.37

Using transistor collector characteristic and dc load line to obtain quiescent operating point (Q-point).

supply voltage will move the load line parallel to that of Fig. 4.35, and the slope remains the same since R_C has not changed.

Since circuit operation depends on both the transistor characteristic and the circuit elements, plotting *both* curves (transistor characteristic and dc load line) on *one* graph allows determination of the circuit Q-point. This is shown in Fig. 4.37. The typical dc bias point shown in Fig. 4.37 is somewhat in the center of the voltage range (0 to V_{CC}) and the center of the current range (0 to V_{CC}/R_C). A large-signal amplifier with output voltage swing near the voltage range set by the voltage supply value would require a centered operating point.

For circuits other than amplifiers, different bias points may be desired. The dc load line describes all the possible values of voltage and current in the output section of the circuit. Figure 4.37 shows a typical bias point set by the amount of base current and the dc load line. Adjusting the base current to higher values moves the operating point toward saturation along the load line, whereas reducing the base current moves the bias point toward transistor cutoff. For the characteristic and load line of Fig. 4.37, base currents in excess of 60 μA will

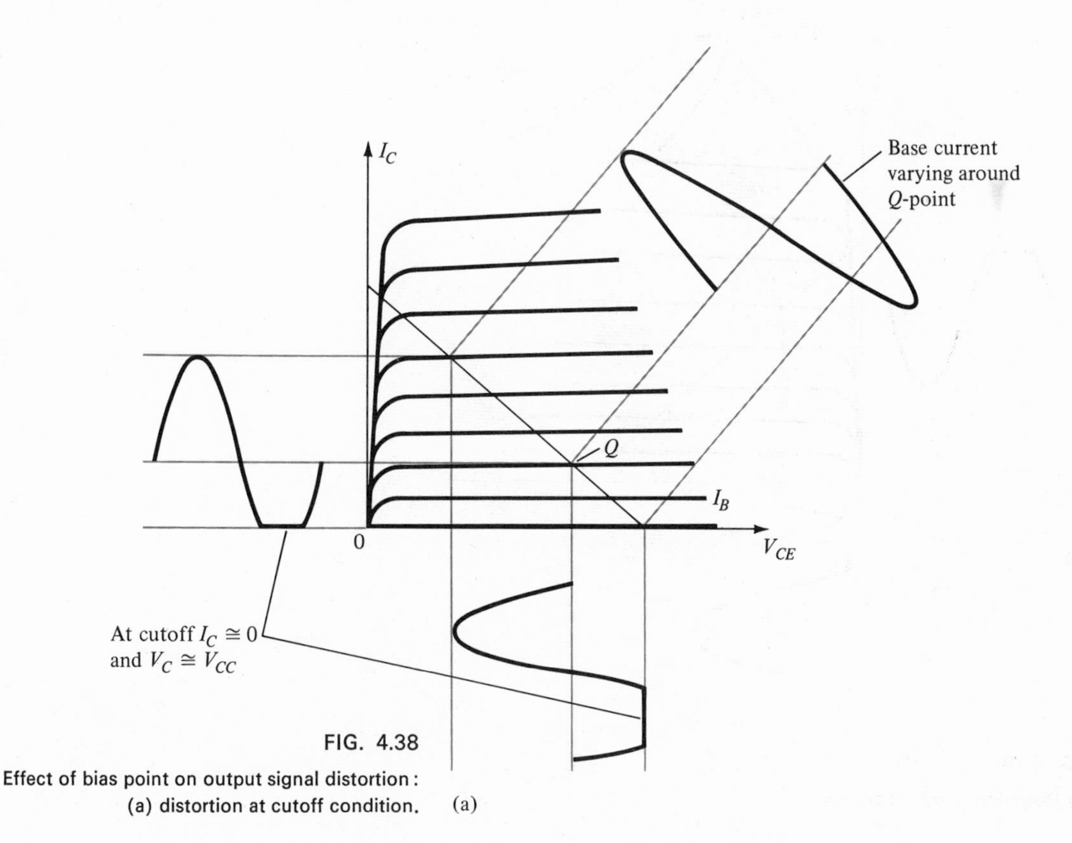

FIG. 4.38
Effect of bias point on output signal distortion: (a) distortion at cutoff condition.

drive the transistor into saturation. Note that for the load line and operating point indicated, the ac input, which adds to the dc base bias current, can go positive only by about 25 μA (from 30 to 55 μA) before the limiting condition (saturation) occurs. The variation of the ac base current, on the other hand, can go negative by 30 μA (30 to 0 μA) before cutoff is reached so that the particular bias point in Fig. 4.37 is not quite centered. For small-signal amplifiers with output voltage swings of less than 1 V, the exact centering of the Q-point is not essential—usually a region of largest transistor gain, or most linear operation, is sought.

To see how the centering of the bias point is important for large-signal amplifiers, Fig. 4.38 shows the resulting distortion for a few different bias points and signal amplitude, which result in the output's distorting because the transistor is driven into saturation, into cutoff, or both (too large an input signal). In the particular case where distortion occurs only in saturation or only in cutoff as shown in Fig. 4.38a and b, the bias point could be adjusted correspondingly to correct

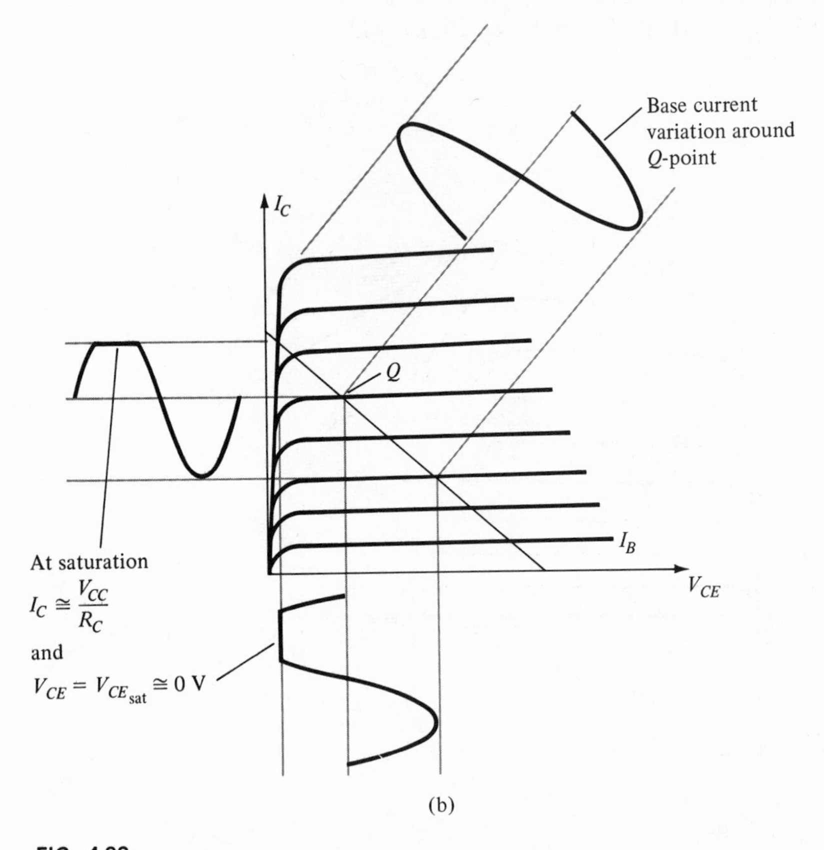

FIG. 4.38
(b) Distortion at saturation condition.

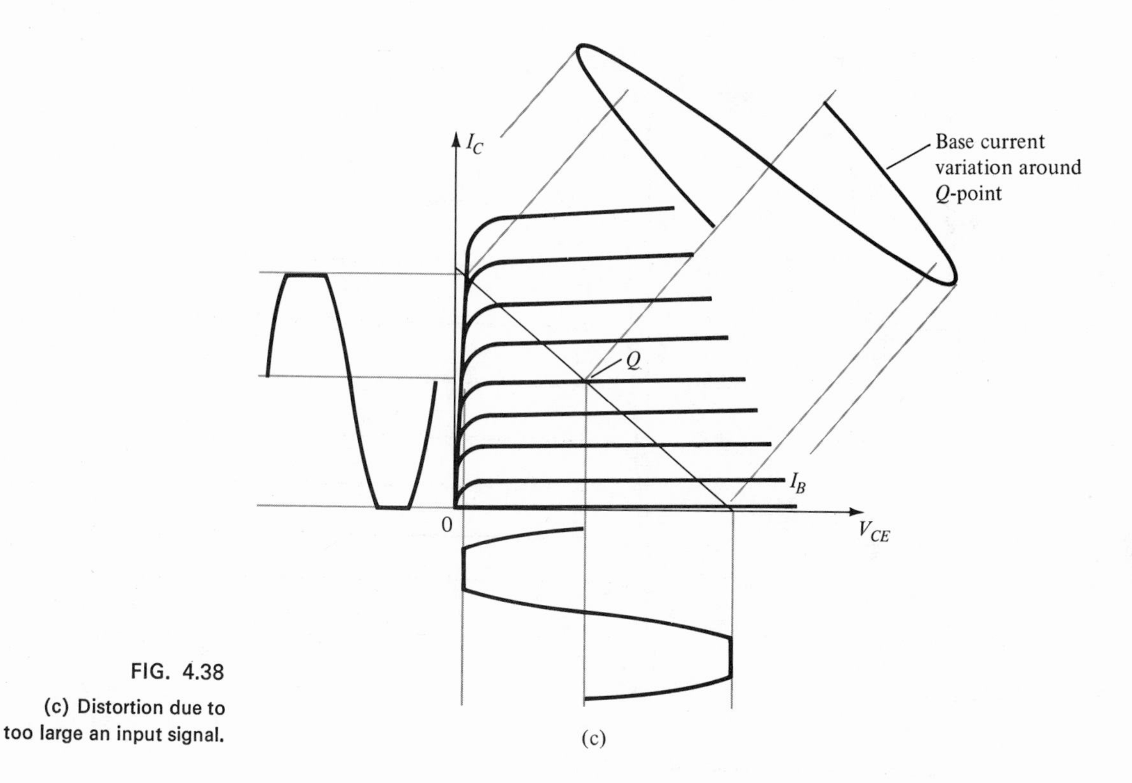

FIG. 4.38
(c) Distortion due to too large an input signal.

the distortion. With too large an input signal as shown in Fig. 4.38c the resulting distortion cannot be helped by operating-point adjustment.

EXAMPLE 4.18 Given the circuit of Fig. 4.39(a) and the transistor collector characteristic of Fig. 4.39b:

(a) plot the dc load line and obtain the Q-point.
(b) find V_{CE}, I_C, $I_C R_C$, and I_E from the graph.

Solution

(a) 1. Draw the dc load line. The two points for the load line are
   a. at $I_C = 0$, $V_{CE} = V_{CC} = 15$ V.
   b. at $V_{CE} = 0$, $I_C = V_{CC}/R_C = 15$ V/3 K $= 5$ mA.

Connect a straight line between these points for the dc load line.

2. Calculate the base current using

$$I_B \cong \frac{V_{CC}}{R_B} = \frac{15\text{ V}}{100\text{ K}} = 150\ \mu\text{A}$$

3. The intersection of the load line and the transistor curve for $I_B = 150\ \mu$A defines the quiescent operating point (see Fig. 4.39c).

(b) From curve $V_{CE} = 7$ V, $I_C R_C = 8$V, $I_C = 2.6$ mA, and $I_E \cong 2.6$ mA.

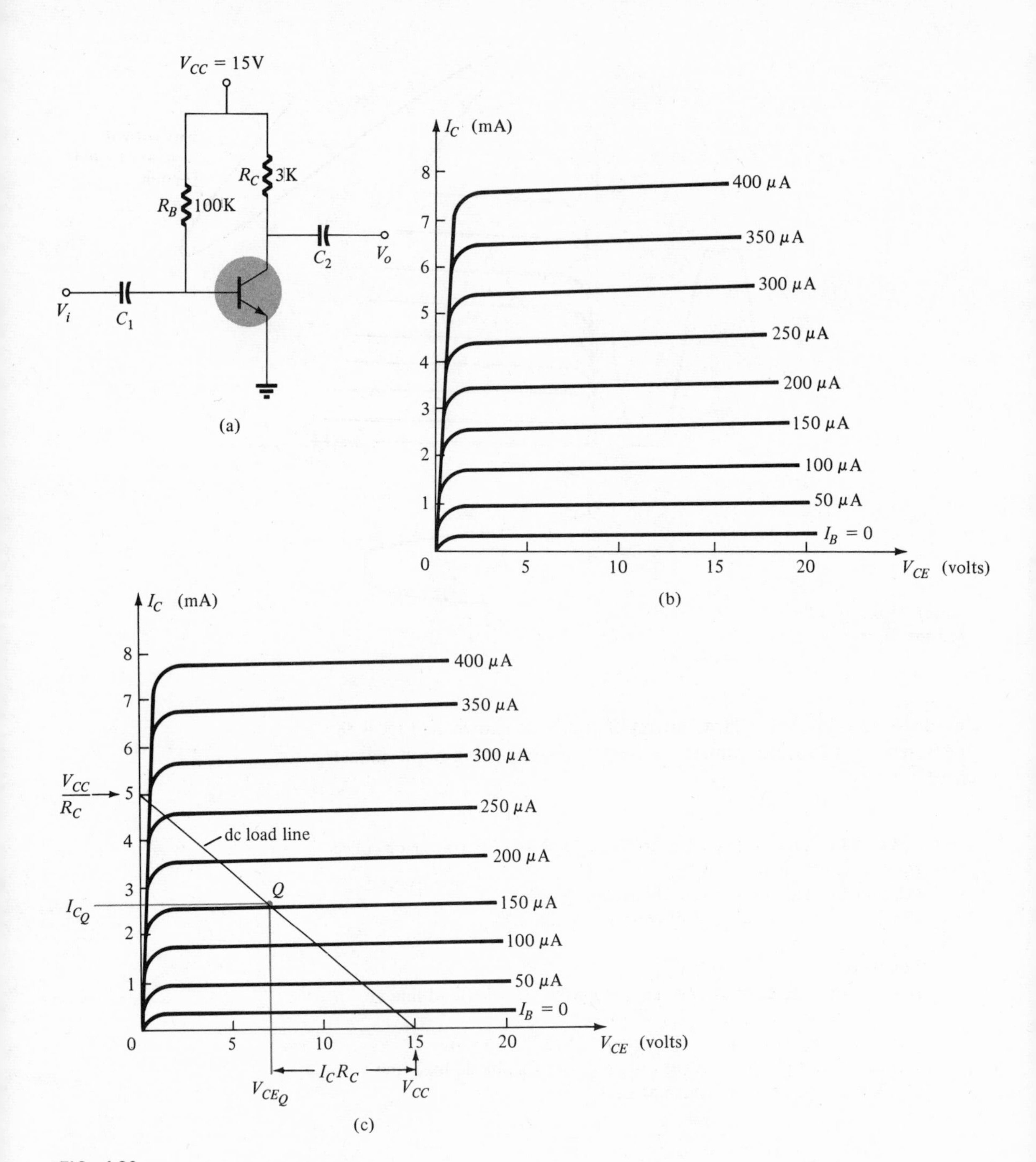

FIG. 4.39

Graphical analysis of fixed-bias circuit for Example 4.18: (a) transistor circuit; (b) transistor collector characteristics; (c) dc load line and Q-point.

EXAMPLE 4.19 For the circuit given in Fig. 4.40a, find the Q-point using the collector characteristic of Fig. 4.40b.

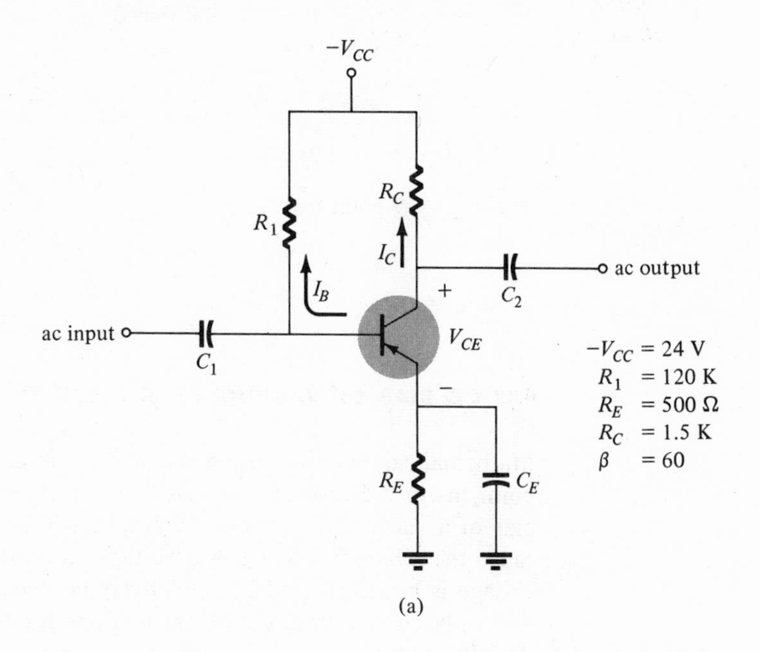

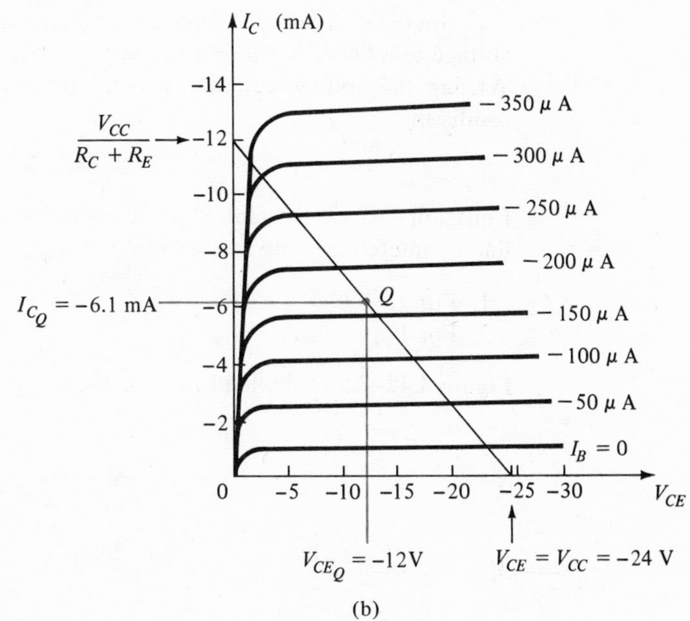

FIG. 4.40

Circuit and characteristics for Example 5.19: (a) emitter-stabilized bias circuit; (b) transistor collector characteristics.

Solution

(a) 1. The two points to obtain the dc load line are

a. for $I_C = 0$, $V_{CE} = V_{CC} = -24$ V.

b. for $V_{CE} = 0$, $I_C = V_{CC}/(R_E + R_C) = -24\text{ V}/(1.5 + 0.5\text{ K})$
$= -12\text{ mA}$.

Draw the load line as in Fig. 4.40b.

2. Calculate I_B:

$$I_B \cong \frac{V_{CC}}{R_1 + (\beta + 1)R_E} = \frac{24\text{ V}}{120\text{ K} + 61(0.5\text{ K})} \cong \frac{24\text{ V}}{150\text{ K}}$$
$$= 160\ \mu\text{A}$$

Q-point from graph

$$V_{CE_Q} = -12\text{ V}$$
$$I_{C_Q} = 6.1\text{ mA}$$

4.13 DC BIAS OF VACUUM-TUBE CIRCUITS

Mathematical methods using equations for calculating the dc bias point (as in the case of a transistor circuit) are not applied to the bias of a vacuum-tube triode. A graphical procedure is necessary because the relation of input grid voltage and output plate current and voltage is best described by a relatively nonlinear plate characteristic. As will be shown the graphical calculations for the vacuum triode are straightforward.

Figure 4.41 shows a fixed-bias circuit using a vacuum triode as a voltage amplifier. A triode plate characteristic is shown in Fig. 4.42. Writing the voltage equation around the output plate-cathode loop results in

$$V_{PK} = V_{PP} - I_P R_P \tag{4.77}$$

Equation 4.77 can be plotted on the plate characteristic as the dc load line by interconnecting the following two points by a straight line:

1. For $I_P = 0$, $V_{PK} = V_{PP}$.
2. For $V_{PK} = 0$, $I_P = V_{PP}/R_P$.

Figure 4.42 shows the load line for the circuit of Fig. 4.41 drawn be-

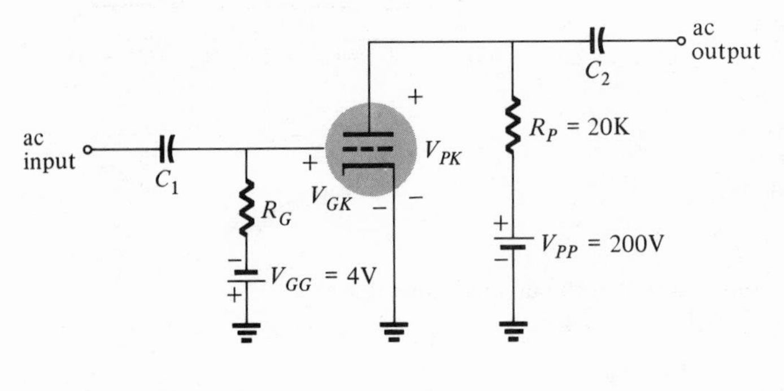

FIG. 4.41
Triode amplifier circuit.

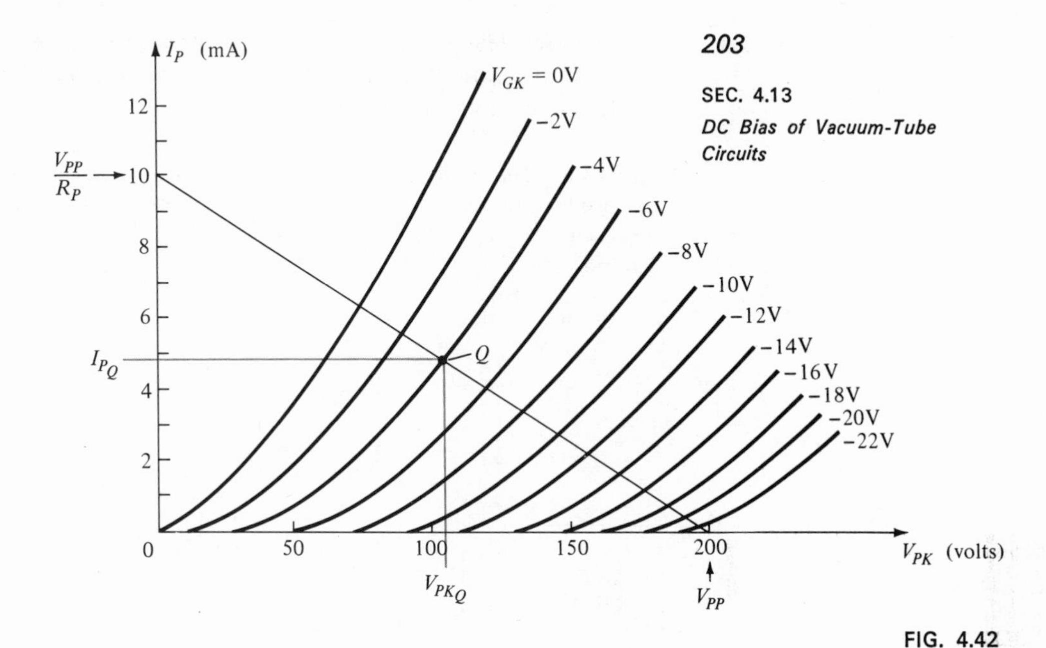

FIG. 4.42
Triode plate characteristics showing dc load line.

tween points 1 and 2 as determined for the values given in the circuit. The grid voltage is set by the grid battery

$$V_{GK} = V_{GG} \tag{4.78}$$

which is −4 V in Fig. 4.41. For the particular grid-cathode voltage and the dc load line set by the plate resistor and plate supply voltage, the quiescent operating point marked on Fig. 4.42 is obtained. The vacuum triode is thus biased into the amplifier region of operation and will provide an output signal that is an amplified version of the input signal applied through the input coupling capacitor to the grid.

Resistor R_G has no effect on the dc bias because there is no dc grid current for negative grid-cathode voltages. The grid resistor is needed for the ac operation of the grid circuit as will be shown in Chapter 5.

To calculate the dc bias (quiescent) plate voltage and current the procedure can be summarized as follows:

1. Draw the dc load line (see Fig. 4.42) on the plate characteristic by interconnecting the following points by a straight line.
   (a) the voltage point V_{PP} along the x-axis.
   (b) the current point V_{PP}/R_P along the y-axis.
2. Find the operating point at the intersection of the dc load line with the fixed value of grid-cathode voltage

$$V_{GK} = V_{GG} \tag{4.79}$$

As demonstrated on Fig. 4.42 this provides the dc bias operating point from which the voltage and current will vary due to the input ac signal.

EXAMPLE 4.20. Calculate the quiescent operating point of the vacuum triode in the circuit of Fig. 4.43a using the plate characteristic of Fig. 4.43b.

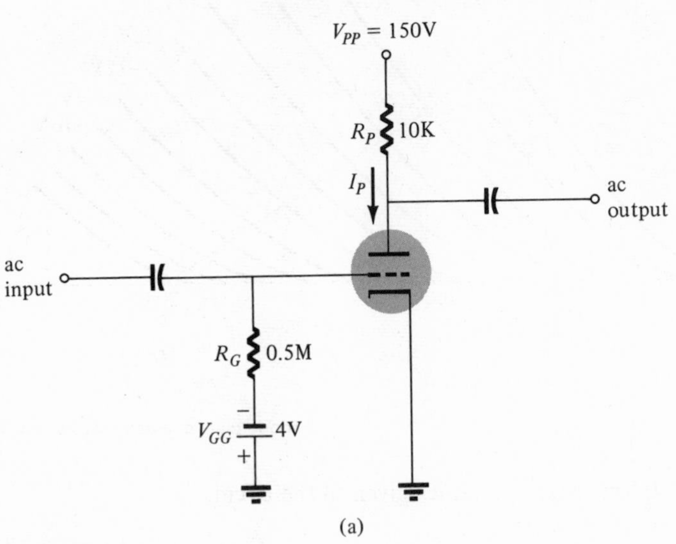

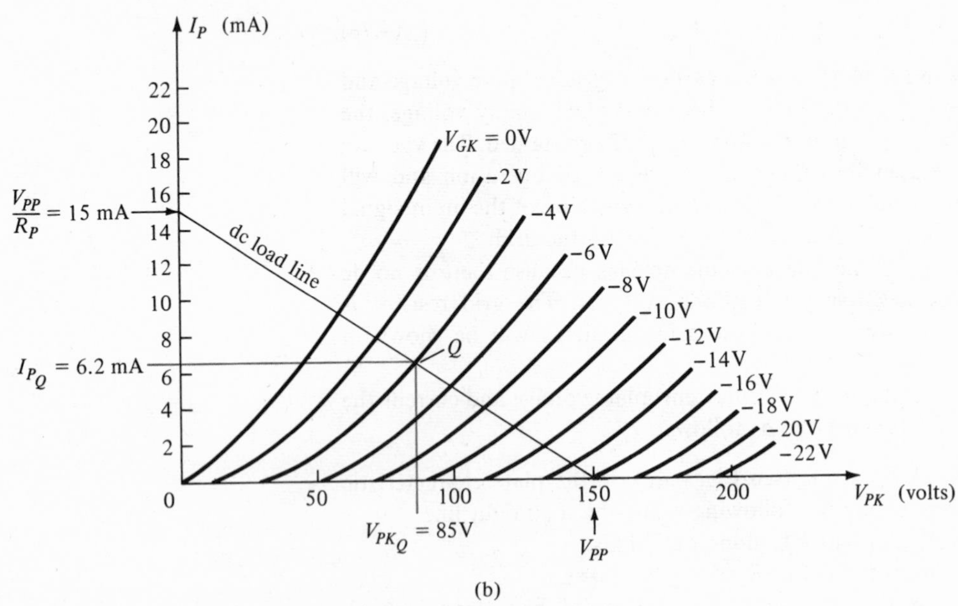

FIG. 4.43

Circuit and characteristics for Example 4.20: (a) triode amplifier circuit; (b) triode characteristics with dc load line and Q-point.

Solution

(a) Obtain the dc load line from the following points:

1. At $V_{PK} = V_{PP} = 150$ V along the x-axis.
2. At $I_P = V_{PP}/R_P = 150$ V/10 K $= 15$ mA along the y-axis. For the load line drawn in Fig. 4.43b:

(b) At the intersection of $V_{GK} = V_{GG} = -4$ V and the dc load line, obtain quiescent operating point Q.

From the graph the quiescent operating point is

$$V_{PKQ} = 85 \text{ V}$$
$$I_{PQ} = 6.2 \text{ mA}$$

Pentode Bias Calculations

A pentode amplifier circuit is shown in Fig. 4.44. As part of the usual bias connection the suppressor grid is connected to ground and the screen grid to a fixed positive dc voltage, V_{SS}, somewhat lower than

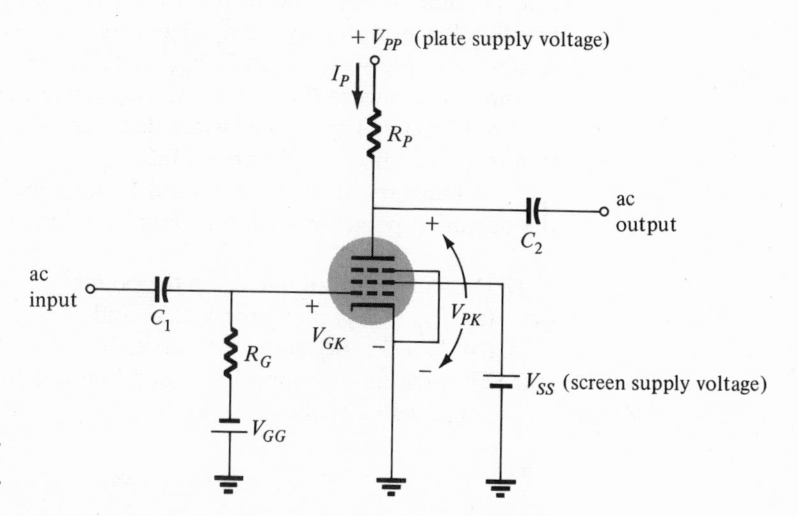

FIG. 4.44
Pentode amplifier circuit.

the plate supply, V_{PP}. The plate supply voltage, V_{PP}, and the plate resistor, R_P, provide much the same operation as in the triode circuit. A dc load line for the circuit operation is obtained using the same method as for the triode circuit. The two points that are used to draw the load line are

1. The point $V_{PK} = V_{PP}$ along the x-axis; and
2. the point $I_P = V_{PP}/R_P$ along the y-axis.

Figure 4.45 shows a pentode plate characteristic and a dc load line drawn on the same graph. The intersection of the load line and the grid-cathode voltage curve defines the operating Q-point of the pentode.

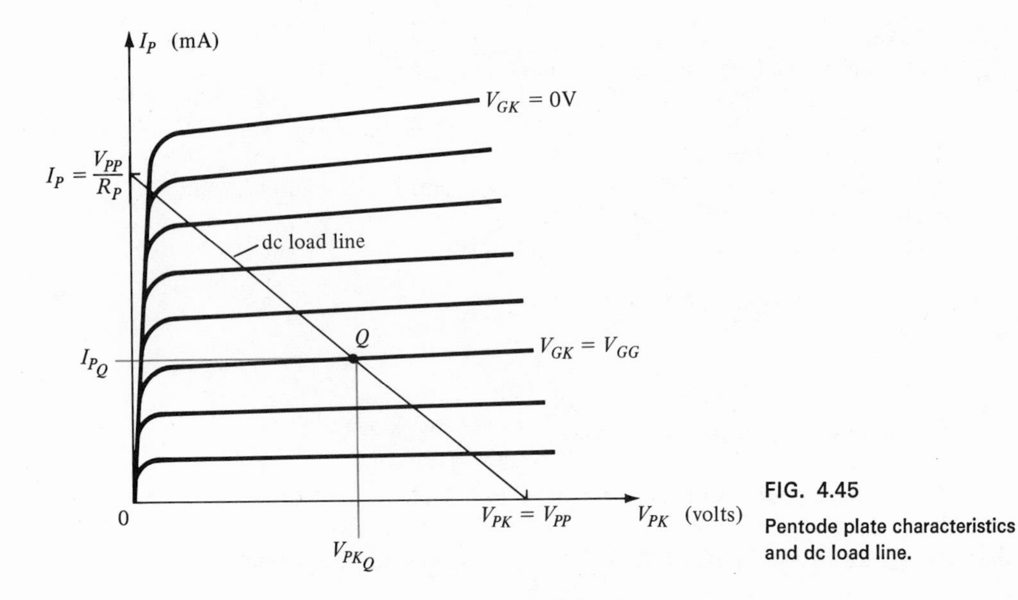

FIG. 4.45
Pentode plate characteristics and dc load line.

The Q-point is thus dependent on the plate supply voltage, the plate resistor, the grid voltage, and, of course, the pentode characteristic. A change in the screen voltage V_{SS} could change the operating point by modifying the pentode characteristic curves. However, the screen voltage is usually kept at a constant dc potential and under this restriction it has no effect on the bias point.

A summary of the steps needed to draw the load line and obtain the operating point for a pentode amplifier circuit are as follows:

1. Draw the dc load line using the points
   (a) $V_{PK} = V_{PP}$ along the x-axis; and
   (b) $I_P = V_{PP}/R_P$ along the y-axis.
2. Find the intersection of the load line and the curve for the particular value of the grid supply

$$V_{GK} = V_{GG} \tag{4.80}$$

EXAMPLE 4.21 Calculate the quiescent operating point of the pentode in the circuit of Fig. 4.46a using the pentode characteristic of Fig. 4.46b.

Solution

(a) 1. $V_{PK} = V_{PP} = 300$ V along the x-axis.
    2. $I_P = V_{PP}/R_P = 300$ V$/10$ K $= 30$ mA along the y-axis.

(b) The intersection of the dc load line and $V_{GK} = V_{GG} = -2$ V gives an operating point of

$$V_{PKQ} = 150 \text{ V}$$
$$I_{PQ} = 15 \text{ mA}$$

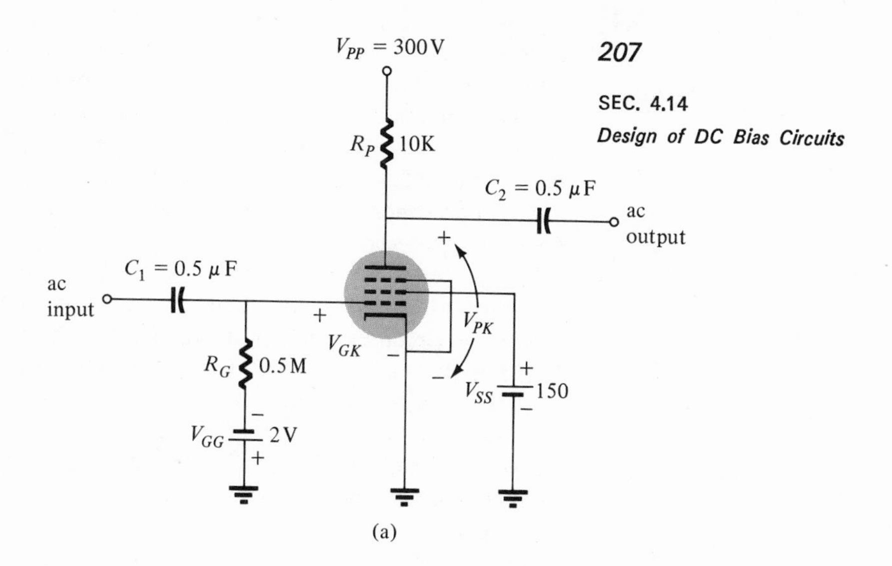

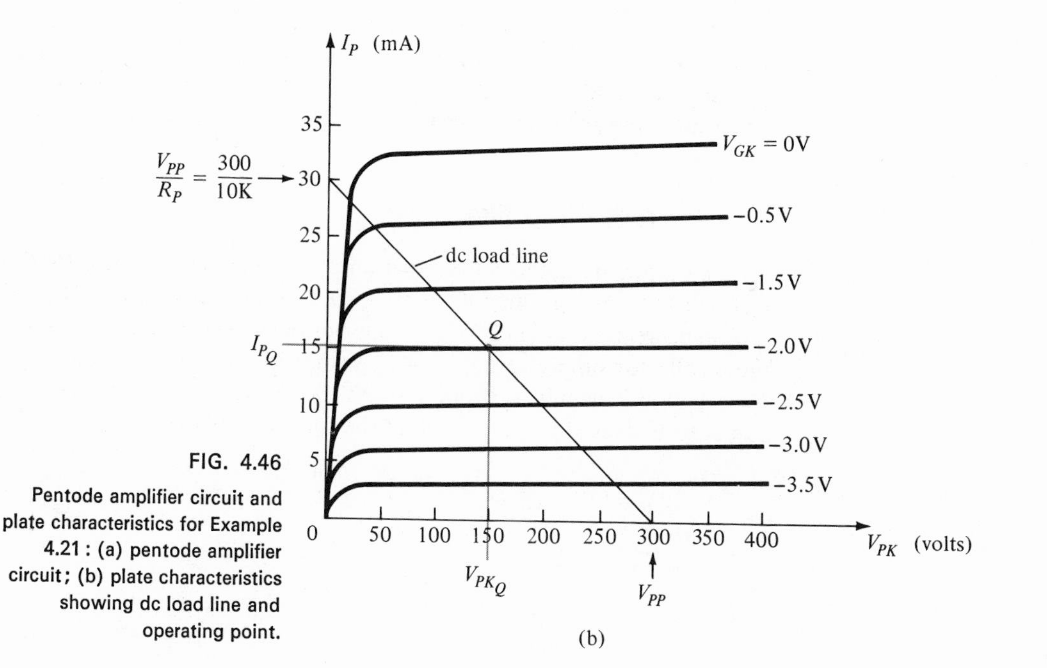

FIG. 4.46
Pentode amplifier circuit and plate characteristics for Example 4.21: (a) pentode amplifier circuit; (b) plate characteristics showing dc load line and operating point.

4.14 DESIGN OF DC BIAS CIRCUITS

Up to now the discussion has been directed to the techniques of analyzing a given transistor or tube circuit to determine the dc operating point. Although it is often necessary to determine the Q-point of a given circuit, it is also quite important to be able to design a circuit to operate at a desired or specified bias point. Quite often the manufacturer's specification (spec) sheets provide information stating a suitable operating point (or operating region) for the particular transistor. In addition, other circuit factors connected with the given amplifier stage may also dictate some conditions of current swing, voltage swing, value of common supply voltage, etc., which can be used in determining the Q-point in a design.

The techniques of synthesis (or design) readily follow from the previous discussions of circuit analysis. In almost all cases, the calculation of the circuit elements proceeds step by step in the reverse to those in the analysis consideration. Basically, the problem of concern in this section can be briefly stated as follows.

Given a desired point or region of operation for a particular transistor, design the bias circuit (resistor and supply voltage values) to obtain the specified operating point.

In actual practice many other factors may have to be considered and may go into the selection of the desired operating point. For the moment we shall concentrate, however, on determining the component values to obtain a specified operating point. Since the basic relations and operation of a number of bias circuits have already been considered no new theory has to be developed.

Design of a Fixed-Bias Circuit

Consider designing a fixed-bias amplifier (Fig. 4.47) using a 2N292 transistor. From the manufacturer's spec sheet the transistor has a typical current gain (β) of 25 at a collector current of 1 mA. The maximum collector supply voltage is +12 V, and an operating point at, say, $V_{CE_Q} = +6$ V would be proper. From the above information the circuit is to be biased for operation at the following bias point:

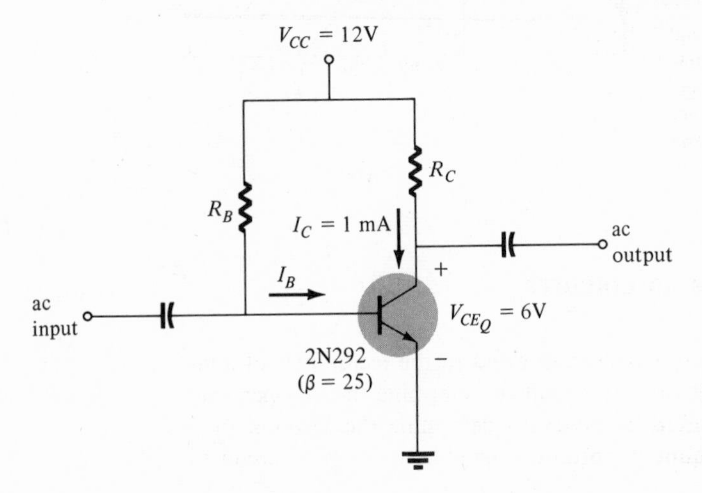

FIG. 4.47
Fixed-bias circuit for design consideration.

$$V_{CEQ} = 6 \text{ V}$$
$$I_{CQ} = 1 \text{ mA}$$

For a fixed-bias circuit, Sec. 4.5 shows the equations to be used. We shall first use Eq. (4.11) relating the voltages around the collector-emitter loop, using the known values of supply voltage, collector current, and collector-emitter voltage to solve for the value of the collector resistor.

Solving for the unknown value of resistance using Eq. (4.11)

$$\boxed{R_C = \frac{V_{CC} - V_{CEQ}}{I_C}} \tag{4.81}$$

where I_C is understood to be I_{CQ}. If units of voltage in volts and current in milliamperes are used the unit of the resistance is kilohms.

We need only calculate the value of the base resistor to complete the design. For this calculation we need the value of the base current, which is

$$\boxed{I_B = \frac{I_C}{\beta}} \tag{4.82}$$

From the equations developed in analyzing the fixed-bias circuit we can solve for the value of base resistance by rewriting Eq. (4.12) in the following form.

$$\boxed{R_B = \frac{V_{CC} - V_{BE}}{I_B}} \tag{4.83}$$

This completes the design of the bias section of the fixed-bias amplifier for the given transistor and operating point. Looking back over the design procedure it should now be obvious that the same equations as in the analysis section were used, but in reverse order. In addition the equations were rewritten to solve for a different unknown quantity—the collector resistance, instead of the collector-emitter voltage, or the base resistance, instead of the base current.

EXAMPLE 4.22 Design a fixed-bias circuit using a 2N188A PNP transistor (see Fig. 4.47). From the manufacturer's spec sheet the transistor has a current gain (β) of around 40 at a collector current of 20 mA. The maximum collector supply voltage is -12 V.

Design Solution: Design of the circuit is required for the operating point $I_{CQ} = 20$ mA and $V_{CEQ} = -6$ V

The emitter resistor is then

$$R_E \cong \frac{V_E}{I_C} = \frac{2\text{ V}}{5\text{ mA}} = \mathbf{400\ \Omega}$$

The collector resistor is calculated to be

$$R_C = \frac{V_{CC} - V_{CEQ} - V_E}{I_C} = \frac{20 - 10 - 2\text{ V}}{5\text{ mA}} = \frac{8\text{ V}}{5\text{ mA}} = \mathbf{1.6\ K}$$

Calculating the base current using

$$I_B = \frac{I_C}{\beta} = \frac{5\text{ mA}}{45} = 110\ \mu\text{A}$$

the base resistor is calculated to be

$$R_B = \frac{V_{CC} - V_{BE} - V_E}{I_B} = \frac{(20 - 0.7 - 2)\text{ V}}{110\ \mu\text{A}} = \frac{17.3\text{ V}}{110\ \mu\text{A}}$$

$$= \mathbf{157\ K} \qquad (\text{use } R_B = 150\text{ K})$$

Design of Current Gain Stabilized Circuit

The circuit of Fig. 4.49 provides stabilization both for leakage current and current gain changes. The values of the four resistors shown must be obtained for a specified operating point. Engineering judgment in selecting a value of emitter voltage, V_E, as in the previous design consideration, leads to a simple straightforward solution for all the resistor values.

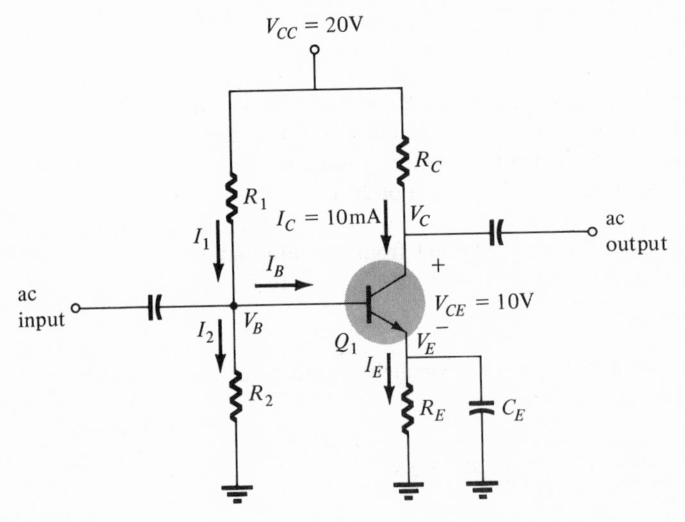

FIG. 4.49
Current gain stabilized circuit for design considerations.

The design steps are as follows:

1. The emitter voltage will be selected to be approximately one-tenth of the supply voltage (V_{CC}).

$$V_E \cong \frac{1}{10} V_{CC} \qquad (4.89)$$

2. Using this value of V_E, the emitter-resistor value is calculated from

$$R_E \cong \frac{V_E}{I_C} \qquad (4.90)$$

3. The collector resistance is then obtained using

$$R_C = \frac{V_{CC} - V_{CEQ} - V_E}{I_C} \qquad (4.91)$$

4. The base voltage is approximately equal to the emitter voltage or, more exactly,

$$V_B = V_E + V_{BE} \qquad (4.92)$$

5. The equation for calculation of the base resistors R_1 and R_2 will require a little thought. Using the values of base voltage calculated in step (4) and the value of the supply voltage will provide one equation—but there are two unknowns, R_1 and R_2. An additional equation can be obtained from an understanding of the operation of these two resistors in providing the base voltage. For the circuit to operate properly, the current through the two resistors should be approximately equal and therefore larger than the base current by an order of magnitude (at least 10 times larger) The two equations that will enable calculating the resistors R_1 and R_2 are

$$V_B = \frac{R_2}{R_1 + R_2}(V_{CC}) \qquad (4.93)$$

$$I_1 \cong I_2 = \frac{V_{CC}}{R_1 + R_2} \geqq 10 I_B \qquad (4.94)$$

Solving these equations results in the following equations, which can be used to calculate R_1 and R_2 directly

$$R_2 = \frac{V_B}{10 I_B} \qquad (4.95)$$

$$R_1 = \frac{V_{CC}}{10 I_B} - R_2 \qquad (4.96)$$

EXAMPLE 4.24 Design a dc bias circuit for an amplifier circuit as in Fig. 4.49 using a 2N635A NPN transistor. From the manufacturer's spec the transistor has a current gain of 100, typical, at a collector current of 10 mA. The supply voltage for the present circuit is 20 V.

Design Solution: From the data provided, the bias point to be obtained is $I_{C_Q} = 10$ mA, $V_{CE_Q} = 10$ V

(a) Select $V_E = \frac{1}{10}(V_{CC}) = \frac{1}{10}(20) = 2$ V
(b) Calculate R_E:

$$R_E \cong \frac{V_E}{I_C} = \frac{2\text{ V}}{10\text{ mA}} = 0.2\text{ K} = \mathbf{200\ \Omega}$$

(c) Calculate R_C:

$$R_C = \frac{V_{CC} - V_{CE_Q} - V_E}{I_C} = \frac{20 - 10 - 2\text{ V}}{10\text{ mA}} = 0.8\text{ K} = \mathbf{800\ \Omega}$$

(d) Calculate V_B:

$$V_B = V_E + V_{BE} = 2 + 0.5 = 2.5\text{ V}$$

(e) Calculate R_1 and R_2:

$$I_B = \frac{I_C}{\beta} = \frac{10\text{ mA}}{100} = 100\ \mu\text{A}$$

$$R_2 = \frac{V_B}{10I_B} = \frac{2.5}{10(100\ \mu\text{A})} = \mathbf{2.5\ K}$$

$$R_1 = \frac{V_{CC}}{10I_B} - R_2 = \frac{20}{10(100\ \mu\text{A})} - 2.5\text{ K} = \mathbf{17.5\ K} \qquad \text{(use 18 K)}$$

4.15 MISCELLANEOUS BIAS CIRCUITS

In practice, one finds that bias circuits do not always conform to the basic forms considered in the present chapter. It should not be difficult to analyze the bias operation of a circuit slightly modified from those considered previously. To show various bias circuits a few examples are included in this section to demonstrate the use of the basic bias concepts discussed in this chapter.

EXAMPLE 4.25 Obtain the bias voltages and currents for the circuit of Fig. 4.50.

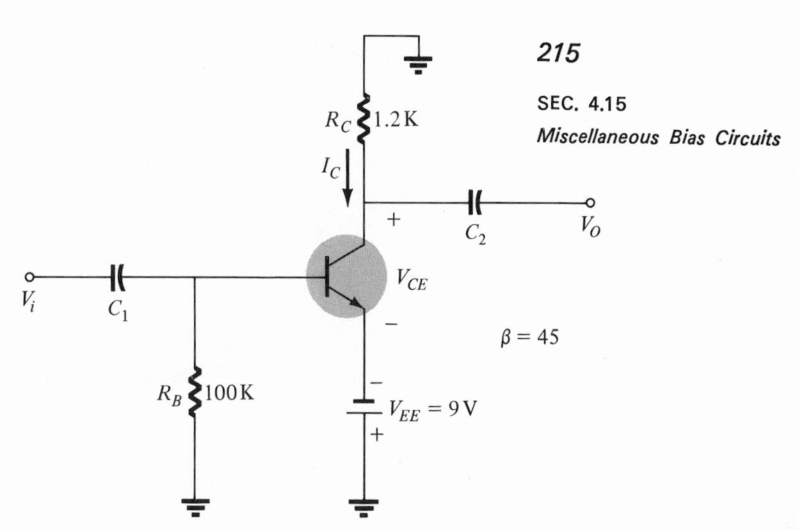

FIG. 4.50
Bias circuit for Example 4.25.

Solution: Input loop:

$$-I_B R_B - V_{BE} + V_{EE} = 0$$

$$I_B = \frac{V_{EE} - V_{BE}}{R_B} = \frac{9 - 0.7\,\text{V}}{100\,\text{K}} = 83\ \mu\text{A}$$

$$I_C = \beta I_B = 45(83\ \mu\text{A}) = 3.74\ \text{mA}$$

Output loop:

$$-I_C R_C - V_{CE} + V_{EE} = 0$$

$$V_{CE} = V_{EE} - I_C R_C = 9 - (3.74\ \text{mA})(1.2\ \text{K})$$
$$= 9 - 4.5 = 4.5\ \text{V}$$

EXAMPLE 4.26 Calculate the bias voltages and currents for the circuit of Fig. 4.51.

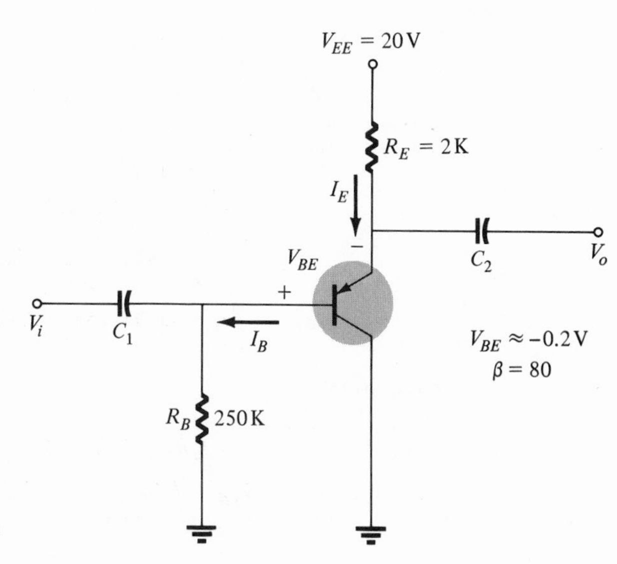

FIG. 4.51
Bias circuit for Example 4.26.

Solution: Writing the voltage loop equation

$$V_{EE} - I_E R_E - V_{EB} - I_B R_B = 0$$

$$(V_{EE} - V_{EB}) = (\beta + 1) I_B R_E + I_B R_B$$

$$I_B = \frac{V_{EE} - V_{EB}}{R_B + (\beta + 1) R_E} = \frac{V_{EE}}{R_B + \beta R_E} = \frac{20\text{ V}}{250\text{ K} + 80(2\text{ K})}$$

$$= \frac{20\text{ V}}{410\text{ K}} = 48.8\ \mu\text{A}$$

$$I_C = \beta I_B = 80(48.8\ \mu\text{A}) = 3.9\text{ mA} \cong I_E$$

$$V_E = V_{EE} - I_E R_E = 20 - (3.9\text{ mA})(2\text{ K}) = 20 - 7.8$$

$$= 12.2\text{ V}$$

EXAMPLE 4.27 Calculate the bias voltages and currents for the circuit of Fig. 4.52.

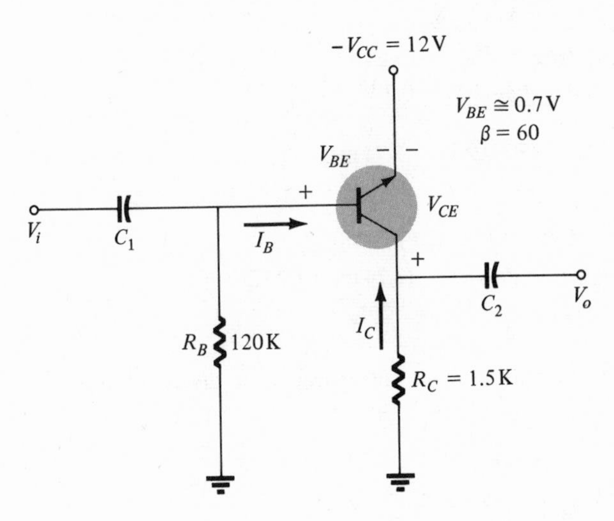

FIG. 4.52
Bias circuit for Example 4.27.

Solution: Input loop:

$$-I_B R_B - V_{BE} + V_{CC} = 0$$

$$I_B = \frac{V_{CC} - V_{BE}}{R_B} = \frac{12 - 0.7\text{ V}}{120\text{ K}} = 94\ \mu\text{A}$$

$$I_C = \beta I_B = 60(94\ \mu\text{A}) = 5.6\text{ mA}$$

Output loop:

$$-I_C R_C - V_{CE} + V_{CC} = 0$$

$$V_{CE} = V_{CC} - I_C R_C = 12 - (5.6\text{ mA})(1.5\text{ K})$$

$$= 12 - 8.5 = 3.5\text{ V}$$

EXAMPLE 4.28 Calculate the bias currents and voltages for the circuit of Fig. 4.53.

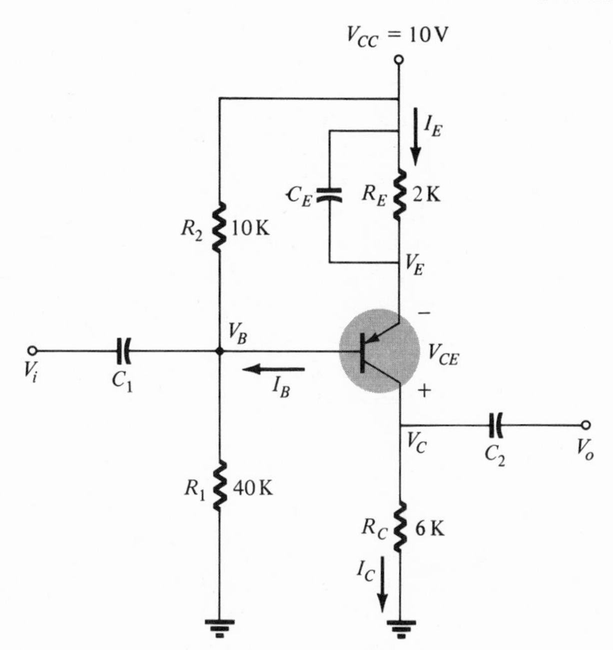

FIG. 4.53
Bias circuit for Example 4.28.

Solution

(a) $V_B = [R_1/(R_1 + R_2)](V_{CC}) = [40/(10 + 40)](10) = 8$ V.
(b) $V_E = V_B - V_{BE} = 8 + 0.2 = 8.2$ V.
(c) $I_E = (V_{CC} - V_E)/R_E = (10 - 8.2)/2 = 0.9 \text{ mA} \cong I_C$.
(d) $V_C = I_C R_C = (0.9 \text{ mA})(6 \text{ K}) = 5.4$ V.
(e) $V_{CE} = V_C - V_E = 5.4 - 8.2 = -2.8$ V.

PROBLEMS

§ 4.3

1. Calculate the dc bias voltages and currents for a PNP common-base bias circuit. Assume transistor values of $\alpha = 0.985$ and $V_{BE} = -0.2$ V. Circuit components are $R_E = 720\ \Omega$ and $R_C = 3.9$ K and supply voltages are $V_{CC} = 9$ V and $V_{EE} = 1.5$ V. (See Fig. 4.2 for circuit connection.)
2. Calculate the collector-base voltage for NPN common-base bias circuit (as in Fig. 4.5) for the following circuit values: $R_E = 1.8$ K, $R_C = 2.7$ K, $V_{EE} = 9$ V, $V_{CC} = 22$ V, $\alpha = 0.995$ and $V_{BE} = +0.7$ V.

§ 4.6

3. For a fixed-bias common-emitter circuit, as in Fig. 4.6, calculate the bias currents and voltages for the following circuit values: $R_B = 150$ K, $R_C = 2.1$ K, $V_{CC} = 9$ V, $V_{BE} = +0.7$ V and $\beta = 45$.

§ 4.13

22. Calculate the quiescent operating point of the vacuum triode in the circuit of Fig. 4.41 using the characteristic of Fig. 4.42. Circuit values are as follows: $R_G = 1$ M, $R_P = 18$ K, $V_{PP} = 200$ V and $V_{GG} = 6$ V.
23. What value of grid supply voltage is needed in Problem 4.22 to obtain an operating point at $V_{PKQ} = 100$ V?
24. Calculate the quiescent operating point of a pentode as in the circuit of Fig. 4.44 for the following circuit values: $R_G = 0.47$ K, $V_{GG} = -3$ V, $V_{PP} = 400$ V and $R_P = 16$ K. Use the tube characteristic of Fig. 4.46b.
25. What value of load resistor is required for a quiescent plate voltage of $V_{PK_Q} = 200$ V in Problem 4.24?

§ 4.14

26. Design a fixed-bias common-emitter circuit using a 2N366 transistor. The NPN transistor has a current gain (β) of 49 and is to be operated at $I_{C_Q} = 1$ mA and $V_{CE_Q} = 10$ V. Use a collector supply of 20 V.
27. Design a fixed-bias common-emitter circuit using a 2N535 PNP transistor rated to have a current gain of 35 at a collector current of 1 mA and collector-emitter voltage of -12 V. Use a supply of 18 V.
28. Calculate the resistor values for an emitter-stabilized amplifier circuit as in Fig. 4.48. Use a 2N1150 silicon NPN transistor having a current gain of 18 at 1 mA and 20 V. Use a 30-V supply voltage.
29. What value of emitter-stabilizing resistor is needed in the circuit designed in Problem 4.28 to obtain a stability factor of 15? How does this value affect the circuit bias?
30. Design a dc bias circuit as in Fig. 4.49 using a 2N1224 PNP transistor having a current gain of 20 at $I_{C_Q} = 1$ mA and $V_{CE_Q} = -12$ V. Use a supply voltage of 22 V.

§ 4.15

31. Obtain the value of dc voltage measured from collector to ground for the circuit of Fig. 4.50 for the following circuit values: $V_{EE} = 15$ V, $R_B = 47$ K, $R_C = 1.2$ K and $\beta = 30$.
32. Calculate the base voltage (with respect to ground) for the circuit of Fig. 4.51 using the following values: $R_B = 120$ K, $R_E = 8.2$ K, $V_{EE} = 12$ V, $V_{BE} = -0.2$ V and $\beta = 20$.
33. Calculate the collector current for the circuit of Fig. 4.52 using the following values: $R_B = 80$ K, $R_C = 1.8$ K, $V_{CC} = 9$ V, $V_{BE} = 0.6$ and $\beta = 35$.
34. Calculate the collector-emitter voltage for a circuit as in Fig. 4.53 for the following circuit values: $R_1 = 120$ K, $R_2 = 15$ K, $R_E = 3.9$ K, $R_C = 12$ K, $V_{CC} = 18$ V, $V_{BE} = -0.3$ V and $\beta = 40$.

5

Small-Signal Analysis

5.1 INTRODUCTION

Chapter 4 was a detailed discussion of transistor and tube circuits purely from a dc viewpoint. We must now begin to investigate the response of these circuits with a sinusoidal ac signal applied.

Of first concern is the magnitude of the input signal. It will determine whether *small-signal* or *large-signal* techniques must be used. There is no set dividing line between the two, but the application, and the magnitude of the variables of interest (i, v) relative to the scales of the device characteristics, will usually make it quite clear which is the case in point. The small-signal technique will be discussed in this case chapter, while large-signal applications will be considered in Chapter 8.

The key to the small-signal approach is an ac equivalent circuit to be derived in this chapter. It is that combination of circuit elements, properly chosen, that will best approximate the actual semiconductor or tube device in a particular operating region. Once the ac equivalent circuit has been determined, the graphic symbol of the device can be replaced in the schematic by this circuit and the basic methods of ac circuit analysis (branch-current analysis, mesh analysis, nodal analysis, and Thévenin's theorem) can be applied to determine the response of the circuit.

There are various methods of introducing and developing the equivalent circuit, one of which includes the use of partial derivatives (calculus). Rather than use this approach we shall first derive an *approximate "piecewise" equivalent* circuit for transistors and tubes from their characteristics. The results will demonstrate that the small-signal ac equivalent circuits derived later (in a more mathematical manner) are valid. For the transistor, the basic two-port equations will then be employed to develop the popular *hybrid equivalent circuit,* while Thévenin's theorem will be used in the derivation of the tube equivalent circuit.

As a means of demonstrating the effect that the ac equivalent circuit will have on the analysis to follow, consider the circuit of Fig. 5.1, discussed in detail in Chapter 4. Let us assume for the present that

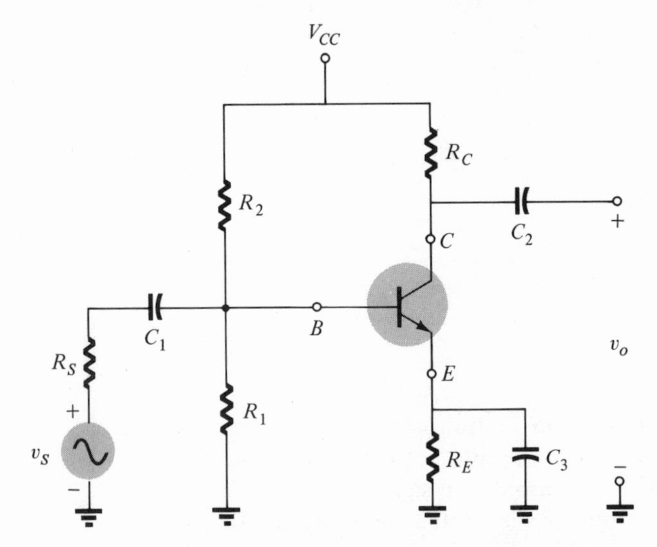

FIG. 5.1
Transistor circuit under examination in this introductory discussion.

the small-signal ac equivalent circuit for the transistor circuit shown in Fig. 5.1 has already been determined. Since we are interested purely in the ac response of the circuit, all the dc supplies can be replaced by short circuits since they will now determine only the dc level of the output voltage and not the magnitude of the swing of the ac output. In addition, the coupling capacitors C_1 and C_2 and bypass capacitor C_3 were chosen to have a very small reactance at the frequency of application. Therefore, they too, may for all practical purposes be replaced by short circuits. Note that this will result in the "shorting out" of the dc biasing resistor R_E. By performing a few fundamental circuit manipulations the circuit of Fig. 5.2 can be obtained.

If we now simply substitute the small-signal ac equivalent circuit for the transistor, the output voltage can be obtained using only those basic laws and methods associated with the fundamental RLC ac circuits.

To ensure completeness, the circuit of Fig. 5.1 will appear as Example 5.3 in the use of the small-signal ac equivalent circuit in Section 5.7

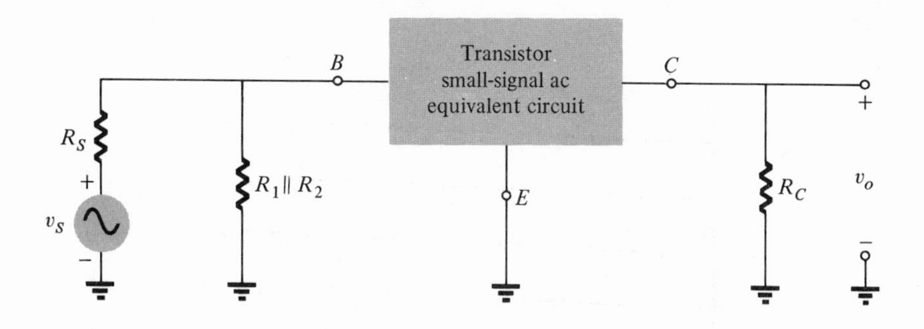

FIG. 5.2
Circuit of Fig. 5.1 redrawn for small-signal ac analysis.

5.2 TRANSISTOR PIECEWISE LINEAR EQUIVALENT CIRCUIT

It must be emphasized and understood before examining this section that we are about to derive a piecewise linear equivalent circuit for the transistor and not the small-signal ac model to be employed in depth later in this chapter. The piecewise linear equivalent circuit formed by representing the device characteristics by piecewise linear (straight lines) curves will be used to establish a circuit for comparison when the more vague, mathematical derivation of the small-signal ac equivalent circuit is considered. The similarities between the two equivalent circuits will serve to reinforce the fact that the small-signal equivalent circuit does, in fact, represent the transistor in the region of operation.

In Fig. 5.3, the collector characteristics of the transistor discussed in Section 3.6 have been approximated by *equally spaced, straight, parallel* lines. The vertical shift in the characteristics is represented by I_o, a dc level of collector current. Since the straight lines are parallel, the slope of each is the same. The slope, defined by g_o (conductance) as shown in Fig. 5.3, is the ratio of a change in current divided by a change in voltage.

The change in collector current due to a change in base current can be found using the forward current gain for the common-emitter configuration. You will recall that

$$\beta = \frac{\Delta I_C}{\Delta I_B}$$

or

$$\Delta I_C = \beta \, \Delta I_B$$

The fact that the lines are all equally spaced, straight, and parallel results in β having the same value in any region of the characteristics.

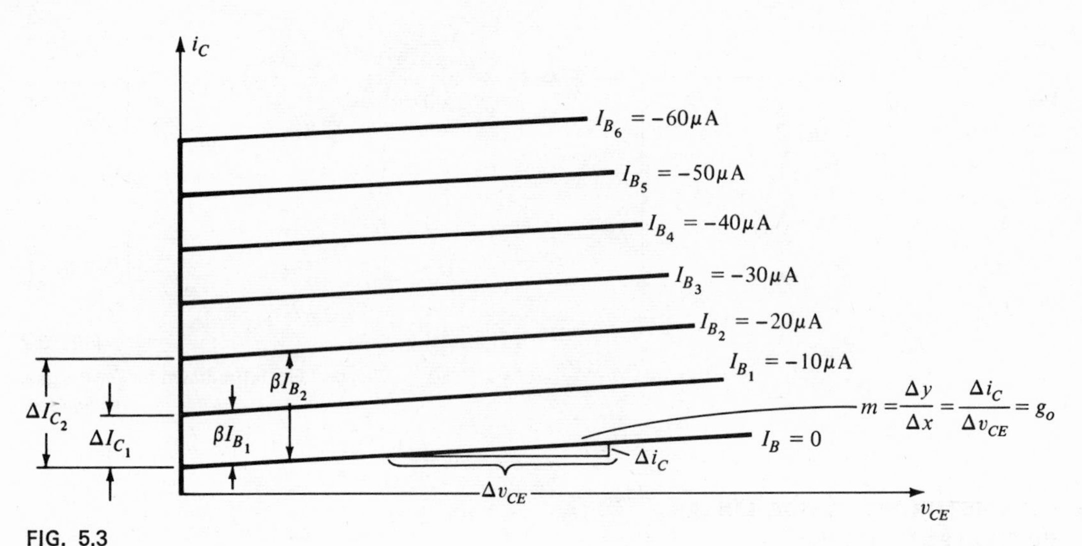

FIG. 5.3
Transistor collector characteristics approximated by equally spaced, straight, parallel lines.

For the straight line defined by $I_{B_2} = -20\ \mu A$,

$$\Delta I_{C_2} = \beta\, \Delta I_B = \beta(I_{B_2} - 0) = \beta I_{B_2}$$

and for the straight line defined by $I_{B_1} = -10\ \mu A$,

$$\Delta I_{C_1} = \beta\, \Delta I_B = \beta(I_{B_1} - 0) = \beta I_{B_1}$$

In general, any straight line in a two-dimensional plane (x, y) can be represented by the following equation:

$$y = mx + b \tag{5.1}$$

For the straight line defined by I_{B_2} (arbitrarily chosen), when $x = v_{CE} = 0$, $y = i_C = I_o + \Delta I_{C_2} = I_o + \beta I_{B_2}$.

Substituting the above conditions into Eq. (5.1),

$$\overbrace{I_o + \beta I_{B_2}}^{y} = \overbrace{g_o \cdot 0}^{mx} + b$$

so that

$$b = I_o + \beta I_{B_2}$$

The equation for the straight line I_{B_2} is, therefore,

$$\begin{aligned} y &= mx + b \\ i_C &= g_o v_{CE} + \overbrace{I_o + \beta I_{B_2}}^{b} \end{aligned}$$

For each straight line, the only quantity that will change is I_B. Substituting i_B to represent any value of base current

$$\boxed{i_C = g_o v_{CE} + I_o + \beta i_B} \tag{5.2}$$

The question now is: What is the significance of Eq. (5.2)? Does it give us some hint as to the type of circuit that will appear in the black box of Fig. 5.4 to represent the transistor collector characteristics on a piecewise approximate basis?

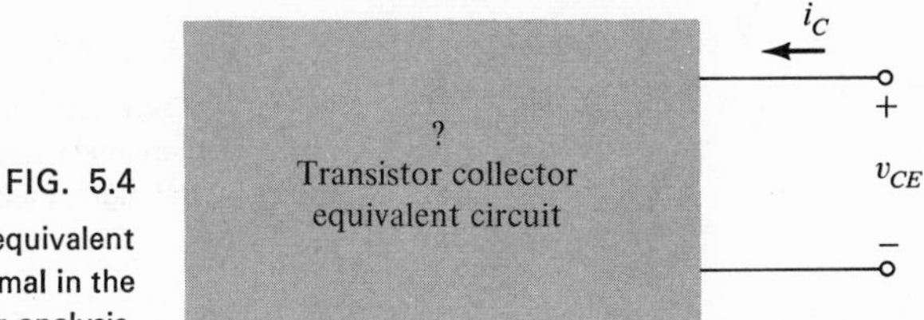

FIG. 5.4
Transistor collector equivalent circuit to be formal in the ensuing analysis.

A second glance at Eq. (5.2) will reveal that each term has the units of current. This being the case, if we simply apply Kirchhoff's current law in reverse; that is, find the circuit corresponding to the equation, the circuit of Fig. 5.5 will result.

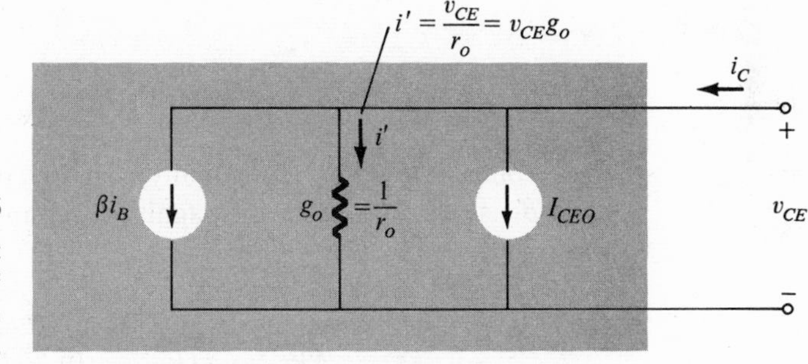

FIG. 5.5
Piecewise linear equivalent circuit for the collector circuit of a transistor, as derived from Eq. (5.2).

We must now turn our attention to the base or input characteristics and determine, using the same procedure, an equivalent circuit for the input characteristics of the common emitter configuration. When this circuit has been obtained, the two can be combined and we shall have a complete piecewise linear equivalent circuit for the transistor.

The input characteristics for the transistor of Section 3.6 have been represented on an approximate basis by the piecewise linear characteristics of Fig. 5.6. The straight lines have been drawn parallel, and equally spaced. The slope, therefore, is the same for each, and for reasons that become obvious in the next few sections, we define it to be $1/r_i$ rather than g_i. The horizontal offset in the characteristics has been

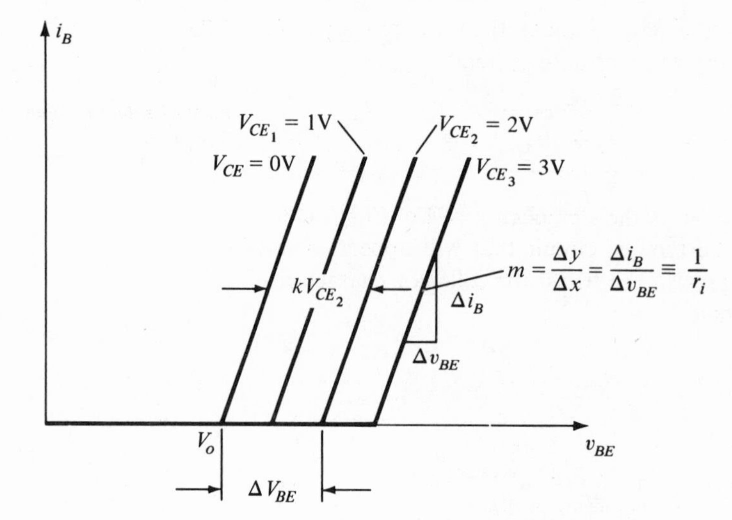

FIG. 5.6
Transistor base characteristics approximated by equally spaced, straight, parallel lines.

included as a dc voltage source V_o. The horizontal spacing can be found in terms of v_{CE} by defining the following ratio to be the constant k.

$$\boxed{k = \frac{\Delta V_{BE}}{\Delta V_{CE}}} \tag{5.3}$$

so that

$$\Delta V_{BE} = k\,\Delta V_{CE}$$

Since the lines are equally spaced, parallel, and straight, k, like β, is also the same in magnitude at any point on the characteristics.

For the curve $V_{CE_2} = 2\text{ V}$,

$$\Delta V_{BE} = k(V_{CE_2} - 0) = kV_{CE_2}$$

and for the same line, when $y = i_B = 0$,

$$x = v_{BE} = V_o + \Delta V_{BE} = V_o + kV_{CE_2}$$

Substituting the above conditions into the basic straight line equation:

$$y = mx + b$$

$$0 = \frac{1}{r_i}(V_o + kV_{CE_2}) + b$$

and

$$b = \frac{-(V_o + kV_{CE_2})}{r_i}$$

The equation for the straight line defined by $V_{CE_2} = 2$ V is, therefore,

$$y = mx + b$$

$$i_B = \frac{1}{r_i} V_{BE} - \frac{V_o + kV_{CE_2}}{r_i}$$

and for any straight line of Fig. 5.6 (substitute v_{CE} for V_{CE}):

$$i_B = \frac{v_{BE}}{r_i} - \frac{(V_o + kv_{CE})}{r_i} \tag{5.4}$$

Rewritten in a slightly different form,

$$i_B r_i = v_{BE} - V_o - kv_{CE}$$

or

$$v_{BE} = i_B r_i + V_o + kv_{CE} \tag{5.5}$$

Again the question arises: will Eq. (5.5) help us find the circuit that will appear in the container of Fig. 5.7 to satisfy the approximate characteristics of Fig. 5.6? Each term of Eq. (5.5) has the units of volts. For this situation, therefore, if we apply Kirchhoff's voltage law in reverse; that is, find the circuit to fit the equation, the circuit of Fig. 5.8 will result.

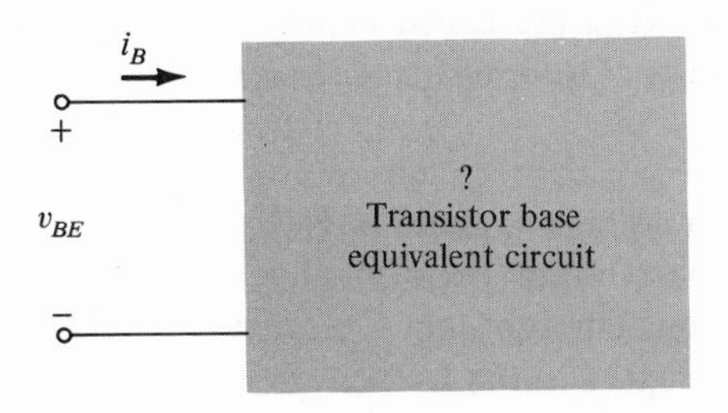

FIG. 5.7

Transistor base equivalent circuit to be found in the ensuing analysis.

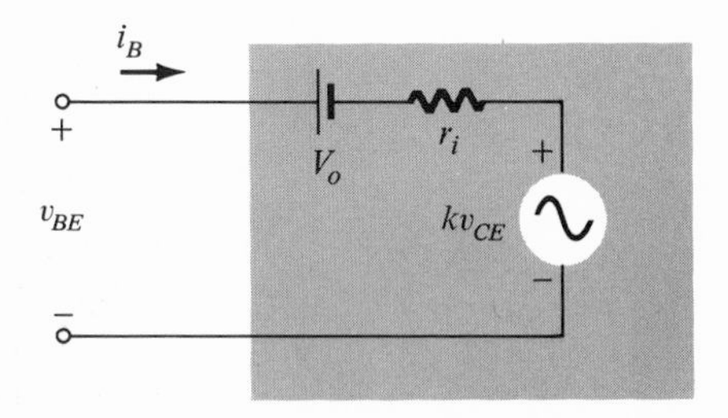

FIG. 5.8

Piecewise linear equivalent circuit for the base circuit of a transistor as derived from Eq. (5.5).

The ratio indicates that the parameter h_{11} is an impedance parameter to be measured in ohms. Since it is the ratio of the *input* voltage to the *input* current with the output terminals *shorted*, it is called the *short-circuit input impedance parameter*.

If I_1 is set equal to zero by opening the input leads, the following will result for h_{12}.

$$h_{12} = \left.\frac{V_1}{V_2}\right|_{I_1=0} \tag{5.8}$$

The parameter h_{12}, therefore, is the ratio of the input voltage to the output voltage with the input current equal to zero. It has no units since it is a ratio of voltage levels. It is called the *open-circuit reverse transfer voltage ratio parameter*. The term reverse is included to indicate that the voltage ratio is an input quantity over an output quantity rather than the reverse, which is usually the ratio of interest.

If in Eq. (5.6b), V_2 is set equal to zero by again shorting the output terminals, the following will result for h_{21}.

$$h_{21} = \left.\frac{I_2}{I_1}\right|_{V_2=0} \tag{5.9}$$

Note that we now have the ratio of an output quantity to an input quantity. The term *forward* will now be used rather than *reverse* as indicated for h_{12}. The parameter h_{21} is the ratio of the output current divided by the input current with the output terminals shorted. It is, for most applications, the parameter of greatest interest. This parameter, like h_{12}, has no units since it is the ratio of current levels. It is formally called the *short-circuit forward transfer current ratio parameter*.

The last parameter, h_{22}, can be found by again opening the input leads to set $I_1 = 0$ and solving for h_{22} in Eq. (5.6b).

$$h_{22} = \left.\frac{I_2}{V_2}\right|_{I_1=0} \quad \text{(mhos)} \tag{5.10}$$

Since it is the ratio of the output current to the output voltage it is the output conductance parameter, and is measured in *mhos*. It is called the *open-circuit output conductance parameter*.

In Section 5.2 we were able to find a circuit that would "fit" a previously derived equation. Since each term of Eq. (5.6a) has the units of volts, let us apply Kirchhoff's voltage law in reverse, as was done in Section 5.2, to find a circuit that "fits" the equation. Performing this operation will result in the circuit of Fig. 5.11. Note the similarities

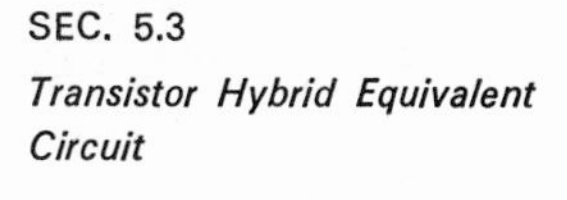

between this circuit and that obtained for the piecewise linear equivalent circuit of Section 5.2. Since the parameter h_{11} has the units of ohms it is represented as a resistor in Fig. 5.11. The dimensionless quantity h_{12} relates itself very nicely to the constant k used in the piecewise equivalent circuit of the transistor. Note that it is a "feedback" of the output voltage to the input circuit.

Each term of Eq. (5.6b) has the units of current. As in Section 5.2, let us now apply Kirchhoff's current law in reverse to obtain the circuit of Fig. 5.12.

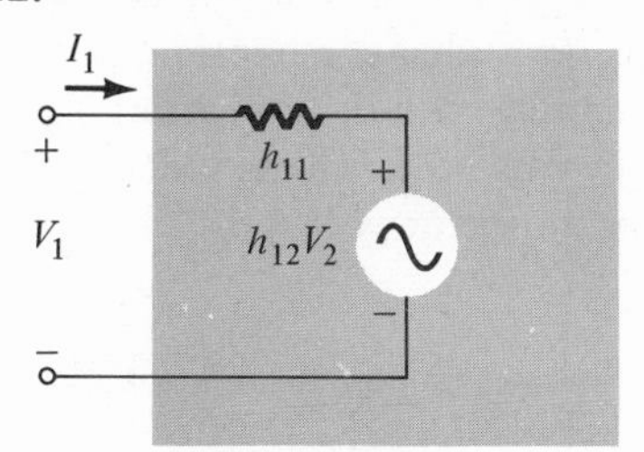

FIG. 5.11
Hybrid input equivalent circuit.

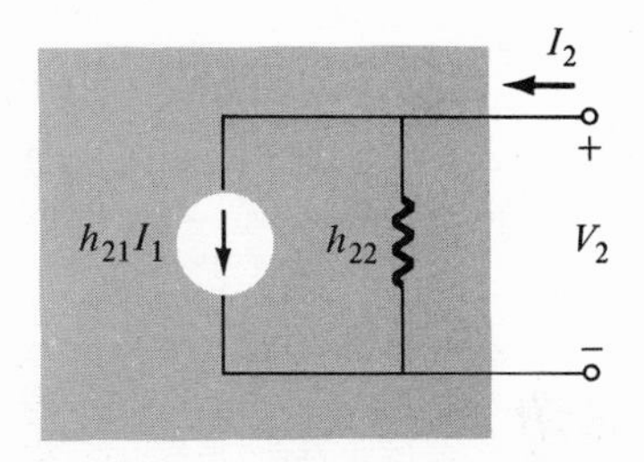

FIG. 5.12
Hybrid output equivalent circuit.

Again, note the similarities between this circuit and that obtained for the output characteristics of a transistor in Section 5.2. Since h_{22} has the units of conductance, it is represented by the resistor symbol. Keep in mind, however, that the resistance in ohms of this resistor is equal to the reciprocal of conductance h_{22} $(1/h_{22})$.

The "total" equivalent circuit for the basic three-terminal linear device is indicated in Fig. 5.13 with a new set of subscripts for the h-parameters.

The notation of Fig. 5.13 is of a more practical nature since it relates the h-parameters to the resulting ratio obtained in the last few paragraphs. The choice of letters is obvious from the following listing:

$h_{11} \longrightarrow$ ***i****nput impedance* $\longrightarrow h_i$
$h_{12} \longrightarrow$ ***r****everse transfer voltage ratio* $\longrightarrow h_r$
$h_{21} \longrightarrow$ ***f****orward transfer current ratio* $\longrightarrow h_f$
$h_{22} \longrightarrow$ ***o****utput conductance* $\longrightarrow h_o$

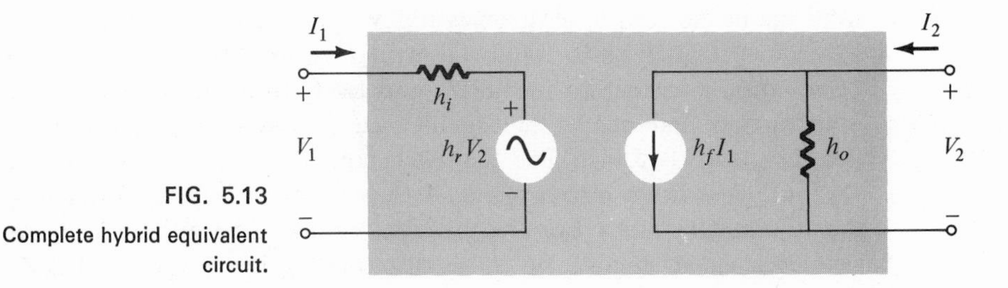

FIG. 5.13
Complete hybrid equivalent circuit.

The circuit of Fig. 5.13 is applicable to any three-terminal device. For the transistor, therefore, even though it has three basic configurations, *they are all three-terminal configurations,* so that the resulting equivalent circuit will have the same format as shown in Fig. 5.13. The *h*-parameters, however, will change with each configuration. To distinguish which parameter has been used or which is available, a second subscript has been added to the *h*-parameter notation. For the common-base configuration the lowercase letter *b* was added, while for the common-emitter and common-collector configurations the letters *e* and *c* were added, respectively. The hybrid equivalent circuit for the common-base and common-emitter configurations with the standard notation is presented in Fig. 5.14. The circuits of Fig. 5.14 are applicable for PNP or NPN transistors.

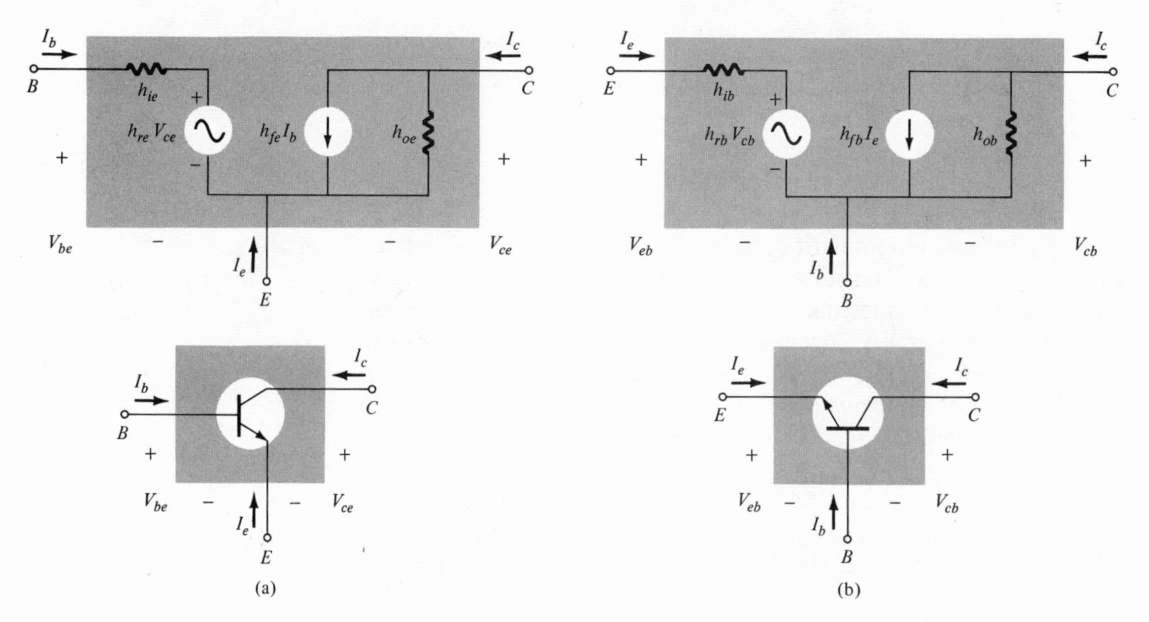

FIG. 5.14

Complete hybrid equivalent circuits: (a) common-emitter configuration; (b) common-base configuration.

The hybrid equivalent circuit of Fig. 5.13 is an extremely important one in the area of electronics today. It will appear over and over again in the analysis to follow. It would be time well spent, at this point, for the reader to memorize and draw from memory its basic construction and define the significance of the various parameters [see Eqs. (5.11)–(5.14)]. The fact that both a Thévenin and Norton circuit appear in the circuit of Fig. 5.13 was further impetus for calling the resultant circuit a *hybrid* equivalent circuit. Two additional transistor equivalent circuits, not to be discussed in this text, called the *Z*-

parameter and Y-parameter equivalent circuits, use either the voltage source or the current source but not both in the same equivalent circuit. In Section 5.4 the magnitude of the various parameters will be found from the transistor characteristics in the region of operation resulting in the desired *small-signal equivalent circuit* for the transistor.

5-4 GRAPHICAL DETERMINATION OF THE *h*-PARAMETERS

Using partial derivatives (calculus) it can be shown that the h-parameters for the small-signal transistor equivalent circuit in the region of operation for the common-emitter configuration can be found using the following equations:

$$h_{ie} = \frac{\partial v_1}{\partial i_1} = \frac{\partial v_B}{\partial i_B} \cong \frac{\Delta v_{BE}}{\Delta i_B}\bigg|_{V_{CE}=\text{constant}} \tag{5.11}$$

$$h_{re} = \frac{\partial v_1}{\partial v_2} = \frac{\partial v_B}{\partial v_C} \cong \frac{\Delta v_{BE}}{\Delta v_{CE}}\bigg|_{I_B=\text{constant}} \tag{5.12}$$

$$h_{fe} = \frac{\partial i_2}{\partial i_1} = \frac{\partial i_C}{\partial i_B} \cong \frac{\Delta i_C}{\Delta i_B}\bigg|_{V_{CE}=\text{constant}} \tag{5.13}$$

$$h_{oe} = \frac{\partial i_2}{\partial v_2} = \frac{\partial i_C}{\partial v_C} \cong \frac{\Delta i_C}{\Delta v_{CE}}\bigg|_{I_B=\text{constant}} \tag{5.14}$$

In each case, the symbol Δ refers to a small change in that quantity around the quiescent point of operation. In other words, the h-parameters are determined in the region of operation for the applied signal, so that the equivalent circuit will be the most accurate available. The constant values of V_{CE} and I_B in each case refer to a condition that must be met when the various parameters are determined from the characteristics of the transistor. For the common-base and common-collector configurations the proper equation can be obtained by simply substituting the proper values of v_1, v_2, i_1, and i_2. *In Appendix A a table has been provided that relates the hybrid parameters of the three basic transistor configurations.* In other words, if the h-parameters for the common-emitter configuration are known the h-parameters for the common-base or common-collector configurations can be found using these tables.

The parameters h_{ie} and h_{re} are determined from the input or base characteristics, while the parameters h_{fe} and h_{oe} are obtained from the output or collector characteristics. Since h_{fe} is usually the parameter of greatest interest we shall discuss the operations involved with equations, such as Eqs. (5.11)–(5.14), for this parameter first. The first step in determining any of the four hybrid parameters is to find the quiescent point of operation as indicated in Fig. 5.15. In Eq. (5.13) the condition V_{CE} = constant requires that the changes in base voltage and current be taken along a vertical straight line drawn through the Q-point representing a fixed collector-to-emitter voltage. Equation (5.13) then requires that a small change in collector current be divided by the corresponding change in base current. For the greatest accuracy these changes should be made as small as possible.

In Fig. 5.15 the change in i_B was chosen to extend from I_{B_1} to I_{B_2} along the perpendicular straight line at V_{CE}. The corresponding change in i_C is then found by drawing the horizontal lines from the intersections of I_{B_1} and I_{B_2} with V_{CE} = constant to the vertical axis. All that remains is to substitute the resultant changes of i_B and i_C into Eq. (5.13); that is,

$$h_{fe} = \left.\frac{\Delta i_C}{\Delta i_B}\right|_{V_{CE}=\text{constant}} = \left.\frac{(2.7-1.7) \times 10^{-3}}{(20-10) \times 10^{-6}}\right|_{V_{CE}=8.4\text{ V}}$$
$$= \frac{10^{-3}}{10 \times 10^{-6}} = \mathbf{100}$$

In Fig. 5.16 a straight line is drawn tangent to the curve I_B through the Q-point to establish a line I_B = constant as required by Eq. (5.14) for h_{oe}. A change in v_{CE} was then chosen and the corresponding change in i_C is determined by drawing the horizontal lines to the vertical axis

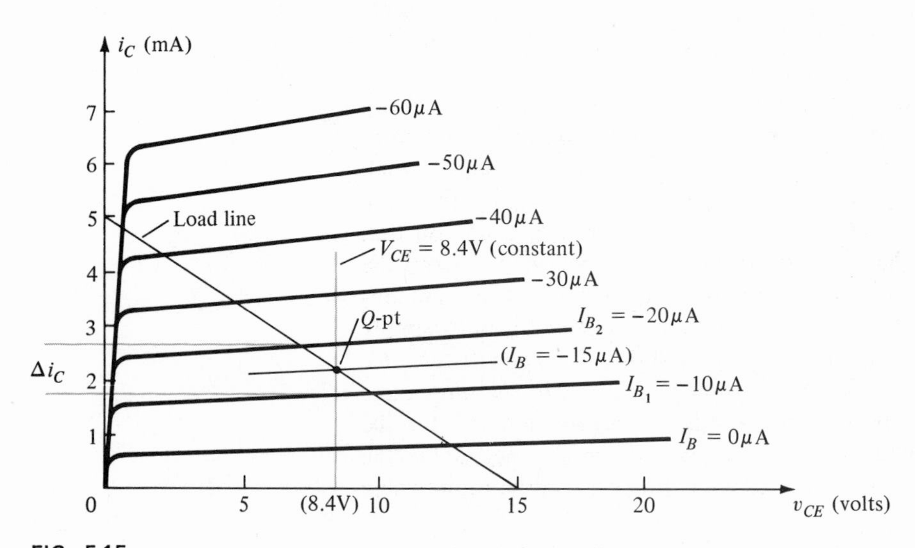

FIG. 5.15
h_{fe} determination.

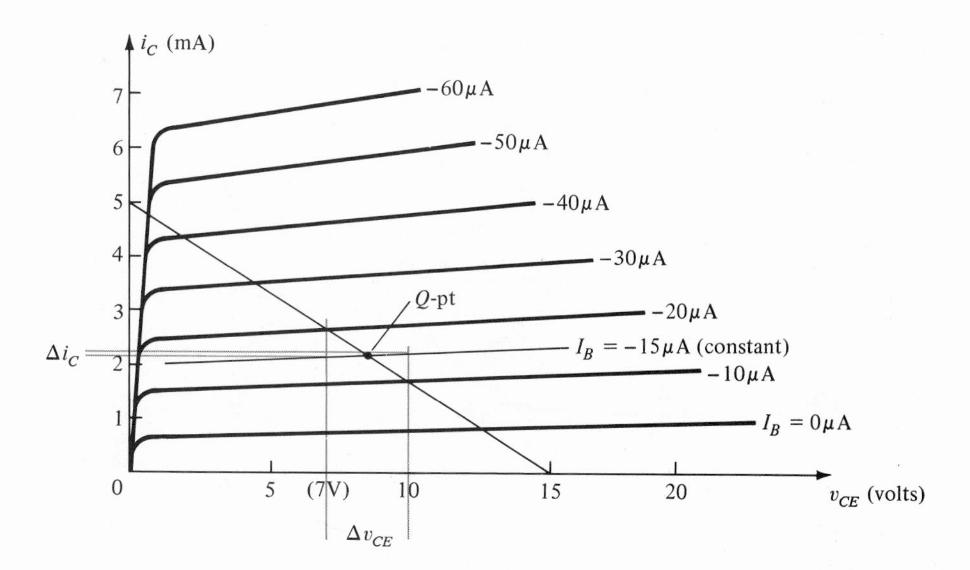

FIG. 5.16
h_{oe} **determination.**

at the intersections on the $I_B =$ constant line. Substituting into Eq. (5.14):

$$h_{oe} = \left.\frac{\Delta i_C}{\Delta v_{CE}}\right|_{I_B=\text{constant}} = \left.\frac{(2.2 - 2.1) \times 10^{-3}}{10 - 7}\right|_{I_B=-15\mu A}$$
$$= \frac{0.1 \times 10^{-3}}{3} = \mathbf{33 \times 10^{-6}\ \mho \text{ or } 33\ \mu A/V}$$

To determine the parameters h_{ie} and h_{re} the Q-point must first be found on the input or base characteristics as indicated in Fig. 5.17. For h_{ie}, a line is drawn tangent to the curve $V_{CE} = 7.5$ V through the Q-point, to establish a line $V_{CE} =$ constant as required by Eq. (5.11). A small change in v_{BE} was then chosen, resulting in a corresponding change in i_B. Substituting into Eq. (5.11):

$$h_{ie} = \left.\frac{\Delta v_{BE}}{\Delta i_B}\right|_{V_{CE}=\text{constant}} = \left.\frac{(187 - 171) \times 10^{-3}}{(20 - 10) \times 10^{-6}}\right|_{V_{CE}=7.5\text{V}}$$
$$= \frac{16 \times 10^{-3}}{10 \times 10^{-6}} = \mathbf{1.6\ K}$$

The last parameter, h_{re}, can be found by first drawing a horizontal line through the Q-point at $I_B = 15\ \mu A$. The natural choice then is to pick a change in v_{CE} and find the resulting change in v_{BE} as shown in Fig. 5.18.

Substituting into Eq. (5.12):

$$h_{re} = \left.\frac{\Delta v_{BE}}{\Delta v_{CE}}\right|_{I_B=\text{constant}} = \frac{(185 - 177) \times 10^{-3}}{20 - 0} = \frac{8 \times 10^{-3}}{20} = \mathbf{4 \times 10^{-4}}$$

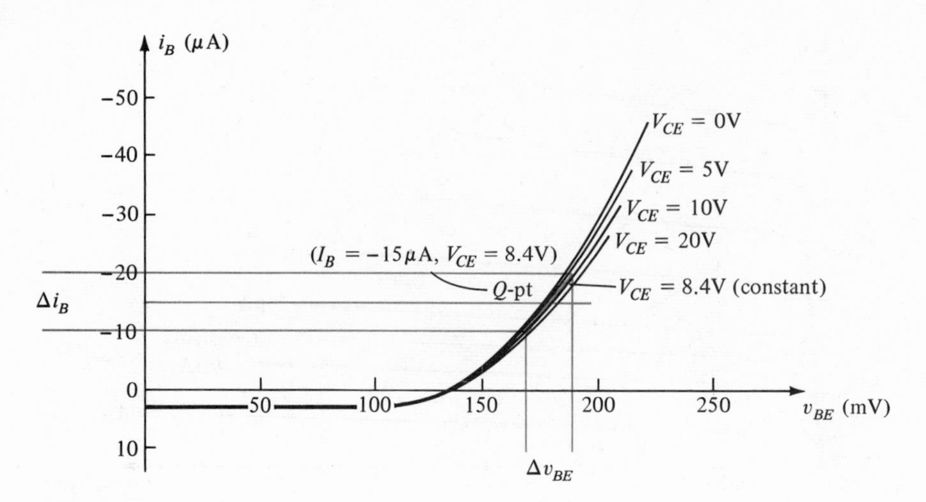

FIG. 5.17
h_{ie} determination.

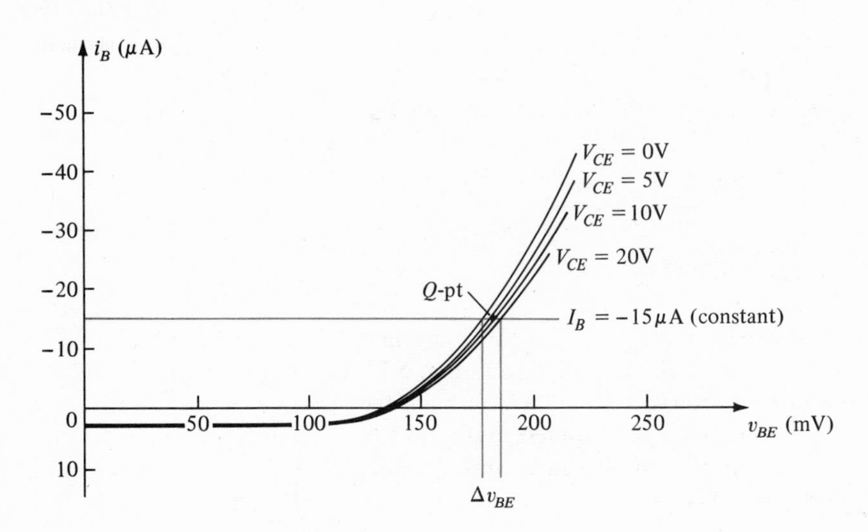

FIG. 5.18
h_{re} determination.

For the transistor whose characteristics have appeared in Figs. 5.15–5.18 the resulting hybrid small-signal equivalent circuit is shown in Fig. 5.19.

As mentioned earlier the hybrid parameters for the common-base and common-collector configurations can be found using the same basic equations with the proper variables and characteristics.

Typical values for each parameter for the broad range of transistors available today in each of its three configurations are provided in Table 5.1. The minus sign indicates that, in Eq. (5.13) as one quantity

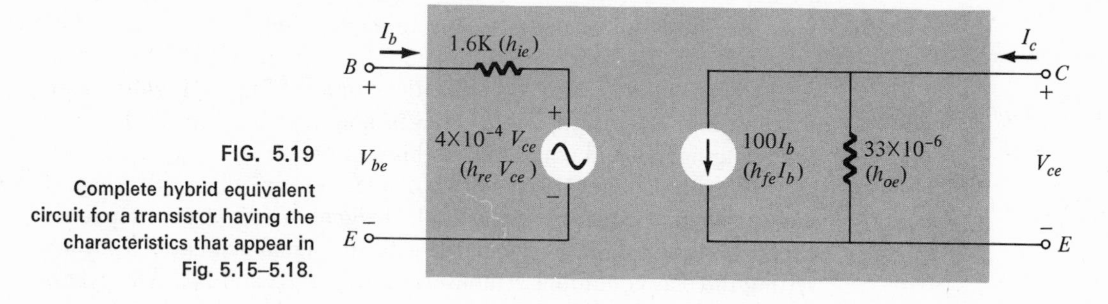

FIG. 5.19
Complete hybrid equivalent circuit for a transistor having the characteristics that appear in Fig. 5.15–5.18.

TABLE 5.1

Typical Parameter Values for the CE, CC, and CB Transistor Configurations

PARAMETER	*CE*	*CC*	*CB*
h_i	1 K	1 K	20 Ω
h_r	2.5×10^{-4}	≅1	3.0×10^{-4}
h_f	50	−50	−0.98
h_o	25 μA/V	25 μA/V	0.5 μA/V
$1/h_o$	40 K	40 K	2 M

increased in magnitude, within the change chosen, the other decreased in magnitude.

Note in retrospect (Section 3.4: Transistor Amplifying Action) that the input resistance of the common-base configuration is low, while the output impedance is high. Consider also that the short-circuit gain is very close to 1. For the common-emitter and common-collector configurations note that the input impedance is much higher than that of the common-base configuration and that the ratio of output to input resistance is about 40 : 1. Consider also, for the common-emitter and common-base configuration, that h_r is very small in magnitude. Transistors are available today with values of h_{fe} that vary from 20 to 200. For any transistor the region of operation and conditions under which it is being used will have an effect on the various *h*-parameters. The effect of temperature and collector current and voltage on the *h*-parameters will be discussed in Section 5.6.

5.5 EXPERIMENTAL DETERMINATION OF THE *h*-PARAMETERS

The *h*-parameters can be determined experimentally by constructing a circuit that will satisfy, under operating conditions, the conditions specified by Eqs. (5.11)–(5.14).

Although four parameters are to be determined only two circuits are needed to determine the parameters. The circuit of Fig. 5.20 can be used to determine h_{ie} and h_{fe}, while the circuit of Fig. 5.21 will be used to determine h_{oe} and h_{re}. All voltages and currents indicated in the following figures and equations of this section are effective values.

The capacitors C_1, C_2, and C_3 of Fig. 5.20 are all effectively short circuits at the frequency of the applied signal. The coupling capacitor C_1 prevents any dc level associated with the signal generator from upsetting the bias conditions established by R_E, V_{EE}, and V_{CC}. The bypass capacitors C_2 and C_3 remove the effects of the biasing circuit from the signal analysis. The resistance R is chosen to be considerably larger than the input resistance R_i so that the variation in R_i with temperature or parameter values will not affect (to any measureable degree) the determination of h_{fe} and h_{ie}. The inductance L is chosen to have a high reactance at the frequency of the applied signal although it is effectively a short circuit for dc (assume $R_{\text{coil}} \cong 0\ \Omega$). In this way it will not appreciably affect the signal response since $X_L \| R_i \cong R_i$, but it will permit the proper dc biasing and the direct measurement of V_{BE}, which, as indicated in Fig. 5.20, is the required V_1 for h_{ie}. The basic equations for h_{ie} [Eq. (5.7)] and h_{fe} [Eq. (5.9)] require that the effective value of the collector-to-emitter voltage (V_2) be 0 V when each

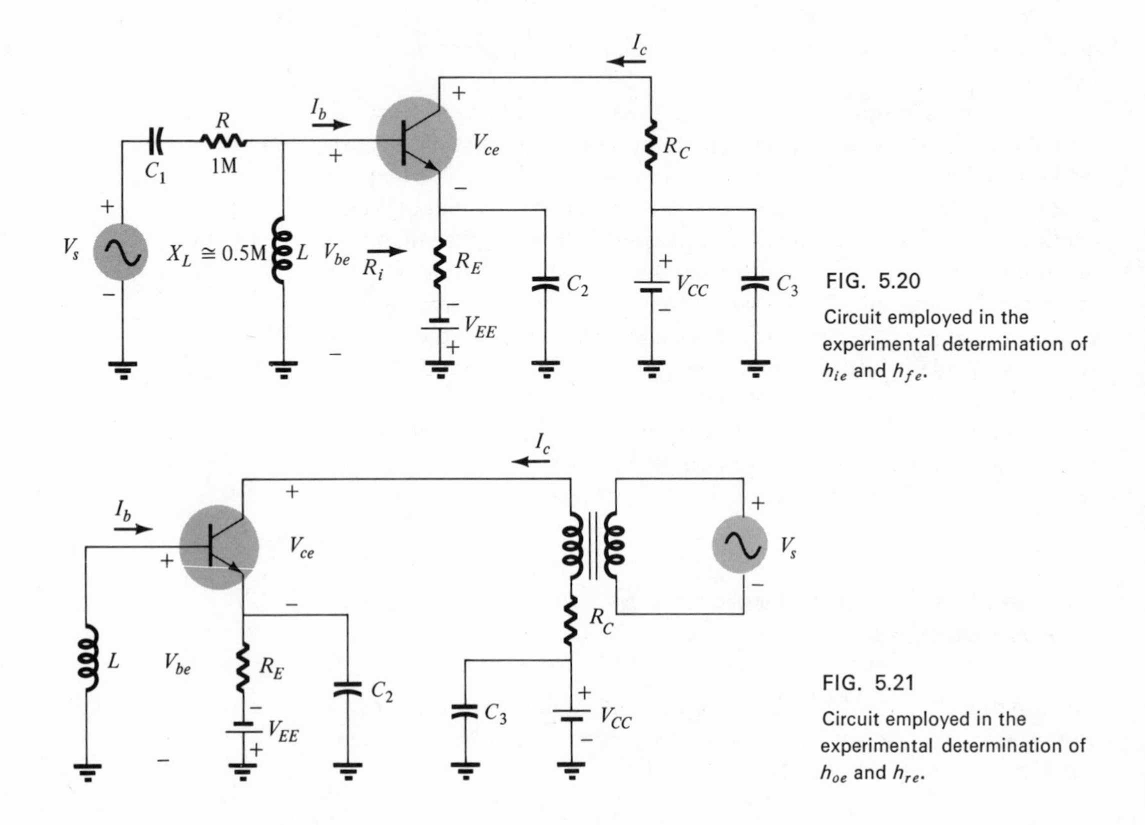

FIG. 5.20 Circuit employed in the experimental determination of h_{ie} and h_{fe}.

FIG. 5.21 Circuit employed in the experimental determination of h_{oe} and h_{re}.

is determined by the indicated ratio. In Fig. 5.20 the resistor R_C is chosen small enough to allow the approximation $V_2 = V_{CE} = V_{R_C} \cong 0$ V, although it is measureable when h_{fe} is to be determined. The resistor R_C is included to permit the necessary measurements for the calculation of h_{fe}.

The equations for h_{ie} [Eq. (5.7)] and h_{fe} [Eq. (5.9)] are repeated below.

$$h_{ie} = \left.\frac{V_1}{I_1}\right|_{V_2=0} = \left.\frac{V_{be}}{I_b}\right|_{V_{ce}=0} \tag{5.15}$$

$$h_{fe} = \left.\frac{I_2}{I_1}\right|_{V_2=0} = \left.\frac{I_c}{I_b}\right|_{V_{ce}=0} \tag{5.16}$$

Substituting the values determined by the circuit of Fig. 5.20:

$$h_{ie} = \left.\frac{V_{be}}{I_b}\right|_{V_{ce}=0} = \left.\frac{V_{be}}{V_s/(R + R_i \cong R)}\right|_{V_{ce}\cong 0} = \left.\frac{V_{be}R}{V_s}\right|_{V_{ce}\cong 0} \tag{5.15a}$$

and

$$h_{fe} = \left.\frac{I_c}{I_b}\right|_{V_{ce}=0} = \left.\frac{V_{ce}/R_C}{V_s/R}\right|_{V_{ce}\cong 0} = \left.\frac{V_{ce}R}{V_sR_C}\right|_{V_{ce}\cong 0} \left(\begin{matrix}\text{although measureable for}\\ \text{the above substitution}\end{matrix}\right) \tag{5.16a}$$

The parameters h_{oe} and h_{re} can be determined using the circuit of Fig. 5.21. The equations for each [Eqs. (5.8) and (5.10), respectively] are indicated below.

$$h_{re} = \left.\frac{V_1}{V_2}\right|_{I_1=0} = \left.\frac{V_{be}}{V_{ce}}\right|_{I_b=0} \tag{5.17}$$

$$h_{oe} = \left.\frac{I_2}{V_2}\right|_{I_1=0} = \left.\frac{I_c}{V_{ce}}\right|_{I_b=0} \tag{5.18}$$

The condition $I_b = 0$ is established by having a high-impedance choke in the base circuit. Capacitors C_2 and C_3 are short circuits to the signal frequency. The dc bias conditions are established by R_E, R_C, V_{EE}, and V_{CC}. The signal source is applied through a transformer to ensure that any dc level associated with the signal V_S does not affect the bias conditions. The resistor R_C is again present to permit the necessary measurements.

Substituting the values determined by the circuit of Fig. 5.21:

$$h_{re} = \frac{V_{be}}{V_{ce}}\bigg|_{I_b=0} = \frac{V_{be}}{V_{ce}}\bigg|_{I_b\cong 0} \tag{5.17a}$$

$$h_{oe} = \frac{I_c}{V_{ce}}\bigg|_{I_b=0} = \frac{I_c}{V_{ce}}\bigg|_{I_b\cong 0} = \frac{V_{Rc}/R_C}{V_{ce}}\bigg|_{I_b\cong 0} = \frac{V_{Rc}}{R_C V_{ce}}\bigg|_{I_b\cong 0} \tag{5.18a}$$

The operation of the circuit can be improved somewhat by replacing the high-impedance coil by a tank circuit (parallel resonant circuit). There will then be a reduction in the resistance R_L associated with the inductor since the required inductance will be reduced. In addition, higher impedances are more readily obtainable with the tank circuit than with increased inductance values at the same frequency.

5.6 VARIATIONS OF TRANSISTOR PARAMETERS

There are a large number of curves that can be drawn to show the variations of the *h*-parameters with temperature, frequency, voltage, and current. The most interesting and useful at this stage of the development include the *h*-parameter variations with junction temperature and collector voltage and current.

In Fig. 5.22, the effect of the collector current on the *h*-parameter has been indicated. Take careful note of the logarithmic scale on the vertical and horizontal axes. The parameters have all been normalized to unity so that the relative change in magnitude with collector current can easily be determined. On every set of curves, such as in Fig. 5.22, the operating point at which the parameters were found is always indicated. For this particular situation the quiescent point is at the intersection of $V_{CE} = 5.0$ V and $I_C = 1.0$ mA. Since the frequency and temperature of operation will also affect the *h*-parameters these quantities are also indicated on the curves. At 0.1 mA, h_{fe} is 0.5 or 50% of its value at 1.0 mA, while at 3 mA, it is 1.5 or 150% of that value. In other words, h_{fe} has changed from a value of $0.5(50) = 25$ to $1.5(50) = 75$. In Section 5.8 we shall find that for the majority of applications it is a fairly good approximation to neglect the effects of h_{re} and h_{oe} in the equivalent circuit. Consider, however, the point of operation as $I_C = 50$ mA. The magnitude of h_{re} is now approximately 11 times that at the defined *Q*-point, a magnitude that may not permit eliminating this parameter from the equivalent circuit. The parameter h_{oe} is approximately 35 times the normalized value. This increase in h_{oe} will decrease the magnitude of the output resistance of the transistor to a point where it may approach the magnitude of the

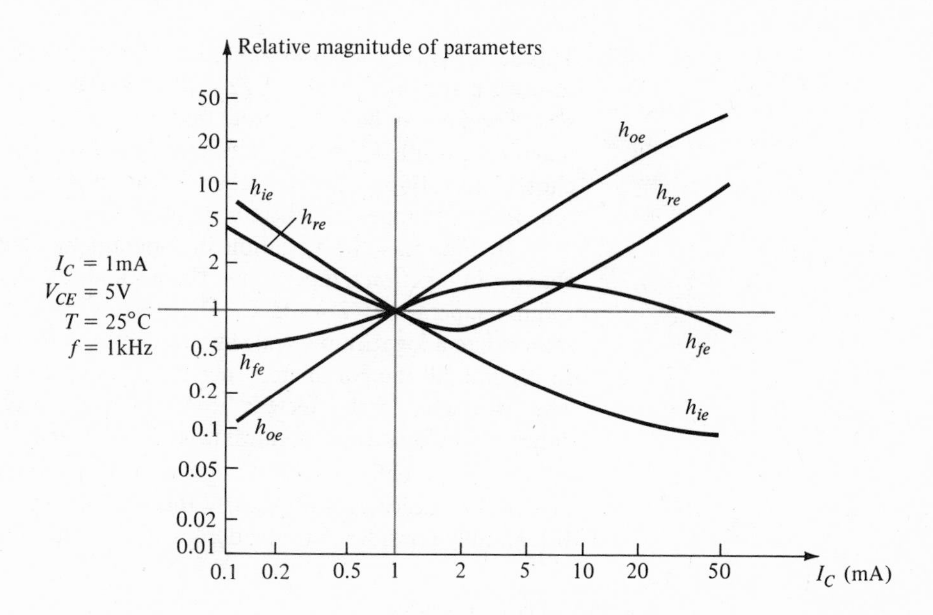

FIG. 5.22
Hybrid parameter variations with collector current.

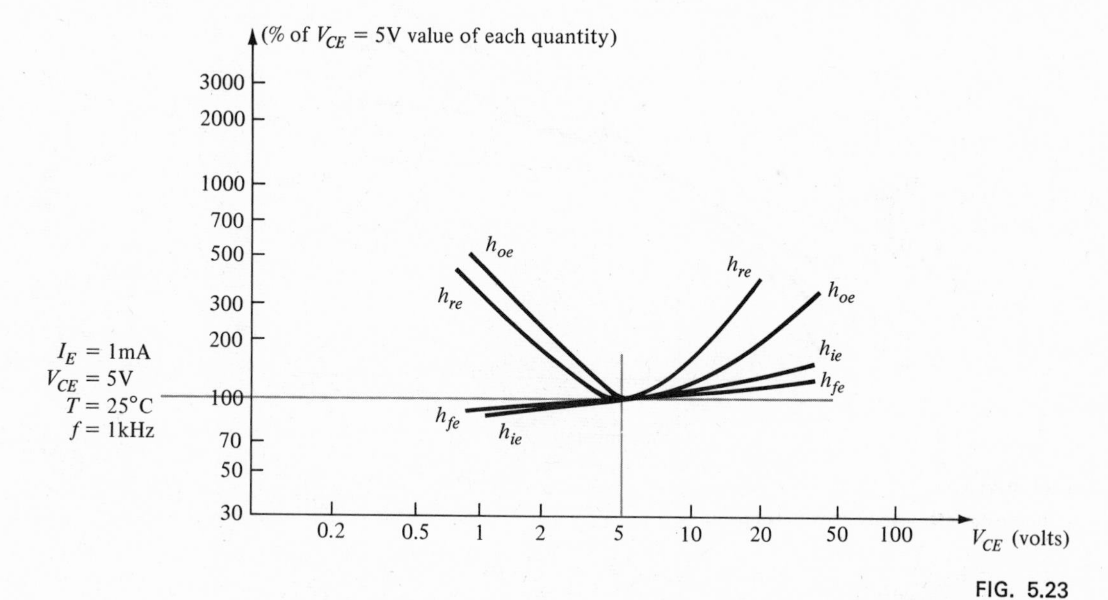

FIG. 5.23
Hybrid parameter variations with collector-emitter potential.

load resistor. There would then be no justification in eliminating h_{oe} from the equivalent circuit on an approximate basis.

In Fig. 5.23 the variation in magnitude of the h-parameters on a normalized basis has been indicated with change in collector voltage.

This set of curves was normalized at the same operating point of the transistor discussed in Fig. 5.22 so that a comparison between the two sets of curves can be made. Note that h_{ie} and h_{fe} are relatively steady in magnitude, while h_{oe} and h_{re} are much larger to the left and right of the chosen operating point. In other words, h_{oe} and h_{re} are much more sensitive to changes in collector voltage than h_{ie} and h_{fe}.

In Fig. 5.24 the variation in h-parameters has been plotted for changes in junction temperature. The normalization value is taken to be room temperature: $T = 25°C$. The horizontal scale is a linear scale rather than a logarithmic scale as was employed for Figs. 5.22 and 5.23. In general, all the parameters increase in magnitude with temperature. The parameter least affected, however, is h_{oe}, while the input impedance h_{ie} changes at the greatest rate. The fact that h_{fe} will change from 50% of its normalized value at $-50°C$ to 150% of its normalized value at $+150°C$ indicates quite clearly that the operating temperature must be carefully considered in the design of transistor circuits.

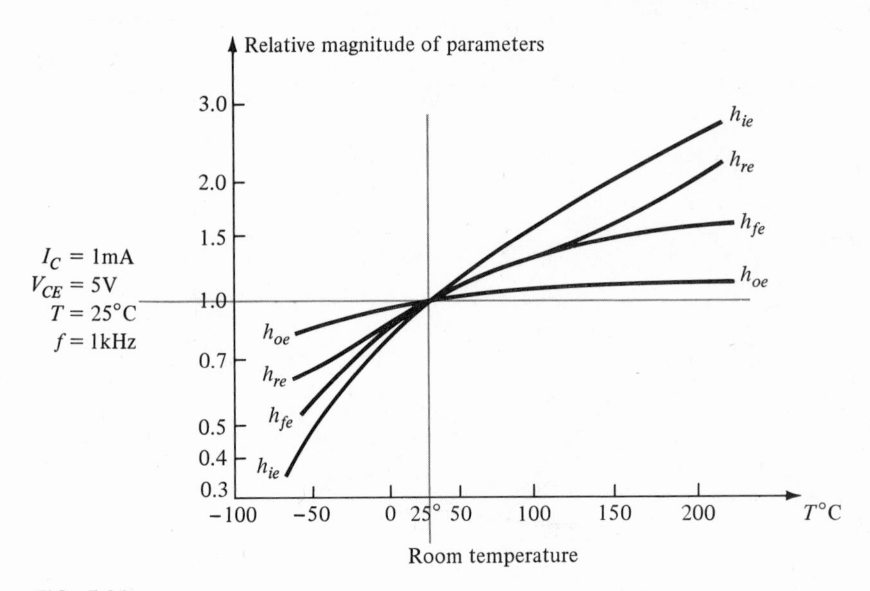

FIG. 5.24
Hybrid parameter variations with temperature.

5.7 SMALL-SIGNAL ANALYSIS OF THE BASIC TRANSISTOR AMPLIFIER USING THE HYBRID EQUIVALENT CIRCUIT

In this section, the basic transistor amplifier will be examined in detail using the hybrid equivalent circuit. It will not be specified whether the transistor is in the common-emitter, base, or collector configuration. The results, therefore, are applicable to each configuration, requiring only that the proper h-parameters be used for the configuration of interest. All amplifiers are basically two-port devices as indicated in Fig. 5.25; that is, there are pair of input terminals and a pair of output terminals. For an amplifier, there are six quantities of general interest: current gain, voltage gain, input impedance, output impedance, power

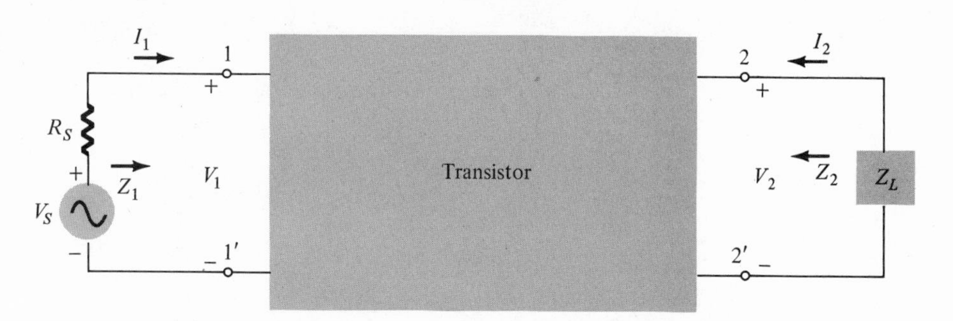

FIG. 5.25
Basic transistor amplifier configuration.

gain, and phase relationships, each of which will be discussed in detail in this section. The load impedance Z_L can be any combination of resistive and reactive elements. In Fig. 5.25, all voltages and currents refer to the effective value of the sinusoidally varying quantities. The resistor R_s represents, in total, the internal resistance of the source and any resistance in series with the source V_s. Before continuing, keep in mind that the analysis to follow is for small-signal inputs. There is no discussion of dc levels and biasing arrangement. For the resulting equations to be useful, however, the quiescent point and resulting h-parameters must be known.

The Current Gain $A_i = (I_2/I_1)$

Substituting the hybrid equivalent circuit for the transistor of Fig. 5.25 will result in the configuration of Fig. 5.26.

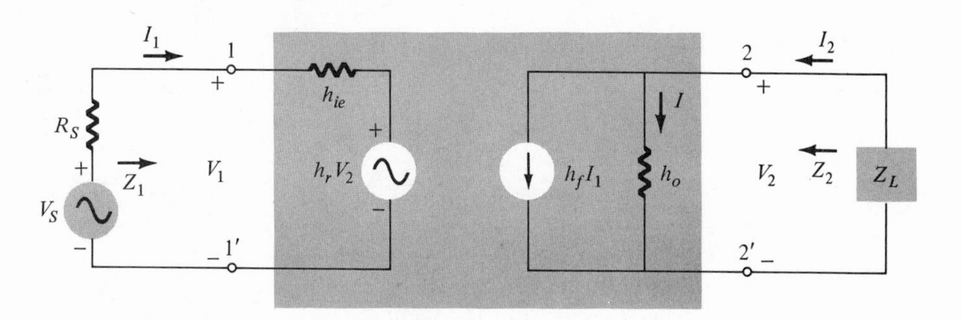

FIG. 5.26
Transistor hybrid equivalent circuit substituted for the transistor of Fig. 5.25.

Applying Kirchhoff's current law to the output circuit:

$$I_2 = h_f I_1 + I = h_f I_1 + h_o V_2$$

Substituting $V_2 = -I_2 Z_L$:

$$I_2 = h_f I_1 - h_o Z_L I_2$$

The minus sign arises because the direction of I_2 as shown in Fig. 5.26 would result in a polarity across the load Z_L opposite to that shown in Fig. 5.26.

Rewriting the above equation

$$I_2 + h_o Z_L I_2 = h_f I_1$$

and

$$I_2(1 + h_o Z_L) = h_f I_1$$

so that

$$\boxed{A_i = \frac{I_2}{I_1} = \frac{h_f}{1 + h_o Z_L}} \tag{5.19}$$

The Voltage Gain $A_v = (V_2/V_1)$

Applying Kirchhoff's voltage law to the input circuit:

$$V_1 = I_1 h_i + h_r V_2$$

Substituting $I_1 = [(1 + h_o Z_L) I_2 / h_f]$ from Eq. (5.19) and $I_2 = -(V_2/Z_L)$ from above:

$$V_1 = \frac{-(1 + h_o Z_L) h_i}{h_f Z_L} V_2 + h_r V_2$$

Solving for the ratio V_2/V_1:

$$\boxed{A_v = \frac{V_2}{V_1} = \frac{-h_f Z_L}{h_i + (h_i h_o - h_f h_r) Z_L}} \tag{5.20}$$

The Input Impedance $Z_1 = (V_1/I_1)$

For the input circuit

$$V_1 = h_i I_1 + h_r V_2$$

Substituting $$V_2 = -I_2 Z_L$$

we have $$V_1 = h_i I_1 - h_r Z_L I_2$$

Since

$$A_i = \frac{I_2}{I_1}$$

$$I_2 = A_i I_1$$

so that the above equation becomes

$$V_1 = h_i I_1 - h_r Z_L A_i I_1$$

Solving for the ratio V_1/I_1

$$Z_1 = \frac{V_1}{I_1} = h_i - h_r Z_L A_i$$

and substituting $$A_i = \frac{h_f}{1 + h_o Z_L}$$

Yields

$$Z_1 = \frac{V_1}{I_1} = h_i - \frac{h_f h_r Z_L}{1 + h_o Z_L} \tag{5.21}$$

The Output Impedance $Z_2 = (V_2/I_2)$

The output impedance of an amplifier is defined to be the ratio of the output voltage to the output current with the signal (V_s) set at zero.

For the input circuit with $V_s = 0$

$$I_1 = \frac{-h_r V_2}{R_s + h_i}$$

Substituting this relationship into the following equation obtained from the output circuit

$$I_2 = h_f I_1 + h_o V_2$$

$$I_2 = \frac{-h_f h_r V_2}{R_s + h_i} + h_o V_2$$

and the ratio

$$Z_2 = \frac{V_2}{I_2}\bigg|_{V_S=0} = \frac{1}{h_o - \left(\dfrac{h_f h_r}{h_i + R_s}\right)} \tag{5.22a}$$

EXAMPLE 5.1 Find the following for the fixed-bias transistor of Fig. 5.27:

(a) Current gain $A_i = I_o/I_i$.
(b) Voltage gain $A_v = V_o/V_i$.
(c) Input impedance Z_i.
(d) Output impedance Z_o.
(e) Power gain A_p.

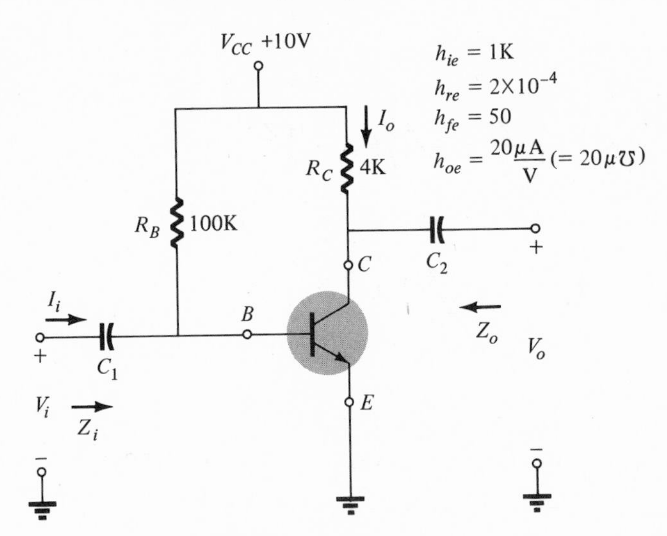

FIG. 5.27
Circuit for Example 5.1.

Solution: Replacing the dc supplies and capacitors by short circuits and substituting the transistor hybrid equivalent circuit will result in the configuration of Fig. 5.28.

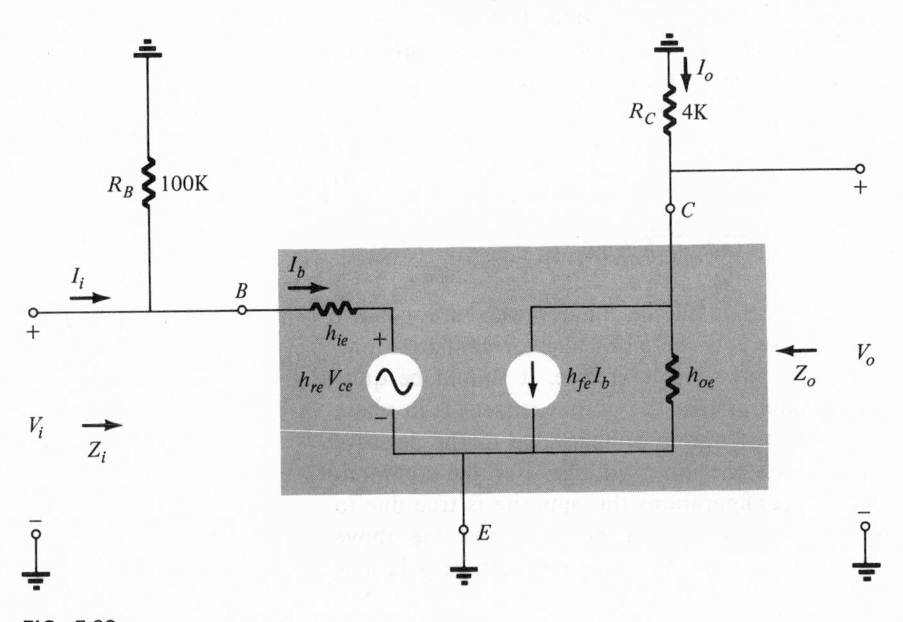

FIG. 5.28
Circuit of Fig. 5.27 following the substitution of the small-signal hybrid equivalent circuit for transistor.

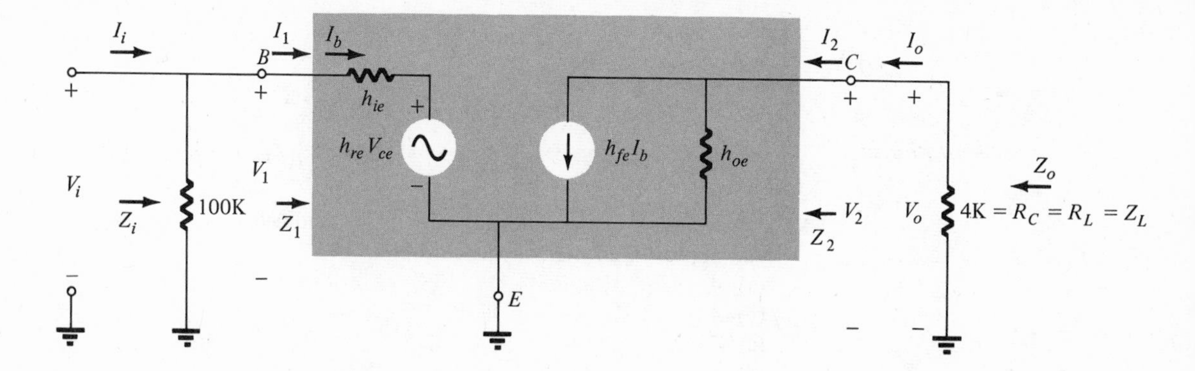

FIG. 5.29
Redrawn circuit of Fig. 5.28.

Redrawing the circuit we obtain Fig. 5.29.

The basic hybrid equations will now be applied to the circuit of Fig. 5.29 to obtain the desired results. Note the similarities in the appearance of Fig. 5.29 as compared to the fundamental configuration of Fig. 5.26. This is a basic requirement if the equations just derived from Fig. 5.26 are to be employed.

(a) To determine A_i, Z_1 must first be found

$$Z_1 = h_{ie} - \frac{h_{fe}h_{re}Z_L}{1 + h_{oe}Z_L} = 1 \times 10^3 - \frac{(50)(2 \times 10^{-4})(4 \times 10^3)}{1 + (20 \times 10^{-6})(4 \times 10^3)}$$

$$= 1 \times 10^3 - \frac{40}{1.08} \cong 963\ \Omega$$

Since $100\text{ K} \gg 0.963\text{ K}$, $I_i \cong I_1$ and

$$A_i = \frac{I_o}{I_i} = \frac{I_o}{I_1} = \frac{I_2}{I_1} = \frac{h_{fe}}{1 + h_{oe}Z_L} = \frac{50}{1 + (20 \times 10^{-6})(4 \times 10^3)}$$

$$= \frac{50}{1.08} = \mathbf{46.3}$$

(b) $$A_v = \frac{V_o}{V_i} = \frac{V_2}{V_1} = \frac{-h_{fe}Z_L}{h_{ie} + (h_{ie}h_{oe} - h_{fe}h_{re})Z_L}$$

$$= \frac{-50(4 \times 10^3)}{1 \times 10^3 + [(1 \times 10^3)(20 \times 10^{-6}) - (50)(2 \times 10^{-4})]4 \times 10^3}$$

$$= \frac{-200 \times 10^3}{1 \times 10^3 + 40} = \mathbf{-192}$$

(c) $$Z_i = 100\text{ K} \parallel Z_1 \cong Z_1 = \mathbf{0.963\ K}$$

(d) $$Z_o = 4\text{ K} \parallel Z_2$$

where $$Z_2 = \frac{1}{h_{oe} - \dfrac{h_{fe}h_{re}}{h_{ie} + R_s}} = \frac{1}{20 \times 10^{-6} - \dfrac{(50)(2 \times 10^{-4})}{1 \times 10^3}}$$

$$= \frac{1}{(20 \times 10^{-6}) - (10 \times 10^{-6})} = \frac{1}{10 \times 10^{-6}} = 100\text{ K}$$

and

$$Z_o = 4\text{ K} \parallel 100\text{ K} \cong \mathbf{3.85\ K}$$

(e) $$A_p = -A_v A_i = -(-192)(46.3) = \mathbf{8.9 \times 10^3}$$

EXAMPLE 5.2. For the following (Fig. 5.30) modified version of the circuit appearing in example 5.1 calculate:

(a) $A_i = I_o/I_i$.
(b) $A_v = V_o/V_i$ and $A_v = V_o/V_s$.
(c) Z_i.
(d) Z_o.

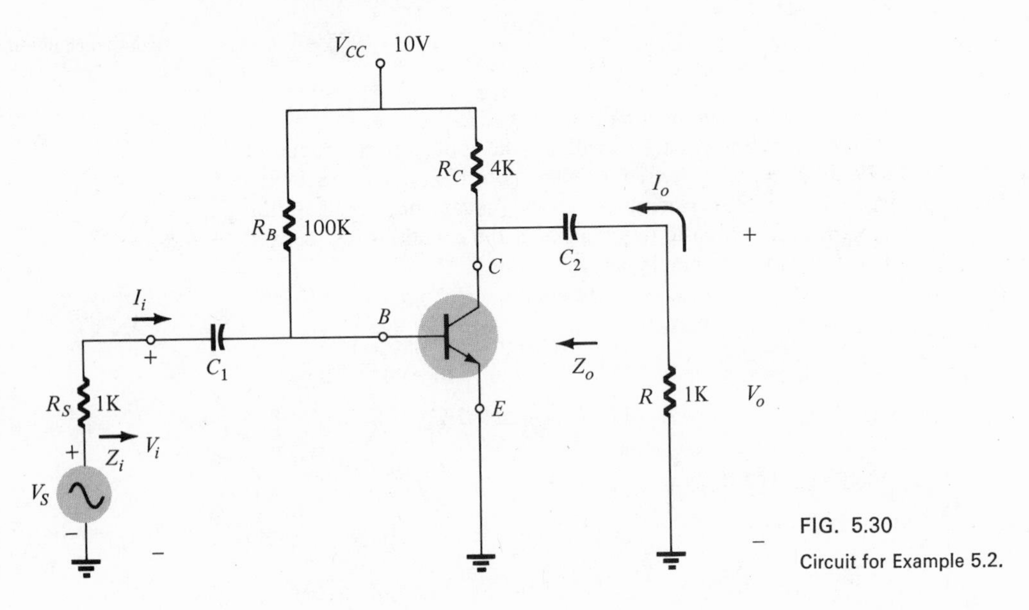

FIG. 5.30
Circuit for Example 5.2.

For small-signal ac analysis the circuit of Fig. 5.30 is redrawn as shown in Fig. 5.31.

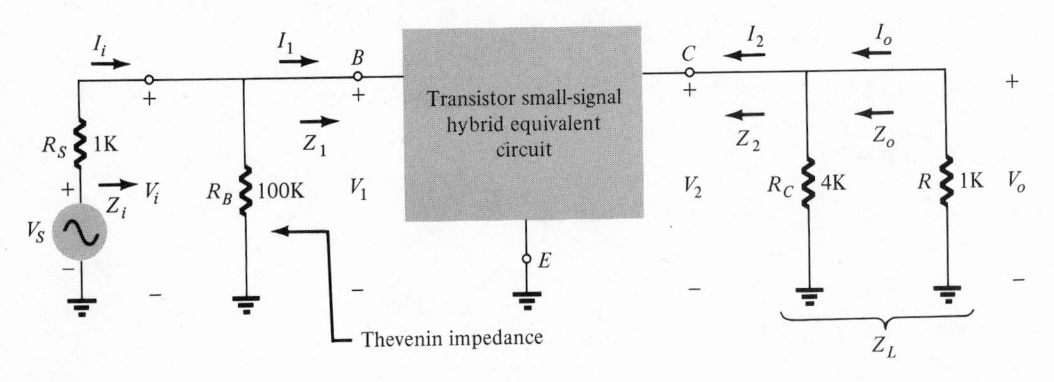

FIG. 5.31
Circuit of Fig. 5.30 redrawn for small-signal ac analysis.

Solution: (a) As in example 5.1, we must first determine Z_1 in order to calculate A_i. However, Z_L is now:

$$Z_L = 4\text{ K} \parallel 1\text{ K} = 0.8\text{ K}$$

so that

$$\begin{aligned} Z_1 &= h_{ie} - \frac{h_{fe}h_{re}Z_L}{1 + h_{oe}Z_L} = 1 \times 10^3 - \frac{(50)(2 \times 10^{-4})(0.8 \times 10^3)}{1 + (20 \times 10^{-6})(0.8 \times 10^3)} \\ &= 1 \times 10^3 - \frac{80 \times 10^{-1}}{1 + 16 \times 10^{-3}} = 1 \times 10^3 - \frac{8}{1.016} \\ &= 1 \times 10^3 - 7.9 = 992\ \Omega \end{aligned}$$

Again, $100\text{ K} \gg Z_1$ with the result $I_i \cong I_1$. Applying the current divider rule:

$$I_o = \frac{4\text{ K}(I_2)}{4\text{ K} + 1\text{ K}} = 0.8\ I_2$$

and

$$\begin{aligned} A_i &= \frac{I_o}{I_i} = \frac{0.8\ I_2}{I_1} = 0.8\left[\frac{I_2}{I_1}\right] = 0.8\left[\frac{h_{fe}}{1 + h_{oe}Z_L}\right] \\ &= 0.8\left[\frac{50}{1 + (20 \times 10^{-6})(0.8 \times 10^3)}\right] = 0.8\left[\frac{50}{1.016}\right] \\ &= 0.8[49.5] = \mathbf{39.6} \end{aligned}$$

(b) $A_v = V_o/V_i$ as calculated in example 5.1 is altered only by the fact that Z_L is now 0.8 K. Therefore,

$$\begin{aligned} A_v &= \frac{V_o}{V_i} = \frac{V_2}{V_1} = \frac{-h_{fe}Z_L}{h_{ie} + (h_{ie}h_{oe} - h_{fe}h_{re})Z_L} \\ &= \frac{-50(0.8 \times 10^3)}{1 \times 10^3 + [(1 \times 10^3)(20 \times 10^{-6}) - (50)(2 \times 10^{-4})]0.8 \times 10^3} \\ &= \frac{-40 \times 10^3}{1 \times 10^3 + 8} \cong \mathbf{-40} \end{aligned}$$

To demonstrate the effect of the source impedance R_s, the voltage gain $A_v = V_o/V_s$ will be found in terms of R_s.

The relationship between V_i and V_s can be determined by applying the voltage divider rule to the circuit of Fig. 5.32.

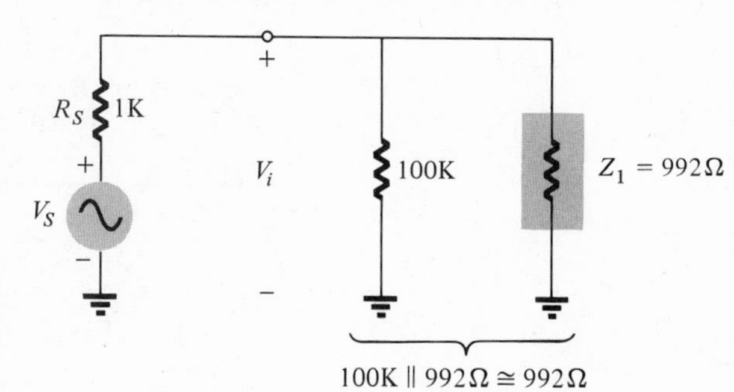

FIG. 5.32
Determining the relationship between V_i and V_s for the network of Fig. 5.31.

That is,

$$V_i = \frac{992(V_s)}{992 + R_s} \quad \text{or} \quad \frac{V_i}{V_s} = \frac{992}{992 + R_s}$$

and

$$A_v = \frac{V_o}{V_s} = \left[\frac{V_o}{V_i}\right]\left[\frac{V_i}{V_s}\right] = [-40]\left[\frac{992}{992 + R_s}\right] = \frac{-39{,}680}{992 + R_s}$$

For $R_s = 1$ K:

$$A_v = \frac{-39{,}680}{992 + 1000} = \frac{-39{,}680}{1{,}992} \cong \mathbf{19.9}$$

For $R_s = 100\ \Omega$

$$A_v = \frac{-39{,}680}{992 + 100} = \frac{-39{,}680}{1092} \cong \mathbf{36.4}$$

A low-impedance source is therefore quite desirable for maximum $A_v = V_o/V_s$.

(c) $$Z_i = 100\text{ K} \parallel Z_1 = 100\text{ K} \parallel 992\ \Omega \cong \mathbf{992\ \Omega}$$

(d) To determine Z_o, we must first determine R_s as defined by Fig. 5.25 so that Z_2 can be obtained. We will refer to this impedance as R_s' to differentiate it from $R_s = 1$ K in this example.

R_s' is simply the Thévenin impedance of the portion of the network indicated in Fig. 5.31 and redrawn in Fig. 5.33

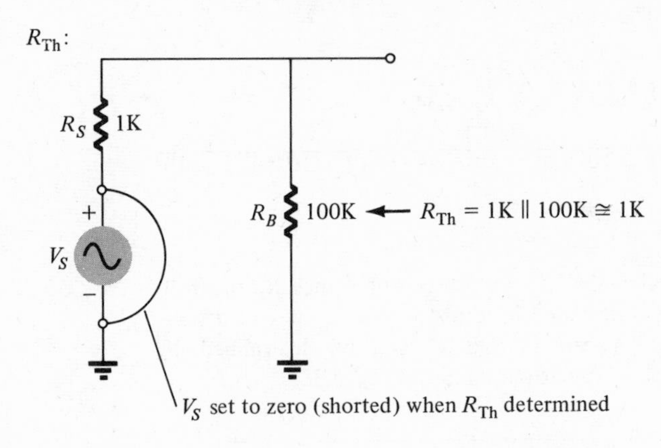

FIG. 5.33
Determining R_{Th} for the portion of the network indicated in Fig. 5.31.

and

$$R_s' = R_{Th} \cong R_s = 1\text{ K}$$

so that

$$Z_2 = \frac{1}{h_{oe} - \dfrac{h_{fe}h_{re}}{h_{ie} + R_s}} = \frac{1}{20 \times 10^{-6} - \dfrac{50(2 \times 10^{-4})}{1 \times 10^3 + 1 \times 10^3}}$$

$$= \frac{1}{20 \times 10^{-6} - \dfrac{100 \times 10^{-4}}{2 \times 10^3}} = \frac{1}{20 \times 10^{-6} - 5 \times 10^{-6}} = 66.7\text{ K}$$

and

$$Z_o = Z_2 \parallel R_C = 66.7\text{ K} \parallel 4\text{ K} \cong \mathbf{3.78\ K}$$

EXAMPLE 5.3 Find the following for the network of Fig. 5.34:

(a) $A_v = V_o/V_i$.
(b) $A_i = I_o/I_i$.
(c) Z_i.
(d) Z_o.

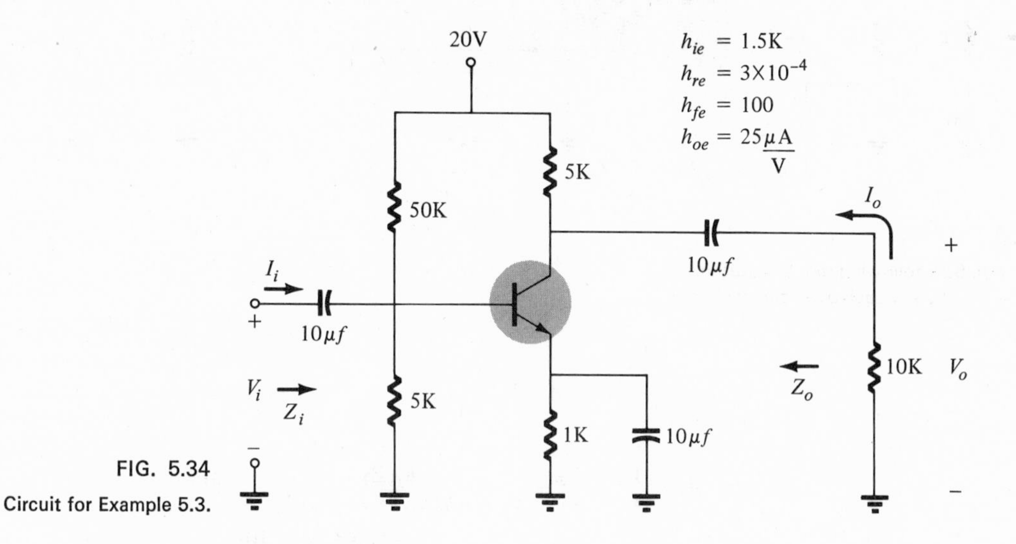

FIG. 5.34
Circuit for Example 5.3.

Solution: Replacing the dc supplies and capacitors by short-circuits will result in the circuit of Fig. 5.35.

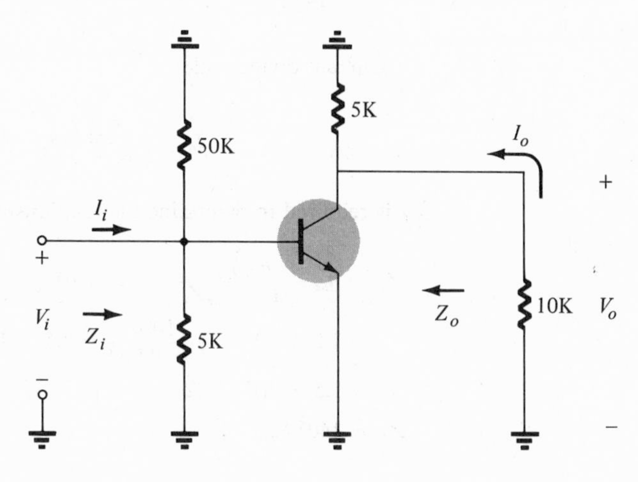

FIG. 5.35
Circuit of Fig. 5.38 redrawn for small-signal ac analysis.

with

$$Z_o = Z_2 \parallel 5\text{ K} = 200\text{ K} \parallel 5\text{ K} \cong \mathbf{4.88\text{ K}}$$

EXAMPLE 5.4 Find the following quantities for the common-base configuration of Fig. 5.39:

(a) $A_i = I_o/I_i$.
(b) $A_{v_1} = V_o/V_i$ and $A_{v_2} = V_o/V_s$.
(c) Z_i.
(d) Z_o.

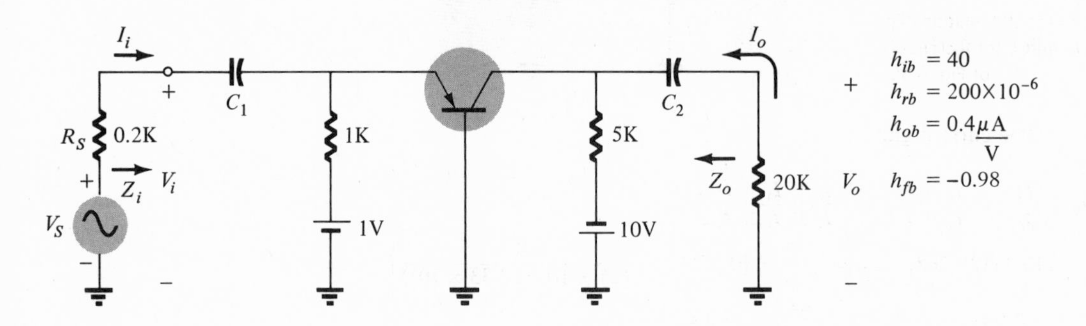

FIG. 5.39
Network for Example 5.4.

Solution: Replacing the dc supplies and capacitors by short-circuits will result in the following configuration (Fig. 5.40).

(a) The equivalent load $Z_L = R_L = 5\text{ K} \parallel 20\text{ K} = 4\text{ K}$

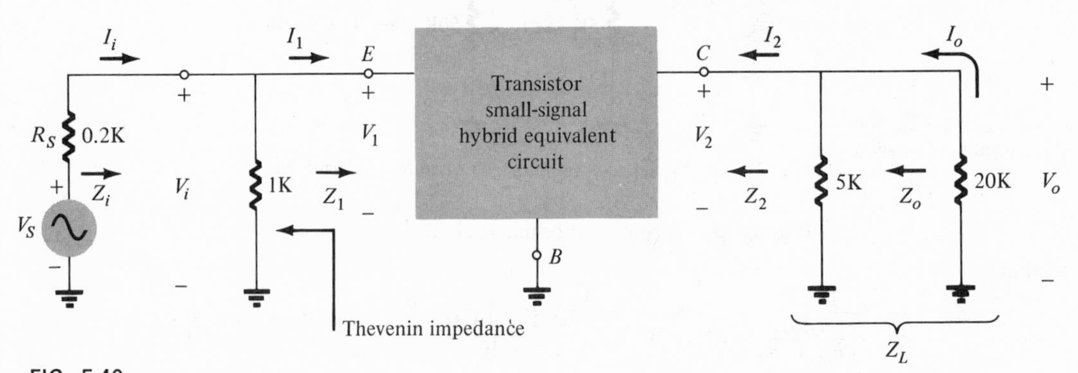

FIG. 5.40
Circuit of Fig. 5.39 redrawn for small-signal ac analysis.

and
$$Z_1 = h_{ib} - \frac{h_{fb}h_{rb}Z_L}{1 + h_{ob}Z_L} = 40 - \frac{(-0.98)(200 \times 10^{-6})(4 \times 10^3)}{1 + (0.4 \times 10^{-6})(4 \times 10^3)}$$
$$= 40 - \frac{(-785 \times 10^{-3})}{1 + 1.6 \times 10^{-3}} = 40 + \frac{785 \times 10^{-3}}{1.0016}$$
$$Z_1 \cong 40.8\ \Omega$$

The relationship between I_1 and I_i can then be determined by applying the current divider to the equivalent circuit of Fig. 5.41.

$$I_1 = \frac{1\text{ K}\,(I_i)}{1\text{ K} + 40.8\ \Omega} = \frac{1\text{ K}(I_i)}{1040.8} \cong 0.96\,I_i$$

And the relationship between I_o and I_2 (Fig. 5.40)

Current divider rule:

$$I_o = \frac{5\text{ K}\,(I_2)}{5\text{ K} + 20\text{ K}} = 0.2\,I_2$$

so that

$$A_i = \frac{I_o}{I_i} = \left[\frac{I_o}{I_2}\right]\left[\frac{I_2}{I_i}\right] = \left[\frac{I_o}{I_2}\right]\left[\frac{I_1}{I_i}\right]\left[\frac{I_2}{I_1}\right]$$
$$= [0.2][0.96]\left[\frac{h_{fb}}{1 + h_{ob}Z_L}\right] = [0.192]\left[\frac{-0.98}{1 + (0.4 \times 10^{-6})(4 \times 10^3)}\right]$$
$$= \frac{-0.188}{1 + 1.6 \times 10^{-3}} = \frac{-0.188}{1.0016} \cong \mathbf{-0.188}$$

(b)
$$A_{v1} = \frac{V_o}{V_i} = \frac{V_2}{V_1} = \frac{-h_{fb}Z_L}{h_{ib} + (h_{ib}h_{ob} - h_{fb}h_{rb})Z_L}$$
$$= \frac{-(-0.98)(4 \times 10^3)}{40 + [(40)(0.4 \times 10^{-6}) - (-0.98)(200 \times 10^{-6})]4 \times 10^3}$$
$$= \frac{3.92 \times 10^3}{40 + (16 \times 10^{-6} + 198 \times 10^{-6})(4 \times 10^3)}$$
$$= \frac{3.92 \times 10^3}{40 + (214 \times 10^{-6})(4 \times 10^3)} = \frac{3.92 \times 10^3}{40 + 0.856} = \mathbf{96}$$

To determine $A_{v_2} = V_o/V_s$, the relationship between V_i and V_s must be found. Applying the voltage divider rule to the equivalent circuit of Fig. 5.42.

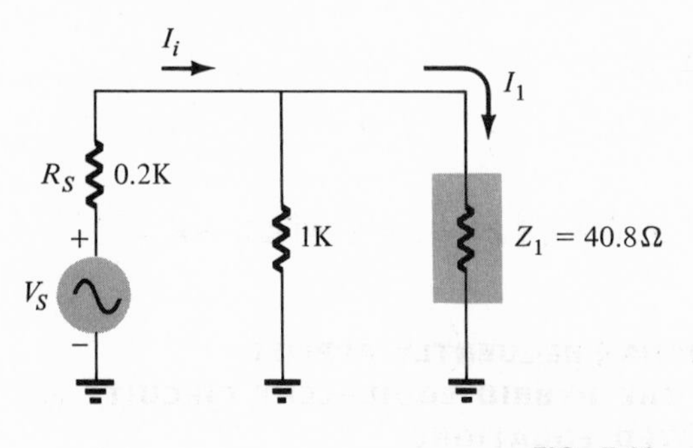

FIG. 5.41

Determining the relationship between I_1 and I_i for the circuit of Fig. 5.40.

R_S 0.2K

V_S

V_i 1K ∥ 40.8Ω ≅ 39.3Ω

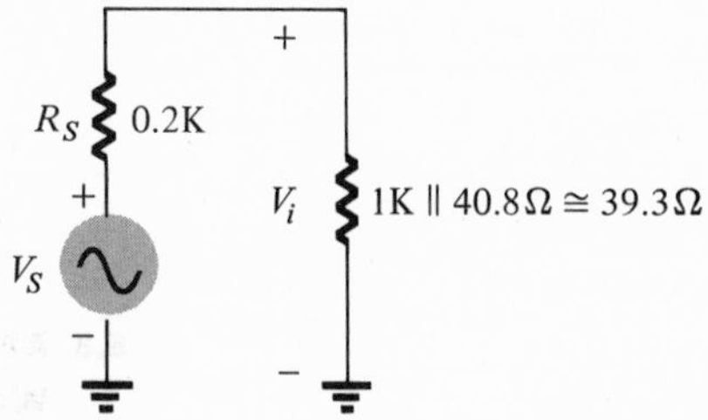

FIG. 5.42

Determining the relationship between V_s and V_i for the circuit of Fig. 5.40.

and

$$h_{ie} + (h_{ie}h_{oe} - h_{fe}h_{re})Z_L = 1000 + (125 \times 10^{-4})(2 \times 10^3)$$
$$= 1000 + 25 \cong 1000 = h_{ie}$$

and

$$A_v \cong \frac{-50(2\times 10^3)}{1 \times 10^3} = -100$$

so that

$$\boxed{A_v \cong \frac{-h_{fe}}{h_{ie}} Z_L} \quad (5.28)$$

due to the fact that the term $Z_L(h_{ie}h_{oe} - h_{fe}h_{re})$ is negligible compared to h_{ie} in Eq. (5.20) because of the small values of h_{oe} and h_{re}.

The Input Impedance Z_1

$$Z_1 = h_{ie} - \frac{h_{fe}h_{re}Z_L}{1 + h_{oe}Z_L}$$

Substituting values in Eq. (5.21)

$$h_{fe}h_{re}Z_L = (50)(2.5 \times 10^{-4})(2 \times 10^3) = 25$$

and

$$\frac{h_{fe}h_{re}Z_L}{(1 + h_{oe}Z_L) \cong 1} \cong \frac{25}{1} = 25$$

and

$$Z_1 = h_{ie} - \frac{h_{fe}h_{re}Z_L}{1 + h_{oe}Z_L} = 1000 - 25 \cong 1000 = h_{ie}$$

so that

$$\boxed{Z_1 \cong h_{ie}} \quad (5.29)$$

The Output Impedance Z_2

$$Z_2 = \frac{1}{h_{oe} - \dfrac{h_{fe}h_{re}}{h_{ie} + R_s}}$$

Equation (5.22a) rewritten in a slightly different form:

$$Z_2 = \frac{h_{ie} + R_s}{h_{oe}(h_{ie} + R_s) - h_{fe}h_{re}}$$

Substituting values:

$$\begin{aligned} h_{oe}(h_{ie} + R_s) - h_{fe}h_{re} &= 25 \times 10^{-6}(1000 + 1000) - 50(2.5 \times 10^{-4}) \\ &= 50 \times 10^{-3} - 12.5 \times 10^{-3} \end{aligned}$$

For Z_2 the order of magnitude of $h_{oe}(h_{ie} + R_s)$ is too close to $h_{fe}h_{re}$ to warrant dropping the smaller quantity. Therefore, there is no clear-cut approximation that can be made for Z_2. Frequently, however, the relationship

$$\boxed{Z_2 > \frac{1}{h_{oe}}} \tag{5.30}$$

is used to obtain some idea of the order of magnitude for the output impedance.

Power Gain A_p

$$A_p = \frac{A_i^2 R_L}{R_i}$$

Through substitution in Eq. (5.26) of the complete expressions for A_i and $Z_i(R_i)$ and the typical values indicated for the parameters at the beginning of this section it can be shown that

$$\boxed{A_p \cong \frac{h_{fe}^2 R_L}{h_{ie}}} \tag{5.31}$$

In summary, the exact equation and the frequently used approximate form for each quantity discussed above are presented in Table 5.2.

For comparison purposes, the exact and approximate values for each quantity using the h-parameters employed in the above derivations have been listed in Table 5.3.

Let us now turn our attention to the hybrid equivalent circuit and note the effect of these approximations on the circuit itself. In all cases, except for Z_2, the hybrid parameter h_{re} does not appear in the

TABLE 5.2

Exact vs. Approximate Equations for the Quantities of Interest for the Basic Transistor Amplifier

QUANTITY	EXACT	APPROXIMATE
A_i	$A_i = \frac{h_{fe}}{1 + h_{oe}Z_L}$	$A_i = h_{fe}$
A_v	$A_v = \frac{-h_{fe}Z_L}{h_{ie} + (h_{ie}h_{oe} - h_{fe}h_{re})Z_L}$	$A_v = \frac{-h_{fe}Z_L}{h_{ie}}$
Z_1	$Z_1 = h_{ie} - \frac{h_{fe}h_{re}Z_L}{1 + h_{oe}Z_L}$	$Z_1 = h_{ie}$
Z_2	$Z_2 = \frac{1}{h_{oe} - \frac{h_{fe}h_{re}}{h_{ie} + R_s}}$	$Z_2 > \frac{1}{h_{oe}}$
A_p	$A_p = \frac{A_i^2 R_L}{R_i}$	$A_p = \frac{h_{fe}^2 R_L}{h_{ie}}$

TABLE 5.3

Exact vs. Approximate Results for a Transistor Amplifier Having the Parameter Values of Section 5.8

QUANTITY	EXACT	APPROXIMATE
A_i	47.62	50
A_v	−97.5	−100
Z_1	975 Ω	1000 Ω
Z_2	53.3 K	$Z_2 >$ 40 K
A_p	4650	5000

approximate equation (Table 5.2) for each quantity. On this basis, the voltage-controlled source $h_{re}V_2$ can be eliminated whenever approximate values of A_i, A_v, and Z_i are desired. As far as Z_2 is concerned, we shall simply keep in mind that whenever $h_{re}V_2$ is eliminated, the resulting Z_2 will be somewhat smaller than the actual value. Removing $h_{re}V_2$ from the complete hybrid equivalent circuit will result in the approximate form of Fig. 5.44. For the circuit just discussed

$$\frac{1}{h_{oe}} = \frac{1}{25 \times 10^{-6}} = 40 \text{ K}$$

and $Z_L = 2$ K. The parallel combination of the two:

$$40 \text{ K} \parallel 2 \text{ K} = \frac{40 \text{ K} \times 2 \text{ K}}{40 \text{ K} + 2 \text{ K}} = \frac{80 \text{ K}}{42 \text{ K}} = 1.91 \text{ K} \cong 2 \text{ K} = Z_L$$

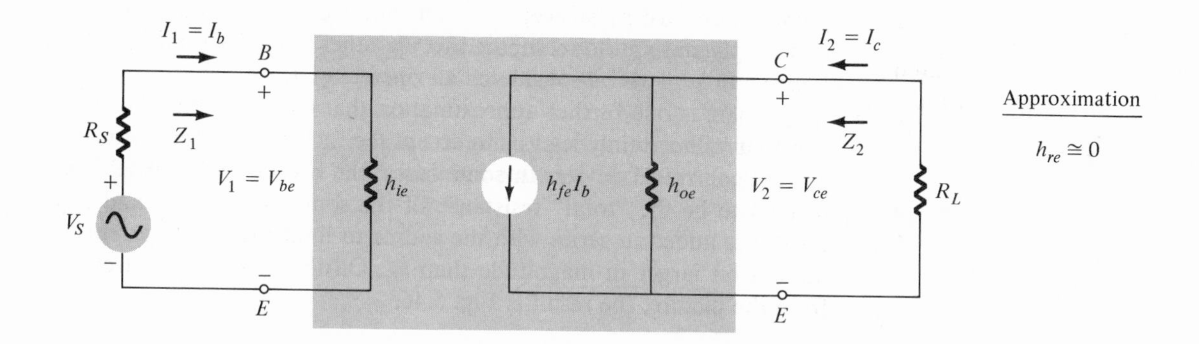

FIG. 5.44
Circuit of Fig. 5.26 following the application of the approximation: $h_{re} \cong 0$.

For a number of applications, the parallel combination of $1/h_{oe}$ and Z_L will result in an equivalent resistance close in magnitude to the load resistance. On this basis, the parameter h_{oe} can be eliminated as a second approximation, from the circuit of Fig. 5.44, resulting in the reduced form of the hybrid equivalent circuit in Fig. 5.45.

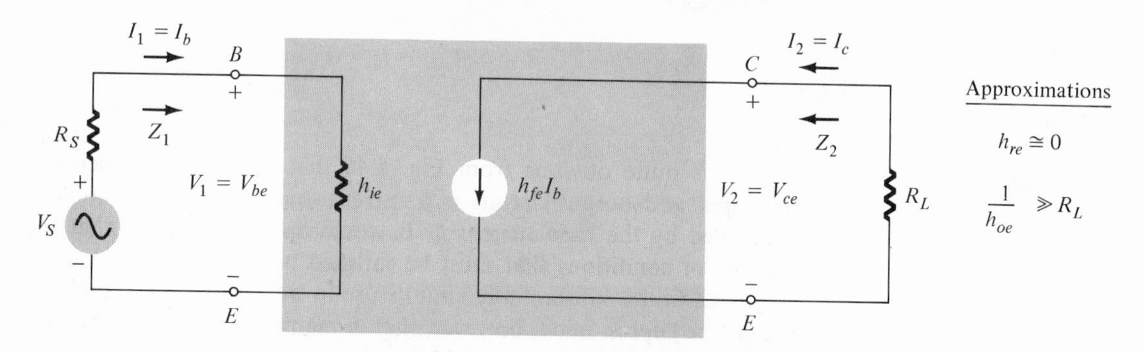

FIG. 5.45
Circuit of Fig. 5.44 following the application of: $\frac{1}{h_{oe}} \gg R_L$.

The circuit of Fig. 5.45 would certainly require fewer calculations to obtain such quantities as the output voltage and current than the complete hybrid equivalent circuit. The conditions associated with the circuit of Fig. 5.45 must always be considered before it can be applied.

In the examples that follow we shall find that, since h_{re} and h_{oe} have been removed from the equivalent circuit, $Z_2 = \infty\ \Omega$ (open circuit). At first glance, this might appear to be a completely unacceptable ap-

Due to the parallel combination of larger and smaller elements, the 100-K and 50-K resistors may be eliminated on an approximate basis as indicated in Fig. 5.48. This will result in the reduced circuit of Fig. 5.49.

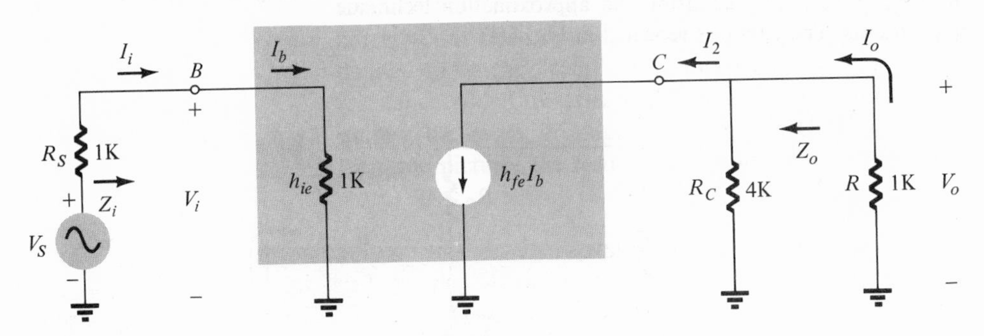

FIG. 5.49
Circuit of Fig. 5.48 following the elimination (on an approximate basis) of certain elements.

(a) A_i:
Current divider rule (Fig. 5.49)

$$I_o = \frac{4\text{ K}(I_2)}{4\text{ K} + 1\text{ K}} = 0.8\,I_2$$

and from Fig. 5.49:

$$I_2 = h_{fe}I_i$$

so that

$$A_i = \frac{I_o}{I_i} = \left[\frac{I_o}{I_2}\right]\left[\frac{I_2}{I_i}\right] = [0.8][h_{fe}] = [0.8][50] = \mathbf{40} \text{ vs. } 39.6$$

obtained in Example 5.2

(b) $A_v = V_o/V_i$:

$$V_o = -(4\text{ K} \,||\, 1\text{ K})I_2 = -0.8\text{ K}I_2 = -0.8\text{ K}[h_{fe}I_b]$$

and

$$I_b = \frac{V_i}{h_{ie}} = \frac{V_i}{1\text{ K}}$$

so that

$$V_o = -(0.8\text{ K})(50)\frac{V_i}{1\text{ K}} = -40\,V_i$$

and

$$A_v = \frac{V_o}{V_i} = \mathbf{-40}$$

as obtained with the move exact method in Example 5.2.

$A_v = V_o/V_s$:

V_o and V_i are clearly related (Fig. 5.49) by the voltage divider rule in the following manner:

$$V_i = \frac{h_{ie}V_s}{h_{ie} + R_s} = \frac{1\text{ K } V_s}{1\text{ K} + 1\text{ K}} = 0.5\ V_s$$

and

$$A_v = \frac{V_o}{V_s} = \left[\frac{V_o}{V_i}\right]\left[\frac{V_i}{V_s}\right] = [-40][0.5] = \mathbf{-20}$$

as compared with 19.9 in Example 5.2.

(c) $Z_i \cong h_{ie} =$ **1 K** vs. 0.992 K obtained in Example 5.2.

(d) Z_o is defined to be the impedance "seen" as the output terminals when V_s is set to zero. If $V_s = 0$, then $I_b = 0$, and $h_{fe}I_b = 0$ resulting the configuration of Fig. 5.50 for the collector circuit, and $Z_o \cong$ **4 K** compared with 3.78 K obtained in Example 5.2.

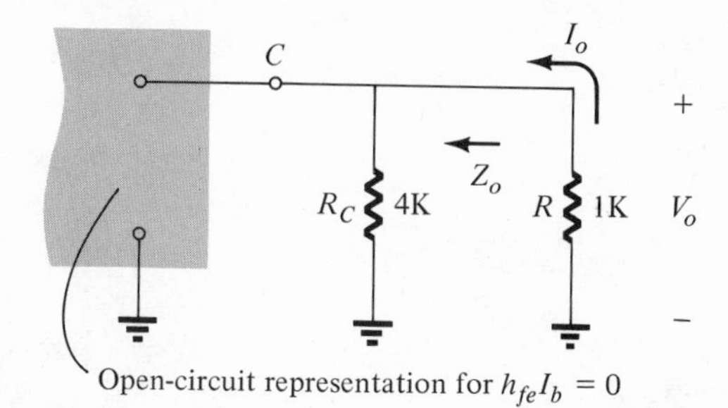

FIG. 5.50
Output circuit configuration for determining Z_o.

Before continuing to the next example, it might be of some interest to determine, at least approximately, how the circuit's response is affected by setting $h_{re} = 0$. For the situation with $h_{re} = 0$, $A_v = V_o/V_i \cong -40$. If we maintain $A_v \cong -40$, but substitute the controlled voltage source $h_{re}V_2$ into the hybrid equivalent circuit the configuration of Fig. 5.51 will result.

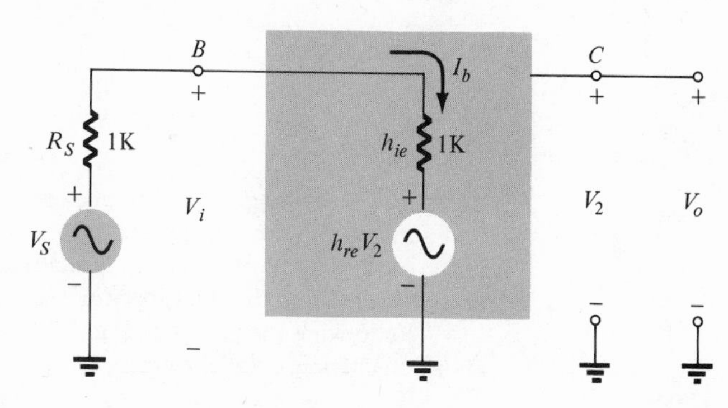

FIG. 5.51
Determining (in an approximate basis) the effect the voltage-controlled voltage source $h_{re}V_2$ will have on the response of the circuit of Fig. 5.47.

Substituting $V_2 = V_o$,

$$h_{re}V_2 = h_{re}V_o = h_{re}(-40V_i)$$
$$= (2 \times 10^{-4})(-40)V_i$$
$$= -0.008V_i$$

The resulting base current $= -0.008V_i$

$$I_b = \frac{V_i + 0.008\ V_i}{2\text{ K}}$$

which is certainly $\cong \frac{V_i}{2\text{ K}}$, substantiating our approximation $h_{re} \cong 0$ in this example.

EXAMPLE 5.7 Find the following for the circuit of Fig. 5.52:

(a) $A_i = I_o/I_i$.
(b) $A_v = V_o/V_i$.
(c) Z_i.
(d) Z_o.

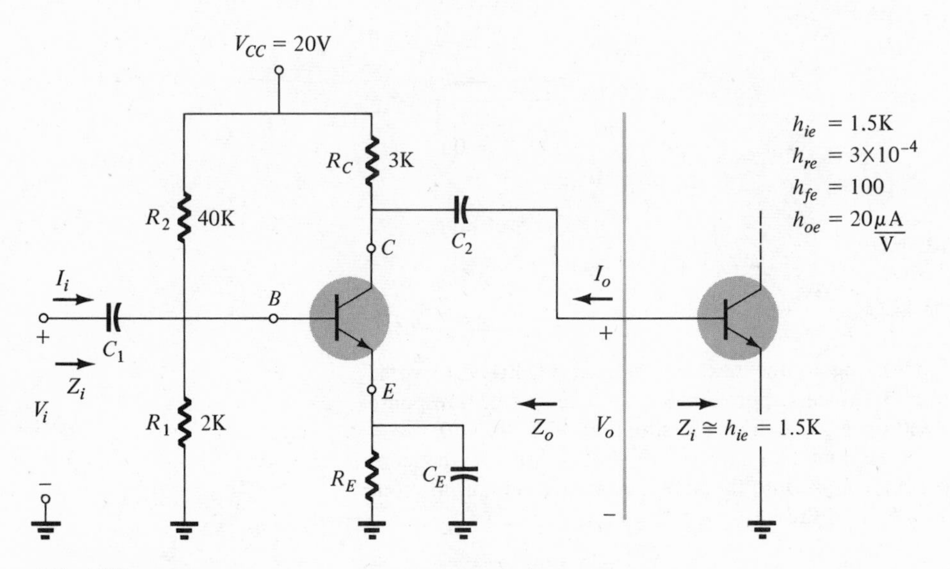

FIG. 5.52
Circuit for Example 5.7.

Solution: Replacing the dc supplies and capacitors by short circuits and substituting the appropriate hybrid equivalent circuit will result in the configuration of Fig. 5.53. Note that the 2-K emitter resistor has been "shorted out" by the capacitor C_E.

Redrawing the circuit of Fig. 5.53 gives us Fig. 5.54 and, finally, after considering parallel elements, Fig. 5.55. Before continuing, note

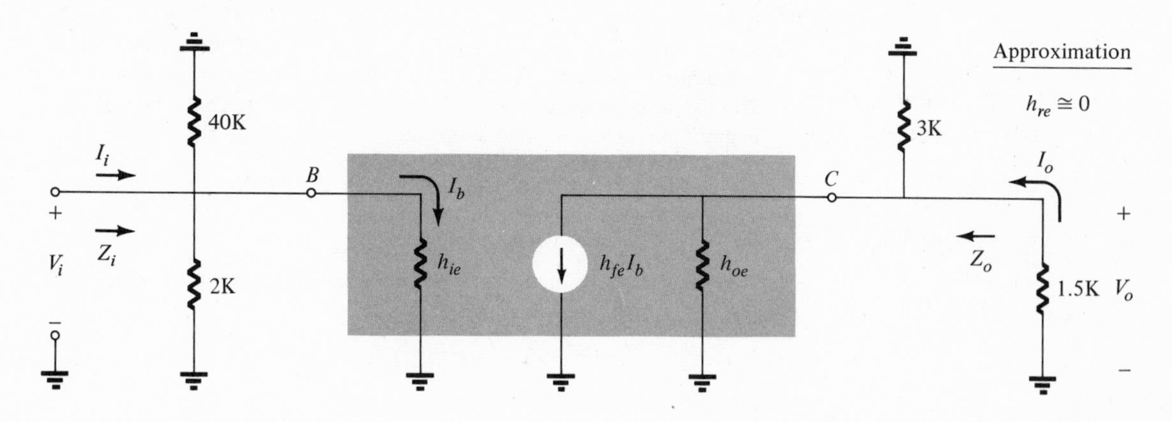

FIG. 5.53

Circuit of Fig. 5.52 following the substitution of the approximate ($h_{re} \cong 0$) hybrid equivalent circuit.

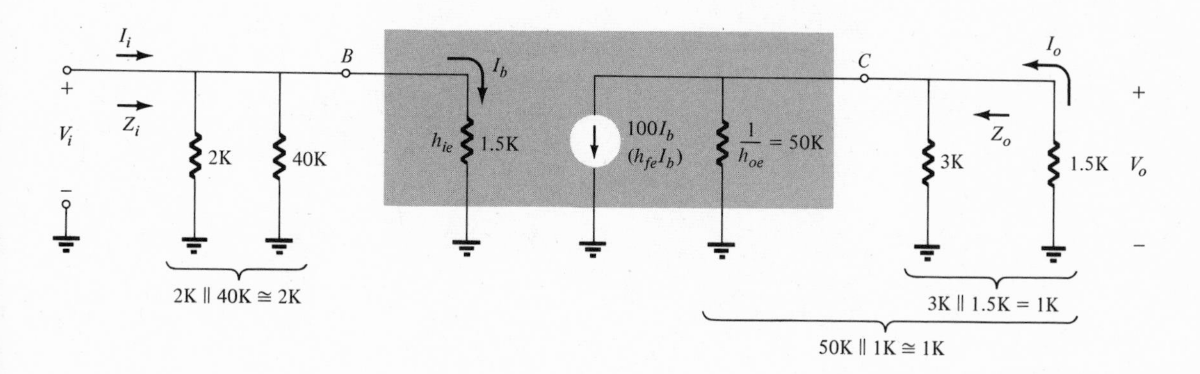

FIG. 5.54

Redrawn circuit of Fig. 5.53.

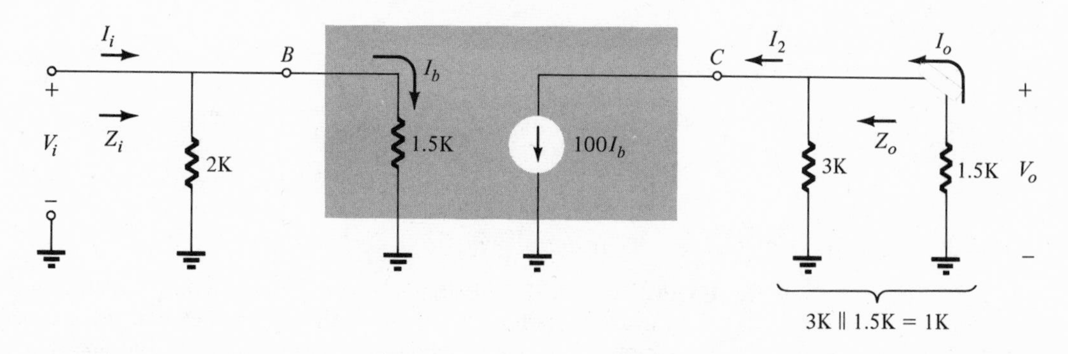

FIG. 5.55

Circuit of Fig. 5.54 following the elimination (on an approximate basis) of certain parallel elements.

that a series of three different figures was required before obtaining the desired result of Fig. 5.55. In time, as the circuits become more familiar, the circuit of Fig. 5.55 can be obtained more directly. In the beginning, however, it is suggested procedure that the intermediate figures be included to preserve the correct unknown quantities and ensure a correct circuit for the remaining numerical calculations.

(a) $A_i = (I_o/I_i)$:

From Fig. 5.55, $I_2 = 100I_b$.

Applying the current divider rule to the input and output circuits:

$$I_b = \frac{2\text{ K}I_i}{2\text{ K} + 1.5\text{ K}} = 0.571I_i$$

and

$$I_o = \frac{3\text{ K}I_2}{3\text{ K} + 1.5\text{ K}} = 0.667I_2$$

Substituting

$$\begin{aligned} A_i = \frac{I_o}{I_i} = \left[\frac{I_o}{I_2}\right]\left[\frac{I_2}{I_i}\right] &= \left[\frac{I_o}{I_2}\right]\left[\frac{I_2}{I_b}\right]\left[\frac{I_2}{I_i}\right] \\ &= [0.667][100][0.571] \\ &= \mathbf{38.1} \end{aligned}$$

(b) $A_v = (V_o/V_i)$

$$V_o = -I_2(3\text{ K} \,||\, 1.5\text{ K}) = -1\text{ K}\,I_2 = -1\text{ K}(100I_b)$$

and

$$I_b = \frac{V_i}{1.5\text{ K}}$$

Substituting:

$$V_o = -1\text{ K}(100)\left(\frac{V_i}{1.5\text{ K}}\right)$$

and

$$A_v = \frac{V_o}{V_i} \cong \frac{1\text{ K } 100}{1.5\text{ K}} = \mathbf{-66.7}$$

(c) $Z_i \cong 2\text{ K} \,||\, 1.5\text{ K} = [(2\text{ K})(1.5\text{ K})/(2\text{ K} + 1.5\text{ K})] = \mathbf{0.86\text{ K}}$.

(d) $Z_o \cong \mathbf{3\text{ K}}$.

EXAMPLE 5.8 In this example, the common-base parameters are made available rather than those of the common-emitter, even though the circuit is in the common-emitter configuration. The first step, in this solution, will be to use Appendix A to obtain the common-emitter values. In Example 5.9, a common-base configuration will be examined with common-base parameters.

Calculate V_o/V_i (see Fig. 5.56).

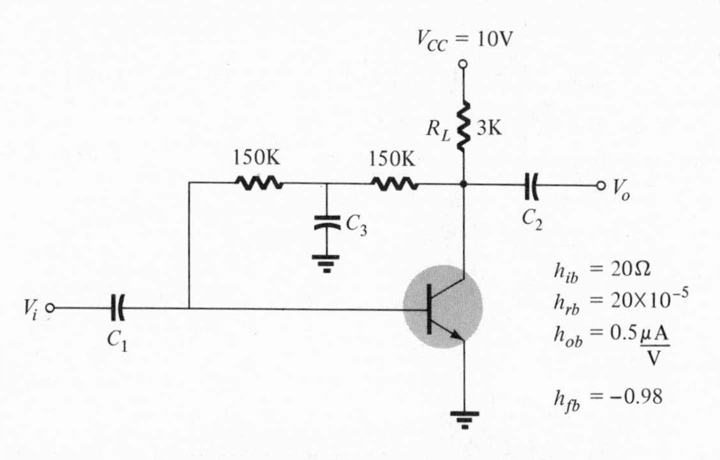

FIG. 5.56
Circuit for Example 5.8.

Solution: Converting to common-emitter parameters (Appendix A):

$$h_{fe} \cong \frac{-h_{fb}}{1 + h_{fb}} = \frac{-(-0.98)}{1 - 0.98} = \frac{0.98}{0.02} = 49$$

$$h_{ie} \cong \frac{h_{ib}}{1 + h_{fb}} = \frac{20}{1 + (-0.98)} = \frac{20}{0.02} = 1 \text{ K}$$

$$h_{oe} \cong \frac{h_{ob}}{1 + h_{fb}} = \frac{0.5 \times 10^{-6}}{1 + (-0.98)} = \frac{0.5 \times 10^{-6}}{0.02} = 25\ \mu\text{A/V}$$

$$h_{re} \cong \frac{(h_{ib}h_{ob} - h_{fb}h_{rb}) - h_{rb}}{1 + h_{fb}}$$

$$= \frac{[(20)(0.5 \times 10^{-6}) - (-0.98)(20 \times 10^{-5})] - 20 \times 10^{-5}}{1 + (-0.98)}$$

$$= \frac{20.6 \times 10^{-5} - 20 \times 10^{-5}}{0.02} = \frac{0.6 \times 10^{-5}}{0.02} = 3 \times 10^{-4}$$

Replacing the dc supplies and capacitors of Fig. 5.56 by short-circuits and substituting the appropriate approximate equivalent circuit will result in the configuration of Fig. 5.57. Redrawing the

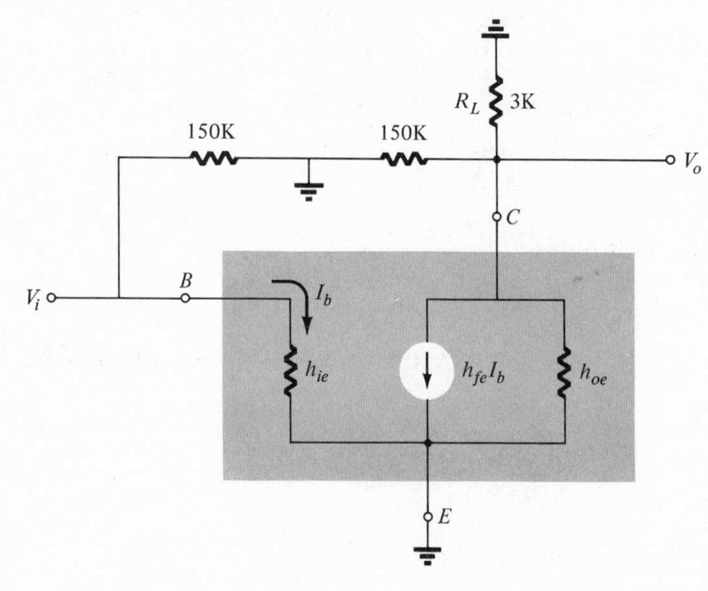

FIG. 5.57
Circuit of Fig. 5.56 following the substitution of the approximate hybrid equivalent circuit.

circuit, we obtain Fig. 5.58 and, finally, Fig. 5.59.

$$V_o = -I_o R_L = -h_{fe} I_b R_L = -h_{fe}\left(\frac{V_i}{h_{ie}}\right) R_L$$

and

$$\frac{V_o}{V_i} \cong -\frac{h_{fe} R_L}{h_{ie}} = \frac{-49(3 \times 10^3)}{1 \times 10^3} = \mathbf{-147}$$

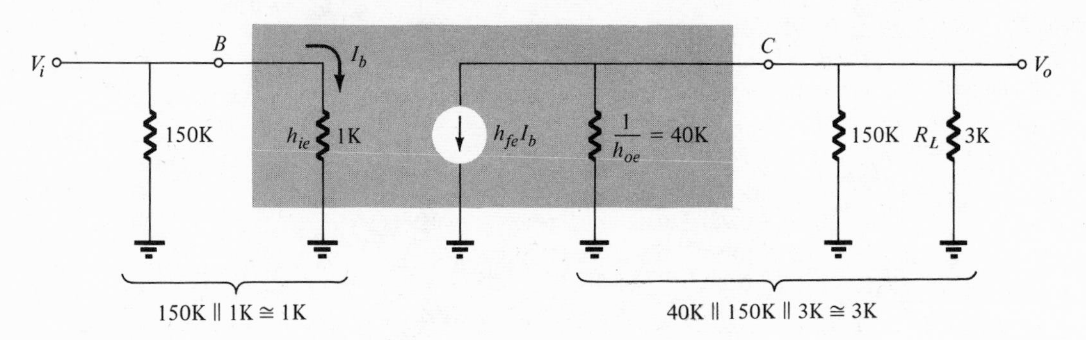

FIG. 5.58
Redrawn circuit of Fig. 5.57.

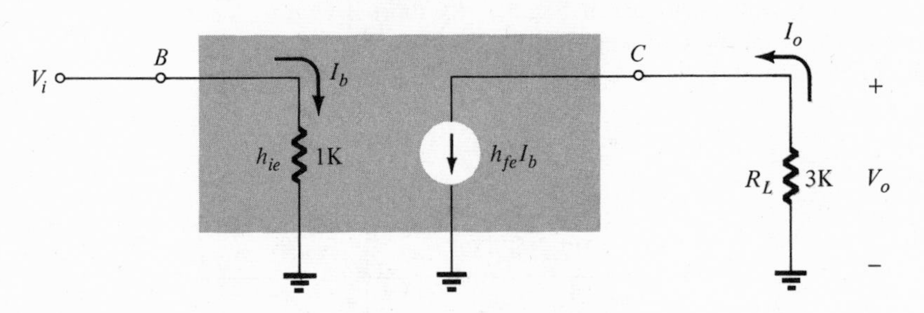

FIG. 5.59
Reduced circuit of Fig. 5.58.

EXAMPLE 5.9 For the common-base circuit of Fig. 5.60 find the following:

(a) $A_i = I_o/I_i$.
(b) $A_v = V_o/V_i$.
(c) Z_i.
(d) Z_o.

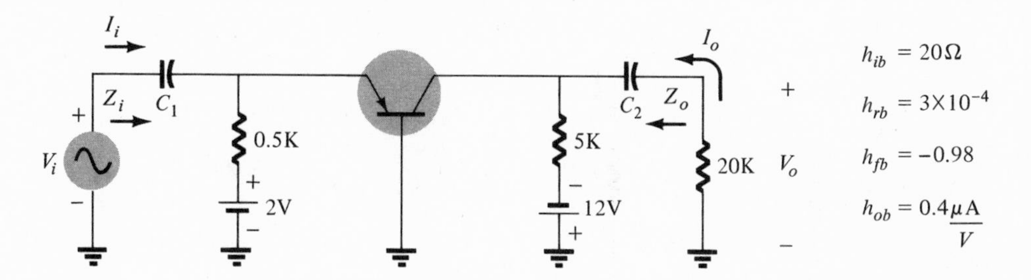

FIG. 5.60
Circuit for Example 5.9.

Solution: Replacing the dc supplies and coupling capacitors by short-circuits will result in the configuration of Fig. 5.61.

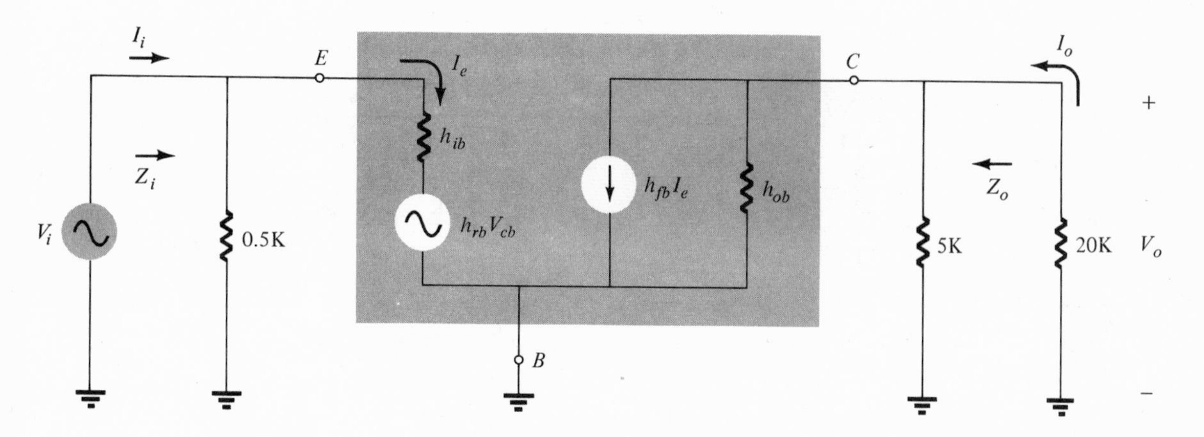

FIG. 5.61
Circuit of Fig. 5.60 following the substitution of the hybrid equivalent circuit.

Note, as mentioned in the above, the common-base equivalent circuit has exactly the same format as the common-emitter configuration but with the appropriate common-base parameters. Consider that now $h_{re}V_{ce}$ is $h_{rb}V_{cb}$ and $h_{fe}I_b$ is now $h_{fb}I_e$. Since the configuration is the same and $1/h_{ob}$ is greater than $1/h_{oe}$ with h_{re} usually close in magnitude to h_{rb}, the approximations ($1/h_{ob} \cong \infty\ \Omega$ and $h_{rb} \cong 0$) made for the common-emitter configuration are applicable here, also.

Applying $h_{rb} \cong 0$, we redraw the circuit (Fig. 5.62). Eliminating the 500-Ω and 500-K ($1/h_{ob}$) resistors due to the lower parallel resistor will result in the circuit of Fig. 5.63.

To determine A_i it will first be necessary to find Z_i as indicated in Fig. 5.65.

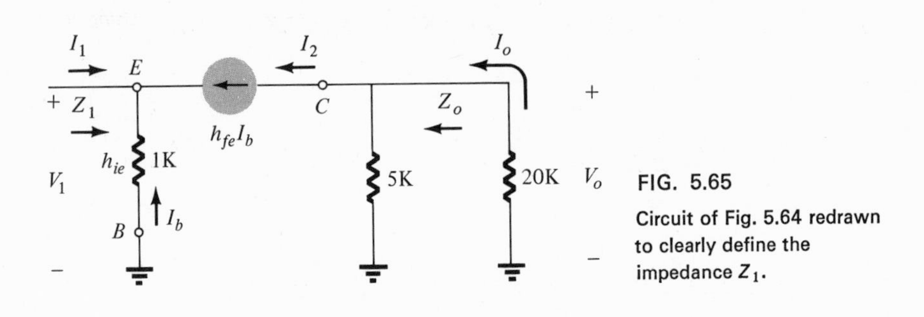

FIG. 5.65
Circuit of Fig. 5.64 redrawn to clearly define the impedance Z_1.

Note that the current-controlled current source ($h_{fe}I_b$) is connected from collector to emitter with h_{ie} between the base and emitter terminals and I_b defined to flow from the base to the emitter terminal.

A_i:

Applying Kirchhoff's current law to node "E" of Fig. 5.65:

$$I_1 + I_b + h_{fe}I_b = 0$$

so that

$$I_1 = -(1 + h_{fe})I_b$$

In addition,

$$I_b = -\frac{V_1}{1\text{ K}}$$

Substituting one into the other:

$$I_1 = -(1 + h_{fe})\left(-\frac{V_1}{1\text{ K}}\right)$$

and

$$Z_1 = \frac{V_1}{I_1} = \frac{1\text{ K}}{1 + h_{fe}} = \frac{1\text{ K}}{50} = 20\ \Omega$$

The parallel combination of the 0.5-K and 20-Ω (Z_1) impedances of the input terminal of the circuit result in

$$0.5\text{ K} \parallel 20\ \Omega \cong 20\ \Omega$$

and $I_i \cong I_1$ (Fig. 5.64)

From the above,

$$I_i = -(1 + h_{fe})I_b$$

and

$$I_o = \frac{5\,\text{K}I_2}{5\,\text{K} + 20\,\text{K}} = 0.2I_2$$

with

$$I_2 = h_{fe}I_b$$

so that

$$A_i = \frac{I_o}{I_i} = \left[\frac{I_o}{I_2}\right]\left[\frac{I_2}{I_i}\right] = \left[\frac{I_o}{I_2}\right]\left[\frac{I_2}{I_b}\right]\left[\frac{I_b}{I_i}\right]$$

$$= [0.2][h_{fe}]\left[-\frac{1}{1 + h_{fe}}\right] = [0.2][49]\left[-\frac{1}{50}\right] = \mathbf{-0.196} \text{ vs. } -0.196$$

obtained in Example 5.9.

A_v:

$$V_o = -I_2(5\,\text{K} \,||\, 20\,\text{K}) = -I_2 4\,\text{K} = -h_{fe}I_b 4\,\text{K}$$

From the input circuit of Fig. 5.65

$$I_b = -\frac{V_1}{1\,\text{K}} = -\frac{V_i}{1\,\text{K}}$$

Substituting into the above equation

$$V_o = -(h_{fe})\left(-\frac{V_i}{1\,\text{K}}\right)(4\,\text{K})$$

and

$$A_v = \frac{V_o}{V_i} = h_{fe}\left(\frac{1}{1\,\text{K}}\right)(4\,\text{K}) = 49(4) = \mathbf{196} \text{ vs. } 196$$

obtained in Example 5.9.

Z_i:
From above:

$$Z_i = 0.5\,\text{K} \,||\, Z_1 \cong Z_1 = \mathbf{20\ \Omega} \text{ vs. } 20\ \Omega$$

for Example 5.9

Z_o:
$Z_o \cong$ **5 K** vs. 5 K obtained in Example 5.9. Z_o is determined solely by the 5-K impedance since $h_{fe}I_b$, when $V_i = 0$, is in its open-circuit state.

Example 5.10 has demonstrated that the parameters of only one hybrid equivalent circuit are necessary to find the quantities of interest for any circuit configuration. It also concludes the examples to be presented in this section. They were chosen to demonstrate the "approximation" technique. It is a technique that permits the development of some "feeling" for the system's response without getting deeply involved in the numerical calculations that will result if the complete hybrid equivalent circuit is employed.

For future reference, note that the approximations $h_{re} \cong 0$ and $(1/h_{oe}) \cong \infty\ \Omega$ (open-circuit) were employed throughout the examples of this section. They are valid approximations in the majority of situations and, therefore, will be used as frequently as possible in the analysis to follow.

5.9 APPROXIMATE BASE, COLLECTOR, AND EMITTER EQUIVALENT CIRCUITS

In the analysis to follow, it will prove very useful to know, at a glance, what the effect of loads and signal sources in another portion of the network will have on the base, collector, or emitter potential and current. In this section we shall find, on an approximate basis, the equivalent circuit "seen" looking into the base, collector, or emitter terminals of a transistor in the common-emitter configuration (Fig. 5.66).

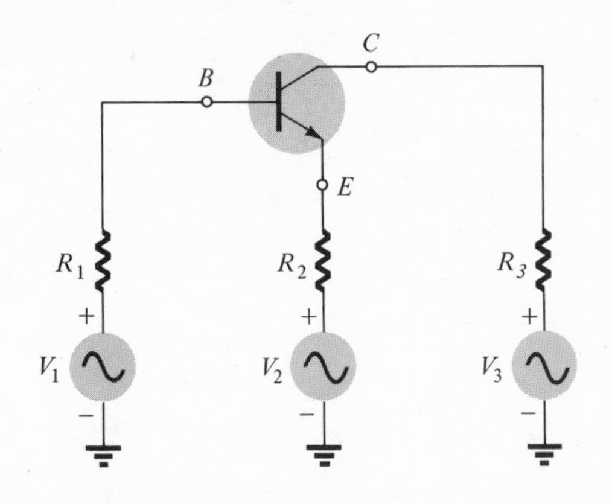

FIG. 5.66
Representative circuit to be employed in the derivation of the base, collector, and emitter approximate equivalent circuits.

The approximations $h_{re} \cong 0$ and $(1/h_{oe}) \cong \infty\ \Omega$ (open circuit), employed continually in Section 5.8, will be used throughout this section (they are valid only if $1/h_{oe} > R_3, R_2$). To include the effects of a supply and a load connected to each terminal the circuit of Fig. 5.66 will be employed. The full benefit of the circuits to be derived will be more apparent when the equivalent circuits have been obtained.

Substituting the equivalent circuit of Fig. 5.45 for the transistor of Fig. 5.66 will result in the circuit of Fig. 5.67.

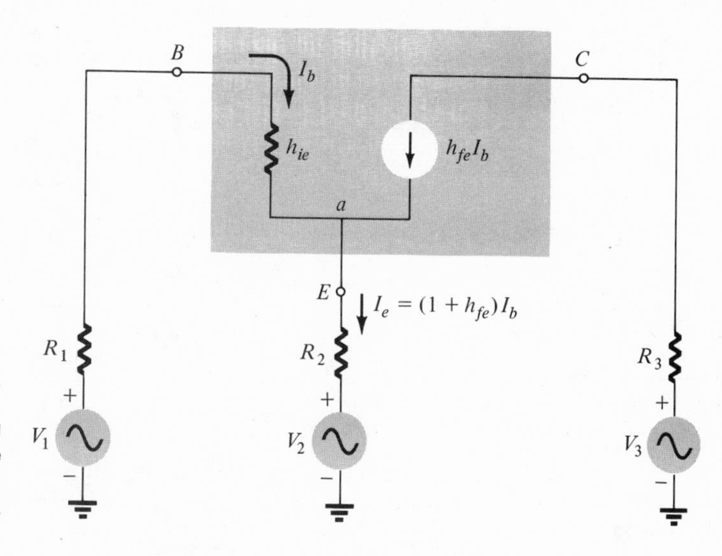

FIG. 5.67
Circuit of Fig. 5.66 following the substitution of the approximate ($h_{re} \cong 0$, $h_{oe} \cong 0$) hybrid equivalent circuit.

Base Circuit

The current through the resistor R_2 can be found by applying Kirchhoff's current law at node a:

$$I_{R_2} = I_b + h_{fe}I_b = (1 + h_{fe})I_b$$

Applying Kirchhoff's voltage law in the loop indicated in Fig. 5.67:

$$V_1 - I_bR_1 - h_{ie}I_b - (1 + h_{fe})I_bR_2 - V_2 = 0$$

Rewritten:

$$I_bR_1 + h_{ie}I_b + (1 + h_{fe})I_bR_2 = V_1 - V_2$$

and solving for I_b:

$$I_b = \frac{V_1 - V_2}{R_1 + h_{ie} + (1 + h_{fe})R_2} \tag{5.32}$$

The circuit that "fits" Eq. (5.32) appears in Fig. 5.68.

We must now take a moment to fully appreciate the benefits of having the base equivalent circuit of Fig. 5.68. It "tells" us that any signal (V_3) or load resistor from collector to ground (R_3) is not reflected to the equivalent "base" circuit of a transistor on an approximate basis. It also indicates quite clearly what effect the source V_2 and load resistor R_2 in the emitter leg will have on the base current I_b. Note that for the common-emitter circuit, when R_2 and $V_2 = 0$, if we substitute these conditions into the circuit of Fig. 5.68 the base circuit of Fig. 5.69 will result, which should, by now, be familiar. A second glance

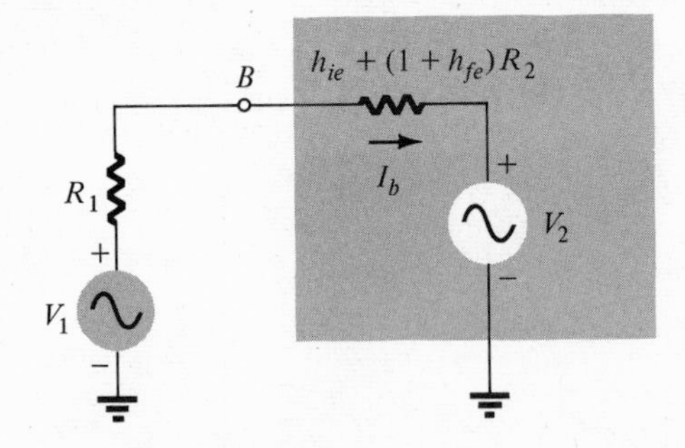

FIG. 5.68

Approximate base equivalent circuit for the transistor.

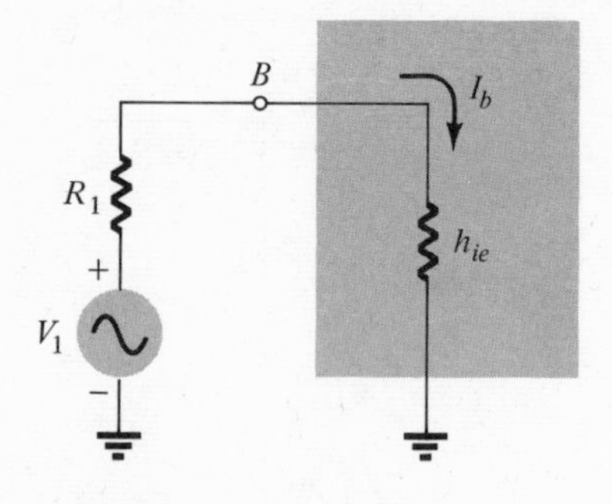

FIG. 5.69

Circuit of Fig. 5.68 with the conditions $R_2 = 0$, $V_2 = 0$.

at Fig. 5.68 will also reveal that any resistor in the emitter leg will appear as a much larger resistance in the base circuit due to the factor $(1 + h_{fe})$.

Emitter Circuit

If Eq. (5.32) is multiplied by $(1 + h_{fe})$ the following equation will result.

$$(1 + h_{fe})I_b = (1 + h_{fe})\left[\frac{V_1 - V_2}{R_1 + h_{ie} + (1 + h_{fe})R_2}\right]$$

Rewritten:

$$(1 + h_{fe})I_b = I_e = \frac{V_1 - V_2}{\dfrac{R_1 + h_{ie}}{1 + h_{fe}} + R_2} \tag{5.33}$$

Constructing a circuit to "fit" Eq. (5.3) we obtain Fig. 5.70.

The "emitter circuit" also indicates quite clearly that a source (V_3) and load resistor (R_3) in the collector circuit are not reflected to the "emitter circuit." Note also that V_1 will affect the current I_e but the reflected resistance is normally small in magnitude due to the division by the factor $(1 + h_{fe})$.

Collector Circuit

The current I_c is $h_{fe}I_b$ independent of the surrounding circuit. The collector voltage,

$$V_c = V_3 - I_cR_3 = V_3 - h_{fe}I_bR_3$$

is also independent of the external circuit. The resulting collector circuit

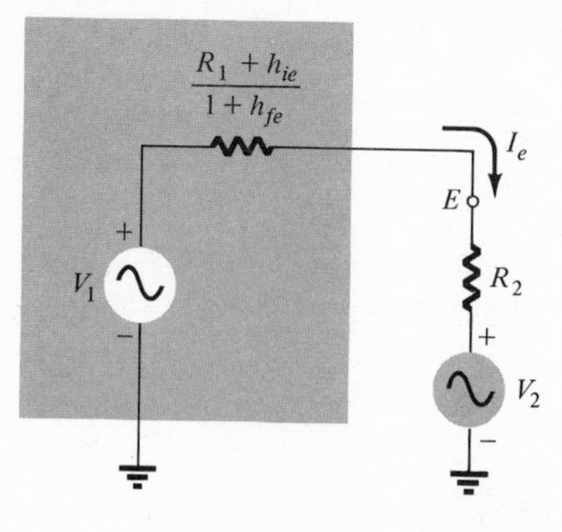

FIG. 5.70
Approximate emitter equivalent circuit for the transistor.

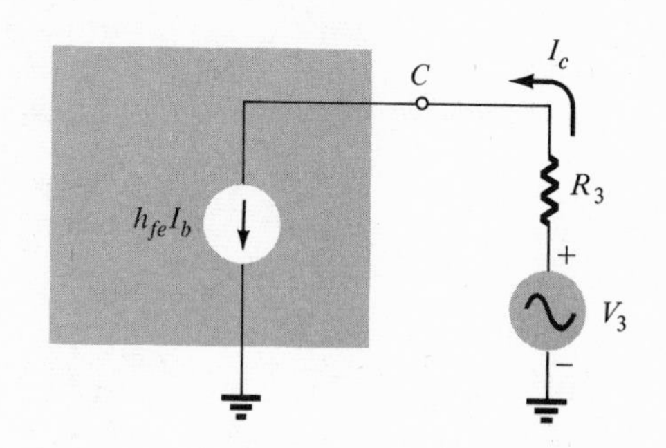

FIG. 5.71
Approximate collector equivalent circuit for the transistor.

is as shown in Fig. 5.71. The collector current or potential, therefore, can be found quite directly after I_b has been determined.

The beneficial aspects of the equivalent circuits just derived will become quite obvious in the examples to follow and in later chapters. It would be wise to memorize these equivalent circuits for future use. They can be very powerful tools in analysis of transistor networks.

EXAMPLE 5.11 The transistor configuration of Fig. 5.72, called the *emitter follower*, is frequently used for impedance matching purposes; that is, it presents a high impedance at the input terminals (Z_i) and a low output impedance (Z_o), rather than the reverse, which is typical of the basic transistor amplifier. The effect of the emitter follower circuit is much like that obtained using a transformer to match a load to the the source impedance for maximum power transfer. The following analysis will reflect that the voltage gain of the emitter follower circuit is always less than one.

For the circuit of Fig. 5.72 calculate the following:

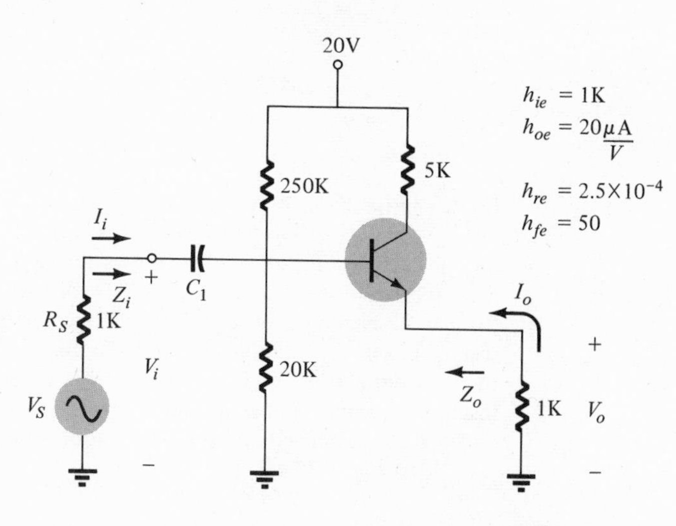

FIG. 5.72
Circuit for Example 5.11.

(a) Z_i.
(b) Z_o.
(c) $A_{v_1} = V_o/V_s$ and $A_{v_2} = V_o/V_i$.
(d) $A_i = I_o/I_i$.

Solution: Eliminating the dc levels and replacing both capacitors by short circuits will result in the circuit of Fig. 5.73.

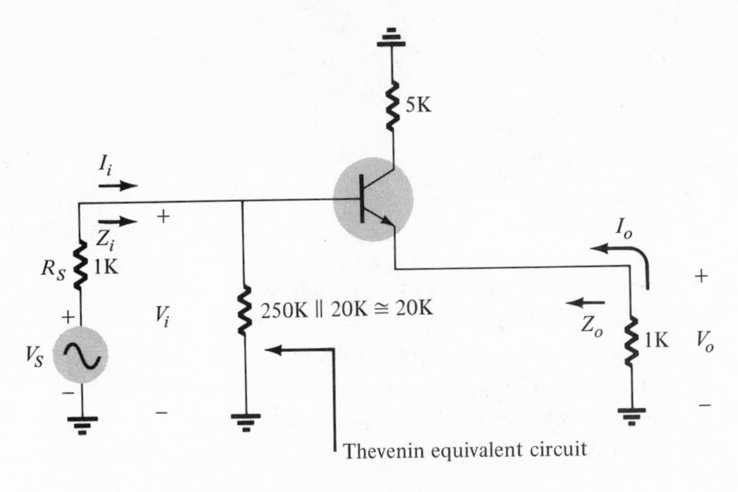

FIG. 5.73
Circuit of Fig. 5.72 redrawn for small-signal ac analysis.

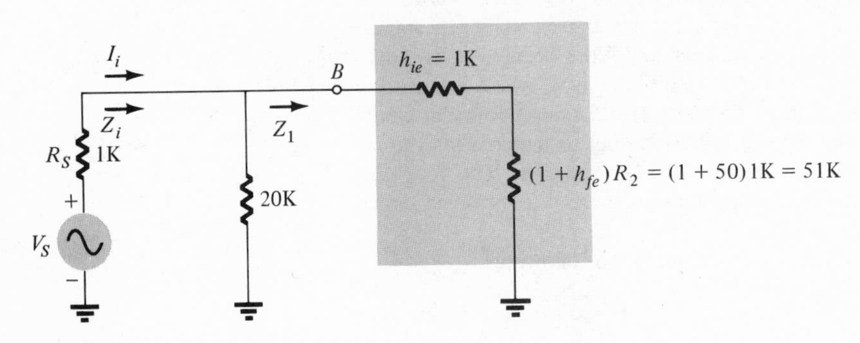

FIG. 5.74
Substitution of the approximate base equivalent circuit into the circuit of Fig. 5.73.

(a) Using the base equivalent circuit (Fig. 5.74):

$Z_1 = h_{ie} + (1 + h_{fe})R_2 = 52$ K (Certainly high compared to the typical input impedance $Z_i \cong h_{ie}$ for the basic transistor amplifier.)

The resulting $Z_i = 20$ K || 52 K = **14.5 K**

(b) The Thévenin equivalent circuit of the portion of the network indicated in Fig 5.73 will now be found so that the input circuit will have the basic configuration of Fig. 5.66. That is, R_1 and V_1 will be found to insure the proper values are substituted into the emitter equivalent circuit to be employed in the determination of Z_o.

R_{Th} (Fig. 5.75)

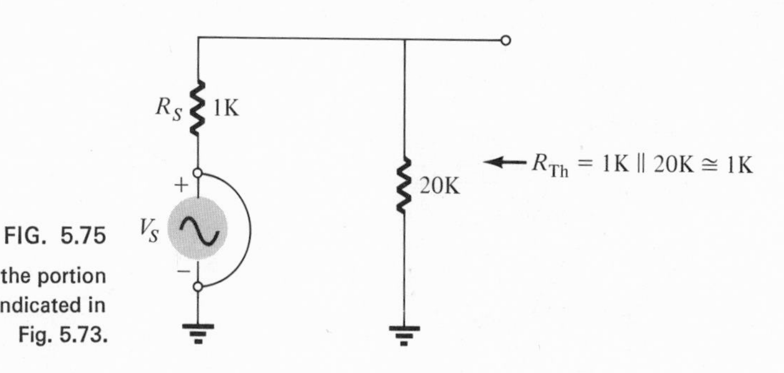

FIG. 5.75
Determining R_{Th} for the portion of the circuit indicated in Fig. 5.73.

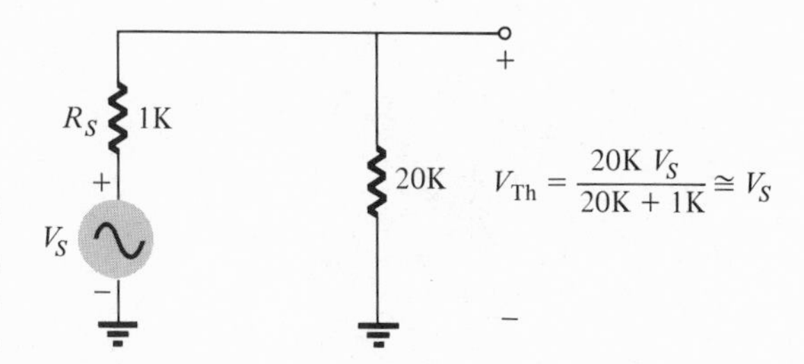

FIG. 5.76
Determining V_{Th} for the portion of the circuit indicated in Fig. 5.73.

V_{Th} (Fig. 5.76)

Substituting the Thévenin equivalent circuit into the circuit of Fig. 5.73 will result in the configuration of Fig. 5.77.

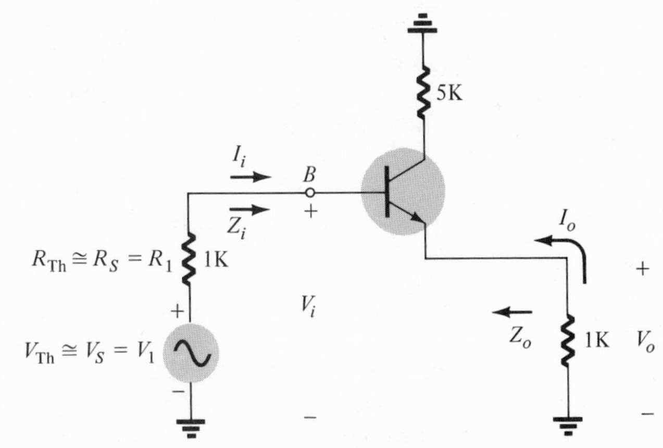

FIG. 5.77
Circuit of Fig. 5.73 following the substitution of the Thévenin equivalent circuit.

Using the emitter equivalent circuit with $V_s = 0$ (as required by definition) to determine Z_o (Fig. 5.78):

$$Z_o = \frac{R_1 + h_{ie}}{1 + h_{fe}} = \frac{1\text{ K} + 1\text{ K}}{51} = \frac{2\text{ K}}{51} \cong \mathbf{39.3\ \Omega}$$

(c) Using the emitter equivalent circuit with the source included (Fig. 5.79):

$$V_o = \frac{1\text{ K}V_s}{1\text{ K} + 39.3} = \frac{V_s}{1.0393} \cong 0.96\ V_s$$

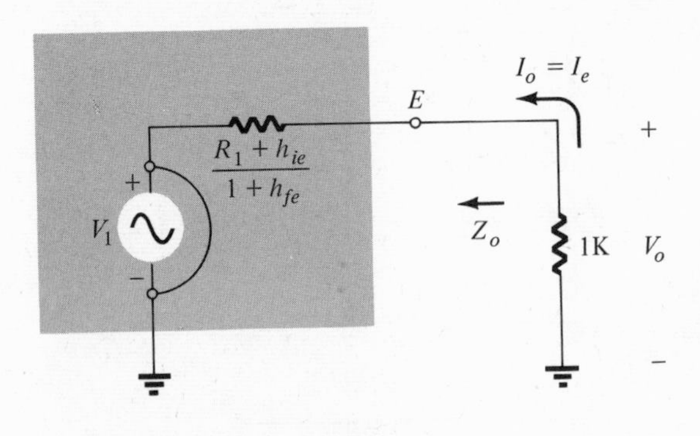

FIG. 5.78

Substitution of the emitter equivalent circuit with V_1 set to zero.

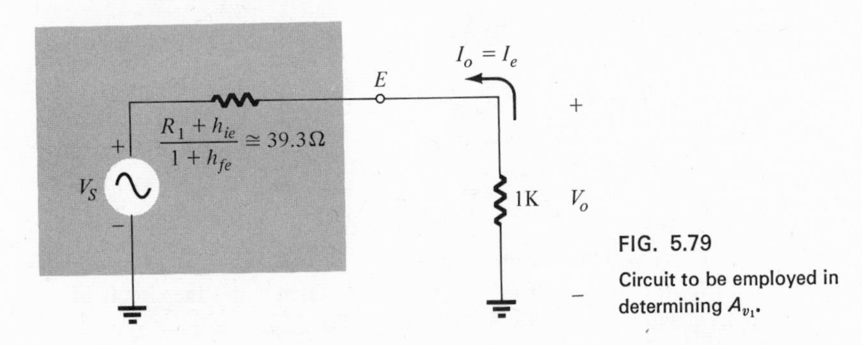

FIG. 5.79

Circuit to be employed in determining A_{v_1}.

and $A_{v_1} = V_o/V_s \cong$ **0.96** which is less than one as indicated in the introduction to this example.

For $A_{v_2} = V_o/V_i$, $R_1 = 0$, since V_i is connected directly to the base of the transistor.

Therefore,

$$\frac{R_1 + h_{ie}}{1 + h_{fe}} = \frac{0 + 1\text{ K}}{1 + 50} = \frac{1\text{ K}}{51} \cong 19.6\ \Omega$$

and

$$V_o = \frac{1\text{ K}V_i}{1\text{ K} + 19.6} = \frac{V_i}{1.0196} \cong 0.98\ V_i$$

and $A_{v_2} = V_o/V_i \cong$ **0.98** which is also less than one but slightly greater than A_{v_1} due to the source impedance R_s.

In addition, note that for the emitter follower circuit, the output and input signals are *in phase* (no minus sign associated with either A_v).

(d) From the emitter equivalent circuit:

$$I_o = -\frac{V_s}{1.0393\text{ K}}$$

and from Fig. 5.74

$$I_i = \frac{V_s}{R_s + Z_i} = \frac{V_s}{1\text{ K} + 14.5\text{ K}} = \frac{V_s}{15.5\text{ K}}$$

or $V_s = I_i\,15.5\text{ K}$

Substituting this result into the above equation:

$$I_o = -\frac{I_i\,15.5\text{ K}}{1.0393\text{ K}}$$

and

$$A_i = \frac{I_o}{I_i} = \frac{-15.5\text{ K}}{1.0393\text{ K}} = \mathbf{-14.9}$$

EXAMPLE 5.12 In Chapter 11 the *difference amplifier* will be discussed in detail. For the present, however, to demonstrate the usefulness of the base, collector, and emitter circuits, we shall consider a simple difference amplifier. In its basic form, a difference amplifier is simply a network that will produce a signal that is the difference of the two applied signals.

Figure 5.80 is such a circuit. Note that a signal has been applied to both the base and emitter leg of the transistor.

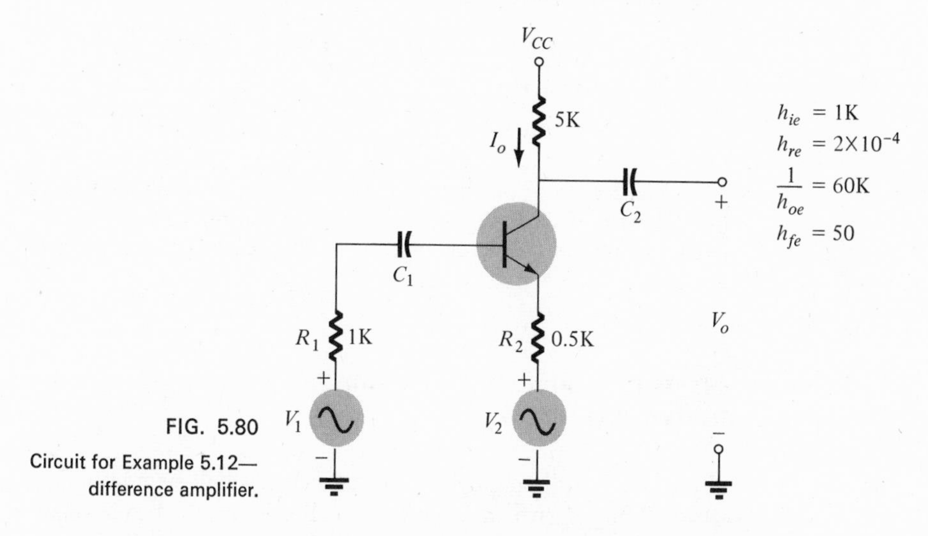

FIG. 5.80
Circuit for Example 5.12—difference amplifier.

Solution: Using the "collector equivalent circuit" after replacing the dc supply and capacitors by short circuits will result in the circuit of Fig. 5.81 where

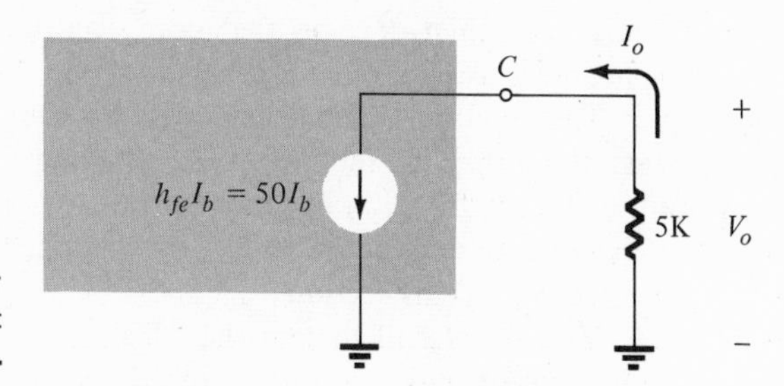

FIG. 5.81
Application of the collector equivalent circuit to the circuit of Fig. 5.80.

$$V_o = -I_o 5\text{ K} = -50\, I_b 5\text{ K}$$

Using the "base equivalent circuit" gives us Fig. 5.82 and

$$I_b = \frac{V_1 - V_2}{2\text{ K} + 25.5\text{ K}} = \frac{V_1 - V_2}{27.5\text{ K}}$$

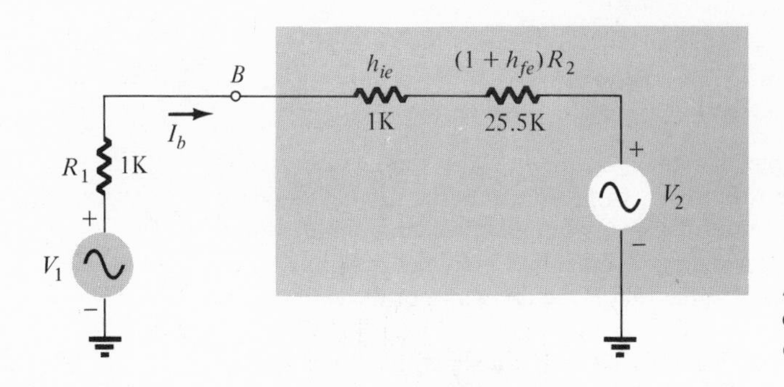

FIG. 5.82
Application of the base equivalent circuit to the circuit of Fig. 5.80.

Substituting into the equation for V_o

$$V_o = -50 I_b 5\text{ K} = -50 \frac{(V_1 - V_2)}{27.5\text{ K}} 5\text{ K} = \frac{-250(V_1 - V_2)}{27.5}$$

and

$$V_o \cong \mathbf{9.1(V_2 - V_1)}$$

The collector potential V_o, therefore, is approximately 9.1 times the difference of the two applied signals. As mentioned earlier, difference amplifiers will be discussed in much greater detail in Chapter 12.

5.10 THE TRIODE PIECEWISE LINEAR EQUIVALENT CIRCUIT

Since the triode is still employed in a number of small-signal applications, we shall now derive, and demonstrate the use of, the triode small-signal equivalent circuit.

The discussion of the triode will proceed in the same order described for the transistor; that is, we shall first consider a graphical derivation of the piecewise linear equivalent circuit from the characteristics and then develop the more accurate model for small-signal applications using a more mathematical approach.

The resulting equivalent circuit for the triode has fewer components than that of the transistor, making it appear somewhat less unwieldy to work with, but there are various facets of its use that must be firmly understood before it can be applied correctly. These will all be discussed in this section.

You will recall that in the *ideal* triode discussed in Chapter 3 there exists an open circuit between the grid and either the plate or the cathode. At high frequencies, however, capacitive effects are introduced that will couple the grid to both the plate and the cathode. Our present discussion, however, is for the open-circuit condition. This being the case, no matter what the grid-to-cathode potential, we will assume that the grid current will be zero, and a simple open circuit will suffice to represent the input, or grid circuit, of a triode.

The output, or plate circuit, however, has the characteristics indicated in Fig. 3.27b, approximated by the straight line segments of Fig. 5.83. The straight line segments are all parallel and equally spaced. The slope of each curve is:

$$m = \frac{\Delta y}{\Delta x} = \frac{\Delta_{i_P}}{\Delta v_{PK}} = \frac{1}{r}$$

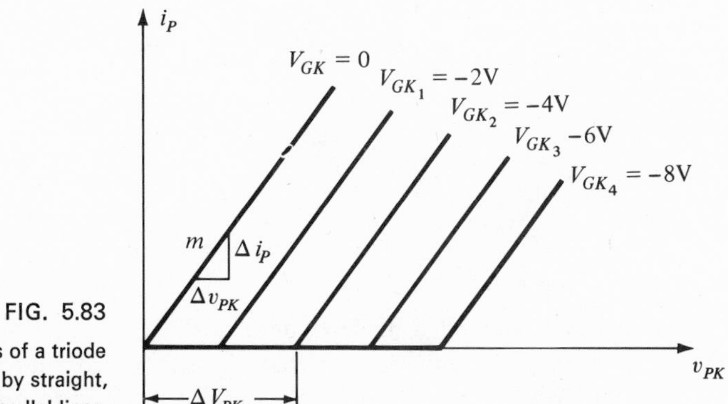

FIG. 5.83
Plate characteristics of a triode approximated by straight, equally spaced, parallel lines.

We will represent the ratio $-(\Delta v_{PK}/\Delta v_{GK})$ by the letter u. The minus sign indicates only that as one quantity becomes more positive the other becomes more negative for a fixed plate current.

For the line defined by V_{GK_2}:

$$\Delta V_{PK_2} = -u\Delta V_{GK_2} = -u(V_{GK_2} - 0) = -uV_{GK_2}$$

Substituting the values obtained for the straight line defined by V_{GK_2} into the basic straight line equation at the intersection of the horizontal axis ($i_P = 0$):

$$y = mx + b$$

$$0 = \frac{1}{r_p}(-uV_{GK_2}) + b$$

$$b = \frac{u}{r_p}V_{GK_2}$$

and the straight line equation for V_{GK_2} becomes

$$i_P = \frac{1}{r}v_{PK} + \frac{u}{r}V_{GK_2}$$

Rewritten:

$$i_p r = v_{PK} + uV_{GK_2}$$

Substituting v_{GK} for V_{GK_2} to include any straight line equation defined by a particular value of v_{GK}:

$$v_{PK} = i_p r - uv_{GK} \tag{5.34}$$

The circuit that "fits" Eq. (5.34) is indicated in Fig. 5.84 with the open-circuit representation for the input circuit. Note that the triode

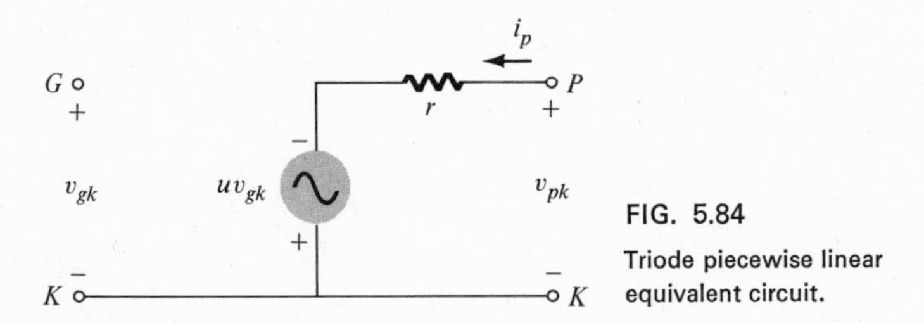

FIG. 5.84
Triode piecewise linear equivalent circuit.

also has a controlled or dependent source in its piecewise linear equivalent circuit. Consider also that the controlled source is dependent on the grid-to-*cathode* potential, and not the grid-to-*ground* potential, an error often made in the use of the triode small-signal equivalent circuit when the cathode is not connected directly to ground.

5.11 SMALL-SIGNAL EQUIVALENT CIRCUIT FOR THE TRIODE

If we assume that the curves of a triode are *linear* in the *region* of the quiescent point of operation, the small-signal equivalent circuit for the output circuit of a tube can be found by applying Thévenin's theorem. The Thévenin impedance can be shown to be the following using partial derivatives (calculus):

$$R_{Th} = r_p = \left.\frac{\Delta v_{PK}}{\Delta i_P}\right|_{V_{GK}=\text{constant}} \tag{5.35}$$

where r_p is called the *plate resistance* of the triode; and $V_{Th} = \mu v_{GK}$, where the Greek letter μ (mu), called the *amplification factor*, has a magnitude defined by

$$\mu = \left.\frac{\Delta v_{PK}}{\Delta v_{GK}}\right|_{I_P=\text{constant}} \tag{5.36}$$

The resulting plate circuit is exactly the same as that shown in Fig. 5.84 with r and μ, however, defined by Eqs. (5.35) and (5.36). The method or procedure used in association with Eqs. (5.35) and (5.36) to determine r_p and μ is exactly the same as that employed for the transistor in the determination of the h-parameters.

A third quantity of interest for the triode is its *transconductance*, defined by Eq. (5.37).

$$g_m = \left.\frac{\Delta_{i_P}}{\Delta v_{GK}}\right|_{V_{PK}=\text{constant}} \tag{5.37}$$

The prefix *trans* from the word transfer is employed because it relates an output quantity (I_P) to an input quantity (V_{GK}).

By simply manipulating eq. (5.36) in the following manner:

$$\mu = \frac{\Delta v_{PK}}{\Delta v_{GK}} = \frac{\Delta v_{PK}}{\Delta_{i_P}} \times \frac{\Delta_{i_P}}{\Delta v_{GK}}$$

and substituting the magnitude of the defined quantities for each ratio:

$$\mu = r_p g_m \tag{5.38}$$

so that if any two quantities are known, the third can be found using Eq. (5.38).

The magnitude of μ, r_p, and g_m will vary as indicated by Table 5.4.

TABLE 5.4

Typical Range of Values for μ, r_p, and g_m for the Triode

μ:	25–100
r_p:	0.5–100 K
g_m:	0.5–10 mA/V

(b) Thévenin equivalent circuit: See Fig. 5.89.
Norton equivalent circuit: See Fig. 5.90.

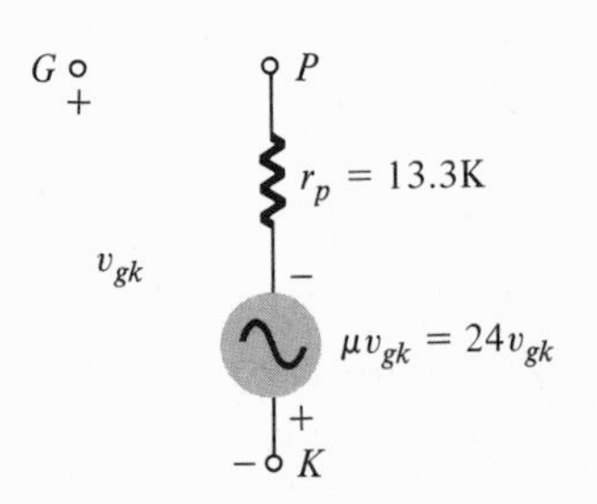

FIG. 5.89

Resulting Thévenin equivalent circuit.

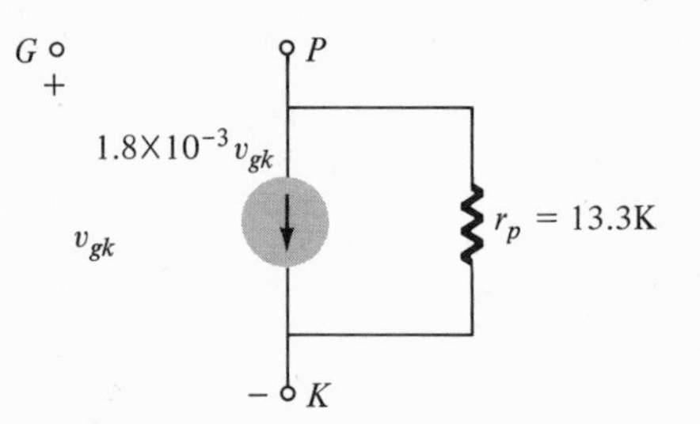

FIG. 5.90

Resulting Norton equivalent circuit.

EXAMPLE 5.14 Find the voltage gain, $A_v = (V_o/V_i)$, for the basic triode amplifier of Fig. 5.91.

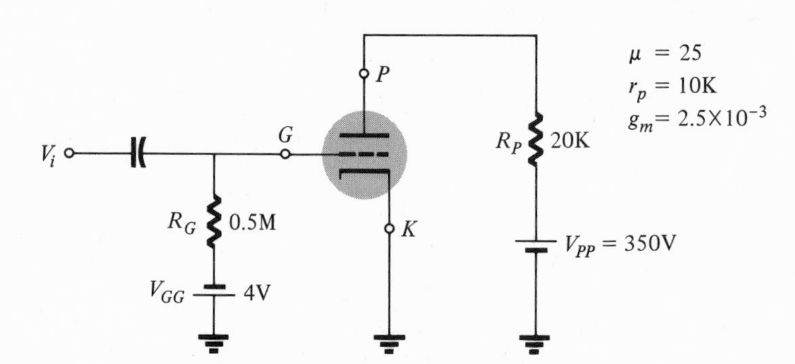

FIG. 5.91

Circuit for Example 5.14.

Solution: Eliminating dc levels and substituting the Thévenin equivalent circuit will result in the configuration of Fig. 5.92. In this case,

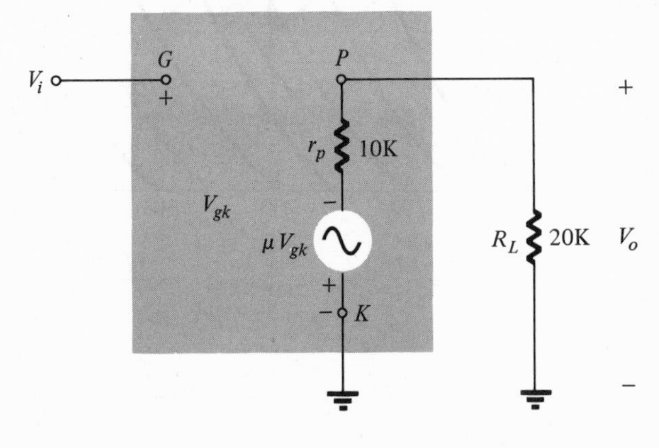

FIG. 5.92

Circuit of Fig. 5.91 following the substitution of the small-signal ac equivalent circuit for the triode.

$$V_{gk} = V_i$$

and

$$V_o = \frac{-R_L(\mu V_{gk})}{R_L + r_p} = \frac{-R_L(\mu V_i)}{R_L + r_p}$$

so that

$$A_v = \frac{V_o}{V_i} = \frac{-\mu R_L}{R_L + r_p} \tag{5.39}$$

Note that the gain is equal to the amplification factor only when $r_p = 0$. For *maximum power transfer* to the load the load impedance R_L must equal the plate resistance (maximum power theorem); that is,

$$R_L = r_{p\ (\text{max. power})} \tag{5.40}$$

for maximum power transfer to R_L.
The negative sign indicates quite clearly that the output (V_o) and input (V_i) voltages are 180° *out of phase*.

Substituting values for this example:

$$A_v = \frac{V_o}{V_i} = \frac{-25(20\text{ K})}{20\text{ K} + 10\text{ K}} = \frac{-500\text{ K}}{30\text{ K}} \cong \mathbf{-16.7}$$

EXAMPLE 5.15 The circuit of Fig. 5.93, called the *cathode follower*, is the tube equivalent of the *emitter follower* circuit for the transistor. It is also used for impedance matching purposes since it has a high input impedance and low output impedance. The input and output voltages are also *in phase* as was true for the emitter follower.

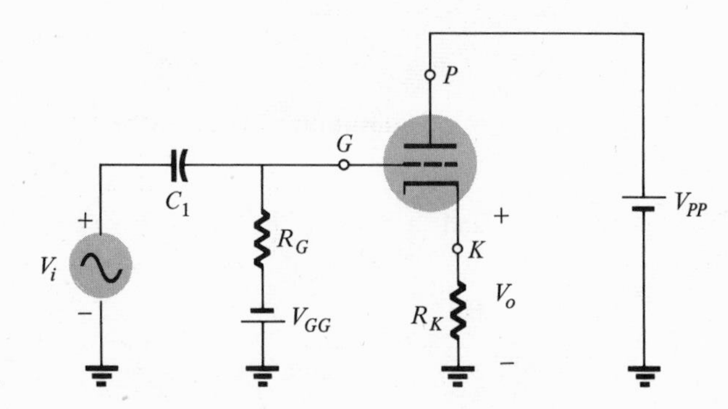

FIG. 5.93
Cathode follower configuration to be explored in Example 5.15.

In this example, the voltage gain $A_v = (V_o/V_i)$, and output impedance Z_o, will be found in terms of the Thévenin equivalent small-signal triode parameters.

Solution: Substituting the equivalent circuit and eliminating dc levels will result in the configuration of Fig. 5.94. Note in this case that V_{gk} is not the potential from grid to ground and therefore is not equal to the input signal V_i.

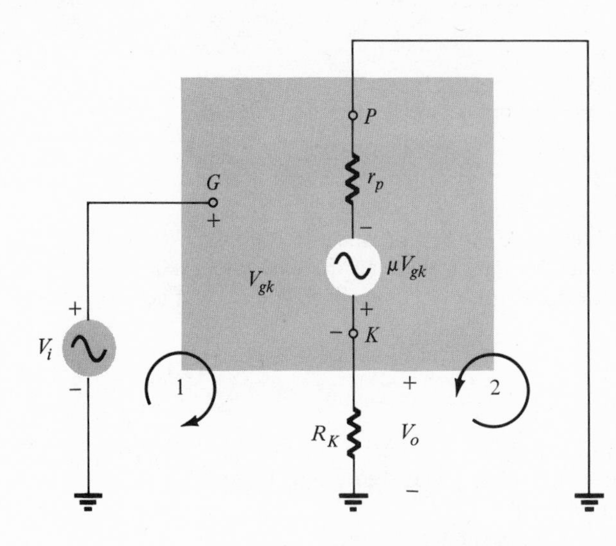

FIG. 5.94
Circuit of Fig. 5.93 following the substitution of the small-signal ac equivalent circuit.

Applying Kirchhoff's voltage law to the grid loop (1):

$$V_i - V_{gk} - I_p R_K = 0$$

and

$$V_{gk} = V_i - I_p R_K$$

Applying Kirchhoff's voltage law to the plate circuit:

$$\mu V_{gk} - V_o - I_p r_p = 0$$

or

$$V_o = \mu V_{gk} - I_p r_p$$

Substituting the above for V_{gk}:

$$V_o = \mu(V_i - I_p R_K) - I_p r_p$$

and

$$V_o = \mu V_i - I_p(\mu R_K + r_p)$$

but

$$I_p = \frac{V_o}{R_K}$$

and

$$V_o = \mu V_i - \frac{V_o}{R_K}(\mu R_K + r_p)$$

$$= \mu V_i - \mu V_o - \frac{r_p}{R_K} V_o$$

or

$$V_o + \mu V_o + \frac{r_p}{R_K} V_o = \mu V_i$$

and

$$\frac{V_o}{V_i} = \frac{\mu}{1 + \mu + r_p/R_K} = \frac{\mu R_K}{(1 + \mu)R_K + r_p} \qquad (5.41)$$

For typical values:

$$\mu = 30, \qquad r_p = 10\text{ K}, \qquad R_K = 20\text{ K}$$

$$A_v = \frac{V_o}{V_i} = \frac{30(20\text{ K})}{(31)(20\text{ K}) + 10\text{ K}} = \frac{600\text{ K}}{620\text{ K} + 10\text{ K}} = \frac{600\text{ K}}{630\text{ K}} \cong \mathbf{0.95}$$

For the cathode follower, like the emitter follower, the gain is always less than one and as indicated above the ratio is always positive so that V_o and V_i are *in phase*.

Rewriting Eq. (5.41) in the following manner:

$$\frac{V_o}{V_i} = \frac{\dfrac{\mu R_K}{\mu + 1}}{R_K + \dfrac{r_p}{\mu + 1}}$$

or

$$V_o = \frac{R_K\left(\dfrac{\mu}{\mu + 1}\right)V_i}{R_K + \dfrac{r_p}{\mu + 1}}$$

and drawing the circuit to "fit" the above equation, we obtain Fig. 5.95.

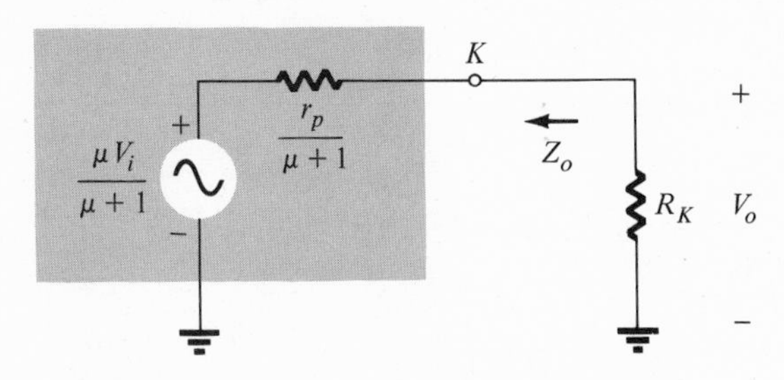

FIG. 5.95
Circuit resulting from the rewriting of the fundamental gain equation (Eq. 5.38).

We can now find the output impedance by setting $V_i = 0$ as defined by the conditions necessary to obtain the output impedance, and

$$Z_o\big|_{V_i=0} = \frac{r_p}{\mu + 1} \qquad (5.42)$$

For the parameters just provided:

$$Z_o = \frac{10\text{ K}}{31} = 322\ \Omega$$

which is considerably less than the 10-K (r_p) output impedance that would result if the tube were used in the basic amplifier configuration of Example 5.14.

5.12 TRIODE PARAMETER VARIATION

The parameters of the small-signal equivalent circuit for the triode will vary with the operating conditions. Figure 5.96 illustrates the variations of μ, r_p, and g_m with plate current for various values of plate-to-cathode potential.

Note that μ remains fairly constant, while r_p and g_m vary considerably with change in plate current (I_p). Consider also that increasing values of plate-to-cathode potentials result in decreasing values of μ and increasing values of r_p.

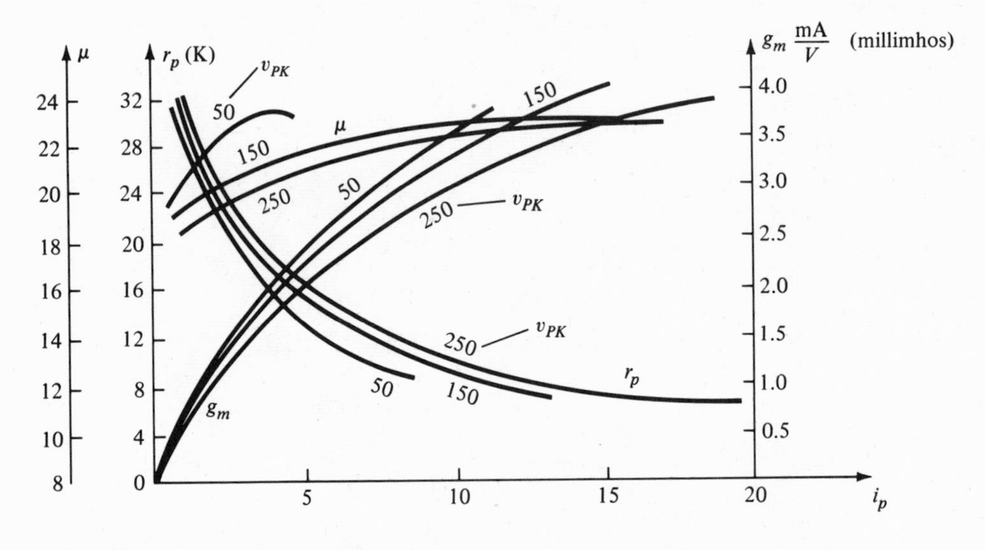

FIG. 5.96
Variations in μ, r_p, and g_m with plate current for various values of plate to cathode potential.

5.13 THE PENTODE SMALL-SIGNAL EQUIVALENT CIRCUIT

The pentode small-signal equivalent circuit is the same in appearance as that discussed for the triode. The magnitude of the parameters, however, is quite different from those of the triode. The pentode small-signal parameters can all be determined using the same equations and procedure described for the triode. The typical range of values for μ, r_p, and g_m are listed in Table 5.5.

TABLE 5.5

μ:	1000–5000
r_p:	0.5–2 M
g_m:	1000–9000 μA/V

Note that the magnitude of μ and r_p has increased considerably over those for the triode while the range of g_m has changed only slightly. For the pentode, the Norton equivalent circuit is the more frequently employed since the current through the load circuit is, for the majority of situations, unaffected by r_p and simply equal to the Norton current: $g_m V_{gk}$. It is a situation similar to that encountered with the transistor, where in general, $(1/h_{oe}) > R_L$. Example 5.16 will demonstrate the validity of the above statement.

EXAMPLE 5.16 Calculate the voltage gain, $A_v = (V_o/V_i)$ for the circuit of Fig. 5.97.

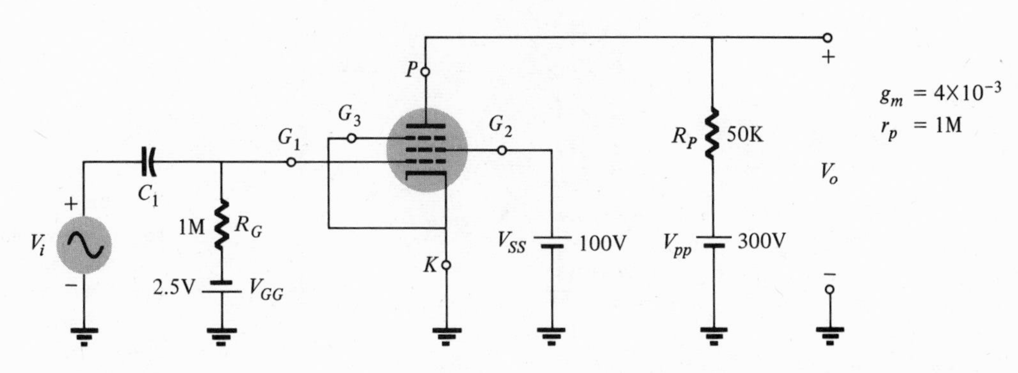

FIG. 5.97
Basic pentode amplifier to be examined in Example 5.16.

Solution: Substituting the Norton small-signal equivalent circuit for the pentode and eliminating dc levels will result in the circuit of Fig. 5.98. Obviously, $V_{gk} = V_i$.

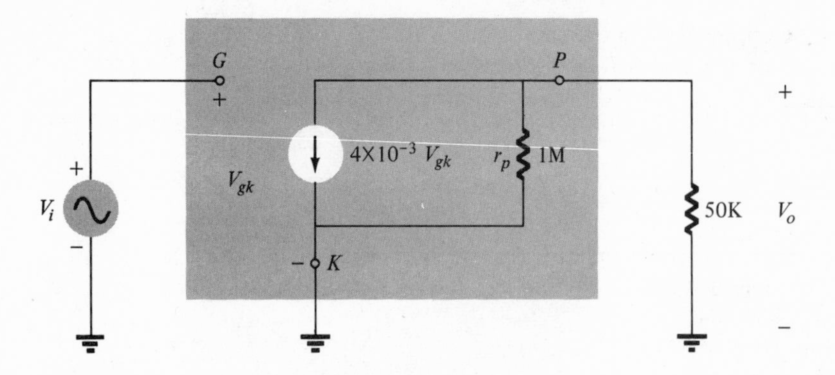

FIG. 5.98

Circuit of Fig. 5.97 following the substitution of the small-signal ac equivalent circuit for the pentode.

The parallel combination of 1 M || 50 K ≅ 50 K (substantiating the discussion of the last few paragraphs) results in the plate circuit of Fig. 5.99 so that

$$\begin{aligned} V_o &= (-4 \times 10^{-3} V_i)(50 \times 10^3) \\ &= -200 V_i \end{aligned}$$

and

$$A_v = \frac{V_o}{V_i} = -200$$

The minus sign again indicates a 180° phase shift between V_o and V_i.

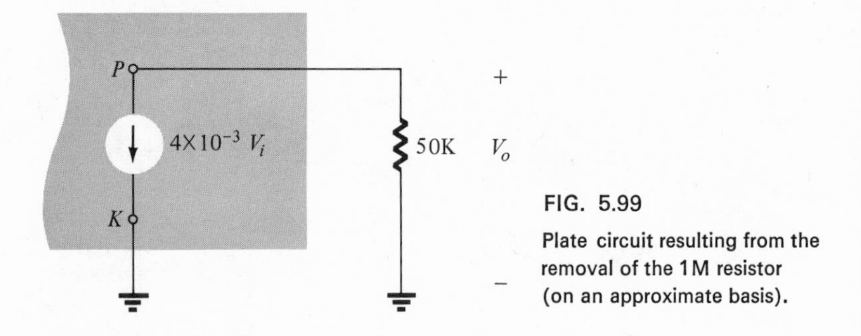

FIG. 5.99

Plate circuit resulting from the removal of the 1 M resistor (on an approximate basis).

PROBLEMS

§ 5.7

1. For the fixed-bias transistor of Fig. 5.100 calculate the circuit
   (a) current gain I_o/I_i.
   (b) voltage gain V_o/V_i.
   (c) power gain A_p.

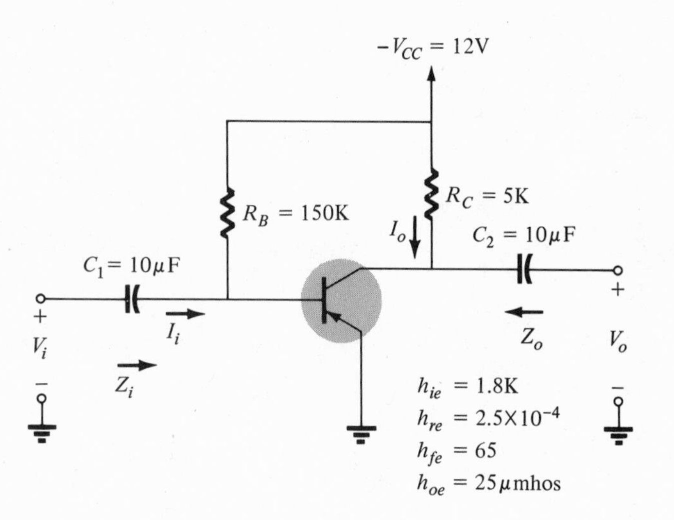

FIG. 5.100
Circuit for Problems 5.1 and 5.2.

2. Using the circuit of Fig. 5.100 calculate the circuit
   (a) input impedance Z_i.
   (b) output impedance Z_o.
3. Using the amplifier circuit of Fig. 5.101 calculate
   (a) $A_i = I_o/I_i$.
   (b) $A_{v_1} = V_o/V_i$.
   (c) $A_{v_2} = V_o/V_s$.

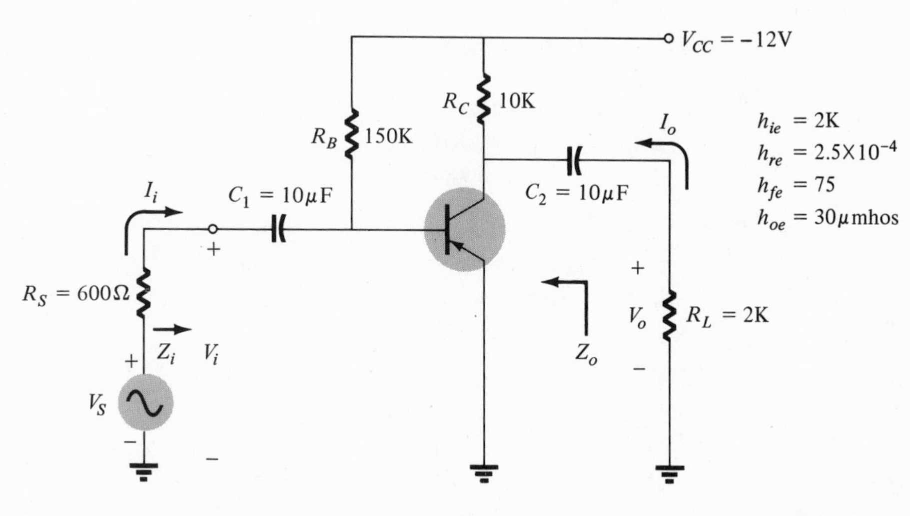

FIG. 5.101
Circuit for Problems 5.3 and 5.4.

4. Using the amplifier circuit of Fig. 5.101 calculate
   (a) Z_i.
   (b) Z_o.

5. Using the amplifier circuit of Fig. 5.102 calculate
   (a) $A_v = V_o/V_i$.
   (b) $A_i = I_o/I_i$.

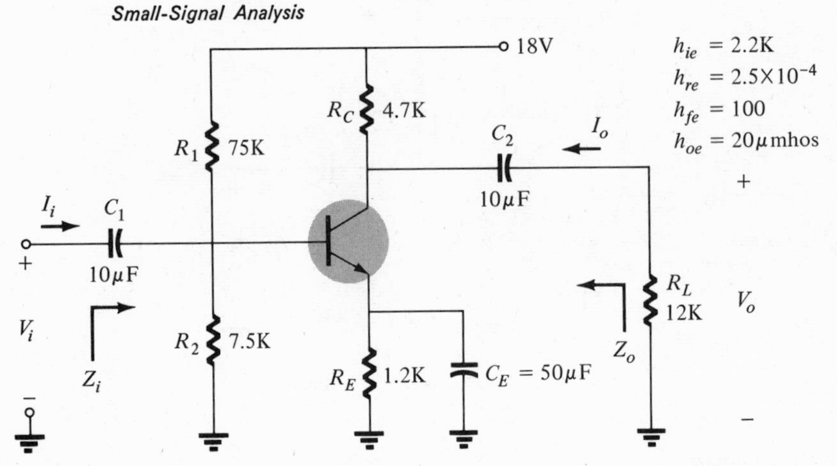

FIG. 5.102
Circuit for Problems 5.5 and 5.6.

6. Using the circuit of Fig. 5.102 calculate
   (a) Z_i.
   (b) Z_o.
7. For the common-base amplifier of Fig. 5.103 calculate
   (a) $A_i = I_o/I_i$.
   (b) $A_v = V_o/V_i$. (c) $A_{v_s} = V_o/V_s$.

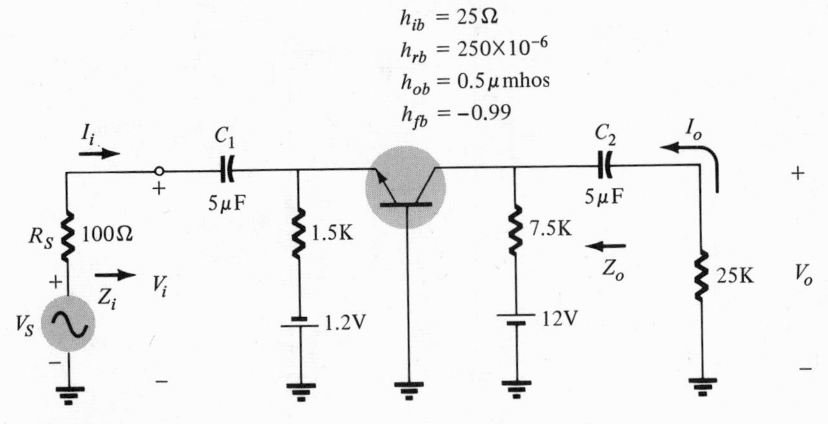

FIG. 5.103
Circuit for Problems 5.7 and 5.8.

8. Using the circuit of Fig. 5.103 calculate
   (a) Z_i.
   (b) Z_o.

9. Using an approximate hybrid equivalent circuit repeat the calculations of Problems 5.1 and 5.2.

10. Using an approximate hybrid equivalent circuit repeat the calculations of Problems 5.3 and 5.4.

11. Using an approximate hybrid equivalent circuit repeat the calculations of Problems 5.5 and 5.6.

12. Using an approximate hybrid equivalent circuit for the transistors of Fig. 5.104 calculate
    (a) $A_i = I_o/I_i$.
    (b) $A_v = V_o/V_i$.
    (c) Z_i.
    (d) Z_o.

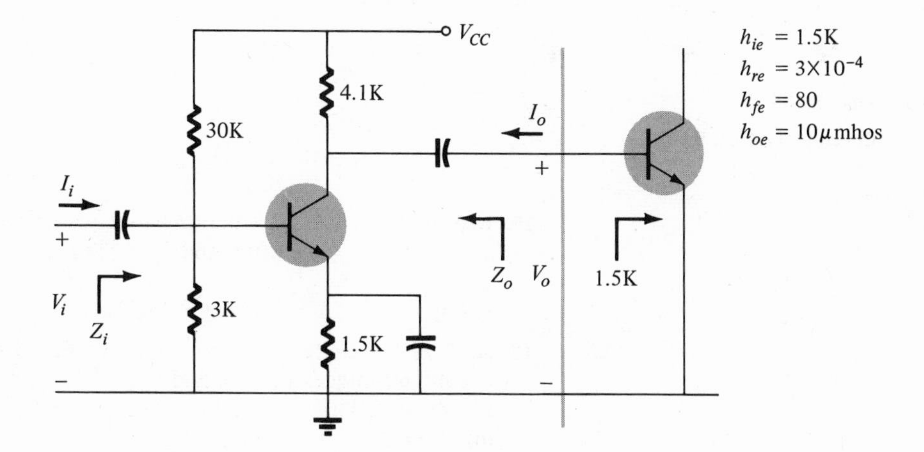

FIG. 5.104
Circuit for Problem 5.12.

13. For the circuit of Fig. 5.105 calculate V_o/V_i, using simplifying approximations where appropriate.

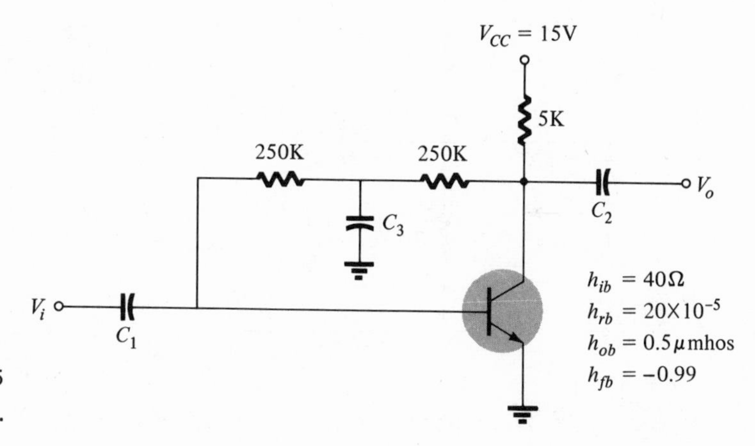

FIG. 5.105
Circuit for Problem 5.13.

14. Using the common-base circuit of Fig. 5.106 calculate the following
(a) $A_v = V_o/V_i$.
(b) $A_i = I_o/I_i$.
(c) Z_i.
(d) Z_o.
Use simplifying approximations where appropriate.

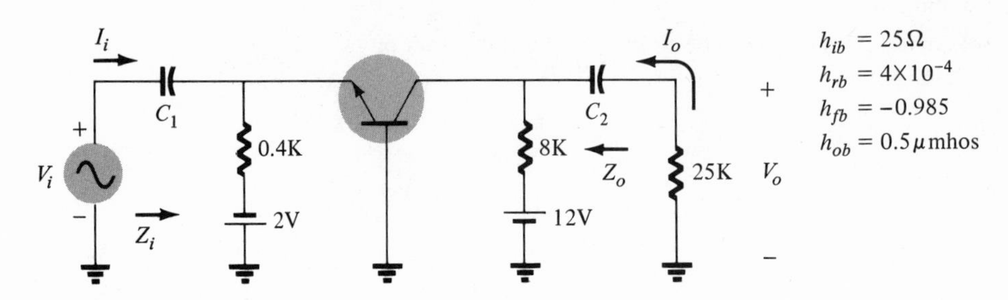

FIG. 5.106
Circuit for Problem 5.14.

15. Repeat Problem 5.14, if the transistor parameters are now $h_{ie} = 2$ K, $h_{fe} = 50$, $h_{re} = 3 \times 10^{-4}$, and $h_{oe} = 20\ \mu$A/V.

§ 5.9

16. Analyze the emitter follower circuit of Fig. 5.107 using "equivalent circuit" techniques; that is, find
(a) $A_v = V_o/V_i$.
(b) $A_i = I_o/I_i$.
(c) Z_i.
(d) Z_o.

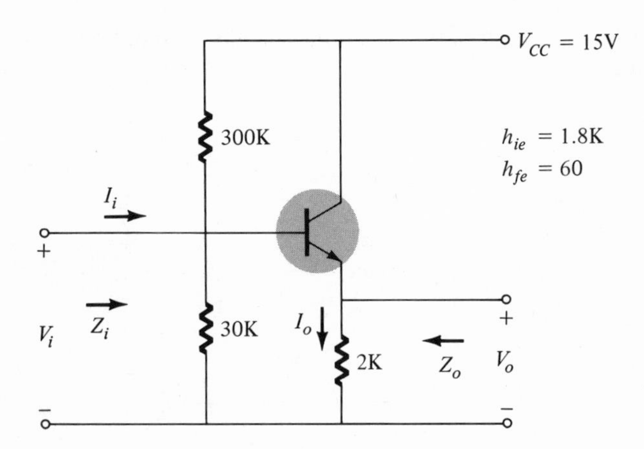

FIG. 5.107
Circuit for Problem 5.16.

17. Using the circuit of Fig. 5.108 calculate V_o.

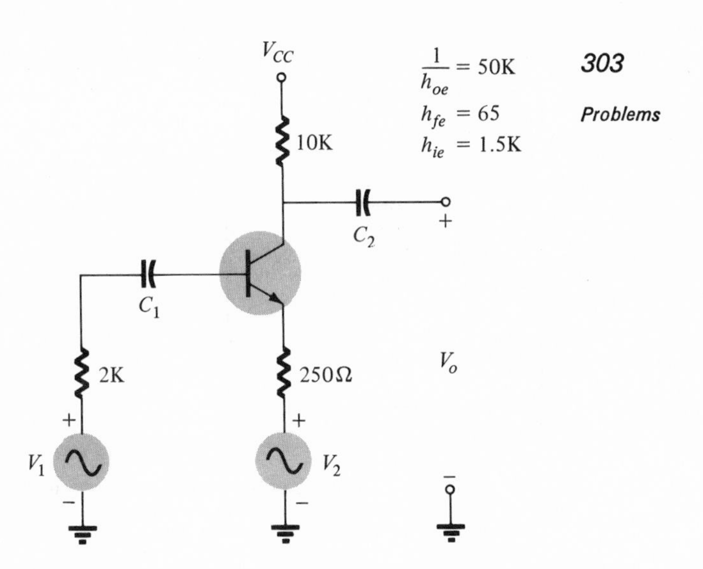

FIG. 5.108
Circuit for Problem 5.17.

§ 5.11

18. Determine the values of μ, r_p, and g_m at an operating point of 200 V and 5 mA on the triode plate characteristic of Fig. 5.109.

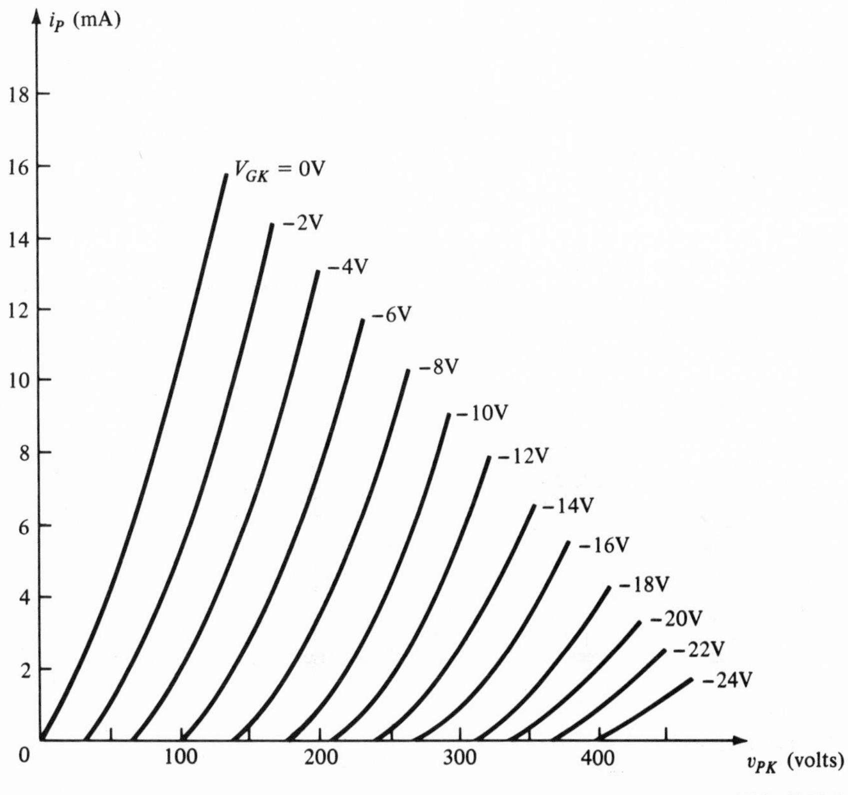

FIG. 5.109

5. It can be operated so as to have greater thermal stability than a conventional transistor.

Counteracting these advantages, it has the disadvantage of a relatively small gain bandwidth which, however, is being improved considerably by manufacturing advances.

6.3 CONSTRUCTION OF FET

A FET is a three-terminal device containing one basic *p-n* junction. Although the FET was one of the earliest solid-state devices proposed* for amplifier operation, the development of a commercially useful device lagged because of manufacturing limitations until the mid 1960s.

The physical structure of an *n-channel* FET is shown in Fig. 6.1a. A bar of *n*-doped material with a small *p*-doped region along one side of the bar is shown. The main part of the device is called the *channel* and is specifically an *n*-channel in Fig. 6.1a. The opposite ends of the channel are called source (*S*) and drain (*D*). The third lead is connected to a small *p*-doped area and is called the *gate* (*G*) lead. As would be expected the channel could also be made *p*-doped and would be a *p-channel* FET (Fig. 6.1b). In that case the remaining lead would connect to a small *n*-doped area for the gate terminal.

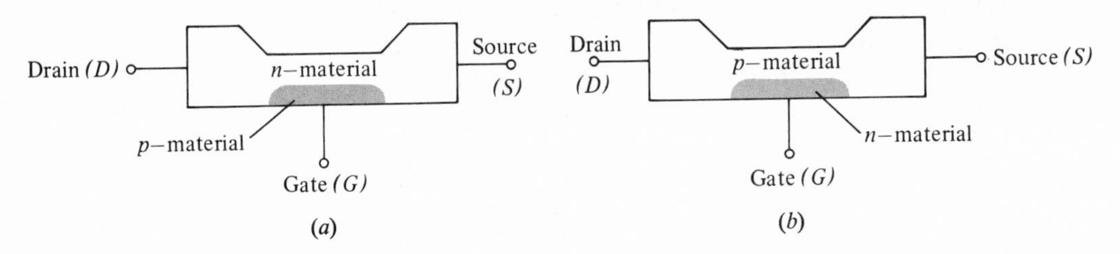

FIG. 6.1
Physical structure of (a) *n*-channel FET; (b) *p*-channel FET.

For the *n*-channel device the current will be carried by the channel majority carriers, namely, electrons. For a *p*-channel device the channel majority carriers are holes. The majority carriers enter the bar of semiconductor material at the source terminal. Having passed through the bar the majority carriers leave at the drain terminal. The *p*-doped gate region forms *p-n* junctions from gate to source and from gate to drain.

* W. Shockly, *Electrons and Holes in Semiconductors*, Van Nostrand Reinhold Co., New York, 1953.

Between points S and D there is an ohmic region whose typical resistance is a few hundred ohms to a few kilohms. A voltage applied across source-drain results in current flow through the channel, which is limited only by the channel resistance. The gate voltage controls this current flow and thereby exerts control over the device operation. Recall that a reversed biased *p-n* junction opposes current flow by forming a region around the *p-n* junction which is devoid of free current carriers. The application of a reverse bias voltage from gate to source (V_{GS}) forms a depletion region (charge-free region) around the gate, as shown in Fig. 6.2. The result of this depletion region is to reduce the

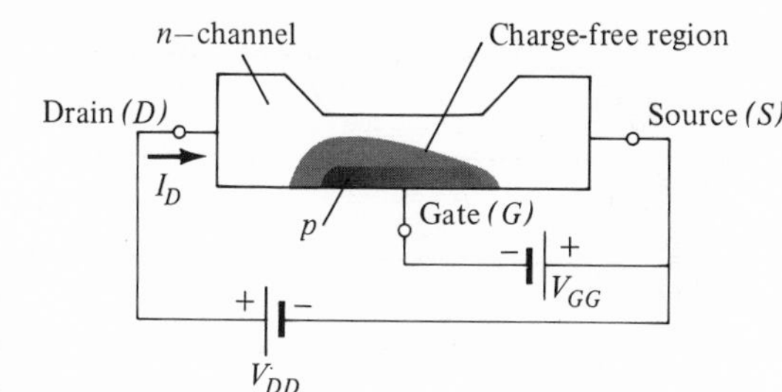

FIG. 6.2
Bias of FET to form depletion region in channel.

width of the channel, forcing the current to move through a narrower channel. As the reverse bias voltage, V_{GS}, is increased by the battery voltage, V_{GG}, the drain-source current I_D will decrease until the depletion region extends across the width of the channel and current can no longer pass. This cutoff value of V_{GS} is called the *pinch-off voltage*, V_P. Typically, V_P is only a few volts. The value of V_P depends on the doping of the *n* and *p* regions of the device and the width of the original channel structure, among other factors.

For the *n*-channel device the majority carrier electrons actually enter the source and leave the drain. Since our desire, as usual, is to use the device in a practical circuit, the mechanism of operation and the details we shall consider are merely to provide some understanding of how the device is made and what the characteristics mean. As far as applications are concerned, conventional current, will be used.

Consider the channel operation for a constant value of gate-source voltage, V_{GS}. For small values of V_{GS} the channel acts as an ohmic resistance whose value is shown by the straight sloped part of the characteristic in Fig. 6.3. As the voltage drop across the drain to source increases due to this ohmic voltage drop, the reverse bias between gate and source increases and the channel depletion region becomes larger and reduces the path for current flow. The channel current will be reduced by this action and will level off to a fairly constant value as shown in Fig. 6.3. As shown in the FET characteristic the current value at which the drain current levels off is less for greater values of reverse-bias voltage, V_{GS}.

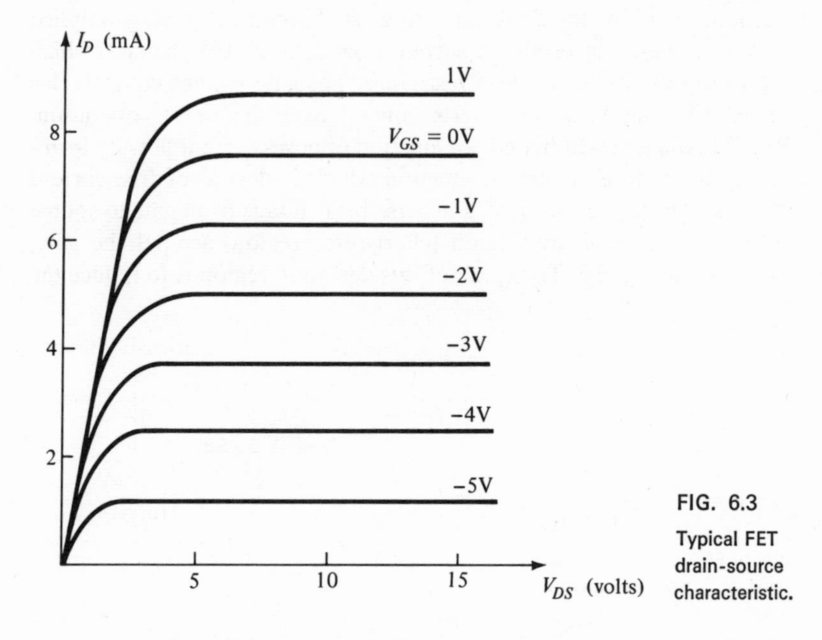

FIG. 6.3
Typical FET drain-source characteristic.

Note in Fig. 6.3 that even for a small positive gate-source voltage the device still operates since the ohmic voltage drop results in increasing reverse bias voltage from gate to source. When the reverse bias across the gate junction exceeds a maximum rated value, avalanche breakdown occurs. This rating limits the maximum voltage supply that may be used. The lowest drain-gate breakdown voltage will occur at the largest gate-source reverse bias voltage since the two add together to produce breakdown. Typical maximum values are 10–30 V across any two terminals in the reverse direction.

6.4 VOLTAGE AND CURRENT CONVENTIONS

Figure 6.4 shows the voltage and current polarity convention for *p*- and *n*-channel FETs. The battery polarities shown will result in reverse biasing the gate-source junction for normal FET operation. Gate current I_G is typically a few nanoamperes and is negligible. This small current indicates a high input impedance for the FET. The drain and source current show the conventional positive current flow. The voltage polarity of V_{DS} indicates the reading obtained on a meter whose positive terminal is connected to the drain and negative terminal to the source—in other words, when properly measuring voltage *from* drain *to* source.

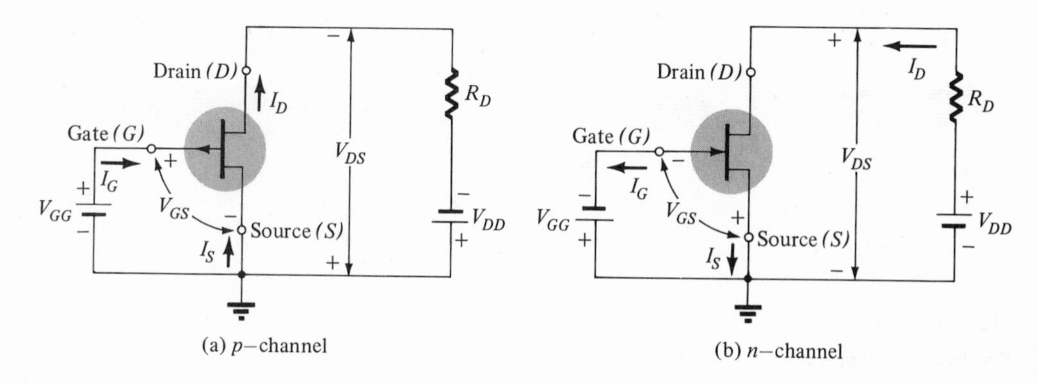

FIG. 6.4
Polarity conventions in FET.

6.5 VARIOUS FET STRUCTURES

There are two popular types of FETs, namely, the *junction FET* or *JFET* and the *metal-oxide-semiconductor FET* or *MOSFET*.

JFET

A practical structure for a JFET is shown in Fig. 6.5. Majority-carrier flow from source to drain takes place through the channel marked in the figure. In terms of manufacturing process the *p*-substrate is doped to produce the *n*-material source-drain slab on top. Then, the *p*-type gate is formed by diffusion. The *p*-type area is made of low-resistivity material so that the depletion region extends mainly into the *n*-type channel.

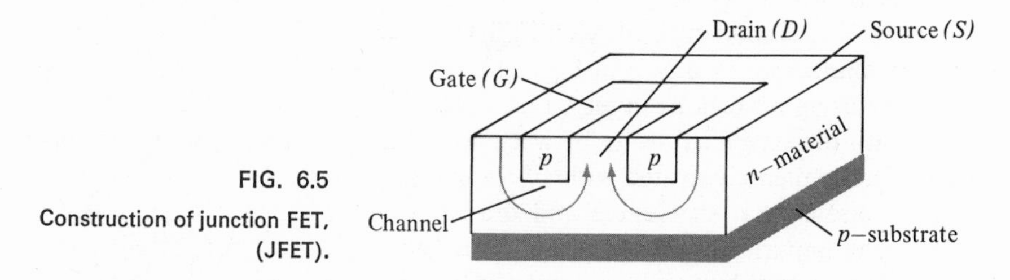

FIG. 6.5
Construction of junction FET, (JFET).

MOSFET

Figure 6.6 shows *channel enhancement* in a MOSFET structure. In the enhancement-type structure *no* channel is physically made during construction of the device. The metalized gate lead and substrate, with

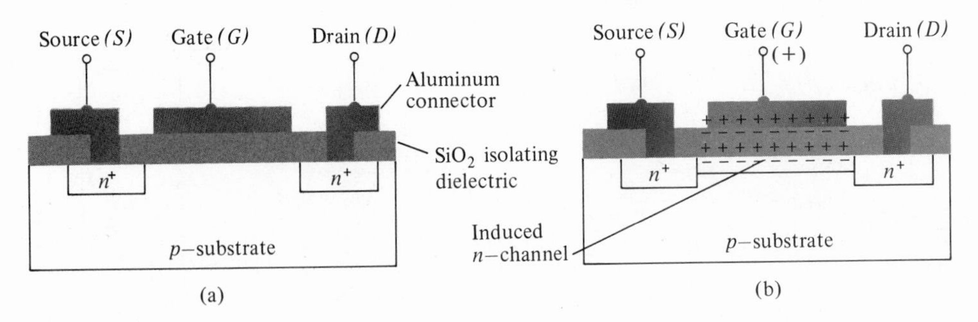

FIG. 6.6

Construction and operation of MOSFET device.

the isolating dielectric (SiO_2) between, form a parallel-plate capacitor.

The source and drain are formed by diffusing n^+ regions in the p-substrate. The n^+ designation indicates an n-doped section which is highly doped (or heavily doped). There are a larger number of free electrons than in the corresponding n-type region near it. With no voltage applied to the gate lead referenced to substrate there exists no connecting n-channel between source and drain; that is, no current will flow from source to drain even with a voltage applied across the source-drain.

A positive voltage applied to the gate will result in a negative charge induced in the p-substrate because of the capacitive action of metal gate, insulating dielectric, and p-substrate. This *induced n-channel* (see Fig. 6.6) will act like the n-channel in the JFET, allowing electron flow from source to drain with suitable drain-source voltage applied. Thus, more positive voltages will result in larger current flow through source and drain, whereas negative voltages will hold the device in cutoff (no current flow). A volt-ampere characteristic of a MOSFET is shown in Fig. 6.7.

A second type of MOSFET provides a physical channel during manufacture as shown in Fig. 6.8a. The source and drain are made by diffusing p^+ regions in an n-type substrate. A p-region is also diffused into the n-type substrate between source and drain and the entire surface is then passivated with silicon dioxide (SiO_2). The p^+ designation indicates that the source and drain are more heavily doped with p-type impurity and have more holes per unit of volume. Finally, the metalization layer is deposited and ohmic leads connected to the FET device. Since a channel already exists a voltage applied from drain to source would result in current flow through the channel (Fig. 6.8b). Negative gate voltages would induce additional holes in the p-channel, thereby *enhancing* the channel operation and the device would operate as an *enhancement-type* MOSFET.

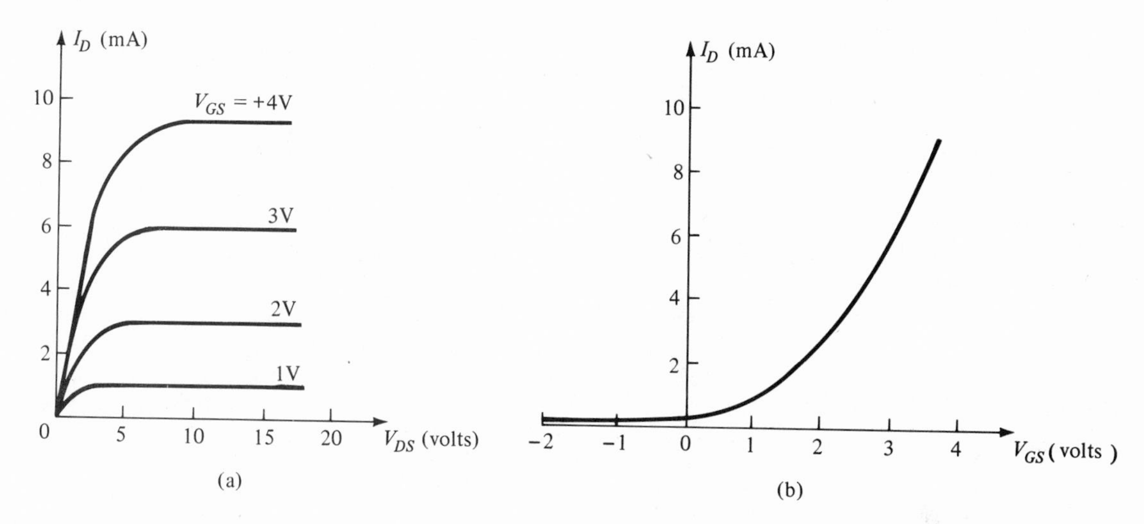

FIG. 6.7

Volt-ampere characteristics of n-channel enhancement-type MOSFET.

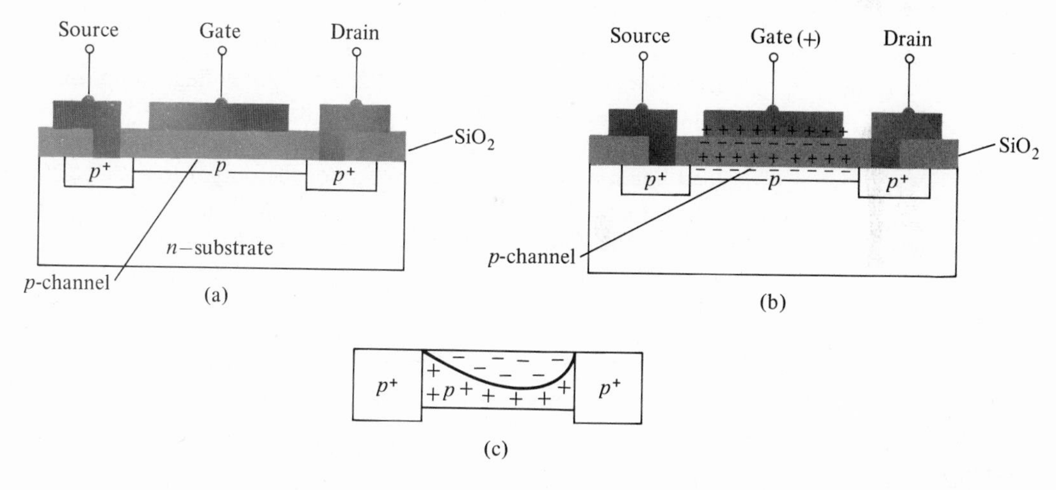

FIG. 6.8

Depletion-type p-channel MOSFET.

In addition, the voltage applied to the gate may also be of positive polarity, resulting in induced electrons in the p-channel and *depleting* the available current carriers (holes) and cutting down the current flow (Fig. 6.8c). A volt-ampere characteristic is shown in Fig. 6.9a and a transfer characteristic in Fig. 6.9b.

For a depletion-type MOSFET with an n-channel, the same description holds, with the voltage and current polarities reversed and election flow used in place of hole flow.

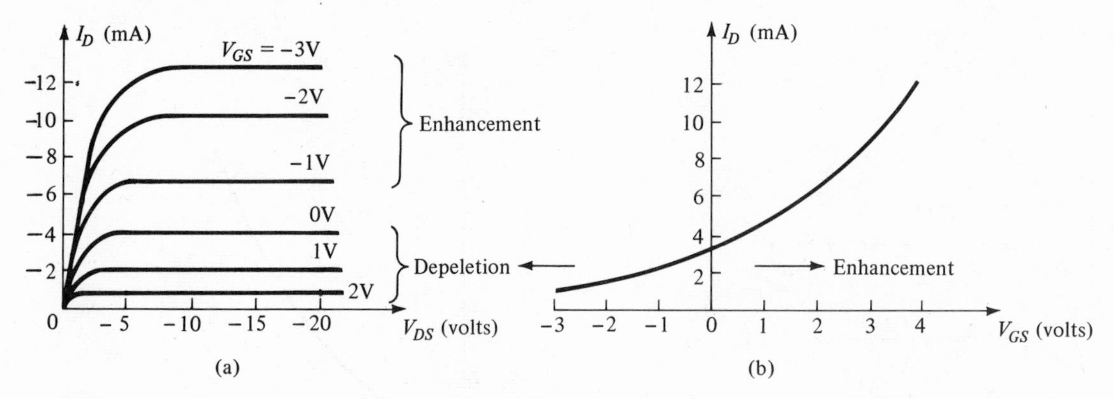

FIG. 6.9

Volt-ampere and transfer characteristics for *p*-channel MOSFET.

6.6 BIASING THE FET

As in all amplifiers, consideration must be given to biasing the active device to place its operating voltage and current within the linear portion of the device active region. A FET bias circuit is shown in Fig. 6.10a. The circuit shown requires two biasing voltage supplies, one to set the gate-source voltage and a second to provide drain-source voltage and drain current. Referring to the device characteristic in Fig. 6.10b. a load line representing the operation of the output loop, V_{DD} battery, R_D resistor, and V_{DS} drain-to-source voltage drop across the FET, is drawn on the volt-ampere graph. It is only necessary to obtain two of

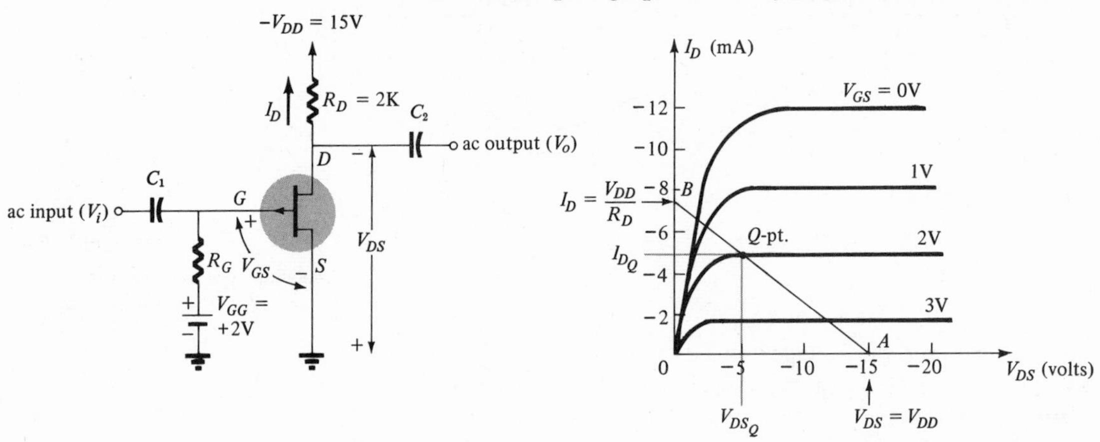

FIG. 6.10

p-channel FET amplifier bias network and drain characteristic (showing bias load line).

the points of the load line since it is a straight line. The two points of the load lines are obtained as follows:

1. With no current flow in the output loop the drain-source voltage is $V_{DS} = V_{DD} = -15$ V (point A).
2. For $V_{DS} = 0$ V the drain current is $V_{DD}/R_D = 15$ V/2 K $= 7.5$ mA in this example. Corresponding to $V_{DS} = 0$ V the point $I_D =$ 7.5 mA is obtained (point B).

Connecting these points by a straight line provides a graphical picture of the output circuit operation. The actual operating voltage and current, the operating point, is now fixed by the gate-source voltage, V_{GS}, and is shown on the graph as the intersecting of the device characteristic with the load line (point C).

The choice of an operating point is usually based on one of the following circuit conditions:

1. Maximum output voltage.
2. Maximum voltage gain.
3. Minimum power dissipation.
4. Minimum drain current drift.
5. Voltage supplies available.

After having decided how to select the bias point, or having chosen a specific bias point, the operating load line is obtained. The following examples demonstrate how a desired bias point is obtained.

EXAMPLE 6.1 Obtain the operating point given a fixed bias circuit as in Fig. 6.11.

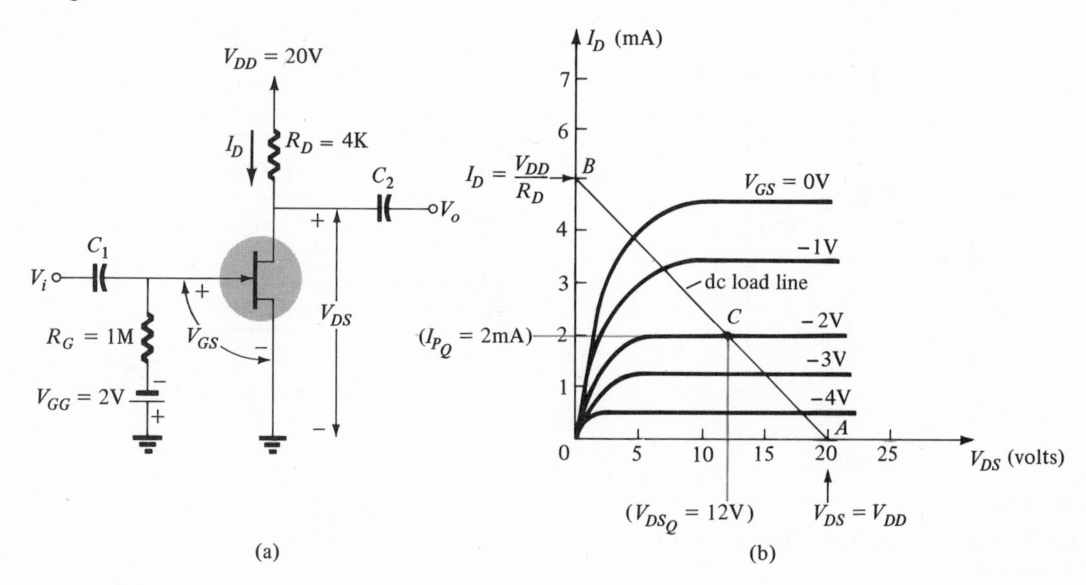

FIG. 6.11
FET bias circuit and characteristic for Example 6.1.

Solution: If the complete circuit (see Fig. 6.11a) and all component values are known the operating point may be directly obtained. Using the typical drain characteristic for the FET device (Fig. 6.11b) we first draw the load line as follows.

From the output loop equation (Kirchhoff's voltage law around drain-source, R_D, power-supply loop),

$$V_{DD} = I_D R_D + V_{DS}$$

We obtain point A: for $I_D = 0$, $V_{DS} = V_{DD}$ (the point corresponding to 20 V on the x-axis) and, point B: for $I_D = V_{DD}/R_D = 20\text{ V}/4\text{ K} = 5\text{ mA}$ (the point corresponding to $+5$ mA on the y-axis). Connecting point A to B forms the dc load line.

The intersection of the load-line curve with the device characteristic curve, for $V_{GS} = 2$ V (fixed by battery V_{GG}) results in point C, the quiescent operating point of the circuit. Reading the bias operating voltage from the graph, V_{DSQ} is 12 V and the current I_{DQ} is then 2 mA. These values signify the voltage and current to be measured with no ac input voltage applied to the circuit and are the values from which the voltage and current will vary when the ac signal to be amplified, is applied.

EXAMPLE 6.2 Obtain the value of the load resistor for desired operating point with fixed bias as in Fig. 6.12.

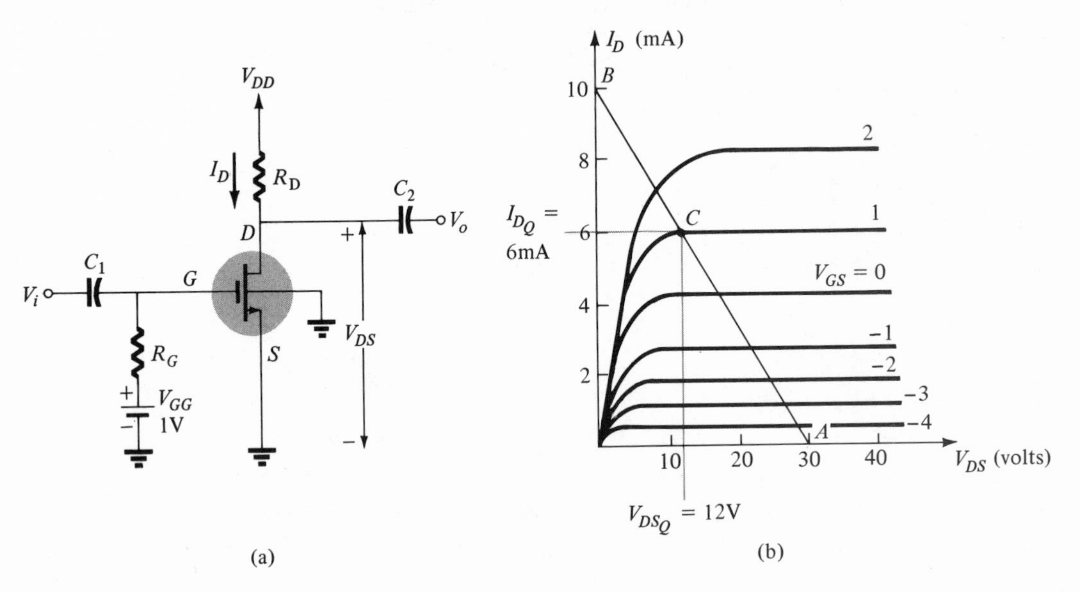

FIG. 6.12
MOSFET bias circuit and drawn characteristic for Example 6.2.

Solution: From the manufacturer's specifications for the MOSFET of Fig. 6.12 we know that this particular FET device operates best at a drain current of about 6 mA and that the maximum transconductance (g_m) occurs around $V_{GS} = +1$ V. It is necessary to determine R_D to obtain the indicated operating condition, namely, $V_{GS} = +1$ V and $I_D = 6$ mA. Using a supply voltage of $V_{DD} = +30$ V (the largest value of V_{DS} is listed as 50 V) we proceed as follows:

The points already specified by the problem are the device curve corresponding to $V_{GS} + 1$ V and the values of $V_{DD} = +30$ V (point *A*) and $I_D = 6$ mA at $V_{GS} = +1$ V (point *C*). Connecting these points to form the load line we obtain a value on the *y*-axis of $I_D = 10$ mA. Since the value of 10 mA is also equal to V_{DD}/R_D, we calculate R_D

$$R_D = \frac{V_{DD}}{I_D} = \frac{30\text{ V}}{10\text{ mA}} = 3\text{ K}$$

EXAMPLE 6.3 Obtain the value of R_D for a given operating point for the circuit of Fig. 6.13.

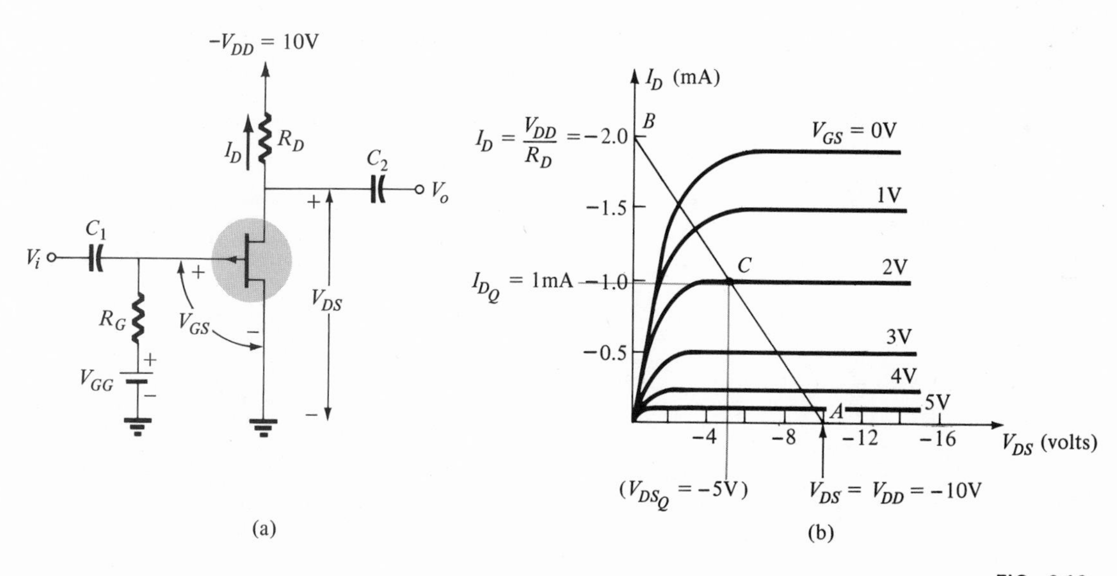

FIG. 6.13
FET circuit and characteristic for Example 6.3.

Solution: If we are given a desired operating point of $V_{DS} = -5$ V and $I_D = 1$ mA for the MOSFET circuit of Fig. 6.13 we must determine both R_D and the value of V_{GG} to obtain proper bias. The desired bias point (point *C*) is marked on the graph, as is point *A*, the supply voltage $V_{DD} = -10$ V. Connecting these we get $I_D = V_{DD}/R_D = 2$ mA, from which $R_D = 10\text{ V}/2\text{ mA} = 5$ K. From the graph the value of V_{GS} is $+2$ V.

6.7 SELF-BIAS OPERATION

It is possible to properly bias a FET using only a single voltage supply. Using the drain supply V_{DD} we can self-bias the gate-source to obtain the desired operating point. Figure 6.14a shows a simple self-bias FET circuit. Notice the addition of source resistor R_S and the elimination of the fixed-bias battery, V_{GG}. The gate voltage (to ground), V_G, is 0 V. The resistor R_G provides a path to ground and since the dc current through R_G is extremely small the voltage drop across R_G is approximately 0 V making the voltage at the gate, $V_G = 0$ V.

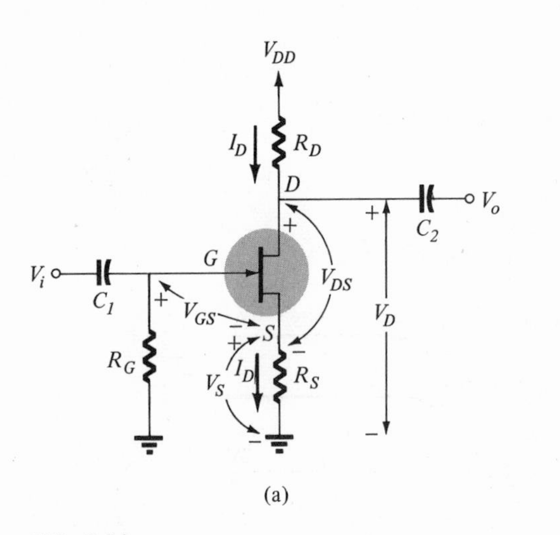

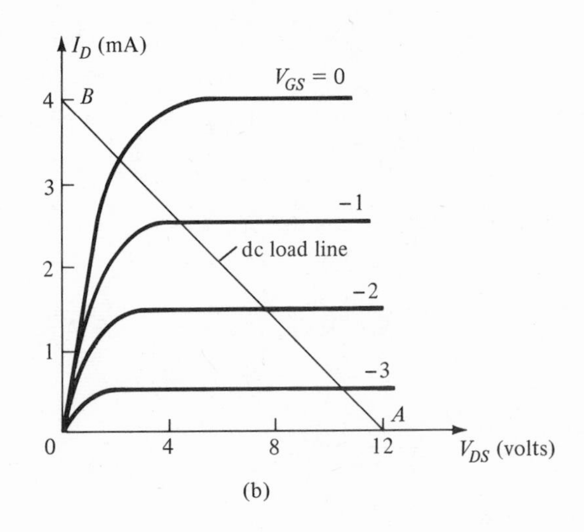

FIG. 6.14
FET self-bias operation.

The presence of resistor R_S from source to ground results in a voltage drop, V_S, due to I_D. The polarity of V_S for the above circuit is positive from source to ground and V_{GS}, which equals $V_G - V_S$, is, $0 - V_S$, or $-V_S$. That is, measuring from 0 V at the gate to, say, $V_S = +2$ V at the source lead, gives a voltage of -2 V from gate to source. Since V_G will be 0 V, the value of V_S will be the gate-source bias voltage and R_S can be selected to set the desired value of V_{GS}.

The dc load line may still be drawn on the device drain characteristic curve. The point along the x-axis, $V_{DS} = V_{DD}$, is still the same. Along the y-axis the point is $I_D = V_{DD}/(R_D + R_S)$, the only difference being the inclusion of R_S in the denominator since R_S and R_D are effectively in series in the output (drain-source) dc loop. Fig. 6.14a shows the load line for the self-bias circuit of Fig. 6.14b.

The load line corresponding to $V_{DD} = 12$ V, $R_D = 2$ K, and

$R_S = 1$ K is

along the x-axis: $\quad V_{DS} = V_{DD} = 12$ V

along the y-axis: $\quad I_D = \dfrac{V_{DD}}{R_D + R_S} = \dfrac{12\text{ V}}{3\text{ K}} = 4\text{ mA}$

Referring to Fig. 6.14b we see the load line describing the operating path for the circuit of Fig. 6.14a. However, we do not yet have the operating point—the values of current and voltage with zero input signal. In order to obtain the alue V_{GS}, we must have the value of V_S. But V_S depends on the current I_D at the quiescent operating point, and in turn, I_D depends on V_{GS}.

What this shows is that we have more unknowns than variables and cannot solve directly for the operating point. If, however, the desired operating current, I_D, is specified (from the device specifications) the bias point can then be easily calculated. The following examples show how self-bias calculations for the operating point are done.

EXAMPLE 6.4 Calculate the self-bias operating point for the FET amplifier circuit shown in Fig. 6.15a. The bias-voltage $V_{GS} = V_G - V_S = -V_S = -I_D R_S$, since $V_G = 0$ V (gate current for reversed-bias gate-source is zero). If the manufacturer's information for the FET specifies an operating point of $I_D = 5$ mA and $V_{DS} = 10$ V, calculate R_D and R_S to obtain this bias condition.

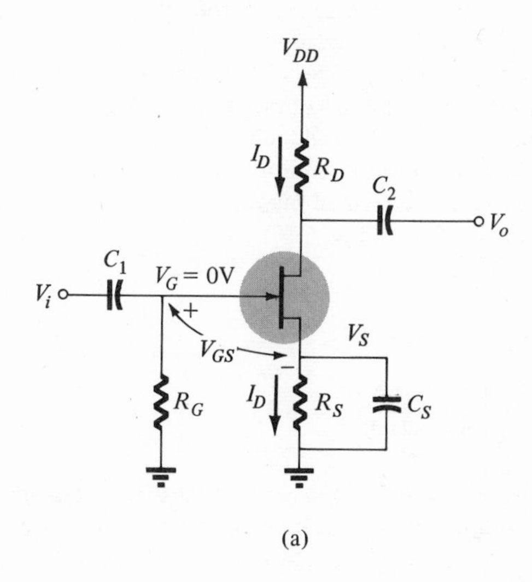

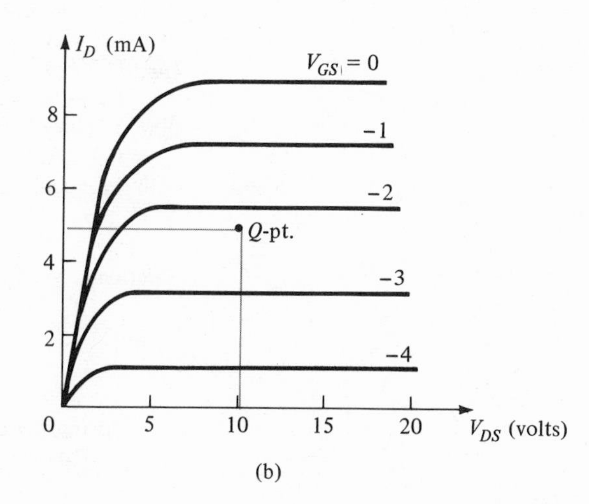

FIG. 6.15
Self-bias circuit and drain characteristic for Example 6.4.

Solution: Neglecting V_S for the moment, we calculate from

$$V_{DS} = V_{DD} - I_D R_D$$
$$10 = 20 - (5 \times 10^{-3}) R_D$$

and

$$R_D = \frac{20 - 10}{5 \times 10^{-3}} = 2\text{ K}$$

From the FET drain characteristic for the desired point, $V_{DS} =$ 10 V, $I_D = 5$ mA (Fig. 6.15b), we find that the required value of V_{GS} is about -2.2 V. Since $V_S = I_D R_S$ we calculate R_S

$$R_S = \frac{V_S}{I_D} = \frac{2.2\text{ V}}{5\text{ mA}} = 0.44\text{ K} = 440\ \Omega$$

Thus, with $R_D = 2$ K and $R_S = 440\ \Omega$ we should obtain the desired operating point. Although we initially neglected V_S, the value of $V_S = 2.2$ V will not appreciably affect our results.

The device characteristic is only typical and we will find in practice that even with these approximations the actual operating point should be within 10% of that desired. If a more exact operating point is needed this would be achieved in practice by adjusting R_S in the operating circuit.

EXAMPLE 6.5 It is possible to select a value of bias voltage, V_{GS}, at which the drain current, I_D, does not vary with temperature. From theory this occurs for $|V_{GS}| = |V_P| - 0.63$ V, where V_P is the specified device pinch-off voltage. If $V_P = 1.63$ V, for example, $|V_{GS}| = 1.63 - 0.63 = 1.0$ V. For the zero-drift condition it can also be shown that $I_D = I_{DSS}(0.63/V_P)^2$, where I_{DSS} is the drain saturation current at $V_{GS} =$ 0 V. If I_{DSS} is specified to be 1.5 mA, I_D is found to be 0.3 mA. Thus we obtain values for the desired operating condition of $V_{GS} = 1.0$ V and $I_D = 0.3$ mA. Using these values as desired bias conditions we proceed with the calculation of R_D and R_S (see Fig. 6.16), using a supply voltage of $V_{DD} = -10$ V.

Solution: We can directly calculate R_S

$$R_S = \frac{V_{GS}}{I_D} = \frac{1.0\text{ V}}{0.3\text{ mA}} \cong 3.3\text{ K}$$

Using the characteristic of Fig. 6.16b we draw a load line from $V_{DS} = V_{DD} = -10$ V along the x-axis through the operating point of $V_{GS} = 1$ V, $I_{DQ} = 0.33$ mA. The resulting load line crosses the y-axis at $I_D = 0.5$ mA, from which we calculate R_D:

$$R_D = \frac{V_{DD}}{I_D} = \frac{10\text{ V}}{0.5\text{ mA}} = 20\text{ K}$$

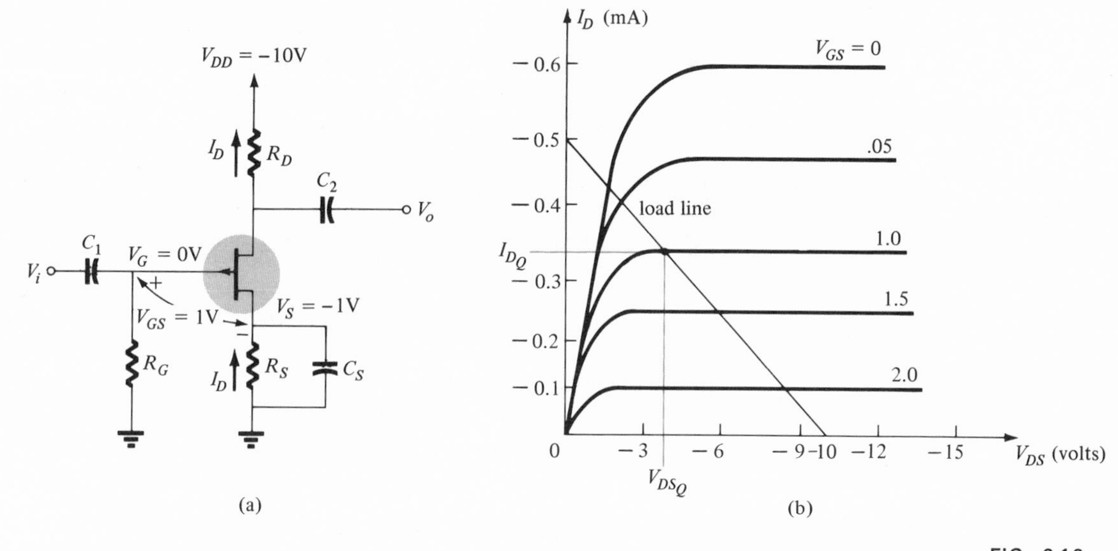

FIG. 6.16
Self-bias FET circuit and characteristic for Example 6.5.

EXAMPLE 6.6 Another variation of a self-bias circuit is shown in Fig. 6.17a. In this circuit the gate-source voltage, V_{GS}, may be set negative, to 0 V, or positive, depending on the resistor values. A MOSFET

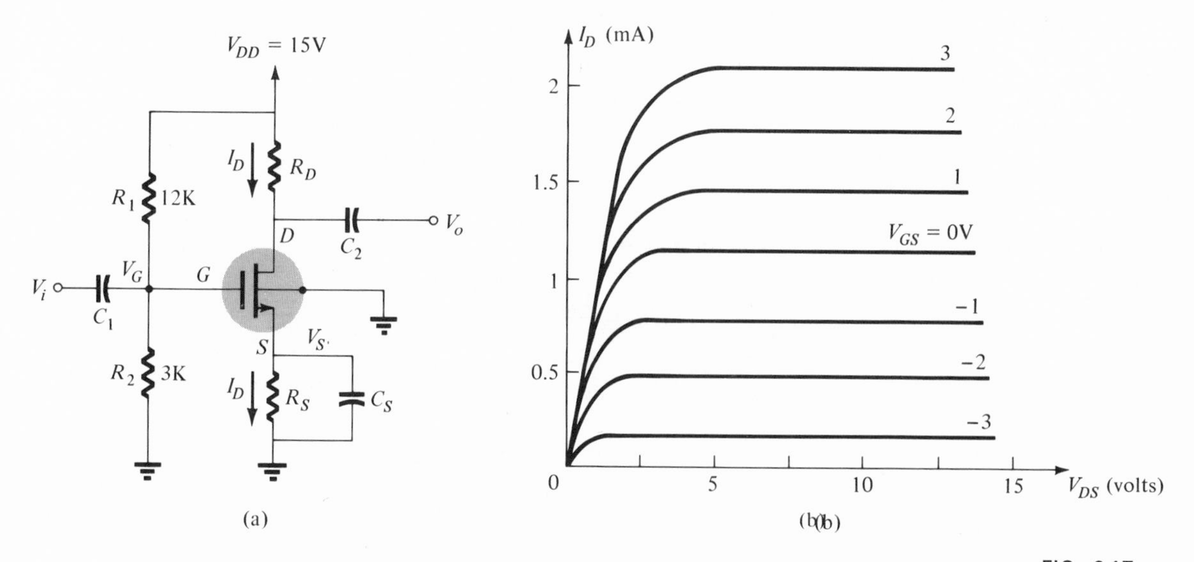

FIG. 6.17
MOSFET bias circuit and characteristic for Example 6.6.

For very large values of R_1, the voltage V_{GS} approaches V_{DS}, whereas for very small values of R_1 the voltage V_{GS} approaches 0 V. The bias voltage V_{GS} can therefore be set from near 0 V to $-V_{DS}$ in value, always negative, as required to bias the FET in the enhancement mode of operation. Reversing the battery polarity will allow the same circuit to properly bias an *n*-channel enhancement MOSFET.

EXAMPLE 6.7 For the circuit of Fig. 6.18c bias the *p*-channel MOSFET to operate at $V_{DS} = -10$ V and $I_D = 2$ mA.

Solution: The MOSFET characteristic of Fig. 6.18b will be used to help determine the circuit resistor values. From the bias point plotted on the device drain characteristic we find $V_{GS} = -2$ V is required.

For $V_{DS} = -10$ V and $I_D = 2$ mA the value of R_D is calculated

$$R_D = \frac{V_{DD} - V_{DS}}{I_D} = \frac{20 - 10\text{ V}}{2\text{ mA}} = 5\text{ K}$$

From the value of $V_{GS} = -2$ V and $V_{DS} = -10$ V we calculate

$$V_{GS} = \frac{R_1}{R_F + R_1}(V_{DS})$$

$$-2\text{ V} = \frac{R_1}{R_F + R_1}(-10\text{ V})$$

$$\frac{R_F + R_1}{R_1} = 5$$

$$1 + \frac{R_F}{R_1} = 5$$

$$\frac{R_F}{R_1} = 4$$

Note that the ratio of R_F to R_1 must be 4. It is necessary (since there are more unknowns than equations to solve for them) to select the value of either resistor—which then fixes the other. Selecting $R_F =$ 400 K, R_1 is then 100 K. Actually, the value of R_F should be determined from stability considerations, the smaller the value of R_F the better the stability (and the lower the amplifier gain).

6.8 LOW-FREQUENCY AC EQUIVALENT CIRCUIT: COMMON SOURCE CONFIGURATION

The low-frequency ac small-signal equivalent circuit of a FET transistor (Fig. 6.19a) is shown in Fig. 6.19b. Notice that the gate to source is represented as an open circuit and no current is drawn by the input terminals of the FET.

Does this mean that the input voltage has no effect on the device?

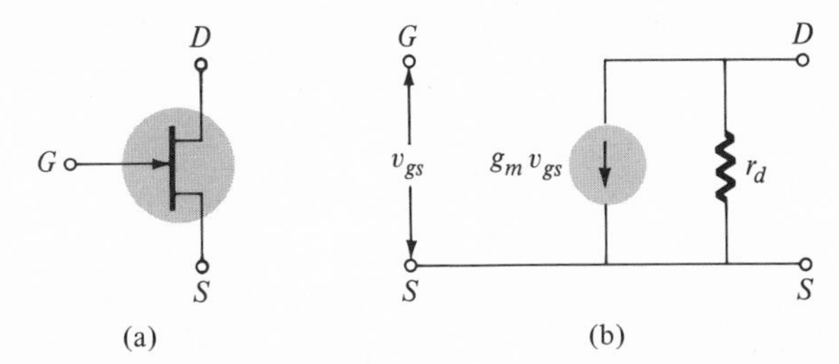

FIG. 6.19
Small-signal ac equivalent circuit for FET at low frequency.

No, for although the gate-source appears as an open circuit in terms of loading considerations in the actual device, the gate-source voltage affects the drain current. This is indicated by the current source, $g_m V_{gs}$, which shows that the theoretical largest current (under output short-circuit) is determined by the FET gain, g_m, and the input voltage, V_{gs}. If, for example, the device has a value of $g_m = 1.5$ mA/V, a 1-V signal ($V_{gs} = 1$ V) will result in a 1.5 mA current in the ideal current generator of the output portion of the equivalent circuit.

The FET transconductance (g_m) is measured in milliamps per volt or millimhos. Typical values of g_m are from 0.5 mA/V to 10 mA/V or, as usually described, 500 μmhos to 10,000 μmhos. The device output resistance, r_d, is measured in kilohms and is typically 100 K to 1 M for JFETs and 1 K to 50 K for MOSFETs.

The FET input resistance at dc is typically 10^8–10^{10} Ω for JFETs and 10^{10}–10^{14} Ω for MOSFET devices. For practical purposes, these values can be considered so large that the input of the FET (from gate to source) can be considered as an open circuit.

Voltage Gain

Consider the small-signal voltage amplifier circuit of Fig. 6.20a. The input voltage, V_i, is coupled through a capacitor to the gate of the FET, amplified, and fed out of the FET through a coupling capacitor as V_o. The circuit voltage gain is $A_v = V_o/V_i$. The ac small-signal equivalent of the amplifier circuit is shown in Fig. 6.20b. The coupling capacitors are large enough that their impedance may be considered negligible for present considerations. In the ac equivalent circuit the dc voltage supplies are replaced by short-circuits. Finally the FET is replaced by an ac equivalent circuit and the complete ac equivalent of the amplifier circuit shown in Fig. 6.20b is obtained. Since the full input voltage is applied across R_G, the value of the bias resistor has no effect on the voltage gain. (If the input signal generator source impedance were considered, then R_G would have an effect.)

For the output circuit loop the current from the current source splits between resistors r_d and R_D; the current through R_D can be calculated using the current divider rule as follows:

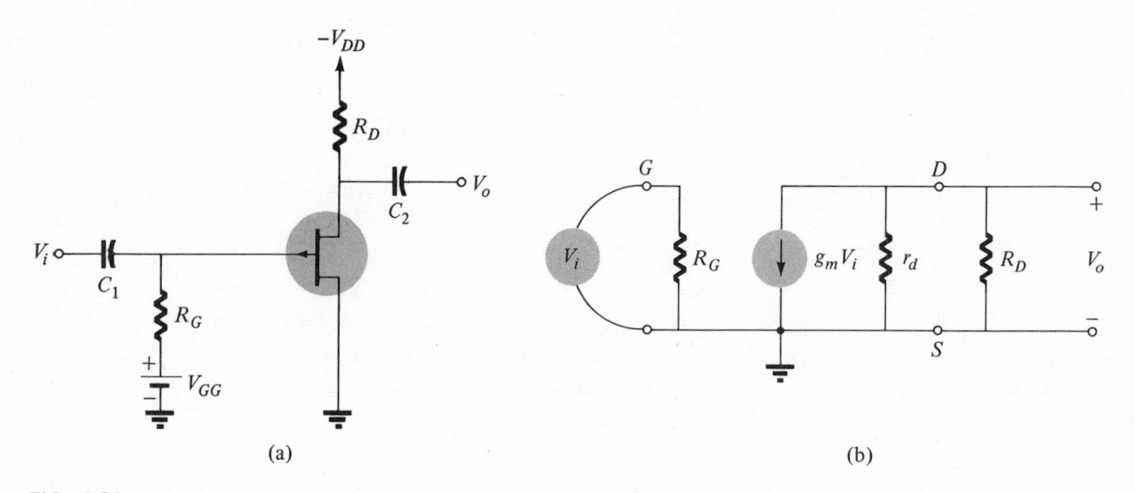

FIG. 6.20
FET amplifier and small-signal equivalent circuit.

$$I_d = \frac{r_d}{R_D + r_d}(g_m V_i)$$

The voltage, V_o is then:

$$V_o = -I_d R_D = -\frac{r_d R_D}{R_D + r_d}(g_m V_i) = -g_m R_L V_i$$

where R_L is the equivalent resistance of resistors r_d and R_D in parallel. Calculating the voltage gain,

$$\boxed{A_v = \frac{V_o}{V_i} = -g_m R_L} \tag{6.1}$$

EXAMPLE 6.8 Calculate the ac voltage gain of the FET amplifier circuit of Fig. 6.21 for the two sets of indicated FET equivalent circuit parameters.

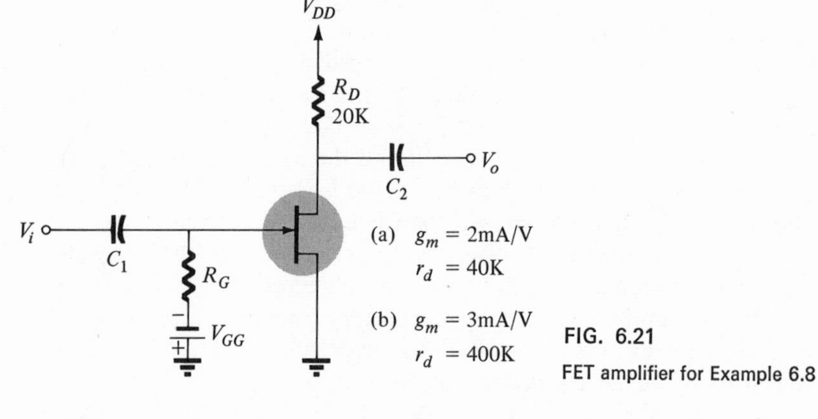

FIG. 6.21
FET amplifier for Example 6.8

Solution: (a) $g_m = 2$ mA/V, $r_d = 40$ K.
Calculating the ac load resistance R_L:

$$R_L = \frac{R_D r_d}{R_D + r_d} = \frac{20 \times 40}{20 + 40} = \frac{800}{60} = \mathbf{13.3\ K}$$

Calculating voltage gain using Eq. (6.1):

$$A_v = -g_m R_L = -(2 \times 10^3)(13.3 \times 10^3) = \mathbf{-26.6}$$

(b) $g_m = 3$ mA/V, $r_d = 400$ K

$$R_L = \frac{R_D r_d}{R_D + r_d} = \frac{20 \times 400}{20 + 400} \cong 19 \text{ K}$$

$$A_v = -g_m R_L = -(3 \times 10^{-3})(19 \times 10^3) = \mathbf{-57}$$

Output Resistance

From the FET equivalent circuit the output resistance is given by resistor r_d. For the amplifier circuit the output impedance is the parallel combination of R_D and r_d.

$$\boxed{R_o = \frac{R_D r_d}{R_D + r_d}} \qquad (6.2)$$

EXAMPLE 6.9 Calcu.ate the output resistance of a FET amplifier circuit having bias resistcr $R_D = 10$ K, for FETs having output resistances of 30 K and 80 K.

Solution

(a) $$R_o = \frac{R_D r_d}{R_D + r_d} = \frac{10 \times 30}{10 + 30} = 7.5 \text{ K}$$

(b) $$R_o = \frac{R_D r_d}{R_D + r_d} = \frac{10 \times 80}{10 + 80} \cong 8.9 \text{ K}$$

Input Resistance

For the equivalent circuit of Fig. 6.19b the FET device is an open circuit across the input terminals, so that $R_i = \infty$. For the amplifier circuit in Fig. 6.20 the input resistance is that of the grid resistor, R_G:

$$\boxed{R_i = R_G} \qquad (6.3)$$

The value of R_G is typically about 1 M, and is limited only by the input leakage current which would cause a dc voltage drop across the gate-source. The value of R_G must not be so large that the resulting voltage drop will affect the circuit bias.

6.9 HIGH-FREQUENCY AC EQUIVALENT CIRCUIT: COMMON SOURCE CONFIGURATION

The ac analysis considered so far applies to dc or low-frequency operation. However, practical operation of the FET requires that additional consideration be given to the effect of higher frequency on device operation. A more complete equivalent circuit is that shown in Fig. 6.22a, which includes additional capacitors between each pair of terminals to account for the frequency dependence of the FET device. Just as there is no actual current source or resistor in the device, there are no actual discrete capacitors, but the use of these in the equivalent circuit includes the frequency effects on the device operation.

Input Capacitance (Miller Effect)

The input capacitance of a FET device is important. Although the typical capacitance values may be a few picofarads (pF), the input

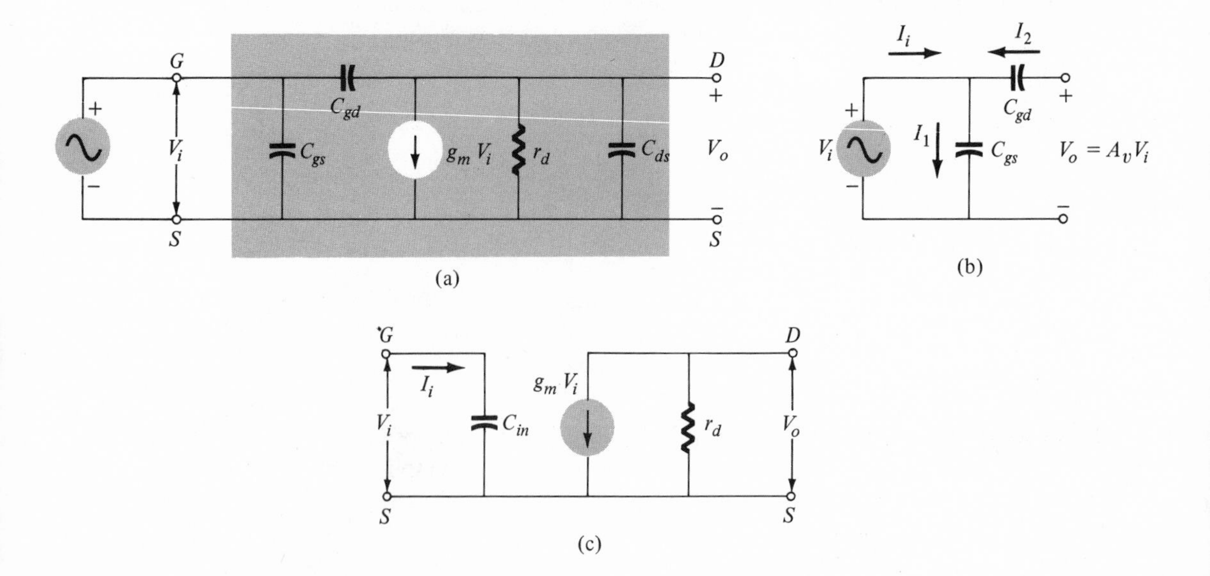

FIG. 6.22

(a) FET high-frequency ac equivalent circuit. (b) Simplified circuit for analysis. (c) Simplified ac equivalent circuit.

capacitance will generally be larger by a significant amount. This is due to the *Miller effect*, as will now be considered.

Figure 6.22b shows a simplified input section of the circuit which will be used to calculate the input capacitance. Using Kirchhoff's current law (KCL) at the input node

$$I_1 = I_i + I_2$$

where

$$I_1 = V_i Y_{gs}$$

$$I_2 = (V_o - V_i) Y_{gd} = (A_v V_i - V_i) Y_{gd} = V_i(A_v - 1) Y_{gd}$$

Using the KCL equation we can write

$$I_i = I_1 - I_2 = V_i Y_{gs} - V_i(A_v - 1) Y_{gd} = V_i[Y_{gs} + (1 - A_v) Y_{gd}]$$

from which we obtain

$$\frac{I_i}{V_i} = Y_i = Y_{gs} + (1 - A_v) Y_{gd}$$

and

$$j\omega C_i = j\omega C_{gs} + (1 - A_v) j\omega C_{gd}$$

so that

$$\boxed{C_i = C_{gs} + (1 - A_v) C_{gd}} \quad (6.4a)$$

Since A_v is $-g_m R_L$ we can write

$$\boxed{C_i = C_{gs} + (1 + g_m R_L) C_{gd}} \quad (6.4b)$$

or, if the load impedance is, in general, Z_L,

$$\boxed{C_i = C_{gs} + (1 + g_m Z_L) C_{gd}} \quad (6.4c)$$

EXAMPLE 6.10 Calculate the input capacitance for the FET amplifier circuit of Fig. 6.21 for the following transistor parameters: $g_m = 2000$ μmhos, $C_{gs} = 2$ pF, $C_{gd} = 1.5$ pF, $C_{ds} = 4$ pF, $r_d = 50$ K.

Solution

(a) Calculate R_L:

$$R_o = R_L = \frac{R_D r_d}{R_D + r_d} = \frac{20 \times 50}{20 + 50} = \mathbf{14.3\ K}$$

$$V_i = \frac{Z_i}{\sqrt{R_s^2 + Z_i^2}}(V_s) = \frac{34.3}{\sqrt{(10)^2 + (34.3)^2}}(1\text{ mV}) = 0.928\text{ mV}$$

$$V_o = A_v V_i = \mathbf{95.8\ mV}$$

(c) At $f = 1$ MHz, $Z_i = 3.43$ K.

$$V_i = \frac{Z_i}{\sqrt{R_s^2 + Z_i^2}}(V_s) = \frac{3.43}{\sqrt{(10)^2 + (3.43)^2}}(1\text{ mV})$$

$$\cong \frac{3.43}{10.85}(1\text{ mV}) \cong 0.32\text{ mV}$$

$$V_o = A_v(V_i) = 100\,(0.32\text{ mV}) = \mathbf{32\ mV}$$

Example 6.12 shows that for frequencies below about 100 kHz the output voltage stays essentially at 100 mV. At 1 MHz, however, the voltage has dropped to only 32 mV, this reduction in gain resulting from the reduced input impedance of the FET device.

Output Impedance

The output impedance of a FET amplifier, as shown in the ac equivalent circuit of Fig. 6.22a, is the parallel combination of the FET output resistance r_d, the load resistance R_D, and the output capacitive impedance due to C_{ds} (and any stray wiring capacitance appearing across the output terminals). For dc considerations the output resistance is calculated as the parallel combination of r_d and R_D [see Eq. (6.2)]. For the additional frequency consideration the output impedance value is obtained as the parallel combination of R_o and the capacitive impedance of C_{ds}, recalling that these components must be combined vectorially. The output impedance can be expressed as

$$\boxed{Z_o = \frac{R_o X_C}{\sqrt{R_o^2 + X_C^2}}} \tag{6.6}$$

where

$$X_C = \frac{1}{\omega C_o}$$

with

$$C_o = C_{ds}$$

EXAMPLE 6.13 Calculate the output impedance of the FET amplifier of Example 6.10 at frequencies of (a) 100 kHz, (b) 1 MHz, and (c) 10 MHz.

Solution: From Example 6.10, $R_o = 14.3$ K.
(a) At $f = 100$ kHz, the output capacitive impedance is

$$X_c = \frac{1}{2\pi f C_{gd}} = \frac{1}{(6.28)(10^5)(4 \times 10^{-12})} \cong 0.4 \text{ M}$$

Since $X_c \gg R_o$.

$$Z_o = R_o \parallel X_c \cong R_o = \mathbf{14.3\ K}$$

(b) At $f = 1$ MHz:

$$X_c = \frac{1}{2\pi(10^6)(4 \times 10^{-12})} \cong 40 \text{ K}$$

and

$$Z_o = \frac{R_o X_c}{\sqrt{R_o^2 + X_c^2}} = \frac{(14.3)(40)}{\sqrt{(14.3)^2 + (40)^2}} \cong \mathbf{13.5\ K}$$

(c) At $f = 10$ MHz:

$$X_c = \frac{1}{2\pi(10^7)(4 \times 10^{-12})} \cong 4 \text{ K}$$

$$Z_o = \frac{(14.3)(4)}{\sqrt{(14.3)^2 + (4)^2}} = \mathbf{3.85\ K}$$

Voltage Gain

For low-frequency considerations Eq. (6.1) gave the amplifier voltage gain as $A_v = -g_m R_L$, where R_L is the load resistance of R_D in parallel with r_d. When considering frequency effects the more general relation would be

$$\boxed{A_v = -g_m Z_o} \tag{6.7}$$

where the output impedance Z_o, is that calculated as the amplifier output impedance of Eq. (6.6).

EXAMPLE 6.14 Calculate the voltage gain of the amplifier of Example 6.10 at frequencies of (a) 100 kHz, (b) 1 MHz, and (c) 10 MHz.

Solution: Referring to the circuit of Fig. 6.21, and the output impedance calculations of Example 6.13:
(a) At $f = 100$ kHz:

$$A_v = -g_m Z_o = -(2 \times 10^{-3})(14.3 \times 10^3) = \mathbf{-28.6}$$

(b) At $f = 1$ MHz

$$A_v = -(2 \times 10^{-3})(13.5 \times 10^3) = \mathbf{-27.0}$$

(c) At $f = 10$ MHz:

$$A_v = -(2 \times 10^{-3})(3.85 \times 10^3) = \mathbf{-6.7}$$

6.10 COMMON-DRAIN (SOURCE-FOLLOWER) AMPLIFIER

Figure 6.23a shows a source-follower circuit with input applied to gate and output voltage taken from source. The circuit is similar in function to a cathode follower or emitter follower. There is a gain of near unity with no phase inversion provided with very high input impedance and low output impedance. Since the input impedance of a FET is already extremely high at dc or low frequency, the impedance consideration of the source follower becomes important only at higher frequencies.

Figure 6.23b shows the ac small-signal equivalent circuit which

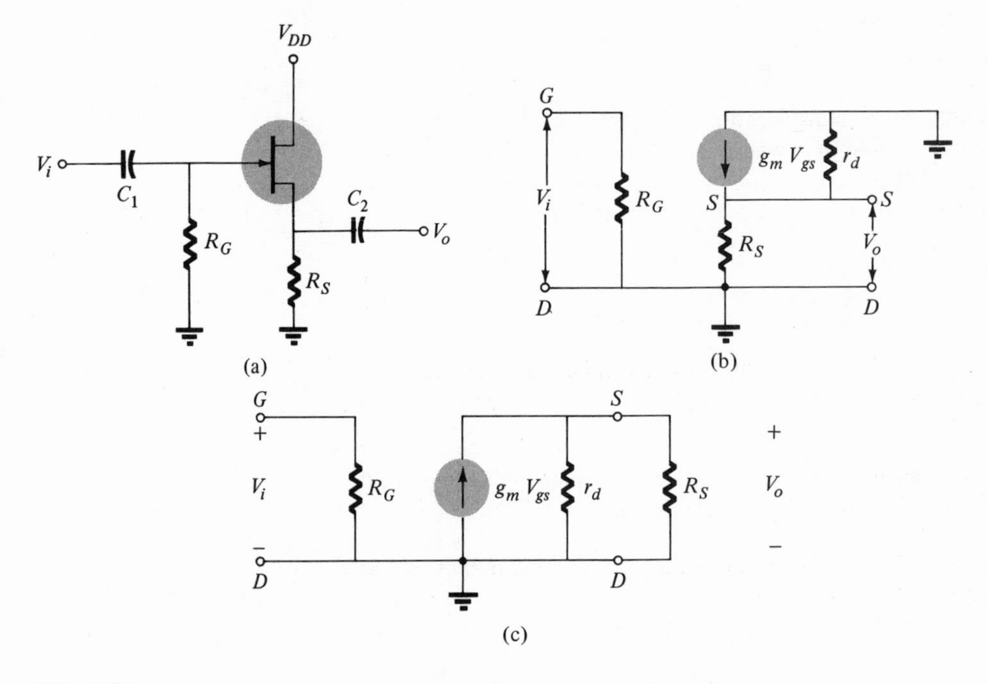

FIG. 6.23
Source follower and ac equivalent circuit.

is redrawn in Fig. 6.23c in simpler form. For dc considerations (and low frequency):

$$\textit{Input impedance:}\ R_i \cong \text{infinity } (\infty) \tag{6.8}$$

$$\textit{Output impedance:}\ R_o \cong \frac{1}{g_m} \tag{6.9}$$

$$\textit{Voltage Gain:}\ A_v \cong 1 \text{ (unity)} \tag{6.10}$$

For high-frequency considerations:

$$\textit{Input admittance:}\ Y_i = j\omega C_i \tag{6.11}$$

where C_i is given by Eq. (6.4).

Output admittance

$$Y_0 = g_m + g_d + j\omega C_T \tag{6.12}$$

where $C_T = C_{ds} + C_{sn}$, with C_{sn} representing any stray capacitance measured from the source to ground (or the capacitance of any device connected across the output), and $g_d =$ FET output conductance $(= 1/r_d)$

Voltage Gain

The voltage gain may be taken as approximately 1, or more exactly as

$$A_v = \frac{(g_m + j\omega C_{gs})R_s}{1 + (g_m + g_d + j\omega C_T)R_s} \tag{6.13a}$$

which simplifies at low frequency to

$$\boxed{A_v = \frac{g_m R_s}{1 + (g_m + g_d)R_s}} \tag{6.13b}$$

(Note that for $g_m \gg g_d$ and also $g_m \gg 1$, the expression reduces to unity, which is approximately the value to expect.)

EXAMPLE 6.15 Calculate the input and output impedances and voltage gain of the FET source follower of Fig. 6.23 for the following circuit and component values at a frequency of (a) 100 kHz and at (b) 1 kHz(low frequency). $C_{gs} = 3$ pF, $C_{gd} = 4$ pF, $C_{ds} = 5$ pF, $g_m = 2$ mmhos, $R_s = 1$ K, $R_G = 1$ M, and $r_d = 50$ K.

Solution: (a) At $f = 100$ KHz:

Input impedance

$$Y_i = \omega C_i \qquad \text{with} \qquad C_i = C_{gs} + (1 - A_v)C_{gd}$$
$$\cong 3 + (1 - 1) \times 4 = 3 \text{ pF}$$
$$Y_i = 2\pi(10^5)(3 \times 10^{-12}) \cong 1.88 \times 10^{-6}$$
$$Z_i = \frac{1}{Y_i} = \frac{1}{1.88 \times 10^{-6}} = \mathbf{532\ K}$$

Output impedance

$$Y_o = g_m + g_d + j\omega C_T \qquad \text{where} \qquad C_T = C_{ds} = 5 \text{ pF}$$
$$= 2 \times 10^{-3} + \frac{1}{50 \times 10^3} + j(2\pi \times 10^5)(5 \times 10^{-12})$$
$$= 2.02 \times 10^{-3} + j3.14 \times 10^{-6} \cong 2 \times 10^{-3}$$
$$Z_o = \frac{1}{Y_o} = \frac{1}{2 \times 10^{-3}} = \mathbf{500\ \Omega}$$

Voltage gain

$$A_v = \frac{(g_m + j\omega C_{gs})R_s}{1 + [g_m + g_d + j\omega C_T]R_s}$$
$$\cong \frac{[(2 \times 10^{-3}) + j(2\pi \times 10^5)(3 \times 10^{-12})](10^3)}{1 + [2 \times 10^{-3}](10^3)}$$
$$= \frac{2 \times 10^{-3} + j1.88 \times 10^{-3}}{3} = \frac{2.74}{3} = \mathbf{0.913}$$

(b) At $f = 1$ kHz (low frequency):

Input impedance $\left(\textit{FET input impedance } X_i = \dfrac{1}{Y_i} = \dfrac{1}{2\pi f C_i}\right)$

$$X_i = \frac{1}{2\pi(10^3)(3 \times 10^{-12})} = 53.2 \text{ M}$$

Circuit input impedance, $Z_i = X_i \,||\, R_G$ which, in this case is $Z_i \cong R_G = \mathbf{1\ M}$.

Output impedance $\left(\textit{FET output impedance } X_o = \dfrac{1}{Y_o} = \dfrac{1}{g_m + g_d + j\omega C_T}\right)$

$$g_m = 2 \times 10^{-3} \text{ mhos}$$
$$g_d = \frac{1}{r_d} = \frac{1}{50 \text{ K}} = 20 \times 10^{-6} \text{ mhos}$$
$$\omega C_T = 2\pi(10^3)(5 \times 10^{-12}) = 31.4 \times 10^{-9} \text{ mhos}$$
$$g_m \gg g_d \gg \omega C_T$$

so that

$$Z_o = R_o \cong \frac{1}{g_m} = \frac{1}{2 \times 10^{-3}} = \mathbf{500\Omega}$$

$$g_m \gg g_d \gg j\omega C_T, \qquad \text{and} \qquad g_m \gg j\omega C_{gs}$$

so that the approximate relation (Eq. (6.13a)), may be written as

$$A_v = \frac{g_m R_s}{1 + g_m R_s}$$

and since $g_m R_s = 2 \times 10^{-3} \times 10^3 = 2$,

$$A_v = \frac{2}{3} = \mathbf{0.67}$$

6.11 DESIGN OF JFET AMPLIFIER CIRCUITS

The design of a few JFET amplifier circuits will provide additional means of presenting the basic characteristics of a JFET. As has been done before, the dc and ac calculations are carried out separately. For dc bias considerations it is necessary to have either a drain or transfer characteristic, or, as will be used here, a mathematical relationship relating input voltages, V_{GS}, and output drain current, I_D. The FET transfer characteristic can be well approximated by

$$\boxed{I_D = I_{DSS}\left(1 - \frac{V_{GS}}{V_P}\right)^2} \tag{6.14}$$

where

I_{DSS} = maximum drain-source current (with $V_{GS} = 0$ V)
V_{GS} = gate-source bias voltage
V_P = gate-source pinch-off voltage (at which $I_D = 0$)
I_D = drain-source bias current

Typically, values of I_{DSS} and V_P are given in the device specifications so that selection of either I_D or V_{GS} provides a means of calculating the other, using Eq. (6.14).

For design purposes consider using the basic amplifier circuit of Fig. 6.24. The circuit design can take a number of forms, depending on what information is given and what is to be determined. The device and circuit relations are the same in any case, as a few examples will show. As a first example consider designing an amplifier as in Fig. 6.24 given the FET device specifications with R_G, R_S, R_D, C_1, and C_2 to be determined.

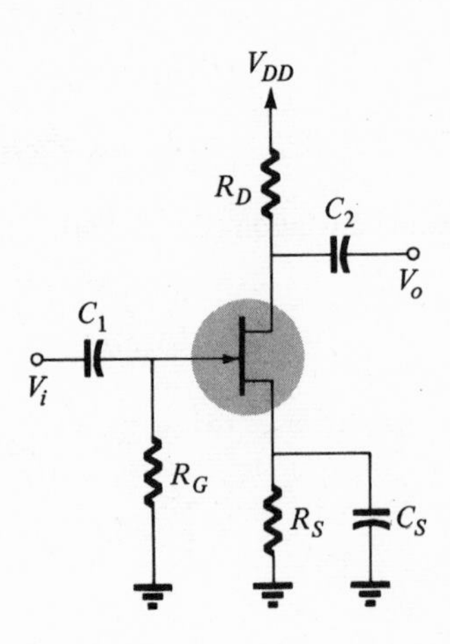

FIG. 6.24

FET self-biased amplifier.

EXAMPLE 6.16 Determine suitable values of R_G, R_S, R_D, C_1, C_2, and C_S for the FET amplifier of Fig. 6.24 using a 2N4220 FET having the following device parameters:

$$V_{DS_{max}} = 30 \text{ V}; \qquad V_{GS_{max}} = -30 \text{ V}$$
$$V_{GS}(\text{off}) = V_P = -4 \text{ V}$$
$$I_{DSS} = 3 \text{ mA}$$
$$I_{GSS} = 0.1 \text{ nA}$$

forward transfer admittance $\equiv |y_{fs}| = g_m = 2500\ \mu$mhos, typical

output admittance $\equiv |y_{os}| = \frac{1}{r_d} = 10\ \mu$mhos, max $= \frac{1}{100}$ K

input capacitance $\equiv C_{iss} = C_{gs} = 4.5$ pF, typical

reverse transfer capacitance $\equiv C_{rss} = C_{gd} = 1.2$ pF, typical

Solution: Consider the dc and then the ac design separately.

dc Design: Since $V_{DS_{max}} = 30$ V we can select a supply of, say $V_{DD} = 20$ V. From the given data we know that the gate-source bias voltage is limited between 0 and -4 V and the drain current will not exceed 3 mA (at $V_{GS} = 0$ V). Let us select $V_{GS} = -2$ V as a suitable bias point (allowing the gate-source to swing ± 2 V, the maximum value). Using Eq. (6.14) we can determine I_D:

$$I_D = I_{DSS}\left(1 - \frac{V_{GS}}{V_P}\right)^2 = 3\left(1 - \frac{-2}{-4}\right)^2 = 3(0.5)^2 = 0.75 \text{ mA}$$

(Note that the factor $(1 - V_{GS}/V_P)$ must always be a fraction since I_D can only be less than the maximum value of I_{DSS}. This is the dc bias current, sometimes listed as I_{DQ} (quiescent current).

Recall that for the FET the gate current is nearly zero (typically nanoamperes) so that no dc current flows either into the gate or through resistor R_G, and

$$V_G = 0 \text{ V}$$

We can select R_G to be 1 M, for which a dc gate current (I_{GSS}) of 0.1 nA would produce only a 1 mV dc voltage drop, which is negligible.

For the bias voltage V_{GS} to be -2 V we have

$$V_S = V_G - V_{GS} = 0 - (-2) = 2\text{ V}$$

We then calculate the required value of R_S:

$$R_S = \frac{V_S}{I_D} = \frac{2\text{ V}}{0.75\text{ mA}} = \mathbf{2.67\text{ K}}$$

(As a suitable value we could choose $R_S = 2.7$ K.) Using the value of R_S calculated for the present design the next step is to select R_D so that the drain-source bias voltage is approximately in the center of its operating range, which is $V_{DD} = +20$ V and $V_S = +2$ V. A choice of $V_D = +11$ V would seem suitable for which we calculate

$$\frac{V_{DD} - V_D}{I_D} = \frac{20 - 11}{0.75\text{ mA}} = \mathbf{12\text{ K}}$$

The bias condition is then

$$V_{GS} = -3\text{ V}$$
$$I_D = 0.75\text{ mA}$$
$$V_{DS} = V_D - V_S = 11 - 2 = 9\text{ V}.$$

ac Design: The input impedance at mid-frequency is basically the value of $R_G = \mathbf{1\text{ M}}$. The value of the impedance of capacitors C_1 and C_2 should be selected so that at the lowest frequency of operation the ac signal attenuation of X_C and R_G acting as a voltage divider is still negligible. Selecting $C_1 = C_2 = C$, and calculating X_C at a frequency $f = 100$ Hz,

$$X_C \leq \frac{1}{10} R_G = \frac{1}{10}(1\text{ M}) = 100\text{ K}$$

$$\frac{1}{2\pi fC} = \frac{1}{6.28(100)C} \leq 100\text{ K}$$

$$C \geq \frac{1}{6.28(100)(100 \times 10^3)} = \frac{10^{-7}}{6.28} = 15.9 \times 10^{-9}$$

$$= 0.0159\ \mu\text{F}$$

(Choose $C_1 = C_2 = \mathbf{0.02\ \mu F}$.)

The value of C_S is selected so that at $f = 100$ Hz we have

$$X_{C_s} \leq \frac{1}{10} R_S = \frac{1}{10}(2.67\text{ K}) = 267\ \Omega$$

$$C_S \geq \frac{1}{2\pi(100)(267)} = \frac{10^{-6}}{0.168} \cong 6\ \mu\text{F (use } C_S = \mathbf{10\ \mu F}\text{)}.$$

The value of V_{GS} is then determined using Eq. (6.14):

$$I_D = I_{DSS}\left(1 - \frac{V_{GS}}{V_P}\right)^2$$

$$1.5 = 3\left(1 - \frac{V_{GS}}{-4}\right)^2$$

$$\left(1 - \frac{V_{GS}}{-4}\right) = 0.707$$

$$V_{GS} = 0.293(-4) = -1.17\text{ V} \cong -1.2\text{ V}$$

The value of V_G is then

$$V_G = V_{GS} + V_S = -1.2 + 10 = 8.8\text{ V}$$

which is set by the voltage divider of R_1 and R_2:

$$V_G = 8.8\text{ V} = \frac{R_2}{R_1 + R_2}(V_{DD}) = \frac{R_2}{R_1 + R_2}(+20\text{ V})$$

Selecting $R_2 = \mathbf{500\ K}$ as a suitably large value,

$$\frac{500\text{ K}}{R_1 + 500\text{ K}} = \frac{8.8}{20} = 0.44$$

$$R_1 = 638\text{ K} \cong \mathbf{650\ K}$$

Using Eq. (6.13b) the amplifier gain is

$$A_v = \frac{g_m R_S}{1 + (g_m + g_d)R_S} = \frac{(2500 + 10^{-6})](6.7 \times 10^3)}{1 + [(2500 + 10)(10^{-6})](6.7 \times 10^3)}$$

$$\cong \frac{16.7}{1 + 16.7} = \mathbf{0.945}$$

At mid-frequency,

$$R_i = \frac{R_1 R_2}{R_1 + R_2} = \frac{500(650)}{500 + 650} = \mathbf{282\ K}$$

At mid-frequency, by Eq. (6.9),

$$R_o \cong \frac{1}{g_m} = \frac{1}{2500 \times 10^{-6}} = \mathbf{40\ \Omega}$$

6.12 DESIGN OF MOSFET AMPLIFIER CIRCUITS

An enhancement MOSFET provides a very simple amplifier circuit (see Fig. 6.26). As the drain characteristic of Fig. 6.26b shows the n-channel enhancement MOSFET requires some positive gate-source

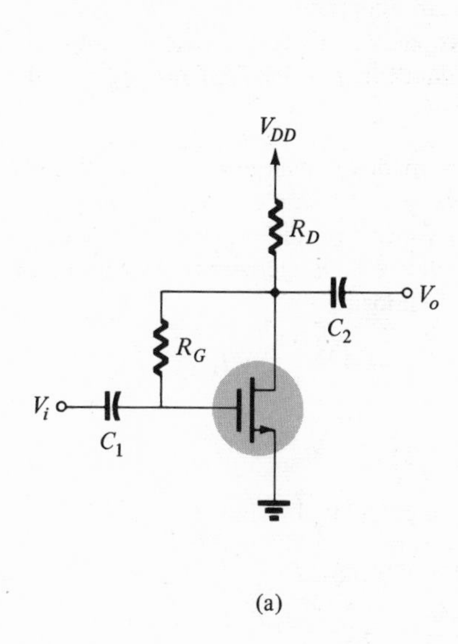

(a)

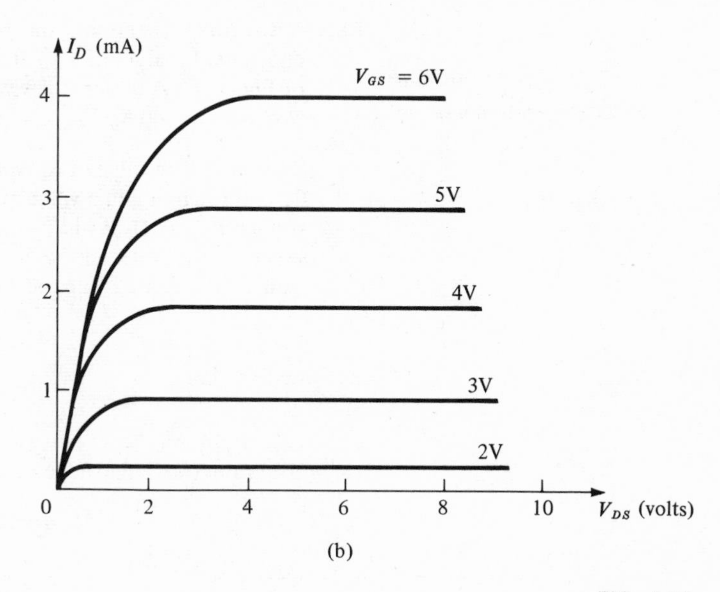

(b)

FIG. 6.26
Practical enhancement MOSFET circuit.

voltage before a channel is created and drain current flows. That is, when

$$V_{GS} < V_1 \qquad I_D = 0$$

which states that when the gate-source voltage (V_{GS}) is below some threshold value (V_T) no drain current flows (because no channel exists). In the circuit of Fig. 6.26a the gate-source voltage is equal to V_{DD} since no dc current flows through R_G. Thus, the only consideration for operation is that

$$V_{DD} < V_T$$

Under this condition the MOSFET is operated in its constant current region (horizontal line section of drain characteristic).

With negligible dc current through R_G the drain-source voltage is given by

$$V_{DS} = V_{DD} - I_D R_D \tag{6.15}$$

The gate-source voltage in the present circuit is the same as the drain-source voltage

$$V_{GS} = V_{DS} \tag{6.16}$$

Consider now the design of an amplifier as given by the circuit of Fig. 6.26.

EXAMPLE 6.19 Determine the resistor component values R_G and R_D, and suitable values of V_{DD} and capacitors C_1 and C_2 for the amplifier of Fig. 6.26. Additional device data for the MOSFET are $g_m = 3000$ μmhos, $r_d = 40$ K, $V_{DS_{max}} = 15$ V.

Solution: The value of R_G can be made quite large since no dc current flows through it and a value of $R_G = \mathbf{10\ M}$ would be typical. A supply voltage $V_{DD} = \mathbf{10\ V}$ will be used. Since V_{DS} and V_{GS} are equal we can select such a point on the given device characteristic (Fig. 6.26b) and read off the desired value of I_{DQ}. Choosing

$$V_{DS} = V_{GS} = 4 \text{ V}.$$

we find that

$$I_D = 2 \text{ mA}$$

Using Eq. (6.15) the value of R_D is calculated to be

$$4 = 10 - (2 \text{ mA})R_D$$

$$R_D = \frac{6 \text{ V}}{2 \text{ mA}} = \mathbf{3\ K}$$

The value of $C_1 = C_2 = C$ must provide a capacitive impedance at the lowest operating frequency (say, $f = 10$ Hz) so that

$$X_C \leq \frac{1}{10} R_G = \frac{1}{10}(10 \text{ M}) = 1 \text{ M}$$

$$C \geq \frac{1}{2\pi(10)(10^6)} = \frac{10^{-6}}{62.8} = 0.016\ \mu\text{F}$$

(Choose $C_1 = C_2 = \mathbf{0.02\ \mu F}$). The circuit gain is calculated as follows:

$$R_L = \frac{R_D r_d}{R_D + r_d} = \frac{3(40)}{3 + 40} = 2.8 \text{ K}$$

$$A_v = -g_m R_L = -(3000 \times 10^{-6})(2.8 \times 10^3) = -8.4$$

For low-frequency and mid-frequency:

$$R_i \cong R_G = 10 \text{ M}$$

$$R_o \cong R_L = 2.8 \text{ K}$$

6.13 FET/BIPOLAR CIRCUITS

Bootstrap Source Follower

If very high input impedance is desired a bootstrap source-follower circuit such as that of Fig. 6.27 might be used. Essentially the circuit couples some of the output back to the gate of the FET via capacitor C_2. Although analysis of this type of signal feedback techni-

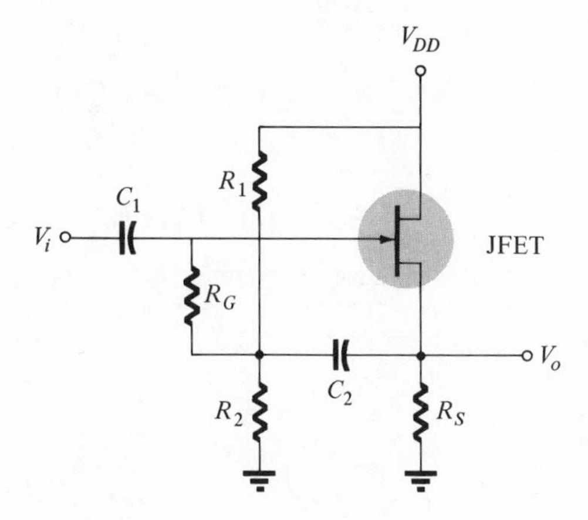

FIG. 6.27
Bootstrap source follower circuit.

que is not covered until Chapter 12 we can note the result that the effective shunt resistance seen looking into the amplifier input is

$$R_i = \frac{R_G}{1 - A_v} \tag{6.17}$$

where A_v for a source follower is typically slightly less than unity. For example, if $A_v = 0.98$ and $R_G = 1$ M we would find that

$$R_i = \frac{R_G}{1 - A_v} = \frac{1 \text{ M}}{1 - 0.98} = \frac{1 \text{ M}}{0.02} = 50 \text{ M}$$

Resistors R_1 and R_2 provide dc bias of the FET as does resistor R_S. Capacitor C_2 is typically chosen so that

$$X_{C_2} \leq \frac{1}{10}(R_1 \| R_2) \tag{6.18}$$

EXAMPLE 6.20 Calculate the dc bias voltages and input impedance for the circuit of Fig. 6.27 for component values below. Determine suitable values of C_1 and C_2. $R_1 = 1.5$ M, $R_2 = 1$ M, $V_{DD} = 18$ V, $R_S = 2.7$ K, $g_m = 5000$ μmhos, $r_d = 50$ K, $R_G = 1$ M, $I_{DSS} = 10$ mA, $V_P = -5$ V

Solution: For dc bias:

$$V_G = \frac{R_2}{R_1 + R_2}(V_{DD}) = \frac{1}{1.5 + 1}(18) = 7.2 \text{ V}$$

(Note, that since no dc current flows through R_G, voltage at the voltage divider junction is gate voltage.)

Using Eq. (6.14),

$$I_D = I_{DSS}\left(1 - \frac{V_{GS}}{V_P}\right)^2 = 10\left(1 - \frac{V_{GS}}{-5}\right)^2$$

and

$$V_{GS} = V_G - V_S = V_G - I_D R_S = 7.2 - I_D(2.7\text{ K})$$

It is possible to determine I_D and V_{GS}, obtaining

$$I_D = 3.5\text{ mA}, \ V_{GS} = 2\text{ V}.$$

For

$$R_1 \,||\, R_2 = \frac{1(1.5)}{1 + 1.5} = 0.6\text{ M}$$

$$X_{C_2} \leq \frac{1}{10}(0.6 \times 10^6) = 60 \times 10^3$$

(at say, $f = 100$ Hz)

$$C_2 \geq \frac{1}{6.28(100)(60 \times 10^3)}$$

$$= \frac{1}{375 \times 10^5} = 0.0265\ \mu\text{F} \quad (\text{Choose } C_2 = 0.05\ \mu\text{F.})$$

The value of C_1 is selected so that at, say, $f = 100$ Hz.

$$X_{C_1} \leq \frac{1}{10} R_G$$

$$C_1 \geq \frac{1}{6.28(100)(10^6)} = 0.0016\ \mu\text{F} \quad (\text{Choose } C_1 = 0.002\ \mu\text{F.})$$

The gain of the circuit is calculated using Eq. (6.13b)

$$A_v = \frac{g_m R_S}{1 + (g_m + g_d)R_S} = \frac{(5 \times 10^{-3})(2.7 \times 10^3)}{1 + (5 \times 10^{-3} + 0.02 \times 10^{-3})(2.7 \times 10^3)}$$

$$\cong \frac{13.5}{14.5} = 0.93$$

Using Eq. (6.17) we have

$$R_i = \frac{R_G}{1 - A_v} = \frac{1\text{ M}}{1 - 0.93} = 14.3\text{ M}$$

Bipolar/FET Bootstrap Source Followers

A circuit such as the FET bootstrap source follower feeding directly into the emitter follower stage shown in Fig. 6.28 provides increased power gain over the circuit of Fig. 6.27. The only change is that the bipolar stage provides additional current gain (still no voltage gain and slightly lower voltage gain which is less than unity) and no

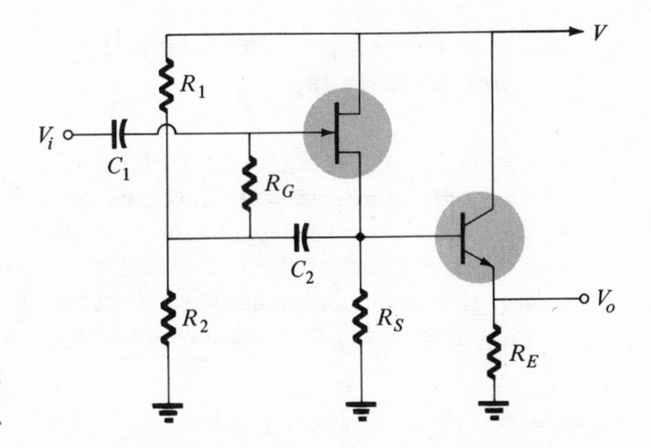

FIG. 6.28

FET/bipolar follower circuit.

phase inversion. The output impedance after the bipolar device is even lower than for the bootstrap stage alone.

Figure 6.29 shows a few additional circuit variations of FET/bipolar follower circuits.

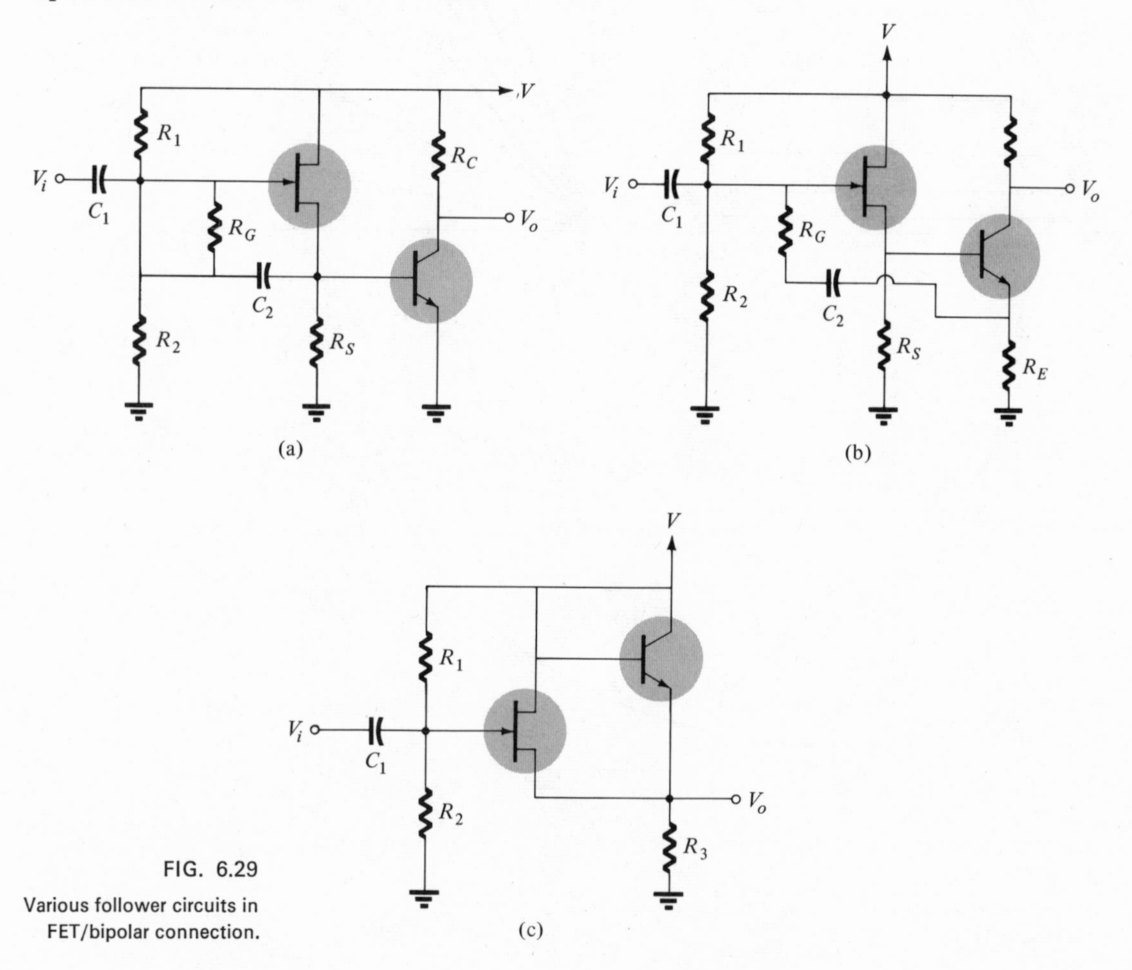

FIG. 6.29

Various follower circuits in FET/bipolar connection.

6.14 THE FET AS A VOLTAGE-VARIABLE RESISTOR (VVR)

The drain-source resistance of a FET can be varied as a function of applied gate-source voltage. This control is fairly linear and applies to the device operating region shown in Fig. 6.30a. Note that this is only a limited part of the FET operating region and is not the linear region of operation as an amplifier. The current range shown is limited to only about 100 μA and a corresponding voltage range of only a few

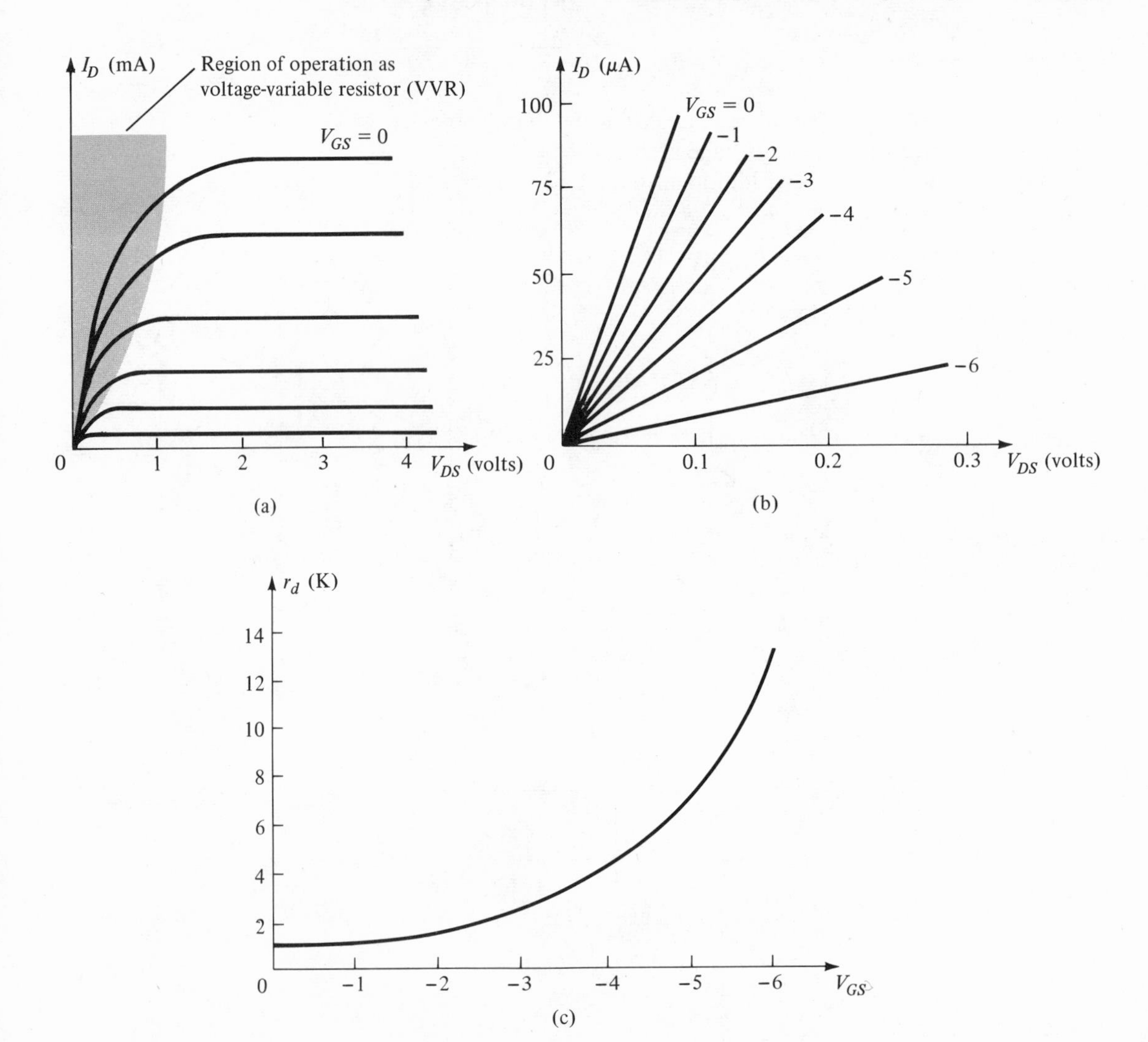

FIG. 6.30
(a) and (b) operating action of FET as VVR;
(c) resistance vs. control voltage.

hundred millivolts. Within this limited operating region the FET can be used as a voltage-variable resistor (VVR).

An enlarged view of the low-level region in which the FET can be used as a VVR is shown in Fig. 6.30b. The slope representing the device resistance is seen to vary as a function of gate-source control voltage. For example, we see that the slope is steepest, and therefore resistance least, for $V_{GS} = 0\ V$ whereas the slope is least, and resistance greatest, for $V_{GS} = -6$ V. A graph of device resistance vs. control voltage obtained from Fig. 6.30b can be made as shown in Fig. 6.30c. Here we also see that the voltage increases with larger control voltage, although not in a linear manner. The change in device resistance is greatest at larger values of gate-source voltage.

Applications of VVR

One common application of a VVR is to vary the gain of an amplifier chain so as to achieve gain control. If this gain control results from a control voltage derived from the output voltage then an automatic gain control (AGC) action is obtained. A simplified circuit diagram for such operation is shown in Fig. 6.31. The ac input signal is applied to an amplifier stage (Q_1) and then to succeeding amplifier stages to obtain an ac output signal. As a means of maintaining the output signal level constant the amplifier gain may be reduced as signal level increases. To achieve this gain control the output ac signal is rectified and filtered, thereby providing a dc voltage whose magnitude increases as the ac signal magnitude increases. This dc voltage is then applied to a FET used as a VVR. Notice that the FET is now used as a resistor, r_d, which effectively is in parallel (for ac operation) with resistor R_e. In this way the resistance of the FET from drain to source acts to vary the effective emitter degeneration resistance seen by ampli-

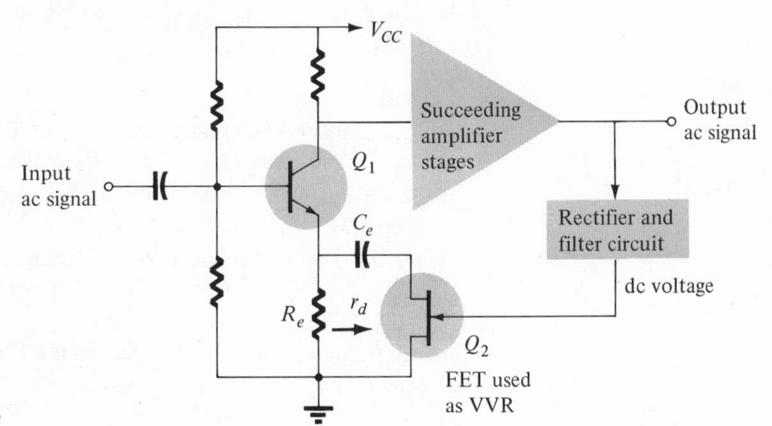

FIG. 6.31
AGC amplifier using FET as VVR.

fier Q_1. Capacitor C_2 serves only to prevent the operation of the VVR from affecting dc bias of stage Q_1.

The gain of transistor Q_1 is then decreased as the output signal level increases. This occurs since the resulting increased dc control voltage to the FET causes its resistance as a VVR to increase, thereby allowing a larger effective emitter degeneration resistance for stage Q_1 and less voltage gain by that stage.

Other applications of the FET as a VVR might include voltage-controlled bandwidth in an LC circuit, electronically tuned RC filter, expander, and compressor circuits in hi-fi, and so on.

PROBLEMS

§ 6.6

1. Calculate the operating point of a FET circuit as in Fig. 6.11a for circuit values $V_{DD} = 12$ V, $R_D = 2$ K, $V_{GG} = 1.5$ V. Use the transistor characteristic of Fig. 6.11b.

2. Calculate the value of R_D needed to obtain a bias operating point $V_{DSQ} = 15$ V, $I_{DQ} = 4$ mA. Use the circuit and characteristic of Fig. 6.12 and select the value of V_{DD}.

3. Using $V_{DD} = 30$ V, and $R_D = 4$ K, determine value of $V_{GG} = V_{GS}$ for operation at $V_{DSQ} = 20$ V. Use circuit and characteristic of Fig. 6.12.

§ 6.7

4. Design a self-bias FET amplifier (as in Fig. 6.15a) using the FET characteristic of Fig. 6.15b for operation at dc bias point. $V_{DSQ} = 12$ V, $I_{DQ} = 4$ mA, using $V_{DD} = 20$ V. Values of R_D and R_S are to be calculated.

5. Using the circuit and characteristic of Fig. 6.17 design a MOSFET circuit (as in Fig. 6.17a) to operate at $I_{DQ} = 1.2$ mA and $V_{DSQ} = 5.0$ V.

6. Using the circuit and characteristic of Fig. 6.18 design a MOSFET circuit (as in Fig. 6.18c) to operate at $I_{DQ} = -2$ mA and $V_{DSQ} = -12$ V.

§ 6.8

7. Calculate the ac voltage gain of a FET amplifier as in Fig. 6.21 for $g_m = 4500$ μmhos, $r_d = 50$ K, $R_D = 40$ K.

8. What value of transistor gain (g_m) is necessary for an amplifier as in Fig. 6.21 to have a gain of $A_v = -40$ if $R_D = 30$ K, $r_d = 60$ K?

9. What value of drain resistor R_D, is required to obtain a circuit gain of $A_v = -25$ for an amplifier as in Fig. 6.21 with $g_m = 6500$ μmhos and $r_d = 35$ K?

10. Calculate the output resistance of the FET amplifiers of Problems 6.7 and 6.8.

11. Calculate the value of drain resistor needed to result in an output impedance of $R_o = 25$ K for a circuit as in Fig. 6.21 having $r_d = 40$ K.

§ 6.9

12. Calculate the input capacitance of a FET amplifier as in Fig. 6.21 having $g_m = 2500$ μmhos, $C_{gs} = 1.5$ pF, $C_{gd} = 1$ pF, $C_{ds} = 5$ pF, $r_d = 40$ K, $R_D = 20$ K, $R_G = 1$ M.
13. Calculate the input impedance of the FET amplifier of Problem 6.12 at frequencies (a) $f = 20$ kHz, and (b) $f = 1$ MHz.
14. Calculate the output voltage of a FET amplifier having voltage gain of $A_v = -80$ at frequencies of (a) 20 kHz, and (b) 1 MHz. Voltage is $V_S = 2$ mV from source impedance $R_S = 7$ K and input impedances are those of Problem 6.13.
15. Calculate the output impedance of the amplifier of Problem 6.12 at frequencies (a) $f = 20$ kHz, and (b) $f = 100$ MHz.

§ 6.10

16. Calculate the input and output impedances and voltage gain of the FET source follower of Fig. 6.23 for circuit values $C_{gs} = 3$ pF, $C_{gd} = 3$ pF, $C_{ds} = 6$ pF, $g_m = 5000$ μmhos, $r_d = 45$ K, $R_G = 1$ M, $R_S = 750\ \Omega$, all at a frequency of $f = 20$ kHz.
17. Calculate input and output impedance and voltage gain of a source follower using the circuit values of Problem 6.16 at $f = 1$ MHz.

§ 6.11

18. Design an amplifier circuit as in Fig. 6.24 using a JFET and circuit values as follows, $V_{DS_{max}} = 20$ V, $V_{GS_{max}} = -20$ V, $V_{GS_{off}} = V_P = -6$ V, $I_{DSS} = 12$ mA, $I_{GSS} = 0.1$ nA, $g_m = 5$ mmhos, $r_d = 40$ K, $C_{iss} = 3.5$ pF, $C_{rss} = 1.3$ pF.
19. Design an amplifier (as in Fig. 6.24) to have a voltage gain of 30 operating from a supply of 18 V. Use the FET specified in Problem 6.18 except for I_{DSS}, which is 4 mA.
20. Select values of R_1, R_2, R_S, C_1, and C_2 for a source-follower circuit (as in Fig. 6.25) using the FET specified in Problem 6.18.

§ 6.12

21. Determine suitable values of resistors R_D and R_G, capacitors C_1, C_2, and voltage supply V_{DD} for a MOSFET circuit as in Fig. 6.26. Some MOSFET data is $g_m = 4500$ μmhos, $r_d = 50$ K, $V_{DS_{max}}$ 12 V. Calculate A_v for the resulting circuit.

§ 6.13

22. Calculate dc bias voltages and input impedance, and select suitable values of C_1 and C_2 for the circuit of Fig. 6.27 for circuit values, $R_1 = 2$ M, $R_2 = 3$ M, $V_{DD} = 15$ V, $R_S = 1.5$ K, $g_m = 4$ mmhos, $r_d = 60$ K, $R_G = 1$ M, $I_{DSS} = 8$ mA, $V_P = -4$ V. Calculate A_v for the resulting circuit.

7

Multistage Systems, Decibels (dB), and Frequency Considerations

7.1 INTRODUCTION

This chapter will include, under the heading of multistage systems, both the *cascaded* and *compound* configurations. The *cascaded* system, for the purposes of this text, is one in which each stage and the connections between each stage are very similar or identical. The *compound* system includes all other possible *multiple* active device configurations, each stage of which can be completely different in appearance with a variety of interconnections.

The first few sections of this chapter examine multistage systems employing the technique of analysis developed in earlier chapters. This is followed by a detailed discussion of decibels (dB) and the effect of frequency on the response of single and multistage systems.

7.2 GENERAL CASCADED SYSTEMS

A discussion of cascaded systems is best initiated by considering the block diagram representation of Fig. 7.1. The quantities of interest are indicated in the figure. The indicated A_v (voltage amplification) and A_i (current amplification) of each stage were determined with all stages connected as indicated in Fig. 7.1. In other words, A_v and A_i of each

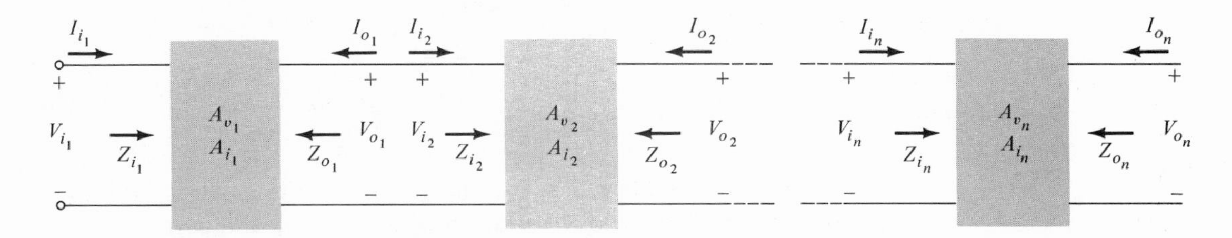

FIG. 7.1
General cascaded system.

stage *do not* represent the gain of each stage on an independent basis. The loading effect of one stage on another was considered when these quantities were determined.

Rather than simply state the result for the overall gain of the system (voltage or current) a simple numerical example will clearly indicate the solution. If $A_{v_1} = -40$ and $A_{v_2} = -50$ with $V_{i_1} = 1$ mV, then $V_{o_1} = A_{v_1} \times V_{i_1} = -40(1 \text{ mV}) = -40$ mV. Since $V_{o_1} = V_{i_2}$,

$$V_{o_2} = A_{v_2} V_{i_2} = -50(-40 \text{ mV}) = 2000 \text{ mV} = 2 \text{ V}$$

The overall gain is $A_{v_T} = 2000 \text{ mV}/1 \text{ mV} = 2000$.

Quite obviously, the total gain of the two stages is simply the product of the individual gains A_{v_1} and A_{v_2}. In general, for n stages,

$$\boxed{A_{v_T} = A_{v_1} A_{v_2} A_{v_3} \cdots A_{v_n}} \tag{7.1}$$

The same is true for the net current gain

$$\boxed{A_{i_T} = A_{i_1} A_{i_2} A_{i_3} \cdots A_{i_n}} \tag{7.2}$$

The input and output impedance of each stage as indicated in Fig. 7.1 are also those values obtained by considering the effects of each and every stage of the system. There is no generally employed equation, such as Eq. (7.2), for the input or output impedances of the system in terms of the individual values. However, in a number of situations (transistor, tube, or FET) the input (or output) impedance can normally be determined to an acceptable degree of accuracy by considering only one, or perhaps two, stages of the system.

The magnitude of the overall voltage gain of the representative system of Fig. 7.1 can be written as

$$|A_{v_T}| = \frac{V_{o_n}}{V_{i_1}} = \frac{I_{o_n} Z_L}{I_{i_1} Z_{i_1}} \tag{7.3a}$$

so that

$$|A_{v_T}| = |A_{i_T}| \frac{Z_L}{Z_{i_1}} \tag{7.3b}$$

Equation (7.3) will prove quite useful in the analysis to follow. To go a step further, if the product of the voltage and current gain is formed,

$$A_{v_T} A_{i_T} = \left(\frac{I_{o_n} Z_L}{I_{i_1} Z_{i_1}}\right)\left(\frac{I_{o_n}}{I_{i_1}}\right) = \frac{I_{o_n}^2 Z_L}{I_{i_1}^2 Z_{i_1}} = \frac{P_o}{P_i}$$

and

$$A_{p_T} = A_{v_T} A_{i_T} \tag{7.4}$$

this being the overall power gain of the system.

There are three types of coupling between stages of a system, such as in Fig. 7.1, that will be considered. The first to be described is the *RC-coupled* amplifier system, which is the most frequently applied of the three. This will be followed by the *transformer* and *direct-coupled* amplifier systems.

7.3 RC-COUPLED AMPLIFIERS

A cascaded RC-coupled transistor amplifier (two-stage) showing typical values and biasing techniques, appears in Fig. 7.2. The terminology "RC-coupled" is derived from the biasing resistors and coupling capacitors employed between stages. Initially, a lengthy, step-by-step procedure of first substituting the equivalent circuit and then writing the

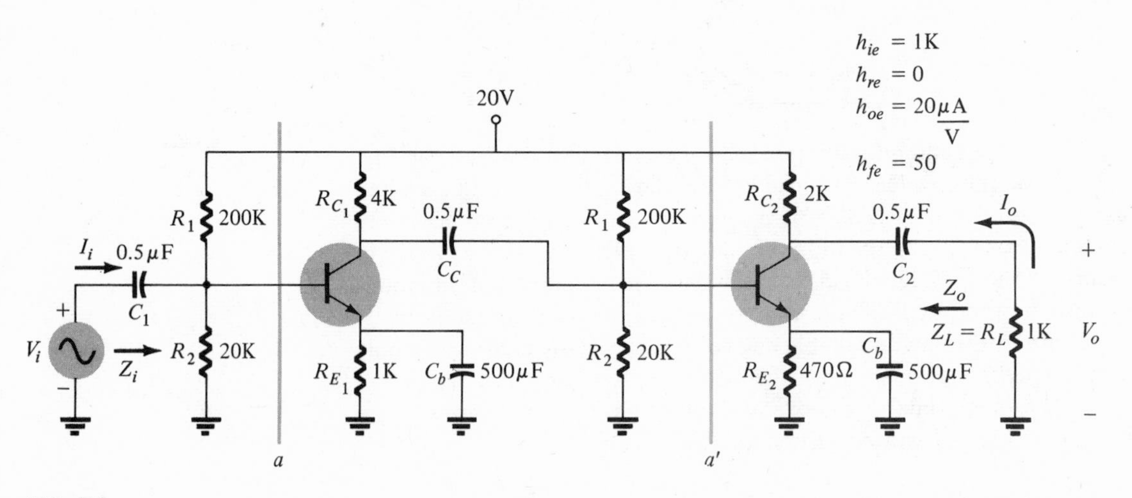

FIG. 7.2
Two-stage RC-coupled amplifier.

necessary equations to determine the required unknowns will be applied. A much shorter approach, employing the approximation techniques of Chapter 5 will then be employed and the results compared. Each approach to the problem is important for reasons described in the text.

If the approximate small-signal ac hybrid equivalent circuit is substituted for each transitor of Fig. 7.2, the network of Fig. 7.3 will

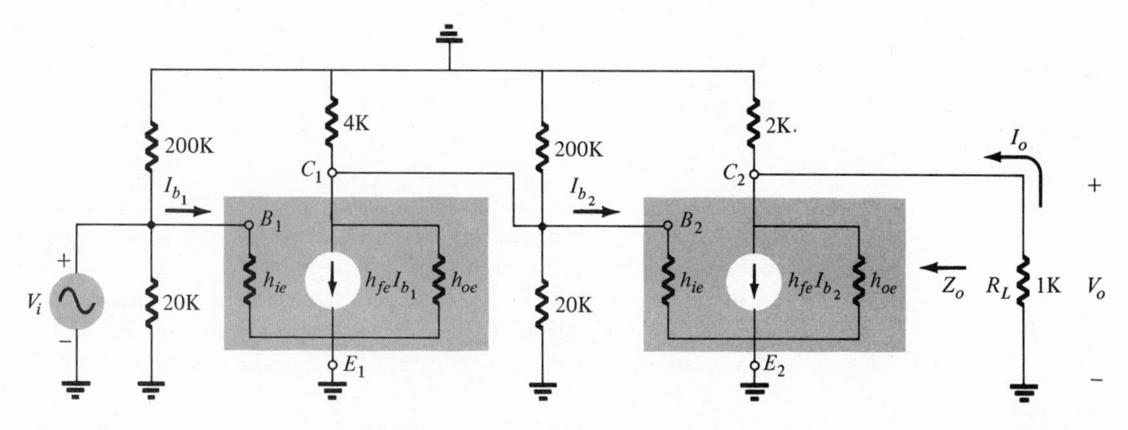

FIG. 7.3
Network of Fig. 7.2 following the substitution of the approximate hybrid equivalent circuit ($h_{re} \cong 0$).

result. Each coupling and bypass capacitor has been replaced by its short-circuit equivalent (very low capacitive impedance at the frequency or frequencies of interest) and the dc levels were eliminated. The circuit is redrawn in Fig. 7.4, and is shown after combining parallel elements in Fig. 7.5.

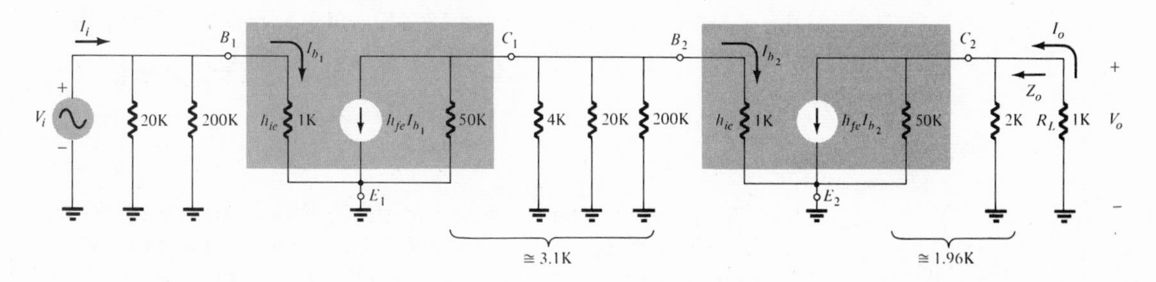

FIG. 7.4
Redrawn network of Fig. 7.3.

Note that some elements were not combined with others to maintain all the unknown quantities and controlling variables (I_{b_1}, I_{b_2}). Although three sketches were required to arrive at the circuit of Fig.

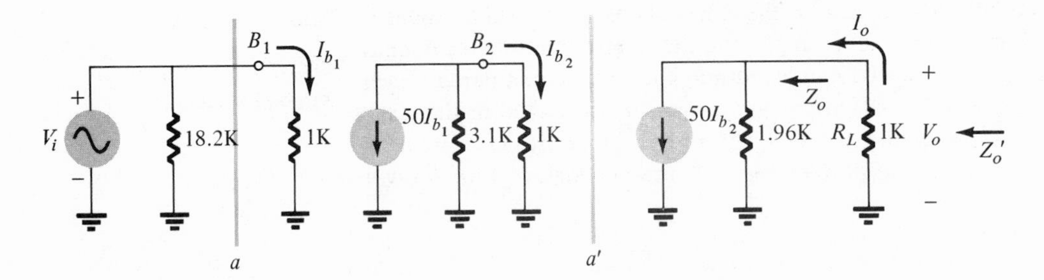

FIG. 7.5

Network of Fig. 7.4 following the combination of parallel elements.

7.5, in time, some of the intermediary steps can be eliminated with little added difficulty or confusion. Initially, however, the student should make every attempt to ensure that the reduced circuit is correct. For an increased number of stages the general appearance of the resulting network is simply an extension of what appears in Fig. 7.5, that is, with additional stages similar to that between the vertical dashed lines *a* and *a'* of Fig. 7.5.

For this system, and those which follow, the voltage and current gain can be found by simply applying one or a combination of the following laws and rules: Kirchhoff's voltage law (KVL), Kirchhoff's current law (KCL), the voltage divider rule (VDR), and the current divider rule (CDR). For the student who feels confident in the use of these four fundamental laws and rules, the subsequent analysis should be relatively easy to follow.

Z_i

The first quantity to be obtained is Z_i. Note that in Fig. 7.5 it is the parallel combination of the 18.2-K and 1-K resistors. Therefore

$$Z_i = 18.2\text{ K}\,||\,1\text{ K} = \frac{(18.2\text{ K})(1\text{ K})}{18.2\text{ K} + 1\text{ K}} = \frac{18.2\text{ K}}{19.2} \cong \mathbf{0.95\text{ K}}$$

Z_o

The output impedance (Z_o) of this system is also relatively easy to determine if it is clearly understood that the output impedance is determined with the input or signal (V_i) set to zero. With $V_i = 0$, I_{b_1}, and hence $h_{fe}I_{b_1}$ and I_{b_2} are also zero. The net result is that $h_{fe}I_{b_2} = 0$, permitting an open-circuit representation for the source and

$$Z_o|_{V_i=0} = \mathbf{2\text{ K}}$$

For the output impedance defined by Z_o' in Fig. 7.5

$$Z_o'|_{V_i=0} = 2\text{ K}\,||\,1\text{ K} = \tfrac{2}{3}\text{ K}$$

A_i

Applying the current divider rule

$$I_{b_1} = \frac{18.2\text{ K} \times I_i}{18.2\text{ K} + 1\text{ K}} = \frac{18.2}{19.2} I_i \cong 0.95 I_i$$

and

$$I_{b_2} = -\frac{3.1\text{ K}(50 I_{b_1})}{3.1\text{ K} + 1\text{ K}} \cong -\frac{155}{4.1} I_{b_1} = -37.9 I_{b_1}$$

and

$$I_o = \frac{1.96\text{ K}(50 I_{b_2})}{1.96\text{ K} + 1\text{ K}} = \frac{98}{2.96} I_{b_2} = 33.1 I_{b_2}$$

The negative sign in the second equation is derived from the fact that the controlled current source and the current desired have conflicting directions. It is important that this sign be properly inserted when A_i has been determined.

Applying the above equations

$$I_o = 33.1 I_{b_2} = 33.1(-37.9 I_{b_1}) = (33.1)(-37.9)(0.95 I_i)$$

and

$$A_i = \frac{I_o}{I_i} \cong \mathbf{-1195}$$

The fact that the result has a negative sign indicates that I_o and I_i *as they appear in Fig. 7.5* are out of phase by 180°. For I_o defined in the opposite direction in Fig. 7.5 a positive sign would result, indicating an *in-phase* relationship.

A_v

The voltage gain can be determined directly from the current gain, since

$$V_o = -I_o R_L = -I_o 1\text{ K}$$

and

$$V_i = I_i Z_i = I_i 0.95\text{ K}$$

so that

$$A_v = \frac{V_o}{V_i} = -\frac{I_o 1\text{ K}}{I_i 0.95\text{ K}} = -A_i \frac{1\text{ K}}{0.95\text{ K}} \quad \text{[by Eq. (7.3b)]}$$

$$= 1195\left(\frac{1}{0.95}\right)$$

and

$$A_v \cong \mathbf{1260} \quad \text{(in phase as defined)}$$

Initially, the analysis of this circuit may appear lengthy, if not difficult. In time, results will develop more quickly with less effort. *Practice, however, is a very necessary element toward this end.*

We will now examine the network in a manner which is theroretically less precise but will, in practice, often lead to results as close to the actual solution as those just obtained. Before we continue, however, a few important facts from Chapter 5 regarding the approximate technique for transistor circuit analysis will be tabulated.

1. $I_c \cong I_e$
2. $I_c \cong h_{fe} I_b$
3. Unless specified otherwise, h_{re}, $h_{oe} \cong 0$, permitting the use of the approximate base-, emitter-, and collector-equivalent circuits as described in Chapter 5.
4. For parallel resistors satisfying a 10:1 ratio or greater, the larger resistor will be eliminated from further investigation unless absolutely required in order to solve for a particular unknown quantity.
5. For grounded emitter transistor configurations,

$$A_v \cong -\frac{h_{fe} R_L}{h_{ie}} \tag{5.28}$$

The primary function of this approximate technique is to obtain a "ball-park" solution with a minimum of time and effort. A reduced time element obviously requires that the network be redrawn a minimum number of times. In fact, let us optimistically say that after the first few examples, many of the problems at the end of this chapter will be solvable working only with the original configuration.

Let us now proceed to find an approximate solution for the quantities required for the circuit of Fig. 7.2, working only with the original network.

Z_i

From past experience with single-stage amplifiers and the analysis just completed, it should be quite clear that for the ac response, both the 20-K and 200-K resistors will appear in parallel if the network is redrawn. Since these two resistors appear in parallel and satisfy the 10:1 ratio the 200-K resistor will no longer be considered a contributing element.

The remaining 20-K resistor is in parallel with the input resistance ($\cong h_{ie}$) of the first transistor since the 1-K resistor (R_{E_1}) is "shorted-out" by the 500-μf (C_b) capacitor.

Quite obviously, then, since the 20-K and 1-K resistors are related by a 20:1 ratio the 20-K resistor can be dropped and

$$Z_i \cong h_{ie} = \mathbf{1\ K}$$

Compared to the 0.95 K value obtained earlier, this is certainly an excellent approximation. Of course, if the parallel resistors did not satisfy the 10:1 ratio the proper equation would have to be applied to determine the combined terminal resistance.

Z_o

Recall that the approximate collector-to-emitter equivalent circuit of a transistor (for h_{re}, $h_{oe} \cong 0$) is simply a current source $h_{fe}I_b$. This being the case, when $V_i = 0$, $h_{fe}I_{b_2} = 0$, and Z_o is simply the 2-K resistor in parallel with the open-circuit representation of the controlled current source. That is,

$$Z_o|_{V_i=0} = \mathbf{2\ K}$$

which equals the value obtained earlier.

A_i

Earlier it was found that both the 200-K and 20-K resistors could be eliminated for the ac response since they appeared in parallel with the input resistance ($\cong h_{ie}$) of the first stage. For this reason $I_{b_1} \cong I_i$.

The collector current of the first stage $I_{c_1} \cong h_{fe}I_{b_1}$. However, I_{c_1} will divide between the 4-K resistor and the *loading* of the second stage (Fig. 7.6). A moment of reflection should indicate that, compared to

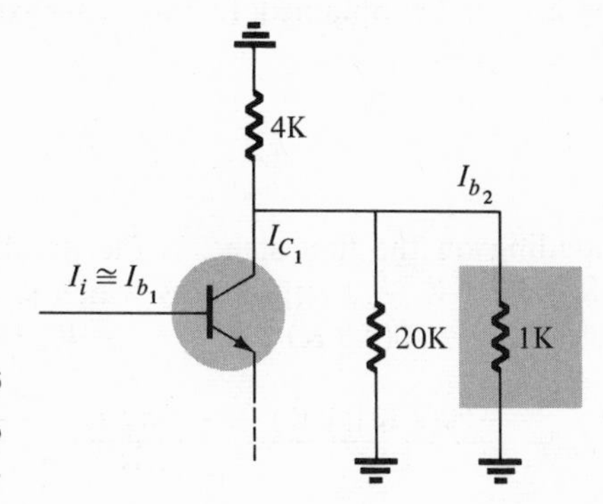

FIG. 7.6

Determining the relationship between I_i and I_{b_2}.

the other parallel elements, the 200-K resistor can be dropped, on an approximate basis, from further consideration. The parallel combination of the 4-K and 20-K resistors will result in an equivalent impedance of $\cong$ 3.33 K.

Applying the current divider rule

$$I_{b_2} = -\frac{3.33\text{ K}I_{c_1}}{3.33\text{ K} + 1\text{ K}} = -\frac{3.33}{4.33}(h_{fe}I_{b_1}) = -0.77(50)I_i$$

and

$$I_{b_2} = -38.5I_i$$

However,

$$I_{c_2} = 50I_{b_2} = 50(-38.5)I_i = -1925I_i$$

and applying the current divider rule once more

$$I_o = \frac{2\text{ K}(I_{c_2})}{2\text{ K} + 1\text{ K}} = \frac{2}{3}(-1925)I_i = -1283I_i$$

and

$$A_i = \frac{I_o}{I_i} \cong \mathbf{-1283}$$

which certainly compares favorably with the value of $A_i = -1195$ obtained earlier.

A_v

A_v can be obtained, as before, using Eq. (7.3). However, to further demonstrate the technique being described it will be obtained assuming A_i is not available.

The direct connection under ac conditions indicates quite clearly in Fig. 7.2, that V_i appears directly at the base of the transistor of the first stage. Since the transistor has a grounded emitter terminal, the ac voltage gain can be obtained (on an approximate basis) using the following equation:

$$A_v \cong \frac{-h_{fe}R_L}{h_{ie}}$$

R_L, the loading on the first stage, is the parallel combination of the 4-K, 20-K, and 1-K (h_{ie}) impedances which is $\cong 4\text{ K} \| 1\text{ K} = 0.8$ K. A_{v_1}, therefore, $= -50(0.8\text{ K})/1$ K or $\cong -40$. For the second stage,

$$A_{v_2} = \frac{-50(2\text{ K} \| 1\text{ K})}{1\text{ K}} = \frac{-50(\frac{2}{3}\text{ K})}{1\text{ K}} = \frac{-100}{3} \cong -33.3$$

The net gain is, therefore,

$$A_{v_T} = A_{v_1}A_{v_2} = (-40)(-33.3)$$

and

$$A_{v_T} \cong \mathbf{1332}$$

which compares very favorably with the previous result, $A_{v_T} = 1260$.

The lengthy discussion required to introduce the approximate technique may tend to raise the question of its validity. Keep in mind, however, that in the approximate method no circuit reductions were necessary, and the number of calculations was minimal. The fact that a great deal of the circuit manipulations and reductions were done mentally requires that a number of problems be approached and completed before the technique can be fully developed and appreciated. The student should become adept at obtaining approximate solutions to single or multistage systems with a minimum of time and effort.

A second example demonstrating each technique follows. It will be less verbal, requiring increased concentration and participation on the part of the student.

EXAMPLE 7.1 We will calculate the input and output impedance, voltage gain, and current gain of the two-stage amplifier of Fig. 7.7. Note that the second stage is an emitter follower configuration. The first approach will be the more lengthy substitution technique, initially applied to the circuit of Fig. 7.2.

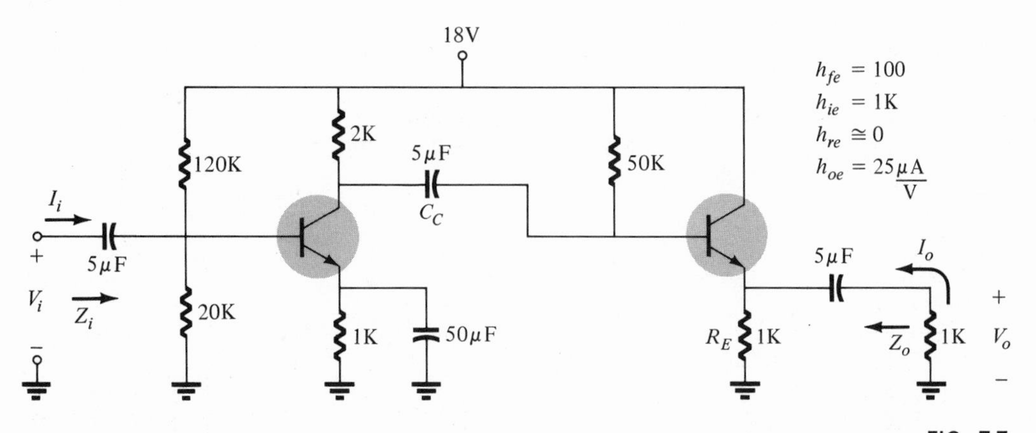

FIG. 7.7
Two-stage transistor network to be examined in detail.

Solution: Z_i**:** Removing the dc level (18 V) and assuming each coupling and bypass capacitor can be replaced by the short-circuit approximation will result in the circuit of Fig. 7.8. The hybrid equivalent

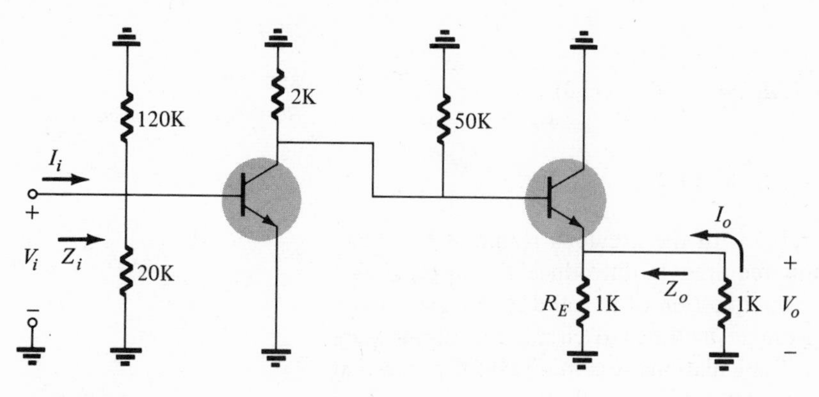

FIG. 7.8

Network of Fig. 7.7 redrawn following the elimination of dc bias levels and replacing all capacitors by the short-circuit equivalent.

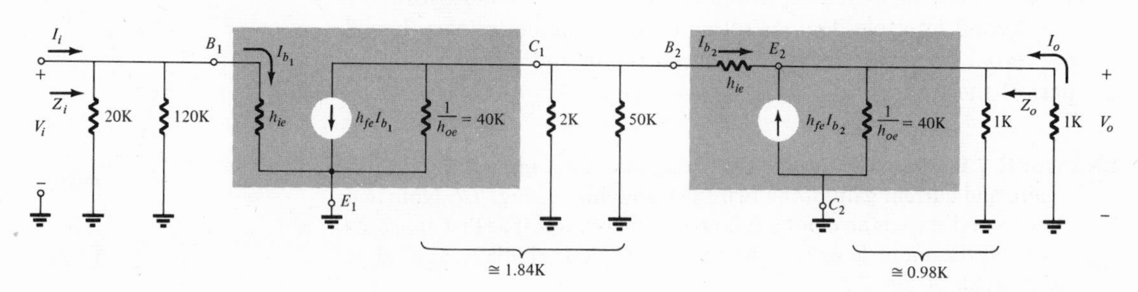

FIG. 7.9

Network of Fig. 7.8 following the substitution of the approximate hybrid equivalent circuit ($h_{re} \cong 0$).

circuit is then substituted as indicated in Fig. 7.9 ($h_{re} \cong 0$). Combining parallel resistors (Fig. 7.10),

$$Z_i = 17.1\text{ K} \,||\, 1\text{ K} = \mathbf{0.945\ K}$$

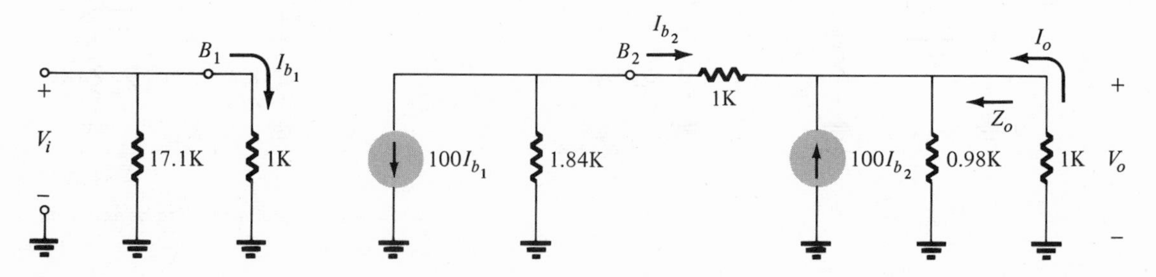

FIG. 7.10

Network of Fig. 7.9 following the combination of parallel elements.

Z_o: With $V_i = 0$, the output circuit of Fig. 7.11 results. It was redrawn as shown to clearly demonstrate that

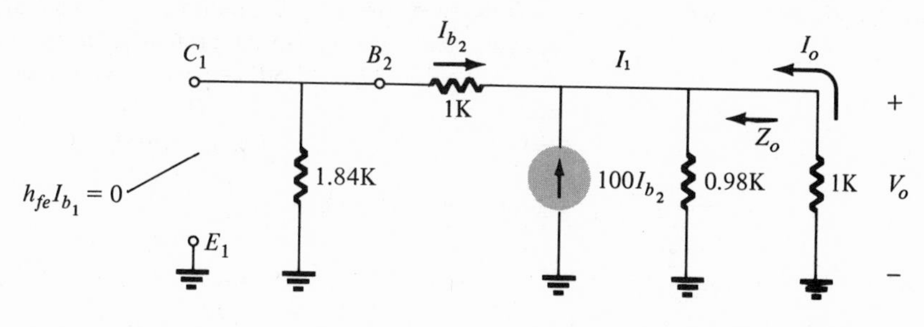

FIG. 7.11
Network of Fig. 7.10 with $V_i = 0$.

$$I_{b_2} = \frac{-V_o}{1\text{ K} + 1.84\text{ K}} = \frac{-V_o}{2.84\text{ K}} \Rightarrow V_o = -I_{b_2}\,2.84\text{ K}$$

In addition, $I_1 = -(I_{b_2} + 100I_{b_2}) = -101I_{b_2}$, and

$$\mathbf{Z}_o' = \frac{V_o}{I_1} = \frac{-I_{b_2}\,2.84\text{ K}}{-101I_{b_2}} \cong 28.1\ \Omega$$

so that

$$Z_o = Z_o' \,||\, 1\text{ K} \cong \mathbf{27.3\ \Omega}$$

This result clearly demonstrates one of the benefits derived from inserting an emitter follower configuration as the last stage of an amplifier: low output impedance.

A_i: From Fig. 7.10,

$$I_{b_1} = \frac{17.1\text{ K}(I_i)}{17.1\text{ K} + 1\text{ K}} = \frac{17.1}{18.1}I_i = 0.945I_i$$

To find the relationship between I_o and I_{b_1}, the output circuit of Fig. 7.10 was redrawn in Fig. 7.12 with the necessary additional

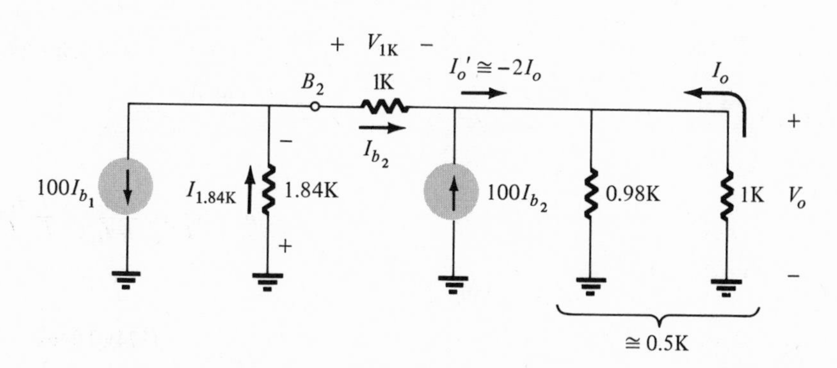

FIG. 7.12
Output circuit of Fig. 7.10 redrawn as an aid in determining A_i.

notation. The solution cannot be obtained by simply applying the current divider rule or Kirchhoff's current law. There must be at least one application of Kirchhoff's voltage law. In general, for networks of this type one *proper* application of Kirchhoff's voltage law *and* current law will result in the desired solution.

Applying Kirchhoff's current law at node a,

$$I_{1.84\text{K}} = 100I_{b_1} + I_{b_2}$$

and Kirchhoff's voltage law,

$$V_{1.84\text{K}} = -(V_{1\text{K}} + V_o)$$

As a general rule, avoid applying this second law around any closed loop including a current source.

Substituting into the above equation for $V_{1.84\text{K}}$

$$(100I_{b_1} + I_{b_2})1.84\text{ K} = -(I_{b_2}\,1\text{ K} + 101\,I_{b_2}\,0.5\text{ K})$$

and

$$184\text{ K}\,I_{b_1} + 1.84\text{ K}I_{b_2} = -I_{b_2}\,1\text{ K} - 50.5\text{ K}I_{b_2}$$

and

$$184I_{b_1} = -(1.84 + 1 + 50.5)I_{b_2}$$

so that

$$I_{b_2} = \frac{-184}{53.34}I_{b_1} \cong -3.45I_{b_1}$$

but

$$-2I_o = 101\,I_{b_2}$$

and

$$I_o = -50.5I_{b_2} = -50.5(-34.5I_{b_1})$$

$$I_o \cong 174\,I_{b_1}$$

so that

$$A_i = \frac{I_o}{I_i} = \frac{I_o}{I_i}\frac{I_{b_1}}{I_{b_1}} = \frac{I_o}{I_{b_1}}\frac{I_{b_1}}{I_i}$$

and

$$A_i = (174)(0.946)$$

$$A_i \cong \mathbf{164}$$

A_v**:**

$$A_v = \frac{V_o}{V_i} = \frac{-I_o Z_L}{I_i Z_i} = -A_i \frac{1\text{ K}}{0.945\text{ K}} = -164(1.06) = \mathbf{-174}$$

Approximate Technique

Z_i**:** The input impedance to the first transistor is approximately $h_{ie} = 1$ K. The 20-K and 200-K resistors in parallel with this impedance can be ignored since they more than satisfy the 10:1 ratio. Therefore,

$$Z_i \cong \mathbf{1\ K} \text{ (compared to 0.945 K obtained earlier)}$$

Z_o**:** Employing the emitter equivalent circuit (Section 5.9) for the second stage will result in the circuit of Fig. 7.13 when $V_i = 0$.

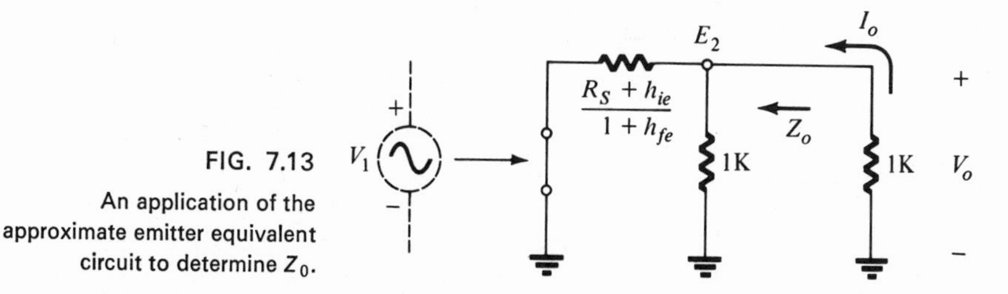

FIG. 7.13
An application of the approximate emitter equivalent circuit to determine Z_0.

To determine Z_o it is only necessary to find R_s. If the concept of the emitter equivalent circuit is the least bit confusing it might be wise to reread the pertinent sections of Chapter 5 before continuing.

When $V_i = 0$ the collector-to-emitter impedance of the first transistor is $\cong \infty\ \Omega$ (open-circuit approximation). The result is the parallel combination of the 2-K and 50-K resistors tied from the base of the second-stage transistor to ground. R_s, therefore, is simply

$$R_s = 50\text{ K} \,||\, 2\text{ K} \cong 2\text{ K}$$

Substituting,

$$Z_o = 1\text{ K} \,||\, \frac{R_s + h_{ie}}{1 + h_{fe}} = \frac{1\text{ K}[(2\text{ K} + 1\text{ K})/101]}{1\text{ K} + [(2\text{ K} + 1\text{ K})/101]} = \frac{1\text{ K}[3\text{K}/101]}{1\text{ K} + [3\text{ K}/101]}$$

$$\cong \frac{0.03\text{ K}}{1.03} \cong \mathbf{29\ \Omega}$$

which compares *very favorably* with the 27.3 Ω obtained earlier.

A_i**:** Since $20\text{ K} \,||\, 200\text{ K} \,||\, (Z_{i_1} \cong h_{ie} = 1\text{ K}) \cong 1$ K, we find that $I_{b_1} \cong I_i$, and $I_{c_1} \cong h_{fe} I_{b_1} \cong h_{fe} I_i = 100\, I_i$.

For the emitter follower stage, the input resistance, as demonstrated in Chapter 5, is $\cong h_{fe} R_E$. This will result in the configuration of Fig. 7.14a to determine the magnitude of I_{b_2}. Applying the current divider rule

$$I_2 = \frac{2\text{ K}(I_{c_1})}{2\text{ K} + 25\text{ K}} = \frac{2}{27}(I_{c_1}) \cong 0.074 I_{c_1}$$

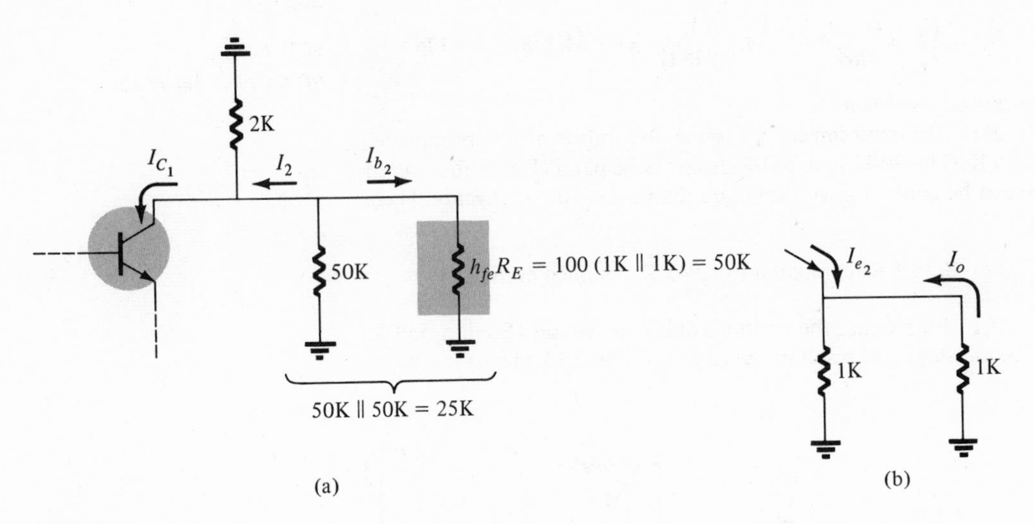

FIG. 7.14

(a) Circuit employed in determining I_{b_2}; (b) determining the relationships between I_{e_2} and I_o.

and I_{b_2}

$$I_{b_2} = -\frac{I_2}{2} = -\frac{0.074I_{c_1}}{2}$$
$$= -0.037I_{c_1} = -0.037(100I_i) = -3.7I_i$$

and

$$I_{e_2} \cong I_{c_2} \cong h_{fe}I_{b_2} = 100(-3.7I_i) = -370I_i$$

Finally

$$I_o = -\frac{I_{e_2}}{2}\text{(Fig. 7.14b)} = \frac{370I_i}{2} = 185I_i$$

so that

$$A_i = \frac{I_o}{I_i} \cong \mathbf{185}$$

which also compares favorably with the value of 164 obtained earlier.

A_v: To further demonstrate the technique, A_v will again be calculated through an analysis of the network rather than by employing Eq. (7.3b) in the first solution.

For the first stage (grounded emitter for the ac response):

$$A_{v_1} \cong -\frac{h_{fe}R_{L_1}}{h_{ie}} = \frac{-100(2\text{ K}\,||\,50\text{ K}\,||\,50\text{ K})}{1\text{ K}}$$
$$\cong \frac{-100(2\text{ K})}{1\text{ K}} = -200$$

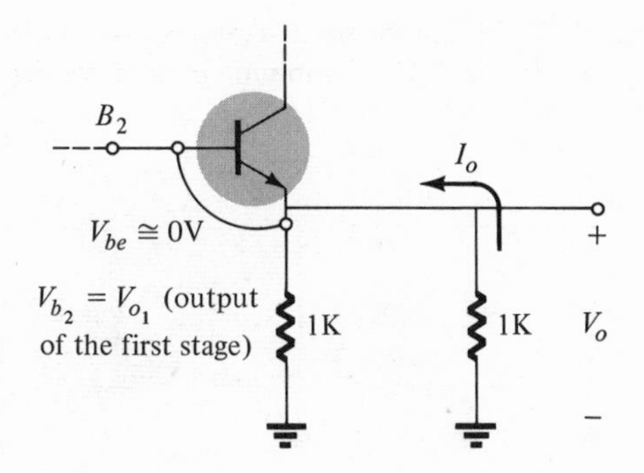

FIG. 7.15
Circuit aiding in the determination of A_v.

For the second stage, if we apply $V_{be_2} \cong 0$ V (Fig. 7.15), then $V_o = V_{e_2} = V_{b_2} = V_{o_1}$, and the overall gain is

$$A_v = \frac{V_o}{V_i} = \frac{V_{o1}}{V_i} \cong -200$$

This also compares reasonably well with the value of −174 obtained earlier.

The student should now take a moment to reflect and compare solutions. It should be quite obvious that the unnecessary redrawing of networks in the second solution, along with the reduced number of calculations, makes this method very useful when quick, approximate, solutions are desired. Keep in mind, however, that the first technique can always be applied if there is some doubt about the validity of an approximation.

EXAMPLE 7.2 *FET RC-Coupled Amplifier.* RC coupling is not limited to transistor stages, as Fig. 7.16 indicates. Priorities require that this circuit be analyzed using only the longer technique.

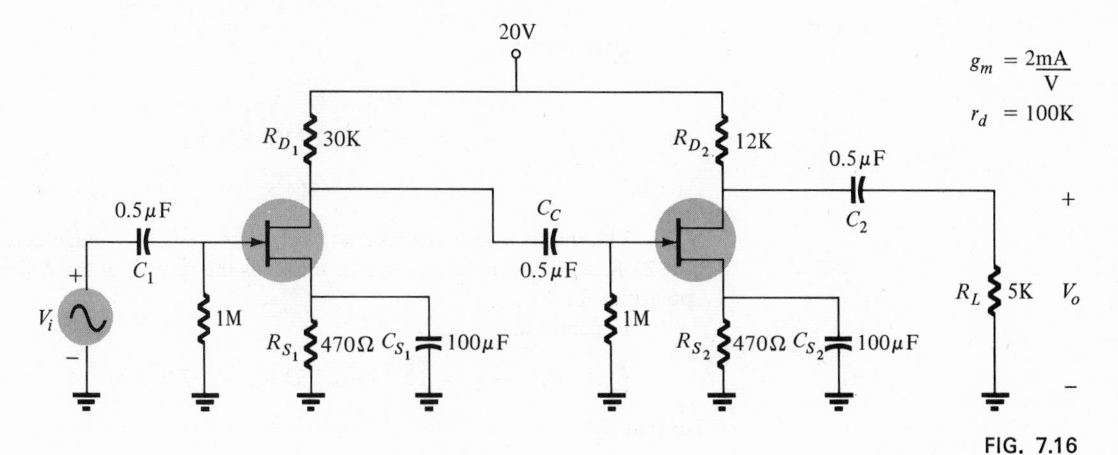

FIG. 7.16
Two-stage FET amplifier.

Solution: Substituting the small-signal equivalent circuit results in the configuration of Fig. 7.17. Combining parallel elements and eliminating

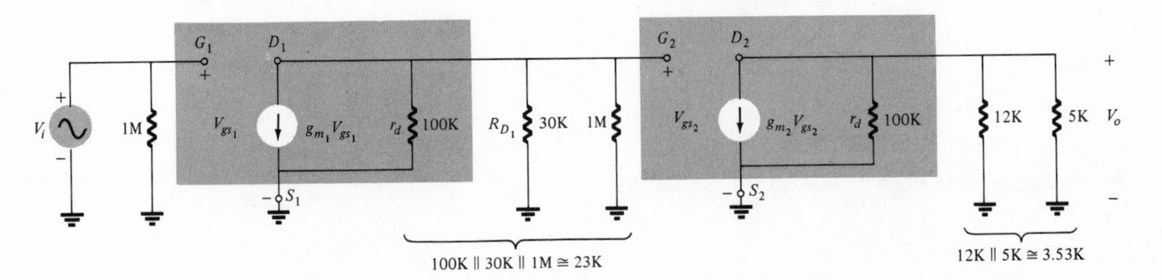

FIG. 7.17
Network of Fig. 7.16 following the substitution of the small-signal ac equivalent circuits.

those having no effect on the desired overall voltage gain will result in Fig. 7.18.

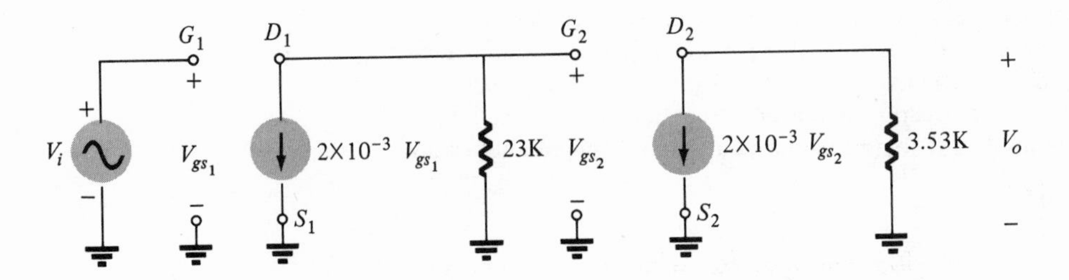

FIG. 7.18
Network of Fig. 7.17 following the combination of parallel elements.

Quite obviously,

$$V_{gs_1} = V_i$$

and
$$V_{gs_2} = -(2 \times 10^{-3} V_{gs_1})(23 \text{ K})$$

so that
$$V_{gs_2} = -46 V_{gs_1}$$

The minus sign indicates that the polarity of the voltage across the 23-K resistor due to the current source is the reverse of the defined polarities for V_{gs_2}.

In conclusion:

$$V_o = -(2 \times 10^{-3} V_{gs_2})(3.53 \text{ K}) = -7.06 V_{gs_2}$$

so that

$$V_o = -7.06 V_{gs_2} = -7.06(-46 V_{gs_1}) = 324.8 V_{gs_1} = 324.8 V_i$$

and

$$A_v = \frac{V_o}{V_i} = \mathbf{32.48}$$

7.4 TRANSFORMER-COUPLED TRANSISTOR AMPLIFIERS

A two-stage transformer-coupled transistor amplifier is shown in Fig. 7.19. Note that step-down transformers are employed between stages while a step-up transformer is connected to the source V_i. The step-up transformer increases the signal level while the step-down transformer

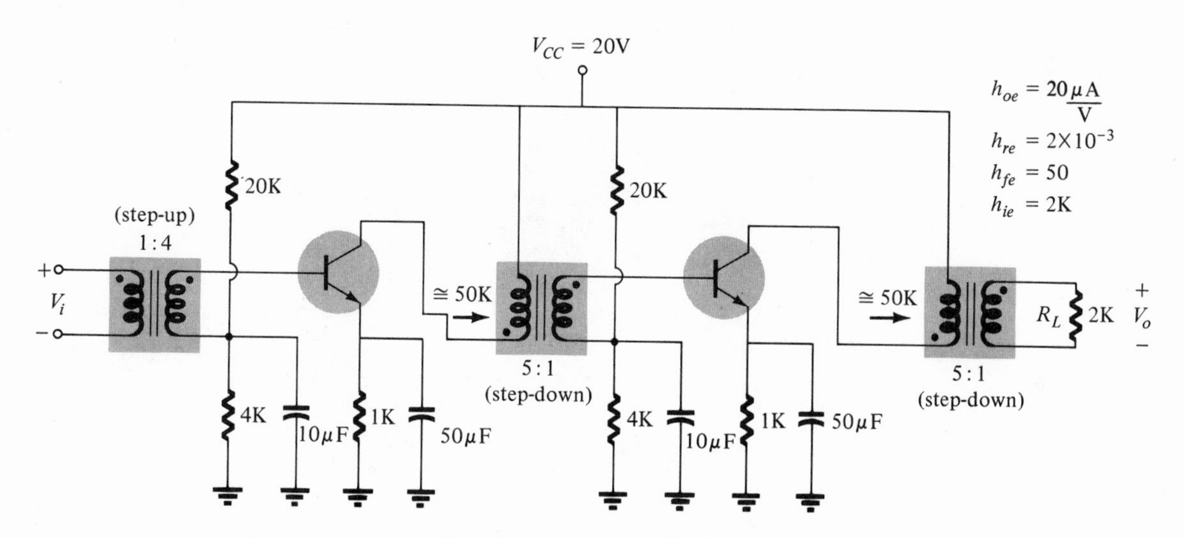

FIG. 7.19
Two-stage transformer-coupled transistor amplifier.

matches, as closely as possible, the loading of each stage to the output impedance of the preceeding stage. This is done in an effort to be as close to maximum power transfer conditions as possible. The effect of this matching technique through the use of transformer coupling will be clearly demonstrated in the following analysis.

Recall that a coupling capacitor was inserted to prevent any dc levels of one stage from affecting the bias conditions of another stage. The transformer provides this dc isolation very nicely.

The basic operation of this circuit is somewhat more efficient than the RC-coupled transistors due to the low dc resistance of the collector circuit of the transformer coupled system. The primary resistance of the transformer is seldom more than a few ohms as compared to the large collector resistance R_C of the RC-coupled system. This lower dc resistance results in a lower dc power loss under operating conditions. The efficiency, as determined by the ratio of the ac power out to the dc power in, is therefore somewhat improved.

There are some decided disadvantages, however, to the transformer-coupled system. The most obvious is the increased size of such a system (due to the transformers) compared to RC-coupled stages. The second is a poorer frequency response due to the newly introduced reactive elements (inductance of coils and capacitance between turns). A third consideration, frequently an important one, is the increased cost of the transformer-coupled, (as compared to the RC-coupled) system.

Before we consider the ac response of the system, the fundamental equations related to transformer action must be reviewed. For the configuration of Fig. 7.20:

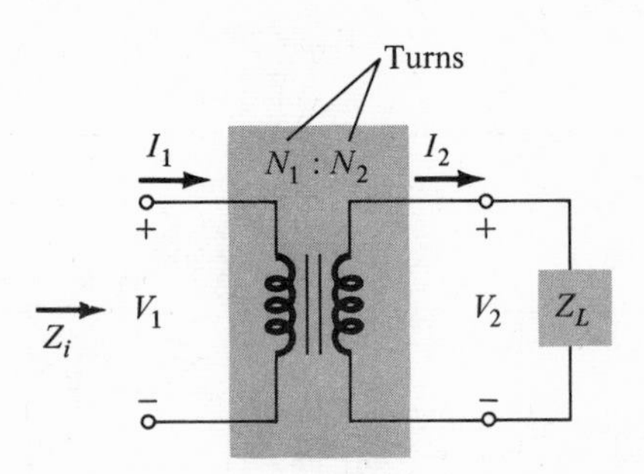

FIG. 7.20
Basic transformer configuration.

$$\frac{V_1}{V_2} = \frac{N_1}{N_2} = a \left(\begin{matrix}\text{transforma-}\\\text{tion ratio}\end{matrix}\right) \tag{7.5a}$$

$$\frac{I_1}{I_2} = \frac{N_2}{N_1} = \frac{1}{a} \tag{7.5b}$$

and

$$Z_i = a^2 Z_L \tag{7.5c}$$

which states, in words, that the input impedance of a transformer is equal to the turns ratio squared times the load impedance.

For the ac response, the circuit of Fig. 7.19 will appear as shown in Fig. 7.21. For maximum power transfer, the impedances Z_2 and Z_4 should be equal to the output impedance of each transistor: $Z_o \cong 1/h_{oe} = 1/20\ \mu$ mhos $= 50$ K. Applying $Z_i = a^2 Z_L$, $Z_4 = a^2 R_L = (5)^2\ 2\text{ K} = 50$ K. Z_2 is also 50 K since the input resistance to each stage

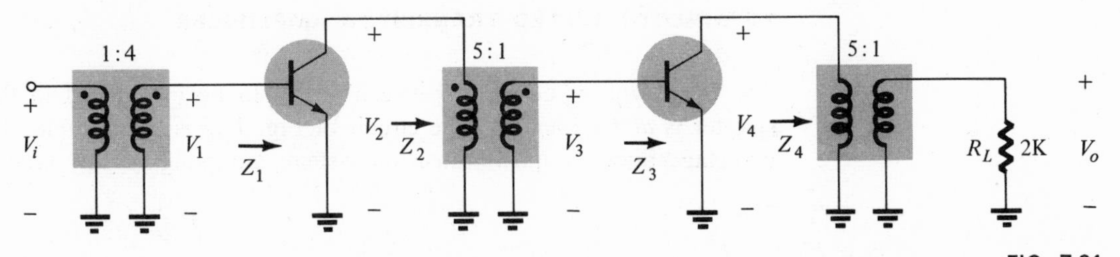

FIG. 7.21

Cascaded transformer-coupled amplifiers of Fig. 7.19 redrawn to determine the small-signal ac response.

(Z_1 and Z_3) is $\cong h_{ie} = 2$ K. Frequency considerations may not always permit Z_2 or Z_4 to be equal to $1/h_{oe}$. For situations of this type Z_2 and Z_4 are usually made as close as possible to $1/h_{oe}$ in magnitude.

Further analysis of the circuit of Fig. 7.21 result, in

$$V_1 = \frac{N_2}{N_1}V_i = 4V_i$$

and

$$A_{v_1} = \frac{-h_{fe}Z_L}{h_{ie}} = \frac{-h_{fe}(\cong 1/h_{oe} \,||\, Z_2)}{h_{ie}} = \frac{-50(50\text{ K} \,||\, 50\text{ K})}{2\text{ K}} = -625$$

so that

$$\begin{aligned} V_2 &= -625V_1 = -625(4V_i) \\ &= -2500V_i \end{aligned}$$

but

$$V_3 = \frac{N_2}{N_1}V_2 = \frac{1}{5}V_2 = \frac{1}{5}(-2500V_i) = -500V_i$$

and

$$A_{v_2} = \frac{-h_{fe}Z_L}{h_{ie}} = \frac{-(50)(25\text{ K})}{2\text{ K}} = -625 = \frac{V_4}{V_3}$$

so

$$\begin{aligned} V_4 &= -625V_3 = -625(-500V_i) \\ &= +312.5 \times 10^3 V_i \end{aligned}$$

with

$$V_L = \frac{1}{5}V_4 = \frac{1}{5}(312.5 \times 10^3 V_i)$$

and

$$A_{v_T} = \frac{V_L}{V_i} = \mathbf{62.5 \times 10^3}$$

7.5 DIRECT-COUPLED TRANSISTOR AMPLIFIERS

The third type of coupling between stages to be introduced in this chapter is *direct coupling*. The circuit of Fig. 7.22 is an example of a two-stage direct-coupled transistor system. Coupling of this type is

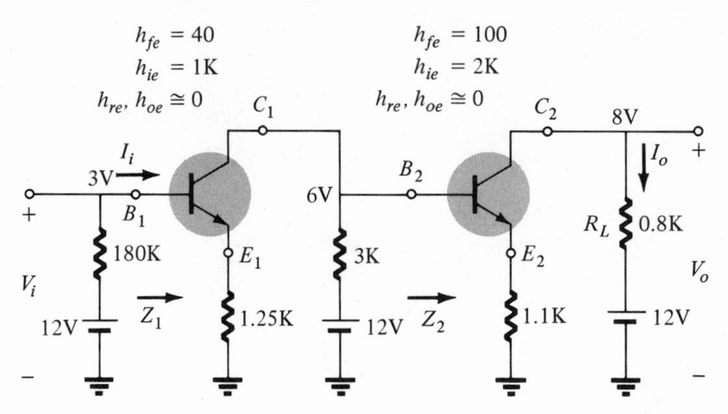

FIG. 7.22
Direct-coupled transistor stages.

necessary for very low-frequency applications. For a configuration of this type the dc levels of one stage are obviously related to the dc levels of the other stages of the system. For this reason the biasing arrangement must be designed for the entire network rather than for each stage independently. Although three separate 12-V supplies are indicated, only one is required if the three terminals of higher (positive) potential for each supply are paralleled.

One of the biggest problems associated with direct-coupled networks is stability. Any variation in dc level in one stage is transmitted on an amplified basis to the other stages. The addition of the emitter resistor aids as a stabilizing element in each stage.

(dc) Bias Conditions

For an output voltage $V_{C_2} = 8$ V as indicated in Fig. 7.22,

$$I_{0.8K} = \frac{4}{0.8\text{ K}} = 5\text{ mA}$$

therefore

$$I_{C_2} \cong I_{E_2} \cong 5\text{ mA}$$

and

$$V_{E_2} = (5\text{ mA})(1.1\text{ K}) = 5.5\text{ V}$$

For

$$V_{BE_2} = 0.5 \text{ V}$$
$$V_{B_2} = V_{C_1} = 5.5 + 0.5 = 6 \text{ V}$$

as indicated. Applying

$$I_C \cong h_{FE} I_B$$
$$I_{B_2} \cong \frac{I_{C_2}}{h_{FE}} = \frac{5 \text{ mA}}{100} = 50 \ \mu\text{A}$$

and

$$I_{3\text{K}} = \frac{6}{3 \text{ K}} = 2 \text{ mA}$$

since

$$I_{3\text{K}} \gg I_{B_2}$$

assume

$$I_{C_1} = I_{3\text{K}} = 2 \text{ mA}$$

and

$$I_{E_1} = 2 \text{ mA}$$

so

$$V_{E_1} = (2 \text{ mA})(1.25 \text{ K}) = 2.5 \text{ V}$$

then

$$V_{B_1} = V_{E_1} + V_{BE_1} = 2.5 + 0.5 = 3 \text{ V}$$

as indicated. Applying

$$I_{C_1} \cong h_{FE} I_{B_1}$$
$$I_{B_1} \cong \frac{I_{C_1}}{h_{FE}} = \frac{2 \text{ mA}}{40} = 50 \ \mu\text{A}$$

and

$$V_{B_1} = 12 - (50 \times 10^{-6})(180 \text{ K}) = 3 \text{ V}$$

as indicated.

The above verification of the potential levels appearing in Fig. 7.22 demonstrates quite clearly the close tie-in required between bias levels of a direct-coupled amplifier.

Now for the ac response. The approach uses the approximate technique introduced earlier in this chapter.

The input impedance to each emitter follower configuration is $\cong h_{fe}R_E$. Therefore,

$$Z_1 \cong h_{fe}R_E = 40(1.25\text{ K}) = 50\text{ K}$$

and

$$Z_2 \cong h_{fe}R_E = 100(1.1\text{ K}) = 110\text{ K}$$

from which

$$I_{b_1} \cong \frac{V_i}{Z_1} = \frac{V_i}{50\text{ K}}$$

and

$$I_{c_1} \cong h_{fe}I_{b_1} = \frac{40}{50\text{ K}}(V_i)$$

Applying the current divider rule

$$I_{b_2} \cong \frac{3\text{ K}(I_{c_1})}{3\text{ K} + Z_2} = \frac{3\text{ K}[(40/50\text{ K})V_i]}{3\text{ K} + 110\text{ K}} = \frac{120V_i}{50(113\text{ K})} = \frac{12}{565\text{ K}}(V_i)$$

so that

$$I_{c_2} \cong h_{fe}I_{b_2} = 100\left[\frac{12}{565\text{ K}}(V_i)\right] = \frac{1200}{565\text{ K}}V_i$$

and

$$V_o = I_{c_2}R_L = \left(\frac{1200V_i}{565\text{ K}}\right)(0.8\text{ K})$$

with

$$A_v = \frac{V_o}{V_i} = \frac{0.8(1200)}{565} \cong \mathbf{1.7}$$

The current gain is

$$A_i = \frac{I_0}{I_i} = \frac{V_o/Z_L}{V_i/Z_1} = \frac{V_oZ_1}{V_iZ_L} = A_v\frac{Z_1}{Z_L} = (1.7)\left(\frac{50\text{ K}}{0.8\text{ K}}\right) = \mathbf{106}$$

The power gain is

$$A_p = A_v A_i = (1.7)(106) = \mathbf{180}$$

The low voltage gain is (to a large part) due to the lack of a bypass capacitor for the emitter resistors. The effect of having these capacitors on Z_1 and Z_2 (both then $\cong h_{ie}$) and the voltage gain is quite obvious. The higher current gain results in a reasonable power gain for the system.

7.6 CASCODE AMPLIFIER

The cascode amplifier is a two-stage compound system having the transistor configuration of Fig. 7.23. Note that the first stage is the common-emitter configuration while the second is the common-base configuration. In practice, the first stage is designed to have a low voltage gain so that the input capacitance (C_{input}) of the first stage has very low value. The interest in C_{input} is derived from high-frequency considerations to be introduced in Section 12 of this chapter. The effect of A_{v_1} the voltage gain, on the input capacitance will also be considered. For now, we will simply find the current gain and voltage gain using the approximate technique discussed in the previous section.

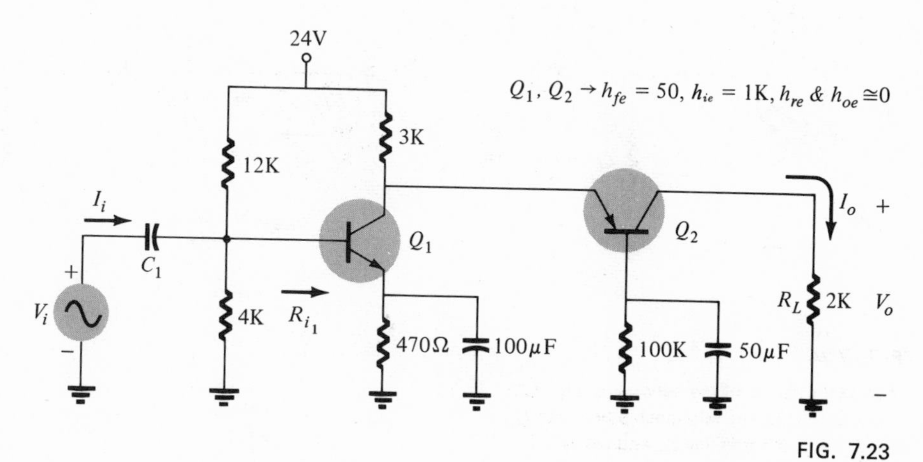

FIG. 7.23
Cascode configuration.

***A*$_i$:**

For the ac response $12\text{ K} \| 4\text{ K} = 3\text{ K}$ and $R_{i_1} \cong h_{ie} = 1\text{ K}$. Therefore,

$$I_{b_1} = \frac{3\,\text{K}(I_i)}{3\,\text{K} + 1\,\text{K}} = 0.75I_i$$

with

$$I_{c_1} \cong h_{fe}I_{b_1} = 50(0.75I_i)$$

and

$$I_{c_1} \cong 37.5I_i$$

In order to find I_{e_2}, the input resistance R_{i_2} of the common-base amplifier must be found (Fig. 7.24a). To find R_{i_2} the common-emitter hybrid parameter equivalent circuit has been substituted for the common-base transistor (Fig. 7.24b). Note in this figure that h_{ie} still exists

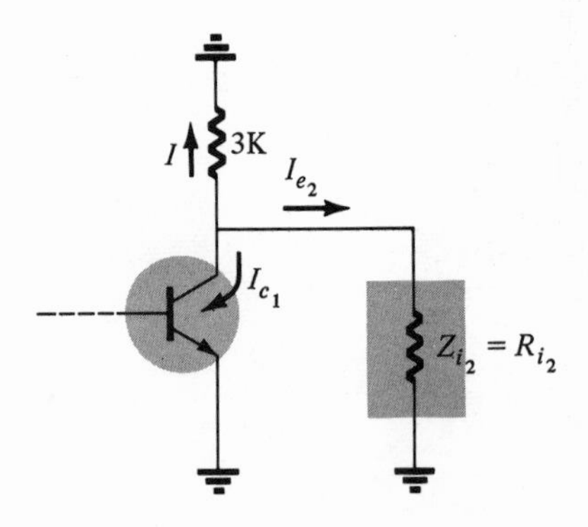

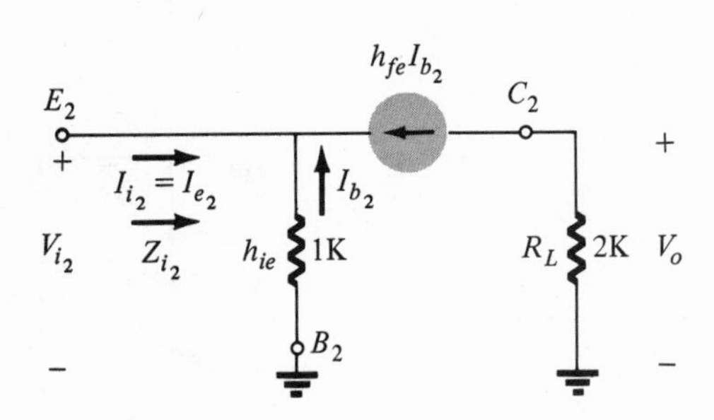

FIG. 7.24
Redrawn portions of the network of Fig. 7.23: (a) to determine the relationship between I_{e_2} and I_{c_1}; (b) to determine Z_{i_2} and aid in calculating the overall voltage gain.

between the base and emitter terminals with I_{b_2} defined from base to emitter. The controlled current source is, as before, between the collector and emitter terminals with the prescribed direction. Quite obviously

$$I_{i_2} = -(I_{b_2} + h_{fe}I_{b_2}) = -(1 + h_{fe})I_{b_2}$$

However,

$$I_{b_2} = -\frac{V_{i_2}}{h_{ie}}$$

Therefore

$$I_{i_2} = -(1 + h_{fe})I_{b_2} = -(1 + h_{fe})\left(-\frac{V_{i_2}}{h_{ie}}\right) = \frac{(1 + h_{fe})}{h_{ie}}V_{i_2}$$

so that

$$Z_{i_2} = \frac{V_{i_2}}{I_{i_2}} \cong \frac{h_{ie}}{1 + h_{fe}} \tag{7.6}$$

which is a general equation for the approximate input impedance of a common-base amplifier.

Substituting numerical values

$$Z_{i_2} = \frac{1\text{ K}}{1 + 50} \cong 20\ \Omega$$

This low input impedance is characteristic of the common-base configuration. It will affect a low voltage gain for the first stage to be determined shortly. To continue (Fig. 7.24a)

$$I_{e_2} = -\frac{3\text{ K}(I_{c_1})}{3\text{ K} + Z_{i_2}} = -\frac{3\text{ K}(I_{c_1})}{3\text{ K} + 20\ \Omega} \cong -I_{c_1}$$

and

$$I_{c_2} \cong I_{e_2} = I_o$$

so that

$$A_i = \frac{I_o}{I_i} = \frac{I_{c2}}{I_i} \cong -\frac{I_{c1}}{I_i} = -\frac{37.5 I_i}{I_i} = \mathbf{-37.5}$$

***A_v*:**

$$A_{v_1} \cong -\frac{h_{fe}R_L}{h_{ie}} = -\frac{h_{fe}(R_{C_1} || R_{i_2})}{h_{ie}} = -\frac{50(\cong 20\ \Omega)}{1\text{ K}} \cong -1$$

(This low voltage gain will result in a low value for C_i, the input capacitance to the system.)

For the gain of the common-base configuration we can make use of the circuit of Fig. 7.24b where

$$V_{i_2} = V_{o_1}$$

and

$$V_o = -h_{fe} I_{b_2} R_L$$

but

$$I_{b_2} = -\frac{V_{i_2}}{h_{ie}}$$

so that

$$V_o = -h_{fe}R_L\left(-\frac{V_{i_2}}{h_{ie}}\right)$$

result in

$$A_{v_2} = \frac{V_o}{V_{i_2}} = \frac{h_{fe}R_L}{h_{ie}} \tag{7.7}$$

which is an equation for the approximate voltage gain for the common-base configuration. The result should appear familiar. It is exactly the same as that obtained earlier for the common-emitter configuration except for the lack of a minus sign, which indicates an in-phase relationship between input and output voltage.

Substituting

$$A_{v_2} = \frac{50(2\text{ K})}{1\text{ K}} = 100$$

and

$$A_v = A_{v_1}A_{v_2} = (-1)(100) = \mathbf{-100}$$

As stated earlier, the common-base configuration will provide the necessary voltage gain.

The cascode amplifier is not limited to transistors. FETs and tubes are also employed in the same configuration to develop a system having an improved frequency response due to the low input capacitance.

7.7 DARLINGTON COMPOUND CONFIGURATION

The Darlington circuit is a compound configuration that results in a set of improved amplifier characteristics. The configuration of Fig. 7.25 has a high input impedance with low output impedance and high current gain, all desirable characteristics for a current amplifier. We shall momentarily see, however, that the voltage gain will be less than one. A variation in the configuration can result in a trade-off between the output impedance and voltage gain.

The description of the biasing arrangement is quite similar to that of a single-stage emitter follower configuration with current feedback (Chapter 4). Note for the Darlington configuration that the emitter current of the first transistor is the base current for the second active device. Obviously the power and current ratings of the second stage must

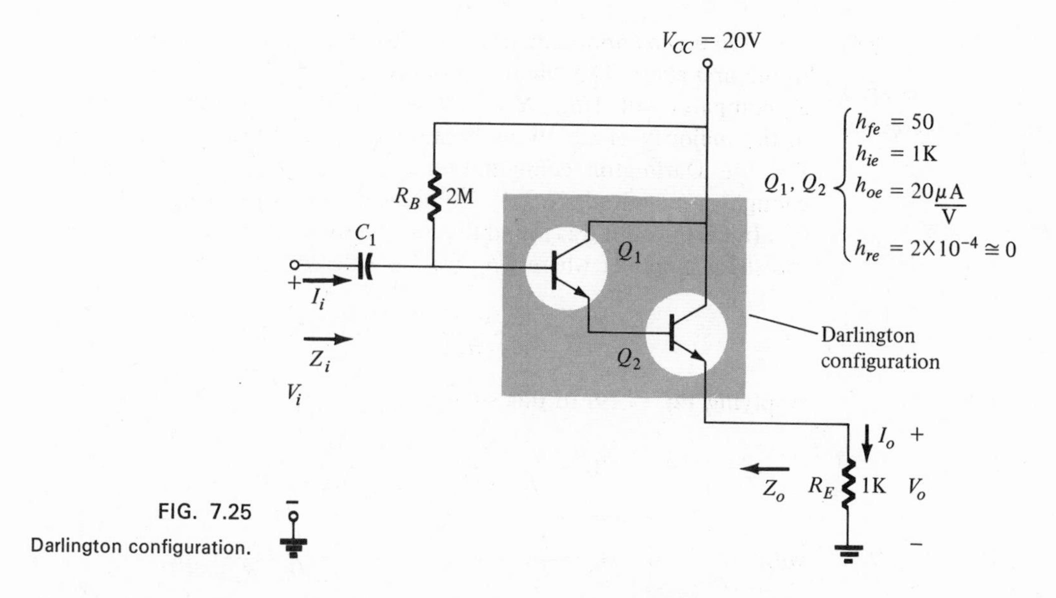

FIG. 7.25
Darlington configuration.

be greater than those of the first stage. There is no reason why each transistor cannot have the same hybrid parameters. This is the chosen case for this example. The capacitor C_1 was introduced to insure that any dc levels associated with V_i do not offset the bias conditions of the network.

In its small-signal ac form, the circuit will appear as shown in Fig. 7.26.

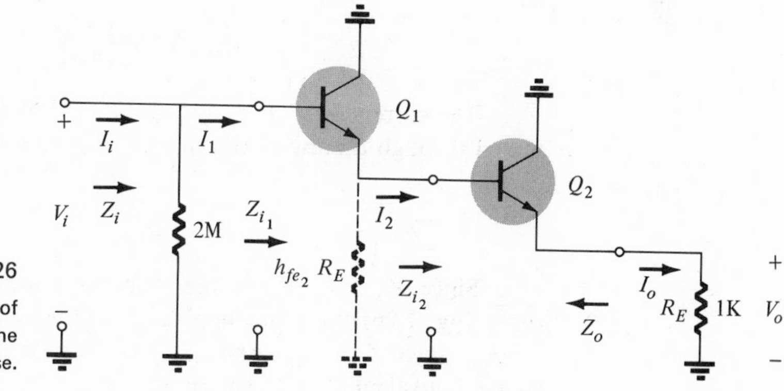

FIG. 7.26
Darlington configuration of Fig. 7.25 redrawn to determine the small-signal ac response.

For the second stage:

$$Z_{i_2} \cong h_{fe} R_E = 50(1\text{ K}) = 50\text{ K}$$

and

$$A_{i_2} = \frac{I_o}{I_2} = \frac{I_{e2}}{I_{b2}} \cong h_{fe2}$$

On a *good* approximate basis, these equations can not be applied to the first stage. The "fly in the ointment" is the closeness with which Z_{i_2} compares with $1/h_{oe_1}$. You will recall that $1/h_{oe_1}$ could be eliminated in the majority of situations because the load impedance $Z_L \ll 1/h_{oe_1}$ For the Darlington configuration the input impedance Z_{i_2} is close enough in magnitude to $1/h_{oe_1}$ to necessitate considering the effects of h_{oe_1}. In Chapter 5 it was found that for the single-stage grounded emitter transistor amplifier where $1/h_{oe}$ was considered,

$$A_i \cong \frac{h_{fe}}{1 + h_{oe}Z_L} \qquad \text{[Eq. (5.19)]}$$

Applying Eq. (5.19) to this situation, $Z_L = Z_{i_2} \cong h_{fe_2}R_E$ and

$$A_{i_1} = \frac{I_2}{I_1} = \frac{I_{c_1}}{I_{b_1}} \cong \frac{h_{fe_1}}{1 + h_{oe_1}(h_{fe_2}R_E)}$$

with

$$A_i = \frac{I_o}{I_1} = A_{i_1}A_{i_2} = \frac{h_{fe_1}h_{fe_2}}{1 + h_{oe_1}(h_{fe_2}R_E)} \tag{7.8}$$

For $h_{fe_1} = h_{fe_2} = h_{fe}$ and $h_{oe_1} = h_{oe_2} = h_{oe}$

$$A_i \cong \frac{h_{fe}^2}{1 + h_{oe}h_{fe}R_E} \tag{7.9}$$

For $h_{oe}h_{fe}R_E < 0.1$ a fairly good approximation (within 10%) is

$$A_i \cong h_{fe}^2 \tag{7.10}$$

The current gain $A_i = I_o/I_i$, as defined by Fig. 7.23, can be determined through the use of the current divider rule

$$I_1 = \frac{R_B I_i}{R_B + Z_{i_1}}$$

Since $Z_{i_2} \cong h_{fe_2}R_E$ is the "emitter resistor" of the first stage (note Fig. 7.26) the input impedance to the first stages is $Z_{i_1} \cong h_{fe_1}(Z_{i_2} \,||\, 1/h_{oe_1})$ since $Z_{i_2} = R_{E_1}$, and $1/h_{oe_1}$ will appear in parallel in the small-signal equivalent circuit. The result is

$$Z_{i_1} \cong h_{fe_1}\left(h_{fe_2}R_E \,\middle\|\, \frac{1}{h_{oe_1}}\right) = \frac{h_{fe_1}h_{fe_2}R_E(1/h_{oe_1})}{h_{fe_2}R_E + 1/h_{oe_1}}$$

and

$$Z_{i_1} = \frac{h_{fe_1}h_{fe_2}R_E}{h_{oe_1}h_{fe_2}R_E + 1} \tag{7.11}$$

which for $h_{fe_1} = h_{fe_2} = h_{fe}$ and $h_{oe_1} = h_{oe_2} = h_{oe}$

$$\boxed{Z_{i_1} \cong \frac{h_{fe}^2 R_E}{1 + h_{oe} h_{fe} R_E}} \tag{7.12}$$

For $h_{ie} h_{fe} R_E < 0.1$

$$\boxed{Z_{i_1} \cong h_{fe}^2 R_E} \tag{7.13}$$

Substituting the parameter values indicated in Fig. 7.25

$$A_i = \frac{I_o}{I_1} \cong \frac{(h_{fe})^2}{1 + h_{oe} h_{fe} R_E} = \frac{(50)^2}{1 + (20 \times 10^{-6})(50)(1\text{ K})}$$
$$= \frac{2500}{1+1} = \mathbf{1250}$$

and

$$Z_{i_1} \cong \frac{h_{fe}^2 R_E}{1 + h_{oe} h_{fe} R_E} = \frac{(50)^2 1\text{K}}{2} = 1250\text{ K} = \mathbf{1.25\text{ M}}$$

so that

$$\frac{I_1}{I_i} = \frac{R_B}{R_B + Z_{i_1}} = \frac{2\text{ M}}{2\text{ M} + 1.25\text{ M}} = \frac{2}{3.25} = 0.615$$

and

$$A_{i_T} = \frac{I_o}{I_i} = \left[\frac{I_o}{I_i}\right]\left[\frac{I_1}{I_1}\right] = \left[\frac{I_o}{I_1}\right]\left[\frac{I_1}{I_i}\right] = A_i \times \frac{I_1}{I_i}$$
$$= (1250)(0.615) = \mathbf{770}$$

with

$$Z_i = 2\text{ M} \,||\, Z_{i_1} = 2\text{ M} \,||\, 1.25\text{ M} = \mathbf{770\text{ K}}$$

Too frequently, the current gain of a Darlington circuit is assumed to be simply $A_i^2 \cong h_{fe}^2$. In this case, $A_i \cong (h_{fe})^2 = 2500$. Certainly, 2500 vs. 1250 is *not* a good approximation. The effect of h_{oe_1} must therefore be considered when the current gain of the first stage is determined.

The output impedance Z_o can be determined, quite directly, from the emitter equivalent circuits (Section 5.9), as follows.

For the first stage

$$Z_{o_1} \cong \frac{R_{s_1} + h_{ie_1}}{1 + h_{fe_1}} \tag{7.14}$$

$$= \frac{0 + 1\text{ K}}{51} \cong \mathbf{19.6\ \Omega}$$

and

$$Z_{o_2} \cong \frac{(Z_{o_1} || 1/h_{oe_1}) + h_{ie_2}}{1 + h_{fe_2}} \tag{7.15}$$

$$= \frac{(19.6\ \Omega || 50\text{ K}) + 1\text{ K}}{51} \cong \frac{19.6\ \Omega + 1\text{ K}}{51} = \frac{1019.6}{51}$$

$$\cong \mathbf{20\ \Omega}$$

Note, as indicated in the introductory discussion, that the input impedance is high, output impedance low, and current gain high. We shall now examine the voltage gain of the system. Applying Kirchhoff's voltage law to the circuit of Fig. 7.25:

$$V_o = V_i - V_{be_1} - V_{be_2}$$

That the output potential is the input *less* the base-to-emitter potential of each transistor indicates quite clearly that $V_o < V_i$. It is closer in magnitude to one than to zero. On an approximate basis ($h_{oe}R_E < 0.1$) it is given by

$$A_v \cong 1 - \frac{h_{ie_2}}{h_{fe_2}R_E} \tag{7.16}$$

Substituting the numerical values of this general example,

$$A_v \cong 1 - \frac{1\text{ K}}{50\text{ K}} = \mathbf{0.98}$$

7.8 CASCADED PENTODE AND TRIODE AMPLIFIERS

Many of the characteristics and general comments made about RC-coupled, transformer-coupled, and direct-coupled transistor amplifiers can also be applied to the pentode or tube amplifier employing the same coupling. Of the three, the RC-coupled pentode or triode amplifier is the most frequently applied, although the transformer-coupled generally result in the highest gain. This latter consideration, however, must be carefully weighed against the cost and desired frequency response.

The RC-coupled pentode amplifier of Fig. 7.27 will now be dis-

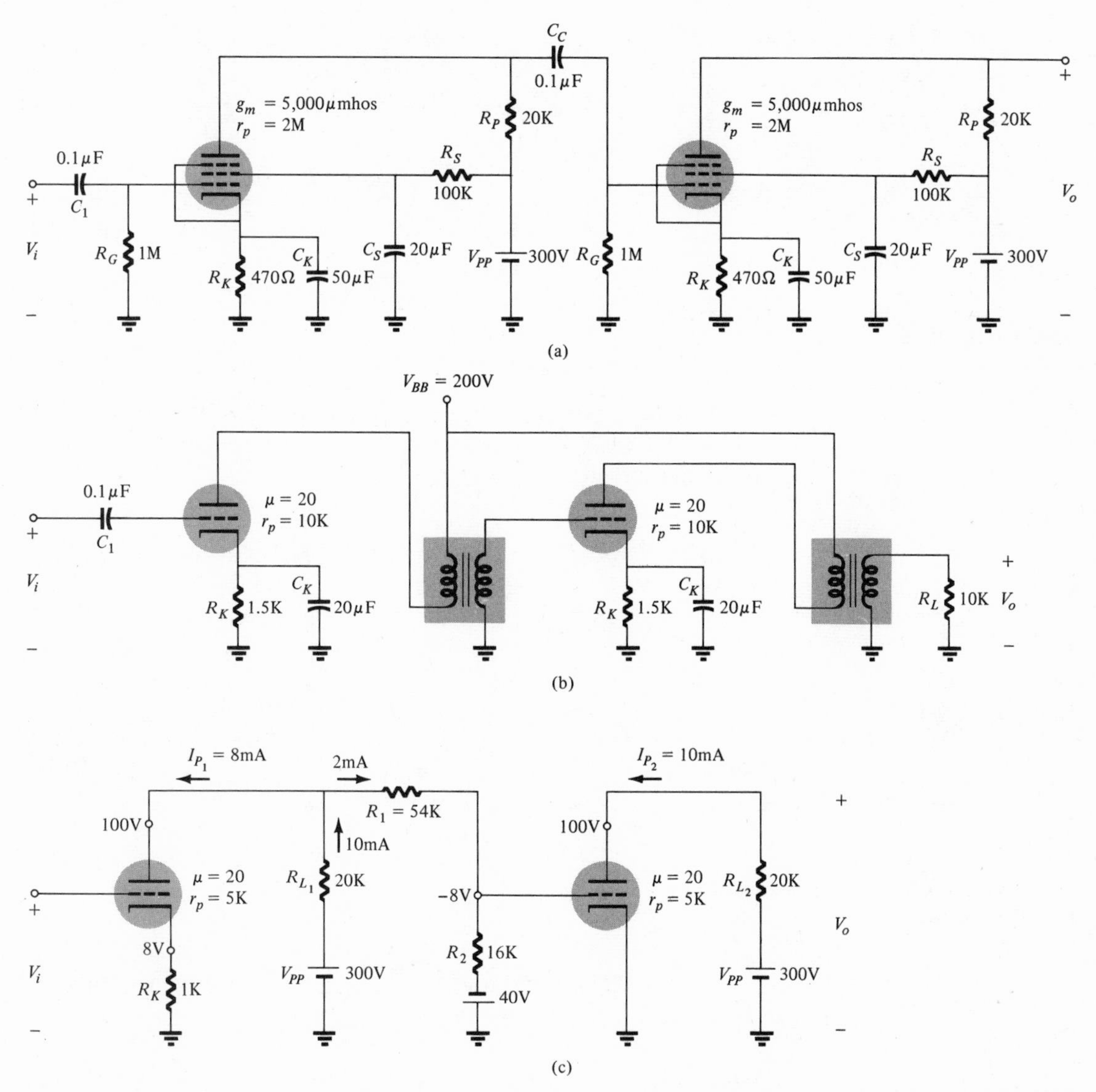

FIG. 7.27
Vacuum-tube multistage systems: (a) RC-coupled; (b) transformer-coupled; (c) direct-coupled.

cussed. The analysis of the remaining transformer and direct-coupled pentode amplifier appearing in the same figure is quite similar to its transistor counterpart and therefore will be left as an exercise for the reader.

Substituting the small-signal ac equivalent circuit for the pentodes of Fig. 7.27a will result in the configuration of Fig. 7.28.

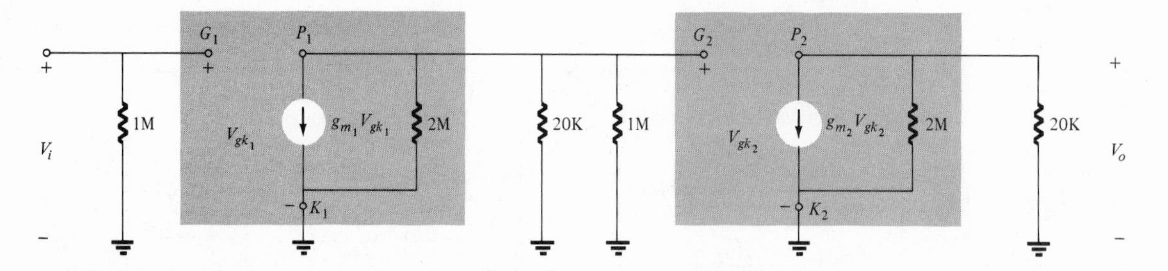

FIG. 7.28

Small signal ac equivalent circuit for the cascaded system of Fig. 7.27a.

Quite obviously, $V_{gk_1} = V_i$, $2\text{ M} \| 1\text{ M} \| 20\text{ K} \cong 20\text{ K}$ and $2\text{ M} \| 20\text{ K} \cong 20\text{ K}$, resulting in the reduced network of Fig. 7.29. From this circuit:

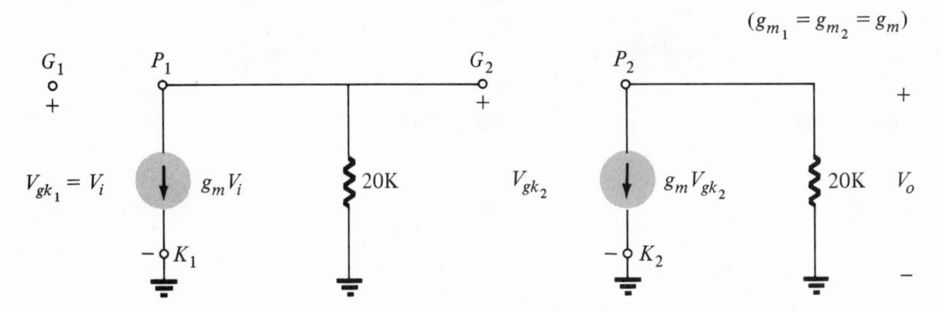

FIG. 7.29

Network of Fig. 7.28 following the combination of parallel elements.

$$V_{gk_2} = V_{20\text{K}} = -(g_m V_i)(20\text{ K}) = -(5000 \times 10^{-6})(20 \times 10^3 V_i)$$
$$= -100V_i$$

and

$$g_m V_{gk_2} = -g_m(100V_i) = -(5000 \times 10^{-6})(10^2 V_i)$$
$$= -0.5V_i$$

so that

$$V_o = -(g_m V_{gk_2})(10\text{ K}) = -(-0.5V_i)(10\text{ K}) = 5000V_i$$

and

$$A_v = \frac{V_o}{V_i} = \mathbf{5000}$$

The concept of the *decibel* (dB) and the associated calculations will become increasingly important in the remaining sections of this chapter. The background surrounding the term decibel has its origin in the old established fact that power and audio levels are related on a logarithmic basis. That is, an increase in power level, say 4 to 16 W, for discussion purposes, does not mean that the audio level will increase by a factor of $16/4 = 4$. It will increase by a factor of 2 as derived from the power of 4 in the following manner: $(4)^2 = 16$. For a change of 4 to 64 W the audio level will increase by a factor of 3 since $(4)^3 = 64$. In logarithmic form, the relationship can be written as

$$\log_4 64 = 3$$

In words, the equation states that the logarithm of 64 to the base 4 is 3. In general: $\log_b a = x$ relates the variables in the same manner as $b^x = a$.

Due to pressures for standardization, the *bel* (B) was defined by the following equation to relate power levels P_1 and P_2:

$$\boxed{\text{bel} = \log_{10}\frac{P_2}{P_1}} \tag{7.17}$$

Note that the common, or base 10, system was chosen to eliminate variability. Although the base is no longer the original power level the equation will result in a basis for comparison of audio levels due to changes in power levels. The term bel was derived from the surname of Alexander Graham Bell.

It was found, however, that the bel was too large a unit of measurement for practical purposes, so the decibel (dB) was defined such that 10 decibels = 1 bel.

Therefore,

$$\#\ \text{dB} = (10)(\#\ \text{bels}) = 10 \log_{10}\frac{P_2}{P_1}$$

and

$$\boxed{\text{dB} = 10 \log_{10}\frac{P_2}{P_1}} \tag{7.18}$$

The terminal rating of electronic communication equipment (amplifiers, microphones, etc.) is commonly rated in decibels. Equation (7.18) indicates quite clearly, however, that the decibel rating is a measure of the difference in magnitude between *two* power levels. For a specified terminal (output) power (P_2) there must be a reference power level (P_1). The reference level is generally accepted to be 1 mW although on occasion the 6 mV standard of earlier years is applied. The resistance to be associated with the 1-mW power level is 600 Ω, chosen because it is the characteristic impedance of audio transmission lines. When the 1-mW level is employed as the reference level the decibel abbreviation frequently appears as dBm. In equation form

$$\text{dBm} = 10 \log_{10} \frac{P_2}{1 \text{ mW}} \bigg|_{600\Omega} \tag{7.19}$$

There exists a second equation for decibels that is applied quite frequently. It can be best described through the circuit of Fig. 7.30a.

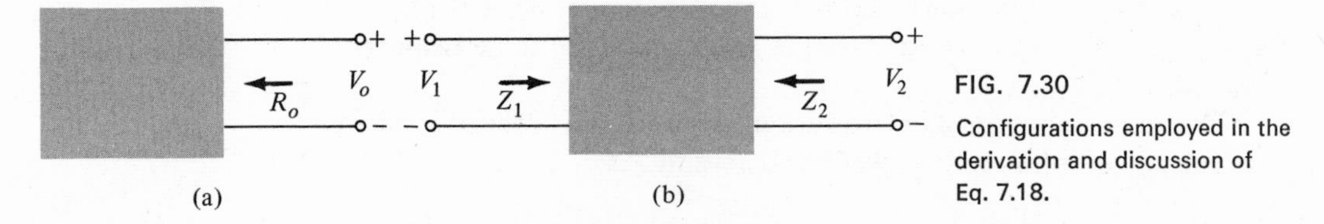

FIG. 7.30
Configurations employed in the derivation and discussion of Eq. 7.18.

For $V_o = V_1$, $P_1 = V_1^2/R_o$ where R_o is the output impedance of the system. If V_o should be increased (or decreased) to V_2, then $P_2 = V_2^2/R_o$. If we substitute into Eq. (7.18) to determine the resulting difference in decibels between the power levels,

$$\text{dB} = 10 \log_{10} \frac{P_2}{P_1} = 10 \log_{10} \frac{V_2^2/R_o}{V_1^2/R_o} = 10 \log_{10} \left(\frac{V_2}{V_1}\right)^2$$

and

$$\text{dB} = 20 \log_{10} \frac{V_2}{V_1} \tag{7.20}$$

Keep in mind, however, that this equation is only correct if the associated resistance for each applied voltage is the same. For the system of Fig. 7.30b, where output and input levels are being compared, $Z_1 \neq Z_2$ and Eq. (7.20) will not result. Equation (7.18) should therefore be employed.

If $\mathbf{Z}_1 = Z_1 \cos \theta_1$ and $\mathbf{Z}_2 = Z_2 \cos \theta_2$ are substituted into Eq. (7.18) such that $P_2 = \dfrac{V_2^2}{Z_2 \cos \theta_2}$, etc., the following general equation will result

$$\text{dB} = 20 \log_{10} \frac{V_2}{V_1} + 10 \log_{10} \frac{Z_1}{Z_2} + 10 \log_{10} \frac{\cos \theta_1}{\cos \theta_2} \tag{7.21}$$

For resistive elements, which are most commonly encountered,

$\cos\theta_1 = \cos\theta_2$, and the last term, $\log_{10}(1) = 0$. In addition, if $Z_1 = Z_2$, the second term will also drop out, resulting in Eq. (7.20).

Frequently the effect of different impedances ($Z_1 \neq Z_2$) is ignored and Eq. (7.20) applied to simply establish a basis of comparison between levels—voltage or current. For situations of this type the decibel gain should more correctly be referred to as the *voltage or current gain in decibels* to differentiate it from the common usage of decibel as applied to power levels.

One of the advantages of the logarithmic relationship is the manner in which it can be applied to cascaded stages. For example, the overall voltage gain of a cascaded system is given by

$$A_{v_T} = A_{v_1} A_{v_2} A_{v_3} \cdots A_{v_n}$$

Applying the proper logarithmic relationship:

$$20\log_{10} A_{v_T} = 20\log_{10} A_{v_1} + 20\log_{10} A_{v_2} + 20\log_{10} A_{v_3} + \cdots + 20\log_{10} A_{v_n} \qquad (7.22)$$

In words, the equation states that the decibel gain of a cascaded system is simply the sum of the decibel gains of each stage, that is,

$$\boxed{\mathrm{dB}_{A_{v_T}} = \mathrm{dB}_{A_{v_1}} + \mathrm{dB}_{A_{v_2}} + \mathrm{dB}_{A_{v_3}} + \cdots + \mathrm{dB}_{A_{v_n}}} \qquad (7.23)$$

The above equations can also be applied to current considerations. For $P_2 = I_2^2 R_o$ and $P_1 = I_1^2 R_o$,

$$\boxed{\mathrm{dB} = 20\log_{10}\frac{I_2}{I_1}} \qquad (7.24)$$

and

$$\boxed{\mathrm{dB}_{A_{i_T}} = \mathrm{dB}_{A_{i_1}} + \mathrm{dB}_{A_{i_2}} + \mathrm{dB}_{A_{i_3}} + \cdots + \mathrm{dB}_{A_{i_n}}} \qquad (7.25)$$

Before considering a few examples, the fundamental operations associated with logarithmic functions will be considered. For many it will be simply a review. For some, an extended amount of time may be required to fully understand the material to follow.

Each equation introduced in this section employs the common or base 10 logarithmic system. As indicated in the introductory discussion, the logarithm of numbers that are powers of the chosen base are quite easily determined. For example,

$$\log_{\underset{b}{\uparrow}10} \overbrace{10{,}000}^{a} = x \Rightarrow (10)^x = 10{,}000$$

and

$$x = 4$$

Similarly,

$$\log_{10} 1000 = \log_{10} (10)^3 = 3$$
$$\log_{10} 100 = \log_{10} (10)^2 = 2$$
$$\log_{10} 10 = \log_{10} (10)^1 = 1$$
$$\log_{10} 1 = \log_{10} (10)^0 = 0$$

For the logarithm of a number such as 24.8

$$\log_{10} 24.8 = x \Rightarrow (10)^x = 24.8$$

The unknown quantity x is obviously between 1 and 2, but a further determination would be purely a trial-and-error process if it were not for the logarithmic function. The procedure for determining the logarithm of a number requires that two components of the result be found separately. These two components are the *characteristic* and *mantissa*. The characteristic is simply the power of 10 associated with the number for which the logarithm is to be determined.

$$24.8 = 2.48 \times 10^1 \Rightarrow 1 = \text{characteristic}$$
$$4860.0 = 4.860 \times 10^3 \Rightarrow 3 = \text{characteristic}$$

The mantissa, or decimal portion of the logarithm must be determined from a set of tables or the D and L scales of a slide rule. The slide-rule procedure is quite direct. Simply find the number for which the logarithm is to be determined on the D scale and the mantissa will appear above (or below) on the L scale (0.3944 as shown in Fig. 7.31),

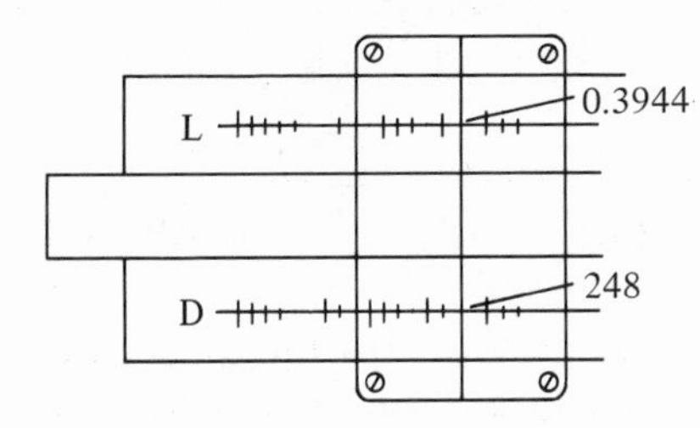

FIG. 7.31 Determining the logarithm of 24.8, to the base 10, using the slide rule.

resulting in

$$\log_{10} 24.8 = 1.3944$$
$$\log_{10} 4860.0 = 3.6870$$

There will be many occasions in which the antilogarithm of a number must be determined; that is, for the example above, determine 24.8 and 4860.0 from the logarithm of these numbers. The process is simply the reverse of that applied to determine the logarithm. For example, find the antilogarithm of 2.140.

$$2.140 \left\{ \begin{array}{l} \text{characteristic} \Rightarrow 10^2 \\ \text{mantissa} \quad \Rightarrow 138 \\ (\text{L} \rightarrow \text{D scale}) \end{array} \right\} 1.38 \times 10^2 = 138$$

For ratios less than 1, the logarithm can be determined by simply inverting the ratio and introducing a negative sign.

$$\log_{10} \frac{16}{24} = -\log_{10} \frac{24}{1.6} = -\log_{10} 15 = -1.176$$

$$\log_{10} 0.788 = -\log_{10} \frac{1}{0.788} = -\log_{10} 1.27 = -0.4315$$

For power ratios, a negative decibel rating simply indicates a reduction in power level as compared to the initial or input power.

EXAMPLE 7.3 Find the magnitude gain corresponding to a decibed gain of 100.

Solution: By Eq. (7.18)

$$100 = 10 \log_{10} \frac{P_2}{P_1} \Rightarrow \log_{10} \frac{P_2}{P_1} = 10$$

so that

$$\frac{P_2}{P_1} = \mathbf{10^{10} = 10{,}000{,}000{,}000}$$

This example clearly demonstrates the range of decibel values to to be expected from practical devices. Certainly a future calculation giving a decibel result in the neighborhood of 100 should be questioned immediately. In fact, a decibel gain of 50 corresponds with a magnitude gain of 100,000, which is still quite large.

EXAMPLE 7.4 The input power to a device is 10,000 W at a voltage of 1000 V. The output power is 500 W while the output impedance is 20 Ω.

(a) Find the power gain in decibels.
(b) Find the voltage gain in delibels.
(c) Explain why (a) and (b) agree or disagree.

Solution

(a) $$dB = 10 \log_{10} \frac{0.5 \times 10^3}{10 \times 10^3} = 10 \log_{10} \frac{1}{20} = -10 \log_{10} 20$$
$$= -10(1.301) = \mathbf{-13.01\ dB}$$

(b) $$dB_v = 20 \log_{10} \frac{V_o}{V_i} = 20 \log_{10} \frac{\sqrt{PR}}{1000} = 20 \log_{10} \frac{\sqrt{500 \times 20}}{1000}$$
$$= 20 \log_{10} \frac{100}{1000} = = 20 \log_{10} \frac{1}{10} = -20 \log_{10} 10 = \mathbf{-20\ dB}$$

(c) $$R_i = \frac{V^2}{P} = \frac{10^6}{10^4} = 10^2 \neq R_o = \mathbf{20\ \Omega}$$

EXAMPLE 7.5 An amplifier rated at 40-W output is connected to a 10-Ω speaker. (a) Calculate the input power required for full power output, if the power gain is 25 dB. (b) Calculate the input voltage for rated output if the amplifier voltage gain is 40 dB.

Solution

(a) Eq. 7.18

$$25 = 10 \log_{10} \frac{40}{P_i} \Rightarrow P_i = \frac{40}{\text{antilog}\,(2.5)} = \frac{40}{3.16 \times 10^2}$$

$$= \frac{40}{316} \cong \mathbf{126\ mW}$$

(b) $$dB_v = 20 \log_{10} \frac{V_o}{V_i} \Rightarrow 40 = 20 \log_{10} \frac{V_o}{V_i}$$

$$\frac{V_o}{V_i} = \text{antilog}\, 2 = 100$$

$$V_o = \sqrt{\text{PR}} = \sqrt{40 \times 10} = 20\ \text{V}$$

$$V_i = \frac{V_o}{100} = \frac{20}{100} = \mathbf{200\ mV}$$

7.10 GENERAL FREQUENCY CONSIDERATIONS

The frequency of the applied signal can have a pronounced effect on the response of a single or multistage network. The analysis thus far has been for the mid-frequency spectrum. At low frequencies we shall find that the coupling and bypass capacitors can no longer be replaced by the short-circuit approximation because of the resulting change in reactance of these elements. The frequency-dependent parameters of the small-signal equivalent circuits and the stray capacitive elements associated with the active device and the network will limit the high-frequency response of the system. An increase in the number of stages of a cascaded system will limit both the high- and low-frequency response.

The magnitude gain of an RC-coupled, direct-coupled, and transformer-coupled amplifier system are provided in Fig. 7.32. Note that the horizontal scale is a logarithmic scale to permit a plot extending from the low- to the high-frequency regions. For each plot, a low-, high-, and mid-frequency region has been defined. In addition, the primary reasons for the drop in gain at low and high frequencies has also been indicated within the parentheses. For the RC-coupled amplifier the drop at low frequencies is due to the increasing reactance of the coupling capacitor, while its upper frequency limit is determined by either the parasitic capacitive elements of the network and active device or the frequency dependence of the gain of the active device. An expla-

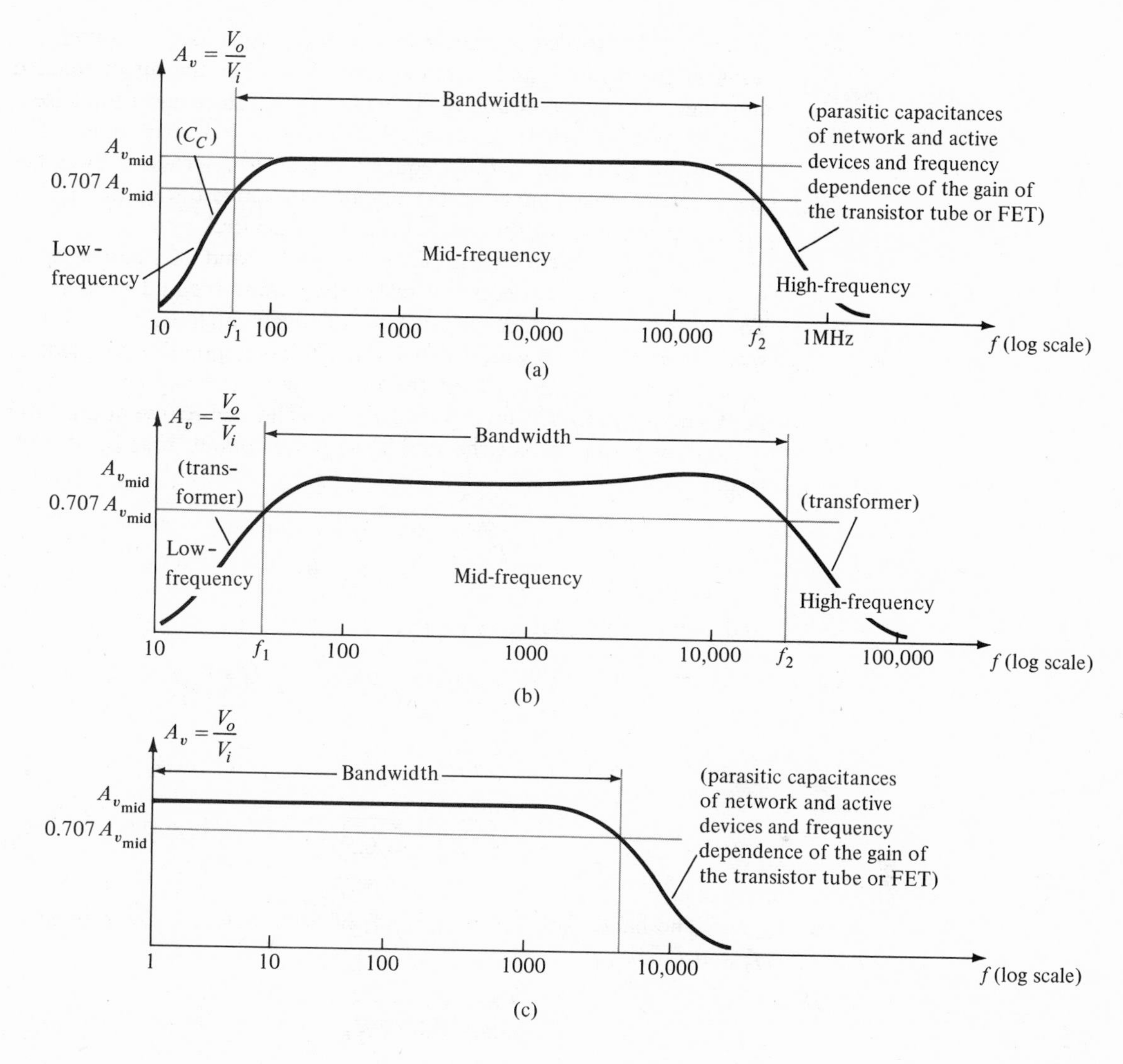

FIG. 7.32

Gain vs. frequency for: (a) RC-coupled amplifiers; (b) transformer-coupled amplifiers; (c) direct-coupled amplifiers.

nation of the drop in gain for the transformer-coupled system requires a basic understanding of "transformer action" and the transformer equivalent circuit. For the moment let us say that it is simply due to the shorting effect (across the input terminals of the transformer) of a magnetizing inductive reactance at low frequencies ($X_L = 2\pi f L$). The gain must obviously be zero at $f = 0$ since at this point there is no longer a changing flux established through the core to induce a secondary or output voltage. As indicated in Fig. 7.32, the high-frequency response is controlled primarily by the stray capacitance between the turns of the primary and secondary windings. For the direct-coupled amplifier, there is no coupling element (C_C or transformer) to cause a drop in gain at low frequencies. As the figure indicates, it is a flat response to the upper cutoff frequency which is determined by either the parasitic capacitances of the circuit and active device or the frequency dependence of the gain of the active device.

For each system of Fig. 7.32 there is a band of frequencies in which the gain is either equal or relatively close to the mid-band value. To fix the frequency boundaries of relatively high gain, $0.707A_{v_{\text{mid}}}$ was chosen to be the gain cutoff level. The corresponding frequencies f_1 and f_2 are generally called the cutoff, band, breakpoint, or half-power frequencies. The multiplier 0.707 was chosen because of this level the output power is half the mid-band power ouput, that is, at mid-frequencies,

$$P_{o_{\text{mid}}} = \frac{V_o^2}{R_o} = \frac{(A_{v_{\text{mid}}} V_i)^2}{R_o}$$

and at the half-power frequencies,

$$P_{o_{HPF}} = \frac{(0.707\, A_{v_{\text{mid}}} V_i)^2}{R_o} = 0.5\left(\frac{A_{v_{\text{mid}}} V_i}{R_o}\right)^2$$

and

$$\boxed{P_{o_{HPF}} = 0.5\, P_{o_{\text{mid}}}} \tag{7.26}$$

The bandwidth (or pass band) of each system is determined by f_1 and f_2, that is,

$$\boxed{\text{bandwidth (BW)} = f_2 - f_1} \tag{7.27}$$

For applications of a communications nature (audio, video), a decibel plot of the voltage gain vs. frequency is more useful than that appearing in Fig. 7.32. Before obtaining the logarithmic plot, however, the curve is generally normalized as shown in Fig. 7.33. In this figure, the gain at each frequency is divided by the mid-band value. Obviously, the mid-band value is then 1 as indicated. At the half-power frequencies the resulting level is $0.707 = 1/\sqrt{2}$. A decibel plot can now be obtained by applying Eq. (7.20) in the following manner:

$$\boxed{\left|\frac{A_v}{A_{v_{\text{mid}}}}\right| = 20 \log_{10} \left|\frac{A_v}{A_{v_{\text{mid}}}}\right|} \tag{7.28}$$

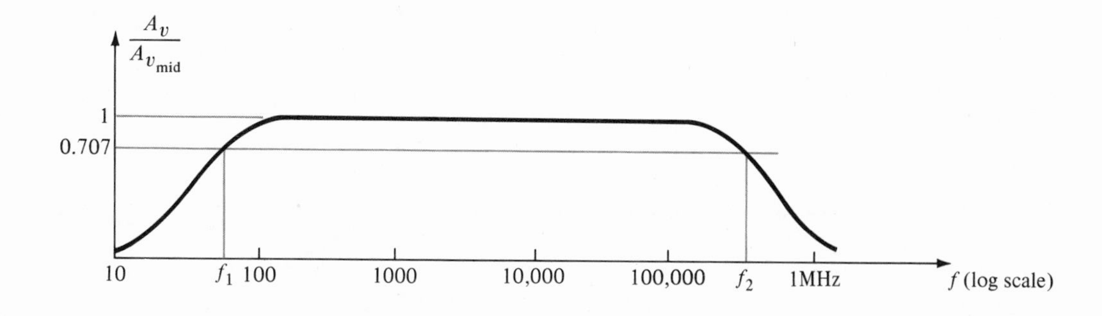

FIG. 7.33

Normalized gain vs. frequency plot

At mid-band frequencies, $20 \log_{10} 1 = 0$, and at the cutoff frequencies, $20 \log_{10} 1/\sqrt{2} = -3$ dB. Both values are clearly indicated in the resulting decibel plot of Fig. 7.34. The smaller the fractional ratio, the more negative the decibel level due to the inversion process discussed in Section 7.9.

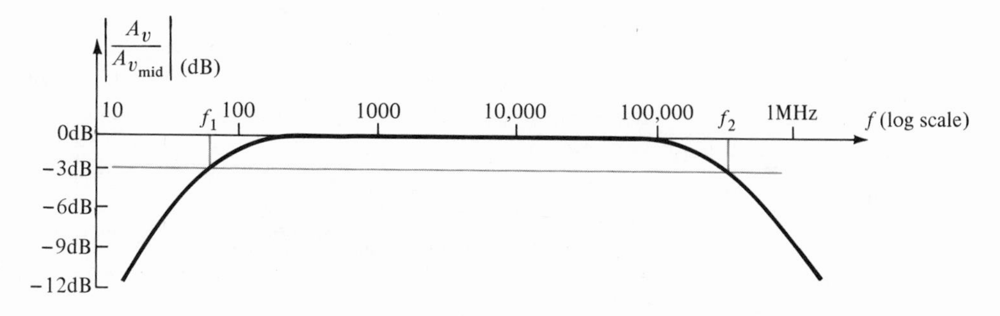

FIG. 7.34

Decibel plot of the normalized gain vs. frequency plot of Fig. 7.34.

For the greater part of the discussion to follow, a decibel plot will be made only for the low- or high-frequency regions. Keep Fig. 7.34 in mind, therefore, to permit a visualization of the broad system response.

It should be understood that an amplifier usually introduces a phase shift of 180° between input and output signals. This fact must now be expanded to indicate that this is only the case in the mid-band region. At low frequencies there is an additional phase shift such that V_o lags V_i by an increased angle. At high frequencies, the phase shift will drop below 180°. Fig. 7.35 is a standard phase plot for an RC-coupled amplifier.

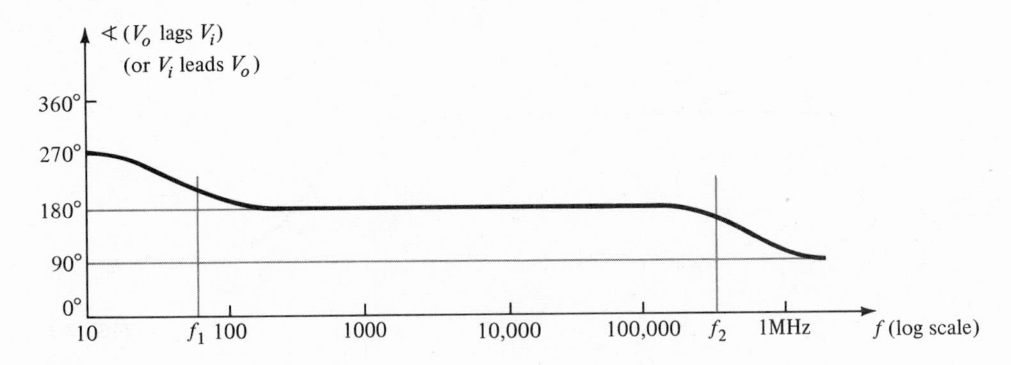

FIG. 7.35

Phase plot for an RC-coupled amplifier system (for each stage).

7.11 LOW-FREQUENCY RESPONSE— RC-COUPLED TRANSISTOR AMPLIFIER

The RC-coupled amplifier is unquestionably the most frequently applied of the three discussed in Section 7.10. This fact, in conjunction with the present-day interest in transistors, necessitates that the greater part of the low-frequency discussion of cascaded systems concentrate its attention on the RC-coupled transistor amplifier. Such a system appears in Fig. 7.36.

The analysis of this section will be for a stage as it appears between a and a' of Fig. 7.36. For the analysis to be complete, the mid-frequency gain as calculated in earlier sections of this chapter must be known.

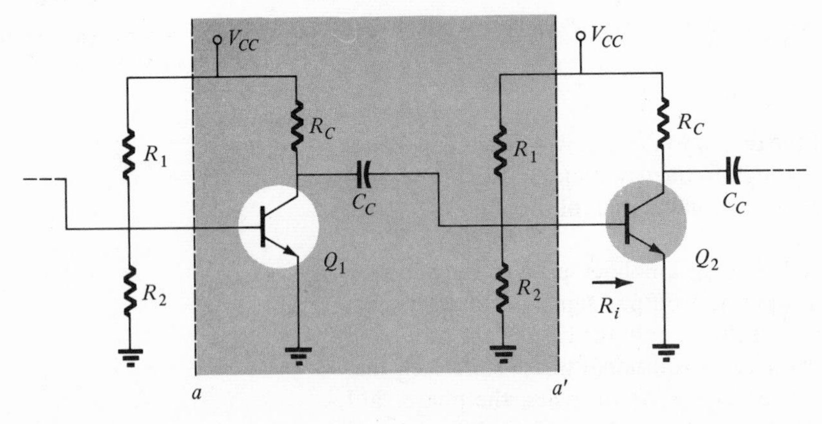

FIG. 7.36

RC-coupled transistor amplifier system.

Substituting the approximate hybrid equivalent circuit will result in the circuit of Fig. 7.37, where it is assumed that $R_1 \,||\, R_2 \,||\, R_i \cong R_i$, an approximation to be applied throughout the following analysis.

FIG. 7.37
Approximate small-signal ac equivalent circuit of section a-a′ of the system of Fig. 7.36 in the mid-frequency region (approximations: h_{re}, $h_{oe} \cong 0$, $R_1 \,||\, R_2 \,||\, R_i \cong R_i$).

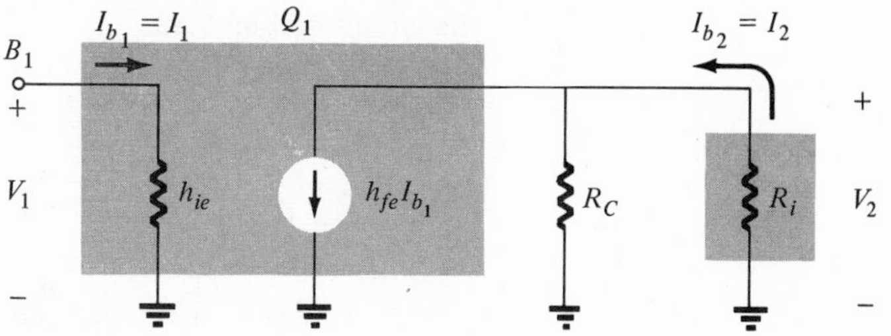

Applying the current divider rule

$$I_2 = \frac{R_C(h_{fe}I_{b_1})}{R_C + R_i} = \frac{h_{fe}R_C I_1}{R_C + R_i}$$

and

$$A_{i_{\text{mid}}} = \frac{I_2}{I_1} = \frac{h_{fe}R_C}{R_C + R_i} = \frac{h_{fe}R_C}{R_C + R_i}\left[\frac{R_i}{R_i}\right]$$

$$= h_{fe}\underbrace{\frac{R_i R_C}{R_i + R_C}}_{R_i \,||\, R_C}\frac{1}{R_i}$$

so that

$$\boxed{A_{i_{\text{mid}}} = \frac{h_{fe}R}{R_i}} \qquad \text{where } R = R_C \,||\, R_i \qquad (7.29)$$

Transistors are fundamentally current amplifiers. For this reason the mid-band current gain rather than voltage gain was obtained. The same will be true in the following analysis of the transistor RC-coupled stage.

The low-frequency behavior of the network of Fig. 7.36 is determined primarily by the coupling capacitor C_C. Inserting this element will result in the modified equivalent circuit of Fig. 7.38. The introduction

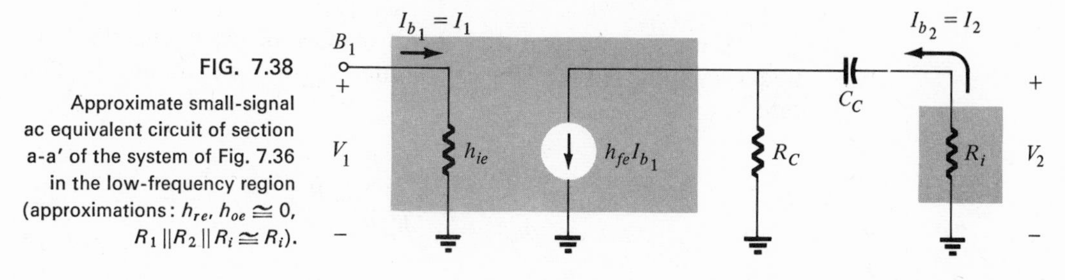

FIG. 7.38
Approximate small-signal ac equivalent circuit of section a-a′ of the system of Fig. 7.36 in the low-frequency region (approximations: h_{re}, $h_{oe} \cong 0$, $R_1 \,||\, R_2 \,||\, R_i \cong R_i$).

of C_C will result in A_i being frequency dependent, since $X_{C_C} = 1/2\pi f C_C$. As the frequency approaches 0 Hz, X_{C_C} becomes increasingly larger, with the net result that I_2, and correspondingly $A_i = I_2/I_1$, will approach zero in the limit ($f = 0$), This low-frequency plot appears in Fig. 7.32.

For the circuit of Fig. 7.38

$$I_2 = \frac{R_C(h_{fe}I_b)}{R_C + R_i + X_C\underline{/-90^\circ}} = \frac{h_{fe}R_C I_1}{R_C + R_i - jX_C}$$

and

$$A_{i_{\text{low}}} = \frac{I_2}{I_1} = \frac{h_{fe}R_C}{R_C + R_i - jX_C}$$

Mathematical manipulations will now be performed to place this result in a more convenient form for further analysis.

Dividing the top and bottom by $R_C + R_i$,

$$A_{i_{\text{low}}} = \frac{h_{fe}R_C/(R_C + R_i)}{1 - jX_C/(R_C + R_i)} = \frac{h_{fe}[R_C/(R_C + R_i)]\overbrace{[R_i/R_i]}^{1}}{1 - jX_C/(R_C + R_i)}$$
$$= \frac{(h_{fe}/R_i)[R_iR_C/(R_C + R_i)]}{1 - jX_C/(R_C + R_i)}$$

and

$$A_{i_{\text{low}}} = \frac{(h_{fe}/R_i)R}{1 - jX_C/(R_C + R_i)}$$

where $R = R_C \,||\, R_i$ as defined earlier, or

$$\boxed{A_{i_{\text{low}}} = \frac{\dfrac{h_{fe}R}{R_i}}{1 - j\dfrac{1}{\omega C_C R_{\text{low}}}}} \tag{7.30}$$

where

$$R_{\text{low}} = R_C + R_i$$

A more convenient and useful manner of examining the low-frequency gain requires that the ratio $A_{i_{\text{low}}}/A_{i_{\text{mid}}}$ be found. In this case

$$\frac{A_{i_{\text{low}}}}{A_{i_{\text{mid}}}} = \frac{\dfrac{h_{fe}R/R_i}{1 - j1/\omega C_C R_{\text{low}}}}{h_{fe}R/R_i}$$

and

$$\frac{A_{i_{\text{low}}}}{A_{i_{\text{mid}}}} = \frac{1}{1 - j\dfrac{1}{\omega C_C R_{\text{low}}}} \tag{7.31}$$

or

$$\frac{A_{i_{\text{low}}}}{A_{i_{\text{mid}}}} = \underbrace{\frac{1}{\sqrt{1^2 + (1/\omega C_C R_{\text{low}})^2}}}_{\text{magnitude}} \underbrace{\angle \tan^{-1} 1/\omega C_C R_{\text{low}}}_{\text{phase angle}} \tag{7.32}$$

The effect of frequency on the gain of the system can now easily be described through Eq. (7.31) or (7.32). As the frequency ($\omega = 2\pi f$) decreases, the magnitude of the second term in the denominator of each equation will increase, resulting in an increase in the magnitude of the denominator and a drop in gain. At increasing frequencies, the second term will become increasingly small in comparison to 1 with the result that the gain approaches the mid-band value.

There is a result of special significance when $1/\omega C_C R_{\text{low}} = 1$. Substituting into Eq. (7.32)

$$\frac{A_{i_{\text{low}}}}{A_{i_{\text{mid}}}} = \frac{1}{\sqrt{1+1}} \angle \tan^{-1} 1 = \frac{1}{\sqrt{2}} \angle 45^\circ = 0.707 \angle 45^\circ$$

The gain, therefore, has dropped to a level defining the lower cutoff frequency.

Applying Eq. (7.20)

$$\left| \frac{A_{i_{\text{low}}}}{A_{i_{\text{mid}}}} \right|_{\text{dB}} = 20 \log_{10} \left| \frac{A_{i_{\text{low}}}}{A_{i_{\text{mid}}}} \right| = 20 \log_{10} \frac{1}{\sqrt{2}}$$
$$= -20 \log_{10} \sqrt{2} = -3 \text{ dB}$$

This result indicates quite clearly that *the gain will drop* **3** *dB from its mid-band level at the cutoff frequencies.*

The frequency at which this occurs can be found in the following manner

$$\frac{1}{\omega C_C R_{\text{low}}} = \frac{1}{2\pi f_1 C_C R_{\text{low}}} = 1 \Rightarrow \boxed{f_1 = \frac{1}{2\pi C_C R_{\text{low}}}} \tag{7.33}$$

To find the lower break frequency it is now only necessary to substitute into Eq. (7.33).

Substituting this value into the general equation will result in

$$\frac{A_{i_{\text{low}}}}{A_{i_{\text{mid}}}} = \frac{1}{1 - jf_1/f} = \frac{1}{\sqrt{1^2 + (f_1/f)^2}} \underline{/\tan^{-1} f_1/f} \qquad (7.34)$$

A plot of this ratio appears in Fig. 7.39a. To ensure that the curve is correctly understood let us examine a few representative points. At $f/f_1 = 1 \Rightarrow f = f_1$, the ratio is 0.707 as determined earlier. At $f/f_1 = 10 \Rightarrow f = 10f_1$, the ratio is, for all practical purposes; 1 and $A_{i_{\text{low}}} = A_{i_{\text{mid}}}$. For $f/f_1 = 0.2$ or $f = 0.2f_1$ the ratio $\cong 0.2$, or $A_{i_{\text{low}}} = 0.2A_{i_{\text{mid}}}$. We will find in the discussions to follow that Eq. (7.34) is representative of the low-frequency response of cascaded FET and tube stages. In each case, f_1 will simply be determined by a different set of parameters.

The phase plot appears in Fig. 7.39b for the same frequency spectrum. The angle to be associated with $A_{i_{\text{mid}}}$ is 180° so that at $f/f_1 = 0.2$

$$\frac{A_{i_{\text{low}}}\underline{/\theta}}{A_{i_{\text{mid}}}\underline{/180^\circ}} \cong \underbrace{0.2\underline{/80^\circ}}_{\text{from Fig. 7.39a}}$$

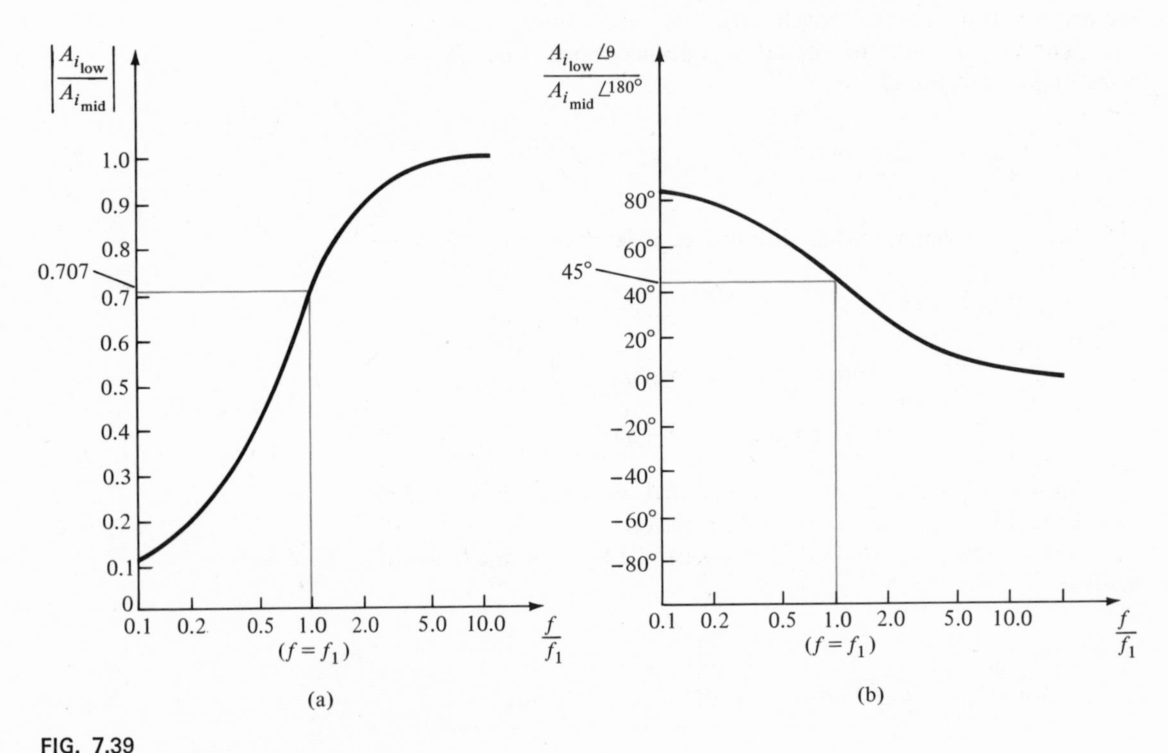

FIG. 7.39
Normalized magnitude and phase plots for the low-frequency region.

and

$$A_{i_{low}}\underline{/\theta} = 0.2\underline{/80^\circ}\, A_{i_{mid}}\underline{/180^\circ}$$

or

$$A_{i_{low}}\underline{/\theta} = 0.2A_{i_{mid}}\underline{/260^\circ}$$

and $\theta = 260^\circ$, the phase shaft between input and output quantities. It states that I_o leads I_i by 260° or I_i leads I_o by $(360 - 260^\circ) = 100^\circ$.

As the frequency increases, the phase shift between input and output quantities approaches 180° (the ratio approaches 0° phase shift.)

A useful logarithmic plot can be obtained by applying Eq. (7.20) directly as indicated below.

$$\left|\frac{A_{i_{low}}}{A_{i_{mid}}}\right|_{dB} = 20 \log_{10}\left|\frac{A_{i_{low}}}{A_{i_{mid}}}\right| = 20 \log_{10}\frac{1}{\sqrt{1 + (f_1/f)^2}}$$
$$= -20 \log_{10}[1 + (f_1/f)^2]^{1/2}$$

and

$$\left|\frac{A_{i_{low}}}{A_{i_{mid}}}\right|_{dB} = -10 \log_{10}[1 + (f_1/f)^2] \qquad (7.35)$$

If we now consider frequencies such that $f \gg f_1$ or $f_1/f \ll 1$,

$$\left|\frac{A_{i_{low}}}{A_{i_{mid}}}\right|_{dB} \cong -10 \log_{10}[1 + \cong 0] = 0 \text{ dB}$$

and for $f \ll f_1$ or $f_1/f \gg 1$, $1 + (f_1/f)^2 \cong (f_1/f)^2$
and

$$\boxed{\left|\frac{A_{i_{low}}}{A_{i_{mid}}}\right|_{dB} \cong -20 \log_{10}\frac{f_1}{f}}_{f_1 \gg f} \qquad (7.36)$$

Ignoring the condition $f_1 \gg f$ for a moment, a plot of Eq. (7.36) on a frequency log scale will yield some results of a useful nature for future decibel plots.

At $f = f_1$, or $\frac{f_1}{f} = 1$, $\left(\frac{f}{f_1} = 1\right)$, $-20 \log_{10}1 = 0$ dB

At $f = 0.5f_1$, or $\frac{f_1}{f} = 2$, $\left(\frac{f}{f_1} = 0.5\right)$, $-20 \log_{10}2 = -6$ dB

At $f = 0.25f_1$, or $\frac{f_1}{f} = 4$, $\left(\frac{f}{f_1} = 0.25\right)$, $-20 \log_{10}4 = -12$ dB

At $f = 0.1f_1$, or $\frac{f_1}{f} = 10$, $\left(\frac{f}{f_1} = 0.1\right)$, $-20 \log_{10}10 = -20$ dB

A plot of these points is indicated in Fig. 7.40 from $f/f_1 = 0.1$ to $f/f_1 = 1$. Note that this results in a straight line when plotted against a log scale. In the same figure a straight line is also drawn for the condition of 0 dB for $f \gg f_1$. As stated earlier, the straight-line segments (asymptotes) are only accurate for 0 db when $f \gg f_1$, and the sloped line when $f_1 \gg f$. We know, however, that when $f = f_1$, there is a 3 dB drop from the mid-band level. Employing this information in association with the straight line segments permits a fairly accurate plot of the frequency response as indicated in the same figure. The piecewise linear plot resulting from the asymtotes and associated breakpoints is called a *Bode plot.*

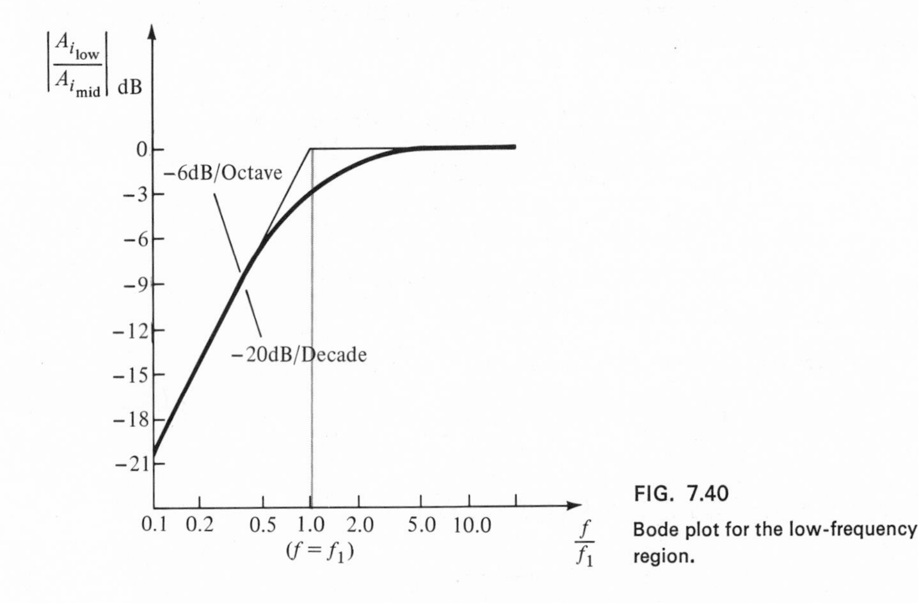

FIG. 7.40
Bode plot for the low-frequency region.

The above calculations and the curve itself demonstrate quite clearly that a change in frequency by a factor of 2 (equivalent to 1 octave) results in a 6 dB change in the ratio. For a 10:1 change in frequency (equivalent to 1 decade) there is a 20 dB change in the ratio. In the future, therefore, a decibel plot can be obtained quite easily for a function having the format of Eq. (7.34). First simply find f_1 from the circuit parameters, then sketch two asymtotes; one along the 0 dB line and the other drawn through f_1 sloped at 6 dB/octave or 20 dB/decade. Then find the 3 dB point corresponding to f_1 and sketch the curve.

EXAMPLE 7.6 For the circuit between a and a' of Fig. 7.2, the breakpoint or lower-band frequency is

$$f_1 = \frac{1}{2\pi C_C R_{low}}$$

where

$$\begin{aligned} R_{\text{low}} &= R_C + R_i \\ &= 4\,\text{K} + 1\,\text{K}\,(\text{note: } R_1 \parallel R_2 \parallel R_i = 200\,\text{K} \parallel 20\,\text{K} \parallel 1\,\text{K} \cong 1\,\text{K} = R_i) \\ &= 5\,\text{K} \end{aligned}$$

and

$$f_1 = \frac{1}{(6.28)(0.5 \times 10^{-6})(5 \times 10^3)} = \frac{1}{283 \times 10^{-3}} = 63.8\ \text{Hz}$$

At $f = 63.8$ Hz the gain of the system has dropped to 0.707 (or -3 dB) of its mid-band value.

The mid-band current gain, as calculated in Section 7.3 is 37.9. For this section (*a-a'*) the current gain I_o/I_i corresponds to I_{b_2}/I_{b_1}. A plot of $A_{i_{\text{low}}}$ vs. frequency (on a log scale) can now be readily determined from the universal curve of Fig. 7.39a. It is provided in Fig. 7.41. The phase plot can be obtained in much the same manner.

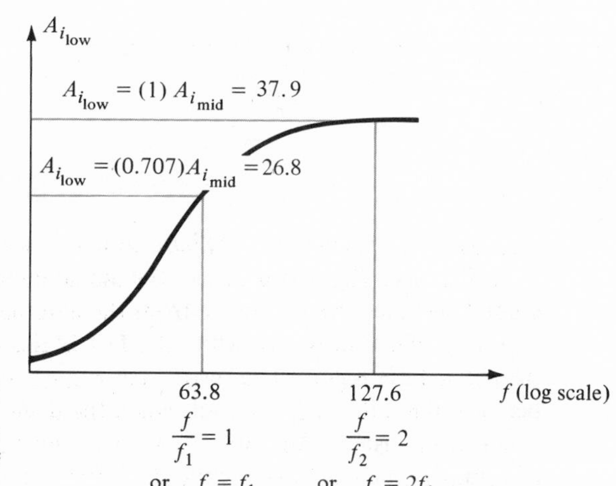

FIG. 7.41
Low-frequency plot for the current gain of the section of Fig. 7.2 between *a* and *a'*.

7.12 HIGH-FREQUENCY RESPONSE—RC-COUPLED TRANSISTOR AMPLIFIER

The high-frequency response is somewhat complicated by the fact that there are two breakpoints to consider. One is determined by the frequency dependence of h_{fe} and the other by the capacitive effects. No general statement can be made about which will determine the upper breakpoint. Both must be determined and compared. We will first consider the frequency dependence of h_{fe}.

The variation of h_{fe} with frequency will approach, with some degree of accuracy, the following relationship:

$$h_{fe} = \frac{h_{fe_{\text{mid}}}}{1 + jf/f_\beta} \tag{7.37}$$

The only undefined quantity, f_β, it determined by a set of parameters employed in the *hybrid π* or *Giacoletto* model frequently applied to best represent the transistor in the high-frequency region. It appears in Fig. 7.42. The various parameters warrant a moment of explanation.

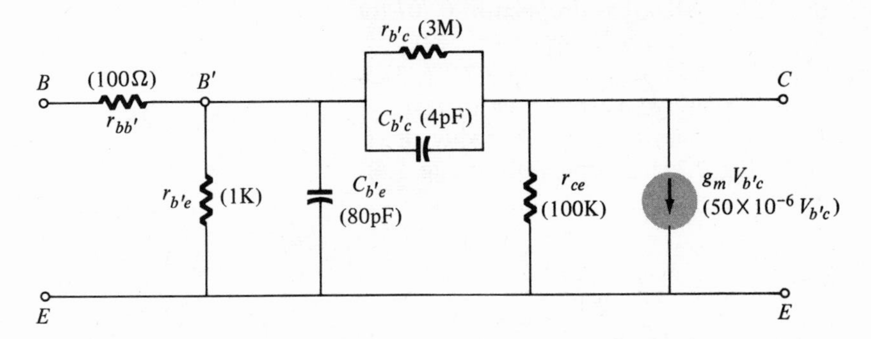

FIG. 7.42
Giacoletto (or hybrid π) high-frequency transistor small-signal ac equivalent circuit.

The resistance $r_{bb'}$ includes the base contact, base bulk, and base spreading resistance. The first is due to the actual connection to the base. The second includes the resistance from the external terminal to the active region of the transistors, while the last is the actual resistance within the active base region. The resistances $r_{b'e}$, r_{ce}, and $r_{b'c}$ are the resistances between the indicated terminals when the device is in the active region. The same is true for the capacitance $C_{b'e}$ and $C_{b'c}$, although the former is a transition capacitance while the latter is a diffusion capacitance. A more detailed explanation of the frequency dependence of each can be found in a number of readily available texts.

In terms of these parameters:

$$f_\beta(\text{sometimes appearing as } f_{h_{fe}}) = \frac{g_{b'e}}{2\pi(C_{b'e} + C_{b'c})} \tag{7.38}$$

or since the hybrid parameter h_{fe} is related to the hybrid parameter by $h_{fe} = g_m/g_{b'c}$

$$f_\beta = \frac{1}{h_{fe}}\left(\frac{g_m}{2\pi(C_{b'e} + C_{b'c})}\right) \tag{7.39}$$

The basic format of Eq. (7.37) should suggest some similarities between it and the curves obtained for the low-frequency response. The most noticeable difference is the fact that f_β appears in the denominator while f_1 appears in the numerator of the frequency ratio. This particular difference will have the effect depicted in Fig. 7.43; the plot

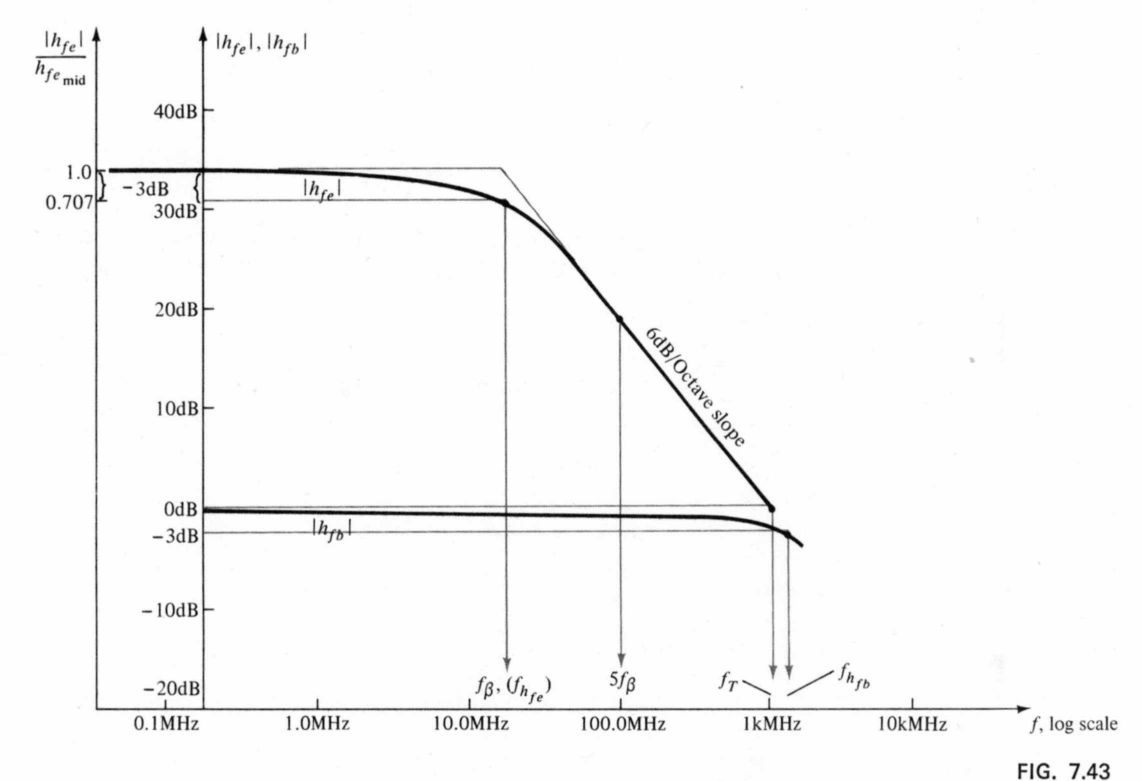

FIG. 7.43

h_{f_e} and h_{f_b} vs. frequency in the high-frequency region.

will drop off from the mid-band value rather than approach it with increase in frequency. The same figure has a plot of h_{fb} vs. frequency. Note that it is almost constant for the frequency range. In general, the common-base configuration displays improved high-frequency characteristics over the common-emitter configuration. For this reason, common-base high-frequency parameters, rather than common-emitter parameters, are often specified for a transistor. The following equation permits a direct conversion for determining f_β if f_α and $|\alpha| = h_{fb}$ are specified.

$$\boxed{f_\beta = f_\alpha(1 - \alpha)} \tag{7.40}$$

For any transistor, FET, or vacuum-tube system, an increase (or decrease) in bandwidth will have the reverse effect on the gain. The product of these two quantities, called the *gain-bandwidth product*, is defined for the transistor by the condition

$$\left| \frac{h_{fe_{mid}}}{1 + jf/f_\beta} \right| = 1$$

so that

$$|h_{fe}|_{dB} = 20 \log_{10} \left| \frac{h_{fe_{mid}}}{1 + jf/f_\beta} \right| = 20 \log_{10} 1 = 0 \text{ dB}$$

The frequency at which $|h_{fe}|_{dB} = 0$ dB is indicated quite clearly by f_T in Fig. 7.43. The magnitude of h_{fe} at the defined condition point is given by

$$\frac{h_{fe_{mid}}}{\sqrt{1 + (f_T/f_\beta)^2}} \cong \frac{h_{fe_{mid}}}{f_T/f_\beta} = 1$$

since

$$f_T \gg f_\beta \text{ or } \frac{f_T}{f_\beta} \gg 1$$

so that

$$f_T \cong \underbrace{h_{fe_{mid}}}_{(\cong \text{ mid-band gain})} \overbrace{f_\beta}^{(\cong \text{ BW})} \text{ (Gain-bandwidth product)} \qquad (7.41)$$

The fact that the above condition results in a gain-bandwidth product is now more obvious. Assuming f_β is the upper cutoff frequency (and not determined by the circuit capacitive elements to be considered next) $f_2 = f_\beta$ and $\text{BW} = f_\beta - f_1 \cong f_\beta$, since $f_\beta \gg f_1$.

Our attention will now be turned toward the capacitive elements and their effect on the high-frequency response.

The high-frequency equivalent circuit for the section *a-a'* of Fig. 7.2 appears in Fig. 7.44. Note that the mid-frequency equivalent circuit has been augmented by the h_{fe} frequency dependence and capacitors C_i and C_W. C_W is simply the wiring capacitance. C_i, however, is the capacitance contributed by the following stage. It can be determined with sufficient accuracy through the use of the following equation

$$C_i = C_{b'e} + (1 + |A_v| C_{b'c}) \qquad (7.42)$$

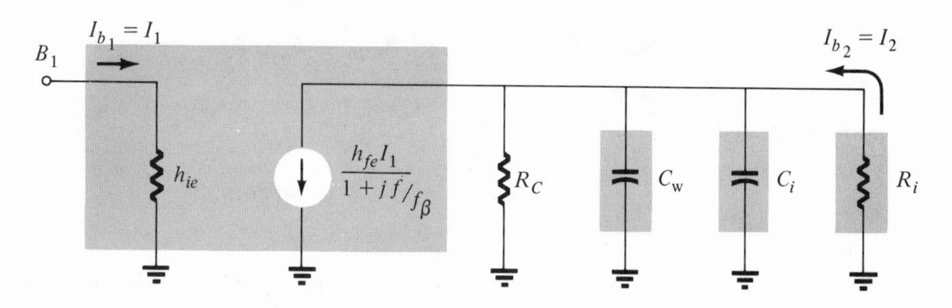

FIG. 7.44
High-frequency small-signal ac equivalent circuit for section *a-a′* of the cascaded system of Fig. 7.36.

where $C_{b'e}$, $C_{b'c}$ are parasitic capacitances of the transistor of the following stage.

The increase in capacitance due to the amplifying factor A_v of the following stage is due to the Miller effect. It was first introduced for the field effect transistor in Chapter 6. The similarities between the circuit employed for that derivation (Fig. 6.22a) and the circuit of Fig. 7.45

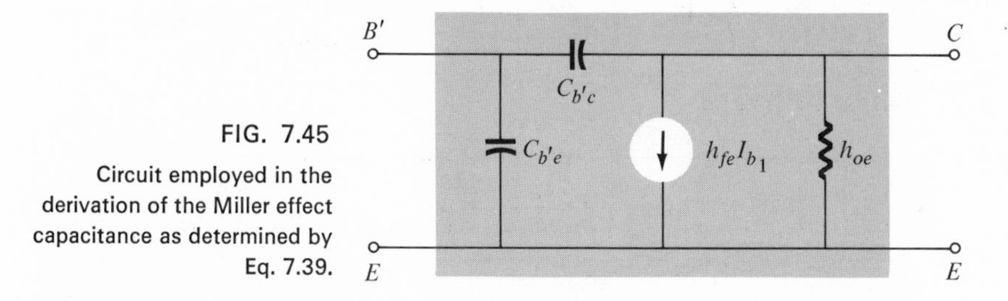

FIG. 7.45
Circuit employed in the derivation of the Miller effect capacitance as determined by Eq. 7.39.

should permit a derivation of Eq. (7.42) with very little difficulty. Since $|A_v|$ is dependent on C_i, $|A_{v_{mid}}|$ is normally employed.

The circuit of Fig. 7.44 can be redrawn as shown in Fig. 7.46.

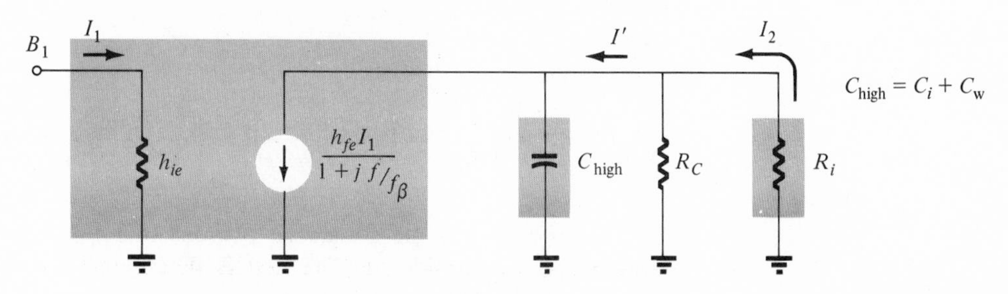

FIG. 7.46
Circuit of Fig. 7.43 redrawn following the combination of parallel capacitive elements and the introduction of required notation.

Defining: $R_{high} = R_C \,||\, R_i$

$$I' = \frac{X_C(h_{fe}I_1/(1 + jf/f_\beta))}{X_C + R_{high}} = \left(\frac{h_{fe}I_1}{1 + jf/f_\beta}\right)\left(\frac{X_C}{X_C + R_{high}}\right)$$

The second term can be written as

$$\frac{1}{1 + (R_{high}/X_C)} = \frac{1}{1 + j\omega C_{high}R_{high}}$$

and

$$I_2 = \frac{R_C}{R_C + R_i} I'$$

or

$$\frac{I_2}{I'} = \frac{R_i}{R_i}\left(\frac{R_C}{R_C + R_i}\right) = \frac{1}{R_i}\left(\frac{R_C R_i}{R_C + R_i}\right) = \frac{1}{R_i}(R_{high})$$

so that

$$\frac{I_2}{I_1} = \frac{I'}{I_1}\frac{I_2}{I'} = \frac{h_{fe}}{1 + j\omega/\omega_\beta}\left(\frac{1}{1 + j\omega C_{high}R_{high}}\right)\frac{1}{R_i}(R_{high})$$

can be written as

$$\boxed{A_{i_{high}} = \frac{I_2}{I_1} = \left(\frac{1}{1 + j\omega/\omega_\beta}\right)\left(\frac{1}{1 + j\omega C_{high}R_{high}}\right)\frac{h_{fe}R_{high}}{R_i}} \tag{7.43}$$

Since $R_{high} = R$ as defined for the mid-frequency region,

$$A_{i_{mid}} = \frac{h_{fe}R}{R_i} = \frac{h_{fe}R_{high}}{R_i}$$

and

$$\boxed{\frac{A_{i_{high}}}{A_{i_{mid}}} = \left(\frac{1}{1 + j\omega/\omega_\beta}\right)\left(\frac{1}{1 + j\omega C_{high}R_{high}}\right)} \tag{7.44}$$

It should now be somewhat obvious that there are two break frequencies to considered. The first is determined by the frequency dependent h_{fe}. The second is determined by the condition employed earlier for the low-frequency response, that is,

$$\omega C_{high}R_{high} = 2\pi f_2 C_{high}R_{high} = 1$$

and

$$f_2 = \frac{1}{2\pi C_{\text{high}} R_{\text{high}}} \tag{7.45}$$

For a given circuit, both f_β and f_2 must be found to determine which will be the predominate factor in determining the upper break frequency. If the two are sufficiently close, the combined effect must be considered as demonstrated in Fig. 7.47. For frequencies greater

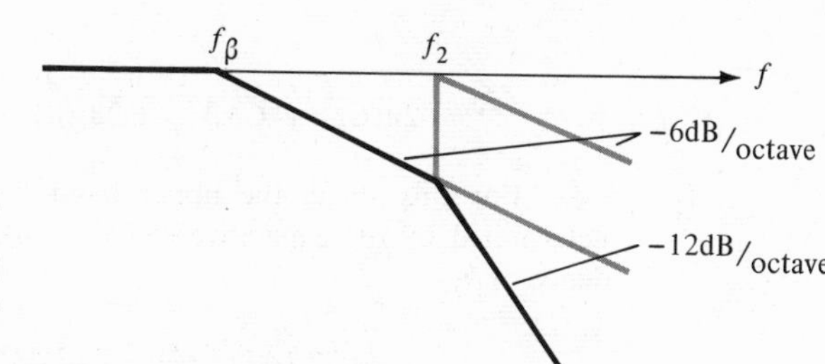

FIG. 7.47
Effect of close values of f_β and f_2 on the high-frequency asymptote (in this case $f_2 > f_\beta$).

than f_2 the slope of the resulting high-frequency asymptote will increase to -12 dB/octave or -40 dB/decade. The situation would be exactly the same if f_2 and f_β were interchanged. In any case the upper break frequency is determined by the point at which the gain has dropped 3 dB from the mid-band value.

EXAMPLE 7.7 For the cascaded system of Fig. 7.2, the transistor parameters are those typical values indicated in Fig. 7.42: $C_{b'c} = 4$ pF, $C_{b'e} = 80$ pF, $g_m = 50$ mA/V, $g_{b'e} = 1/1$ K, and $C_W = 2$ pF. For the section a-a',

$$|A_i| = \frac{I_o}{I_i} = \frac{I_{b_2}}{I_{b_1}} = 37.9$$

and

$$|V_2| = I_{b_2} R_{i_2} \cong I_{b_2} h_{ie_2}$$

$$|V_1| \cong I_{b_1} h_{ie_1}$$

so that

$$|A_{v_{\text{mid}}}| = \frac{V_2}{V_1} \cong \frac{I_2 h_{ie_2}}{I_1 h_{ie_1}} = A_{i_{\text{mid}}} \frac{h_{ie_2}}{h_{ie_1}}$$

But

$$h_{ie_2} = h_{ie_1}$$

so that

$$|A_{v_{\text{mid}}}| = |A_{i_{\text{mid}}}| = 37.9$$

$$C_i = C_{b'e} + (1 + |A_v|\, C_{b'c}) = 80 \text{ pF} + [1 + (37.9)(4)] = 232.6 \text{ pF}$$

and

$$C_{\text{high}} = C_i + C_W = 234.6 \text{ pF}$$

$$R_{\text{high}} = R_C \parallel R_i = 4 \text{ K} \parallel 1 \text{ K} = \frac{4}{5} \text{K}$$

so that

$$f_2 = \frac{1}{2\pi C_{\text{high}} R_{\text{high}}} = \frac{1}{(6.28)(234.6 \times 10^{-12})(0.800 \times 10^3)} \cong \mathbf{0.7MHz}$$

with

$$f_\beta = \frac{g_{b'e}}{2\pi(C_{b'e} + C_{b'c})} = \frac{10^{-3}}{(6.28)(84 \times 10^{-12})} \cong \mathbf{1.9\ MHz}$$

For this stage, the upper band frequency is quite obviously determined by the capacitive elements and not the frequency dependence of h_{fe}.

$$f_T \cong h_{fe} f_\beta = 50(1.9 \text{ MHz}) = \mathbf{95\ MHz}$$

Employing the results of an earlier example for low frequency, the band-pass plot for this stage will appear as shown in Fig. 7.48.

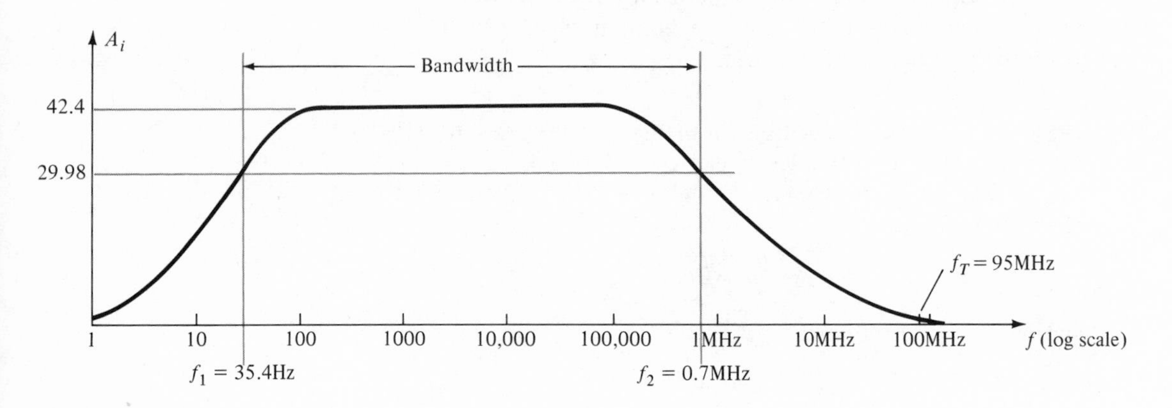

FIG. 7.48
A_i vs. log f for the network of Fig. 7.2.

7.13 MULTISTAGE FREQUENCY EFFECTS

An increase in the number of stages of a cascaded system can have a pronounced effect on the frequency response. For each additional stage the upper cutoff frequency will be determined primarily by that stage having the lowest cutoff frequency. The low-frequency cutoff is primarily determined by that stage having the highest cutoff frequency. Obviously, therefore, one poorly designed stage can offset an otherwise well-designed cascaded system.

The effect of increasing the number of *identical* stages can be clearly demonstrated by considering the situations indicated in Fig. 7.49. In each case the upper and lower cutoff frequencies of each of the cascaded stages is identical. For a single stage the cutoff frequencies are f_1 and f_2 as indicated. For two identical stages in cascade the drop-off

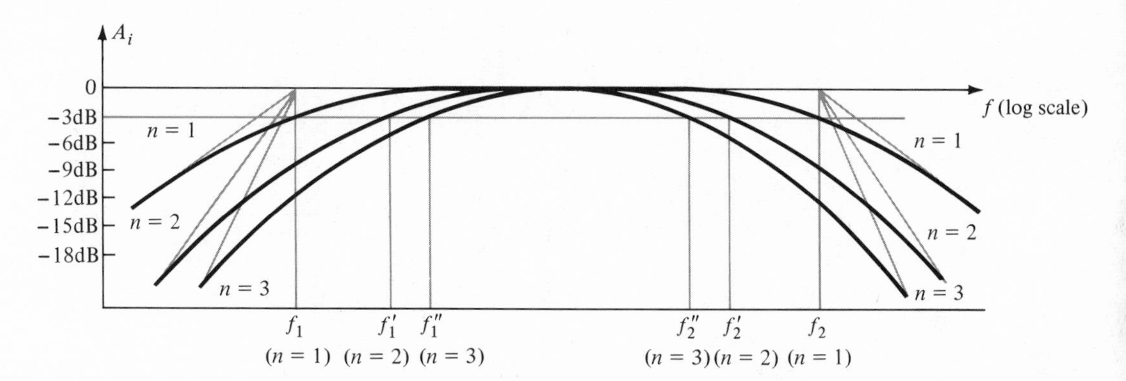

FIG. 7.49
Effect of an increased number of stages on the cutoff frequencies and the bandwidth.

rate in the high- and low-frequency regions has increased to -12 dB/octave or -40 dB/decade. At f_1 and f_2, therefore, the decibel drop is now -6 dB rather than the defined band frequency gain level of -3 dB. The -3 dB point has shifted to f_1' and f_2' as indicated with a resulting drop in the bandwidth. A -18 dB/octave or -60 dB/decade slope will result for a three-stage system of identical stages with the indicated reduction in bandwidth (f_1'' and f_2'').

Assuming identical stages, an equation for each band frequency as a function of the number of stages (n) can be determined in the following manner:

For the low-frequency region,

$$A_{i_{\text{low, (overall)}}} = A_{i_{1\,\text{low}}} A_{i_{2\,\text{low}}} A_{i_{3\,\text{low}}} \cdots A_{i_{n\,\text{low}}}$$

but since each stage is identical, $A_{i_{1\,\text{low}}} = A_{i_{2\,\text{low}}} =$ etc.

and
$$A_{i_{\text{low, (overall)}}} = (A_{i_{1\,\text{low}}})^n$$

or

$$\frac{A_{i_{\text{low}}}}{A_{i_{\text{mid}}}}(\text{overall}) = \left(\frac{A_{i_{1\,\text{low}}}}{A_{i_{\text{mid}}}}\right)^n = \frac{1}{(1 + jf_1/f)^n}$$

Setting the magnitude of this result equal to $1/\sqrt{2}$ (−3 db level)

$$\frac{1}{[\sqrt{1+(f_1/f_1')^2}]^n} = \frac{1}{\sqrt{2}}$$

or

$$\left\{\left[1+\left(\frac{f_1}{f_1'}\right)^2\right]^{1/2}\right\}^n = \left\{\left[1+\left(\frac{f_1}{f_1'}\right)^2\right]^n\right\}^{1/2} = \{2\}^{1/2}$$

so that

$$\left[1+\left(\frac{f_1}{f_1'}\right)^2\right]^n = 2$$

and

$$1+\left(\frac{f_1}{f_1'}\right)^2 = 2^{1/n}$$

with the result

$$\boxed{f_1' = \frac{f_1}{\sqrt{2^{1/n}-1}}} \qquad (7.46)$$

In a similar manner, it can be shown that for the high-frequency region,

$$\boxed{f_2' = \sqrt{2^{1/n}-1}\, f_2} \qquad (7.47)$$

Note the presence of the same factor $\sqrt{2^{1/n}-1}$ in each equation. The magnitude of this factor for various values of n is listed below.

n	$\sqrt{2^{1/n}-1}$
1	1
2	0.64
3	0.51
4	0.44
5	0.39

For $n = 2$, consider that the upper cutoff frequency $f_2' = 0.64f_2$ or 64% of the value obtained for a single stage, while $f_1' = (1/0.64)f_1 =$

$1.56f_1$. For $n = 3, f_2' = 0.51f_2$ or approximately $\frac{1}{2}$ the value of a single stage with $f_1' = (1/0.51)f_1 = 1.96f_1$ or approximately *twice* the single-stage value.

Consider the example of the past few sections, where $f_2 = 0.7$ MHz and $f_1 = 35.4$ Hz. For $n = 2$,

$$f_2' = 0.64f_2 = 0.64(0.7 \text{ MHz}) = 0.448 \text{ MHz}$$

and

$$f_1' = 1.56f_1 = 1.56(35.4) = 55.2 \text{ Hz}$$

The bandwidth is now $0.448 \text{ MHz} - 55.2 \text{ Hz} \cong f_2' \Rightarrow 64\%$ of its single-stage value, a drop of some significance.

For the RC-coupled transistor amplifier, if $f_2 = f_\beta$, or if they are close enough in magnitude for both to affect the upper 3-dB frequency the number of stages must be increased by a factor of 2 when determining f_2', due to the increased number of factors $1/(1 + jf/f_x)$.

A decrease in bandwidth is not always associated with an increase in the number of stages if the mid-band gain can remain fixed independent of the number of stages. For instance, if a single-stage amplifier produces a gain of 100 with a bandwidth of 10,000 Hz, the resulting gain-bandwidth product is $10^2 \times 10^4 = 10^6$. For a two-stage system, the same gain can be obtained by having two stages with a gain of 10 since $(10 \times 10 = 100)$. The bandwidth of each stage would then increase by a factor of 10 to 100,000 due to the lower gain requirement and fixed gain-bandwidth product of 10^6. Of course the design must be such as to permit the increased bandwidth and establish the lower gain level.

This discussion of the effects of an increased number of stages on the frequency response was included here, rather than at the conclusion of this chapter, to add a note of completion to the analysis of cascaded transistor amplifier systems. The results, however, can be applied directly to the discussion of FET and vacuum-tube cascaded systems to follow.

7.14 FREQUENCY RESPONSE OF CASCADED FET AMPLIFIERS

A representative cascaded system employing FET amplifiers appears in Fig. 7.50. A similar system was analyzed in depth (for the mid-frequency region) in section 7.3 of this chapter.

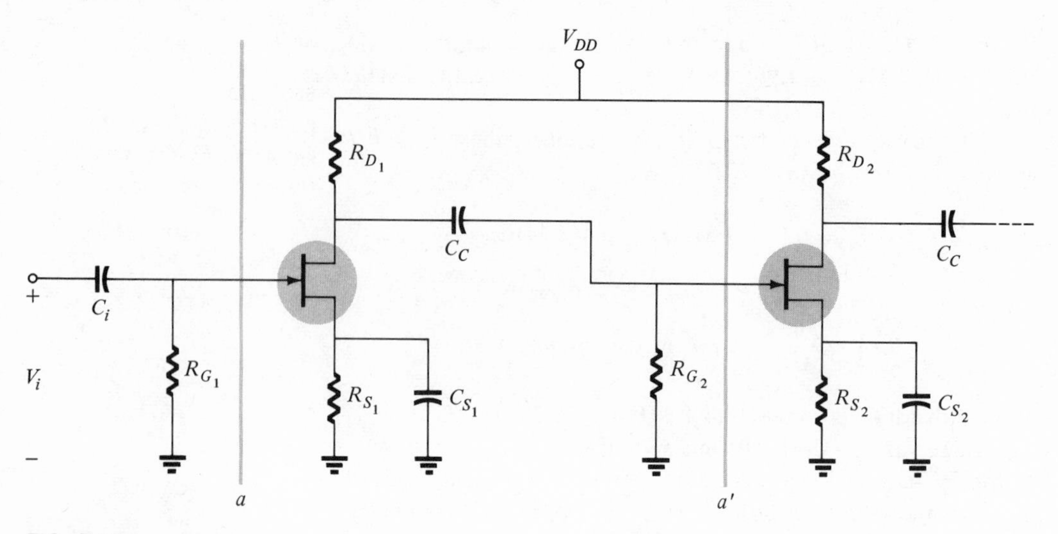

FIG. 7.50
Cascaded FET amplifiers.

The equivalent circuit for the section *a-a′* indicated in Fig. 7.50 is presented in Fig. 7.51 for both the high- and low-frequency regions. It is assumed for the low-frequency response that the breakpoint frequency due to C_s is sufficiently less than that due to C_C, to result in C_C determining the lower-band frequency. For this reason C_s does not appear in the low-frequency model. For future reference, the break-

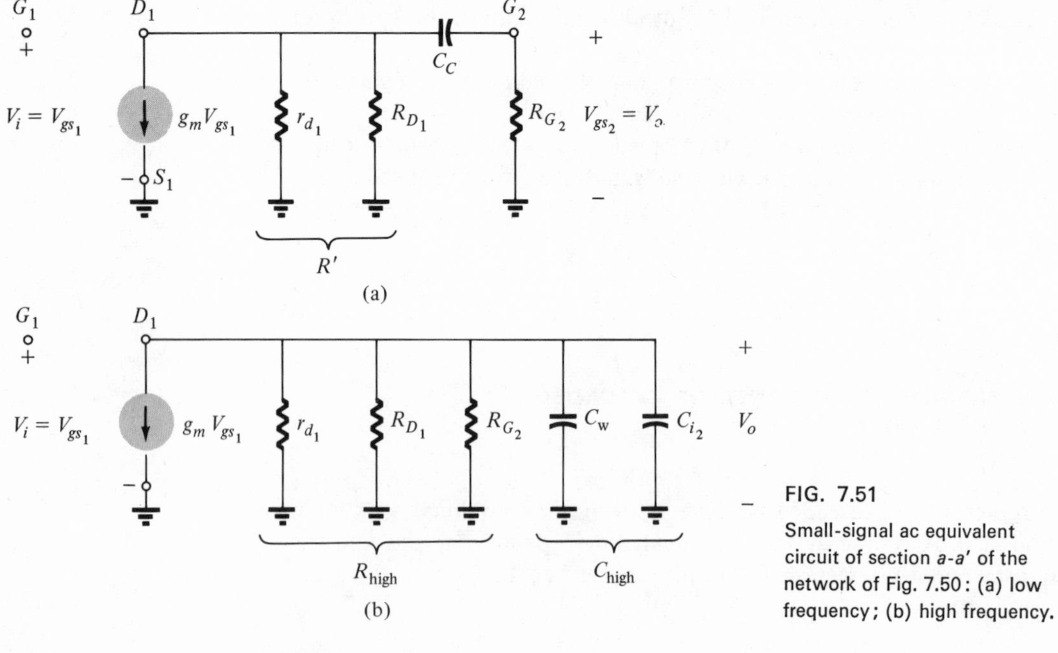

FIG. 7.51
Small-signal ac equivalent circuit of section *a-a′* of the network of Fig. 7.50: (a) low frequency; (b) high frequency.

point frequency determined by C_s is given by

$$f_s = \frac{1 + R_s(1 + g_m r_d)/(r_d + R_D)}{2\pi C_s R_s} \tag{7.48}$$

It was demonstrated in this and Chapter 6 that the mid-band voltage gain is given by

$$A_{v_{\text{mid}}} = \frac{V_o}{V_i} = -g_m R \qquad \text{where } R = r_{d_1} || R_{D_1} || R_{G_2}$$

The similarities between the circuits of Fig. 7.51 and those introduced in the last few sections on the transistor cascaded system are quite obvious. With this in mind, the steps leading to the following results should also be obvious;

$$\frac{A_{v_{\text{low}}}}{A_{v_{\text{mid}}}} = \frac{1}{1 + jf_1/f} \tag{7.49}$$

where

$$f_1 = \frac{1}{2\pi C_C(R' + R_{G_2})}$$

and

$$R' = r_{d_1} || R_{D_1}$$

In addition,

$$\frac{A_{v_{\text{high}}}}{A_{v_{\text{mid}}}} = \frac{1}{1 + jf/f_2} \tag{7.50}$$

where

$$f_2 = \frac{1}{2\pi C_{\text{high}} R_{\text{high}}}$$

and

$$C_{\text{high}} = C_W + C_{i_2}$$

with

$$R_{\text{high}} = R = r_{d_1} || R_{D_1} || R_{G_2}$$

For FET amplifiers, the input capacitance of the succeeding stage, as derived in Chapter 6 is given by

$$C_i = C_{gs} + C_{gd}(1 + |A_v|) \tag{7.51}$$

The gain-bandwidth product is

$$[\text{gain}][\text{BW}] = [A_{v_{\text{mid}}}][f_2 - f_1 \cong f_2] = [g_m R]\left[\frac{1}{2\pi R(C_W + C_i)}\right]$$

and

$$\boxed{\text{GBW} = \frac{g_m}{2\pi(C_W + C_i)} = f_T = A_{v_{\text{mid}}} f_2} \qquad (7.52)$$

The similarities between Eqs. (7.49) and (7.50) with those obtained for the transistor network permit the use of the universal curves of Sections 7.10 and 7.11 for magnitude and phase variations with frequency. The discussion of the previous section is also directly applicable to cascaded FET amplifiers.

7.15 FREQUENCY RESPONSE OF CASCADED TRIODE AND PENTODE AMPLIFIERS

A cascaded vacuum triode amplifier system appears in Fig. 7.52. The high- and low-frequency equivalent circuits for the section *a-a'* of Fig. 7.52 appear in Fig. 7.53. The similarities between the circuits of Fig. 7.53 and those of Fig. 7.51 should be immediately obvious. The following discussion of cascaded vacuum-tube triode amplifiers will be a

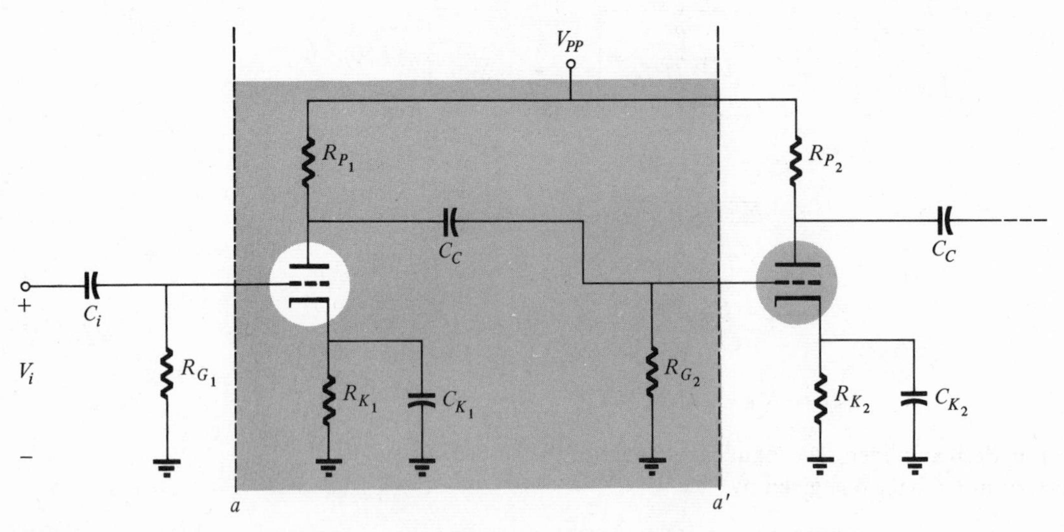

FIG. 7.52

Cascaded vacuum tube triode stages.

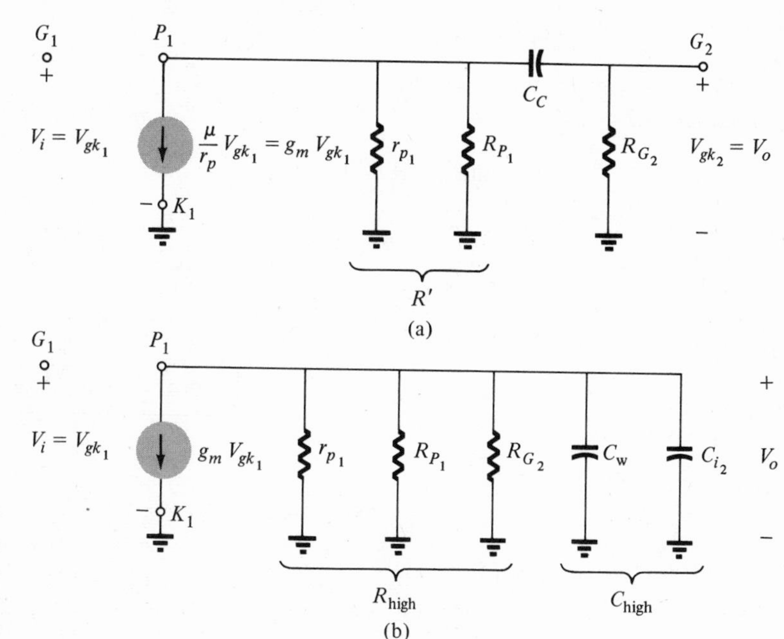

FIG. 7.53
Small-signal ac equivalent circuit of section *a-a'* of the network of Fig. 7.52: (a) low frequency; (b) high frequency.

replica of that presented for the FET stages of the Section 7.14. The breakpoint due to C_K is again assumed to be sufficiently less than that due to C_C that it may be neglected in the low-frequency models. The breakpoint due to C_K is given by

$$f_K = \frac{1 + R_K(1+\mu)/(r_p + R_P)}{2\pi C_K R_K} \tag{7.53}$$

The mid-band gain is

$$A_{v_{\text{mid}}} = \frac{V_o}{V_i} = -g_m R \tag{7.54}$$

where

$$R = r_p \| R_P \| R_G$$

and

$$\frac{A_{v_{\text{low}}}}{A_{v_{\text{mid}}}} = \frac{1}{1 + jf_1/f} \tag{7.55}$$

where

$$f_1 = \frac{1}{2\pi C_C(R' + R_G)}$$

and

$$R' = r_p || R_P$$

with

$$\boxed{\frac{A_{v_{\text{high}}}}{A_{v_{\text{mid}}}} = \frac{1}{1 + jf/f_2}} \tag{7.56}$$

where

$$f_2 = \frac{1}{2\pi C_{\text{high}} R_{\text{high}}}$$

$$C_{\text{high}} = C_W + C_{pk_1} + C_{i_2}$$

$$R_{\text{high}} = R = r_p || R_P || R_G$$

For triode amplifiers, the input capacitance of the suceeding stage is given by

$$\boxed{C_i = C_{gk} + G_{gp}(1 + |A_v|)} \tag{7.57}$$

Through similarities with the FET circuit of Fig. 6.22b. Eq. (7.57) can be derived quite directly through the use of the following triode equivalent circuit (Fig. 7.54). The gain-bandwidth product is

$$\boxed{\text{GBW} = \frac{g_m}{2\pi C_{\text{high}}} = f_T = |A_{v_{\text{mid}}}| f_2} \tag{7.58}$$

As for the FET stages, the universal curves of Sections 7.10 and 7.11 and the discussion of Section 7.12 are also applicable to triode stages.

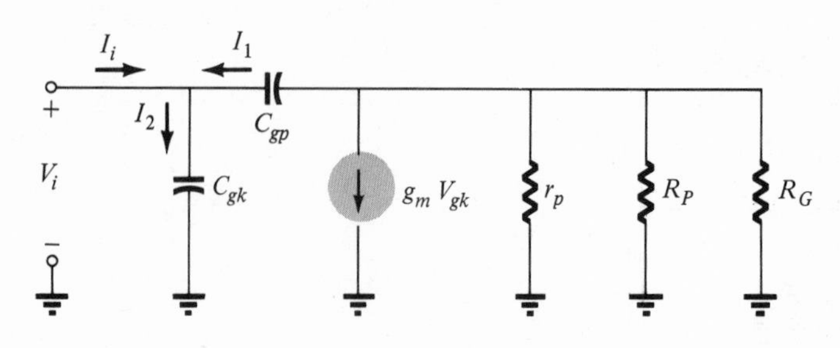

FIG. 7.54

Small-signal ac equivalent circuit employed in the derivation of the input Miller capacitance of a triode. It has the same basic structure as the circuit of Fig. 6.22a employed in the derivation of the Miller capacitance for the FET amplifiers.

For pentode stages, the circuits and equations are exactly the same as those obtained for the triode system except for C_{high}. The action of the screen grid between plate and grid permits the approximation $C_{gp} \cong 0$, eliminating the high Miller feedback capacitance so that

$$C_{\text{high}_{(\text{pentode})}} \cong C_W + C_{pk_1} + C_{gk_2} \quad (7.59)$$

PROBLEMS

§ 7.3

1. A two-stage RC-coupled amplifier is shown in Figure 7.55. Calculate Z_i snd Z_o for the amplifier.

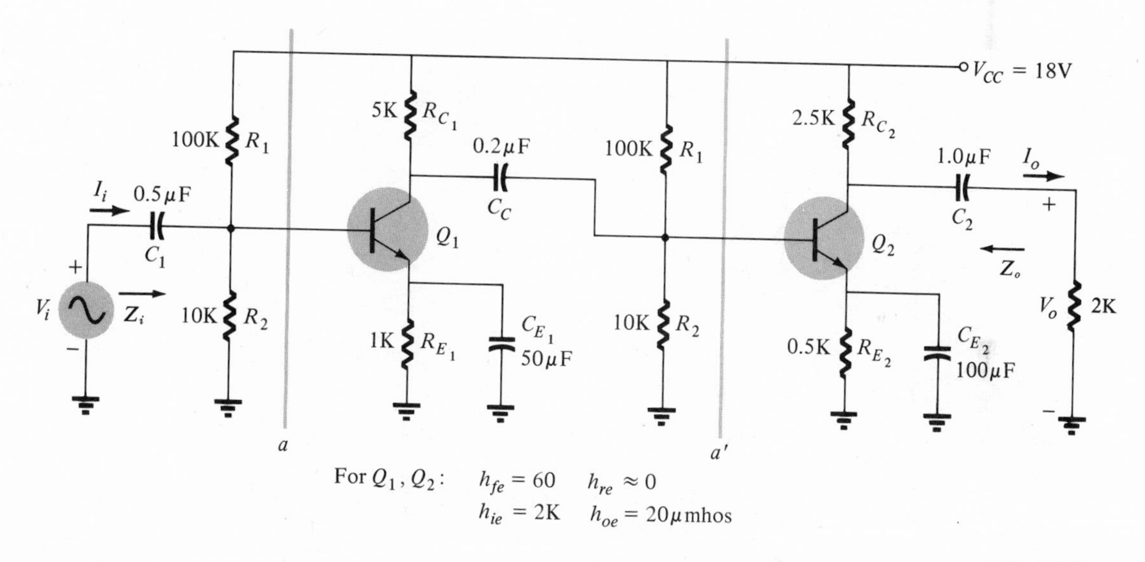

FIG. 7.55
Two-stage RC-coupled amplifier.

2. Calculate the voltage gain ($A_v = V_o/V_i$) for the amplifier of Fig. 7.55.
3. Calculate the current gain ($A_i = I_o/I_i$) for the amplifier of Fig. 7.55.
4. If the collector resistors in the amplifier of Fig. 7.55 are both made 15 K, calculate the voltage and current gain of the overall amplifier.
5. If the transistors of Fig. 7.55 are replaced by two having the hybrid parameters specified below, calculate the voltage and current gain of the overall amplifier.

$$h_{ie} = 2\text{ K},\ h_{re} \approx 0,\ h_{fe} = 75,\ h_{oe} = 40\ \mu\text{mhos}$$

6. Design a two-stage RC-coupled amplifier as in Fig. 7.55 to provide an overall voltage gain of 2000. The circuit is to operate into a load of 10 K, while the signal is supplied from a perfect voltage source. Show typical component values for each element and calculate the voltage gain of the resulting circuit as a check.

§ 7.4

7. Calculate the impedance seen looking into the primary of a 5:1 step-down transformer connected to a load of 20 Ω.
8. Calculate the necessary transformer turns ratio to match a 50-Ω load to a 10-K source impedance.
9. (a) Calculate the voltage gain (V_o/V_i) of the transformer-coupled amplifier of Fig. 7.56.
   (b) What is the voltage gain of the circuit of Fig. 7.56 if the load is reduced to 0.5 K?

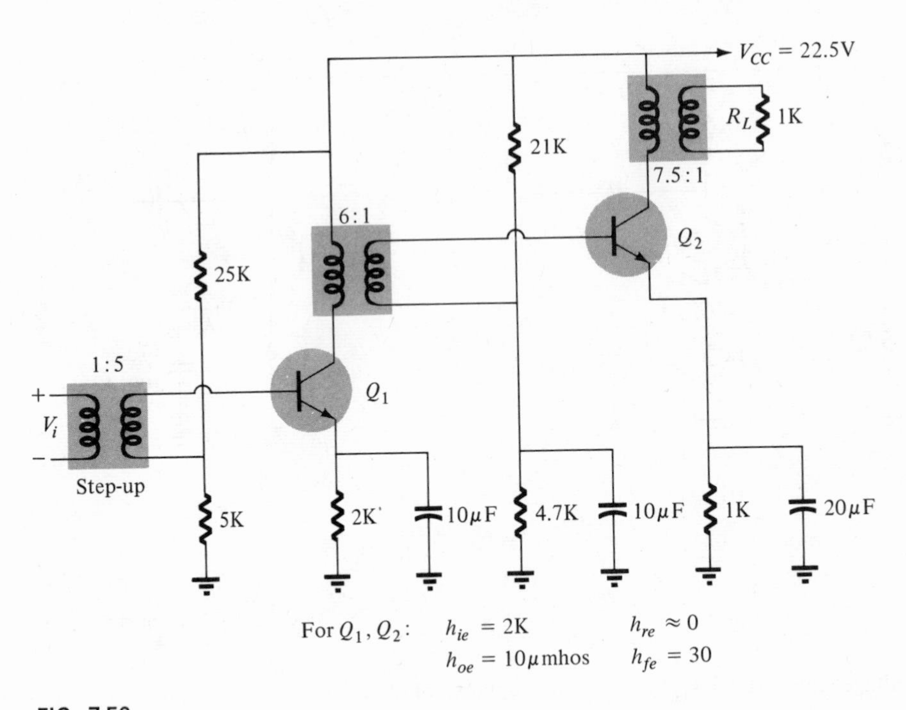

FIG. 7.56

Two-stage transformer-coupled amplifier.

§ 7.6

10. Calculate the voltage gain (V_o/V_i) of the two-stage amplifier of Fig. 7.57.
11. For the circuit of Fig. 7.57 calculate the overall current gain (I_o/I_i).

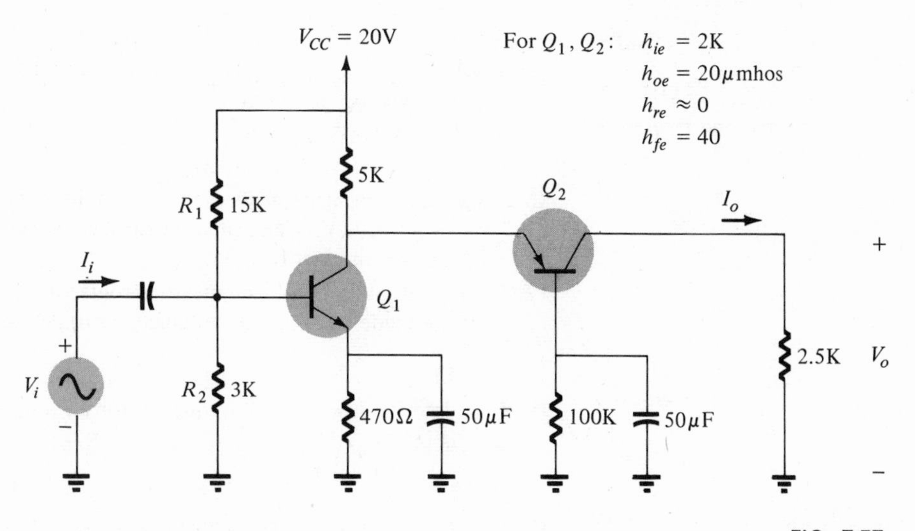

FIG. 7.57
Two-stage amplifier for Problem 7.10.

§ 7.7

12. Calculate the current gain of the amplifier of Figure 7.58.

13. For the amplifier of Fig. 7.58 calculate Z_i and Z_o.

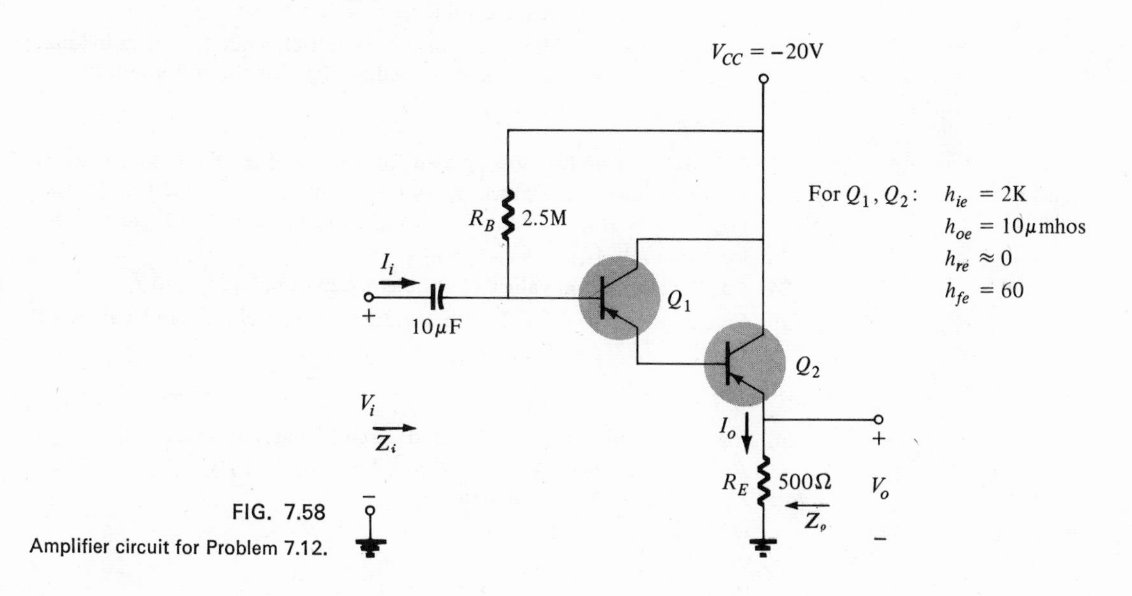

FIG. 7.58
Amplifier circuit for Problem 7.12.

§ 7.8

14. Calculate the voltage gain of a pentode amplifier as in Fig. 7.27a for tube parameters $g_m = 8000$ μmhos, $r_p = 2$ M.

§ 7.9

15. Calculate the decibel power gain for
(a) $P_o = 100$ W, $P_i = 5$ W.
(b) $P_o = 100$ mW, $P_i = 5$ mW.
(c) $P_o = 100\ \mu$W, $P_i = 20\ \mu$W.

16. Two voltage measurements made across the same resistance are $V_1 = 25$ V, $V_2 = 100$ V. Calculate the decibel power gain of the second reading over the first reading.

17. Input and output voltage measurements of $V_i = 10$ mV and $V_o = 25$ V are made. What is the voltage gain in decibels?

§ 7.11

18. Calculate the lower 3-dB frequency for an amplifier circuit between *a-a'* as in Fig. 7.55.

§ 7.12

19. For the circuit in Fig. 7.55 calculate f_2, f_β, and f_T using: $C_{b'c} = 2$ pF, $C_{b'e} = 10$ pF, $g_m = 40\ \mu$mhos, $g_{b'e} = 1/1$ K, $C_W = 7$ pF, $|A_{\text{mid}}| = = 40$, $R_L = 10$ K, $R_i = 2$ K, $h_{fe} = 30$.

§ 7.13

20. Calculate the overall voltage gain of four identical stages of an amplifier, each having a gain of 20.

21. Calculate the overall upper 3-dB frequency for a four-stage amplifier having an individual stage value of $f_2 = 2.5$ MHz.

22. A four-stage amplifier has a lower 3-dB frequency for an individual stage of $f_1 = 40$ Hz. What is the value of f_1 for the full amplifier.

§ 7.14

23. Calculate the mid-frequency gain for a stage of a FET amplifier as in Fig. 7.50 for circuit values: $g_m = 6000\ \mu$mhos, $r_d = 50$ K (for Q_1 and Q_2); $R_{D_1} = R_{D_2} = 10$ K; $R_{G_1} = R_{G_2} = 1$ M; $C_s = 10\ \mu$F, $R_s = 1$ K, $C_C = 0.1\ \mu$F, $C_{gs} = C_{gd} = 4$ pF.

24. For the circuit and values of Problem 7.23 calculate f_1 and f_2.

25. Calculate the gain-bandwidth product for the circuit and values of Problem 7.23.

§ 7.15

26. For the circuit of Fig. 7.52 and circuit values $r_p = 25$ K, $\mu = 50$, $R_P = 10$ K, $R_G = 0.5$ M, $R_K = 2$ K $C_K = 1\ \mu$F, $C_C = 0.5\ \mu$F, $C_{gk} = C_{gp} = 5$ pF, calculate
(a) $A_{v,\text{mid}}$.
(b) f_K.
(c) f_T.

8

Large-Signal Amplifiers

8.1 GENERAL

An amplifier system generally consists of a signal pickup transducer, followed by a small-signal amplifier, a large-signal amplifier, and an output transducer device. The input transducer signal is generally quite small and must be amplified sufficiently to be used to operate some output device. The factors of prime interest in small-signal amplifiers are usually linearity and gain. Since the signal voltage and current from the input transducer is usually small, the amount of power handling capacity and power efficiency are of slight concern. Voltage amplifiers provide a large enough voltage signal to the large-signal amplifier stages to operate such output devices as speakers and motors. A large-signal amplifier must operate efficiently and be capable of handling large amounts of power—typically, a few watts to hundreds of watts. This chapter concentrates on the amplifier stage used to handle large signals, typically a few volts to tens of volts. The amplifier factors of greatest concern are the power efficiency of the circuit, the maximum amount of power that the circuit is capable of handling, and impedance matching to the output device.

A class-A series-fed amplifier stage is considered first to show some of the limitations in using such a simple circuit connection. The single-ended transformer-coupled stage is then discussed to show how impedance matching between driver stage and load (output transducer) is accomplished. The push-pull connection, a very popular connection for low distortion and efficient coupling of the signal to a speaker or motor device, is discussed next. Finally, complementary transistors, for push-pull operation without a transformer are presented.

8.2 SERIES-FED CLASS-A AMPLIFIER

The simple fixed-bias circuit connection can be used as a large-signal class-A amplifier as shown in Fig 8.1. The only difference between this circuit and the small-signal version considered previously is that the signals handled by the large-signal circuit are in the range of volts and the transistor used is a power transistor capable of operating in the range of a few watts. As will be shown, this circuit is not the best to use for a large-signal amplifier.

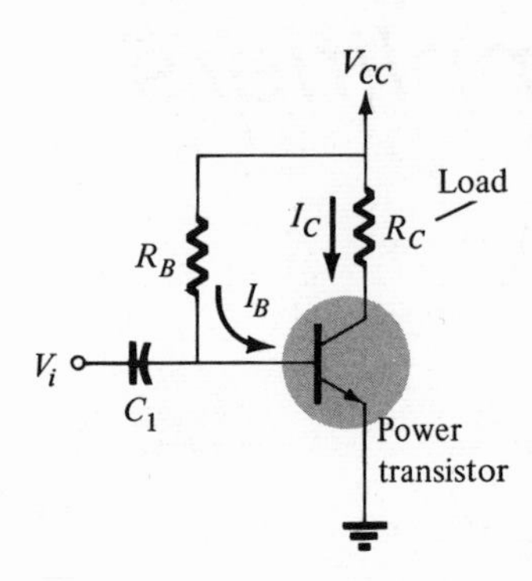

FIG. 8.1
Series-fed class-A large signal amplifier.

Figure 8.2a shows a typical power transistor circuit with appropriate circuit values. The transistor-collector characteristic of Fig. 8.2b shows the load line and input and output signals. The input signal is an ac voltage of 0.5 V, peak amplitude. The circuit has an ac input resistance of 50 Ω. The output ac signal is to be developed across the 20-Ω load resistor. For the bias value of base current shown, the transistor is operated at a collector-emitter voltage of 10 V and a collector current of about 500 mA.

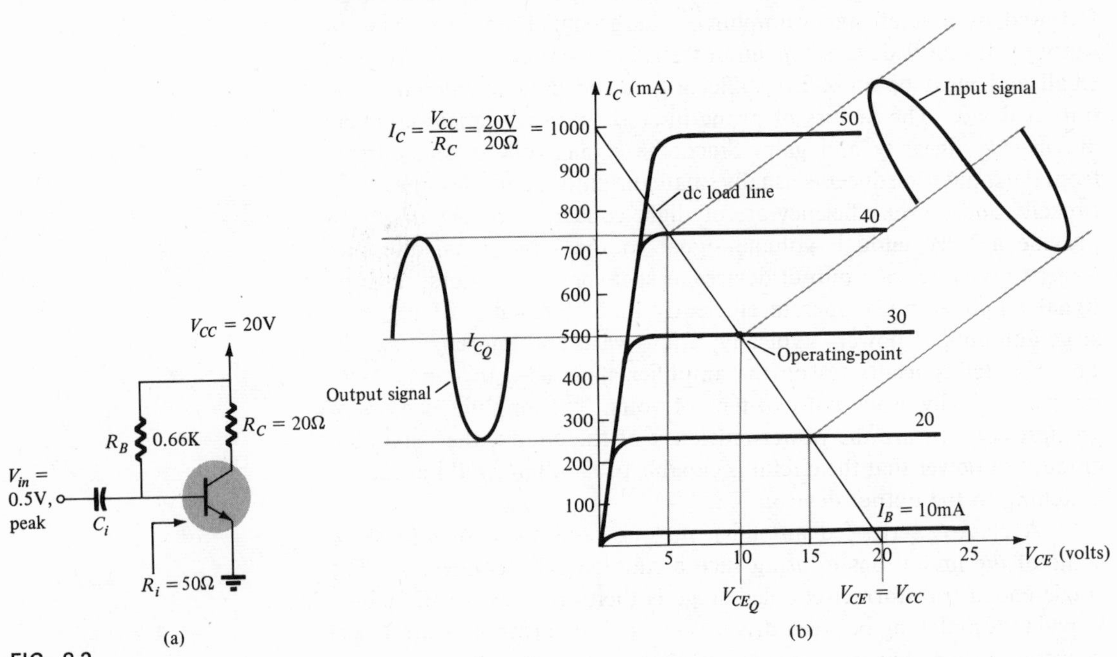

FIG. 8.2
Operation of a series-fed circuit.

The input ac current variation is calculated to be

$$I_i = \frac{V_i}{R_i} = \frac{0.5\ \text{V}}{50\ \Omega} = 10\ \text{mA, peak}$$

As shown on the transistor-collector characteristic (Fig. 8.2b) the corresponding current swing in the output (across the load resistor) is about 250 mA, peak. Corresponding to the indicated current swing is a voltage variation of 5 V, peak. From the present information we can calculate the current gain, voltage gain, and power gain of the circuit as follows:

$$A_v = \frac{V_o}{V_i} = \frac{5\ \text{V, peak}}{0.5\ \text{V, peak}} = 10$$

$$A_i = \frac{I_o}{I_1} = \frac{250\ \text{mA, peak}}{10\ \text{mA, peak}} = 25$$

$$A_p = \text{power gain} = A_v A_i = (10)(25) = 250$$

The overall operation of the circuit is poor with small voltage gain, a current gain of 25, and a power gain of 250. So far the circuit operation seems poor and worse is yet to come.

So far now no consideration of the power distribution in the amplifier has been considered since the amount of power handled by small-signal circuits is quite small. In the large-signal circuit the amount of power in different parts of the amplifier is considerable and the efficiency of the amplifier circuit is of great interest. The dc battery is the source of power in a transistor amplifier. The dc power drawn from the battery is dissipated as heat lost in the load resistor and the transistor. The output power (ac signal developed across the load) is a part of the power taken from the dc battery. In fact, the operation of the circuit of Fig. 8.2a is to convert as much of the dc power drawn from the battery into ac power (output power) across the load (the 20-Ω resistor in this case). If the load were a speaker the power delivered would result in audible sound.

In the present circuit we can calculate separately a number of different power terms as follows:

The average power taken from the dc battery is

$$\boxed{P_i(\text{dc}) = V_{CC} I_{CQ}} \tag{8.1}$$

the product of the dc battery voltage times the *average* current drawn from the battery—the quiescent or average current for the nondistorted output signal, as shown in Fig. 8.2b.

The ac signal power developed across the load is

$$\boxed{P_o(\text{ac}) = I_C^2\,(\text{rms})\, R_C} \tag{8.2}$$

the square of the root-mean square value of the ac current *through* the load resistor times the value of load resistance.

The dc power dissipated by the load resistor is

$$P_L(\text{dc}) = I_{C_Q}^2 R_L \tag{8.3}$$

The dc power dissipated as heat by the load resistor is equal to the average or dc current through the resistor times the value of the load resistance ($R_L = R_C$ in the present circuit).

The dc power dissipated by the transistor is

$$P_t = P_i - P_o - P_L \tag{8.4}$$

the product of the average voltage across the transistor times the average current flowing through the transistor. This dc power is dissipated as heat with the transistor power rating indicating the maximum amount of power the particular device is capable of handling. (Transistor ratings and specifications are discussed in Section 8.7.)

The efficiency of the circuit in converting the dc power drawn from the battery into ac signal power across the load is calculated to be

$$\%\text{ efficiency} = \eta = \frac{P_o}{P_i} \times 100 \tag{8.5}$$

The numerical calculations for the circuit of Fig. 8.2 are carried out in Example 8.1.

EXAMPLE 8.1 Calculate the various power terms for the circuit of Fig. 8.2a and transistor characteristic of Fig. 8.2b. Obtain the circuit power efficiency from these calculated values.

Solution

(a) $$P_i(\text{dc}) = V_{CC} I_{CQ} = (20\text{ V})(500\text{ mA}) = 10\text{ W}.$$

(b) $$P_o(\text{ac}) = I_C^2(\text{rms})\, R_C = \left(\frac{0.25}{\sqrt{2}}\right)^2 20 = 0.625\text{ W}$$

(c) $$P_L(\text{dc}) = I_{CQ}^2 R_L = (0.5\text{ A})^2(20\ \Omega) = 5\text{ W}$$

(d) $$P_t(\text{dc}) = P_i - P_o - P_L = 10 - 0.625 - 5 = 4.375\text{ W}$$

(e) $$\eta = \frac{P_o}{P_i} \times 100 = \frac{0.625}{10} \times 100 = 6.25\%.$$

The power efficiency of 6.25% is extremely poor and the circuit of Fig. 8.2 is not a good amplifier for handling large amounts of power.

Of the 10 W of power drawn from the battery, about half is dissipated in heat by the transistor and almost the same amount is wasted as heat in the load resistor, with only a very small amount of ac power being developed across the load. Although a 20-Ω speaker could be directly connected in the circuit as the load resistor the operation of the circuit would be quite inefficient.

8.3 TRANSFORMER-COUPLED AUDIO POWER AMPLIFIER

A more reasonable class-A amplifier connection uses a transformer to couple the load to the amplifier stage as shown in Fig. 8.3a. This is a simple version of the circuit for the presentation of a few basic concepts. More practical circuit versions will be covered shortly. Fig. 8.3b shows the output coupling transformer with voltage, current, and impedances indicated.

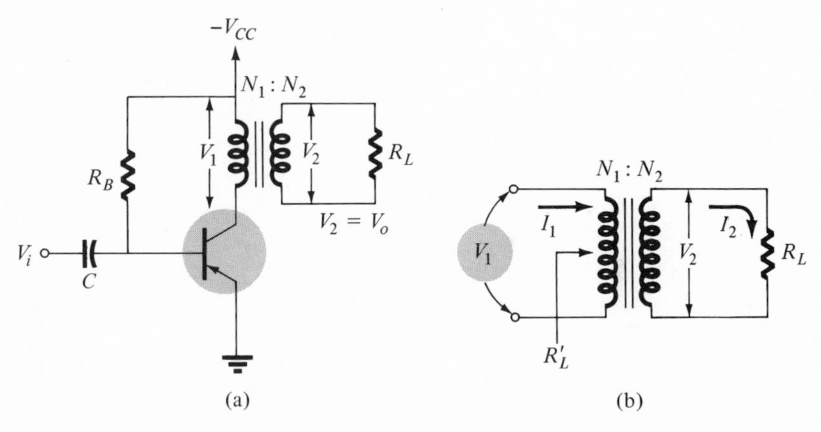

FIG. 8.3
Transformer-coupled audio power amplifier.

Consider first the operation of the transformer as an impedance device. Typically, a low-impedance speaker is connected as the load. We have seen the result of connecting this speaker directly as the load in the circuit of Fig. 8.2. With the transformer used to connect the speaker load to the transistor-collector circuit, there results an impedance transformation.

Transformer Impedance Matching

The resistance seen looking into the primary of the transformer is related to the resistance connected across the secondary. The ratio of secondary resistance to primary resistance may be expressed as follows:

$$\frac{R'_L = V_1/I_1}{R_L = V_2/I_2} = \frac{R'_L}{R_L} = \frac{V_1}{I_1}\frac{I_2}{V_2} = \frac{V_1}{V_2}\frac{I_2}{I_1} = \frac{N_1}{N_2}\frac{N_1}{N_2} = \left(\frac{N_1}{N_2}\right)^2$$

where $V_1/V_2 = N_1/N_2$ and $I_2/I_1 = N_1/N_2$. Hence the ratio of the transformer input and output resistance varies directly as the *square* of the transformer turns ratio:

$$\boxed{\frac{R'_L}{R_L} = \left(\frac{N_1}{N_2}\right)^2 = a^2} \tag{8.6}$$

and

$$\boxed{R'_L = a^2 R_L = \left(\frac{N_1}{N_2}\right)^2 R_L} \tag{8.7}$$

where

R_L = resistance of load connected across the transformer secondary,
R'_L = effective resistance seen looking into primary of transformer,
$a = N_1/N_2$ is the step-down turns ratio needed to make the load resistance appear as a larger effective resistance seen from the transformer primary.

EXAMPLE 8.2 Calculate the effective resistance (R'_L) seen looking into the primary of a 15:1 transformer connected to an output load of 8 Ω.

Solution: Using Eq. (8.7)

$$R'_L = \left(\frac{N_1}{N_2}\right)^2 R_L = (15)^2 8 = 1800\ \Omega = \mathbf{1.8\ K}$$

EXAMPLE 8.3 What transformer turns ratio is required to match a 16-Ω speaker load to an amplifier so that the effective load resistance is 10 K?

Solution: Using Eq. (8.6)

$$\frac{R'_L}{R_L} = \left(\frac{N_1}{N_2}\right)^2 = \frac{10{,}000}{16} = 625$$

$$\left(\frac{N_1}{N_2}\right) = \sqrt{625} = \mathbf{25:1}$$

For the transformer-coupled amplifier of Fig. 8.3 the various power losses and output power developed across the load resistor will allow calculation of the circuit power efficiency. Figure 8.4 shows a

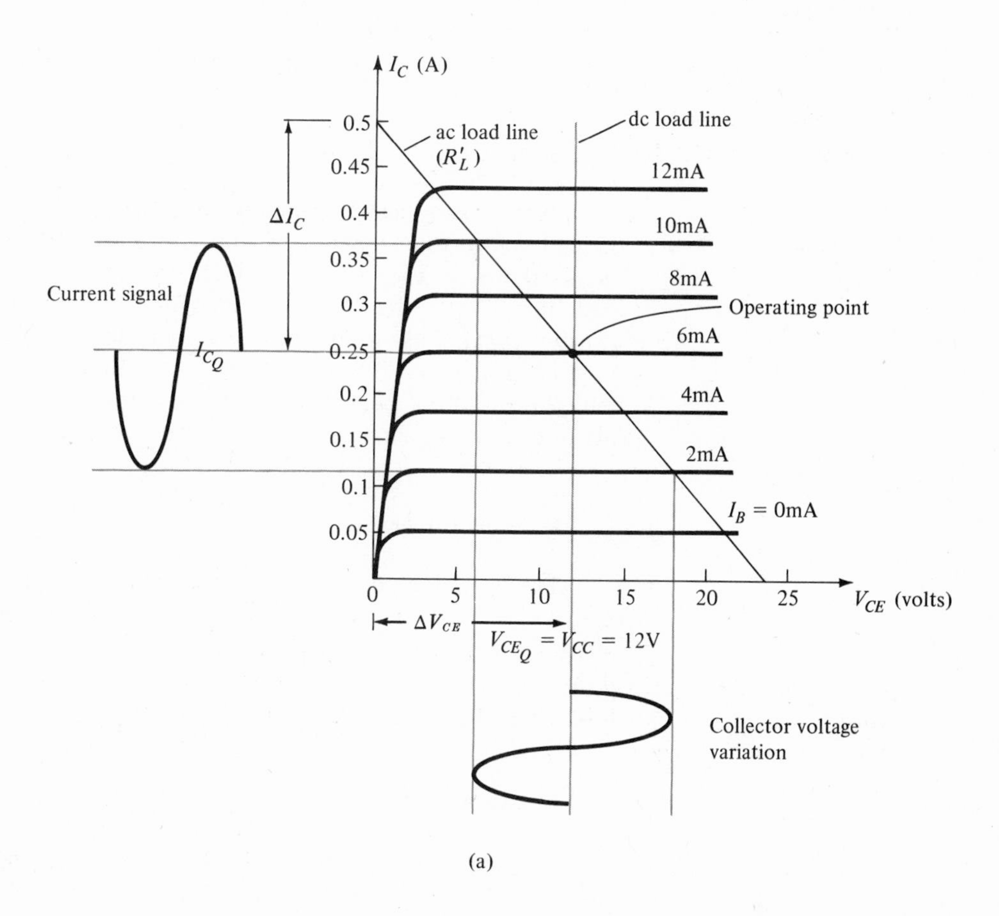

(a)

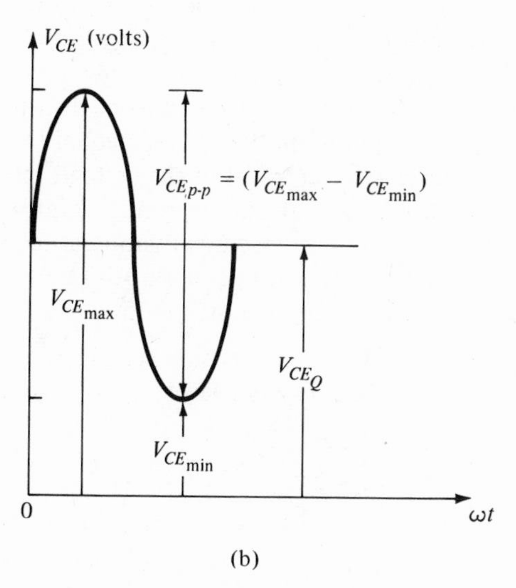

(b)

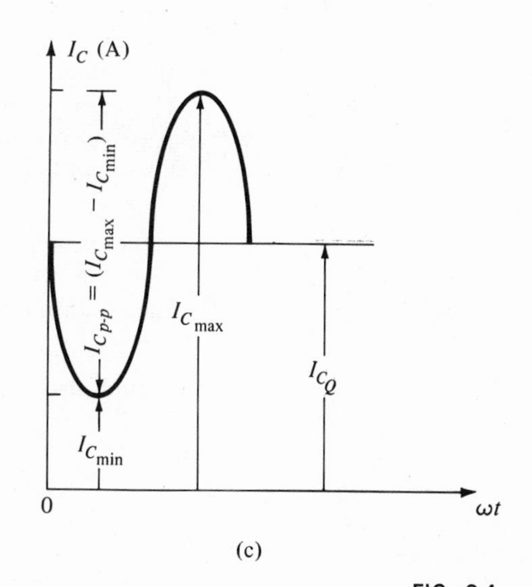

(c)

FIG. 8.4
Graphical operation of transformer-coupled audio power amplifier.

transistor-collector characteristic, load lines, input, and output signals.

dc Load Line

The transformer dc (winding) resistance is used to obtain the dc load line for the circuit. Typically, this dc resistance is quite small and is shown in Fig. 8.4a to be 0 Ω, providing a straight (vertical) load line. This is the ideal load line for the transformer. Practical transformer windings would provide a slight slope for the load line, but only the ideal case will be considered in this discussion. There is no dc voltage drop across the dc load resistance in the ideal case and the load line is drawn straight vertically from the voltage point, $V_{CE_Q} = V_{CC}$.

Quiescent Operating Point

The operating point is obtained graphically as the point of intersection of the dc load line and the transistor base current curve. From the operating point the quiescent collector current I_{C_Q} is read. The value of the base current is calculated separately from the circuit as considered in the dc bias calculations of Chapter 4.

ac Load Line

In order to obtain the ac signal operation it is necessary to first calculate the ac load resistance seen looking into the primary side of the transformer, and then to draw the ac load line on the transistor characteristic. The effective load resistance is calculated using Eq. (8.7) from the values of secondary load resistance and transformer turns ratio. Having obtained the value of R'_L the ac load line must be drawn so that it passes through the operating point and has a slope equal to $-1/R'_L$. The load line slope is the negative reciprocal of the ac load resistance. Since the collector signal passes through the operating point when no signal is applied the load line must pass through the operating point.

In order to simplify drawing a load line of slope $-1/R'_L$ through the operating point the following technique may be used:

If the ac signal were to vary from the quiescent level to 0 V, it would also vary from the quiescent current level, I_{C_Q}, by an amount

$$\Delta I_C = \frac{\Delta V_{CE}}{R'_L} \qquad (8.8)$$

Mark a point on the y-axis of the transistor characteristic ΔI_C units above the quiescent level and connect this point through the operating point to draw the ac load line desired (see Fig. 8.4a). Notice that the ac load line shows that the output signal swing can exceed the value of V_{CC}, the supply voltage. In fact the voltage developed across the transformer primary can be quite large. One of the maximum operating values that will have to be checked carefully is that of $V_{CE_{max}}$, as specified by the transistor manufacturer, to see that the value of maximum voltage obtained after drawing the ac load line does not exceed the transistor rated maximum value. Assuming for the moment that maximum power and voltage ratings are not exceeded (the consideration of these maximum values is deferred until Section 8.7), ac signal swings of current and voltage are obtained as shown in Fig. 8.4a and redrawn in detail in Figs. 8.4b and 8.4c.

Signal Swing and Output ac Power

From the signal variations as shown in Figs. 8.4b and 8.4c the values of the peak-to-peak signal swings are obtained

$$V_{\text{swing}} = V_{CE}\,(\text{peak-to-peak}) = (V_{CE_{max}} - V_{CE_{min}}) \tag{8.9}$$

$$I_{\text{swing}} = I_C\,(\text{peak-to-peak}) = (I_{C_{max}} - I_{C_{min}}) \tag{8.10}$$

using the values as indicated (and defined) in Fig. 8.4. The ac power developed across the transformer primary can be calculated to be

$$P_o(\text{ac}) = V_{CE}\,(\text{rms})\,I_C\,(\text{rms})$$

$$= \frac{V_{CE}\,(\text{peak})}{\sqrt{2}}\,\frac{I_C\,(\text{peak})}{\sqrt{2}}$$

$$= \frac{V_{CE}\,(\text{peak-to-peak})/2}{\sqrt{2}} \times \frac{I_C\,(\text{peak-to-peak})/2}{\sqrt{2}}$$

$$\boxed{P_o(\text{ac}) = \frac{(V_{CE_{max}} - V_{CE_{min}})(I_{C_{max}} - I_{C_{min}})}{8}} \tag{8.11}$$

The ac power calculated is that developed across the primary of the transformer. Assuming a highly efficient transformer the power across the speaker is approximately equal to that calculated by Eq. (8.11). For our purposes an ideal transformer will be assumed so that the ac power calculated using Eq. (8.11) is also the ac power delivered to the load.

For the ideal transformer considered, the voltage across the secondary of the speaker can be calculated from

$$V_2 = V_S = V_L = \left(\frac{N_2}{N_1}\right) V_1 \tag{8.12}$$

where the secondary voltage (V_2) equals the speaker (V_S) or load voltage

(V_L). The load voltage is related by the transformer turns ratio, N_2/N_1, to the voltage developed across the transformer primary (V_1). The voltage across the primary was previously labelled V_{CE} (rms), and for power considerations the rms values of voltage are usually used [unless otherwise stated, as in Eq. (8.11)].

From the calculated value of the secondary rms voltage the power across the load can be obtained

$$P_S = P_L = \frac{V_L{}^2}{R_L} \tag{8.13}$$

and equals the power calculated using Eq. (8.11). Thus the ac power can be calculated in a number of ways, including the following:

$$I_L\ (\text{rms}) = \frac{N_1}{N_2} I_C\ (\text{rms}) \tag{8.14}$$

where I_L is the rms value of the current through the load resistor (or speaker resistance) and the load current is related to the rms value of the ac component of collector current by the transformer turns ratio.

The ac power is then calculated from

$$P_S = P_L = I_L^2 R_L \tag{8.15}$$

EXAMPLE 8.4 The circuit of Fig. 8.5a shows a transformer-coupled class-A audio power amplifier driving an 8-Ω speaker. The coupling transformer has a 3:1 step-down turns ratio. If the circuit component values result in a dc base current of 6 mA and the input signal (V_i) results in a peak base current swing of 4 mA, calculate the following circuit values using the transistor characteristic of Fig. 8.5b: $V_{CE_{\max}}$, $V_{CE_{\min}}$, $I_{C_{\max}}$, $I_{C_{\min}}$, the rms values of load current and voltage, and the ac power developed across the load. Caclulate the ac power using different equations as a check; that is, Eq. (8.11), Eq. (8.13), and Eq. (8.15). Draw the dc and ac load lines to obtain the voltages and currents in the collector side of the transformer.

Solution

(a) The dc load line can be drawn vertically from the voltage point $V_{CEQ} = V_{CC} = 10$ V (see Fig. 8.5c).

(b) For $I_B = 6$ mA the operating point on Fig. 8.5c is

$$V_{CEQ} = 10\ \text{V} \qquad I_{CQ} = 140\ \text{mA}$$

(c) The effective ac resistance R_L' is [use Eq. (8-7)]

$$R_L' = \left(\frac{N_1}{N_2}\right)^2 R_L = (3)^2 8 = 72\ \Omega.$$

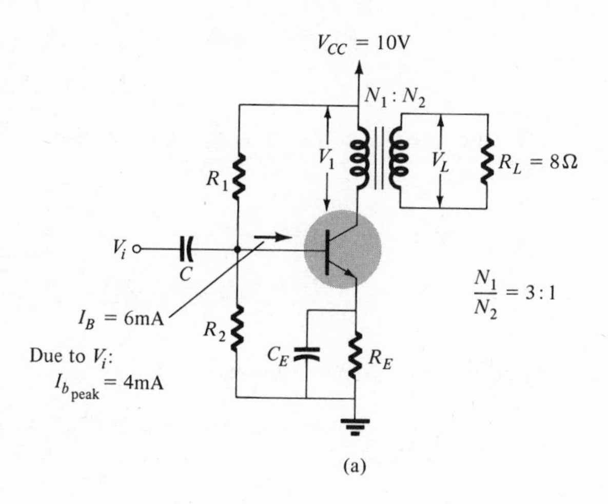

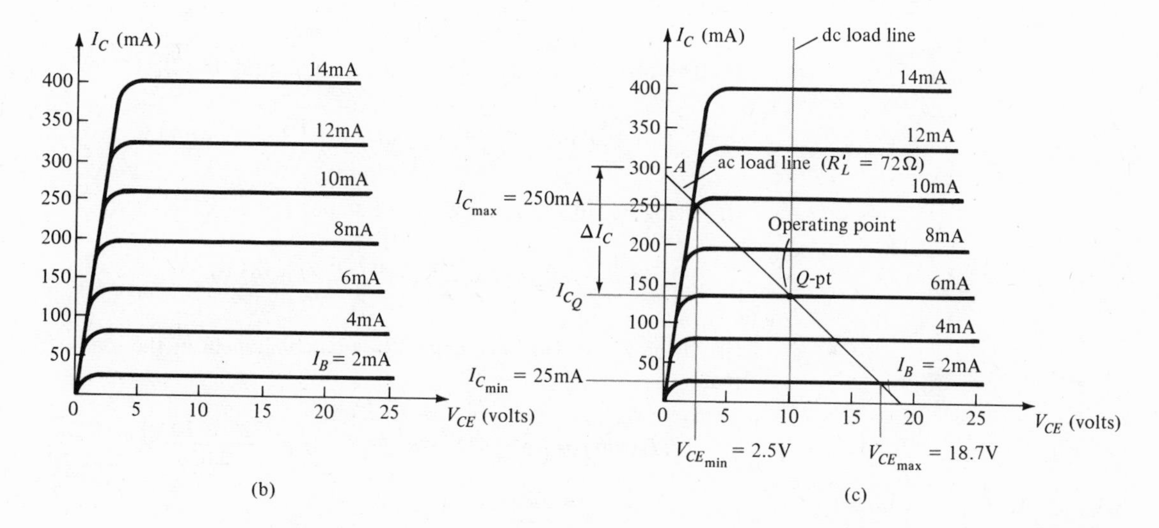

FIG. 8.5
Transformer-coupled audio power amplifier and transistor characteristic for Example 8.4.

(d) Draw the ac load line as follows: Use Eq. (8.8) to calculate the current swing above the operating current

$$\Delta I_C = \frac{\Delta V_{CE}}{R'_L} = \frac{10\text{ V}}{72\ \Omega} = 139\text{ mA}$$

Mark point A (Fig. 8.5c) $= I_{C_{EQ}} + \Delta I_C = 150 + 139 = 289$ mA along the y-axis. Connect point A to Q-point to draw ac load line.

(e) For the given peak base current swing of 4 mA, the maximum and minimum values of collector current and voltage obtained from Fig. 8.5c are

$$V_{CE_{min}} = 2.5 \text{ V} \qquad I_{C_{min}} = 25 \text{ mA}$$

$$V_{CE_{max}} = 17.5 \text{ V} \qquad I_{C_{max}} = 250 \text{ mA}$$

(f) Calculate the ac power across the transformer primary using Eq. (8.11)

$$P_o(\text{ac}) = \frac{(V_{CE_{max}} - V_{CE_{min}})(I_{C_{max}} - I_{C_{min}})}{8}$$

$$= \frac{(17.5 - 2.5)(250 - 25) \times 10^{-3}}{8} = \mathbf{0.423\ W}$$

(g) Calculate the rms voltage across the primary

$$V_1\,(\text{rms}) = \frac{V_1(\text{peak-to-peak})}{2\sqrt{2}} = \frac{V_{CE_{max}} - V_{CE_{min}}}{2\sqrt{2}}$$

$$= \frac{15}{2.828} = 5.3 \text{ V}$$

(h) The rms value of the load voltage is [using Eq. (8.12)]

$$V_L\,(\text{rms}) = \left(\frac{N_2}{N_1}\right) V_1\,(\text{rms}) = \left(\frac{1}{3}\right)(5.3) = 1.77 \text{ V}$$

(i) Using Eq. (8.13) to calculate the ac power:

$$P_L(\text{ac}) = \frac{V_L^2}{R_L} = \frac{(1.77)^2}{8} = \mathbf{0.392\ W}$$

(j) Using Eq. (8.14) to calculate the rms component of the load current

$$I_L(\text{rms}) = \left(\frac{N_1}{N_2}\right) \frac{(I_{C_{max}} - I_{C_{min}})/2}{\sqrt{2}}\,(3) = \frac{(225 \times 10^{-3})}{2.828}$$

$$= 238 \text{ mA}$$

(k) Calculating the ac power using Eq. (8.15)

$$P_L(\text{ac}) = I_L^2 R_L = (238 \times 10^{-3})^2 8 = \mathbf{0.454\ W}$$

Considering the inaccuracies of drawing the load lines and obtaining the values of minimum and maximum currents and voltages graphically, the power calculated by the various methods is in good agreement.

Power and Efficiency Calculations

So far we have considered calculating the ac power delivered to the load (the output ac power). We next consider the input power from the battery, power losses in the amplifier, and the overall power efficiency of the transformer-coupled class-A amplifier. The input dc power obtained from the battery is calculated from the values of dc battery voltage and average current drawn from the battery

$$P_i(\text{dc}) = V_{CC} I_{C_Q} \tag{8.16}$$

For the transformer-coupled amplifier, as shown in Fig. 8.3, the power dissipated by the transformer is quite small and will be ignored in the present calculations. Thus for the transformer-coupled amplifier the only lost power is that dissipated by the power transistor as calculated by the following equation

$$P_t = P_D = P_i - P_o \tag{8.17}$$

where P_t is the power dissipated as heat by the active device (transistor in this case), also labelled P_D. The equation seems quite simple but is significant in operating a power amplifier. The amount of power dissipated by the transistor (which then sets the transistor power rating) is the difference between the average dc input power from the battery (which is a constant for a fixed battery and operating point) and the output ac power drawn by the load. If the output power is zero then the transistor must handle the maximum amount, that set by the battery voltage and bias current. If the load does draw some of the power then the transistor has to handle that much less (for the moment). In other words, the transistor has to work hardest (dissipate the most power) when the load is disconnected from the amplifier circuit and the transistor dissipates least power when the load is drawing maximum power from the circuit. Obviously, the safest rating of the transistor used is the maximum set when the load is disconnected. Since normal operation with the load connected requires the transistor to dissipate less power, it is always preferable to keep the load connected as long as the amplifier unit is turned on.

EXAMPLE 8.5. Calculate the efficiency of the amplifier circuit of Example 8.4. Calculate, also, the power dissipated by the transistor.

Solution: Using Eq. (8.16) to calculate the input power

$$P_i = V_{CC} I_{C_Q} = (10)(150 \times 10^{-3}) = 1.5 \text{ W}$$

Using Eq. (8.17) the power dissipated by the transistor is

$$P_D = P_i - P_o = 1.5 - 0.423 \cong 1.06 \text{ W}$$

$$\%\eta = \frac{P_o}{P_i} \times 100 = \frac{0.423}{1.5} \times 100 = 28.2\%$$

For a class-A amplifier the maximum theoretical efficiency for the series-fed circuit is 25% and for the transformer-coupled circuit it is 50%. From analysis of the operating range for a series-fed amplifier circuit the efficiency can be stated in the following form:

$$\%\eta = 25\frac{V_{CE_{max}} - V_{CE_{min}}}{V_{CE_{max}}} \tag{8.18}$$

In practical operation the minimum value of collector-emitter voltage is not 0 V and the efficiency is less than 25%. In the circuit of Example 8.1 the efficiency was only 6.25%, indicating a poorly designed series-fed circuit.

The efficiency of a transformer-coupled class-A amplifier can be expressed by

$$\%\eta = 50\frac{(V_{CE_{max}} - V_{CE_{min}})}{(V_{CE_{max}} + V_{CE_{min}})} \tag{8.19}$$

The larger the value of $V_{CE_{max}}$ and the smaller the value of $V_{CE_{min}}$ the closer the efficiency approaches the theoretical limit of 50%. In the circuit of Example 8.4 the value obtained was 28.2%. Well-designed circuits can approach the limit of 50% quite closely so that the circuit of Fig. 8.5a would be considered average in operation. The larger the amount of power handled by the amplifier the more critical the efficiency becomes. For a few watts of power a simpler, cheaper circuit with less than maximum efficiency is acceptable (and sometimes quite desirable). For power levels in the tens to hundreds of watts, efficiency as close as possible to the theoretical maximum would be desired.

The value of 50% considered as the maximum for transformer-coupled amplifiers is only for class-A operation. There are additional ways of operating (biasing) the amplifier to obtain even higher efficiency, as will now be considered.

8.4 CLASSES OF AMPLIFIER OPERATION AND DISTORTION

Operating Classes

The only class of amplifier operation so far considered has been class A. By definition class-A operation provides collector (output) current flow during the complete cycle of the input signal (over a 360° interval). Figure 8.6a shows the output for class-A circuit operation. The bias

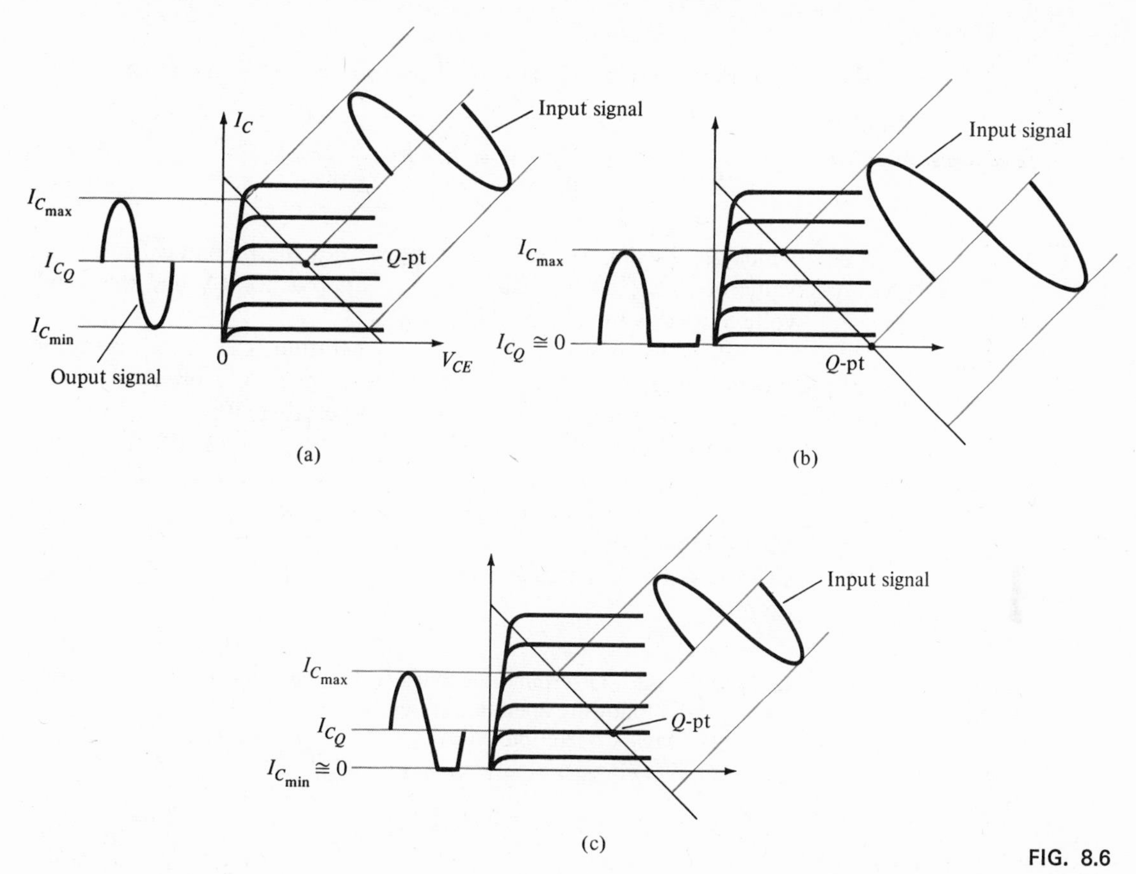

FIG. 8.6

Various amplifier operating classes.

level of current is I_{C_Q} and for the load line shown, the output signal does not exceed values of $I_{C_{max}}$ or $I_{C_{min}}$, which would take the operation out of the linear region of device operation. Figure 8.6b shows *class-B* operation. The bias point is set at cutoff so that input signal variations driving the transistor into conduction will cause the output current to vary although input signal variations which tend to cut down the device conduction merely result in the device remaining in cutoff. As shown in the figure, the output current flows for only about 180° of the cycle, this being the definition of class-B operation. Note that with no input signal the device is biased with no collector current flow and therefore no power dissipated by the transistor. Only when signal is applied does the transistor handle an average current which increases for larger input signals. Contrary to class-A operation, in which the worst condition occurs with no input signal and the least power is dissipated by the transistor for maximum input signal, the operation of a class-B circuit is to increase transistor dissipation for increased input signal. Since the average current in class-B operation is less than in class A the amount of power dissipated by the transistor is less in class B.

A relation for the circuit efficiency of a class-B operating circuit is

$$\% \eta = 78.5\left(1 - \frac{V_{CE_{\min}}}{V_{CC}}\right) \quad (8.20)$$

which shows that the efficiency approaches the theoretical maximum value of 78.5% as the value of $V_{CE_{\min}}$ approaches 0 V and the supply voltage used is made larger.

In between class-A and class-B operation is *class-AB* operation, shown in Fig. 8.6c. The collector current flows for more than 180° of the signal cycle but less than 360°. The operating efficiency of class AB is between that of class A and class B, that is, between 50 and 78.5%, theoretic maximum.

Operation with the output conducting for less than 180° is called *class-C* operation and is not found in audio-type amplifier circuits. Class C is used in tuned amplifier circuits in radio or television, for example.

EXAMPLE 8.6 Determine the efficiency of the types of circuits indicated for the following operating conditions
(a) Class-A operation with $V_{CE_{\max}} = 30$ V and $V_{CE_{\min}} = 3$ V.
(b) Class-B operation with $V_{CE_{\min}} = 3$ V and $V_{CC} = 20$ V.

Solution: (a) Using Eq. (8.19)

$$\% \eta = 50 \frac{V_{CE_{\max}} - V_{CE_{\min}}}{V_{CE_{\max}} + V_{CE_{\min}}} = 50\left(\frac{30 - 3}{30 + 3}\right) = 40.9\%$$

(b) Using Eq. (8.20)

$$\% \eta = 78.5\left(1 - \frac{V_{CE_{\min}}}{V_{CC}}\right) = 78.5\left(1 - \frac{3}{20}\right) = 66.8\%$$

Distortion

Output signal variations of less than 360° of the signal cycle are considered to have *distortion.* This means that the output signal is no longer just an amplified version of the input signal but in some ways is distorted or changed from that of the input. One common example of distortion is the poor quality of music coming from a radio of hi-fi system with distortion, the music or voice no longer sounding like that which was originally recorded or transmitted. Distortion can come from a number of different places in any audio system.

Distortion can occur because the device characteristic is not linear: *nonlinear or amplitude distortion.* This can occur with all classes of operation. In addition the circuit elements and the amplifying device can respond to the signal differently at various frequency ranges of operation: *frequency distortion.*

When distortion occurs as amplitude distortion in class-A, class-AB, or class-B operation the output signal no longer represents the input signal exactly. One technique of accounting for this change in the output signal is the method of *Fourier analysis,* which provides a means for describing a periodic signal in terms of its fundamental frequency component and frequency components at integer multiples—components called *harmonic components or harmonics.* For example, a signal which is originally 1000 Hz could result, after distortion, in a frequency component at 1000 Hz, and harmonic components at 2 kHz (2×1000 Hz), at 3 kHz (3×1000 Hz), 4 kHz (4×1000 Hz), and so on. The original frequency of 1000 Hz is called the fundamental frequency and those at integer multiples are the harmonics—that at 2 kHz is the second harmonic, the component at 3 kHz is the third harmonic, and so on. The fundamental signal is considered the first harmonic. (No harmonics at fractional amounts of the fundamental frequency exist using this technique.)

An instrument such as spectrum analyzer would allow measurement of the harmonics present in the signal by providing display of the fundamental component of the signal and a number of its harmonics on a CRT screen as in Fig. 8.7a. Similarly, a wave analyzer instrument allows more precise measurement of the harmonic components of a distorted signal by filtering out each of these components and providing a calibrated dial reading of these components, one at a time (see Fig. 8.7b).

In any case the technique of considering any distorted signal as containing a fundamental component and harmonic components is quite practical and useful. For a signal occurring in class AB or class B the distortion may be mainly even harmonics, of which the second harmonic component is greatest. Thus, although the distorted signal contains all harmonic components from second harmonic on up, the most important in terms of the amount of distortion for the classes of operation we will consider is the second harmonic.

A current output waveform is shown in Fig. 8.8 with the quiescent, minimum and maximum signal levels, and the times they occur, marked on the waveform. The signal shown indicates some distortion is present. An equation which approximately describes the distorted signal waveform is

$$i_C \cong I_{CQ} + I_O + I_1 \cos \omega t + I_2 \cos 2\omega t \qquad (8.21)$$

The current waveform contains the original quiescent current I_{CQ}, which occurs with zero input signal, an additional dc current I_O, due to the nonzero average of the distorted signal, the fundamental component of the distorted ac signal I_1, and a second harmonic component

(a)

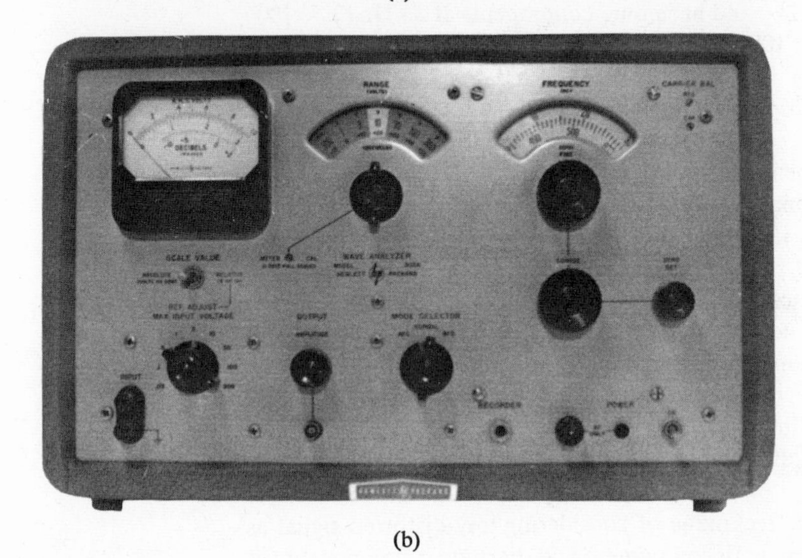

(b)

FIG. 8.7
(a) Spectrum analyzer; (b) wave analyzer.

I_2, at twice the fundamental frequency. Although other harmonics are also present only the second is considered here. Equating the resulting current from Eq. (8.21) at a few points in the cycle to that shown on the current waveform provides the following three relations:

At $\omega t = 0$

$$i_C = I_{C\max} = I_{CQ} + I_O + I_1 \cos(0) + I_2 \cos(0)$$

$$I_{C\max} = I_{CQ} + I_O + I_1 + I_2 \tag{8.22}$$

At $\omega t = \pi/2$

$$i_C = I_{CQ} = I_{CQ} + I_O + I_1 \cos\left(\frac{\pi}{2}\right) + I_2 \cos\left(\frac{2\pi}{2}\right)$$

$$I_{CQ} = I_{CQ} + I_O - I_2 \tag{8.23}$$

At $\omega t = 3\pi/2$

$$i_C = I_{C\min} = I_{CQ} + I_O + I_1 \cos\left(\frac{3\pi}{2}\right) + I_2 \cos\left(2\frac{3\pi}{2}\right)$$

$$I_{C\min} = I_{CQ} + I_O - I_1 + I_2 \tag{8.24}$$

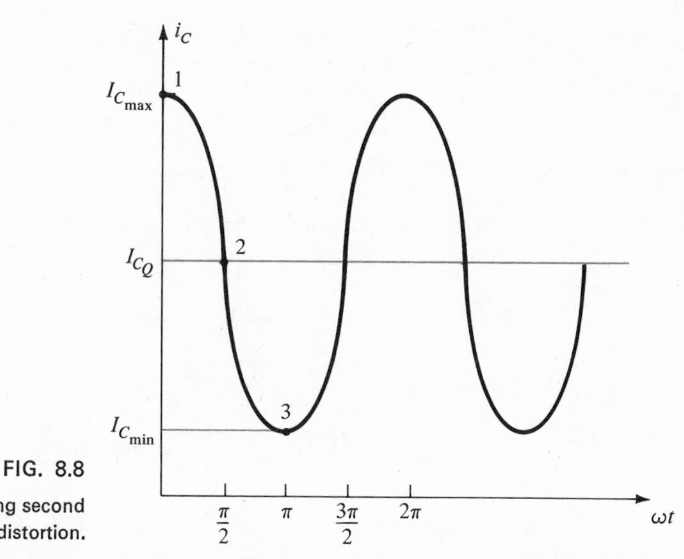

FIG. 8.8
Waveform for obtaining second harmonic distortion.

Solving Eq. (8.22), (8.23), and (8.24) simultaneously gives the following results

$$I_O = I_2 = \frac{I_{C_{\max}} + I_{C_{\min}} - 2I_{CQ}}{4}$$

$$I_1 = \frac{I_{C_{\max}} - I_{C_{\min}}}{2} \tag{8.25}$$

By definition the percent of the second harmonic distortion is given by

$$\% D_2 \equiv \left|\frac{I_2}{I_1}\right| \times 100 \tag{8.26}$$

The second harmonic distortion is the percent of the second harmonic component present in the output current waveform with respect to the amount of the fundamental component. Obviously, 0% distortion is the ideal condition of no distortion.

Using the results in Eq. (8.25) to express the second harmonic distortion defined by Eq. (8.26)

$$\% D_2 = \left|\frac{\frac{1}{2}(I_{C_{\max}} + I_{C_{\min}}) - I_{CQ}}{I_{C_{\max}} - I_{C_{\min}}}\right| \times 100 \tag{8.27}$$

In a similar manner, the amount of second harmonic distortion can be related to the measured values of the distorted output voltage waveform

$$\% D_2 = \left|\frac{\frac{1}{2}(V_{CE_{\max}} + V_{CE_{\min}}) - V_{CEQ}}{V_{CE_{\max}} - V_{CE_{\min}}}\right| \times 100 \tag{8.28}$$

EXAMPLE 8.7 An output waveform displayed on a scope provides the following measured values

(a) $V_{CE_{\min}} = 1\text{ V}, \quad V_{CE_{\max}} = 22\text{ V}, \quad V_{CE_Q} = 12\text{ V}$

(b) $V_{CE_{\min}} = 4\text{ V}, \quad V_{CE_{\max}} = 20\text{ V}, \quad V_{CE_Q} = 12\text{ V}$

For each set of values calculate the amount of the second harmonic distortion.

Solution: Using Eq. (8.28)

(a) $$\% D_2 = \left| \frac{\frac{1}{2}(22 + 1) - 12}{22 - 1} \right| \times 100 = 2.38\%$$

(b) $$\% D_2 = \left| \frac{\frac{1}{2}(20 + 4) - 12}{20 + 4} \right| \times 100 = 0\% \qquad \text{(no distortion)}$$

The method used to obtain the amount of second harmonic distortion was called the three-point method since it involved equating the assumed form of the output voltage to the measured voltage at three points in the signal cycle. Using an assumed output signal equation containing more harmonic terms, along with choosing more points in the waveform, results in obtaining relations for the magnitude of the harmonic components at higher harmonic frequencies. Using a five-point method provides the dc component, first harmonic (fundamental), second harmonic, third harmonic, and fourth harmonic components. The harmonic distortion for each of these components is then defined as

$$D_2 = \left| \frac{I_2}{I_1} \right|, \qquad D_3 = \left| \frac{I_3}{I_1} \right|, \qquad D_3 = \left| \frac{I_4}{I_1} \right| \tag{8.29}$$

The total distortion may be defined, in general, using the individual distortion components

$$D = \sqrt{D_2^2 + D_3^2 + D_4^2 + \cdots} \tag{8.30}$$

When distortion does occur the output power calculated by the undistorted case is no longer correct. Equation (8.11), for example, is true *only* for the nondistorted case. When distortion is present the output power due to the fundamental component of the distorted signal is

$$P_1 = \frac{I_1^2 R_L}{2} \tag{8.31}$$

The total output power due to all the harmonic components of the distorted signal is

$$P = (I_1^2 + I_2^2 + I_3^2 + \cdots)\frac{R_L}{2} \tag{8.32}$$

The total power can also be expressed in terms of the total distortion

$$P = (1 + D_2^2 + D_3^2 + \cdots)I_1^2 \frac{R_L}{2} = (1 + D^2)P_1 \qquad (8.33)$$

EXAMPLE 8.8 Using a five-point method to calculate harmonic components gives the following results: $D_2 = 0.1$, $D_3 = 0.02$, $D_4 = 0.01$, with $I_1 = 4$ A and $R_L = 8\ \Omega$. Calculate the total distortion, fundamental power component and total power.

Solution: Total distortion, using Eq. (8.30)

$$D = \sqrt{D_2^2 + D_3^2 + D_4^2} = \sqrt{(0.1)^2 + (0.02)^2 + (0.01)^2} \cong 0.1$$

Fundamental power, using Eq. (8.31)

$$P_1 = \frac{I_1^2 R_L}{2} = \frac{(4)^2 8}{2} = 64 \text{ W}$$

Total power, using Eq. (8.33)

$$P = (1 + D^2)P_1 = [1 + (0.1)^2]64 = (1.01)64 = 64.64 \text{ W}$$

(Total power mainly due to fundamental component even with 10% second harmonic distortion.)

GRAPHICAL DESCRIPTION OF HARMONIC COMPONENTS OF DISTORTED SIGNAL

A demonstration of the use of harmonic components to represent a distorted signal is provided to help clarify the concept. As an example, a distorted waveform such as that resulting from class-B operation is shown in Fig. 8.9a. The signal is clipped on the negative half-cycle so that only the positive sinusoidal half-cycle provides an output signal.

Using Fourier analysis techniques a fundamental component of the distorted signal is calculated as shown in Fig. 8.9b. Figure 8.9b does not show the distorted waveform, only the fundamental component (which is a perfectly sinusoidal signal itself). Similarly, the second and third harmonic components can be obtained and are shown in Figs. 8.9c and 8.9d, respectively.

We now wish to check whether these components, each a purely undistorted sinusoidal signal, add up approximately to the original distorted signal. Figure 8.9e shows the resulting waveform when adding the fundamental and second harmonic components together. Note the flattening of the second half of the cycle. In Fig. 8.9f the third harmonic component is added to give a resulting waveform that comes fairly close to the original distorted signal. The addition of higher harmonic components of the right amplitude and correct phase will further alter the

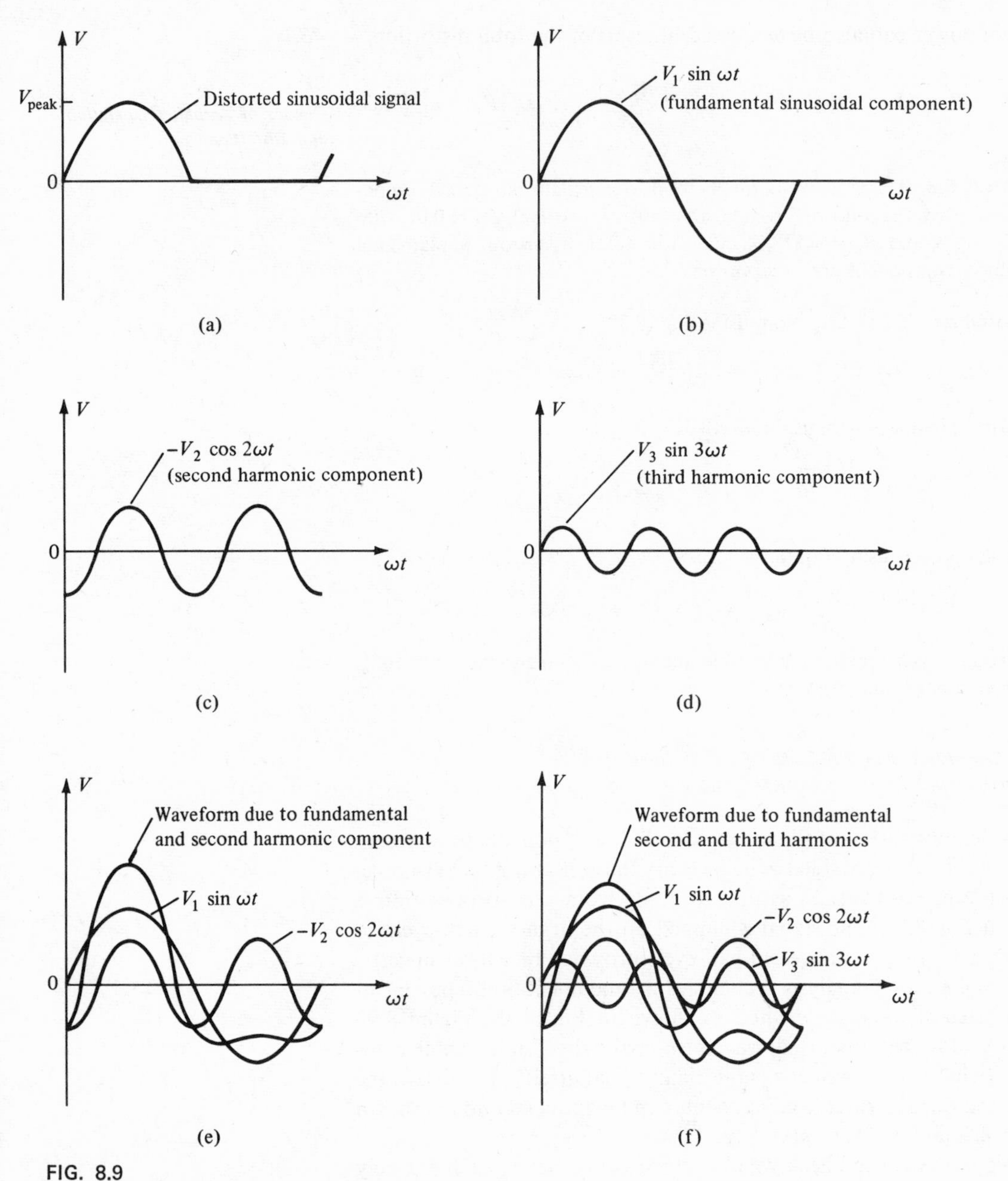

FIG. 8.9

Graphical representation of a distorted signal through the use of harmonic components.

resulting waveform to approximate the original distorted signal. In a relatively simple manner we can observe that addition of a fundamental component and harmonic components can result in the original distorted waveform. In general, any periodic waveform can be represented by a fundamental component and harmonic components, each of varying amplitudes and at various phase angles.

The concept of harmonics is quite useful is both analyzing distorted (nonsinusoidal) waveforms and in providing a means of working with such signals. Since all the harmonic components are sinusoidal signals, we can separately consider the effect of each component on the circuit and obtain the total effect using superposition—adding together the voltages or currents being considered. Particular mention of harmonic content of a waveform will be made in the discussion of push-pull amplifiers, where it will be shown that the particular circuit connection eliminates the even harmonic components and leaves only the fundamental component and the odd harmonic components. Since the largest distortion component is the second harmonic, the elimination of that component will considerably reduce the total amount of distortion.

8.5 PUSH-PULL AMPLIFIER CIRCUIT

Previous discussion has shown that efficient operating conditions occur for class-AB or class-B operation. On the other hand class-AB or class-B operation, as discussed so far, result in considerable distortion. To be forced to choose one or the other—more efficient operation or less distortion— is not the best of choices. Ideally, one wants to have the efficient operation of class-B but also the low distortion of class-A operation. This is not quite realistic, but a surprisingly low-distortion, high-efficiency operation can be obtained using the circuit connection known as *push-pull*. A typical transistor push-pull circuit connection is shown in Fig. 8.10. Similar circuits are also made using vacuum tubes or FETs.

The circuit of Fig. 8.10 requires an input transformer to produce opposite polarity signals to the two transistor inputs, and an output transformer to drive the load in a push-pull mode of operation to be described. Figure 8.11a shows the input transformer with center tap at ground. As an example, the voltage across the secondary from the plus (+) to the minus (−) terminal is 100 V, peak. As shown in Fig. 8.11a, the voltages across each half of the transformer are 50 V, peak, adding up to the total of 100 V, peak across the transformer. With the center tap of the transformer connected to ground (0-V potential) the signal observed from the plus terminal to ground is, as shown in Fig. 8.11a, in phase with the voltage across the full transformer. However, since the voltage measured across the bottom half of the transformer is also in phase with the total signal when measured from center tap to minus, it is opposite when measured from minus to center tap (or from minus to ground). The signals at the plus and minus terminals measured with

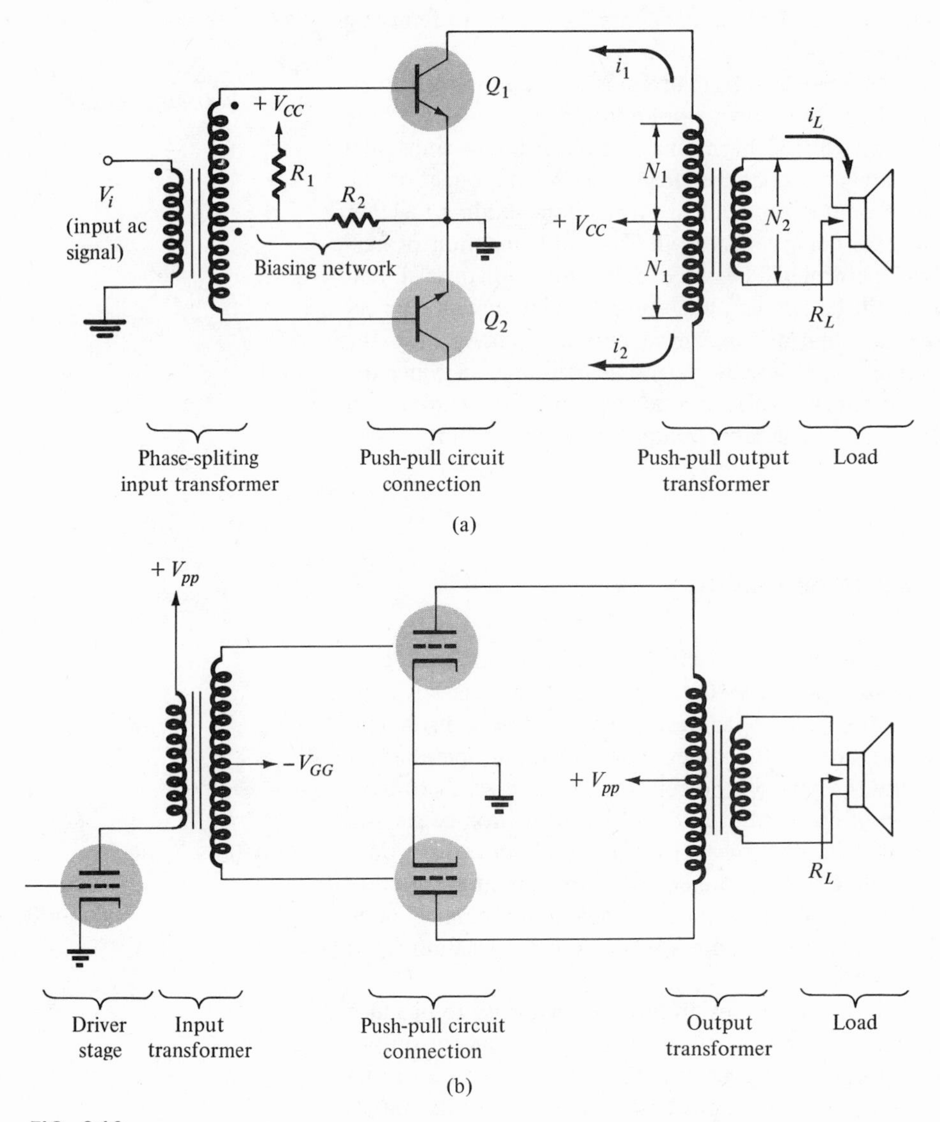

FIG. 8.10

Push-pull circuit.

respect to ground are therefore opposite in phase as shown in Fig. 8.11a.

Having obtained opposite-phased input to the two transistor units the push-pull nature of the circuit operation can be considered as shown in the partial circuit diagram of Fig. 8.11b. Consider first the dc bias current (I_{CQ}) for each transformer. Figure 8.11b shows that the bias currents for each transistor flow in opposite directions through the transformer winding. The magnetic flux set up by each of these

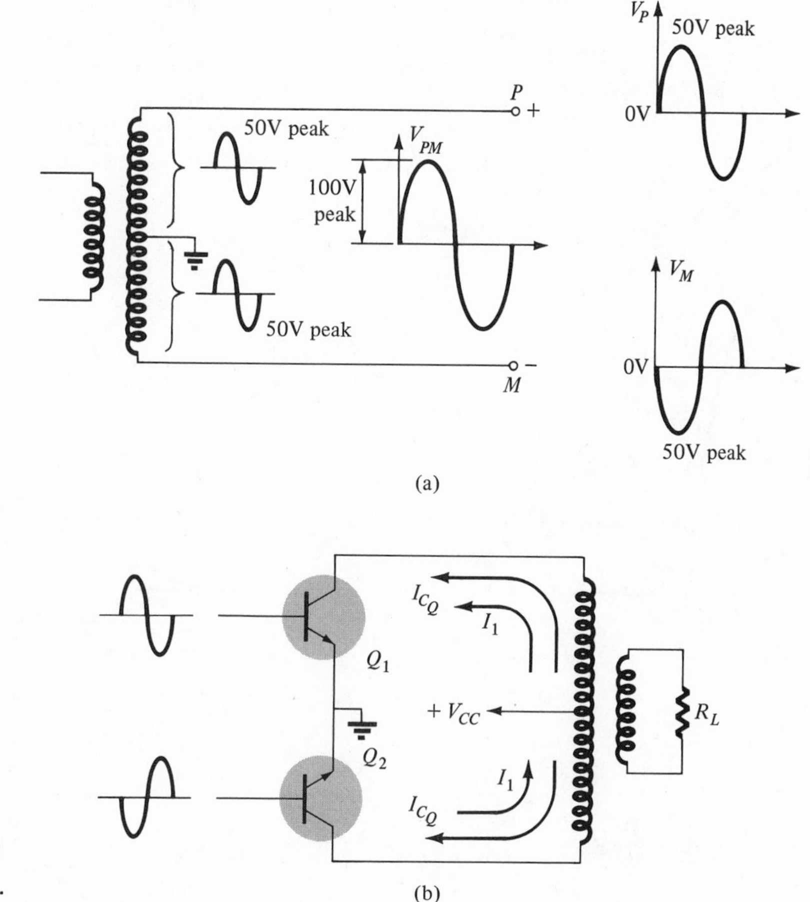

FIG. 8.11
Details of push-pull operation.

currents results in opposite flux through the magnetic core so that the net flux, in the perfectly matched case, is zero. Thus the transformer need not handle a large flux due to the dc bias currents, resulting in a smaller size core, biased to operate near zero flux.

Considering only the first half-cycle of operation, transistor Q_1 is driven further into conduction whereas transistor Q_2 is driven less into conduction. The varying component of current for each transformer is marked I_1 in Fig. 8.11b. Since Q_1 is driven further into conduction the varying component of current flows in the same direction as the dc bias current, resulting in a larger total current. The varying component of current flow in Q_2, however, flows in an opposite direction to the bias current for that transistor, resulting in a net decrease in current flow for transistor Q_2. Note from Fig. 8.11b that the overall operation results in a net current flow through the transformer. The input signals to each stage being of opposite phase results in a net output signal across the transformer. Had in-phase signals been applied to the transistor inputs the net output signal for the varying components would have been zero. We now need to consider how the push-pull connection with class-AB or class-B operation of the transistors will provide low-distortion output.

Figure 8.12a shows the output signal for each section of the push-pull circuit when operated in class B. Note that since the two transistors

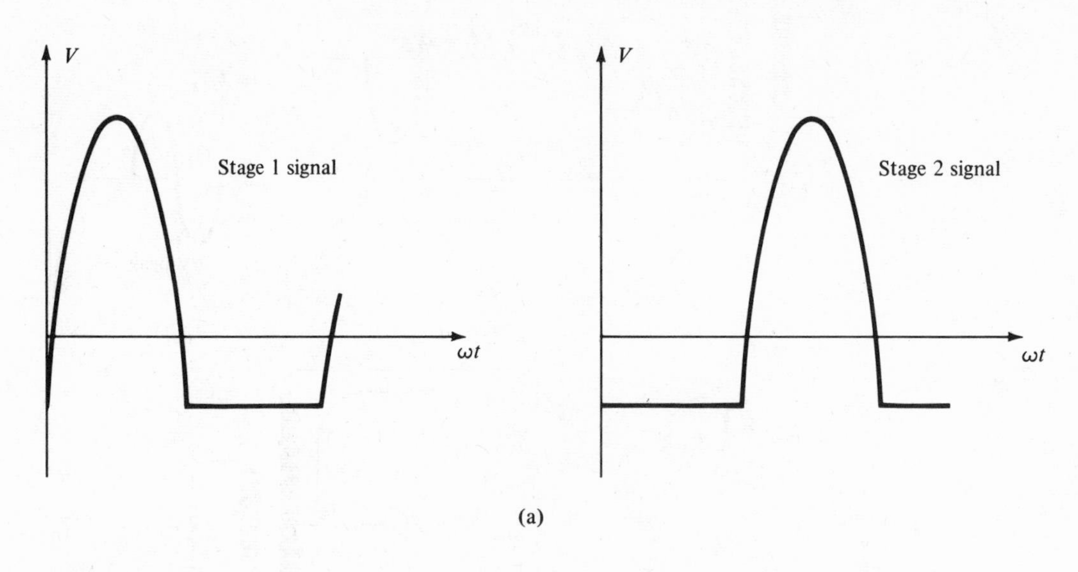

(a)

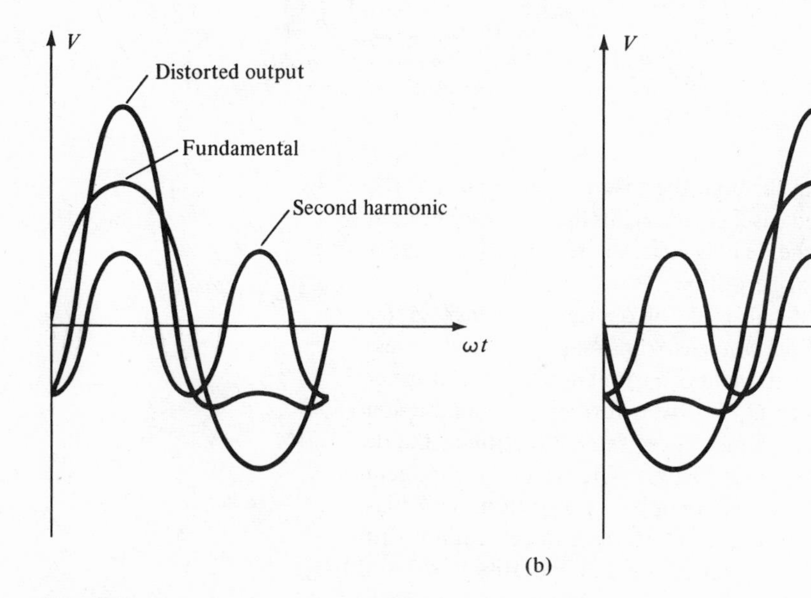

(b)

FIG. 8.12

Distorted output waveforms for each section of the push-pull circuit operation.

are identically biased they distort the output waveform on the negative half-cycle for each signal—the second half-cycle for stage 1 and the first half-cycle for stage 2, as shown. Intuitively, one might say that the positive half-cycle due to stage 1 provides the output signal during the first half-cycle and that the positive half-cycle signal operates stage 2 during the second half-cycle. Since the two stages would cause current flow in opposite directions the two half-cycles, as indicated in Fig. 8.12a, would result in opposite half-cycles of signal in the transformer—thus providing a full or complete cycle of signal flow even with half-cycles of operation of each stage.

For a more detailed and meaningful discussion of the circuit operation the approximate distorted output waveforms in each stage of the circuit are shown in Fig. 8.12b as being made up of a fundamental component and a second harmonic component. Although other harmonic components would be necessary to describe more fully the actual distorted waveforms of Fig. 8.12a, consideration of only the fundamental and second harmonic components will provide the necessary information here.

Notice in Fig. 8.12b that the fundamental components of each stage are opposite in phase. For the push-pull circuit connection, opposite phase output signals will result in a net voltage in the secondary which is the sum of the two component signals. The second harmonic components, however, are in phase and therefore cancel, providing a net voltage in the secondary of 0 V due to these signal components. In other words, the action of the push-pull connection would be to cancel out the second harmonic components of the signals applied to the primary of the output transformer, and the same cancellation can be shown to result for all other even harmonic components. From our previous consideration of harmonics we can now appreciate that a distorted signal applied to the push-pull transformer connection, such as those shown in Fig. 8.12a, would result in cancellation of all even harmonic components so that the resulting output signal across the secondary can be considered to be made up of the fundamental component, third harmonic component, and all odd harmonic components of the distorted signal. Since the second harmonic is the largest component for the distorted signal of Fig. 8.12a, elimination of that component will result in an output signal having considerably less distortion. In calculating the distortion of the output signal we need to consider the amount of the third, fifth, and so on, harmonic components, which we expect will not be too large, so that the total distortion as calculated by Eq. (8.30) will not be much more than the small amount of the third harmonic distortion.

Of course, the above conclusion of even harmonic cancellation is an ideal condition which will occur only if the circuit is perfectly balanced—transistors exactly matched, transformer center-tap connection perfect, and input signals exactly equal and opposite. This will never practically occur, but still the circuit can be operated class AB or class B for improved power efficiency with quite low distortion occurring (although more than for class-A operation).

To summarize, then, the advantages of the push-pull circuit connection are

1. The dc components of the collector currents oppose each other in the transformer resulting in no net flux due to the bias current and hence, smaller size cores can be used.
2. Class-AB or class-B operation is possible with small resulting distortion due to the cancellation of all the even harmonic components.
3. Ripple voltage in the voltage supply is cancelled out by the push-pull operation of the circuit.
4. High efficiency operation is possible, using class-AB or class-B biasing.

To complete the picture it must be pointed out that the power supply used in class-B operation must have good voltage regulation since the amount of current drawn is about zero for no signal operation and rises in value for larger amounts of signal level. Since the current drawn from the supply ranges considerably from no load to full load the supply voltage regulation must be good. Also, hum voltages (60-Hz pickup), which are picked up on the input lines of the circuit, are *not* eliminated by the push-pull connection since these are brought along with the input signals to the driver transformer and connected out-of-phase as input to the push-pull circuit, thereby acting as proper input signals to drive the output device.

8.6 VARIOUS PUSH-PULL CIRCUITS INCLUDING TRANSFORMERLESS CIRCUITS

Although the circuit shown in Fig. 8.10 is the most common form of the push-pull circuit connection a number of other circuit arrangements are possible. We will consider a few of these and their various advantages and disadvantages. It is important to keep in mind the overall operation of the circuit in order to appreciate the different methods of obtaining the advantages of push-pull operation. For the push-pull circuit it is necessary to develop the output voltage across the load in such a manner that two stages operating in class B will still provide a full cycle of signal by conducting on alternate half-cycles.

Starting with an input signal obtained from a driver amplifier stage, it is necessary to operate the two-stage push-pull circuit on alternate half-cycles for class-B operation. The opposite phase input signals to the two stages of the push-pull circuit can be obtained in a number of ways. Figure 8.10 shows the use of an input transformer to provide the phase inversion between the two push-pull input signals. Another means of obtaining opposite phase input signals is to use the paraphase circuit of Fig. 8.13. The input signal applied to the base appears at the collector, 180° out of phase. The output from the emitter is in phase with the input so that the two output signals are out of phase as shown in Fig. 8.13. Choosing values of R_C, R_E, and h_{fe} the voltage gain for the collector output signal can be set to 1. The gain for the signal taken from the emitter is 1 for the emitter follower operation. Thus, the circuit provides no net gain but results in opposite phase signals to drive the push-pull amplifier stage. The advantage of this driver connection is the savings on the use of a center-tapped transformer which is

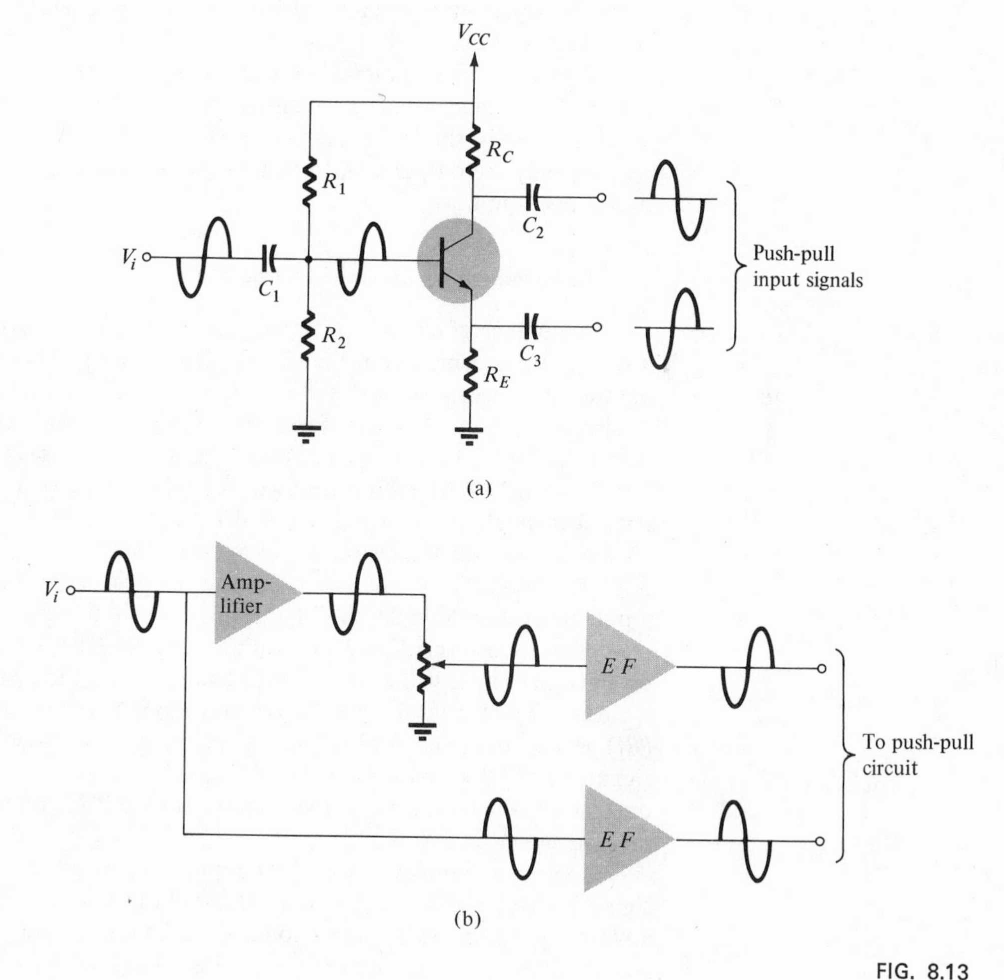

FIG. 8.13
Phase-splitter circuits.

expensive, bulky, and has a limited frequency operating range. A disadvantage is that the two signals do not come from similar impedance sources. The signal from the emitter provides a good driver connection since the source resistance viewed from the emitter is low. The collector circuit resistance, however, is relatively high and although the unloaded output signals are equal they are quite different under load conditions. One possible improvement would be to add an additional emitter follower stage to connect the output to the load since such stage would provide no additional voltage gain or phase inversion and would drive the push-pull stage from a low-resistance source.

Another means of obtaining opposite phase signals to drive the push-pull stage is illustrated by the block diagram of Fig. 8.13b. The input signal is amplified and inverted by one amplifier stage and then attenuated for an overall gain of unity. The use of two emitter followers (possibly Darlington circuits) drive the push-pull stage from low-impedance sources.

One even more practical and popular circuit for obtaining the out-of-phase signals is using the difference amplifier discussed in Chapter 11. As will be shown in that chapter the use of a difference amplifier requires only the output transformer to provide the complete push-pull circuit operation.

Complementary Symmetry Circuits

A number of circuits go beyond eliminating only the input phase inverting transformer from the circuit. These circuits also remove the output transformer so that the circuit is completely transformerless. A simple version of a transformerless push-pull amplifier circuit is shown in Fig. 8.14. Complementary transistors are used; that is, an NPN and a PNP transistor are used instead of using two of the same type. The single input signal required is applied to both base inputs. However, since the transistors are of opposite type they will conduct on opposite half-cycles of the input. During the positive half-cycle of the input signal, for example, the PNP transistor will be reverse biased by the positive half-cycle signal and will not conduct. The NPN transistor, on the other hand, will be biased into conduction by the positive half-cycle signal with a resulting half-cycle of output across the load resistor (R_L) as shown in Fig. 8.14b. During the negative half-cycle of input signal the NPN transistor is biased off and the output half-cycle developed across the load is due to the operation of the PNP transistor at this time, as shown in Fig. 8.14c.

During a complete cycle of the input, a complete cycle of output signal is developed across the load. It should be obvious that one disadvantage of this circuit connection is the need for two supply voltages. Another, less obvious, but important, disadvantage with the complementary circuit as shown is the resulting *crossover* distortion in the

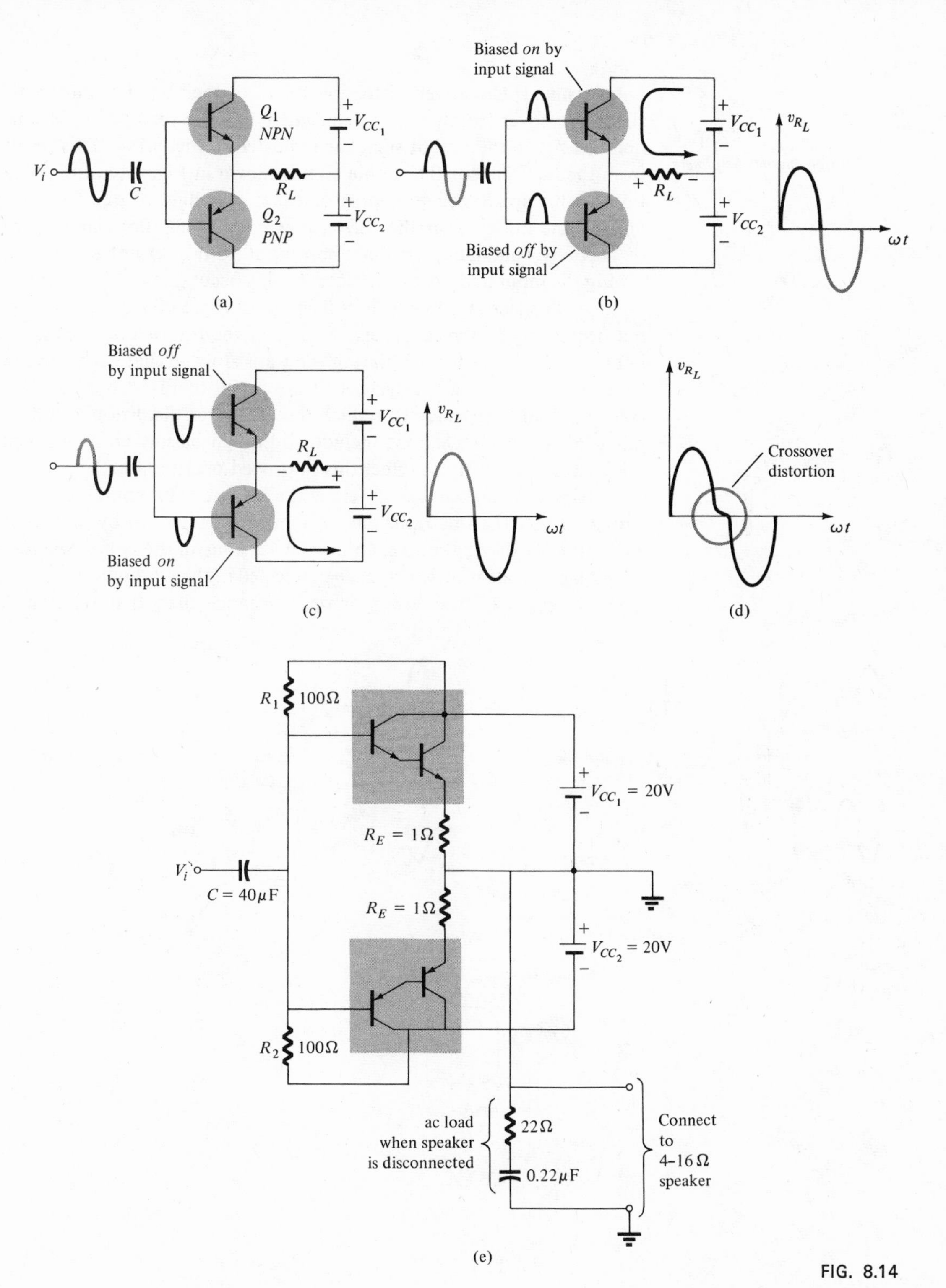

FIG. 8.14

Complementary symmetry push-pull circuit.

output signal. Crossover distortion refers to the fact that during the signal crossover from positive to negative (or vice-versa) there is some nonlinearity in the output signal as indicated in Fig. 8.14d. This results from the fact that for the simple circuit shown in Fig. 8.14a the operation of the circuit does not provide exact switching of one transistor *off* and the other *on* at the zero voltage condition. Both may be off or partially conducting so that the output voltage is not exactly following the input and distortion occurs. This occurance at the crossover point is of concern for the push-pull circuit of Fig. 8.10 as well, although not necessarily to the same degree. Bias of the transistors in class AB improves the operation by biasing the transistors so that each stays on for more than half of the cycle. For the circuit of Fig. 8.14a considerable effort must be made to reduce the crossover distortion and more practical circuit connections include additional biasing components in the base circuit to try to effect this improved operation.

Note that the load is driven as the output of an emitter follower circuit so that the low resistance of the load is matched by low resistance from the driving source. Improved versions of the complementary circuit include the transistors, each connected in the Darlington arrangement, to provide even lower driver resistance than that with single

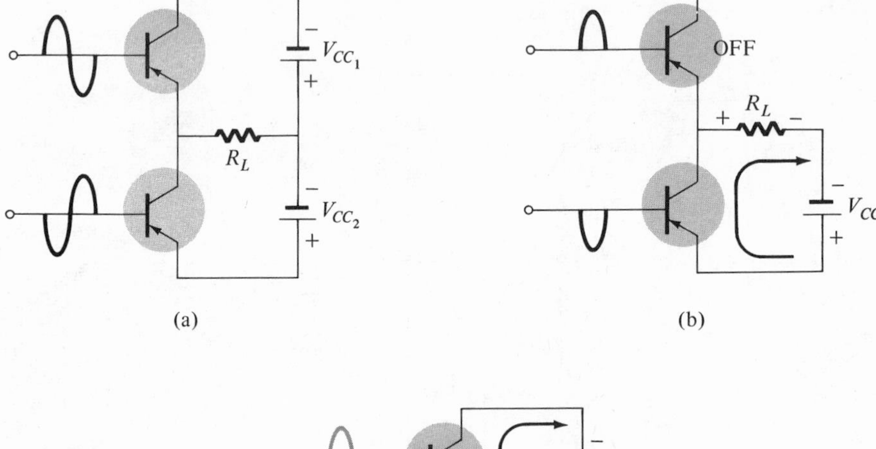

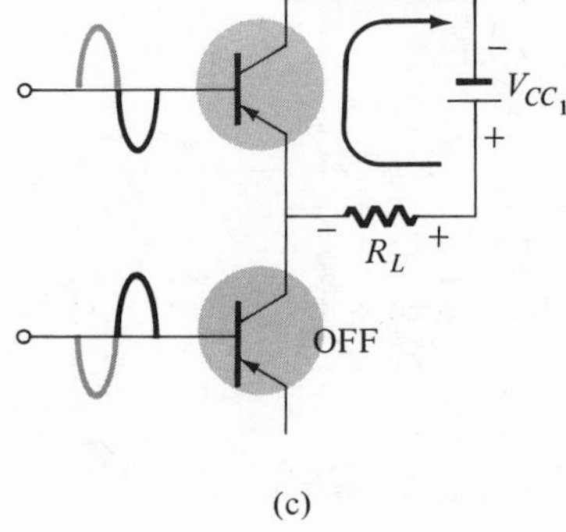

FIG. 8.15
Transformerless push-pull circuit.

transistors. The circuit of Fig. 8.14e shows a practical circuit connection using the Darlington connection of the transistors and additional emitter resistors for temperature bias stabilization.

Another transformerless push-pull circuit, shown in Fig. 8.15a, uses the same type of transistors rather than complementary transistors. However, the inputs applied to the circuit must be opposite in phase so that some phase-splitter circuit is required. Figures 8.15b and 8.15c show the operation of the transistors for alternate halves of the cycle and the resulting polarity of the signal across the load resistor.

Figure 8.16 shows a more practical version of a circuit using complementary driver transistors to provide the phase-splitting operation. In addition the circuit requires only one voltage supply which is an attractive feature of this circuit.*

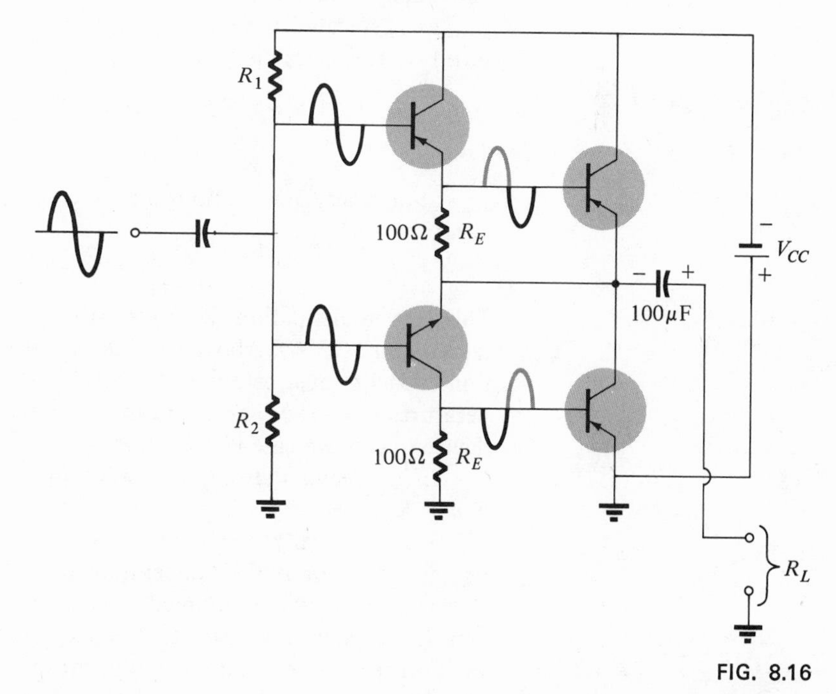

FIG. 8.16
Transformerless push-pull circuit using single power supply.

8.7 POWER TRANSISTOR HEAT SINKING

Recent trends in electronics have been to replace individual transistors by complete integrated circuits for small-signal and low-power applications. Most high-power applications, however, still require individual

* A few practical versions of the circuit type of Fig. 8.16 can be found in the GE transistor manual, 7th edition.

power transistors. Improvements in production techniques have provided higher power ratings in smaller-sized packaging cases, have increased the maximum transistor breakdown voltage, and provided faster switching power transistors.

Some discussion of power transistor rating occurred in Chapter 3. The maximum power handled by a particular device and the temperature of the transistor junctions are related since the power dissipated by the device causes an increase in temperature at the junctions of the device. Obviously, a 100-W transistor will provide more power capability than a 10-W transistor. On the other hand, proper heat sinking techniques will allow operation of a device closer to its maximum power rating.

We should note that of the two types of transistors—germanium and silicon—silicon transistors provide greater maximum temperature ratings. Typically, the maximum junction temperature of these types of power transistors is

germanium	100 — 110°C
silicon	150 — 200°C

For many applications the average power dissipated may be approximated by

$$P_D = V_{CE}I_C$$

This power dissipation, however, is only allowed up to about room temperature (25°C). Above this temperature the device power dissipation capacity must be reduced (or *derated*) so that at higher case temperatures the power handling capacity is reduced—down to 0 W at the device maximum case temperature.

The greater the power handled by the transistor (dependent on power level set by the circuit) the higher the case temperature of the transistor. Actually, the limiting factor in power handled by a particular transistor is the temperature of the device collector junction. Power transistors are mounted in large metal cases to provide a large area from which the heat generated by the device may radiate. Even so, operating a transistor directly into air (mounting it on a plastic board, for example) severely limits the device power rating. If, instead (as is usual practice), the device is mounted on some form of *heat sink* its power handling capacity can approach the rated maximum value more closely. A few heat sinks are shown in Fig. 8.17. When the heat sink is used, the heat produced by the transistor dissipating power has a larger area from which to radiate the heat into the air, thereby holding the case temperature to a much lower value than would result without the heat sink. Even with an infinite heat sink (which, of course, is not available), for which the case temperature is held at the *ambient* (air) temperature, the junction will be heated above the case temperature and a maximum power rating must be considered.

FIG. 8.17

Typical power heat sinks.

Since even a good heat sink cannot hold the transistor case temperature at ambient (which, by the way could be more than 25°C if the transistor circuit is in a confined area where other devices are also radiating a good deal of heat) it is necessary to derate the amount of *maximum power* allowed for a particular transistor as a function of increased case temperature.

Figure 8.18 shows typical power derating curves for silicon and germanium transistors. The curves show that the manufacturer will specify an upper temperature point (not necessarily 25°C) after which a linear derating takes place. For silicon the maximum power that should be handled by the device does not reduce to 0 W until a case temperature of 200°C (or 150° in some devices), whereas for germanium the derating goes to a temperature of 100°C (110°C in some devices). For the germanium transistor, for example, the curve shows that at a case temperature of 75°C the maximum device dissipation should be reduced from 100 W to about 50 W. A silicon device rated at 100 W might still be able to handle the full power at this temperature as shown in Fig. 8.18.

It is not necessary to provide a derating curve since the same information could be given simply as a listed derating factor on the device specification sheet. For example, a derating factor for a germanium transistor may be stated as follows:

Derate linearly to 100°C case temperature at the rate of 1 watt per degree above 50°C.

For a 50-W transistor (rated below 50°C) this means that at 80°C the maximum power rating of the device must be reduced by

$$(80°\text{C} - 50°\text{C})\left(1\frac{\text{W}}{°\text{C}}\right) = 30\text{ W}$$

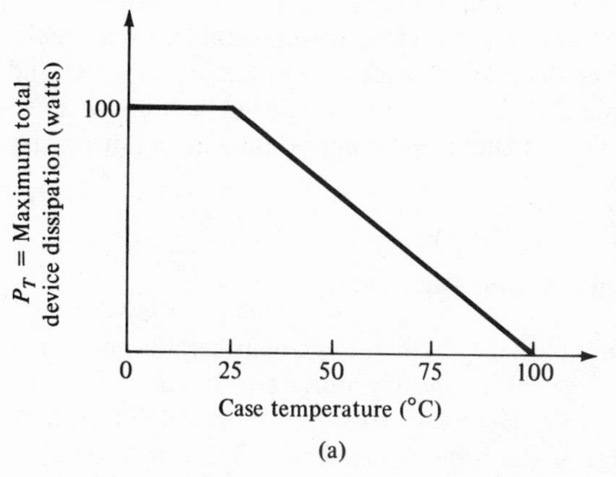

(a)

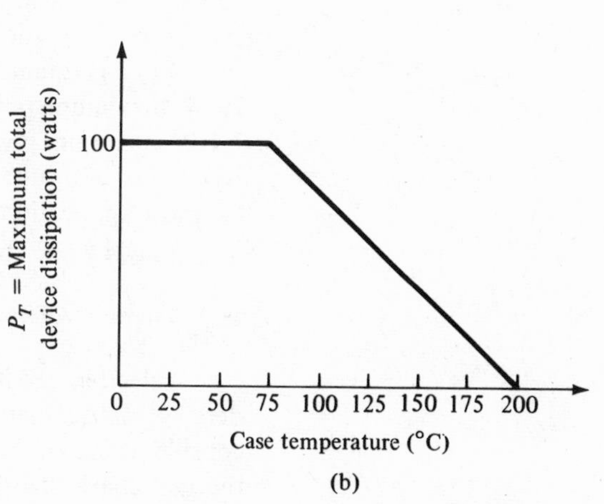

(b)

FIG. 8.18

Typical power derating curves for silicon and germanium transistors.

so that the rated power dissipated should only be

$$50 \text{ W} - 30 \text{ W} = 20 \text{ W}$$

at 80° case temperature. Stated mathematically,

$$P_D(\text{temp}_1) = P_D(\text{temp}_0) - (\text{Temp}_1 - \text{Temp}_0)(\text{Derating factor})$$

where the value of Temp_0 is the temperature at which derating should begin, the value of Temp_1 is the particular temperature of interest (above the value temp_0), P_D (temp_0) and $P_D(\text{temp}_1)$ are the maximum power dissipations at the temperatures specified, and the derating factor is the value given by the manufacturer in units of watts (or milliwatts) per degree of temperature.

EXAMPLE 8.9 Determine what maximum dissipation will be allowed for an 80-W silicon transistor (rated at 25°C) if derating is required above 25°C by a derating factor of 0.5 W/°C at a case temperature of 125°C.

Solution

$$P_D(125°\text{C}) = P_D(25°\text{C}) - (125°\text{C} - 25°\text{C})(0.5 \text{ W}/°\text{C})$$
$$= 80 \text{ W} - 100(0.5) = 30 \text{ W}$$

It is interesting to note what power rating results using a power transistor without a heat sink. For example, a silicon transistor rated at 100 W at (or below) 100°C is rated at only 4 W at (or below) 25°C, free-air temperature. Thus, operated without a heat sink the device can handle a maximum of only 4 W at a room temperature of 25°C. Using a heat sink large enough to hold the case temperature to 100°C at 100 W allows operation at the maximum power rating.

For a germanium transistor the comparison might be, for example, 20-W maximum power dissipation at case temperature of 25°C and 1.4-W maximum power dissipation into a free-air temperature of 25°C. Again, use of a suitable heat sink is quite important—in fact, necessary for most applications.

Thermal Analogy of Power Transistor

Selection of a suitable heat sink requires a considerable amount of detailed determination which is not appropriate to our present basic considerations of the power transistor. However, more detail about the thermal characteristics of the transistor and its relation to the power dissipation of the transistor may help provide a clearer understanding of power as limited by temperature. The following discussion should provide some background information.

A picture of how the junction temperature (T_J), case temperature (T_C), and ambient (air) temperature (T_A) are related by the device heat handling capacity—a temperature coefficient usually called thermal resistance—is presented in the thermal-electrical analogy shown in Fig. 8.19.

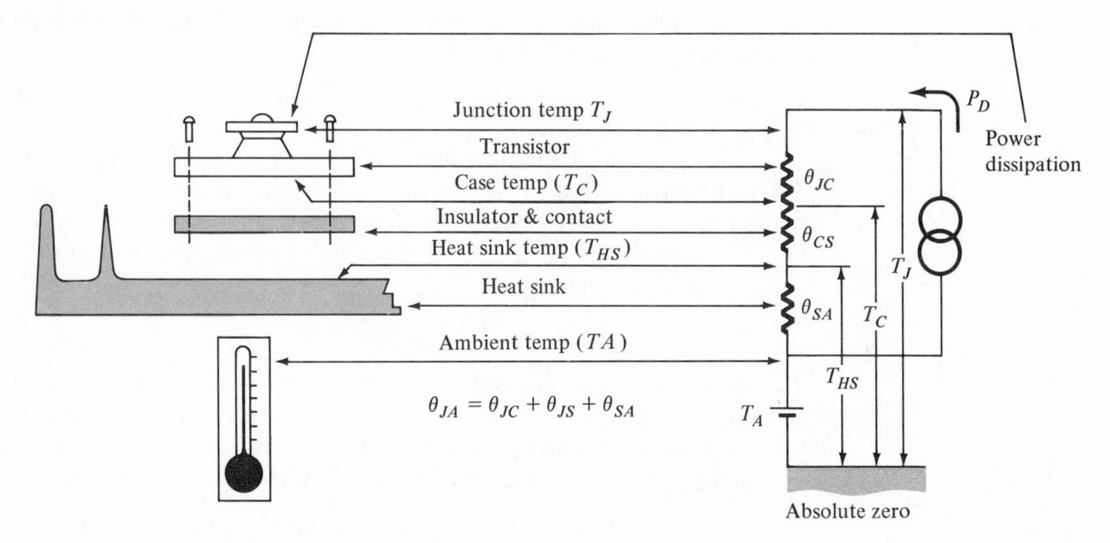

FIG. 8.19
Thermal-to-electrical analogy.

In providing a thermal-electrical analogy the term thermal resistance is used to describe heat effects by an electrical term. The terms in Fig. 8.19 are defined as follows:

θ_{JA} = total thermal resistance (junction to ambient)
θ_{JC} = transistor thermal resistance (junction to case)
θ_{CS} = insulator thermal resistance (case to heat sink)
θ_{SA} = heat sink thermal resistance (heat sink to ambient)

Using the electrical analogy for thermal resistances we can write

$$\theta_{JA} = \theta_{JC} + \theta_{CS} + \theta_{SA} \tag{8.34}$$

The analogy can also be used in applying Kirchhoff's law to obtain

$$T_J = P_D\theta_{JA} + T_A \tag{8.35}$$

The last relation shows that the junction temperature "floats" on the ambient temperature and that the higher the ambient temperature the lower the allowed value of device power dissipation.

The thermal factor θ provides information about how much temperature drop (or rise) results for a given amount of power dissipation. For example, the value of θ_{JC} is usually about 0.5°C/W. This means that for a power dissipation of 50 W the difference in temperature between case temperature (as measured by a thermocouple) and the inside junction temperature is only

$$T_J - T_C = \theta_{JC} P_D = (0.5°\text{C/W})(50\text{ W}) = 25°\text{C}$$

Thus, if the heat sink can hold the case at, say 50°C, the junction is then only at 75°C. This is a relatively small temperature difference, especially at lower power dissipation levels.

The value of thermal resistance from junction to free air (using no heat sink) is, typically

$$\theta_{JA} = 40°\text{C/W}$$

Note, that in this case only 1 W of power dissipation results in a junction temperature 40°C greater than the ambient. A germanium transistor operating into an ambient temperature of 25°C could not dissipate 2 W without exceeding the junction temperature limit of 100°C.

$$T_J = T_A + \theta_{JA} P_D = 25°\text{C} + (40°\text{C/W})(2\text{ W}) = 105°\text{C}$$

A heat sink can now be seen to provide a low thermal resistance between case and air—much less than the 40°C/W value of the transistor case alone. Using a heat sink having

$$\theta_{SA} = 2°\text{C/W}$$

and, with an insulating thermal resistance (from case to heat sink) of

$$\theta_{CS} = 0.8°\text{C/W}$$

and, finally for the transistor,

$$\theta_{CJ} = 0.5°\text{C/W}$$

we can obtain

$$\begin{aligned}\theta_{JA} &= \theta_{SA} + \theta_{CS} + \theta_{CJ} \\ &= 2.0 + 0.8 + 0.5 = 3.3\ °\text{C/W}\end{aligned}$$

So, with a heat sink, the thermal resistance between air and junction is only 3.3°C/W as compared to, say 40°C/W for the transistor operating directly into free air. Using the value of θ_{JA} above for a transistor operated at, say, 2 W we calculate

$$(T_J - T_A) = \theta_{JA} P_D = (3.3°\text{C/W})(2\text{ W}) = 6.6°\text{C}$$

In other words, the use of a heat sink in this example provided only a 6.6°C increase in junction temperature as compared to an 80°C rise without a heat sink.

EXAMPLE 8.10 A germanium power transistor is operated with a heat sink ($\theta_{SA} = 3.4$°C/W). The transistor, rated at 50 W (25°C) has $\theta_{JC} = 0.5$ °C/W and the mounting insulation has $\theta_{CS} = 0.6$°C/W. What maximum power can be dissipated if the ambient temperature is 40°C and $T_{J_{\max}} = 100$°C.

Solution

$$P_D = \frac{T_J - T_A}{\theta_{JC} + \theta_{CS} + \theta_{SA}} = \frac{100 - 40}{0.5 + 0.6 + 3.4} = \frac{60°\text{C}}{4.5°\text{C/W}}$$

$$= 13.3 \text{ W}$$

PROBLEMS

§ 8.3

1. A class-A transformer-coupled amplifier uses a 25:1 transformer to drive a 4-Ω load. Calculate the effective ac load (seen by the transistor connected to the larger turns side of the transformer).
2. What turns ratio transformer is needed to couple to an 8-Ω load so that it appears as a 10-K effective load.
3. Calculate the transformer turns ratio required to connect four parallel 16 Ω speakers so that they appear as an 8-K effective load.
4. A transformer-coupled class-A amplifier drives a 16-Ω speaker through a 15:1 transformer. Using a power supply of 36 V (V_{CC}) the circuit can deliver 2 W to the load. Calculate
   (a) The ac power across the transformer primary.
   (b) The rms value of load voltage.
   (c) The rms value of primary voltage.
   (d) The rms values of load and primary current.
5. Calculate the efficiency of the circuit of Problem 8.4 if the bias current is $I_{CQ} = 150$ mA.
6. Draw the circuit diagram of a class-B transformer-coupled amplifier using an NPN transistor.

§ 8.4

7. Calculate the efficiency of the following amplifier classes and voltages
   (a) Class-A operation with $V_{CE_{\max}} = 24$ V and $V_{CE_{\min}} = 2$ V.
   (b) Class-B operation with $V_{CE_{\min}} = 2$ V, and $V_{CC} = 28$ V.
8. For the following voltage values measured on a scope calculate the amount of the second harmonic distortion: $V_{CE_{\max}} = 27$ V, $V_{CE_{\min}} = 14$ V, $V_{CEQ} = 20$ V.

§ 8.5

9. Draw the circuit diagram of a PNP push-pull power amplifier.

10. Sketch the waveforms of the ac signal in the circuit of Problem 8.9 at each point in the circuit for a sinusoidal input signal.

11. Draw the circuit diagram of an NPN push-pull power amplifier operated class AB. Show the driver stage before the push-pull stage.

§ 8.6

12. Sketch the circuit diagram of a complementary symmetry transformerless amplifier.

13. List any advantages or disadvantages of a transformerless circuit over a transformer-coupled circuit.

§ 8.7

14. Determine the maximum dissipation allowed for a 100-W silicon transistor (rated at 25°C) for a derating factor of 0.6 W/°C, at a case temperature of 150°C.

15. A 60-W germanium power transistor operated with a heat sink (θ_{SA} = 5.6 °C/W) has θ_{JC} = 0.5 °C/W and mounting insulation of θ_{CS} = 0.8 °C/W. What maximum power can be handled by the transistor at an ambient temperature of 60°C? Junction temperature should not exceed 90°C.

PNPN and Other Devices

9.1 INTRODUCTION

In this chapter we will consider other important devices not discussed in detail in the previous chapters. Recall from Chapter 3 that the two-element vacuum tube led the way to the three-element triode, the four-element tetrode and the five-element pentode. A similar natural sequence paved the way for a number of the three-layer and four-layer semiconductor devices to be introduced in this chapter. The family of four-layer PNPN devices will first be considered (SCR, SCS, GTO, LASCR, Shockley diode, DIAC, and TRIAC), followed by an increasingly important device—the UJT (unijunction transistor). The chapter will close with a brief discussion of the phototransistor and the thyratron and beam power tubes.

PNPN DEVICES

9.2 SILICON CONTROLLED RECTIFIER (SCR)

Within the family of PNPN devices the silicon controlled rectifier (SCR) SCR is unquestionably of the greatest interest today. It was first introduced in 1956 by a group of Bell Telephone Laboratory engineers. A few of the more common areas of application for SCRs include relay controls, time delay circuits, regulated power supplies, static switches, motor controls, choppers, invertors, cycloconverters, battery chargers, protective circuits, heater controls, and phase controls.

In recent years, SCR's have been designed to control powers as high as 10 mW with ratings as high as 1200 A at 1800 V (water-cooled) and 475 A at 1200 V (air-cooled). Its frequency range of application has also been extended to about 50 kHz, permitting some high-frequency applications such as induction heating and ultrasonic cleaning.

In some manuals and texts the SCR is called a *thyristor*, derived from the tube equivalent—the thyratron. The term thyristor is not limited only to the SCR but refers, in general, to all members of the PNPN family that have a control mechanism.

9.3 BASIC SILICON CONTROLLED RECTIFIER (SCR) OPERATION

As the terminology indicates the SCR is a rectifier constructed of silicon material which has a third terminal for control purposes. Silicon was chosen because of its high temperature and power capabilities. The basic operation of the SCR is quite different from the fundamental two-layer semiconductor diode in that a third terminal, called a *gate*, determines when the rectifier switches from the open-circuit to short-circuit state. It is not enough simply to forward bias the anode-to-cathode region of the device. In the conduction region the dynamic resistance of the SCR is typically 0.01–0.1 Ω. The reverse resistance is typically 100 K or more.

The graphic symbol for the SCR is shown in Fig. 9.1 with the corresponding connections to the four-layer semiconductor structure. As indicated in Fig. 9.1a, if forward conduction is to be established, the anode must be positive with respect to the cathode. This is not, however, a sufficient criterion for turning the device on. A pulse of sufficient magnitude must also be applied to the gate to establish a turn-on gate current, represented symbolically by I_{GT}.

A more detailed examination of the basic operation of an SCR is best effected by splitting the four-layer PNPN structure of Fig. 9.1b into two three-layer transistor structures as shown in Fig. 9.2a and then considering the resultant circuit of Fig. 9.2b.

Note that one transistor for Fig. 9.2 is a NPN device while the

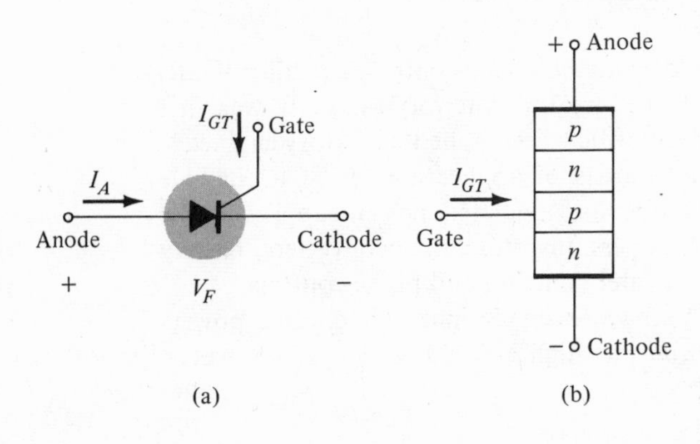

FIG. 9.1
(a) SCR symbol and (b) basic construction.

other is a PNP transistor. For discussion purposes, the signal shown in Fig. 9.3a will be applied to the gate of the circuit of Fig. 9.2b. During the interval $0 \longrightarrow t_1$, $V_{gate} = 0$ V, the circuit of Fig. 9.2b will appear as shown in Fig. 9.3b ($V_{gate} = 0$ V is equivalent to the gate terminal being grounded as shown in the figure). For $V_{BE_2} = V_{gate} = 0$ V, the base current $I_{B_2} = 0$ and I_{C_2} will be approximately I_{CO}. The base current of Q_1, $I_{B_1} = I_{C_2} = I_{CO}$ is too small to turn Q_1 on. Both transistors are therefore in the OFF state, resulting in a high impedance between the collector and emitter of each transistor and the open circuit representation for the controlled rectifier as shown in Fig. 9.3c.

At $t = t_1$ a pulse of V_G volts will appear at the SCR gate. The circuit conditions established with this input are shown in Fig. 9.4a. The potential V_G was chosen sufficiently large to turn Q_2 on ($V_{BE_2} = V_G$). The collector current of Q_2 will then rise to a value sufficiently large to turn Q_1 on ($I_{B_1} = I_{C_2}$). As Q_1 turns on, I_{C_1} will increase, resulting

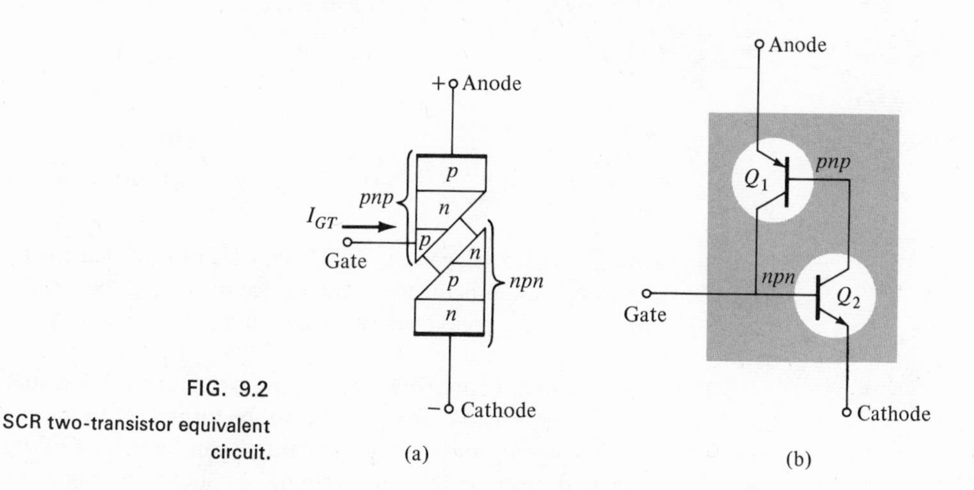

FIG. 9.2
SCR two-transistor equivalent circuit.

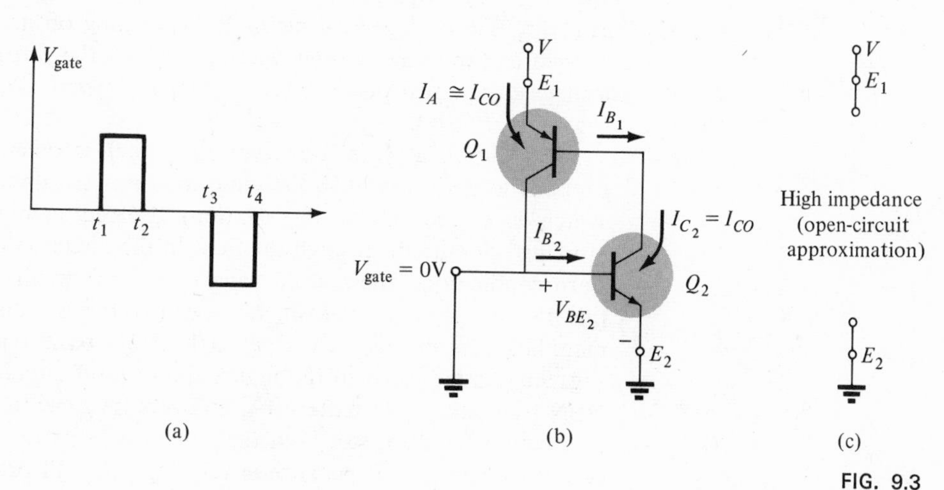

FIG. 9.3
OFF state of the SCR.

in a corresponding increase in I_{B_2}. The increase in base current for Q_2 will result in a further increase in I_{C_2}. The net result is a regenerative increase in the collector current of each transistor. The resulting anode to cathode resistance ($R_{SCR} = V/I_A$ — large) is then very small, resulting in the short-circuit representation for the SCR as indicated in Fig. 9.4b. The regenerative action described above results in SCRs having typical turn-on-times of 0.1 to 1 μsec.

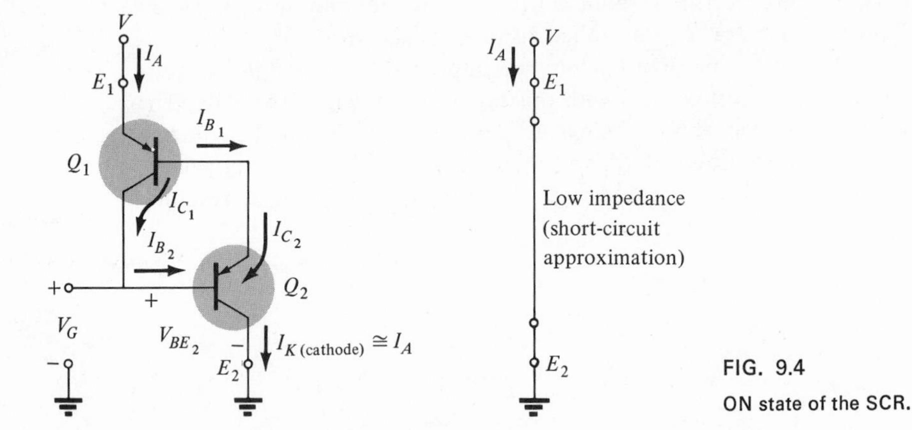

FIG. 9.4
ON state of the SCR.

In addition to gate triggering, SCR's can also be turned on by significantly raising the temperature of the device or raising the anode-to-cathode voltage to the breakover value shown on the characteristics of Fig. 9.7.

The next question of concern is "How long is the turn-off time and how is turn-off accomplished?" An SCR *cannot* be turned off by simply removing the gate signal, and only a special few can be turned off by applying a negative pulse to the gate terminal as shown in Fig. 9.3a at $t = t_3$. The two general methods for turning off an SCR are categorized as the *anode current interruption* and the *forced commutation technique*. The two possibilities for current interruption are shown in Fig. 9.5.

In Fig. 9.5a, I_A is zero when the switch is opened (series interruption) while in Fig. 9.5b the same condition is established when the switch is closed (shunt interruption). Forced commutation is the "forcing" of current through the SCR in the direction opposite to forward conduction. There are a wide variety of circuits for performing this function, a number of which can be found in the manuals of major manufacturers in this area. One of the more basic types is shown in Fig. 9.6. As indicated in the figure, the turn-off circuit consists of an NPN transistor, a dc battery V_B, and a pulse generator. During SCR conduction the transistor is in the "off state", that is, $I_B = 0$ and the collector-to-emitter impedance is very high (for all practical purposes an open-circuit). This high impedance will isolate the turn-off circuitry

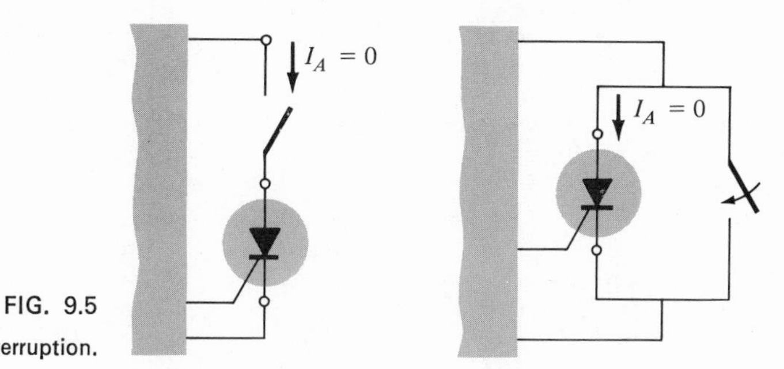

FIG. 9.5
Anode current interruption.

from affecting the operation of the SCR. For turn-off conditions, a positive pulse is applied to the base of the transistor, turning it heavily on, resulting in a very low impedance from collector to emitter (short-circuit representation). The battery potential will then appear directly across the SCR as shown in Fig. 9.6b, forcing current through it in the reverse direction for turn-off. Turn-off time of SCR's are typically 5–30 μsec.

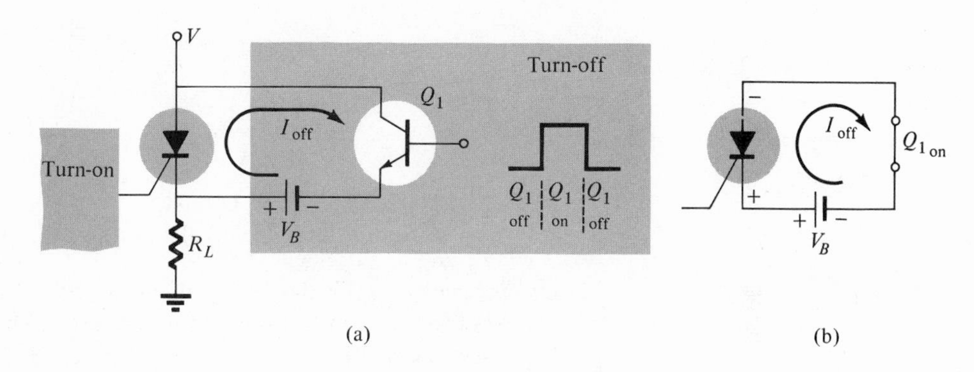

FIG. 9.6
Forced commutation technique.

9.4 SCR CHARACTERISTICS AND RATINGS

The characteristics of an SCR are provided in Fig. 9.7 for various values of gate current. The currents and voltages of usual interest are indicated on the characteristic. A brief description of each is listed below.

1. *Forward breakover voltage* $V_{(BR)F^*}$ is that voltage above which the SCR enters the conduction region. The asterisk (*) is a letter to be added that is dependent on the condition of the gate terminal as listed below

   O = open-circuit from G to K.
   S = short-circuit from G to K.

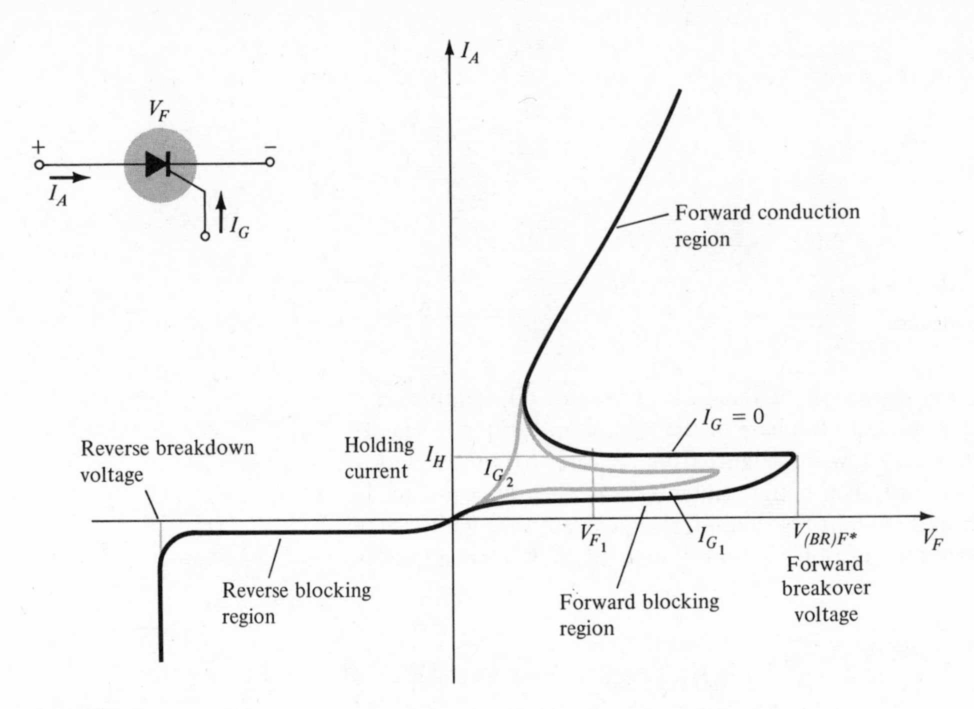

FIG. 9.7
SCR characteristics.

R = resistor from G to K.
V = fixed bias (voltage) from G to K.

2. *Holding current* (I_H) is that value of current below which the SCR switches from the conduction state to the forward blocking region under stated conditions.
3. *Forward and reverse blocking regions* are the regions corresponding to the open-circuit condition for the controlled rectifier which *block* the flow of charge (current) from anode to cathode.
4. *Reverse breakdown voltage* is equivalent to the Zener or avalanche region of the fundamental two-layer semiconductor diode.

It should be immediately obvious that the SCR characteristics of Fig. 9.7 are very similar to those of the basic two-layer semiconductor diode except for the horizontal offshoot before entering the conduction region. It is this horizontal jutting region that gives the gate control over the response of the SCR. For the characteristic having the solid line in Fig. 9.7 ($I_G = 0$) V_F must reach the largest required breakover voltage before the "collapsing" effect will result and the SCR can enter the conduction region corresponding to the *on* state. If the gate current is increased to I_{G_1}, as shown in the same figure, by applying a bias voltage to the gate terminal the value of V_F required for the conduction is considerably less. Note also that I_H drops with increase in I_G. If increased to I_{G_2} the SCR will fire at very low values of voltage and the characteristics begin to approach those of the basic *p-n* junction diode. Looking at the characteristics in a completely different sense, for a particular V_F voltage, say V_{F_1} (Fig. 9.7); if the gate current is increased from $I_G = 0$ to $I_G = I_{G_1}$ the SCR will fire.

The gate characteristics are provided in Fig. 9.8. The characteristics of Fig. 9.8b are an expanded version of the shaded region of Fig. 9.8a. In Fig. 9.8a the three gate ratings of greatest interest, P_{GFM}, I_{GFM}, and V_{GFM} are indicated. Each is included on the characteristics in the same manner employed for the triode and transistor in Chapter 3. Except for portions of the shaded region any combination of gate current and voltage that falls within this region will fire any SCR in the series of components for which these characteristics are provided. Temperature will determine which sections of the shaded region must be avoided. At $-65°C$ the minimum current that will trigger the series of SCR's is 80 mA while at $+150°C$ only 20 mA are required. The effect of temperature on the minimum gate voltage is usually not indicated on curves of this type since gate potentials of 3 V or more are usually obtained easily. As indicated on Fig. 9.8b a minimum of 3 V is simply indicated for all units for the temperature range of interest.

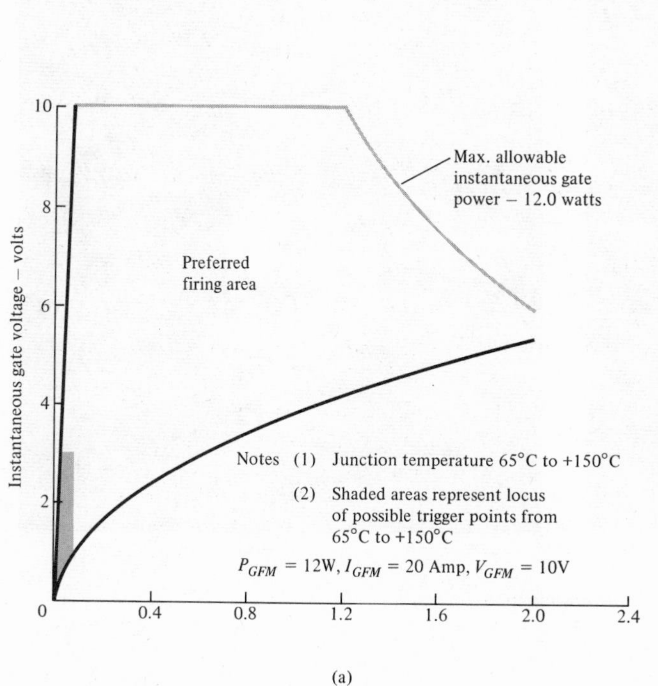

(a)

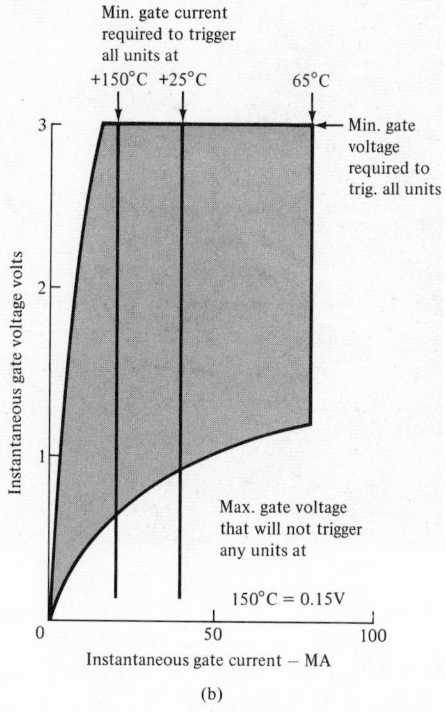

(b)

FIG. 9.8
SCR gate characteristics (GE series—C38).

As an aid in the design of systems that will not trigger prematurely, an additional piece of information is usually provided as indicated in Fig. 9.8b. It states that at 150°C, any gate voltages 0.15 V or less will not trigger the device into the ON state.

Other parameters usually included on the specification sheet of an SCR are the turn-on-time (t_{on}), turn-off time (t_{off}), junction temperature (T_j) and case temperature (T_C) all of which should by now be, to some extent, self-explanatory.

9.5 SCR CONSTRUCTION AND TERMINAL IDENTIFICATION

The basic construction of the four-layer pellet of an SCR is shown in Fig. 9.9a. The complete construction of a thermal-fatigue-free, high-current SCR is shown in Fig. 9.9b. Note the position of the gate,

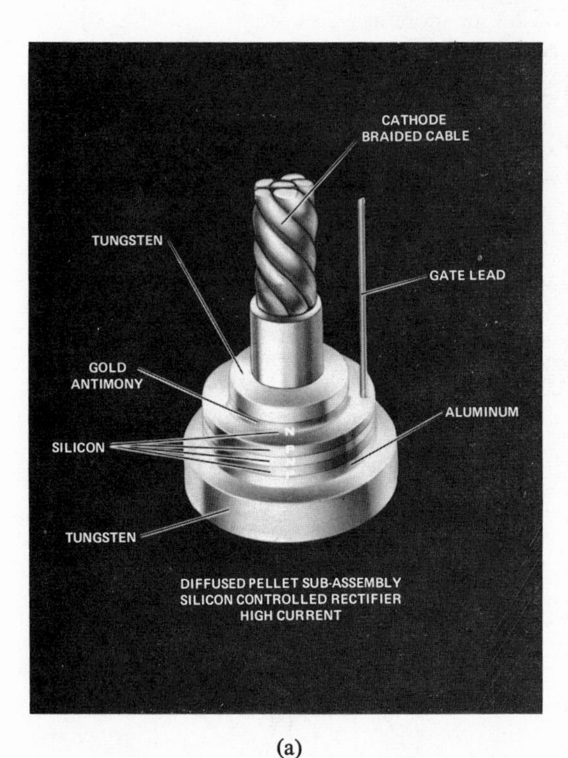

(a)

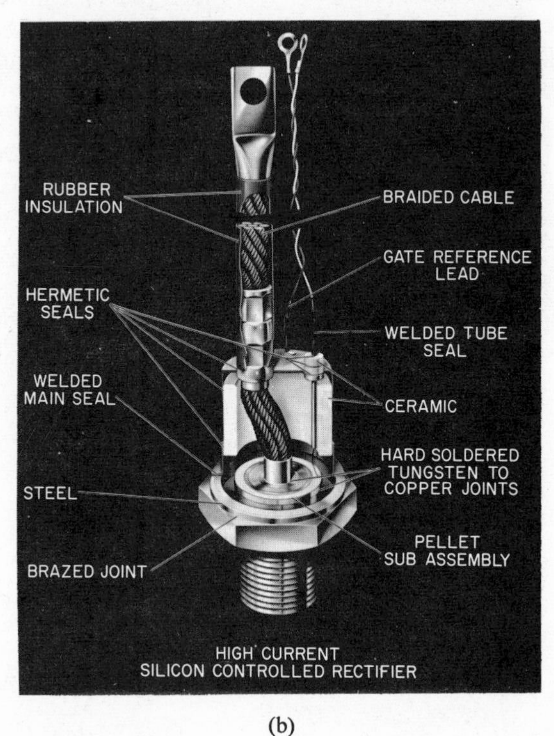

(b)

(Courtesy General Electric Company)

FIG. 9.9
(a) Alloy-diffused SCR pellet; (b) thermal fatigue-free SCR construction.

(Courtesy General Electric Company)

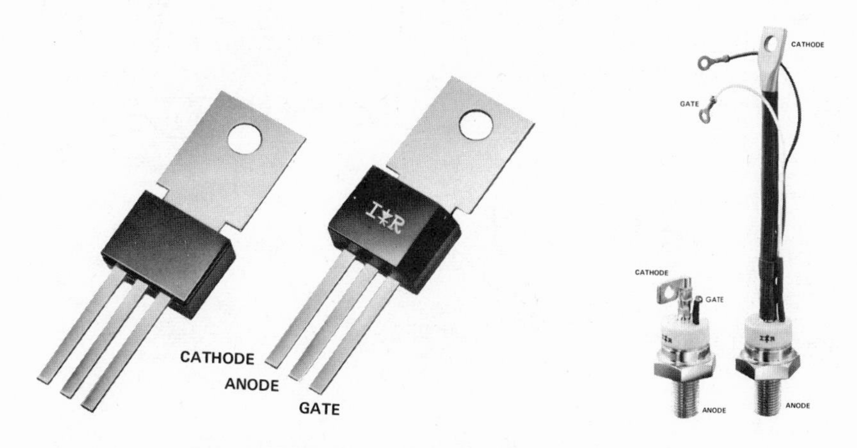

(Courtesy International Rectifier Corp. Inc.)

FIG. 9.10

SCR case construction and terminal identification.

cathode, and anode terminals. The pedestal acts as a heat sink by transferring the heat developed to the chassis on which the SCR is mounted. The case construction and terminal identification of SCRs will vary with the application. Other case-construction techniques and the terminal identification of each are indicated in Fig. 9.10.

9.6 SCR APPLICATIONS

A few of the possible applications for the SCR are listed in the introduction to the SCR (Section 9.2). In this section we will consider three: a static switch, a phase control system, and a battery charger.

A half-wave *series static switch* is shown in Fig. 9.11a. If the switch is closed as shown in Fig. 9.11b, a gate current will flow during the positive portion of the input signal turning the SCR on. Resistor R_L limits the magnitude of the gate current. When the SCR turns on, the anode-to-cathode voltage (V_F) will drop to the conduction value resulting in a greatly reduced gate current and very little loss in the gate circuitry. For the negative region of the input signal the SCR will turn off since the anode is negative with respect to the cathode. The diode D_1 is included to prevent a reversal in gate current.

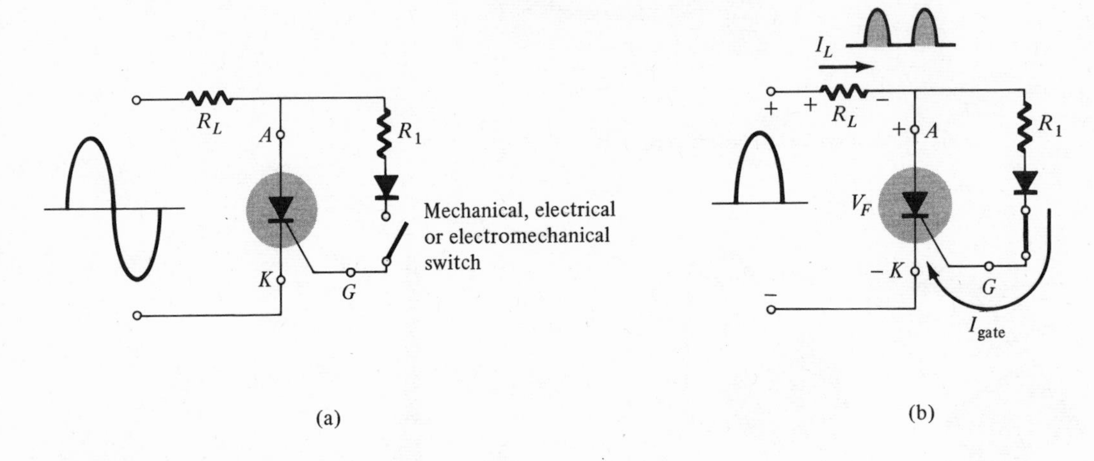

FIG. 9.11
Half-wave series static switch.

The waveforms for the resulting load current and voltage are shown in Fig. 9.11b. The result is a half-wave rectified signal through the load. If less than 180° conduction is desired, the switch can be closed at any phase displacement during the positive portion of the input signal. The switch can be electronic, electromagnetic, or mechanical, depending on the application.

A circuit capable of establishing a conduction angle between 90° and 180° is shown in Fig. 9.12a. The circuit is similar to that of Fig. 9.11a except for the addition of a variable resistor and the elimination of the switch. The combination of the resistors R and R_1 will limit the gate current during the positive portion of the input signal. If R_1 is set to its maximum value the gate current may never reach turn-on magnitude. As R_1 is decreased from the maximum the gate current will increase for the same input voltage. In this way, the required turn-on gate current can be established in any point between 0° and 90° as shown in Fig. 9.12b. If R_1 is low the SCR will fire almost immediately, resulting in the same action as that obtained from the circuit of Fig. 9.11a (180° conduction). However, as indicated above, if R_1 is increased a larger input votage (positive) will be required to fire the SCR. The control as shown in Fig. 9.12b cannot be extended past a

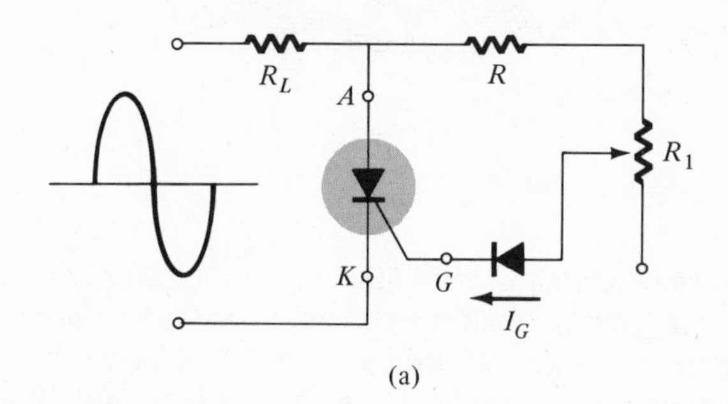

(a)

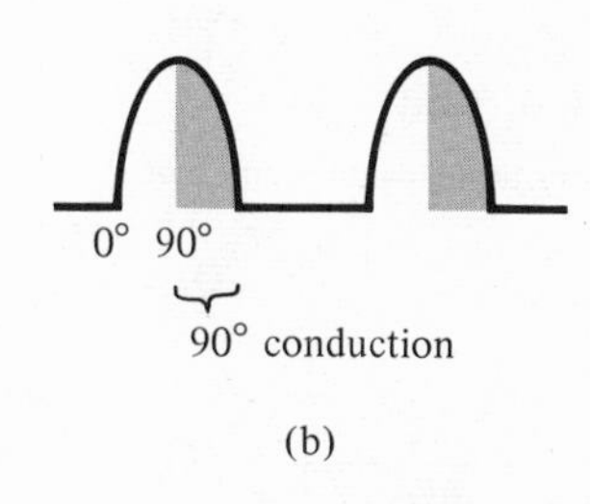

(b)

FIG. 9.12
Half-wave variable-resistance phase control.

90° phase displacement since the input is its maximum at this point. If it fails to fire at this and lesser values of input voltage on the positive slope of the input the same response must be expected from the negatively sloped portion of the signal waveform. The operation here is normally referred to in technical terms as half-wave variable resistance phase control. It is an effective method of controlling the rms current and therefore power to the load.

A second popular application of the SCR is in a *battery charging regulator*. The fundamental components of the circuit are shown in Fig. 9.13. You will note that the control circuit has been blocked off for discussion purposes.

As indicated in the figure, D_1 and D_2 establish a full-wave rectified signal across SCR_1 and the 12-V battery to be charged. At low battery voltages SCR_2 is in the off state for reasons to be explained shortly. With SCR_2 open, the SCR_1 controlling circuit is exactly the same as

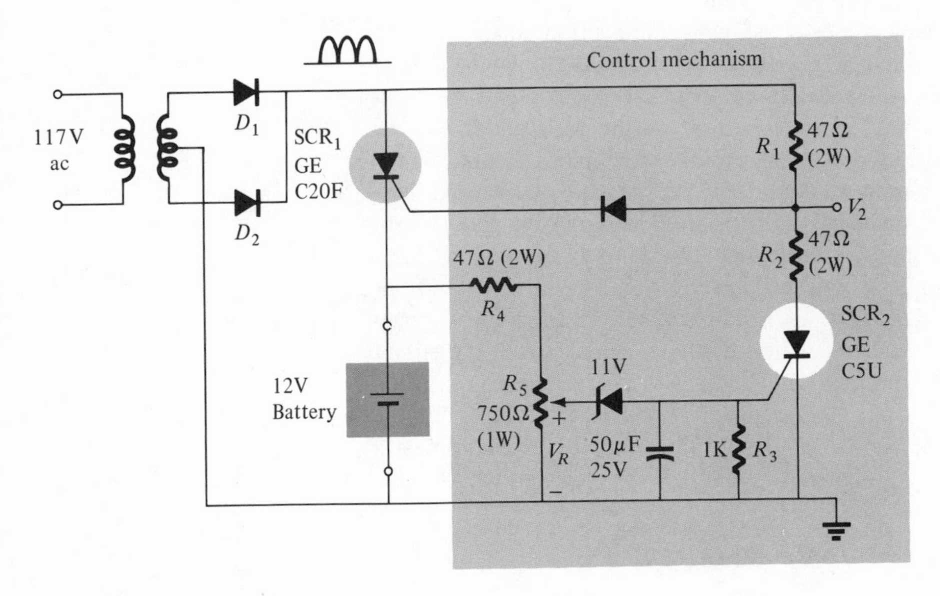

FIG. 9.13

Battery charging regulator.

the series static switch control discussed earlier in this section. When the full-wave rectified input is sufficiently large to produce the required turn-on gate current (controlled by R_1), SCR_1 will turn on and charging of the battery will commence. At the start of charging, the low battery voltage will result in a low voltage V_1 as determined by the simple voltage divider circuit. Voltage V_1 is in turn too small to cause 11.0-V Zener conduction. In the off state, the Zener is effectively an open-circuit maintaining SCR_2 in the off state since the gate current is zero. The capacitor C_1 is included to prevent any voltage transients in the circuit from accidently turning on SCR_2. Recall from your fundamental study of circuit analysis that the voltage cannot instantaneously change across a capacitor. In this way C_1 prevents transient effects from affecting the SCR.

As charging continues, the battery voltage rises to a point where V_1 is sufficiently high to turn the 11.0-V Zener on and fire SCR_2. Once SCR_2 has fired, the short-circuit representation for SCR_2 will result in a voltage divider circuit determined by R_1 and R_2 that will maintain V_2 at a level too small to turn SCR_1 on. When this occurs the battery is fully charged and the open-circuit state of SCR_1 will cut off the charging current. Thus, the regulator recharges the battery whenever the voltage drops, and prevents overcharging when fully charged.

The silicon controlled switch (SCS), like the silicon controlled rectifier is a four-layer PNPN device. All four semiconductor layers of the SCS are available due to the addition of an anode gate as shown in Fig. 9.14a. The graphic symbol and transistor equivalent circuit are shown in the same figure.

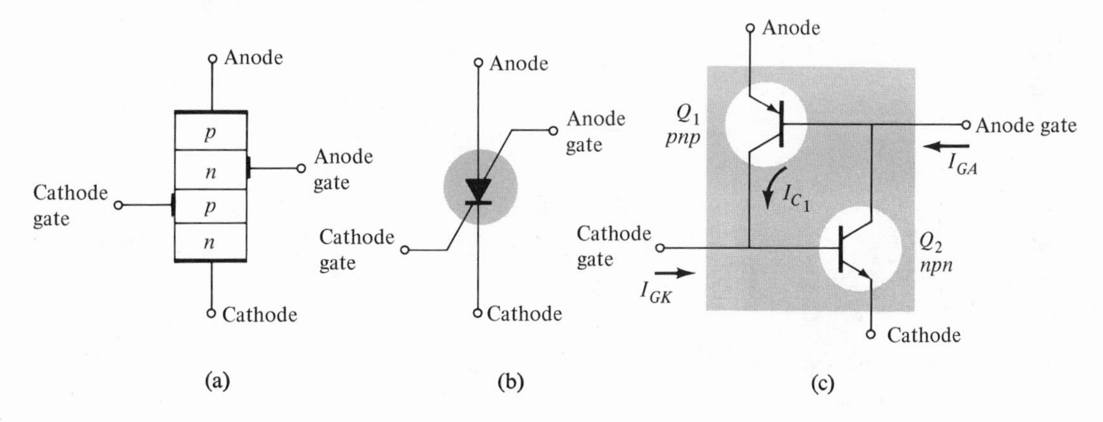

FIG. 9.14
Silicon controlled switch (SCS): (a) basic construction; (b) graphic symbol; (c) equivalent transistor circuit.

The characteristics of the device are essentially the same as those for the SCR. The effect of an anode gate current is very similar to that demonstrated by the gate current in Fig. 9.7. The higher the anode gate current the lower the required anode-to-cathode voltage to turn the device on.

The anode gate connection can be used to either turn on or turn off the device. To turn on the device, a negative pulse must be applied to the anode gate terminal, while a positive pulse is required to turn off the device. The need for the type pulse indicated above can be demonstrated using the circuit of Fig. 9.14c. A negative pulse at the anode gate will forward bias the base-to-emitter junction of Q_1, turning it on. The resulting heavy collector current I_{C_1} will turn on Q_2, resulting in a regenerative action and the on state for the SCS device. A positive pulse at the anode gate will reverse bias the base-to-emitter junction of Q_1, turning it off, resulting in the open-circuit off state of the device. In general, the triggering (turn-on) anode gate current is larger in magnitude than the required cathode gate current. For one representative SCS device, the triggering anode gate current is 1.5 mA while the required cathode gate current is 1 μA. The required turn-on gate current at either terminal is affected by many factors. A few include the operating temperature, anode-to-cathode voltage, load placement, and type of cathode, gate-to-cathode or anode gate-to-anode connection (short-circuit, open-circuit, bias, load, etc.). Tables, graphs, and curves are normally available for each device to provide the type of information indicated above.

Three of the more fundamental types of turn-off circuits for the SCS are shown in Fig. 9.15. When a pulse is applied to the circuit of

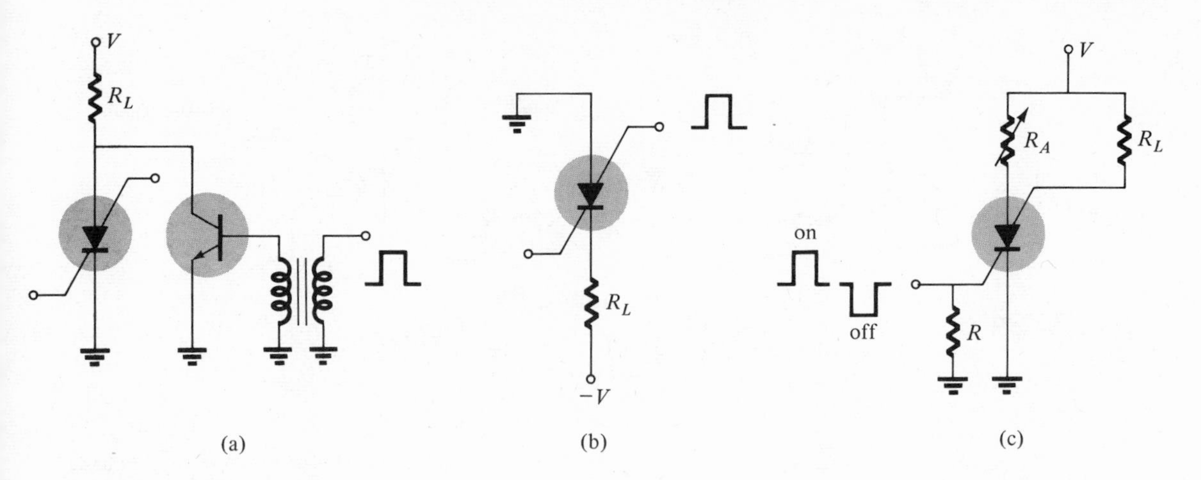

FIG. 9.15
SCS turn-off techniques.

Fig. 9.15a the transistor conducts heavily, resulting in a low impedance (≅ short-circuit) characteristic between collector and emitter. This low impedance branch diverts anode current away from the SCS, dropping it below the holding value and consequently turning it off. Similarly, the positive pulse at the anode gate of Fig. 9.15b will turn the SCS off by the mechanism described earlier in this section. The circuit of Fig. 9.15c can be turned either off *or* on by a pulse of the proper magnitude at the cathode gate. The turn-off characteristic is only possible if the correct value of R_A is employed. It will control the amount of regenerative feedback, the magnitude of which is critical for this type of operation. Note the variety of positions in which the load resistor R_L can be placed. There are a number of other possibilities that can be found in any comprehensive semiconductor handbook or manual.

An advantage of the SCS over a corresponding SCR is the reduced turn-off time, typically within the range 1–10 μsec for the SCS and 5–30 μsec for the SCR.

Some of the remaining advantages of the SCS over an SCR include increased control and triggering sensitivity and a more predictable firing situation. At present, however, the SCS is limited to low power, current, and voltage ratings. Typical maximum anode currents range from 100 to 300 mA with dissipation (power) ratings of 100 to 500 mW.

A few of the more common areas of application include a wide variety of computer circuits (counters, registers, and timing circuits) pulse generators, voltage sensors, and oscillators. One simple application for an SCS as a voltage-sensing device is shown in Fig. 9.16. It is an alarm system with n inputs from various stations. Any single input will turn that particular SCS on, resulting in an energized alarm relay and light in the anode gate circuit to indicate the location of the input (disturbance). The terminal identification of an SCS is shown in Fig. 9.17 with a packaged SCS.

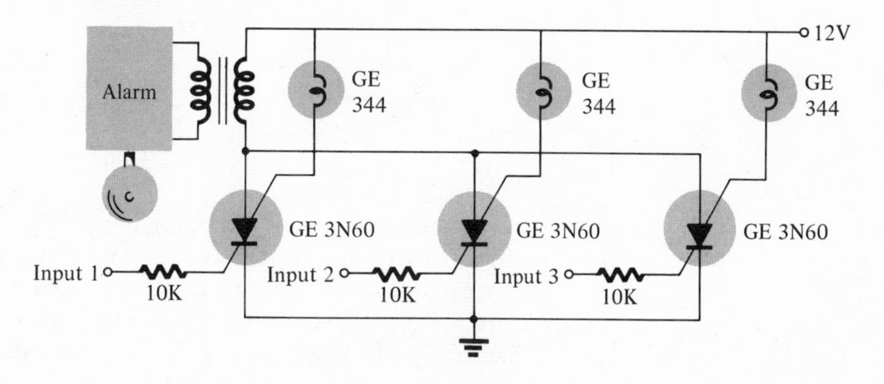

FIG. 9.16
SCS alarm circuit.

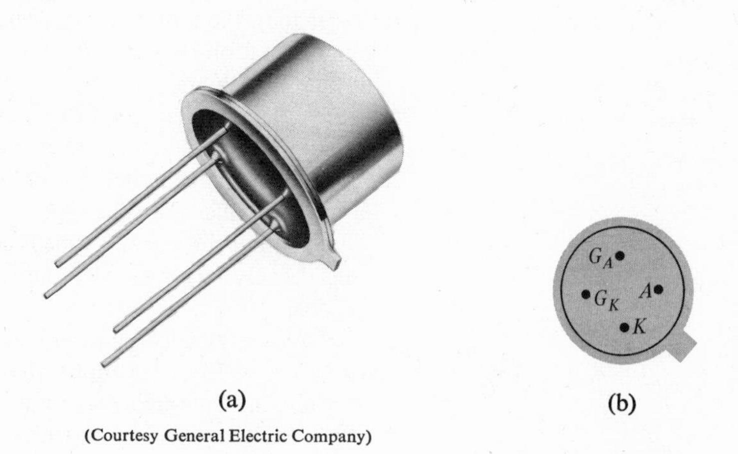

(a) (b)

(Courtesy General Electric Company)

FIG. 9.17
Silicon controlled switch (SCS): (a) device and (b) terminal identification.

The gate turn-off switch is the third PNPN device to be introduced in this chapter. Like the SCR, however, it has only three external terminals as indicated in Fig. 9.18a. Its graphic symbol is also shown in the same figure (Fig. 9.18b). Although the graphic symbol is quite different from either the SCR or SCS the transistor equivalent is exactly the same and the characteristics are quite similar.

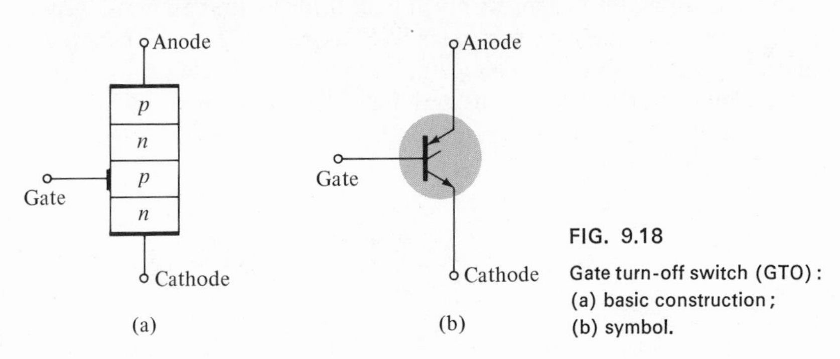

FIG. 9.18
Gate turn-off switch (GTO): (a) basic construction; (b) symbol.

The most obvious advantage of the GTO over the SCR or SCS is the fact that it can be turned on *or* off by applying the proper pulse to the cathode gate (without the anode gate and associated circuitry required for the SCS). A consequence of this turn-off capability is an increase in the magnitude of the required gate current for triggering. For an SCR and GTO of similar maximum rms current ratings, the gate triggering current of a particular SCR is 30 μA while the triggering current of the GTO is 20 mA. The turn-off current of a GTO is slightly larger than the required triggering current. The maximum rms current and dissipation ratings of GTOs manufactured today is limited to about 3 A and 20 W, respectively.

A second very important characteristic of the GTO is improved switching characteristics. The turn-on time is quite similar to the SCR (typically 1 μsec) but the turn-off time of about the *same* duration (1 μsec) is much smaller than the typical turn-off time of an SCR (5–30 μsec). The fact that the turn-off time is similar to the turn-on time rather than considerably larger permits the use of this device in high-speed applications.

A typical GTO and its terminal identification are shown in Fig. 9.19. The GTO gate input characteristics and turn-off circuits can be found in a comprehensive manual or specification sheet. The majority of the SCR turn-off circuits can also be used for GTOs.

Some of the areas of application for the GTO include counters, pulse generators, multivibrators, and voltage regulators. Figure 9.20 is an illustration of a simple sawtooth generator employing a GTO and Zener diode.

When the supply is energized, the GTO will turn on, resulting

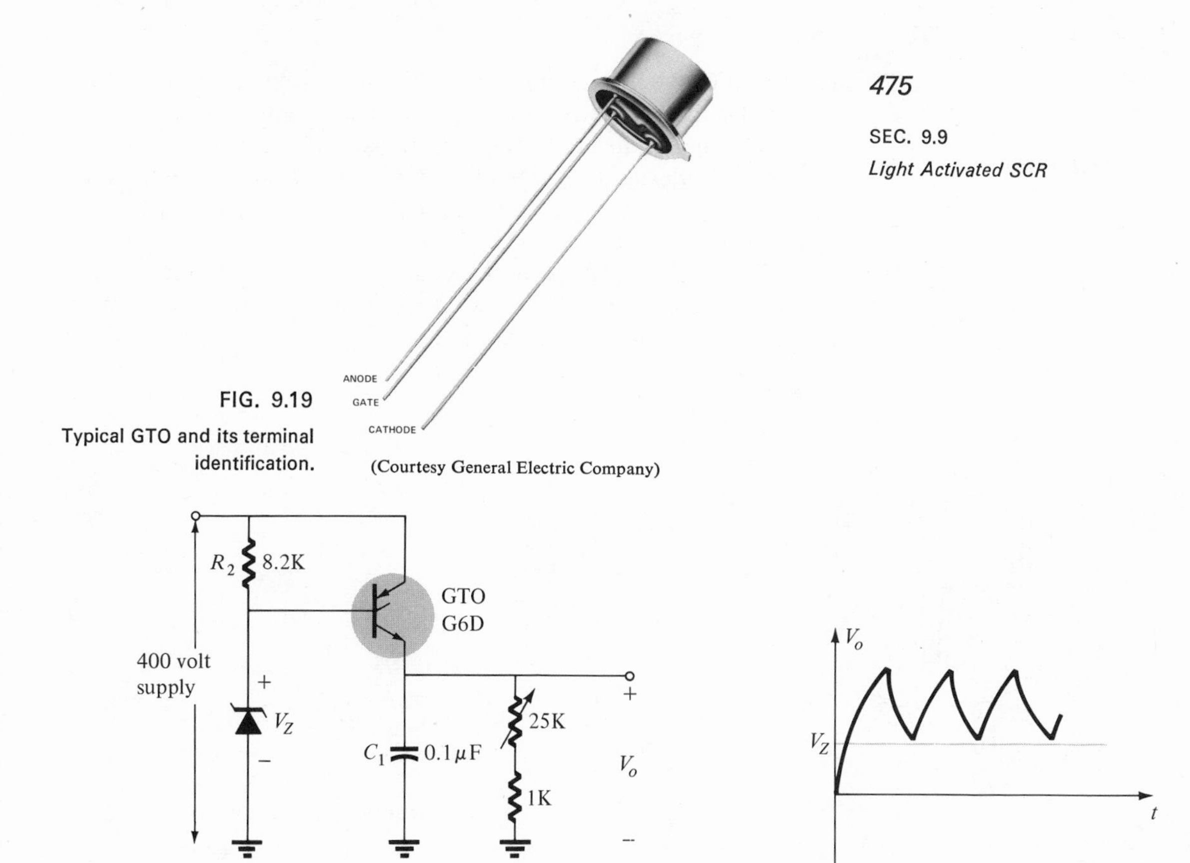

FIG. 9.19
Typical GTO and its terminal identification.

FIG. 9.20
GTO sawtooth generator.

in the short-circuit equivalent from anode to cathode. The capacitor C_1 will then begin to charge toward the supply voltage as shown in Fig. 9.20. As the voltage across the capacitor C_1 charges above the Zener potential a reversal in gate-to-cathode voltage will result, establishing a reversal in gate current. Eventually the negative gate current will be large enough to turn the GTO off. Once the GTO turns off, resulting in the open-circuit representation, the capacitor C_1 will discharge through the resistor R_3. The discharge time will be determined by the circuit time constant $\tau = R_3C_1$. The proper choice of R_3 and C_1 will result in the sawtooth waveform of Fig. 9.20. Once the output potential V_o drops below V_Z, the GTO will turn on and the process repeated.

9.9 LIGHT ACTIVATED SCR (LASCR)

The next in the series of PNPN devices is the light activated SCR (LASCR). As indicated by the terminology, it is an SCR whose state is controlled by the light falling upon a silicon semiconductor layer of the device. The basic construction of an LASCR is shown in Fig. 9.21a.

As indicated in Fig. 9.21a, a gate lead is also provided to permit triggering the device using typical SCR methods. Note also, in the same figure, that the mounting surface for the silicon pellet is the anode connection for the device.

The graphic symbols most commonly employed for the LASCR are provided in Fig. 9.21b. The terminal identification and typical LASCRs are shown in Fig. 9.22a.

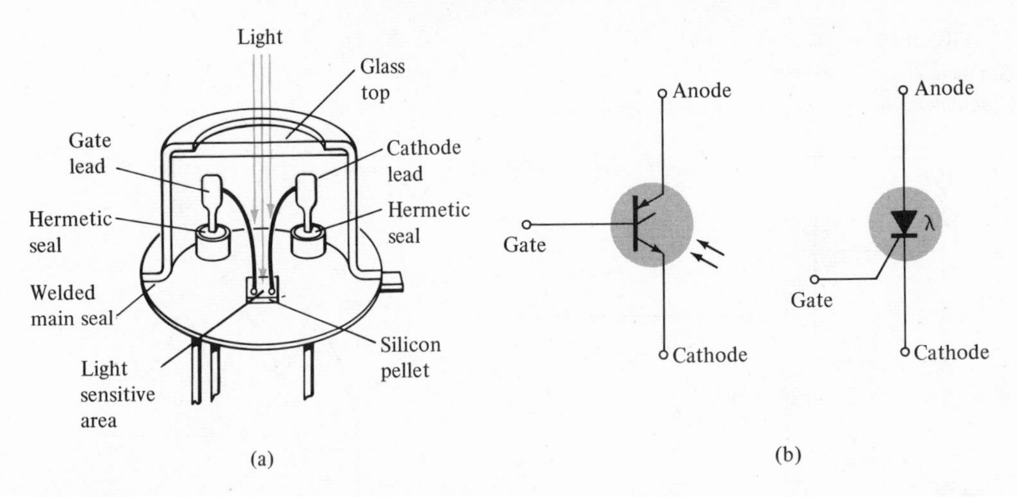

FIG. 9.21
Light activated SCR (LASCR) : (a) basic construction ; (b) symbols.

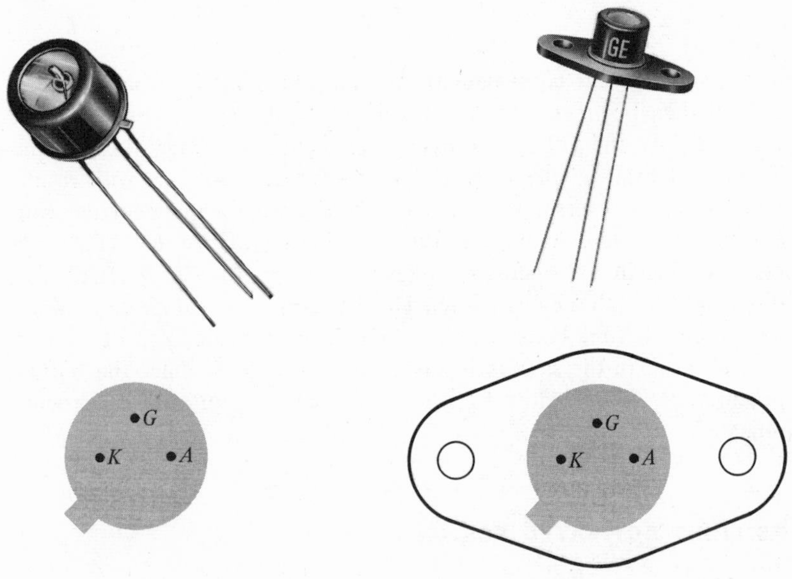

(a) (Courtesy General Electric Company)

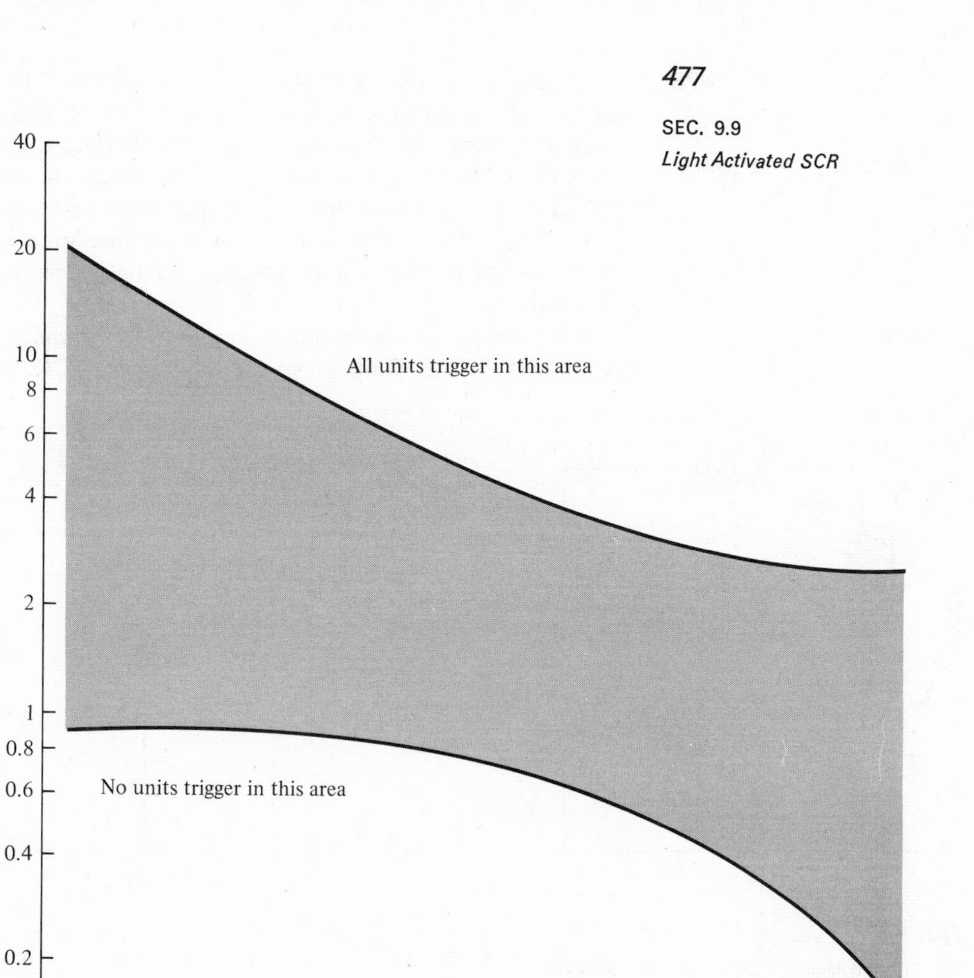

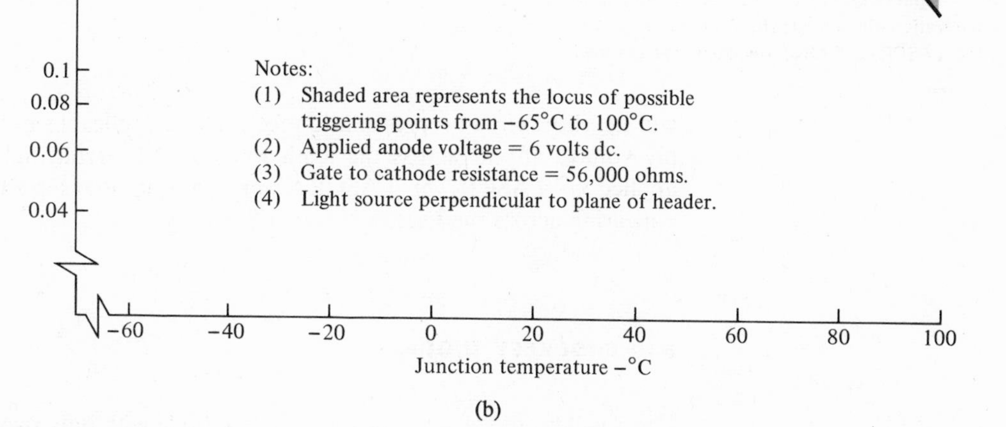

(b)

FIG. 9.22

LASCR: (a) appearance and terminal identification; (b) light triggering characteristics.

Some of the areas of application for the LASCR include optical light controls, relays, phase control, motor control, and a variety of computer applications. The maximum current (rms) and power (gate) ratings for LASCRs commercially available today are about 3 A and 0.1 W. The characteristics (light-triggering) of a representative LASCR are provided in Fig. 9.22b. Note in this figure that an increase in junction temperature results in a reduction in light energy required to activate the device.

One interesting application of an LASCR is in the AND and OR circuits of Fig. 9.23. Only when light falls on $LASCR_1$ *and* $LASCR_2$

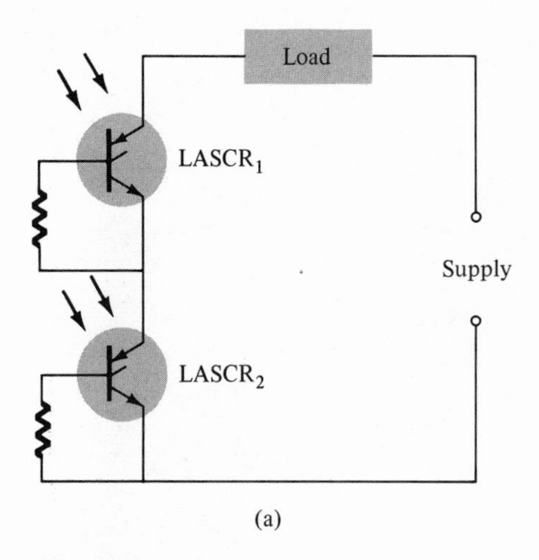

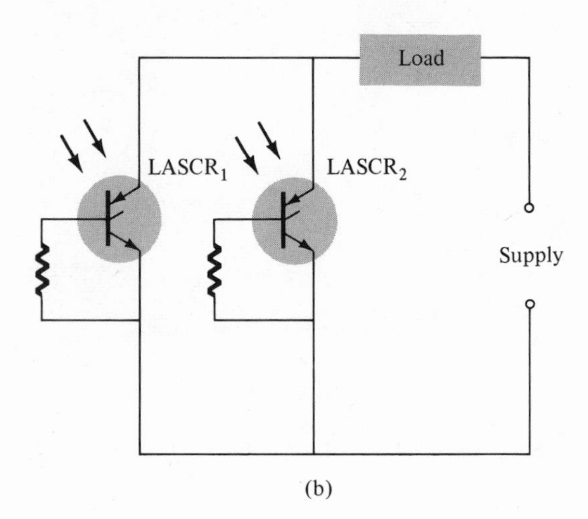

FIG. 9.23

LASCR optoelectronic logic circuitry: (a) AND gate—input to $LASCR_1$ *and* $LASCR_2$ required for energization of the load; (b) OR gate—input to either $LASCR_1$ *or* $LASCR_2$ will energize the load.

will the short-circuit representation for each be applicable and the supply voltage appear across the load. For the OR circuit, light energy applied to $LASCR_1$ *or* $LASCR_2$ will result in the supply voltage appearing across the load.

9.10 SHOCKLEY DIODE

The Shockley diode is a four-layer PNPN diode with only two external terminals as shown in Fig. 9.24a, with its graphic symbol. The characteristics (Fig. 9.24b) of the device are exactly the same as those

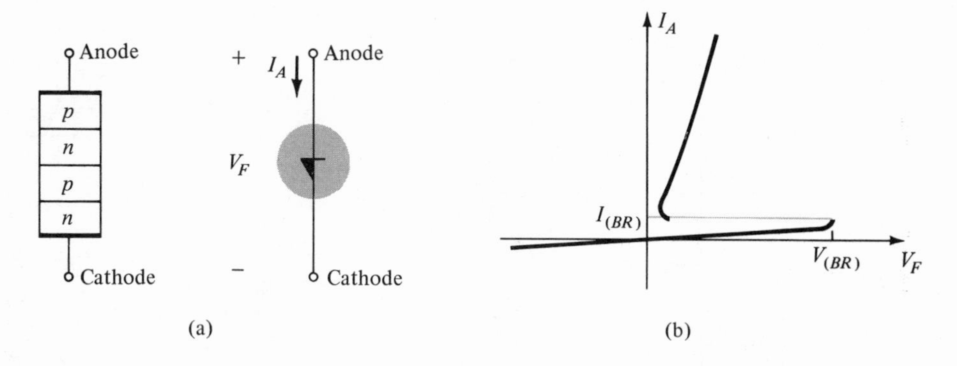

FIG. 9.24
Shockley diode: (a) basic construction and symbol; (b) characteristics.

encountered for the SCR with $I_G = 0$. As indicated by the characteristics the device is in the off state (open-circuit representation) until the breakover voltage is reached, at which time avalanche conditions develop and the device turns on (short-circuit representation).

One common application of the Shockley diode is shown in Fig. 9.25 where it is employed as a trigger switch for an SCR. When the circuit is energized, the voltage across the capacitor will begin to change toward the supply voltage. Eventually the voltage across the capacitor will be sufficiently high to first turn on the Shockley diode and then the SCR.

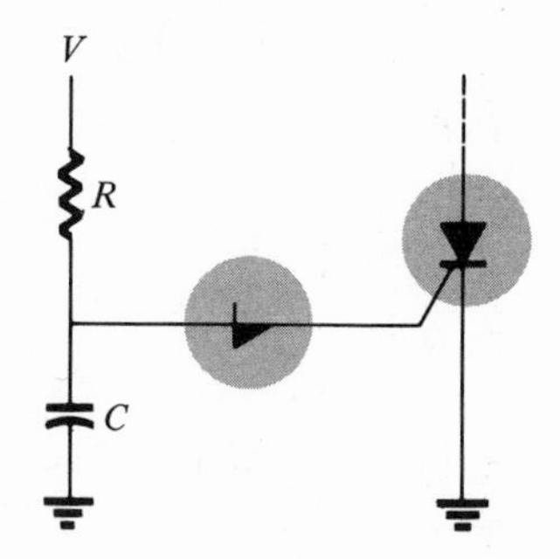

FIG. 9.25
Shockley diode application—trigger switch for an SCR.

9.11 DIAC

The DIAC is basically a parallel-inverse combination of semiconductor layers that permits triggering in either direction. The characteristics of the device, presented in Fig. 9.26a, demonstrate quite clearly that there is a breakover voltage in either direction. This possibility of an on condition in either direction can be used to its fullest advantage in ac applications.

The basic arrangement of the semiconductor layers of the DIAC is shown in Fig. 9.26b, along with its graphic symbol. Note that neither terminal is referred to as the cathode. Instead there is an anode 1 (or electrode 1) and an anode 2 (or electrode 2). When anode 1 is positive with respect to anode 2 the semiconductor layers of particular interest are $p_1n_2p_3$ and n_3. For anode 2 positive with respect to anode 1 the applicable layers are $p_2n_2p_1$ and n_1.

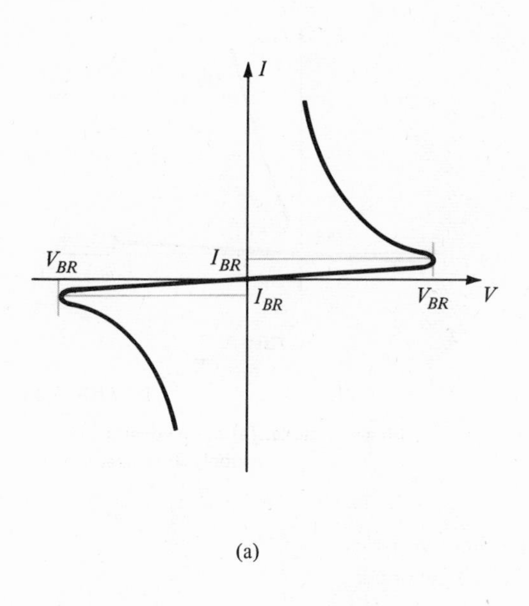

(a)

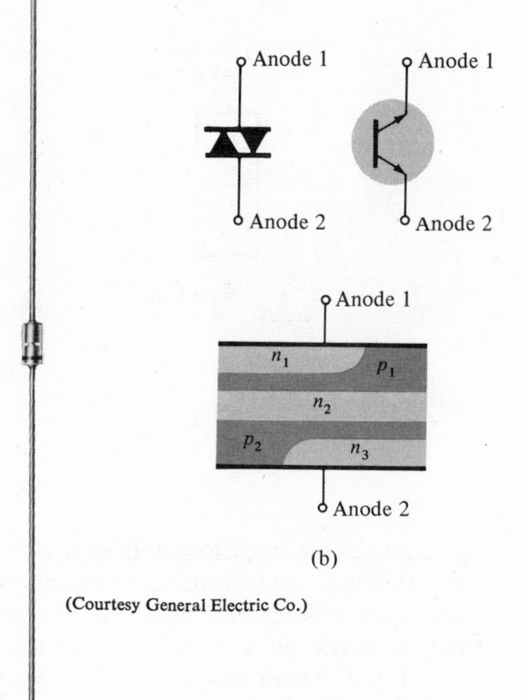

(b)

(Courtesy General Electric Co.)

FIG. 9.26

DIAC: (a) characteristics; (b) symbols and basic construction.

9.12 TRIAC

The TRIAC is fundamentally a DIAC with a gate terminal for controlling the turn-on conditions of the bilateral device in either direction. In other words, for either direction, the gate current can control the action of the device in a manner very similar to that demonstrated for an SCR. The characteristics, however, of the TRIAC in the first and third quadrants are somewhat different from those of the DIAC as shown in Fig. 9.27c. Note the holding current in each direction not present in the characteristics of the DIAC.

The graphic symbol for the device and the distribution of the semiconductor layers is provided in Fig. 9.27 with photographs of the device. For each possible direction of conduction there is a combination of semiconductor layers whose state will be controlled by the signal applied to the gate terminal.

One fundamental application of the TRIAC is presented in Fig. 9.28. In this capacity, it is controlling the ac power to the load by switching on and off during the positive and negative regions of input sinusoidal signal. The action of this circuit during the positive portion of the input signal is very similar to that encountered for the Shockley

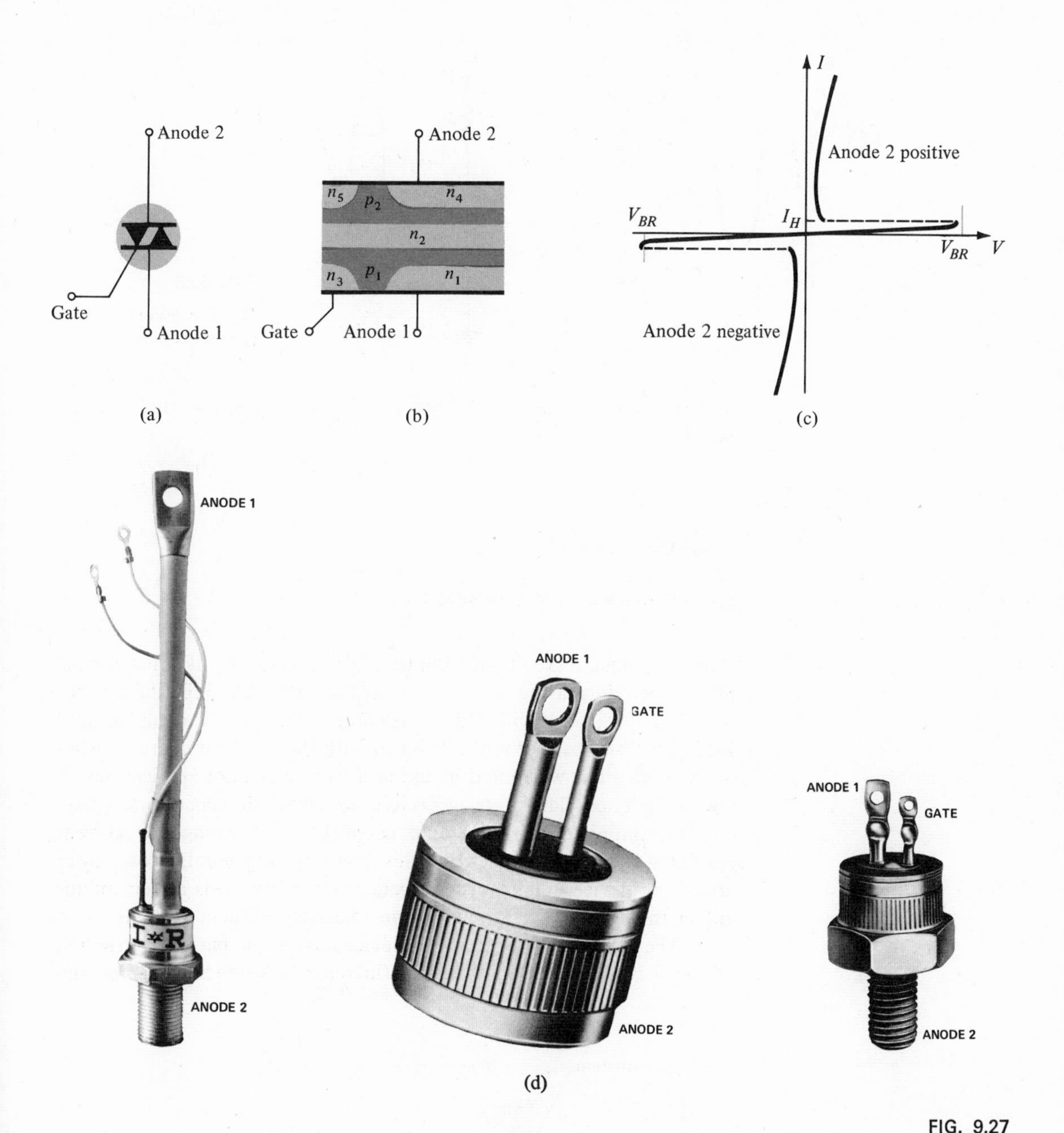

FIG. 9.27
TRIAC: (a) symbol; (b) basic construction; (c) characteristics; (d) photographs.

diode in Fig. 9.25. The advantage of this configuration is that during the negative portion of the input signal the same type of response will result since both the DIAC and TRIAC can fire in the reverse direction. The resulting waveform for the current through the load is provided

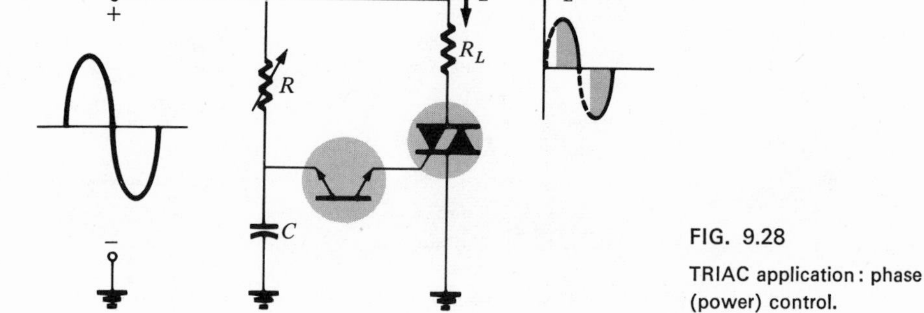

FIG. 9.28
TRIAC application: phase (power) control.

in Fig. 9.28. By varying the resistor R the conduction angle can be controlled.

OTHER DEVICES

9.13 UNIJUNCTION TRANSISTOR

Recent interest in the unijunction transistor (UJT) has, like that for the SCR, been increasing at an exponential rate. Although first introduced in 1948, the device did not become commercially available until 1952. The low cost per unit combined with the excellent characteristics of the device has warranted its use in a wide variety of applications. A few include oscillators, trigger circuits, sawtooth generators, phase control, timing circuits, bistable networks, and voltage or current regulated supplies. The fact that this device is, in general, a low power absorbing device under normal operating conditions, is a tremendous aid in the continual effort to design relatively efficient systems.

The UJT is a three-terminal device having the basic construction of Fig. 9.29. A slab of lightly doped (increased resistance characteristic)

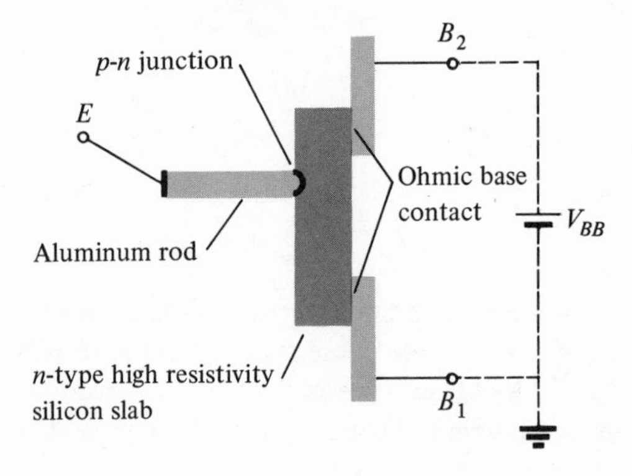

FIG. 9.29
Unijunction transistor (UJT): basic construction.

n-type silicon material has two base contacts attached to both ends of one surface and an aluminum rod alloyed to the opposite surface. The p-n junction of the device is formed at the boundary of the aluminum rod and the n-type silicon slab. The single p-n junction accounts for the terminology unijunction. It was originally called a duo (double) base diode due to the presence of two base contacts. Note in Fig. 9.29 that the aluminum rod is alloyed to the silicon slab at a point closer to the base 1 contact than the base 2 contact and that the base 1 terminal is made positive with respect to the base 2 terminal by V_{BB} volts. The effect of each will become evident in the paragraphs to follow.

The symbol for the unijunction transistor is provided in Fig. 9.30. Note that the emitter leg is drawn at an angle to the vertical line representing the slab of n-type material. The arrowhead is pointing in the direction of conventional current (hole) flow when the device is in the forward-biased, active, or conducting state.

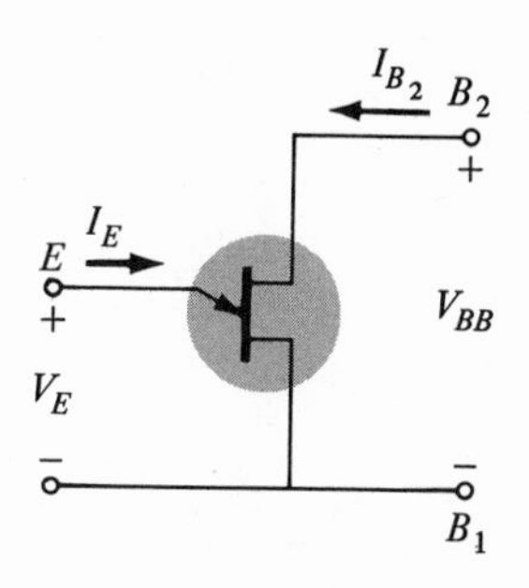

FIG. 9.30
Symbol and basic biasing arrangement for the unijunction transistor.

The circuit equivalent of the UJT is shown in Fig. 9.31. Note the relative simplicity of this equivalent circuit: two resistors (one fixed, one variable) and a single diode. The resistance R_{B_1} is shown as a variable resistor since its magnitude will vary with the current I_E.

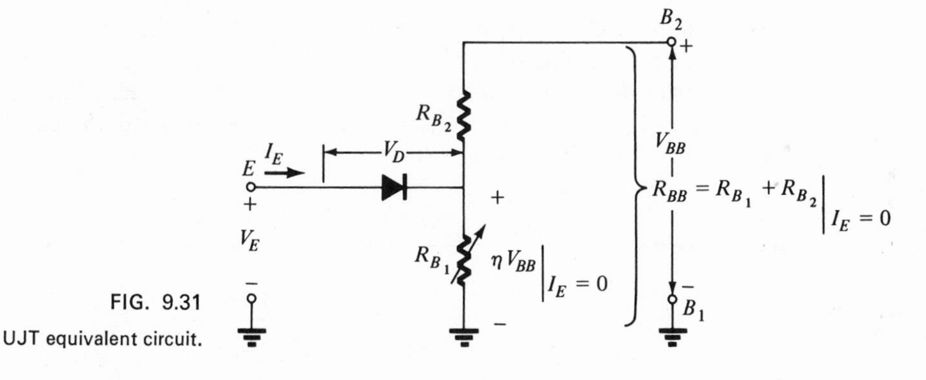

FIG. 9.31
UJT equivalent circuit.

In fact, for a representative unijunction transistor, R_{B_1} may vary from 5 K down to 50 Ω for a corresponding change of I_E from 0 to 50 μA. The interbase resistance R_{BB} is the resistance of the device between terminals B_1 and B_2 when $I_E = 0$. In equation form

$$\boxed{R_{BB} = R_{B_1} + R_{B_2}\big|_{I_E=0}} \qquad (9.1)$$

R_{BB} is typically within the range 4–10 K. The position of the aluminum rod of Fig. 9.29 will determine the relative values of R_{B_1} and R_{B_2} with $I_E = 0$. The magnitude of V_{RB_1} (with $I_E = 0$) is determined by the voltage divider rule in the following manner

$$V_{RB_1} = \frac{R_{B_1}V_{BB}}{R_{B_1} + R_{B_2}} = \eta V_{BB}\Big|_{I_E=0} \tag{9.2}$$

The Greek letter η (eta) is called the *intrinsic stand-off* ratio of the device and is given by

$$\eta = \frac{R_{B_1}}{R_{B_1} + R_{B_2}}\Big|_{I_E=0} \tag{9.3}$$

For applied emitter potentials (V_E) greater than $V_{RB_1} = \eta V_{BB}$ by the forward voltage drop of the diode, $V_D(0.35 \longrightarrow 0.70\text{ V})$ the diode will fire, assume the short-circuit representation (on an ideal basis), and I_E will begin to flow through R_{B_1}. In equation form the emitter firing potential is given by

$$V_P = \eta V_{BB} + V_D \tag{9.4}$$

The characteristics of a representative unijunction transistor are shown for $V_{BB} = 10$ V in Fig. 9.32. Note that for emitter potentials to the left of the peak point, the magnitude of I_E is never greater than I_{EO} (measured in microamperes). The current I_{EO} corresponds very closely with the reverse leakage current I_{CO} of the conventional bipolar transistor. This region, as indicated in the figure, is called the cutoff

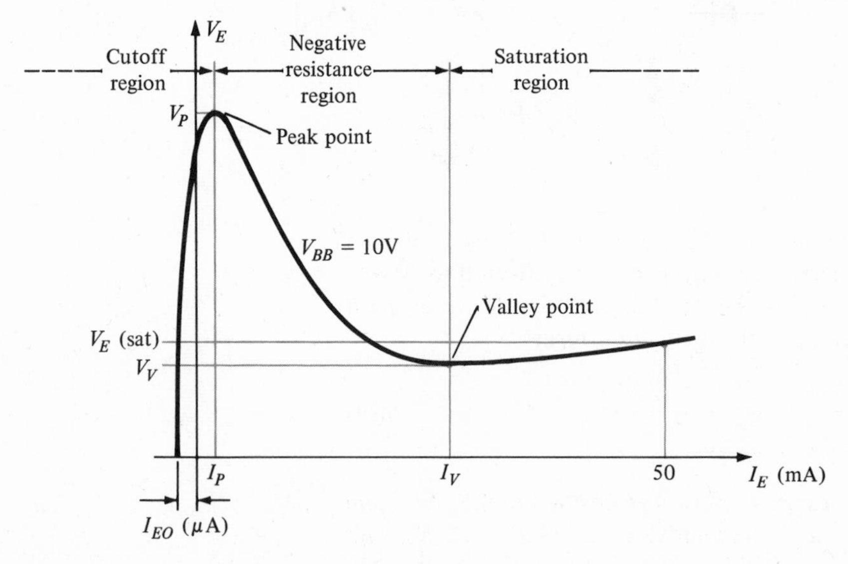

FIG. 9.32
UJT static emitter characteristic curve.

region. Once conduction is established at $V_E = V_P$, the emitter potential V_E will drop with increase in I_E. This corresponds exactly with the decreasing resistance R_{B_1} for increasing current I_E as discussed earlier. This device, therefore, has a *negative resistance* region which is stable enough to be used with a great deal of reliability in the areas of application listed earlier. Eventually, the valley point will be reached, and any further increase in I_E will place the device in the saturation region. In this region the characteristics approach that of the semiconductor diode in the equivalent circuit of Fig. 9.31.

The decrease in resistance in the active region is due to the holes injected into the n-type slab from the aluminum p-type rod when conduction is established. The increased hole content in the n-type material will result in an increase in the number of free electrons in the slab producing an increase in conducitivity (G) and a corresponding drop in resistance ($R\downarrow = 1/G\uparrow$). Three other important parameters for the unijunction transistor are I_P, V_V and I_V. Each is indicated on Fig. 9.32. They are all self-explanatory.

The emitter characteristics as they normally appear are provided in Fig. 9.33. Note that I_{EO} (μA) is not in evidence since the horizontal scale is in milliamperes. The intersection of each curve with the vertical

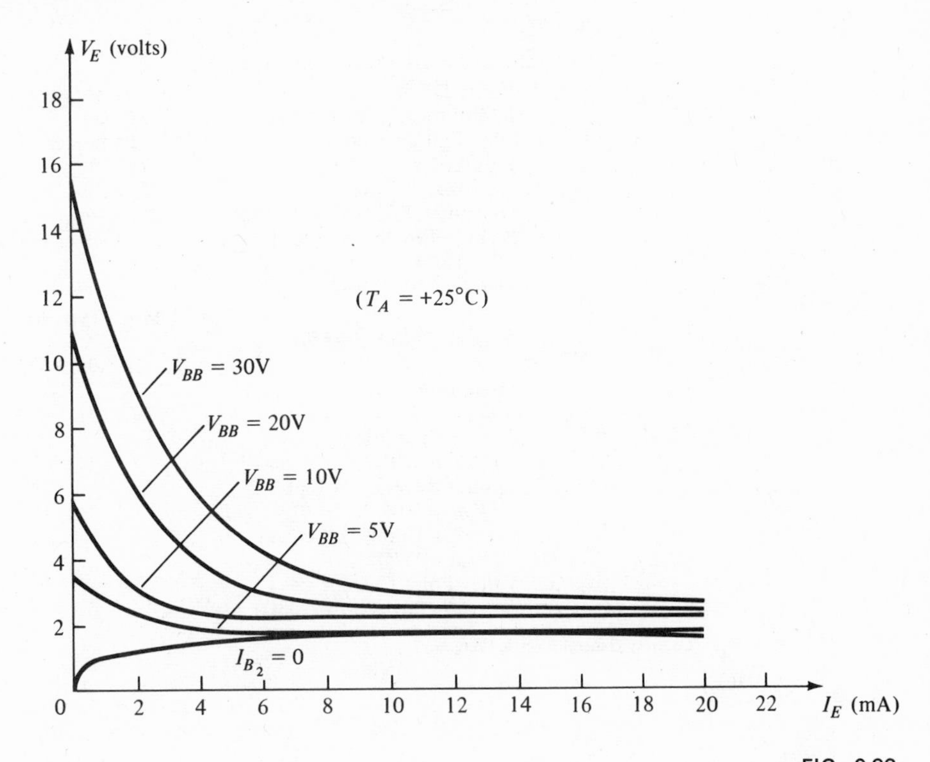

FIG. 9.33
Typical static emitter characteristic curves for a UJT.

axis is the corresponding value of V_P. For fixed values of η and V_D, the magnitude of V_P will vary as V_{BB}, that is,

$$V_P\uparrow = \underbrace{\eta V_{BB}\uparrow + V_D}_{\text{fixed}}$$

A typical set of specifications for the UJT is provided in Fig. 9.34b. The discussion of the last few paragraphs should make each quantity readily recognizable. The terminal identification is provided in the same figure with a photograph of a representative UJT. Note that the base terminals are opposite each other while the emitter terminal is between the two. In addition, the base terminal to be tied to the higher potential is closer to the extension on the lip of the casing.

One rather common application of the UJT is in the triggering of other devices such as the SCR. The basic elements of such a triggering circuit are shown in Fig. 9.35. The resistor R_1 must be chosen to insure the load line determined by R_1 passes through the device characteristics to the right of the peak point but to the left of the valley point. If the load line fails to pass to the right of the peak point the device cannot turn on. An equation for R_1 which will ensure a

(a)

absolute maximum ratings: (25°C)

Power Dissipation	300 mw
RMS Emitter Current	50 ma
Peak Emitter Current	2 amperes
Emitter Reverse Voltage	30 volts
Interbase Voltage	35 volts
Operating Temperature Range	−65°C to +125°C
Storage Temperature Range	−65°C to +150°C

electrical characteristics: (25°C)

Characteristic	Symbol	Min.	Typ.	Max.
Intrinsic Standoff Ratio ($V_{BB} = 10V$)	η	0.56	0.65	0.75
Interbase Resistance ($V_{BB} = 3V, I_E = 0$)	R_{BB}	4.7	7	9.1
Emitter Saturation Voltage ($V_{BB} = 10V, I_E = 50$ ma)	$V_{E(SAT)}$		2	
Emitter Reverse Current ($V_{BB} = 30V, I_{B1} = 0$)	I_{EO}		0.05	12
Peak Point Emitter Current ($V_{BB} = 25V$)	I_P		0.4	5
Valley Point Current ($V_{BB} = 20V, R_{B2} = 100\Omega$)	I_V	4	6	

(Courtesy General Electric Company)

(b)

B_1 B_2 E

(c)

FIG. 9.34

UJT: (a) appearance; (b) specification sheet; (c) terminal identification.

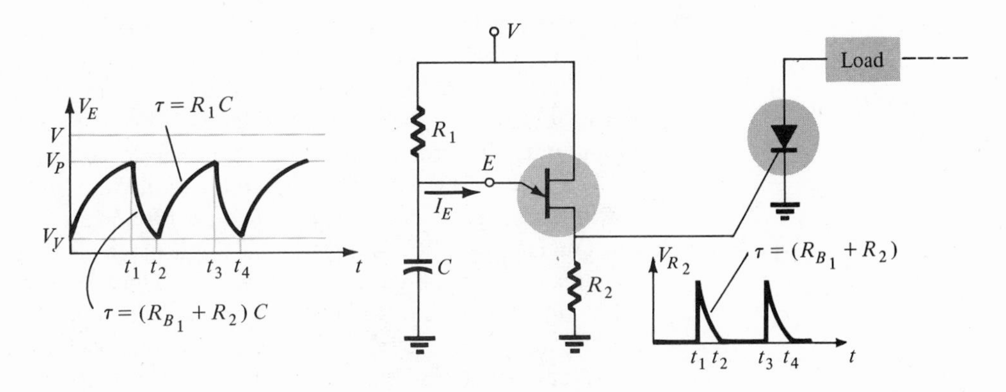

FIG. 9.35
UJT triggering of an SCR.

turn-on condition can be established if we consider the peak point at which $I_P = I_{R_1}$ and $V_E = V_P$. (The equality $I_P = I_{R_1}$ is valid since the charging current of the capacitor, at this instant, is zero; that is, the capacitor is at this particular instant changing from a charging to discharging state.) Then $V - I_P R_1 = V_P$ or $(V - V_P)/I_P = R_1$.

To ensure firing

$$\frac{V - V_P}{I_P} > R_1 \tag{9.5}$$

At the valley point $I_E = I_V$ and $V_E = V_V$ so that to ensure turning off

$$\frac{V - V_V}{I_V} < R_1 \tag{9.6}$$

For the typical values of $V = 30$ V, $\eta = 0.5$, $V_V = 1$ V, $I_V = 10$ mA, $I_P = 10\ \mu$A, and $R_{BB} = 5$ K.

$$\frac{V - V_P}{I_P} = \frac{30 - [0.5(30) + 0.5]}{10 \times 10^{-6}} = \frac{14.5}{10 \times 10^{-6}} = 1.45\ \text{M} > R_1$$

and

$$\frac{V - V_V}{I_V} = \frac{30 - 1}{10 \times 10^{-3}} = 2.9\ \text{K} < R_1$$

Therefore 1.45 M $> R_1 >$ 2.9 K.

The range for R_1 is therefore quite extensive. The resistor R_2 must be chosen small enough to insure that the SCR is not turned on by the interbase current I_{BB} that will flow through R_2 when $I_E = 0$.

The capacitor C will determine, as we shall see, the time interval between triggering pulses and the time span of each pulse.

At the instant the dc supply voltage V is applied, the voltage V_E will charge toward V volts since the emitter circuit of the UJT is in the open-circuit state. The time constant of the charging circuit is R_1C. When $V_E = V_P$, the UJT will enter the conduction state and the capacitor C will discharge through R_{B_1} and R_2 at a rate determined by the time constant $(R_{B_1} + R_2)C$. This time constant is much smaller than the former, resulting in the patterns of Fig. 9.35. Once V_E decays to V_V the UJT will turn off and the charging phase will repeat itself. Since I_{R_2} and V_{R_2} are related by Ohm's law (linear relationship) the waveform for I_{R_2} appears the same as for V_{R_2}. The positive pulse of V_{R_2} is designed to be sufficiently large to turn the SCR on. The operation of a UJT in an oscillator will be discussed further in Chapter 12.

9.14 PHOTOTRANSISTORS

The fundamental behavior of photoelectric devices was introduced in Chapter 1 with the description of the photodiode. This discussion will now be extended to include the phototransistor, which has a photo-

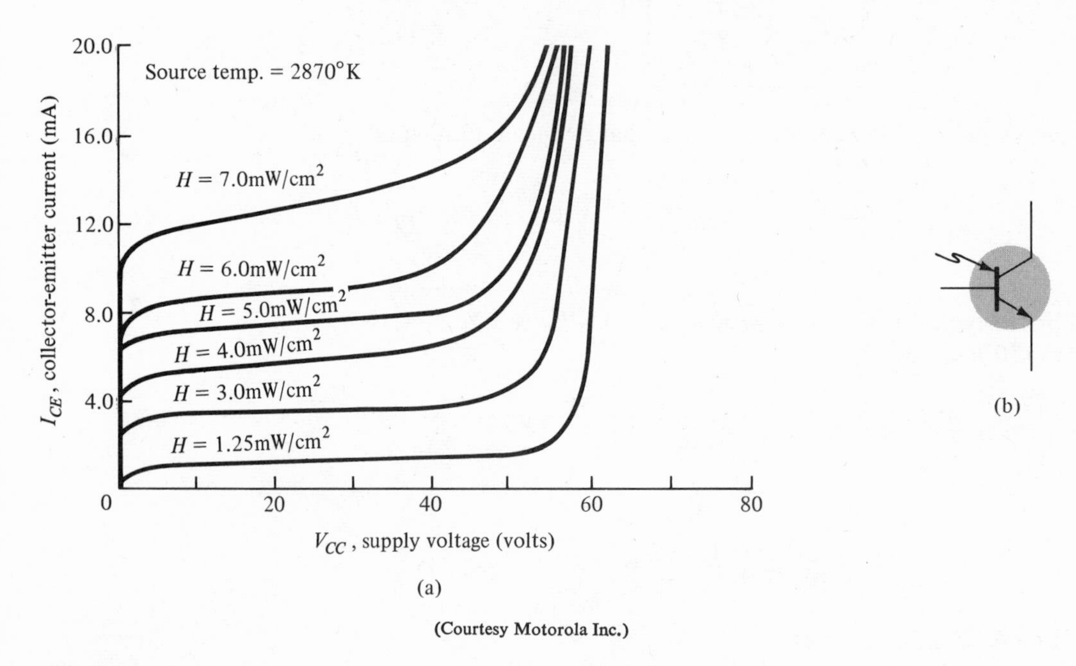

FIG. 9.36

Phototransistor: (a) collector characteristics (MRD300); (b) symbol.

sensitive collector-base *p-n* junction. The current induced by photoelectric effects is the base current of the transistor. If we assign the notation I_λ for the photoinduced base current, the resulting collector current, on an approximate basis, is

$$\boxed{I_C \cong h_{fe} I_\lambda} \tag{9.7}$$

A representative set of characteristics for a phototransistor is provided in Fig. 9.36 with the symbolic representation of the device.

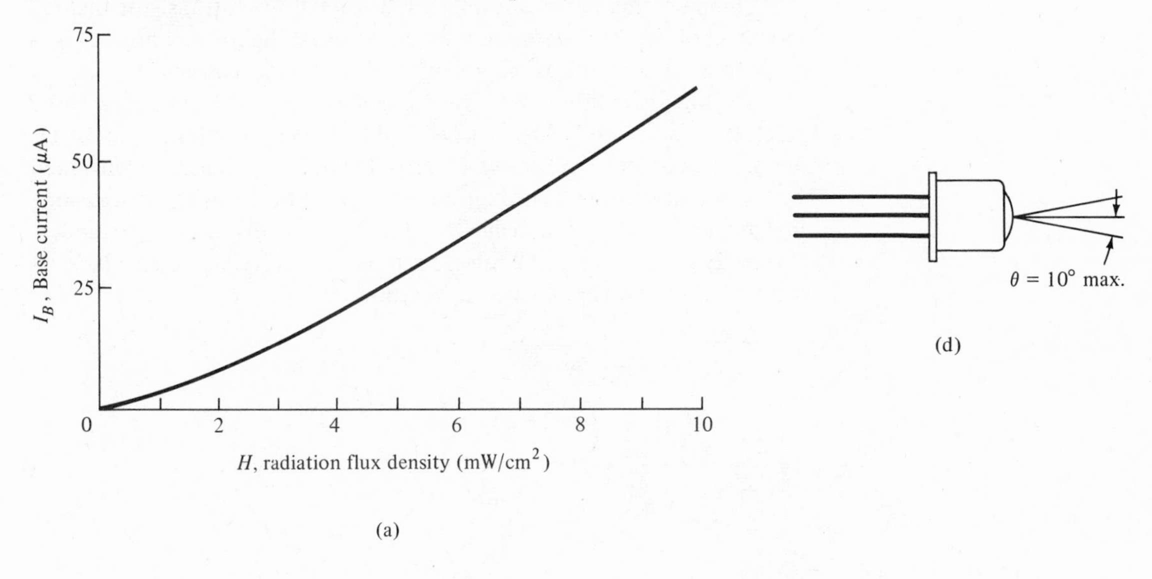

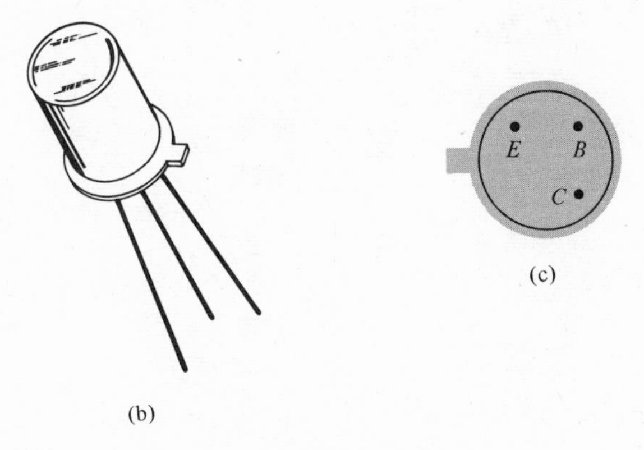

(Courtesy Motorola Inc.)

FIG. 9.37

Phototransistor: (a) base current vs. flux density; (b) device; (c) terminal identification; (d) angular alignment.

Note the similarities between these curves and those of a typical bipolar transistor. As expected, an increase in light intensity corresponds with an increase in collector current. To develop a greater degree of familiarity with the light intensity unit of measurement, milliwatts per square centimeter, a curve of base current vs. flux density appears in Fig. 9.37a. Note the exponential increase in base current with increasing flux density. In the same figure a sketch of the phototransistor is provided with the terminal identification and the angular alignment.

Some of the areas of application for the phototransistor include punch card readers, computer logic circuitry, lighting control (highways, etc.), level indication, relays, and counting systems.

A high-isolation AND gate is shown in Fig. 9.38 using three phototransistors and three LEDs (light emitting diodes). The LEDs are semiconductor devices which emit light at an intensity determined by the forward current through the device. With the aid of discussions in Chapter 1 the circuit behavior should be relatively easy to understand. The terminology "high-isolation" simply refers to the lack of a physical connection between the input and output circuits.

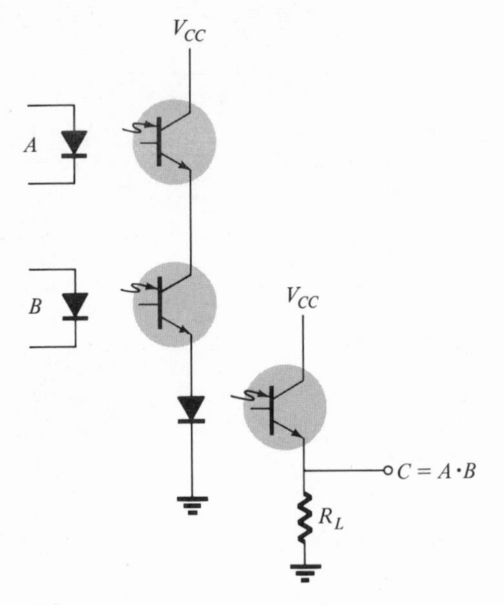

FIG. 9.38
High-isolation AND gate employing phototransistors and light emitting diodes (LEDs).

9.15 THYRATRON

The thyratron is the tube equivalent of the semiconductor thyristor in the sense that it also has a control terminal for determining the state of the device (conduction or nonconduction). Although the semiconductor devices are, in general, more efficient, they have yet to reach the high PIV ratings of the tube equivalent. Fundamentally, the thyraton is nothing more than a gas diode with an additional grid structure for controlling the conditions for ionization of the gas in the tube. The basic structure of the device and its graphic symbol appear in Fig. 9.39a. Of the three structures (plate, cathode, and grid) the grid differs most in appearance between the high-vacuum triode and the thyratron. It is much larger in the latter device, usually being the largest visible element (Fig. 9.39b). The larger size usually requires that the plate appear above (Fig. 9.39) rather than around the grid structure as was true for most high-vacuum triodes. The plate of the tube is connected directly to the cap on top of the tube as shown in Fig. 9.39a and b or internally as for the 2050 tube in Fig. 9.39b.

At low forward potentials (below the ionization level), the control grid has negligible effect on the shape of the characteristics of the device. In fact, up to the ionization level, the characteristics of a thyratron are, for all practical purposes, the same as those obtained for the gas or high-vacuum diode. The more negative the potential applied to the grid the greater will be the plate-to-cathode voltage necessary to initiate the ionization process. This is a direct result of the negative grid potential effectively "slowing down" the "free" electrons traveling toward the plate for a particular plate to cathode potential. A reduction in speed will reduce the kinetic energy associated with the moving charged particles and forestall the ionization process.

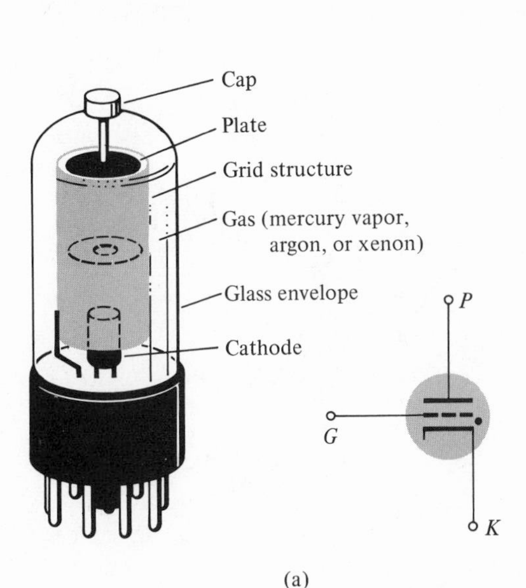

(a)

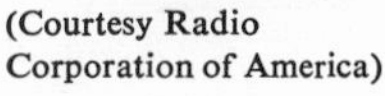
(Courtesy Radio Corporation of America)

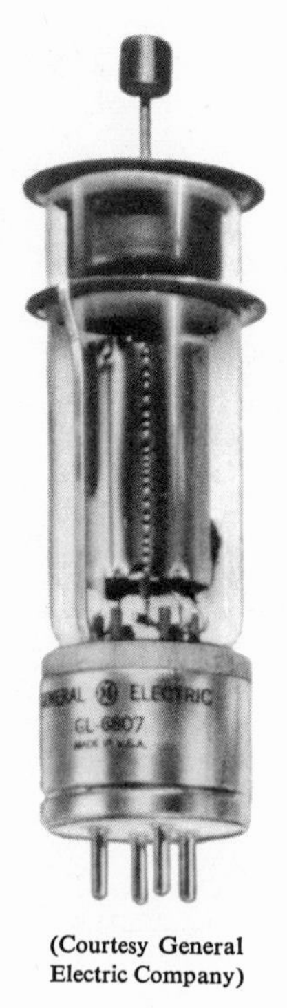

(Courtesy General Electric Company)

(b)

FIG. 9.39
Thyratron: (a) basic structure and symbol; (b) photographs.

Once the ionization process is initiated, the grid *loses control* of the tube's behavior and the characteristics appear much like those of a hot-cathode gas diode; that is, there is a sharp rise at the ionization potential. These then are three ways in which the device can be turned off. The first is to drop the plate-to-cathode potential below the ionization level. The second is to reverse the polarity of the voltage across the plate to cathode terminals, such as with an ac signal. The last, and most obvious, is to open the plate circuit. The control action of the grid is best described by examining the circuit of Fig. 9.40 employing a thyratron having the characteristics of Fig. 9.41. For the circuit

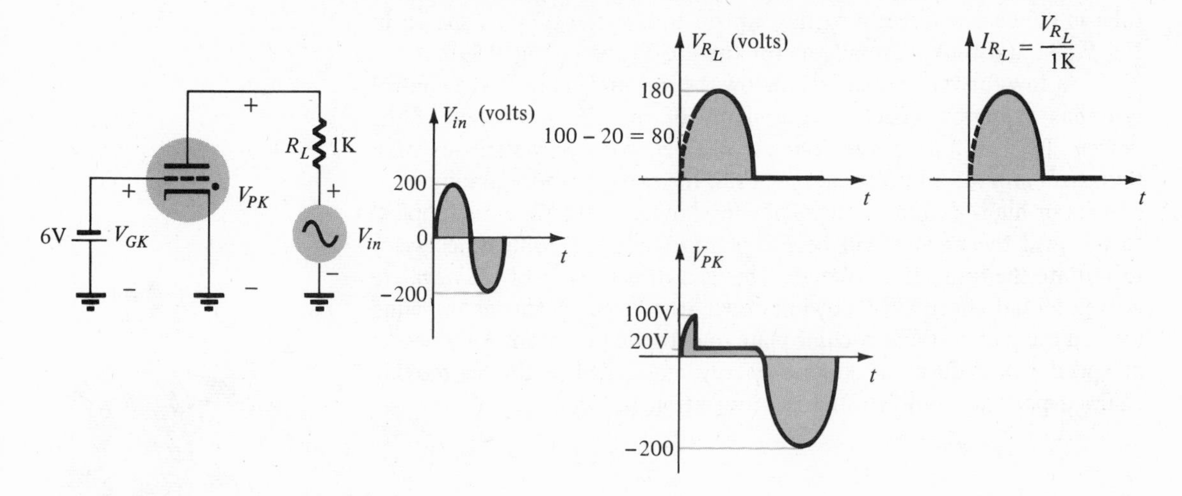

FIG. 9.40
Thyratron conduction control circuit.

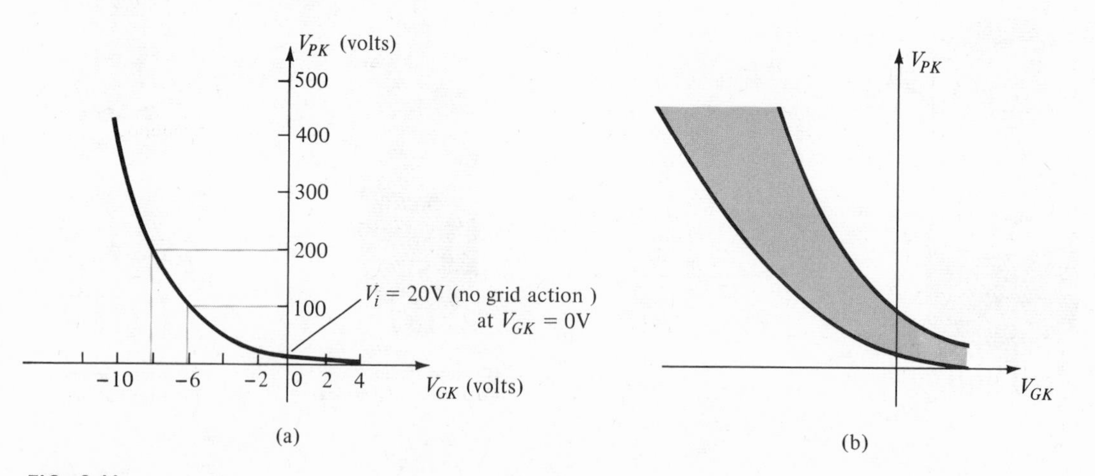

FIG. 9.41
Thyratron control characteristics: (a) single tube; (b) any tube of a particular production series.

of Fig. 9.40, $V_{GK} = -6$ V as indicated on the curve of Fig. 9.37a. The horizontal line to the V_{PK} axis indicates, quite clearly, that for this applied grid voltage, the tube will not fire for plate-to-cathode voltages below 100 V. When $V_{in} = V_f = 100$ V, the firing potential for the indicated V_{GK}, the device will enter the conducting state and V_{PK} will drop, almost instantaneously, to the ionization potential (V_i) of the gas. You will recall that for the gas diode $V_f = V_i$. For the conditions above, the resulting voltage across the load is

$$\begin{aligned} V_L &= V_{in} - V_i \\ &= 100 - 20 \\ &= 80 \text{ V} \end{aligned}$$

and

$$I_L = \frac{80 \text{ V}}{1 \text{ K}} = 80 \text{ mA}$$

These calculated values are indicated on the curves of Fig. 9.40. As V_{in} continues to increase and then decrease in a sinusoidal manner the curves of V_L and I_L will trace the same shape as shown in the same figure. Only after V_{in} drops below V_i will the ionization process collapse and the tube return to its nonconducting state. The curves for V_{PK}, V_L and I_L for a complete cycle of V_{in} are indicated in Fig. 9.40.

If $V_{GK} = -8$ V, the tube would not fire until $V_{in} = 200$ V and only the latter half of the positive portion of the sinusoidal ac waveform would appear. For $V_{GK} = -9$ V, the tube will not fire and both V_L and I_L will be zero.

Manufacturers do not generally provide a single curve for a particular thyratron as shown in Fig. 9.41a, but rather a shaded region as shown in Fig. 9.41b. The shaded region includes all the thyratrons of a particular series that will exhibit different operating characteristics due to age, heater variations, and operating temperatures.

A few general areas of application for the thyratron include rectification and control (motor, phase, etc.), oscillators, relays, and sawtooth generators.

9.16 BEAM POWER TUBE

The last of the tube devices to be introduced in this chapter is the beam power tube. This is by no means to indicate that all the high-vacuum or gas tubes in use today have been considered. To merely name all those available would require a number of pages of this text.

It is the consensus of the authors, however, that the more important and more frequently applied devices have been introduced.

The beam power tube is a four-element device having plate characteristics very similar to those of the pentode. The primary difference in the characteristics is that the beam power tube has a sharper knee for each grid line at low plate-to-cathode voltages. This results in a reduced nonlinear region and an ability to handle larger signal swings. The fact that large voltage swings are synonomous with power applications accounts for the term "power" in the nomenclature. We shall find in the discussion to follow that the term "beam" is derived from the fundamental operation of the device.

The basic construction and symbol for the beam power tube is provided in Fig. 9.42. Note in Fig. 9.42b that the beam forming plates

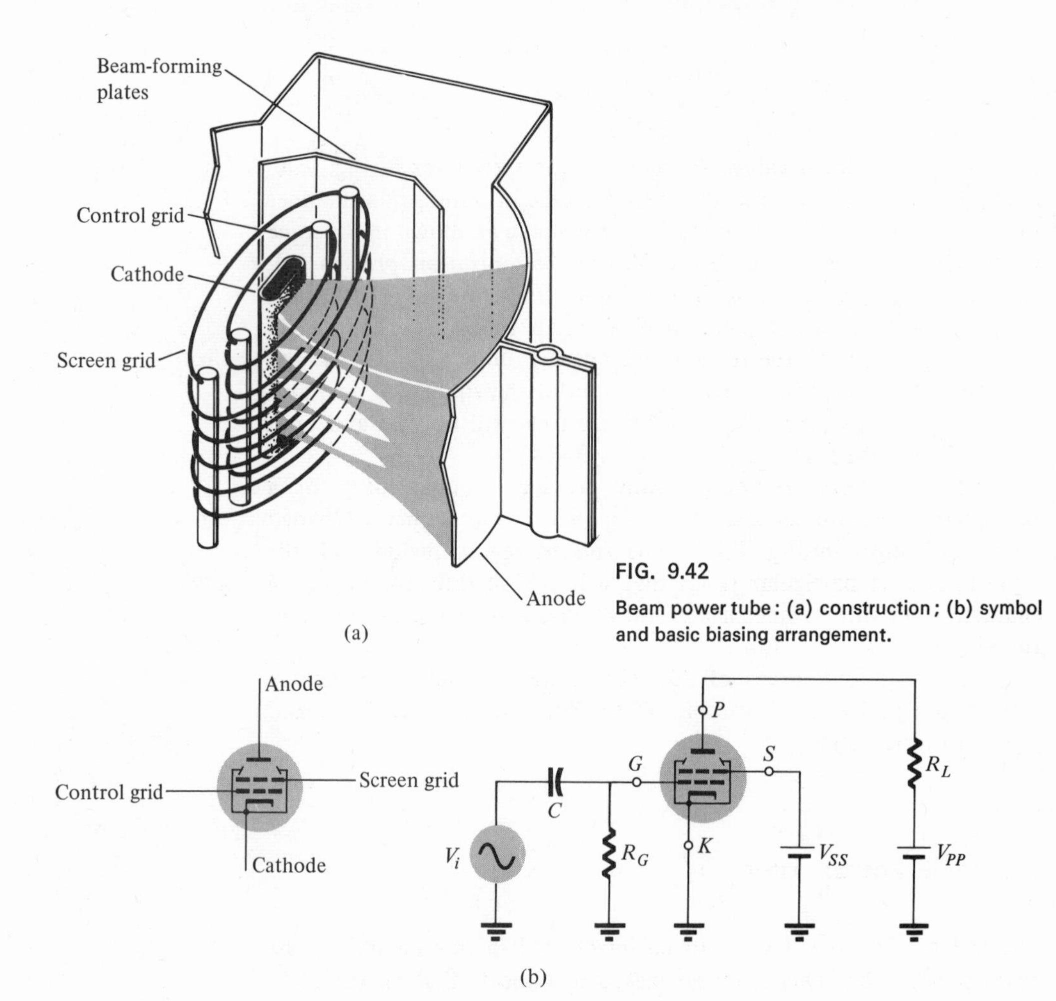

FIG. 9.42
Beam power tube: (a) construction; (b) symbol and basic biasing arrangement.

(a)

FIG. 9.43

Beam power tube: (a) photograph; (b) plate characteristics.

(Courtesy Radio Corporation of America)

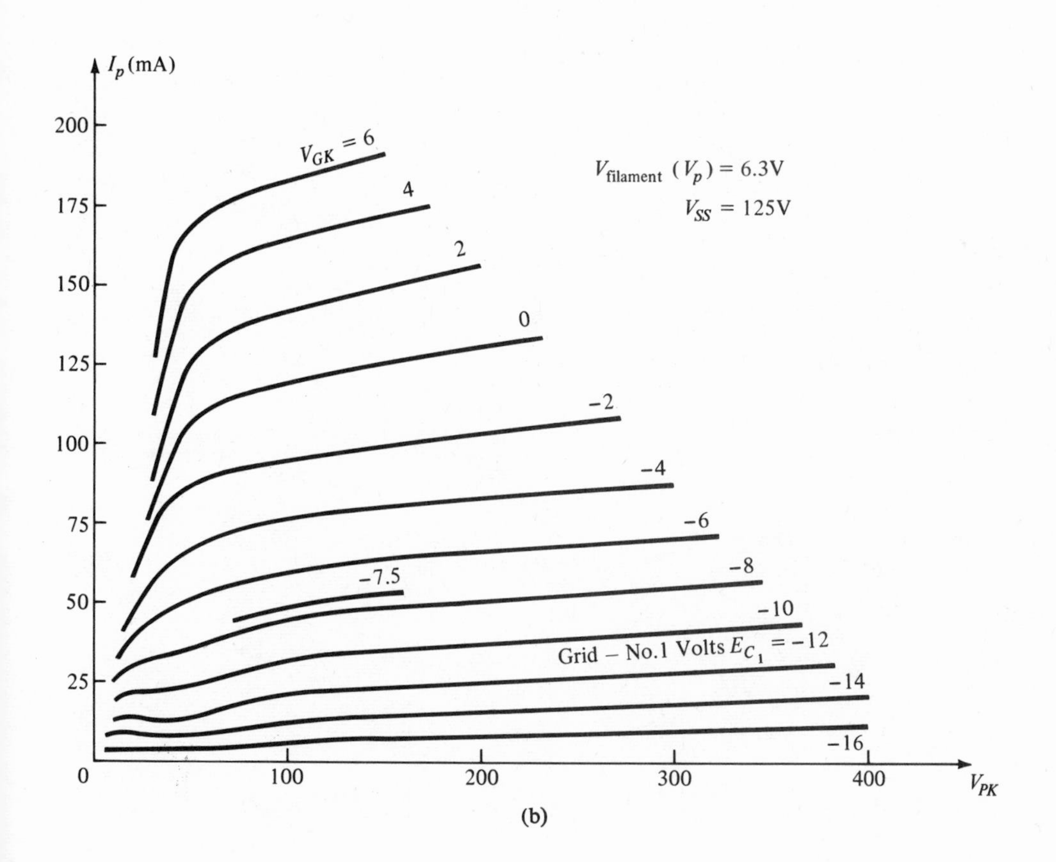

(b)

are connected to the cathode terminal which under normal operating conditions is connected to ground potential. In Fig. 9.42a it is quite obvious that the screen and grid wires are directly in line with each other. This will result in a shielding action on the part of the control grid that will substantially reduce the screen current and resulting losses. The appearance of the electron beam in Fig. 9.42a is due to the effect of both the control grid and beam forming plates on the negatively charged electrons. A dense region of electrons is thereby established between the screen and plate that will have an effect much like that of the suppressor grid of the pentode. This dense region of charge, however, is spaced over a distance between the plate and screen rather than at an abrupt point between the two as was true for the suppressor grid of the pentode. This accounts for the improved characteristics for the beam power tube at low plate potentials.

A beam power tube and its plate characteristics are presented in Fig. 9.43. Note the similarities with those of a typical pentode and the sharper knee of each grid bias line.

10

Integrated Circuits (ICs)

10.1 INTRODUCTION

During the last few years, the excellent characteristics of (and advantages associated with) a relatively new type of electronic package have substantially altered the course of many areas of research and development. This product, called an integrated circuit (IC), has, through expanded usage and the various media of advertising, become a product whose basic function and purpose are now understood by the layman. The most noticeable characteristic of an IC is its size. It is typically hundreds and even thousands of times smaller than a semiconductor structure built in the usual manner with discrete components. In Fig. 10.1a, all the circuit elements appearing in the circular pattern can be found in the IC appearing in the center of the figure. Figure 10.1b is an indication of the reduction in size for a 120 gate computer package.

Integrated circuits are seldom, if ever, repaired; that is, if a single component within an IC should fail, the entire structure (complete circuit) is replaced—a more economical approach. There are three types of ICs commercially available on a large scale today. They include the *monolithic*, *thin* (*or thick*) *film*, and *hybrid* integrated circuits.

10.2 MONOLITHIC INTEGRATED CIRCUIT

The term *monolithic* is derived from a combination of the Greek words *monos*, meaning single, and *lithos*, meaning stone, which in combination

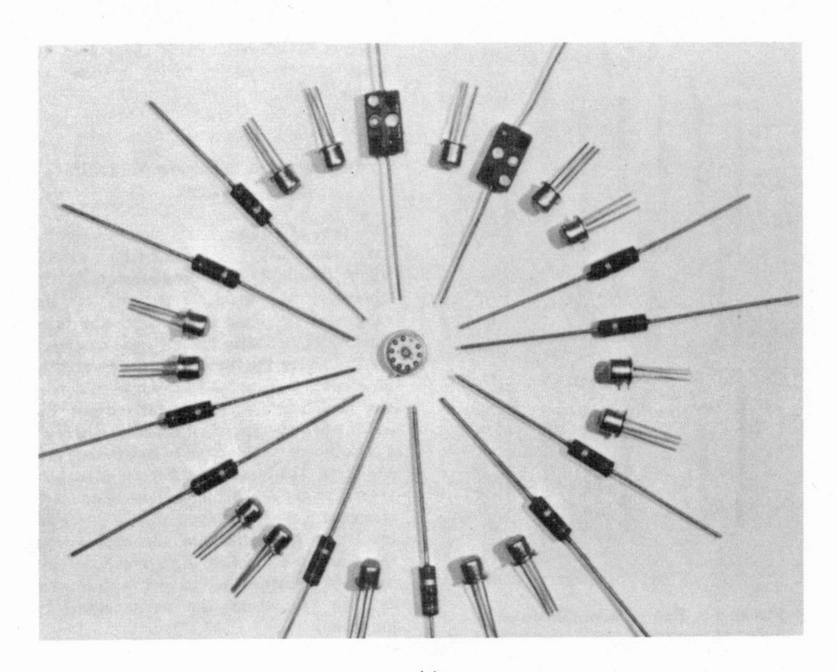

(a)

(Courtesy Motorola, Inc.)

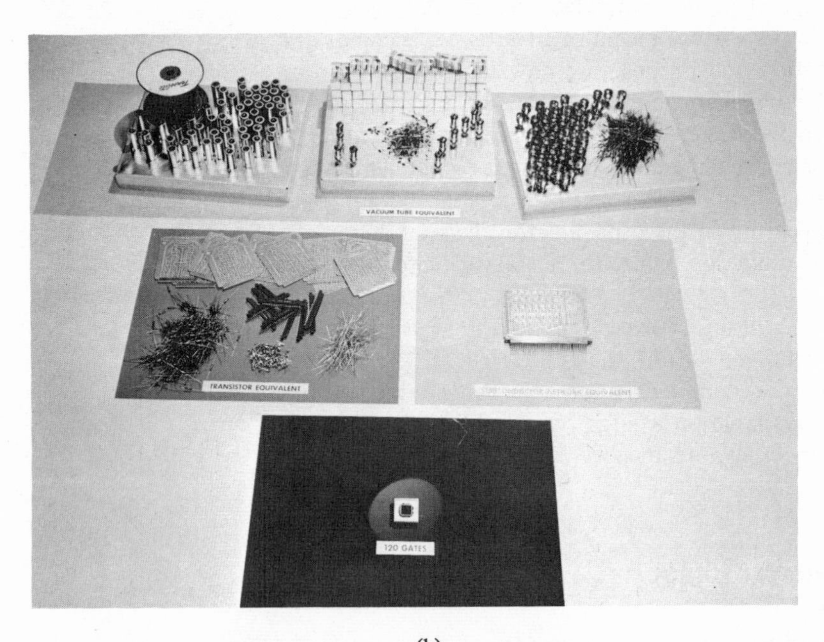

(b)

(Courtesy Texas Instruments, Inc.)

FIG. 10.1

(a) An integrated circuit and the discrete elements required to build a circuit to perform the same function; (b) comparison of the resulting size of a 120 gate computer package manufactured by the indicated methods.

result in the literal translation, single-stone, or more appropriately, single-solid-structure. As this descriptive term implies, the monolithic IC is constructed within a *single* wafer of semiconductor material. The greater portion of the wafer will simply act as a supporting structure for the very thin resulting IC. An overall view of the stages involved in the fabrication of monolithic ICs is provided in Fig. 10.2. The actual number of steps leading to a finished product is many times that appearing in Fig. 10.2. The figure does, however, point out the major production phases of forming a monolithic IC. The initial preparation of the semiconductor wafer of Fig. 10.2 was discussed in Chapter 3 in association with the fabrication of transistors. As indicated in the figure, it is first necessary to design a circuit that will meet the specifications. The circuit must then be laid out so as to ensure optimum use of available space and a minimum of difficulty in performing the diffusion processes to follow. The appearance of the mask and its function in the sequence of stages indicated will be introduced in Section 10.4. For the moment, let it suffice to say that a mask has the appearance of a negative through which impurities may be diffused (through the light

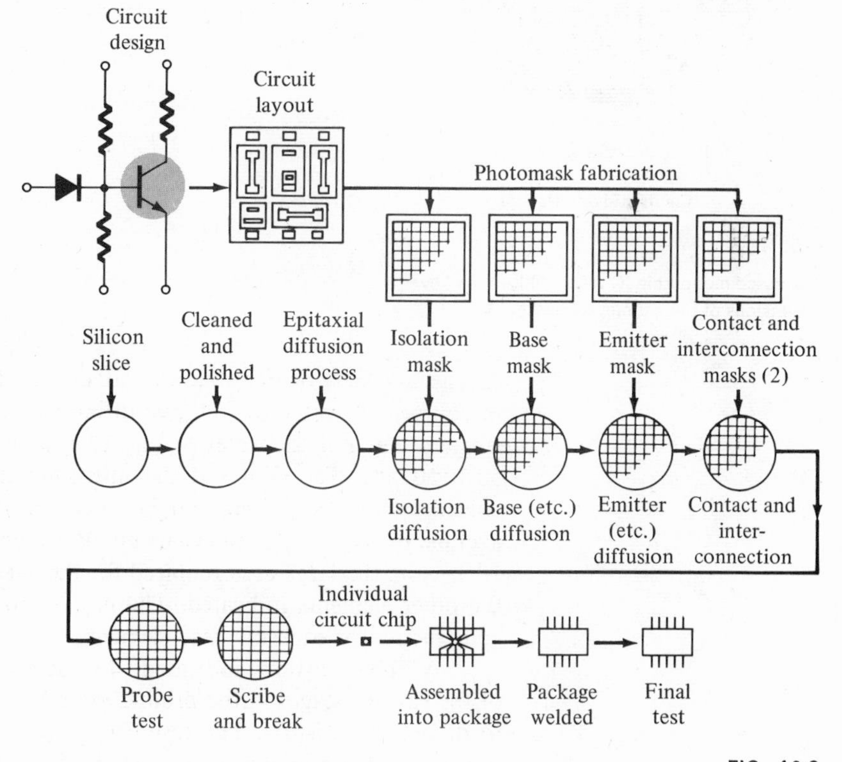

FIG. 10.2

Monolithic integrated circuit fabrication.
(Courtesy of Robert Hibberd)

areas) into the silicon slice. The actual diffusion process for each phase is quite similar to that applied in the fabrication of diffused transistors in Chapter 3. The last mask of the series will control the placement of the interconnecting conducting pattern between the various elements. The wafer then goes through various testing procedures, is scribed and broken into individual chips, packaged and assembled as indicated. A processed silicon wafer appears in Fig. 10.3. The original wafer

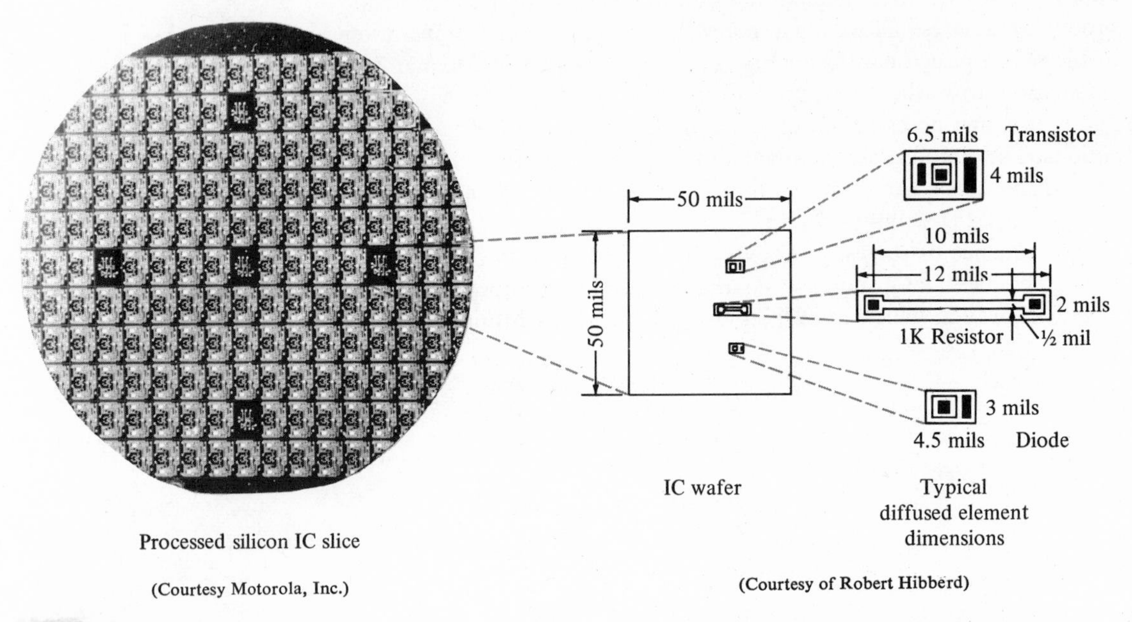

FIG. 10.3

Processed monolithic IC wafer with the relative dimensions of the various elements.

can be anywhere from $\frac{1}{2}$ to 2 in. in diameter. The size of each chip will, of course, determine the number of individual circuits resulting from a single wafer. For the wafer of Fig. 10.3, each chip has the dimensions-50×50 mils. To point out the microminiature size of these chips, consider that 20 of them can be lined up along a 1-in. length. The average relative size of the elements of a monolithic IC appear in Fig. 10.3. Note the large area required for the 1-K resistor as compared to the other elements indicated. The next section will examine the basic construction of each of these elements.

A recent article indicated, by percentage, the relative costs of the various stages in the production of monolithic ICs as compared to discrete transistors. The resulting graphs appear in Fig. 10.4. The processing phase includes all stages leading up to the individual chips of Fig. 10.3. Note the high cost of packaging the integrated circuits and of testing the complex silicon integrated circuits (SIC). The cost

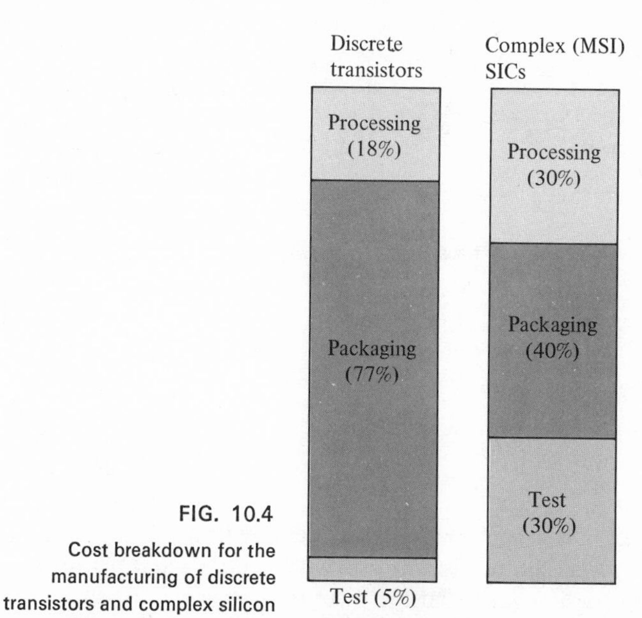

FIG. 10.4
Cost breakdown for the manufacturing of discrete transistors and complex silicon integrated circuits (SICs).

(Courtesy of J. J. Suran, General Electric Co.)

of packaging has resulted in an increase (wherever feasible) in the number of IC chips within a single package. This multichip, hybrid type of integrated circuit will be considered in a Section 10.7.

10.3 MONOLITHIC CIRCUIT ELEMENTS

The surface appearance of the transistor, diode, and resistor appear in Fig. 10.3. We will now examine the basic construction of each in more detail.

Resistor

You will recall that the resistance of a material is determined by the resistivity, length, area, and temperature of the material. For the integrated circuit, each necessary element is present in the sheet of semiconductor material appearing in Fig. 10.5. As indicated in the figure, the semiconductor material can be either *p*- or *n*-type although the *p*-type is most frequently employed.

The resistance of any bulk material is determined by

$$R = \rho \frac{l}{A}$$

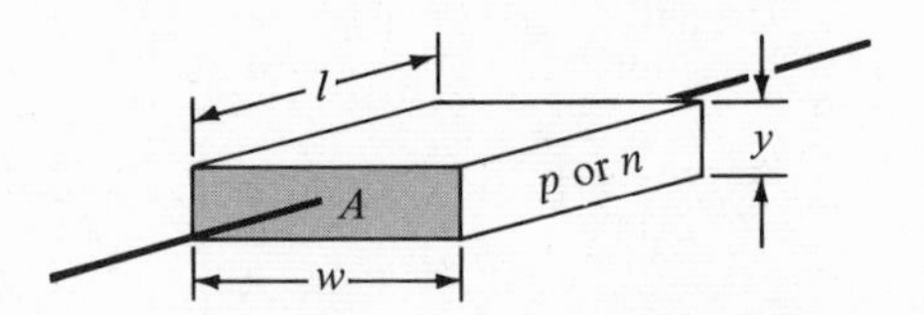

FIG. 10.5
Parameters determining the resistance of a sheet of semiconductor material.

For $l = w$, resulting in a square sheet,

$$R = \frac{\rho l}{yw} = \frac{\rho l}{yl}$$

and

$$R_s = \frac{\rho}{y} \quad \text{(ohms)} \tag{10.1}$$

where ρ is in ohm-centimeters and y is in centimeters.

R_s is called the sheet resistance and has the units ohms per square. The equation clearly reveals that the sheet resistance is independent of the size of the square.

In general, where $l \neq w$,

$$R = R_s \frac{l}{w} \quad \text{(ohms)} \tag{10.2}$$

For the resistor appearing in Fig. 10.3, $w = \frac{1}{2}$ mil, $l = 10$ mils, and $R_s = 100\ \Omega$/square:

$$R = R_s \frac{l}{w} = 100 \times \frac{10}{\frac{1}{2}} = 2\text{ K}$$

A cross-sectional view of a monolithic resistor appears in Fig. 10.6 along with the surface appearence of two monolithic resistors. In Fig. 10.6a, the sheet resistive material (p) is indicated with its aluminum terminal connections. The n-isolation region performs exactly that function indicated by its name; that is, it isolates the monolithic resistive elements from the other elements of the chip. Note in Fig. 10.6b the method employed to obtain a maximum l in a limited area. The resistors of Fig. 10.6 are called base-diffusion resistors since the p-material is diffused into the p-type substrate during the base-diffusion process indicated in Fig. 10.2.

Capacitor

Monolithic capacitive elements are formed by making use of the transition capacitance of a reverse-biased p-n junction. At increasing

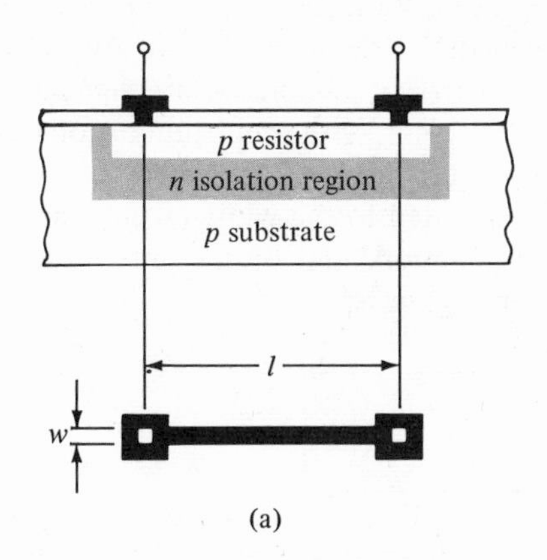

(a)

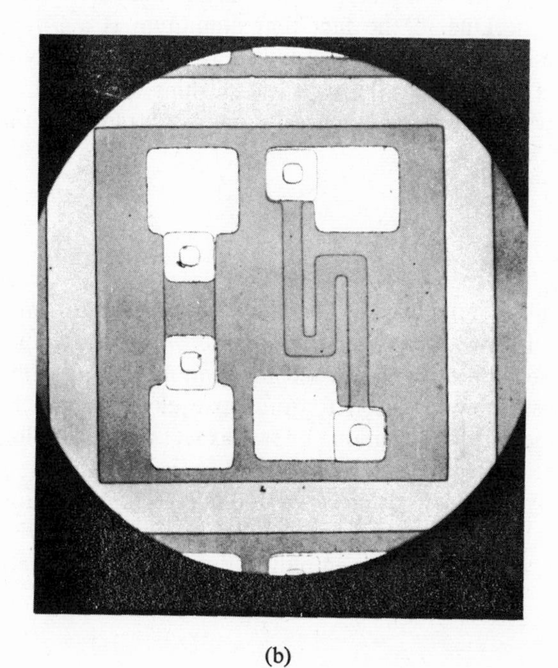

(b)

FIG. 10.6

Monolithic resistors: (a) cross section and determining dimensions; (b) surface view of two monolithic resistances in a single die.
(Courtesy Motorola, Inc.)

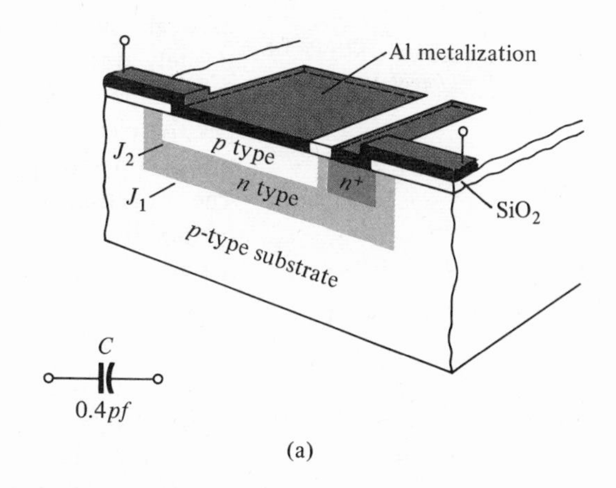

(a)

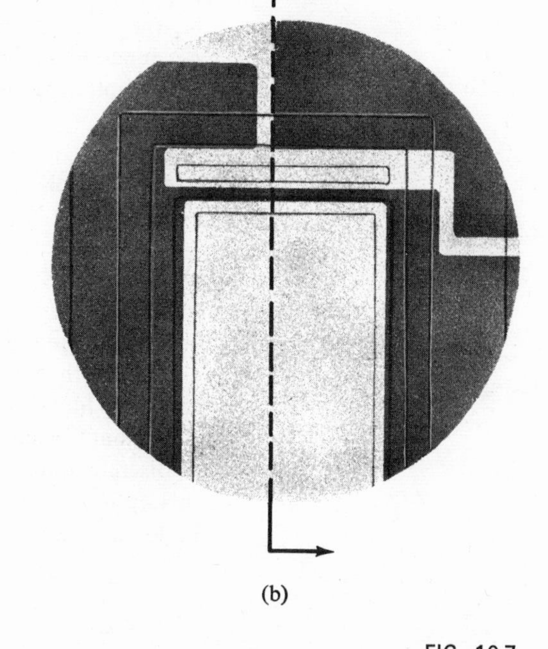

(b)

FIG. 10.7

Monolithic capacitor: (a) cross section; (b) photograph.
(Courtesy Motorola, Inc.)

reverse-bias potentials, there is an increasing distance at the junction between the *p*- and *n*-type impurities. The region between these oppositely doped layers is called the depletion region (see Chapter 3) due to the absence of "free" carriers. The necessary elements of a capacitive element are therefore present,—the depletion region has insulating characteristics that separates the two oppositely charged layers. The transition capacitance is related to the width (W) of the depletion region, the area (A) of the junction, and the permittivitiy (ϵ) of the material within the depletion region by

$$C_T = \frac{\epsilon A}{W} \tag{10.3}$$

The cross section and surface appearence of a monolithic capacitive element appear in Fig. 10.7. The reverse-biased junction of interest is J_2. The undesirable parasitic capacitance at junction J_1 is minimized through careful design. Due to the fact that aluminum is a *p*-type impurity in silicon, a heavily doped n^+ region is diffused into the *n*-type region as shown to avoid the possibility of establishing an undesired *p-n* junction at the boundary between the aluminum contact and the *n*-type impurity region.

Inductor

Whenever possible, inductors are avoided in the design of integrated circuits. An effective technique for obtaining nominal values of inductances has so far not been devised for monolithic integrated circuits. In many instances, the need for inductive elements can be eliminated through the use of a technique known as RC synthesis. Thin (or thick) film or hybrid integrated circuits have an option open to them that cannot be employed in monolithic integrated circuits: the addition of discrete inductive elements to the surface of the structure. Even with this option, however, they are seldom employed due to their relatively bulky nature.

Transistors

The cross section of a monolithic transitor appears in Fig. 10.8a. Note again the presense of the n^+ region in the *n*-type epitaxial collector region. The vast majority of monolitic IC transistors are NPN rather than PNP for reasons to be found in more advanced texts on the subject. Keep in mind when examining Fig. 10.8 that the *p*-substrate is only a supporting and isolating structure forming no part of the

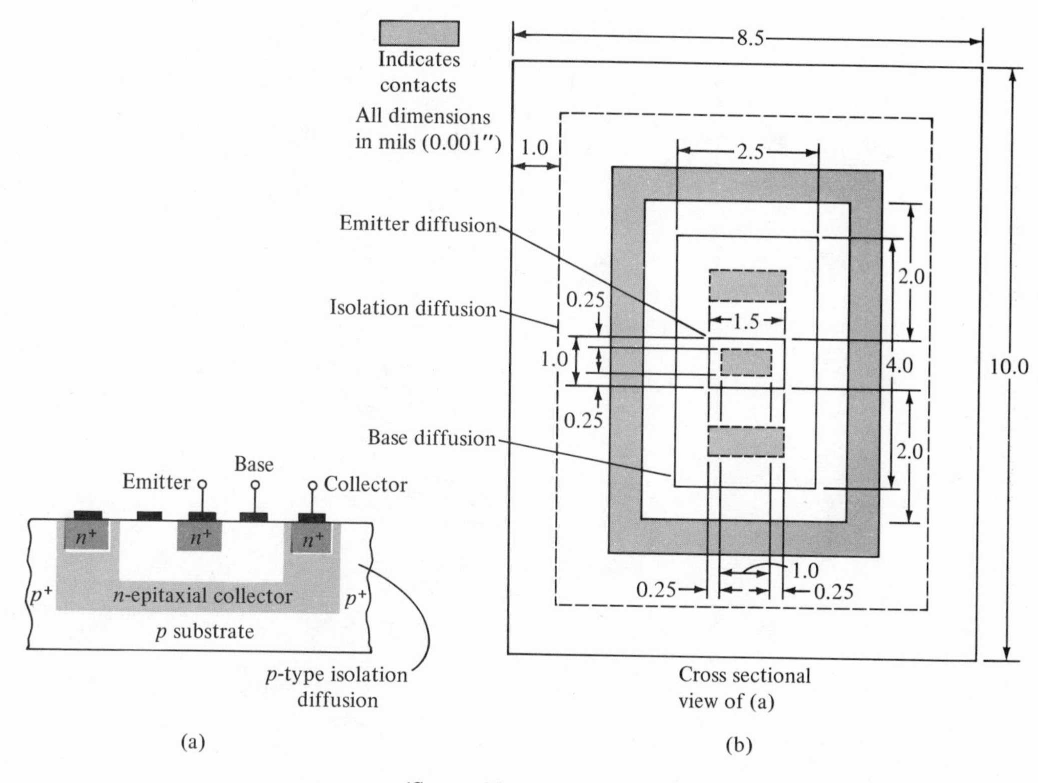

FIG. 10.8
Monolithic transistor: (a) cross section; (b) surface appearance and dimension for a typical monolithic transistor.

active device itself. The base, emitter, and collector regions are formed during the corresponding diffusion processes of Fig. 10.2.

The top view of a typical monolithic transistor appears in Fig. 10.8b. Note that two base terminals are provided while the collector has the outer rectangular aluminum contact surface.

Diodes

The diodes of a monolithic integrated circuit are formed by first diffusing the required regions of a transistor and then masking the diode rather than transistor terminal connections. There is, however, more than one way of hooking up a transistor to perform a basic diode action. The two most common methods, applied to monolithic integrated circuits appear in Fig. 10.9. The structure of a *BC-E* diode appears in Fig. 10.10. Note that the only difference between the cross-sectional

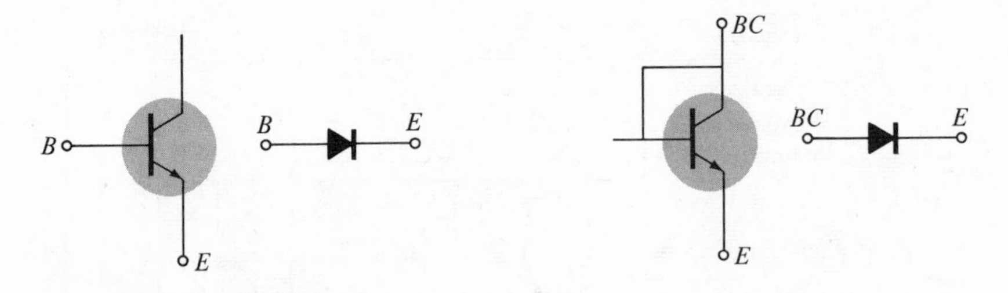

FIG. 10.9

Transistor structure and connections employed in the formation of monolithic diodes.

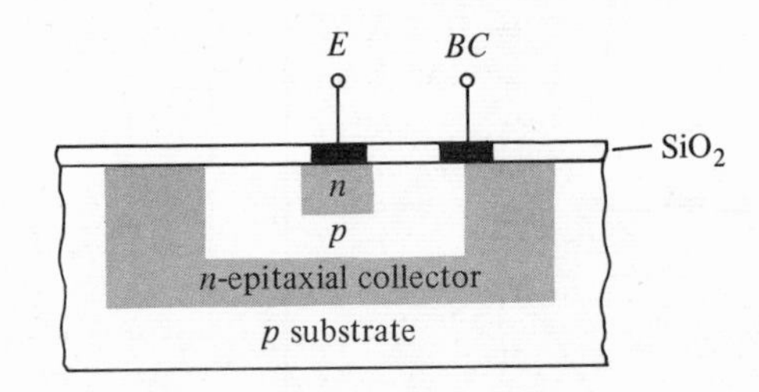

FIG. 10.10

The cross-sectional view of a BC-E monolithic diode.

view of Fig. 10.10 and that of the transistor of Fig. 10.8a is the position of the ohmic aluminum contacts.

10.4 MASKS

The selective diffusion required in the formation of the various active and passive elements of an integrated circuit is accomplished through the use of masks such as those appearing in Fig. 10.11. This figure depicts and describes those major steps envolved in the production of these masks. We will find in the next section that the light areas are the only areas through which donor and acceptor impurities can pass. The dark areas will block the diffusion of impurities somewhat as a shade will prevent sunlight from changing the pigment of the skin. The next section will demonstrate the use of these masks in the formation of a computer logic circuit.

10.5 MONOLITHIC INTEGRATED CIRCUIT–THE NAND GATE

This section is devoted to the sequence of production stages leading to a monolithic NAND-gate circuit (the operation of which is covered in Chapter 13). A detailed examination of each process would

(a)

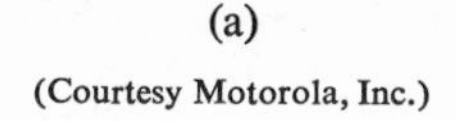

(Courtesy Motorola, Inc.)

(b)

(c)

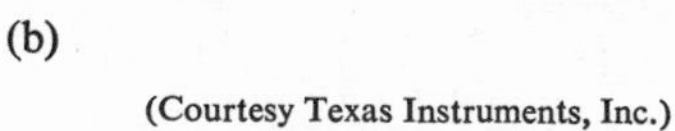

(Courtesy Texas Instruments, Inc.)

(d)

(Courtesy Motorola, Inc.)

FIG. 10.11

Mask preparation: (a) aristo handcutting of the mask pattern; (b) photo-reduction of the mask pattern; (c) step and repeat machine for the placement of a large number of the reduced mask pattern or a single production mask; (d) final mask.

require many more pages than it is possible to include in this text. The description, however, should be sufficiently complete and informative to aid the reader in any future contact with this highly volatile area. The circuit to be prepared appears in Fig. 10.12a. The criteria of space allocation, placement of pin connection, and so on require that the elements be situated in the relative positions indicated in Fig. 10.12b. The regions to be isolated from one another appear within the solid

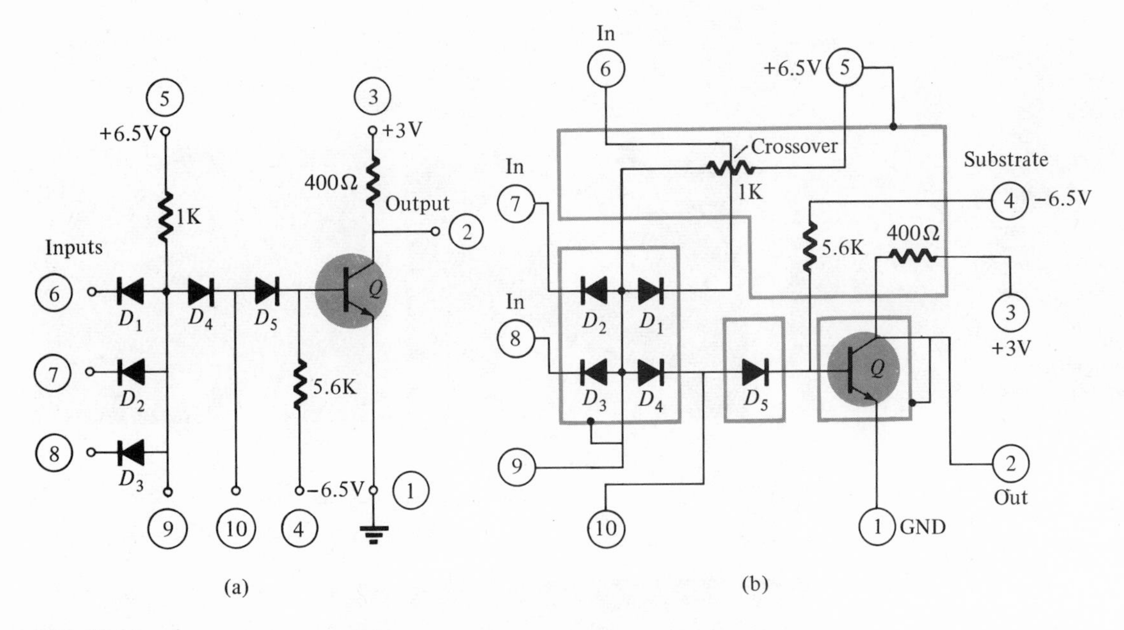

FIG. 10.12

NAND gate: (a) circuit; (b) layout for monolithic fabrication.

heavy lines. A set of masks for the various diffusion processes must then be made up for the circuit as it appears in Fig. 10.12b.

We will now slowly proceed through the first diffusion process to demonstrate the natural sequence of steps that must be followed through each diffusion process indicated in Fig. 10.2.

p-Type Silicon Wafer Preparation

After slicing from the grown ingot, a *p*-type silicon wafer is polished and cleaned to produce the structure of Fig. 10.13.

n-Type Epitaxial Region

An *n*-type epitaxial region is then diffused into the *p*-type substrate as shown in Fig. 10.14. It is *in* this thin epitaxial layer that the active and passive elements will be diffused. The *p*-type area remaining is simply adding some thickness to the structure to give it increased strength and permit easier handling.

Silicon Oxidation (SiO_2)

The resulting wafer is then subjected to an oxidation process resulting in a surface layer of SiO_2 (silicon dioxide) as shown in Fig. 10.15.

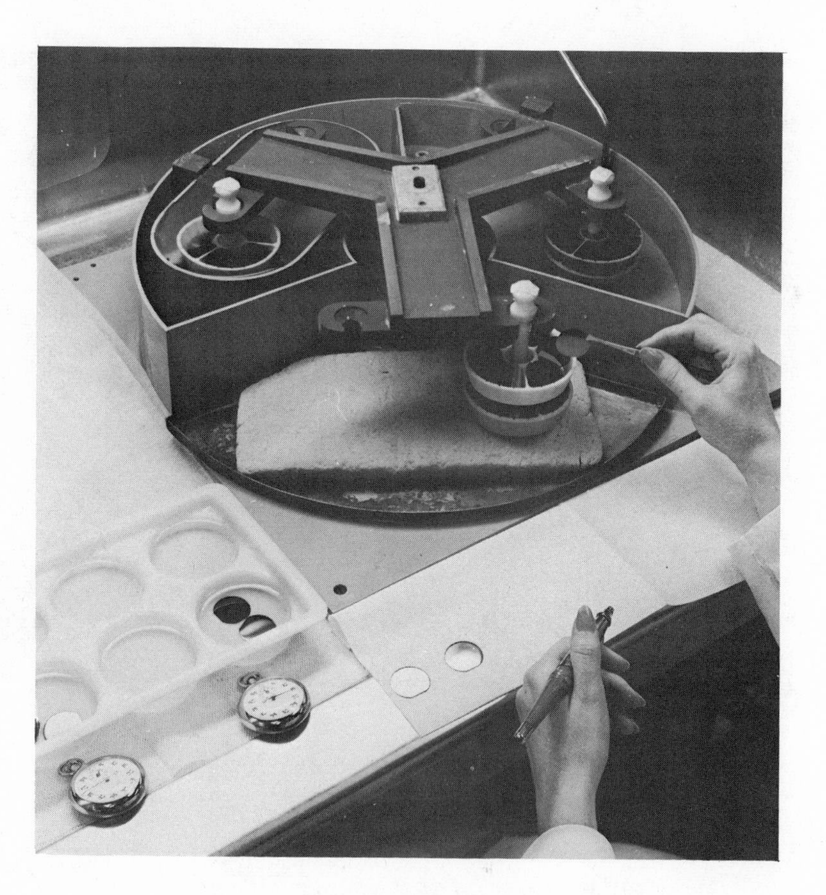

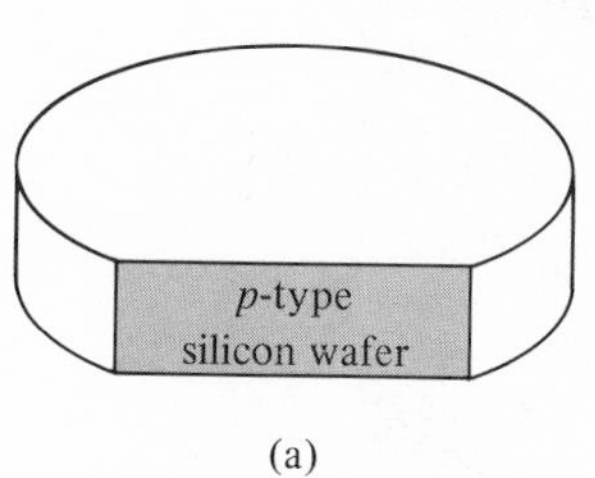

(a)

(Courtesy Texas Instruments, Inc.)

FIG. 10.13

(a) *p*-Type silicon wafer;
(b) polishing apparatus.

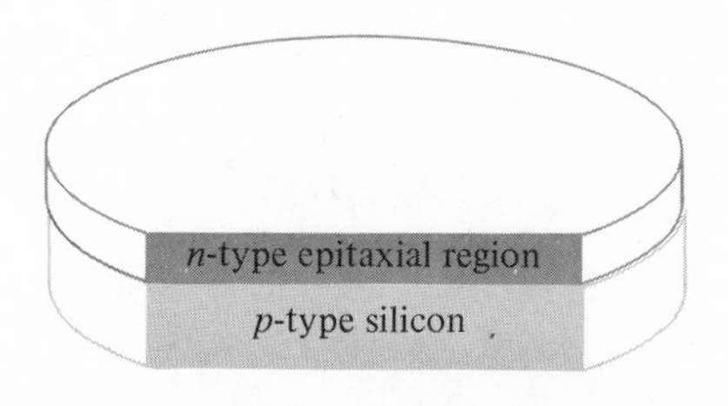

FIG. 10.14

p-Type silicon wafer after the *n*-type epitaxial diffusion process.

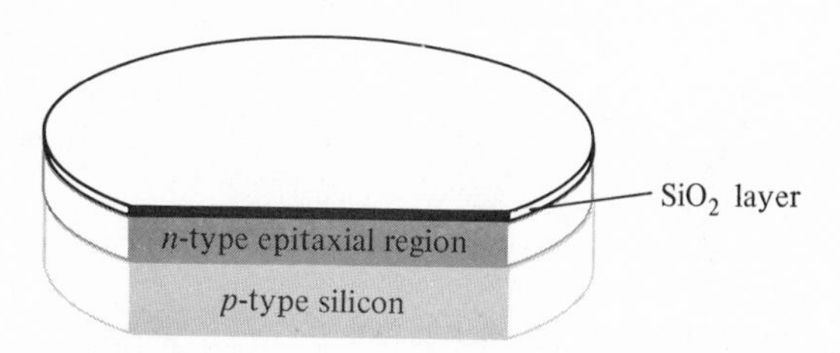

FIG. 10.15

Wafer of Fig. 10.14 following the deposit of the SiO_2 layer.

(a)

(Courtesy Motorola, Inc.)

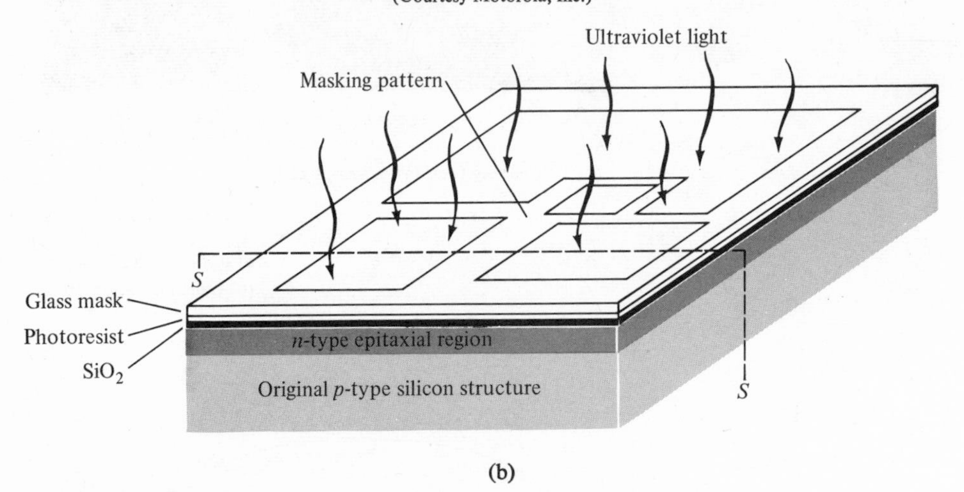

(b)

FIG. 10.16

Photolithographic process: (a) Applying the photoresist; the wafer is spun at a high speed to insure an even distribution of the photoresist. (b) The application of ultraviolet light after the mask is properly set; the structure is only one of the 200, 400, or even 500 individual NAND gate circuits being formed on the wafer of Figs. 10.13 through 10.15.

This surface layer will prevent any impurities from entering the n-type epitaxial layer. However, selective etching of this layer will permit the diffusion of the proper impurity into designated areas of the n-type epitaxial region of the silicon wafer.

Photolithographic Process

The selective etching of the SiO_2 layer is accomplished through the use of a photolithographic process. A mask is first prepared on a glass plate as explained in Section 10.4. The first mask will determine those areas of the SiO_2 layer to be removed in preparation for the isolation diffusion process. The wafer is first coated with a thin layer of photosensitive material, commonly called photoresist, as demonstrated in Fig. 10.16a. This new layer is then covered by the mask and an ultraviolet light is applied that will expose those regions of the photosensitive material not covered by the masking pattern (Fig. 10.16b). The resulting wafer is then subjected to a chemical solution that will remove the unexposed photosensitive material. A cross section of a chip (S-S) will then appear as indicated in Fig. 10.17. A second solution will then etch away the SiO_2 layer from any region not covered by the photoresist material (Fig. 10.18). The final step before the diffusion process is the removal, by solution, of the remaining photosensitive material. The structure will then appear as shown in Fig. 10.19.

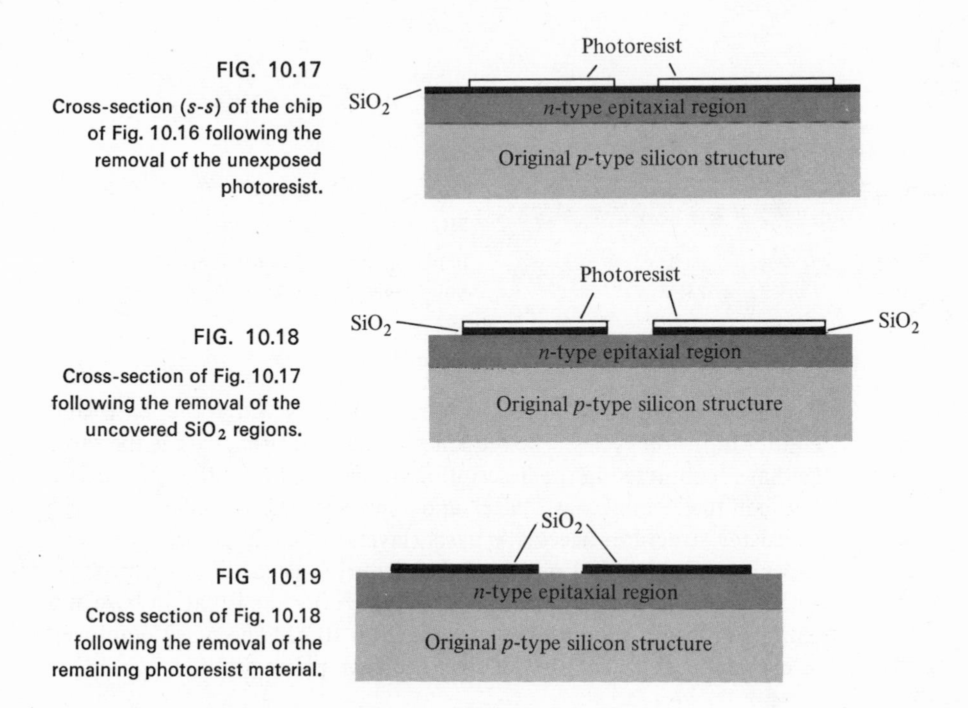

FIG. 10.17

Cross-section (s-s) of the chip of Fig. 10.16 following the removal of the unexposed photoresist.

FIG. 10.18

Cross-section of Fig. 10.17 following the removal of the uncovered SiO_2 regions.

FIG 10.19

Cross section of Fig. 10.18 following the removal of the remaining photoresist material.

The structure of Fig. 10.19 is then subjected to a p-type diffusion process resulting in the islands of n-type regions indicated in Fig. 10.20.

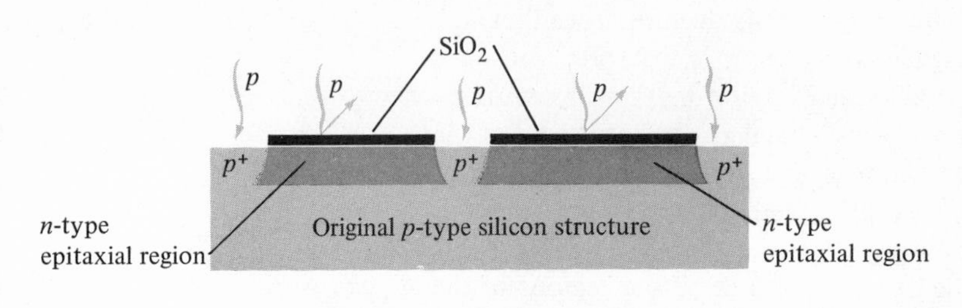

FIG. 10.20

Cross section of Fig. 10.19 following the isolation diffusion process.

The diffusion process is such as to insure a heavily doped p-type region (indicated by p^+) between the n-type islands. The p^+ regions will result in improved *isolation* properties between the active and passive components to be formed in the n-type islands. In preperation for the next masking and diffusion process, the entire surface of the wafer is coated with a SiO_2 layer as indicated in Fig. 10.21

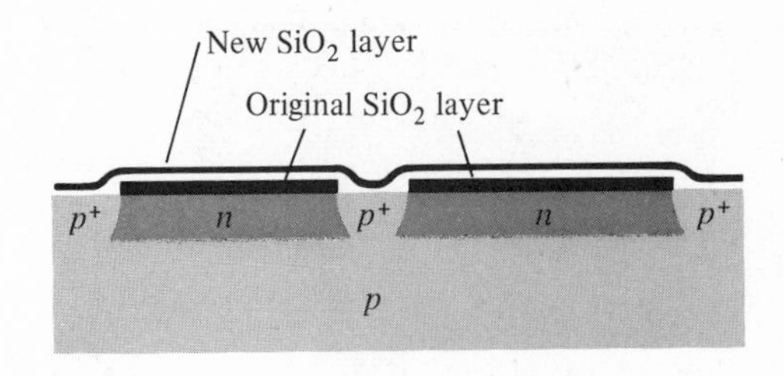

FIG. 10.21

In preparation for the next diffusion process the entire wafer is coated with a SiO_2 layer.

Base and Emitter Diffusion Processes

The isolation diffusion process is followed by the base and emitter diffusion cycles. The sequence of steps in either case is the same as that encountered in the description of the isolation diffusion process. Altough the terminology "base" and "emitter" refer specifically to the transistor structure, necessary parts (layers) of each element (resistor, capacitor, and diodes) will be formed during each diffusion process. The surface appearance of the NAND gate after the isolation base and emitter diffusion processes appears in Fig. 10.22. The mask employed in each process is also provided next to each photograph.

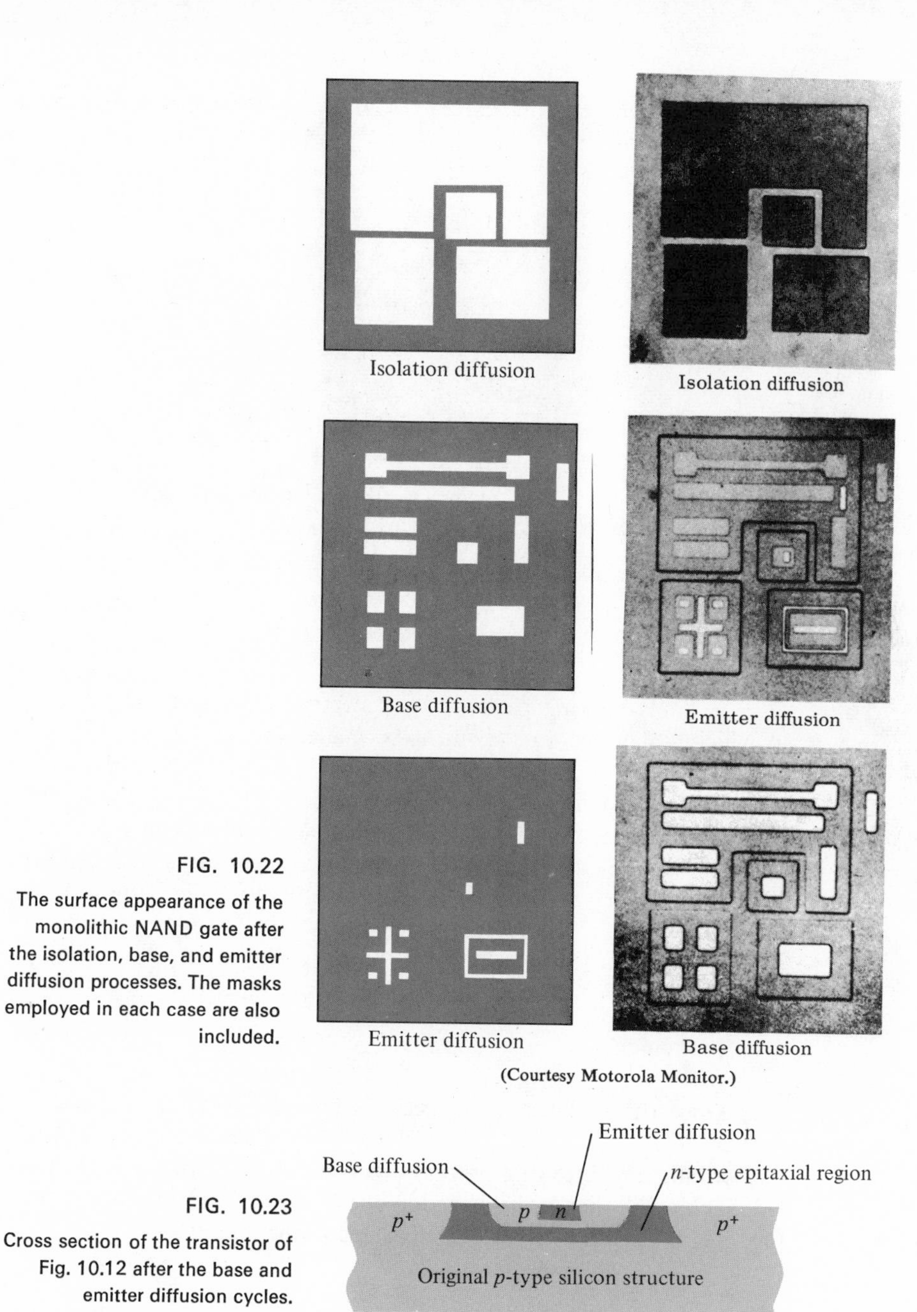

FIG. 10.22
The surface appearance of the monolithic NAND gate after the isolation, base, and emitter diffusion processes. The masks employed in each case are also included.

FIG. 10.23
Cross section of the transistor of Fig. 10.12 after the base and emitter diffusion cycles.

The cross section of the transistor of Fig. 10.12 will appear as shown in Fig. 10.23 after the base and emitter diffusion cycles.

Preohmic Etch

In preparation for a good ohmic contact, n^+ regions (see Section 10.3) are diffused into the structure as clearly indicated by the light areas of Fig. 10.24. Note the correspondence between the light areas and the mask pattern.

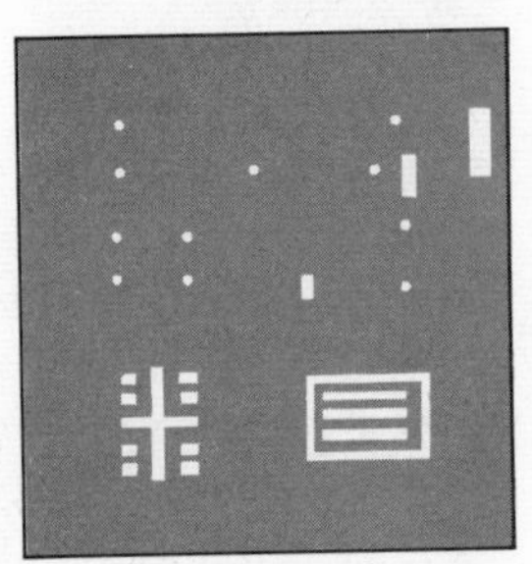

Preohmic etch

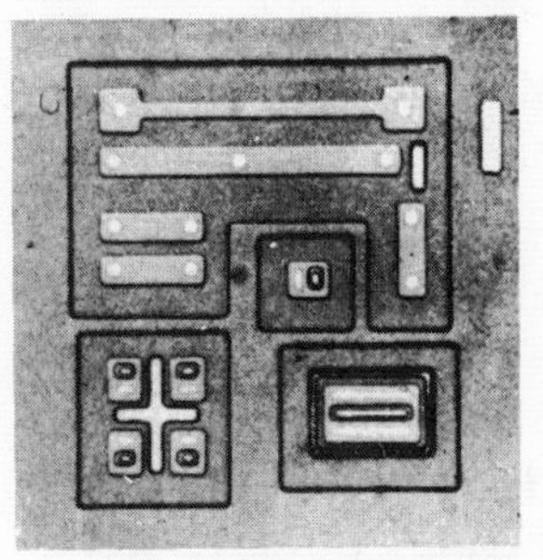

Preohmic etch

(Courtesy Motorola Monitor.)

FIG. 10.24

Surface appearance of the chip of Fig. 10.22 after the preohmic etch cycle. The mask employed is also included.

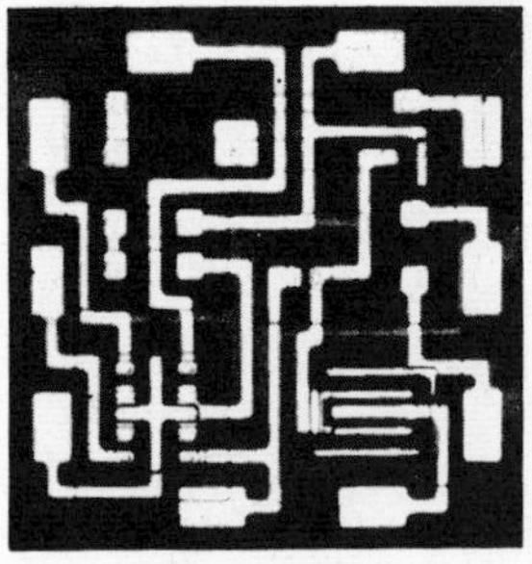

Metalization

(Courtesy Motorola Monitor.)

FIG. 10.25

Completed metalization process.

Metalization

A final masking pattern exposes those regions of each element to which a metallic contact must be made. The entire wafer is then coated with a thin layer of aluminum that after being properly etched will result in the desired interconnecting conduction pattern. A photograph of the completed metalization process appears in Fig. 10.25.

The complete structure with each element indicated appears in Fig. 10.26. Try to relate the interconnecting metallic pattern to the original circuit of Fig. 10.12a.

Packaging

Once the metalization process is complete, the wafer must be broken down into its individual chips. This is accomplished through the scribing and breaking processes depicted in Fig. 10.27. Each individual chip will then be packaged in one of the three forms indicated in Fig. 10.28. The name of each is provided in the figure.

Testing

The final production stage, as with every commercial electronic package, is the testing of the system. As indicated in Fig. 10.4 this can demand a good percentage of the manufacturing costs. Photographs of various testing procedures appear in Fig. 10.29.

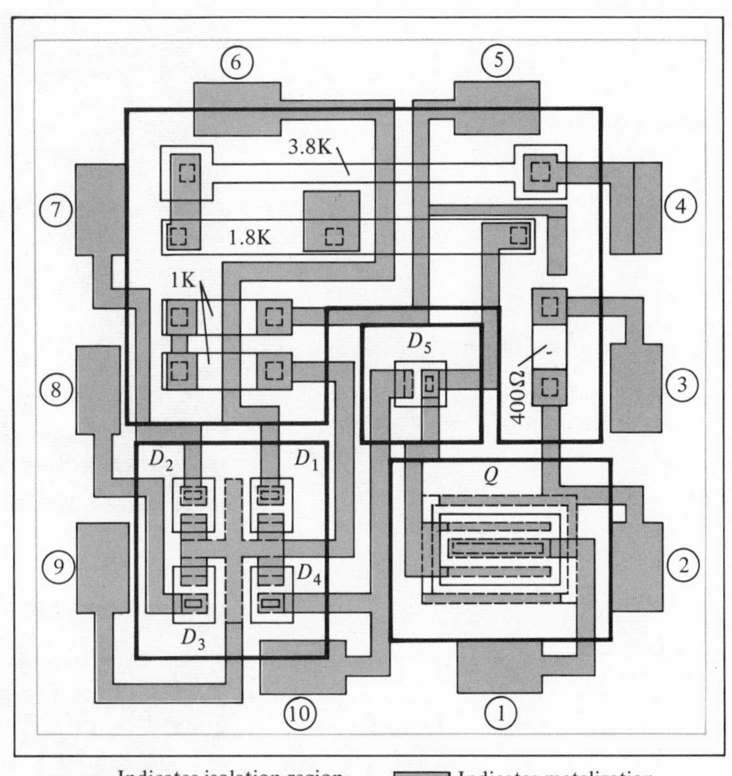

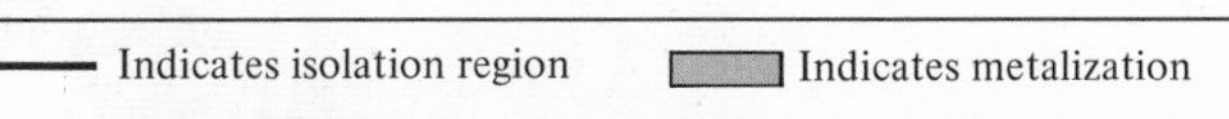

FIG. 10.26

Monolithic structure for the NAND gate of Fig. 10.12.

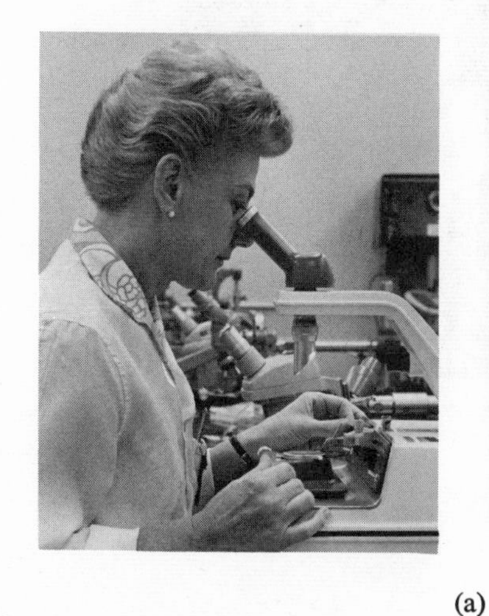

(a)

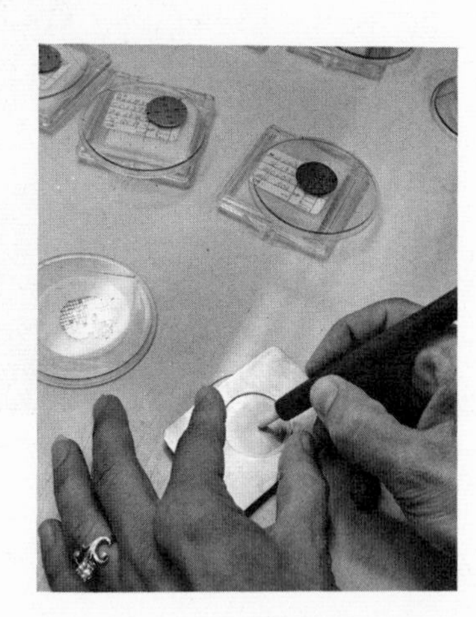

(b)

(Courtesy Autonetics, North American Rockwell Corp.)

(Courtesy Texas Instruments, Inc.)

(Courtesy Motorola, Inc.)

FIG. 10.27

(a) Scribing and (b) breaking of the monolithic wafer into individual chips.

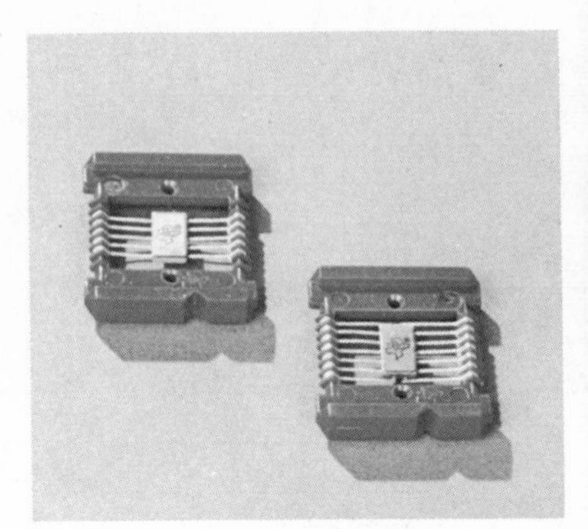

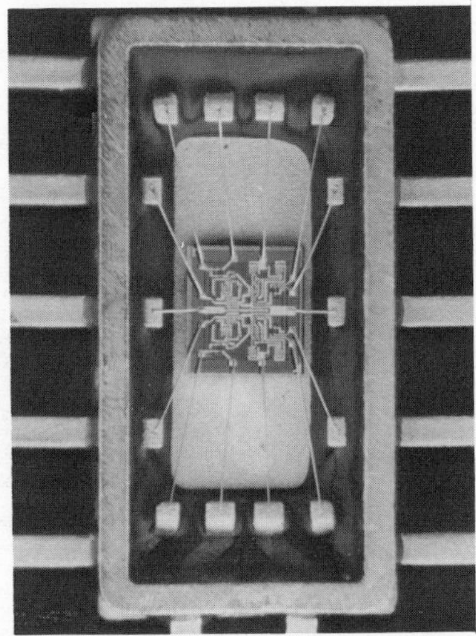

(a)

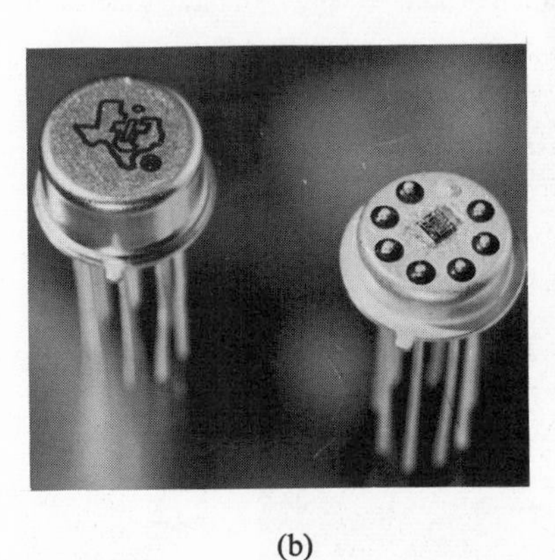

(b)

(c)

(Courtesy Texas Instruments, Inc.)

FIG. 10.28

Monolithic packaging techniques: (a) flat package; (b) TO (top-hat)-type package; (c) dual in-line plastic package.

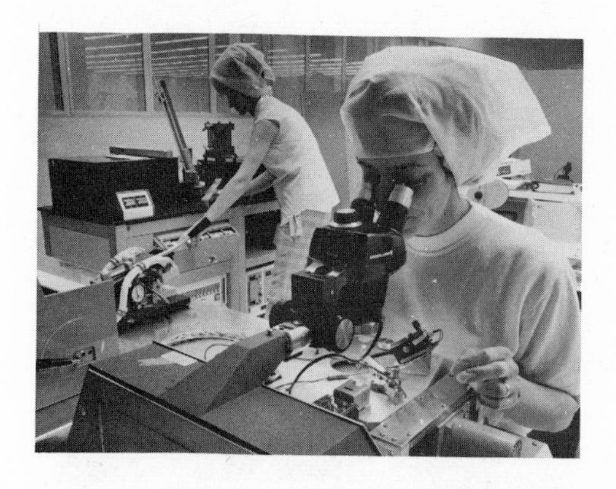

(a)
(Courtesy Autonetics, North American Rockwell Corp.)

(b)
(Courtesy Texas Instruments, Inc.)

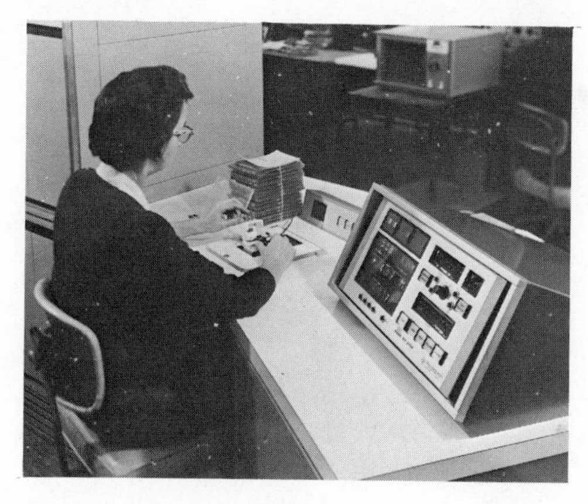

(c)
(Courtesy Texas Instruments, Inc.)

FIG. 10.29
Production testing.

10.6 THIN AND THICK FILM INTEGRATED CIRCUITS

The general characteristics, properties, and appearance of thin and thick film integrated circuits are quite similar although they both differ in many respects from the monolithic integrated circuit. They are not formed within a semiconductor wafer but *on* the surface of an insulating substrate such as glass or an appropriate ceramic material. In addition, *only* passive elements (resistors, capacitors) are formed through thin or thick film techniques on the insulating surface. The active elements (transistors, diodes) are added as *discrete* elements to the surface of the structure after the passive elements have been formed. The discrete active devices are frequently produced using the monolithic process.

(Courtesy Autonetics, North American Rockwell Corp.)

FIG. 10.30

Monolithic transistors to be employed in thin or thick film integrated circuits.

A number of thin film integrated circuits appear in Fig. 10.30. Note the active elements added on the surface between the proper aluminum contacts. The interconnecting conduction pattern and the passive elements are prepared through masking techniques.

The primary difference between the thin and thick film techniques is the process employed for forming the passive components and the metallic conduction pattern. The thin film circuit employs an evaporation or cathode-sputtering technique while the thick film resorts to silk-screen techniques. Priorities do not permit a detailed description of these processes here.

In general, the passive components of film circuits can be formed with a broader range of values and reduced tolerances as compared to the monolithic IC. The use of discrete elements also increases the flexibility of design of film circuits although, quite obviously, the resulting circuit will be that much larger. The cost of film circuits with a larger number of elements is also, in general, considerably higher than that of monolithic integrated circuits.

10.7 HYBRID INTEGRATED CIRCUITS

The terminology *hybrid integrated circuit* is applied to the wide variety of multichip integrated circuits and also those formed by a combination of the film and monolithic IC techniques. The multichip integrated circuit employs either the monolithic or film technique to form the various components, or set of individual circuits, which are then interconnected on an insulating substrate and packaged in the same container. Integrated circuits of this type appear in Fig. 10.31. In a more sophisticated type of hybrid integrated circuit, the active devices are first formed within a semiconductor wafer which is subsequently covered with an insulating layer such as SiO_2. Film techniques are then employed to form the passive elements on the SiO_2 surface. Connections are made from the film to the monolithic structure through "windows" cut in the SiO_2 layer.

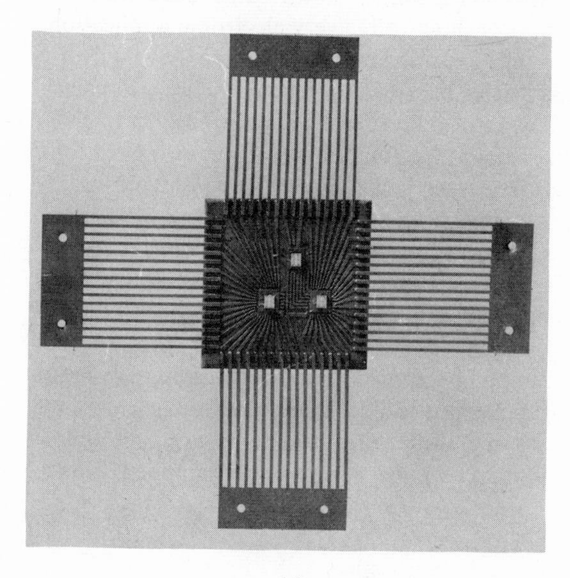

(a)

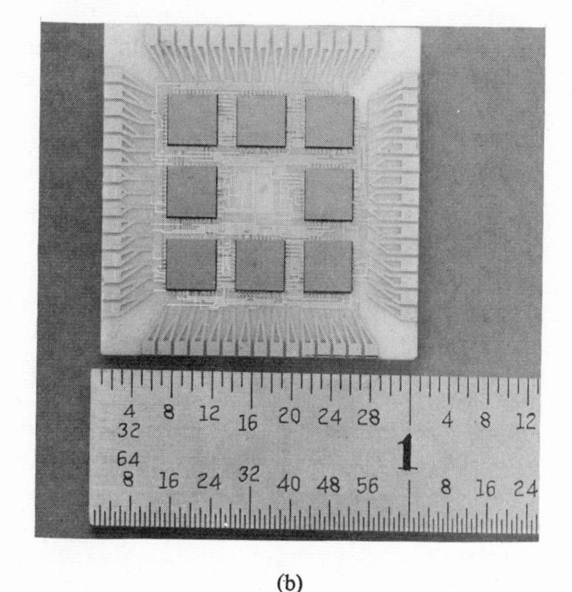

(b)

(Courtesy Texas Instruments, Inc.)

FIG. 10.31
Thin film integrated circuit.

11

Differential and Operational Amplifiers

11.1 BASIC DIFFERENTIAL AMPLIFIER

An amplifier is an electronic circuit containing transistors, may tubes, FETs, or IC circuits, and so on, that provides voltage gain. It may also provide current gain, or power gain, or allow impedance transformation. As it is a basic part of practically every electronic application, the amplifier is an essential circuit. Amplifiers, as we have already discovered, may be classified in many ways. There are low-frequency amplifiers, audio amplifiers, ultrasonic amplifiers, radio-frequency (RF) amplifiers, wide-band amplifiers, video amplifiers, and so on; each type operating in a prescribed frequency range. We have considered small-signal and large-signal amplifiers, and amplifiers that may be interconnected as RC-coupled, transformer-coupled, and so on.

The *differential amplifier* is a special type of circuit that is used in a wide variety of applications. Let us consider a number of basic properties of differential amplifiers. Figure 11.1 shows a block symbol of a differential amplifier unit. As shown, there are two separate input (1 and 2) and two separate output (3 and 4) terminals. We must first consider the relation between these terminals to obtain an understanding of how the differential amplifier (D.A.) may be applied. Notice

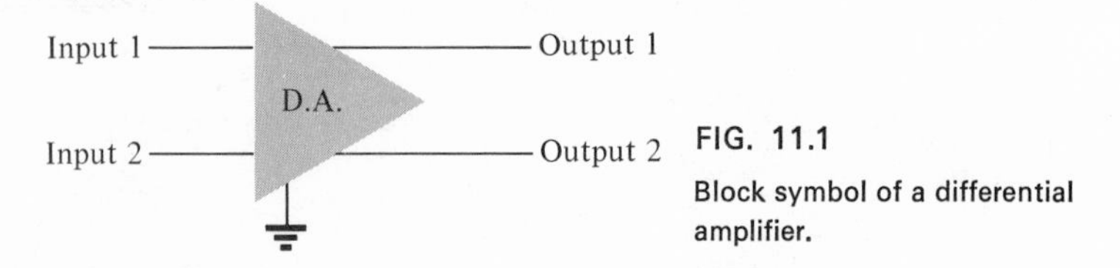

FIG. 11.1
Block symbol of a differential amplifier.

that, in Fig. 11.1, a ground connection is shown separately to make clear that both input or output terminals are different from ground. Voltages may be applied to either or both input terminals and output voltages may appear at both output terminals. However, there are some very specific phase relations between both input and both output terminals.

Figure 11.2 shows the block and circuit diagrams of a basic differential amplifier to be used in the following discussion. There are two inputs and outputs shown in the block diagram. Inputs are applied essentially to each base of the two separate transistors. As shown, however, the transistor emitters are connected to a common-emitter resistor so that the two output terminals V_{o1} and V_{o2} are affected by either or both input signals. The outputs are taken from the collector terminals of each transistor. The input and output terminals are also numbered to facilitate reference. There are two supply voltages shown in the circuit diagram and it should be carefully noted that no ground terminal is indicated within the circuit although the opposite points of both positive- and negative-voltage supplies are understood to be connected to ground.

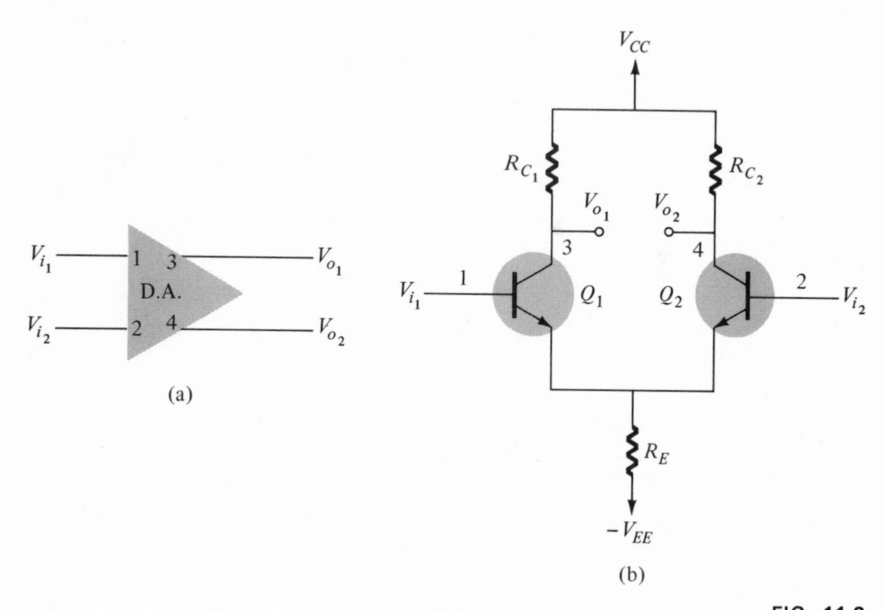

FIG. 11.2
Basic differential amplifier: (a) block diagram; (b) circuit diagram.

Single-Ended Input D.A.

Consider first the operation of the differential amplifier with a single input signal applied to terminal 1, with terminal 2 connected to ground (0 V). Only one output signal at either terminal 3 or terminal

4 is considered. Figure 11.3 shows the block and circuit diagrams for input signal V_{i_1} at terminal 1 and output V_{o_1} at terminal 3. The block diagram shows a sinusoidal input and an amplified, inverted output. The circuit diagram shows the sinusoidal input applied to the base of a transistor with the amplified output at the collector inverted, as we would expect from past knowledge of a single-stage transistor amplifier.

With input 2 grounded it might seem that there is no output at terminal 4—but this is incorrect. The block diagram of Fig. 11.4 shows the operation of the differential amplifier with the V_{o_2} output at terminal 4 resulting from an input V_{i_1} at terminal 1. The input at terminal 1 is shown as V_{i_1}, which is a small sinusoidal voltage measured with respect to ground. Since an emitter resistor is connected in common with both emitters, a voltage that is developed by V_{i_1} appears at the common-emitter point. This sinusoidal voltage, measured with respect to ground, is approximately the same magnitude and is in phase with V_{i_1} because it results from emitter follower action of the circuit.

To be sure that this is clearly understood, the part of the circuit, acting as an emitter follower is shown in Fig. 11.4c. An input applied to the base of Q_1 appears in phase and about the same magnitude at the emitter of Q_1 for the emitter follower part of the circuit shown. Recall that for an emitter follower the gain is nearly unity (with no phase reversal). This emitter signal is measured with respect to ground. Figure 11.4d shows the part of the circuit with the emitter voltage affecting the operation of transistor Q_2. The voltage at the emitter

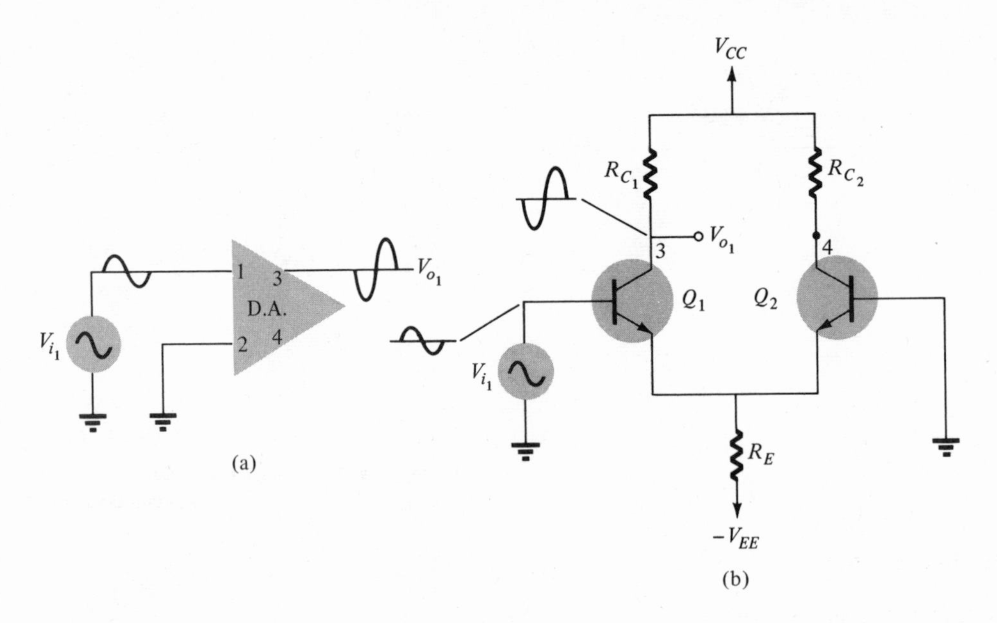

FIG. 11.3

Single-ended operation of differential amplifier: (a) block diagram; (b) circuit diagram.

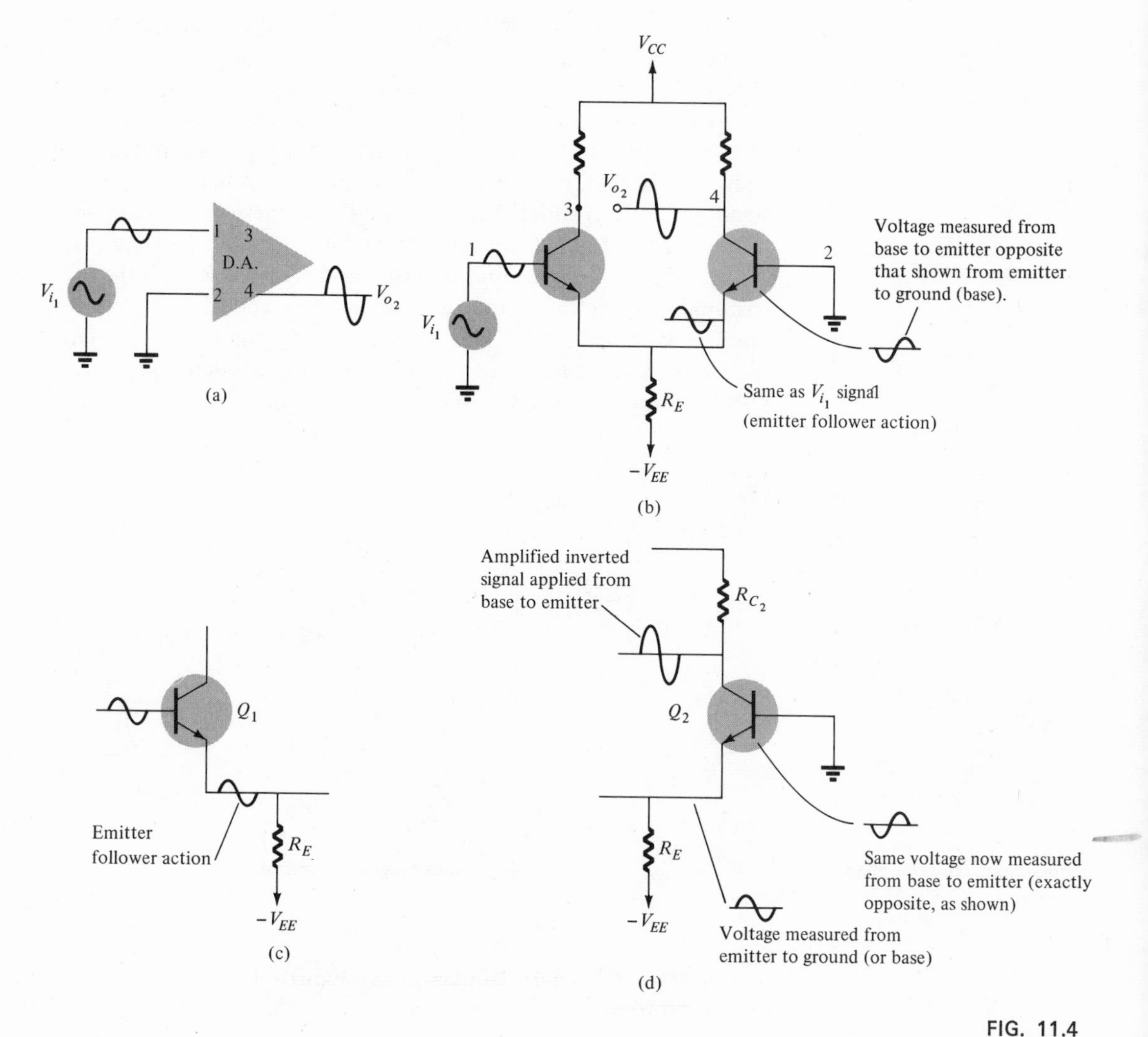

FIG. 11.4
Single-ended operation of differential amplifier.

of Q_2 is the same as that of Q_1 (since the emitters are connected together) and appears from the emitter of Q_2 to ground or to the base of Q_2 (since that is connected to ground). If the voltage measured from the emitter to base of Q_2 is in phase with input V_{i_1} as shown, the voltage measured from base to emitter of Q_2 is the same signal with opposite polarity. Thus, measuring from base to emitter of Q_2 a voltage of about the same magnitude as V_{i_1} is obtained but the signal is opposite in polarity to that of V_{i_1}. The amplifier action of transistor Q_2 and load resistor R_{C_2} provides an output at the collector of Q_2 that is amplified and inverted from the signal developed across base to emitter of Q_2.

In summary, an input V_{i_1} is applied to input 1 and an amplified, in-phase signal V_{o_2} results at output terminal 4. Because the input at terminal 2 is grounded this does not mean that no output occurs at terminal 4. It should be clear that internal connection (of common emitters) results in the input at terminal 1 causing an output at terminal 4. In fact we can now see that the input at terminal 1 causes output signals at both terminals 3 and 4. In addition, these outputs are opposite in phase and of about the same magnitude. Finally, we should see (as in Fig. 11.5) that the output at terminal 4 is in phase with the input at terminal 1, while the output at terminal 3 is opposite in phase to the input at terminal 1. It should be understood from the previous discussion that an input applied to terminal 2 with terminal 1 grounded will result in output voltages as shown in Fig. 11.6.

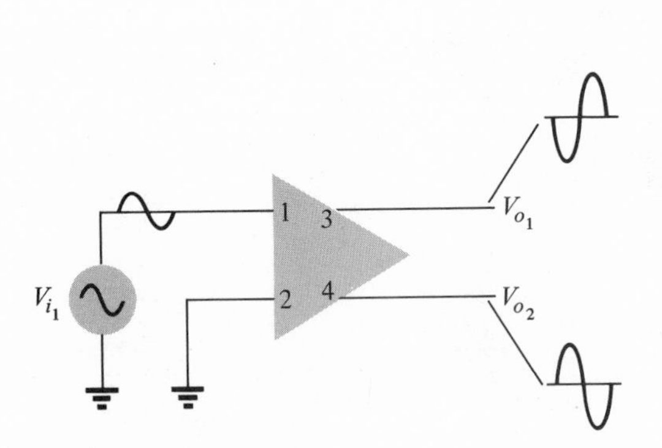

FIG. 11.5
Single-input, opposite-phase outputs.

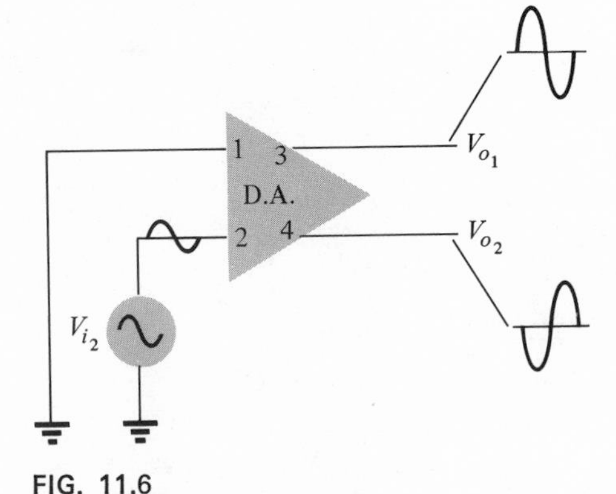

FIG. 11.6
Single-ended input to terminal.

Differential Input (Double-Ended Input) Operation

In addition to using only one input to operate the differential amplifier circuitry it is possible to apply signals to each input terminal, with opposite outputs appearing at the two output terminals. The usual

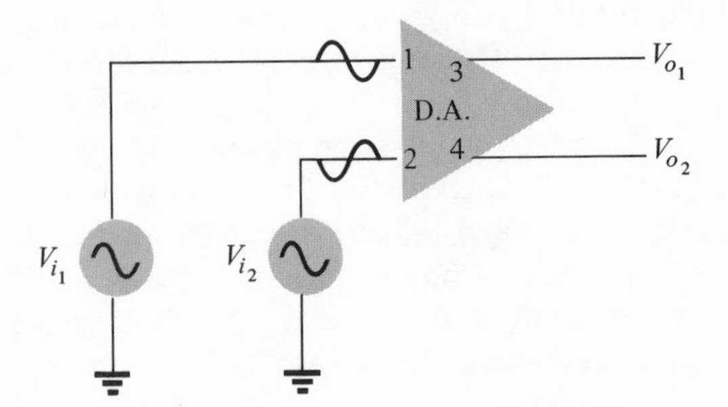

FIG. 11.7
Operation with differential input signals.

use of the *double-ended* or *differential* mode of input is when the two input signals are themselves opposite in phase (180° out of phase), and about the same magnitude. Figure 11.7 shows such a situation.

We now must consider how each input affects the outputs and what the resulting output signal looks like. This can be done using the *superposition principle*, considering each input applied separately with the other at 0 V and summing the resulting output voltages at each terminal. Figures 11.8a and 11.8b show the result of each input acting alone and Fig. 11.8c shows the resulting overall operation. The input applied to terminal 1 results in an opposite-phase, amplified output at terminal 3 *and* an in-phase, amplified output at terminal 4. Assume that the inputs are about equal in magnitude and that the output magnitudes are about equal, of value V, for discussion purposes.

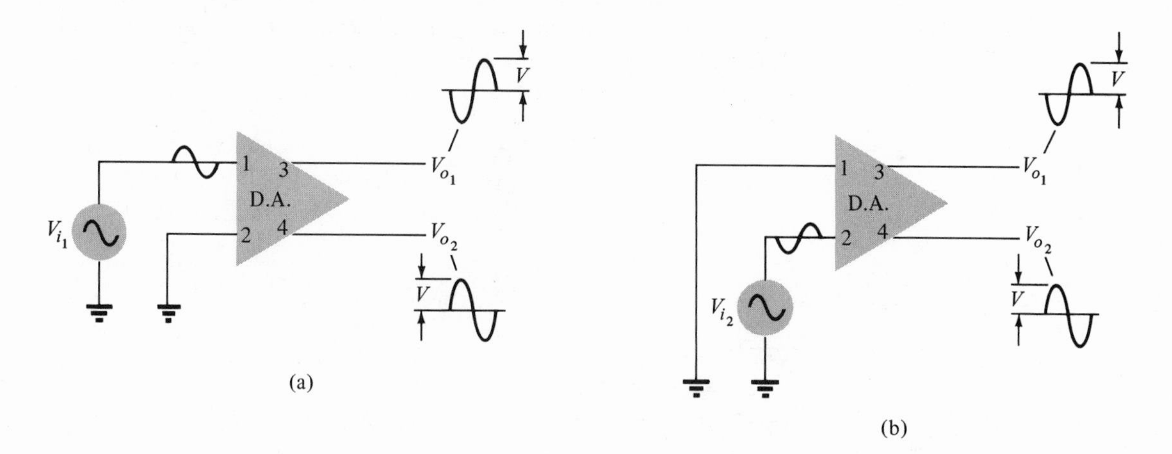

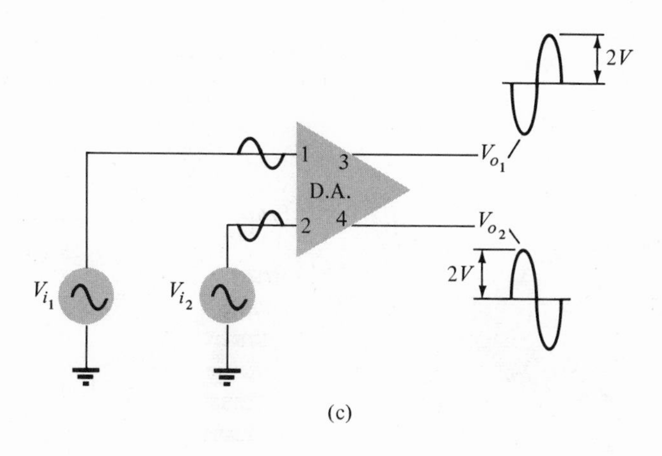

FIG. 11.8
Differential operation of amplifier: (a) $V_{i_2} = 0$; (b) $V_{i_1} = 0$; (c) both inputs present.

The input applied to terminal 2 results in an opposite-phase, amplified output at terminal 4 *and* an in-phase, amplified output at terminal 3. The magnitudes of the outputs will both be V since the input magnitudes were assumed to be about the same. It is important to note that the outputs in each case are of the same phase at each output terminal. By superposition, the resulting signals at each output terminal are added, and we obtain the full operation of the circuit shown in Fig. 11.8c. The output at each terminal is twice the resulting from single-ended operation because the outputs due to each input are in phase. If the inputs applied were both in phase (or if the same input were applied to both input terminals) the resulting signals due to each input acting alone would be opposite in phase at each output and the resulting output would be about 0 V, as shown in Fig. 11.9.

To bring the operation as single- and double-ended differential amplifier states into full perspective consider the connection of two differential amplifiers shown in Fig. 11.10. From the previous discussion, if the amplifiers had identical single-ended gains then the outputs

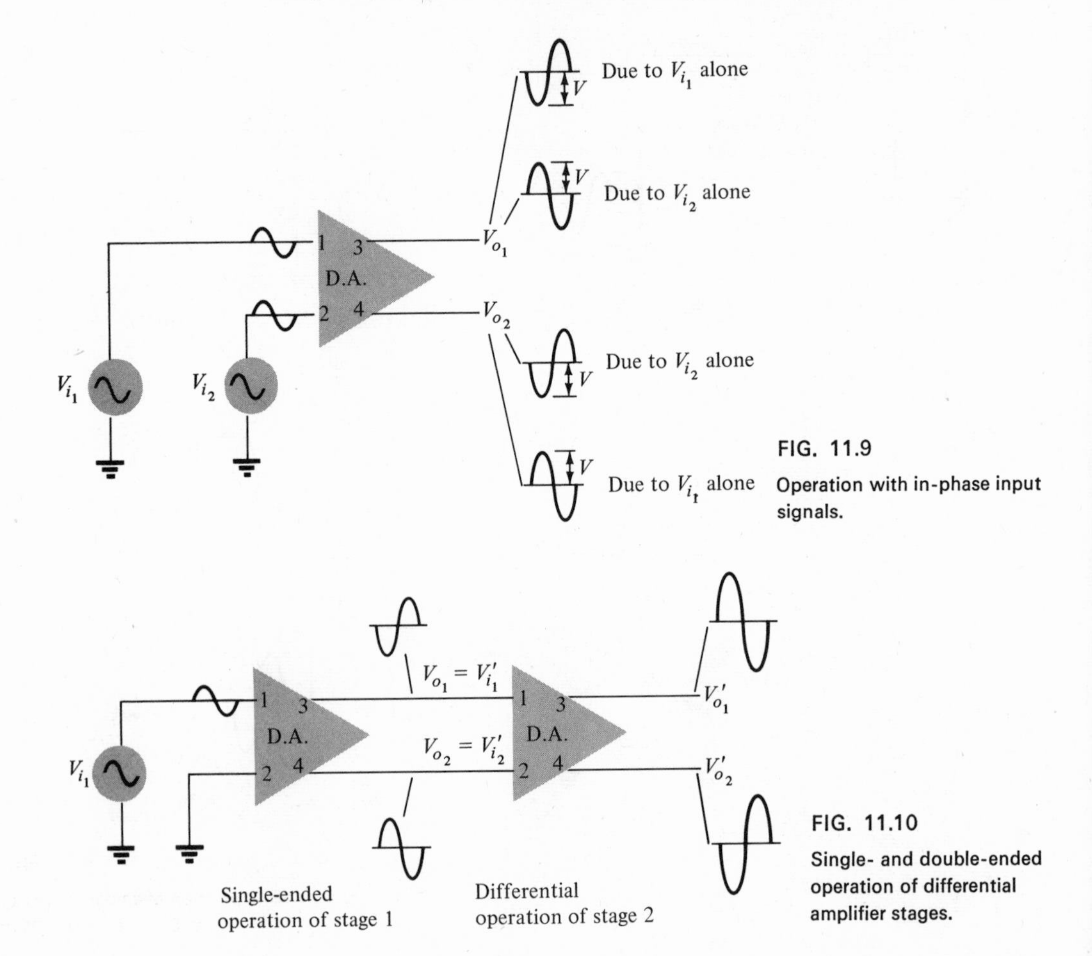

FIG. 11.9
Operation with in-phase input signals.

FIG. 11.10
Single- and double-ended operation of differential amplifier stages.

of stage 1 would be larger than the inputs by the amount of amplifier gain, while the outputs of stage 2 would be larger than the inputs to stage 2 by twice the amplifier gain. The initial signal, from an antenna of a radio, or a phonograph pickup cartridge, etc., is single-ended and is used as such. The second differential amplifier stage, however could be operated double-ended to obtain twice the stage gain. Either output of stage 2 (or both) could then be used as amplified signals to the next section of the system. Although differential operation requires about equal and opposite phase signals, this is often available, especially after one single-ended stage of gain.

11.2 DIFFERENTIAL AMPLIFIER CIRCUITS

Having considered some features of use and operation of a differential amplifier stage we now look into some details of a differential amplifier circuit. In particular, we shall consider the voltage gain of the stage and its input and output impedance. A basic circuit of a differential amplifier is shown in Fig. 11.11. Input sources are shown as a voltage source and source resistance in the general case.

dc Bias Action of Circuit

Before considering the main action of the circuit as a voltage amplifier let us see how the circuit is biased to operate. A circuit diagram

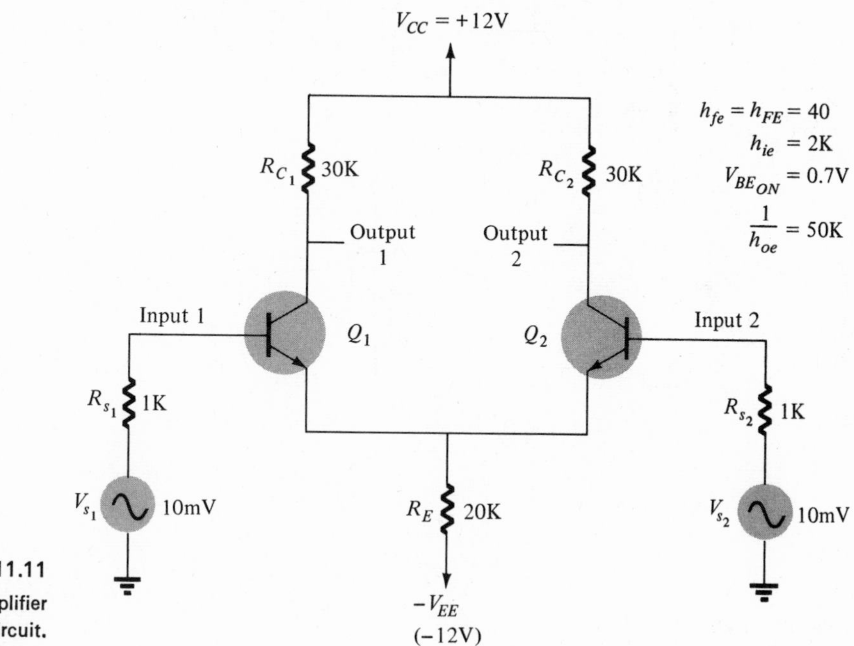

FIG. 11.11
Basic differential amplifier circuit.

(Fig. 11.12) shows the main voltage and current features of the circuit for dc operation. No ac signal sources are present, the sources having been set to 0 V with only the source resistances present. The base-emitter of Q_1 is forward biased by the $-V_{EE}$ battery from ground through resistor R_{S_1}, through the base-emitter, through resistor R_E to $-V_{EE}$ (see Fig. 11.13a). The voltage drop across the forward-biased base emitter is about

$$V_{BE_{ON}} = 0.7 \text{ V}$$

We would have to write a number of equations to solve for the dc voltages and currents. However, it is possible to use good approximations to make the calculations more direct. For example, the dc voltage drop across source resistor R_{S_1} will be quite small as the following calculation indicates (assuming a typical base current in the order of microamperes)

$$I_{B_1}R_{S_1} = (100\ \mu\text{A})(1\text{ K}) = 100\text{ mV} = 0.1\text{ V}$$

If the base current were only 10 μA then the dc voltage drop across R_{S_1} would be 10 mV, which is quite negligible. On the other hand, a source resistance of 10 K with base current of 100 μA would result in a voltage drop of 1 V, which is not negligible. For our purposes we shall as-

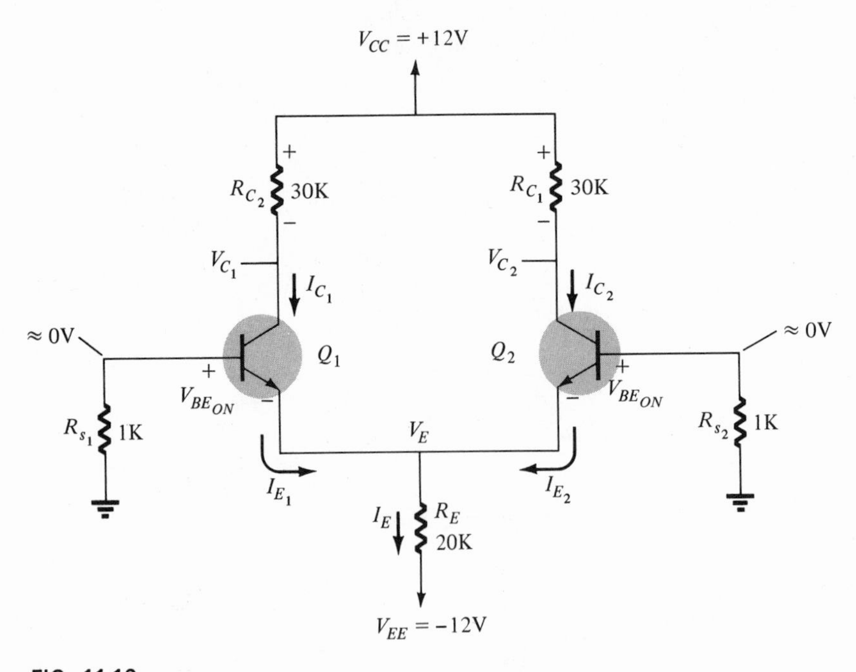

FIG. 11.12
dc bias action of circuit.

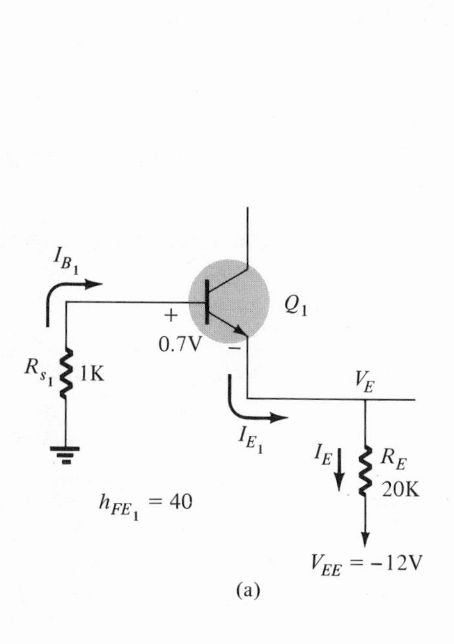

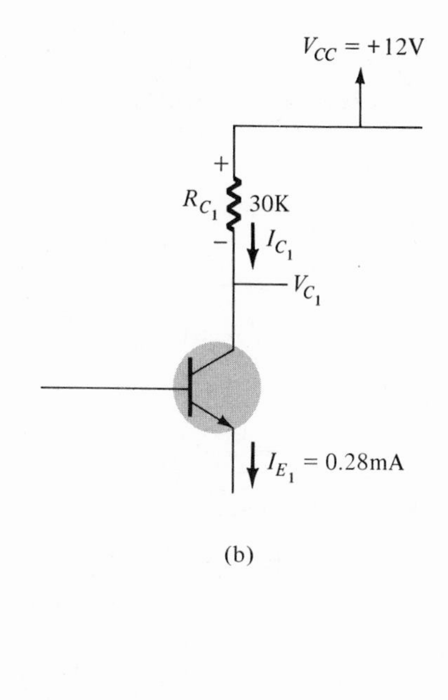

FIG. 11.13
Partial circuits of differential amplifier: (a) input section; (b) output section.

sume the voltage drop to be quite small (which is often correct) and check later in the calculations to be sure we were able to make such an assumption.

If we assume that

$$V_{B_1} = 0 \text{ V}$$

then the emitter voltage is directly calculated to be

$$\begin{aligned} V_E = V_{E_1} &= V_{B_1} - V_{BE_1} \\ &= 0 - 0.7 \text{ V} \end{aligned} \tag{11.1}$$

and

$$V_E = -0.7 \text{ V}$$

The current through resistor R_E is then calculated directly to be

$$\begin{aligned} I_E &= \frac{V_E - V_{EE}}{R_E} \\ &= \frac{-0.7 - (-12)}{20\ K} = \frac{11.3}{20} = 0.565 \text{ mA} \end{aligned} \tag{11.2}$$

The current through resistor R_E is made up of the emitter currents from each transistor. Assuming that the transistors are well matched (which should always be the case) the emitter current of each transistor is one-half the total current through R_E.

$$I_{E_1} = I_{E_2} = \frac{I_E}{2} \tag{11.3}$$
$$= \frac{0.565\text{ mA}}{2} = 0.2825\text{ mA} \cong 0.28\text{ mA}$$

We can now check our assumption of V_{B_1} by calculating I_{B_1} as follows:

$$I_{B_1} = \frac{I_{E_1}}{1 + h_{FE_1}} \tag{11.4}$$
$$= \frac{0.28\text{ mA}}{1 + 40} = 0.068\text{ mA} = 68\ \mu\text{A}$$
$$V_{B_1} = I_{B_1}R_{S_1} = 68\ \mu\text{A} \times 1\text{ K} = 68\text{ mV} = 0.068\text{ V}$$

which is negligible compared to the other voltage drops in the circuit.

The output section of the circuit can now be directly handled. Figure 11.13b shows a partial circuit diagram. The collector current is obtained from the calculation of emitter current.

$$I_{C_1} \cong I_{E_1} \tag{11.5}$$
$$= 0.28\text{ mA}$$

and the collector voltage is

$$V_{C_1} = V_{CC} - I_{C_1}R_{C_1} \tag{11.6}$$
$$= 12 - (0.28\text{ mA})(30\text{ K})$$
$$= 12 - 8.4 = +3.6\text{ V}$$

In summary, the base voltage being about 0 V, the emitter voltage is then fixed at about -0.7 V. The emitter currents of each transistor are then set by resistor R_E, this also setting the collector current, which is approximately equal to the emitter current of either transistor. For the collector current resulting, the output dc voltage V_{C_1} is adjusted by choosing the value of R_{C_1}. These considerations hold true for the currents and voltages of transistor Q_2. The output voltage at V_{C_1} will probably be set about in the center of the potential voltage swing from 0 V to $+V_{CC}$. In the present circuit a lower value of R_{C_1} would result in a lower voltage drop with the value of V_{C_1} then higher, a value of about 6 V being desirable. It should be clearly noted that the value of R_{C_1} has almost no effect on the current I_{C_1}, which depends on the value of resistor R_E in setting I_{E_1}.

If an input ac signal causes the transistor to approach cutoff the output voltage will approach V_{CC} in value. If, however, the input ac signal causes the transistor to turn on, the lowest the output voltage can go is about -0.7 V, the fixed dc voltage at the emitter. Thus, the largest possible voltage swing at V_{C_1} would be from near 0 to $+12$ V, in the present circuit. Biasing the circuit in the center of this range will allow the largest voltage swing from dc bias before distortion (clipping) occurs.

ac Operation of Differential Amplifier Circuit

To consider the ac operation of the circuit all dc voltage supplies are set at zero and the transistors are replaced by small-signal ac equivalent circuits. Figure 11.14 shows the resulting ac equivalent circuit, with the transistors replaced by hybrid equivalent circuits. The circuit obviously appears quite complex and analyzing the total circuit would become quite involved. Again we can break up the calculations by using some simplifying approximations so that smaller parts of the circuit can be analyzed separately. In addition,

$h_{ie_1} = h_{ie_2} = h_{ie}$, $h_{fe_1} = h_{fe_2} = h_{fe}$, $h_{oe_1} = h_{oe_2} = h_{oe}$, and

$R_{C_1} = R_{C_2} = R_C$, $R_{s_1} = R_{s_2} = R_s$

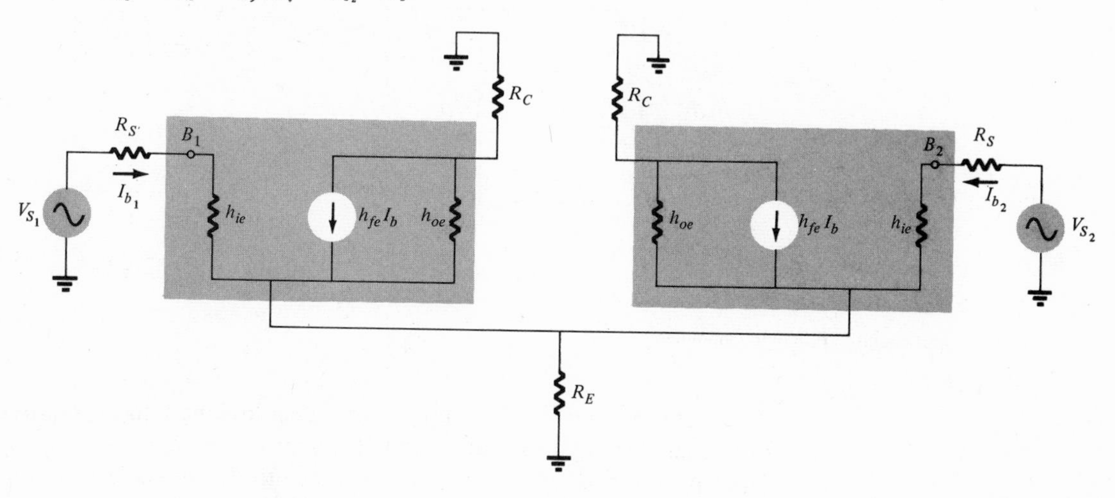

FIG. 11.14
ac equivalent circuit of differential amplifier.

INPUT AC SECTION

Figure 11.15a shows the partial ac equivalent circuit of the input for transistor Q_1. Looking into the emitter of transistor Q_2 a small ac equivalent resistance is present, equal in value to

$$R_{e_2} = \frac{R_s + h_{ie}}{1 + h_{fe}} \tag{11.7}$$

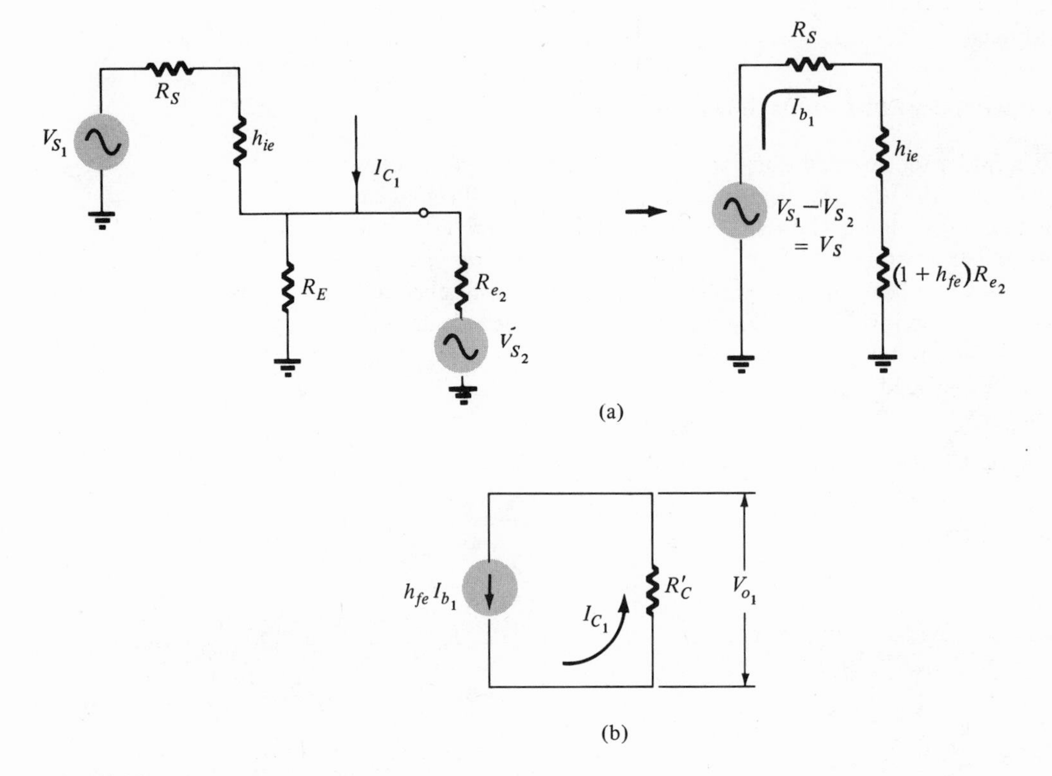

FIG. 11.15
Partial ac equivalent circuit of difference amplifier.

As a general statement, the ac resistance seen looking into the emitter of a transistor circuit is approximately equal to the value of the transistor plus source resistances divided by transistor current gain. For the values of Fig. 11.11

$$R_{e_2} = \frac{1\text{ K} + 2\text{ K}}{1 + 40} = 73\ \Omega$$

The parallel combination of resistors R_E and R_{e_2} give an equivalent ac resistance of

$$\frac{R_{e_2} R_E}{R_{e_2} + R_E} = \frac{73 \times 20{,}000}{73 + 20{,}000} \cong 73\ \Omega$$

Since the differential amplifier circuit generally has an R_E of large value we can make the approximate statement that, if

$$R_E \gg R_{e_2}$$

the parallel combination is approximately R_{e_2} in value, as shown in Fig. 11.15a. Using the resulting ac equivalent circuit the value of the ac base current is calculated to be

$$I_{b_1} = \frac{V_{s_1} - V_{s_2}}{R_s + h_{ie} + (1 + h_{fe}) R_{e_2}} = \frac{V_{s_1} - V_{s_2}}{2(R_s + h_{ie})} \tag{11.8a}$$

and defining $V_s \equiv V_{s_1} - V_{s_2}$ as the difference input voltage

$$I_{b_1} = \frac{V_s}{2(R_s + h_{ie})} \tag{11.8b}$$

Using the given circuit values

$$I_{b_1} = \frac{20 \text{ mV}}{2(1000 + 2000)} = 3.3 \ \mu\text{A}$$

OUTPUT AC SECTION

Figure 11.15b shows the output collector section of the ac equivalent circuit. The base current has already been calculated from consideration of only the input ac section of the circuit. An ac equivalent resistance is obtained

$$R'_C = \frac{R_C \times 1/h_{oe}}{R_C + 1/h_{oe}} \tag{11.9a}$$

$$= \frac{R_C}{1 + h_{oe} R_C} \tag{11.9b}$$

where h_{oe} is the hybrid parameter of the output ac conductance of transistor Q_1, which may be expressed as the output ac resistance $1/h_{oe}$.

For the circuit values of Fig. 11.11

$$R_C = \frac{30 \times 50}{30 + 50} = \frac{1500}{80} \cong 8.8 \text{ K}$$

From the resulting ac equivalent of current source and resistance R_C in parallel, the output ac voltage is calculated

$$V_{o_1} = -I_{C_1} R'_C \tag{11.10}$$

where

$$I_{C_1} = h_{fe} I_{b_1} \tag{11.11}$$

Calculating the numerical values

$$I_{C_1} = 40 \times 3.3\ \mu\text{A} = 132\ \mu\text{A}$$

$$V_{o_1} = -(132\ \mu\text{A})(8.8\ \text{K}) = -1160\ \text{mV} = -1.16\ \text{V}$$

where the minus sign shows that there is 180° phase inversion between input and output ac voltages.

The circuit ac difference voltage gain is

$$A_{v_1} = \frac{V_{o_1}}{V_s} \tag{11.12}$$

$$= \frac{-1.16\ \text{V}}{20\ \text{mV}} = -0.058 \times 10^3 = -58$$

With an ac circuit gain of $A_{v_1} = -58$, and an input difference voltage of 20 mV, rms is amplified (and inverted) to an ac value of 1.16 V. This value of gain could be obtained from a single expression as follows:

$$\text{A}_{v_1} = \frac{V_{o_1}}{V_s} = \frac{-I_C R'_C}{V_s} = \frac{-(h_{fe} I_{b_1}) R'_C}{V_s}$$

$$\cong \frac{-h_{fe} R'_C [V_s / 2(R_s + h_{ie})]}{V_s} = \frac{-h_{fe} R'_C}{2(R_s + h_{ie})}$$

$$\boxed{A_{v_1} \cong \frac{-h_{fe} R'_C}{2(R_s + h_{ie})}} \tag{11.13}$$

Using the values of the circuit of Fig. 11.11

$$A_{v_1} \cong -\frac{40 \times 8.8\ \text{K}}{2(1\ \text{K} + 2\ \text{K})} = -\frac{40 \times 8.8}{6} = -58.5$$

which agrees quite closely with the gain calculated in the step-by-step derivation above.

INPUT RESISTANCE

From the ac equivalent circuit of Fig. 11.15a the input resistance of the circuit seen from the source is

$$\boxed{R_{i_1} = h_{ie} + (1 + h_{fe}) R_{e_2}} \tag{11.14}$$

For the circuit of Fig. 11.1 and the previous calculation of $R_{e_2} = 73\ \Omega$,

$$R_{i_1} = 2{,}000 + 73(1 + 40) \cong 5\text{K}$$

From the ac equivalent circuit of Fig. 11.15b the resulting approximate output resistance is

$$R_{o_1} = R'_C \cong \frac{R_C \times 1/h_{oe}}{R_C + 1/h_{oe}} = \frac{R_C}{1 + h_{oe} R_C} \tag{11.15}$$

which has already been determined to be 8.8 K.

Input and output equations that hold for the one half of the circuit should be the same for the other half of the circuit.

EXAMPLE 11.1 Calculate the input and output resistance of a difference amplifier circuit as in Fig. 11.11 for the following circuit values: $R_{C_1} = R_{C_2} = 15$ K, $R_E = 10$ K, $h_{fe} = 60$, $h_{ie} = 2.5$ K, $h_{oe} = 12.5$ $\mu℧$, and $R_{s_1} = R_{s_2} = 600\ \Omega$.

Solution

$$R_{e_2} = \frac{R_{s_2} + h_{ie_2}}{1 + h_{fe_2}} = \frac{600 + 2500}{1 + 60} = 50.8\ \Omega$$

$$R_{i_1} = h_{ie_1} + (1 + h_{fe_1}) R_{e_2} = 2500 + (1 + 60)\, 50.8 \cong 6.5\ \text{K}$$

$$R_{o_1} = \frac{R_{c_1}}{1 + h_{oe_1} R_{c_1}} = \frac{15 \times 10^3}{1 + 12.5 \times 10^{-6} \times 15 \times 10^3} = \frac{15 \times 10^3}{1 + 0.188} \cong 8\ \text{K}$$

Difference Amplifier Circuit with Constant-Current Source

One important thing to note in the previous circuit considerations was that with $R_{e_2} \ll R_E$, the value of R_E was very large and therefore negligible. In fact, the larger the value of R_E, the better certain desirable aspects of a difference amplifier circuit. The main reason for R_E being very large is a circuit factor called *common-mode rejection*, which will be discussed in detail in Section 11.3.

However, dc bias calculations showed that the emitter (and thus the collector) current is determined partly by the value of R_E. For a fixed negative-voltage supply of, say, $V_{EE} = -20$ V a value of R_E of 10 K would limit the emitter resistor current to about

$$I_E \cong \frac{V_{EE}}{R_E} = \frac{20\ \text{V}}{10\ \text{K}} = 2\ \text{mA}$$

If a preferably larger value of $R_E = 100$ K were used, the value of dc emitter resistor current would then be

$$I_E \cong \frac{V_{EE}}{R_E} = \frac{20\ \text{V}}{100\ \text{K}} = 0.2\ \text{mA} = 200\ \mu\text{A}$$

and if a very large value of $R_E = 1$ M were used,

$$I_E = \frac{20 \text{ V}}{1 \text{ M}} = 20 \ \mu\text{A}$$

We see that as larger values of R_E are used the resulting dc emitter current becomes much too small for proper operation of the transistors since the emitter and collector current of each transistor is half the already very small emitter resistor current.

One way to achieve high ac resistance while still allowing reasonable dc emitter currents is to use a constant-current source as shown in Fig. 11.16. The value of I_E could be set by the constant-current circuit to any desired value—1, 10, 20 mA, and so on. The ac resistance of a constant-current source is ideally infinite and practically from 100 K to about 1 M.

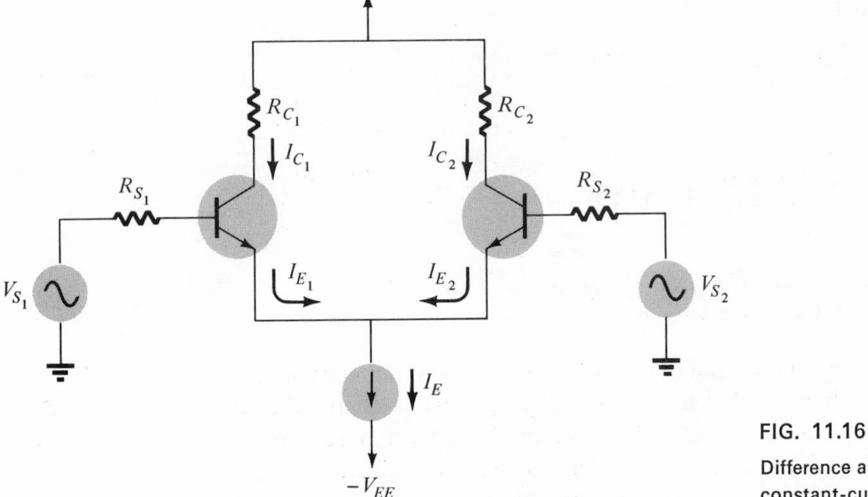

FIG. 11.16
Difference amplifier with constant-current source.

A practical difference amplifier circuit containing a constant-current source is shown in Fig. 11.17. To determine the dc currents and voltages let us first consider the details of the constant-current circuit.

DC OPERATION

The constant-current section of the difference amplifier is shown in Fig. 11.18a. No connection to the collector is shown since the collector current is determined by the value of emitter current set by the base-emitter section of the circuit. To a great degree the amount of

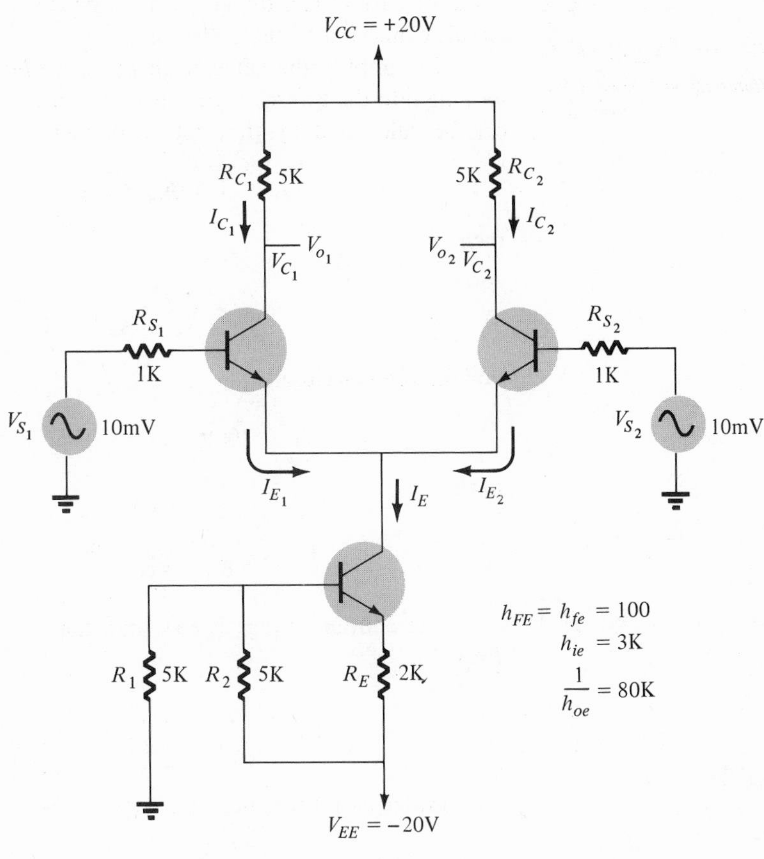

FIG. 11.17

Practical difference amplifier circuit with constant-current source.

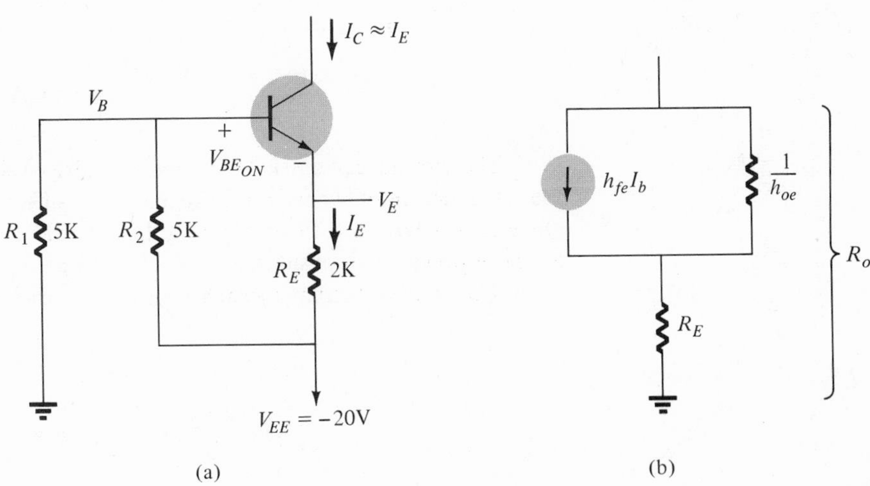

FIG. 11.18

Constant-current circuit: (a) dc considerations; (b) ac considerations.

collector current can be set at any desired value without regard to the circuit connected to the collector.

To simplify the calculation of I_E we note that if the resistance looking into the transistor base is much larger than R_2 the base voltage can be calculated by the voltage divider of R_1 and R_2. That is, if

$$(1 + h_{FE})R_E \gg R_2$$

then

$$V_B = \frac{R_1}{R_1 + R_2}(-V_{EE}) \tag{11.16}$$

For the present circuit

$$(1 + 100)(2\text{ K}) = 202\text{ K} \gg 5\text{ K}$$

so that

$$V_B = \frac{5\text{ K}}{5\text{ K} + 5\text{ K}}(-20) = -10\text{ V}$$

The emitter voltage is less than the base voltage by the voltage drop $V_{BE_{ON}}$

$$\begin{aligned} V_E &= V_B - V_{BE_{ON}} \\ &= -10 - (0.7) = -10.7\text{ V} \end{aligned} \tag{11.17}$$

The emitter current is then calculated to be

$$\begin{aligned} I_E &= \frac{V_E - V_{EE}}{R_E} \\ &= \frac{-10.7 - (-20)\text{ V}}{2\text{ K}} = 4.65\text{ mA} \end{aligned} \tag{11.18}$$

The emitter current is held constant fairly well and variations in the difference amplifier section have almost no effect on the value of I_E. Once I_E is determined the remaining dc bias calculations are the same as those previously considered.

The emitter current of each transistor is then

$$\begin{aligned} I_{E_1} = I_{E_2} &= \frac{I_E}{2} \\ &= \frac{4.65}{2}\text{ mA} \cong 2.3\text{ mA} \end{aligned} \tag{11.19}$$

and

$$\begin{aligned} I_{C_1} = I_{C_2} &\cong I_{E_1} = I_{E_2} \\ &= 2.3\text{ mA} \end{aligned} \tag{11.20}$$

The collector voltage is, as before,

$$V_C = V_{CC} - I_C R_C \tag{11.21}$$
$$= 20 - (2.3 \text{ mA})(5 \text{ K}) = 20 - 11.5 = 8.5 \text{ V}$$

AC OPERATION

The ac action of the constant-current source is that of a very high resistance—ideally infinite. An ac equivalent of the constant-current circuit is shown in Fig. 11.18b. From the equivalent circuit

$$R_o = 1/h_{oe} + R_E \cong 1/h_{oe} \tag{11.22}$$
$$= 80 \text{ K} + 2 \text{ K} \cong 80 \text{ K}$$

Thus, with as high a resistance as 80 K for ac operation, we still obtained a dc bias current near 5 mA.

IMPROVED CONSTANT-CURRENT CIRCUIT

An improved version of the constant-current circuit is shown in Fig. 11.19. A Zener diode is used to ensure that the current remains constant. The Zener diode conducts when the reverse-bias voltage exceeds the Zener breakdown voltage, V_Z. The Zener diode will then conduct keeping the voltage across the diode fixed at V_Z, for a wide range of current values. In the circuit, the emitter current is calculated from the voltage drops around the loop containing the Zener diode and emitter resistor

$$+V_Z - V_{BE_{ON}} - I_E R_E = 0$$
$$I_E = \frac{V_Z - V_{BE_{ON}}}{R_E} \tag{11.23}$$

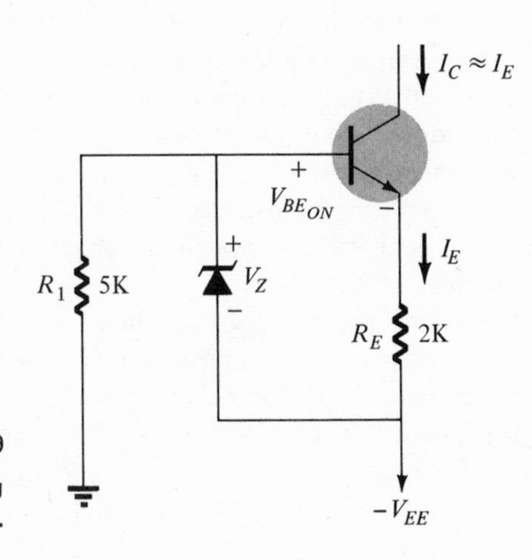

FIG. 11.19
Constant-current circuit using Zener diode.

For a Zener voltage of $V_Z = 10$ V, we calculate

$$V_E = -10.0 - 0.7 = -10.7 \text{ V}$$

and, as in the previous constant-current circuit,

$$I_E = \frac{V_{EE} - V_E}{R_E} = \frac{20 - 10.7}{2 \text{ K}} = 4.65 \text{ mA}$$

All other dc calculations are the same.

We can thus bias the circuit with either bias resistors or Zener diode. The improvement with the Zener diode comes from the independence of I_E from all other circuit factors except V_Z and R_E. From Eq. (11.23) it can be seen that changes in V_{CC}, V_{EE}, or V_S, and so on, have no effect on I_E.

11.3 COMMON-MODE REJECTION

One of the more important features of a difference amplifier is its ability to cancel out or reject certain types of unwanted voltage signals. These unwanted signals are referred to as "noise" and can occur as voltages induced by stray magnetic fields in the ground or signal wires, as voltage variations in the voltage supply, and so on. What is important in this consideration is that these noise signals are not the signals that are desired to be amplified in the difference amplifier. Their distinguishing feature is that the noise signal appears equally at both inputs of the circuit.

We can say then that any unwanted (noise) signals that appear in phase (are common) to both input terminals will be greatly rejected (cancelled out) at the output of the difference amplifier. The signal that is to be amplified appears at only one input or opposite in phase at both inputs. What we wish to consider in this section is, if undesirable noise does occur (is picked up), how much can the amplifier reject or cancel out this noise? A measure of this rejection of signals common to both inputs is called the amplifier's *common-mode rejection* and a numerical value is assigned, which is called the *common-mode rejection ratio* (CMRR).

Figure 11.20a shows an amplifier with two input signals. These signals can, in general, be considered to contain components that are exactly opposite in phase *and* components that are exactly in phase. For ideal operation we would want the difference amplifier to provide high gain for the out-of-phase components of the signals and zero gain for the in-phase components of the signals.

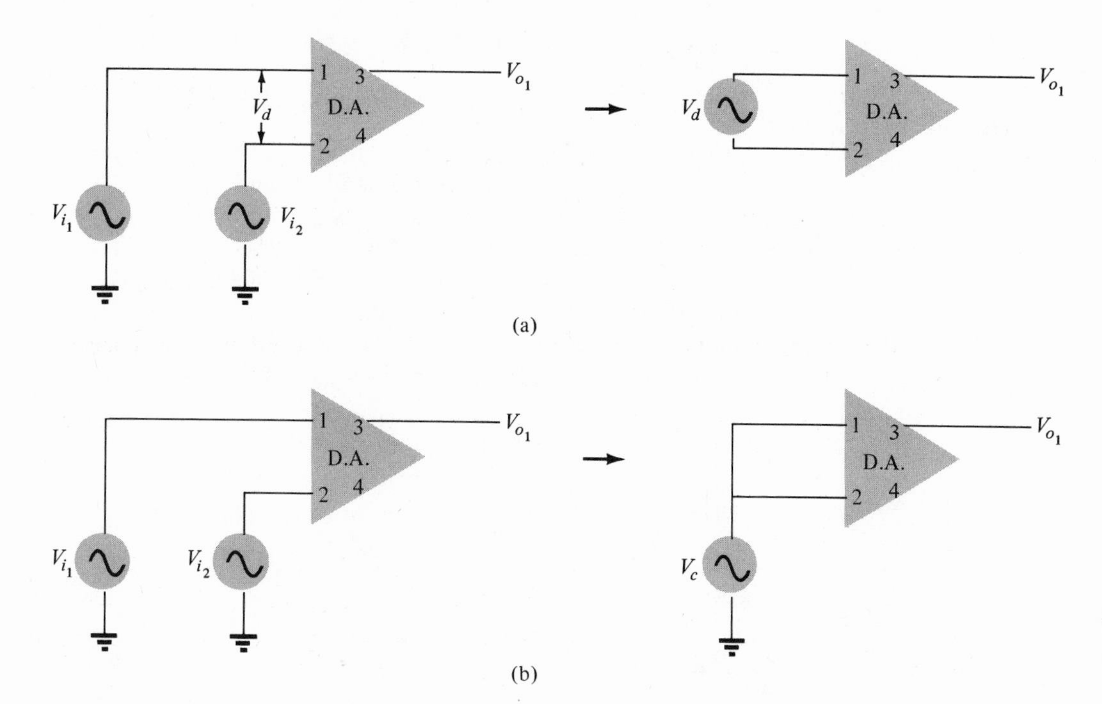

FIG. 11.20
Difference and common operation: (a) ideal difference-mode operation; (b) ideal common-mode operation.

The voltage measured *from* terminal 1 *to* terminal 2 can be considered as a difference voltage

$$\boxed{V_d = V_{i_1} - V_{i_2}} \qquad (11.24)$$

If, as in the ideal case, $V_{i_1} = -V_{i_2}$, we note that

$$V_d = V_{i_1} - (-V_{i_1}) = 2V_{i_1} = -2V_{i_2}$$

In general, there may also exist common components to the input signals. We can define a common input as

$$\boxed{V_c = \tfrac{1}{2}(V_{i_1} + V_{i_2})} \qquad (11.25)$$

The ideal case (shown in Fig. 11.20b) is $V_{i_1} = V_{i_2}$, for which

$$V_c = \tfrac{1}{2}(V_{i_1} + V_{i_2}) = V_{i_1} = V_{i_2}$$

The output voltage could be expressed as

$$V_{o_1} = A_1 V_{i_1} + A_2 V_{i_2} \tag{11.26}$$

where A_1 = gain from input terminal 1 to output terminal 3 (with input terminal 2 grounded), and
A_2 = gain from input terminal 2 to output terminal 3 (with input terminal 1 grounded).

It is more important to consider the difference and common-mode operation of the amplifier since this allows determining the common-mode rejection of the circuit. This second way of considering the operation of the amplifier provides an output voltage as

$$\boxed{V_{o_1} = A_d V_d + A_c V_c} \tag{11.27}$$

where A_d = difference-mode gain of the amplifier,
A_c = common-mode gain of the amplifier, and
V_d and V_c are defined in Eqs. (11.24) and (11.25), respectively.

Opposite-Phase Inputs: If the inputs are equal and opposite, $V_{i_1} = V_s$ and $V_{i_2} = -V_s$, then from Eq. (11.24)

$$V_d = V_{i_1} - V_{i_2} = V_s - (-V_s) = 2V_s$$

and from Eq. (11.25)

$$V_c = \tfrac{1}{2}(V_{i_1} + V_{i_2}) = \tfrac{1}{2}[V_s + (-V_s)] = 0$$

so that in Eq. (11.27)

$$V_{o_1} = A_d V_d + A_c V_c = A_d(2V_s) + A_c(0)$$
$$V_{o_1} = 2A_d V_s$$

which shows that only differential-mode operation occurs (and that the overall gain is twice the value of A_d).

In-Phase Inputs: If the inputs are equal and in phase, $V_{i_1} = V_s = V_{i_2}$, then from Eq. (11.24)

$$V_d = V_{i_1} - V_{i_2} = V_s - V_s = 0$$

and from Eq. (11.25)

$$V_d = \tfrac{1}{2}(V_{i_1} + V_{i_2}) = \tfrac{1}{2}(V_s + V_s) = V_s$$

so that in Eq. (11.27)

$$V_{o_1} = A_d V_d + A_c V_c = A_d(0) + A_c V_s$$
$$= A_c V_s$$

which shows that only common-mode operation occurs.

The above calculations indicate how A_d and A_c can be measured in a difference amplifier circuit.

To measure A_d: Set $V_{i_1} = -V_{i_2} = 0.5$ V so that $V_d = 1$ V and $V_c = 0$ V. Under these conditions the output voltage is $A_d \times (1)$ so that the output voltage equals A_d.

To measure A_c: Set $V_{i_1} = V_{i_2} = 0.5$ V so that $V_d = 0$ V and $V_c = 1$ V. Then the output voltage measured equals A_c.

Having measured A_d and A_c for the amplifier we can now calculate a common-mode rejection ratio, which is defined as

$$\boxed{\text{CMRR} = \frac{A_d}{A_c}} \tag{11.28}$$

It should be clear that the desired operation will have A_d very large with A_c very small. That is, the signals appearing opposite in phase will appear greatly amplified at the output terminal, whereas the in-phase signals will mostly cancel out so that the common-mode gain A_c is very small. Ideally, A_d is very large and A_c is zero so that the value of CMRR is infinite. The larger the value of CMRR the better the common-mode rejection of the circuit.

It is possible to obtain an expression for the output voltage as follows:

$$\boxed{V_o = A_d V_d\left(1 + \frac{1}{\text{CMRR}}\frac{V_c}{V_d}\right)} \tag{11.29}$$

Even if both V_c and V_d components of voltage exist at the inputs the value of $(1/\text{CMRR})(V_c/V_d)$ will be very small, for CMRR very large, and the output voltage will be approximately $A_d V_d$. In other words the output will be almost completely due to the difference signal with the common-mode input signals rejected (or cancelled out). Some practical examples should help clarify these ideas.

EXAMPLE 11.2 Determine the output voltage of a difference amplifier for input voltages of $V_{i_1} = 150\ \mu$V and $V_{i_2} = 100\ \mu$V. The amplifier has a difference-mode gain of $A_d = 1000$ and the value of CMRR is

(a) 100

(b) 10^5

Solution

$$V_d = V_{i_1} - V_{i_2} = 150 - 100 = 50\ \mu\text{V}$$

$$V_c = \tfrac{1}{2}(V_{i_1} + V_{i_2}) = \frac{150 + 100}{2} = 125\ \mu\text{V}$$

Note that the common signal is more than twice as large as the difference signal.

(a)

$$V_o = A_d V_d \left(1 + \frac{1}{\text{CMRR}} \frac{V_c}{V_d}\right)$$
$$= A_d V_d \left(1 + \frac{1}{100} \times \frac{125}{50}\right)$$
$$= A_d V_d (1.025)$$
$$= (1000)(50\ \mu\text{V})(1.025) = \mathbf{51.25\ mV}$$

The output is only 0.025 or 2.5% more than the output, due only to a difference signal of 50 μV.

The common-mode signal, even larger than the difference component, has been rejected so that only 1.25 mV appear as output.

(b)

$$V_o = A_d V_d \left(1 + \frac{1}{10^5} \times \frac{125}{50}\right)$$
$$= A_d V_d (1.00025)$$
$$\cong 1000 \times 50\ \mu\text{V} = \mathbf{50\ mV}$$

The output in this case is larger than that due to only the difference signal by 0.025%.

Example 11.2 shows that the larger the value of CMRR the better the circuit rejects common-input signals. Thus, one of the important difference amplifier factors to consider is the circuit's common-mode rejection ratio.

11.4 PRACTICAL DIFFERENTIAL AMPLIFIER UNITS—IC CIRCUITS

Differential amplifiers are quite versatile and useful in many areas of electronic operation. They are some of the most widely used linear IC devices. Since it is often easier, cheaper, and thus more desirable to use an IC circuit than to build an equivalent circuit using discrete components, we shall consider some typical IC units in this section.

As an example, Fig. 11.21 shows the schematic and block diagrams of an RCA CA3000 IC differential amplifier. The manufacturer lists some of the possible uses of this unit as communications, telemetry, instrumentation, and data processing. Some of the specific applications include RC-coupled feedback amplifier, crystal oscillator, sense amplifier, comparator, and modulator. The manufacturer also lists a number of specifications for the unit, including

Input impedance: 195 K, typical
Voltage gain: 37 dB, typical
CMRR: 98 dB, typical
Frequency capability: dc to 30 MHz
Push-pull input and output

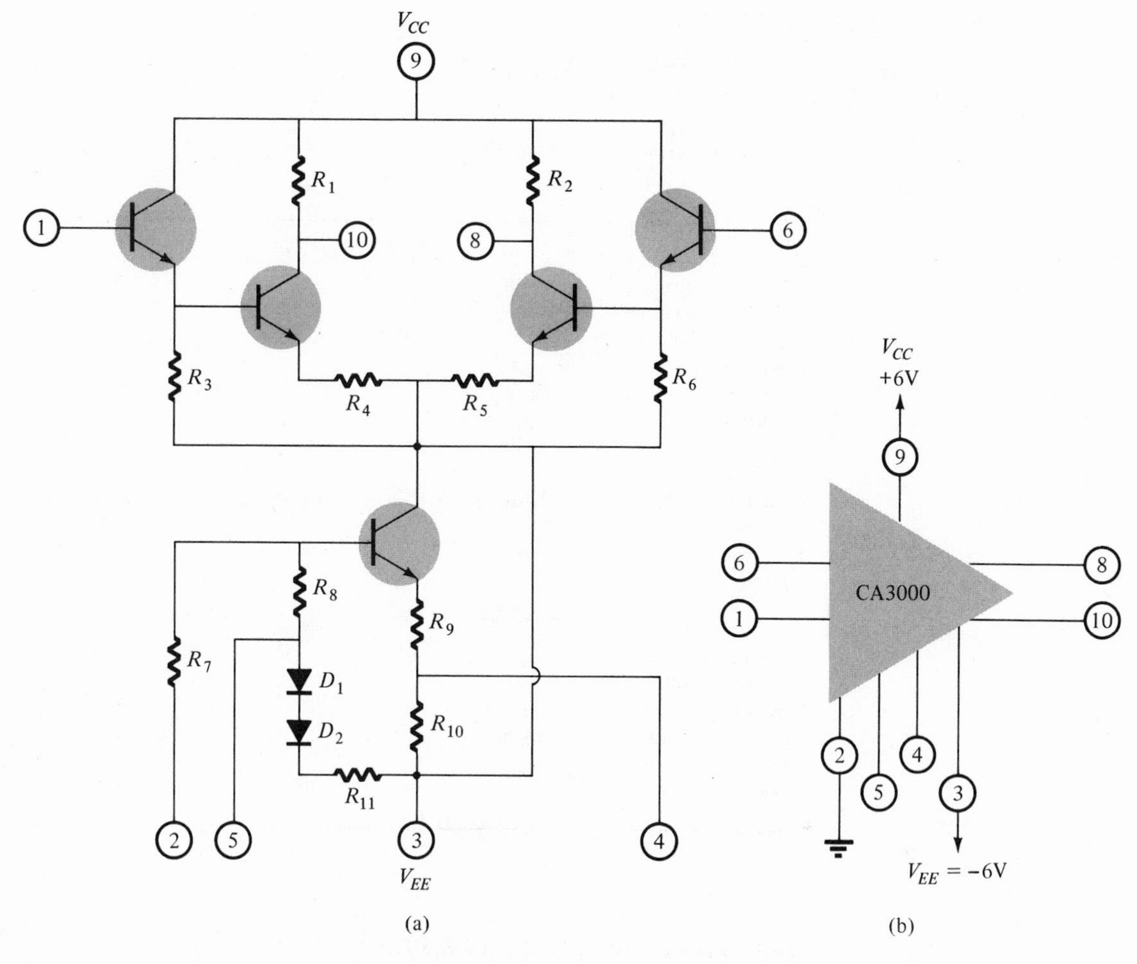

FIG. 11.21
IC differential amplifier: (a) schematic diagram; (b) block diagram.

Notice how much more complex the circuit appears. However, we need not consider the details of the circuit at all in order to use it. The external features that will be considered next are sufficient to allow using the amplifier. It is helpful to consider the operation of a basic version of the circuit so that the external operation of the unit makes some sense. The additional complexity in most IC units improves certain of these circuit features but the basic operation remains the same.

Table 11.1 is part of the manufacturer's listing of electrical characteristics and definitions of terms. The following discussion will elaborate on a number of the more important characteristics and give some examples.

TABLE 11.1

Electrical Characteristics, at T_{FA} = 25°C, V_{CC} = +6 V, V_{EE} = −6 V, f = 1 KHz

CHARACTERISTICS	SYMBOLS	LIMITS			UNITS
		Min.	*Typ.*	*Max.*	
Dynamic Characteristics					
Differential voltage gain, single-ended input	A_{diff}	28	32		dB
Single-ended input impedance	Z_i	70 K	195 K		Ω
Single-ended output impedance	Z_o	5.5 K	8 K	10.5 K	Ω
Common-mode rejection ratio	CMR	80	98		dB
Maximum output voltage swing	V_o(p–p)		6.4		V (p–p)
Bandwidth at −3-dB point	BW		650		kHz
Static Characteristics					
Input offset voltage	V_{IO}		1.4	8	mV
Input offset current	I_{IO}		1.2	10	μA
Quiescent operating voltage	V_8 or V_{10}		2.6		V
Device dissipation	P_T		30		mW

DEFINITIONS OF TERMS FOR CA3000

INPUT OFFSET VOLTAGE: The difference in the dc voltages that must be applied to the input terminals to obtain equal quiescent operating voltages (zero-output offset voltage) at the output terminals

INPUT OFFSET CURRENT: The difference in the currents at the two input terminals

QUIESCENT OPERATING VOLTAGE: The dc voltage at either output terminal, with respect to ground

DC DEVICE DISSIPATION: The total power drain of the device with no signal applied and no external load current

COMMON-MODE VOLTAGE GAIN: The ratio of the signal voltages developed between the two output terminals to the signal voltage applied to the two input terminals connected in parallel for ac

DIFFERENTIAL VOLTAGE GAIN—SINGLE-ENDED INPUT-OUTPUT: The ratio of the change in output voltage at either output terminal with respect to ground, to a change in input voltage at either input terminal with respect to ground

COMMON-MODE REJECTION RATIO: The ratio of the full differential voltage gain to the common-mode voltage gain

BANDWIDTH AT—3-dB POINT (BW): The frequency at which the voltage gain of the device is 3 dB below the voltage gain at a specified lower frequency

MAXIMUM OUTPUT VOLTAGE V_o(p–p): The maximum peak-to-peak output voltage swing, measured with respect to ground, that can be achieved without clipping of the signal waveform

SINGLE-ENDED INPUT IMPEDANCE (Z_i): The ratio of the change in input voltage to the change in input current measured at either input terminal with respect to ground

SINGLE-ENDED OUTPUT IMPEDANCE (Z_o): The ratio of the change in output voltage to the change in output current measured at either output terminal with respect to ground

Differential Voltage Gain—Single-Ended Input-Output

The typical value of 32 dB is the gain from one input terminal to either output terminal. This was considered the gain A_1 or A_2 in Sections 11.1–11.3. The manufacturer lists the gain in units of decibels (dB). The relation of decibels and the gain as numerical ratio of output voltage (V_o) to input voltage (V_i) is

$$A_{\text{dB}} = 20 \log A_v = 20 \log \frac{V_o}{V_i} \tag{11.30}$$

As an example, a gain of $A_v = 100$ is the same as

$$A_{\text{dB}} = 20 \log 100 = 20\,(2) = 40 \text{ dB}$$

and a gain of $A_v = 10$ is the same as

$$A_{\text{dB}} = 20 \log 10 = 20\,(1) = 20 \text{ dB}$$

A gain of 32 dB is then the same as a gain between 10 and 100 and can be calculated exactly as

$$32 = 20 \log A_v$$

$$1.6 = \log A_v$$

$$A_v = \text{antilog } 1.6 \cong 2 \times 10 = 20$$

The input impedance is measured at either input terminal. A listed value of 195 K indicates a relatively high value. It should be recalled from Chapter 7 how important input impedance values are when interconnecting amplifier stages or driving the amplifier with a practical voltage source. If the input impedance is not much larger than the source impedance loading will cause the input voltage to be less than that of the unloaded source signal, resulting in less output voltage.

Figure 11.22 shows how the input impedance can be measured. As shown, a voltage (V_A) is applied through a resistor of fixed value (20 K), and due to the amplifier input impedance, the voltage, V_B, is less than V_A. Supply voltages are properly connected and the second input is grounded through a matching 20-K resistor. Output voltage is present although it is of no concern for the present measurements. Note that although an impedance calculation would require measurements of input voltage and current, the present method avoids having to measure current. From the equivalent circuit showing the input as an impedance Z_i we can express the voltage V_B from the voltage divider rule as

$$V_B = \frac{Z_i}{Z_i + 20\text{ K}} V_A \tag{11.31}$$

Solving for Z_i

$$Z_i = \frac{20\text{ K} \times V_B}{V_A - V_B} = \frac{20\text{ K}}{(V_A/V_B) - 1}$$

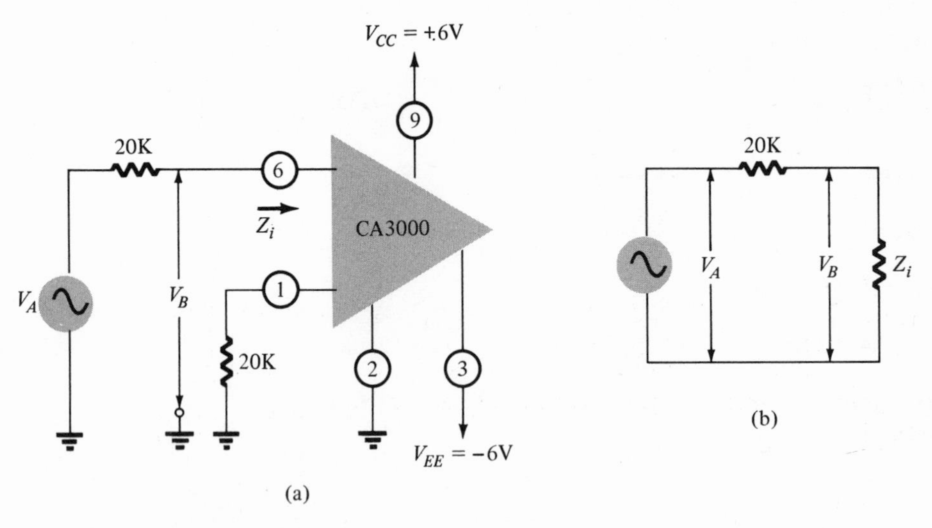

FIG. 11.22
Measurement of input impedance (Z_i).

From the manufacturer's measurements a typical value of $Z_i =$ 195 K was obtained. Since this value is only typical it might be desired to repeat the measurement for the particular unit on hand. If, then, the measurements gave $V_A = 1$ V and $V_B = 0.90$ V, then

$$Z_i = \frac{R_x}{(V_A/V_B) - 1} = \frac{20 \text{ K}}{(1/0.9) - 1} = \frac{20 \text{ K}}{1.11 - 1} = \frac{2\text{K}0}{0.11} = 198 \text{ K}$$

A more direct measurement can be obtained by using a variable resistance (R_x) in place of the fixed 20-K resistor. If this resistance is adjusted until $V_B = \frac{1}{2}V_A$ we note that

$$Z_i = \frac{R_x}{(V_A/V_B) - 1} = \frac{R_x}{(V_A/\frac{1}{2}V_A) - 1} = R_x$$

That is, the value of R_x is then the same value as Z_i. If R_x is a calibrated resistor box then the value of Z_i can be read directly with no additional calculation required. Thus, with the amplifier properly biased and operating, the use of a single variable resistor unit, and a voltmeter to measure V_A and V_B, the value of the input impedance is directly obtained.

Single-Ended Output Impedance (Z_o)

The output impedance of either output terminal with respect to ground is listed at, typically, 8 K. If a measurement is to be made on a particular unit, the technique shown in Fig. 11.23 can be used. First the output voltage, unloaded, is measured. Then the switch is closed, connecting R_o as a load on the amplifier and R_o is adjusted until the output voltage is $V_o/2$. The value of R_o is then equal to the output impedance of the amplifier ($Z_o = R_o$). Note that if CA3000s are connected in cascade (series) the loading of one stage on another is quite small; that is, if a source voltage of V_o and source resistance of $Z_o = 8$ K from the output of one stage is connected as input to a second stage ($Z_i = 195$ K) the resulting loaded input signal will be

$$V_i = \frac{Z_i}{Z_i + Z_o}(V_o) = \frac{195}{195 + 8}(V_o) = 0.96\, V_o$$

so that almost no loading takes place.

Common-Mode Rejection Ratio (CMRR)

The common-mode rejection ratio defined as

$$\text{CMRR} = \frac{A_d}{A_c}$$

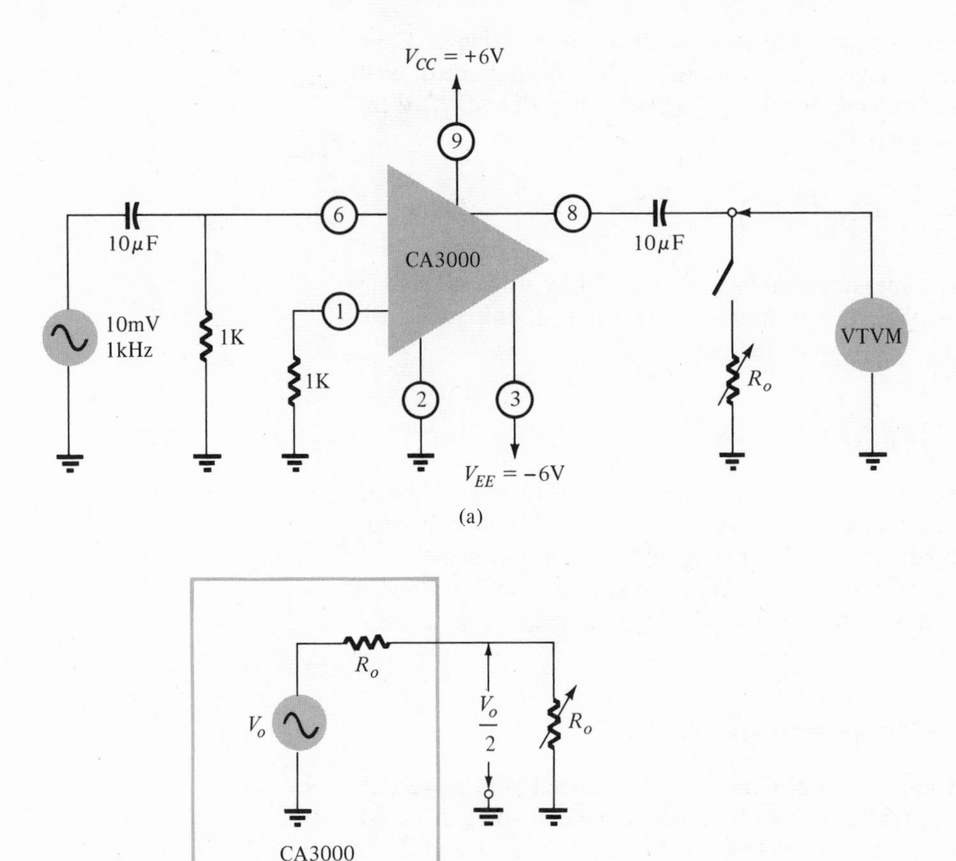

FIG. 11.23
Measurement of output impedance.

may also be calculated in decibel units as

$$\text{CMRR} = 20 \log \frac{(2A_1)}{V_o/V_i}$$

where V_i and V_o are measured in the circuit of Fig. 11.24 and A_1 was previously measured as the single-ended voltage gain. The ratio of V_o/V_i is the value of A_c, the gain of the amplifier with common-input signal (ideally zero). If the measurements are $V_i = 0.3$ V and $V_o = 0.12$ mV, and if A_1 previously was 20, the value of CMRR is calculated to be

$$\text{CMRR} = 20 \log = \frac{2A_1}{V_o/V_i} = 20 \log \frac{2(20)}{0.12\ \text{m}V/0.3\ \text{m}V}$$

$$= 20 \log \frac{40}{0.4 \times 10^{-3}} = 20 \log 10^5 = 20(5) = 100\ \text{dB}$$

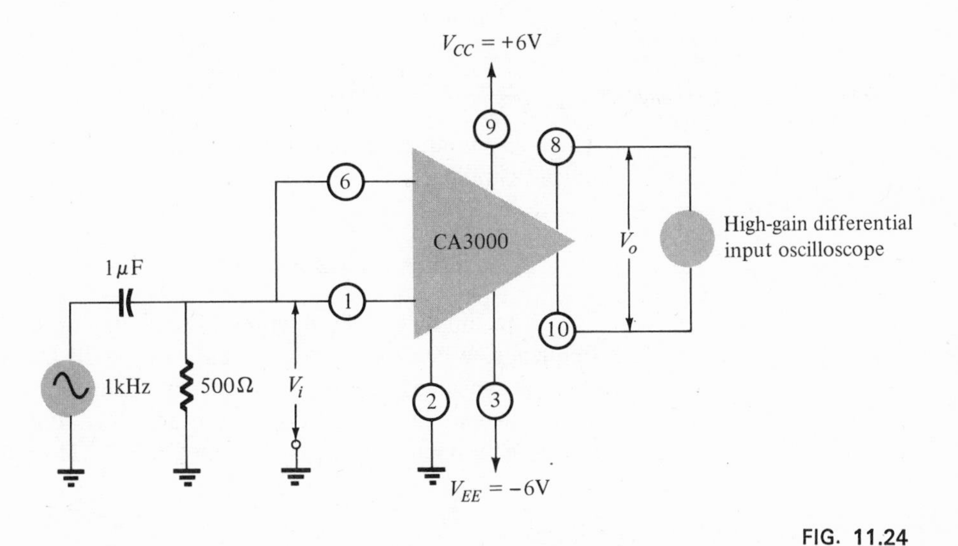

FIG. 11.24
Measurement of common-mode gain (A_c).

Notice that the value of A_c is 0.4×10^{-3}, which is quite small—the smaller the better.

Maximum Output Voltage Swing, V_o(p-p)

The voltage gain of the amplifier is meaningful only within the linear operating range of the amplifier. Within the linear range the output is an amplified representation of the input signal. It should be clear that the overall voltage swing at the amplifier output is limited at least by the power-supply voltages. The manufacturer lists the maximum output voltage swing as V_o(p-p) = 6.4 V. Thus, the output can vary from the quiescent operating point by a peak-to-peak swing of 6.4 V, at most.

Referring to the static characteristics, the typical quiescent operating point is + 2.6 V. If the output voltage is then caused to rise by 3.2 V due to the input signal the output will rise to + 5.8 V, which is about the largest value of supply voltage. If the output also varies by 3.2 V down from the bias point, it goes to −0.6 V, which is about the value of the emitter voltage of the amplifier. So we see that the overall swing of 6.4 V (p-p) can take place but that any larger voltage swing will result in clipping of the output signal and thereby large-signal distortion.

If we use the typical gain of $A_1 = 20$ we can determine the largest input signal (peak to peak) that can be applied without causing the output to be clipped. Since

$$V_o = A_1 V_i$$

$$V_i = \frac{V_o}{A_1} = \frac{6.4\text{ V(p-p)}}{20} = 0.32\text{ V(p-p)}$$

Thus, any input larger than 0.32 V or 320 mV (p-p) will result in the output clipping.

Bandwidth at −3-dB Point (BW)

The amplifier can operate from 0 Hz, or dc, to some upper frequency. When the gain has dropped by 3 dB the frequency at that point is considered the upper frequency of the amplifier and is, in this case, equal to the bandwidth of the amplifier. Figure 11.25 shows a typical gain-bandwidth curve for the CA3000. The gain has been nor-

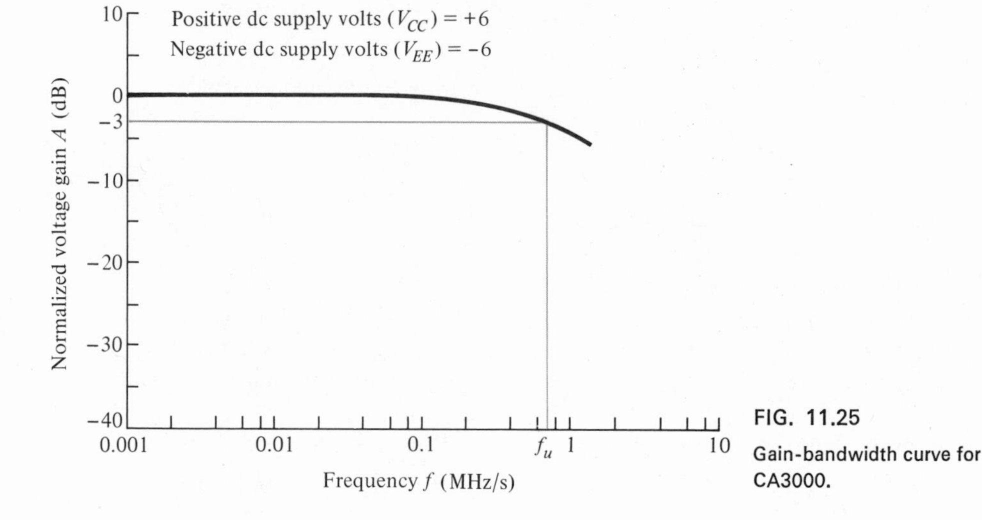

FIG. 11.25 Gain-bandwidth curve for CA3000.

malized so that the gain of 32 dB is now 0 dB. When the gain drops by 3 to 29 dB, or on the curve to −3 dB, then the upper frequency (f_u) can be read (not too easily from this graph) as 650 kHz, so that the value of bandwidth is BW = 650 kHz. The frequency is plotted on a logarithmic scale because of its very wide range.

Input Offset Voltage (V_{IO})

The input offset voltage (V_{IO}) provides a measure of how much the input voltage must be offset from 0 V to result in both output voltages being biased at exactly the same voltage point. Figure 11.26

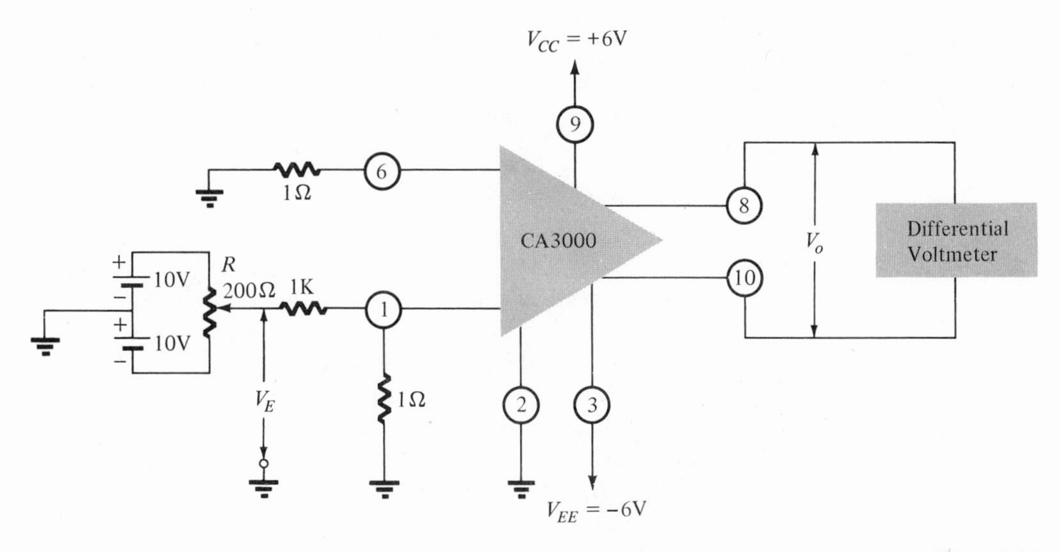

FIG. 11.26

Test circuit to measure V_{IO}: (1) adjust R for $V_o(\text{dc}) = V0$; (2) measure V_E and record V_{IO} in millivolts.

shows a test circuit to apply an offset voltage until the output voltages are exactly the same. Notice that the typical value is only 1.4 mV so that reasonably matched circuits can be expected in IC form with the components being made in the same process. A voltage divider of 1 Ω and 1 K provides attenuation of 1000 so that the actual voltage applied to terminal 1 is 1000 times smaller than V_E, or is in millivolts.

11.5 BASICS OF OPERATIONAL AMPLIFIERS (OPAMP)

An operational amplifier is a very-high-gain differential amplifier that uses voltage feedback to provide a stabilized voltage gain. The basic amplifier used is essentially a difference amplifier having very high open-loop gain (no signal feedback condition) as well as high input impedance and low output impedance. Typical uses of the operational amplifier (OPAMP) are scale changing; analog computer operations, such as addition and integration; and a great variety of phase shift, oscillator and instrumentation circuits.

Figure 11.27 shows an OPAMP unit having two inputs and a single output. Recalling how the two inputs affect an output in a difference amplifier the inputs are marked here with *plus* (+) and *minus* (−) to indicate noninverting and inverting inputs, respectively. A

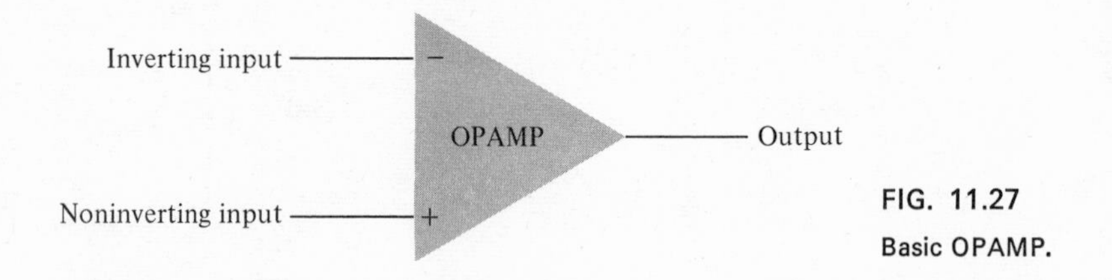

FIG. 11.27
Basic OPAMP.

signal applied to the *plus* input will appear in phase and amplified at the output, whereas an input applied to the *minus* (−) terminal will appear amplified but inverted at the output.

The basic circuit connection of an operational amplifier is shown in Fig. 11.28a. As shown the circuit operates as a scale changer or constant-gain multiplier. An input signal V_1 is applied through a resistor R_1 to the minus input terminal. The output voltage is fed back, through resistor R_f to the same input terminal. The plus input terminal is connected to ground. We now wish to determine the overall gain of the circuit (V_o/V_1). To do this we must consider some more details of the OPAMP unit.

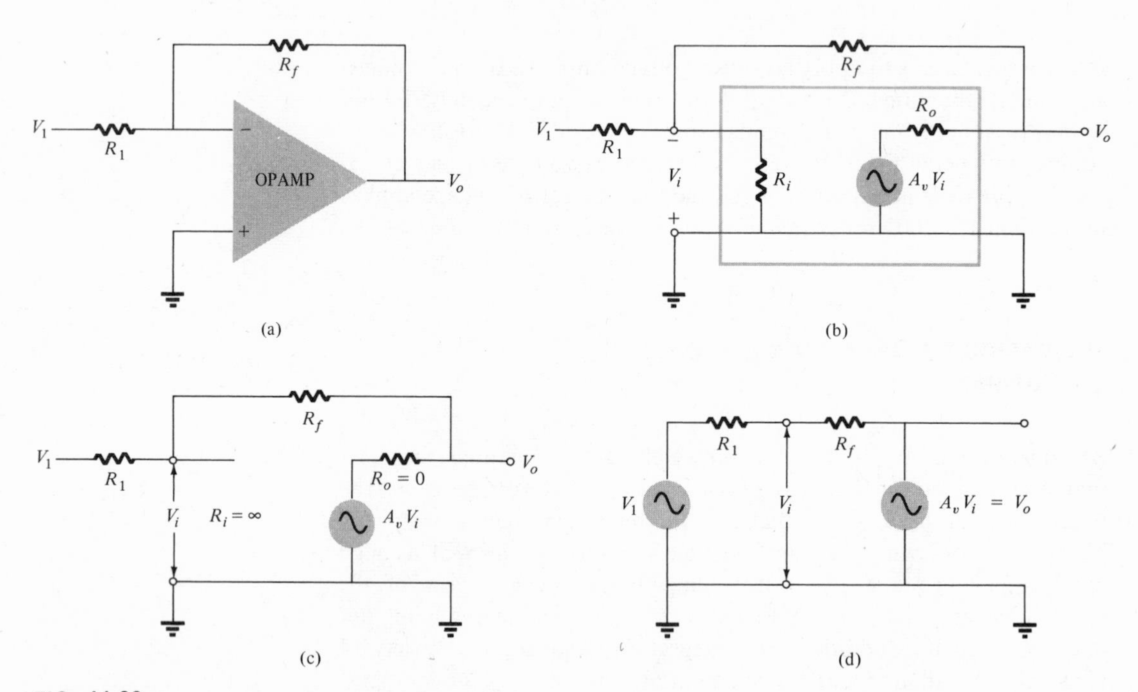

FIG. 11.28
Operation of OPAMP as scale changer: (a) basic connection (constant-gain multiplier); (b) effect of OPAMP circuit; (c) ideal OPAMP; (d) ideal equivalent circuit.

Figure 11.28b shows the OPAMP replaced by an equivalent circuit of input resistance R_i, and output voltage source and resistance. An ideal OPAMP as shown in Fig. 11.28c has infinite resistance ($R_i = \infty$), zero output resistance ($R_o = 0$), and very high voltage gain ($A_v \gg 1$). The connection for the ideal amplifier is shown redrawn in Fig. 11.28d. Using the resulting equivalent circuit we can determine the overall gain of the circuit.

Using superposition we can solve for the voltage V_i, in terms of the components due to each of the sources. For source V_1 only ($-A_v V_i$ set to zero)

$$V_{i_1} = \frac{R_f}{R_1 + R_f} V_1$$

For source $-A_v V_i$ only (V_1 set at zero)

$$V_{i_2} = \frac{R_1}{R_1 + R_f}(-A_v V_i)$$

The total voltage of V_i is then

$$V_i = V_{i_1} + V_{i_2} = \frac{R_f}{R_1 + R_f} V_1 + \frac{R_1}{R_1 + R_f}(-A_v V_i)$$

which can be solved for V_i as

$$V_i = \frac{R_f}{R_f + (1 + A_v)R_1} V_1$$

if $A_v \gg 1$ and $A_v R_1 \gg R_f$ as is usually true, then

$$V_i \cong \frac{R_f}{A_v R_1} V_1$$

Solving for V_o/V_1:

$$\frac{V_o}{V_1} = \frac{-A_v V_i}{V_1} = \frac{-A_v}{V_1}\left(\frac{R_f V_1}{A_v R_1}\right) = -\frac{R_f}{R_1}$$

$$\boxed{\frac{V_o}{V_1} = -\frac{R_f}{R_1}} \tag{11.32}$$

The result shows that the ratio of overall output to input voltage is dependent only on the values of resistors R_1 and R_f—provided that A_v is very large.

If $R_f = R_1$ the gain is

$$A_v = -\frac{R_1}{R_1} = -1$$

and the circuit provides a sign change with no magnitude change.

If $R_f = 2R_1$, then

$$A_v = \frac{-2R_1}{R_1} = -2$$

and the circuit provides a gain of 2 along with 180° phase inversion of the input signal.

Selecting precise resistor values for R_f and R_1, a wide range of gains can be obtained, the gain being as accurate as the resistors used and quite independent of temperature and other circuit factors.

VIRTUAL GROUND

The output voltage is limited by the supply voltage of, typically, a few volts. Voltage gains as stated before are very high. If, for example, $V_o = -10$ V and $A_v = 10{,}000$ the input voltage is

$$V_i = -\frac{V_o}{A_v} = -\frac{(-10)}{10{,}000} = 1 \text{ mV}$$

If the circuit had an overall gain (V_o/V_1) of, say, 1, the value of V_1 would be 10 V. The value of V_i, compared to all the other voltages, is then quite small, and may be considered 0 V. Note that although $V_i \cong 0$ V it is not exactly 0 V, since the output is the value of V_i times the gain of the amplifier $(-A_v)$.

The fact that $V_i \cong 0$ V leads to the concept that at the input to the amplifier there exists a virtual short circuit or *virtual ground.* The concept of a virtual short implies that although the voltage is nearly 0 V, no current flows through the amplifier input to ground. Figure 11.29 depicts the virtual ground concept. The heavy line is used

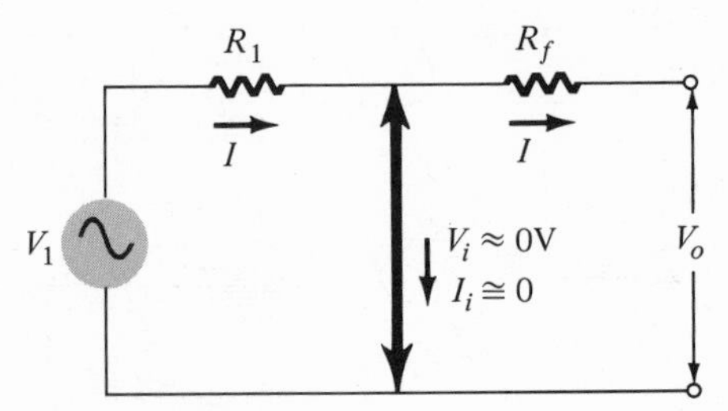

FIG. 11.29
Virtual ground in an OPAMP.

to indicate that we may consider that a short exists with $V_i \cong 0$ V, but that this is a virtual short in that no current flows through the short to ground. Current flows through resistor R_1 and through R_f as shown.

If we use the virtual ground concept we can write equations for the current I as follows:

$$I = \frac{V_1}{R_1} = -\frac{V_o}{R_f}$$

which can be solved for V_o/V_1

$$\frac{V_o}{V_1} = -\frac{R_f}{R_1}$$

The virtual ground concept, which depended on A_v being very large, allowed simple solution of overall voltage gain. It should be understood that although the circuit of Fig. 11.29 is not a physical circuit it does allow an easy means for solving the overall circuit gain.

11.6 OPAMP CIRCUITS

Constant-Gain Multiplier

An inverting constant-multiplier circuit has already been considered but is repeated here to provide a fuller listing of basic OPAMP circuits. Figure 11.30 shows an inverting constant-gain-multiplier circuit for which we know

$$\frac{V_o}{V_1} = -\frac{R_f}{R_1} \tag{11.33}$$

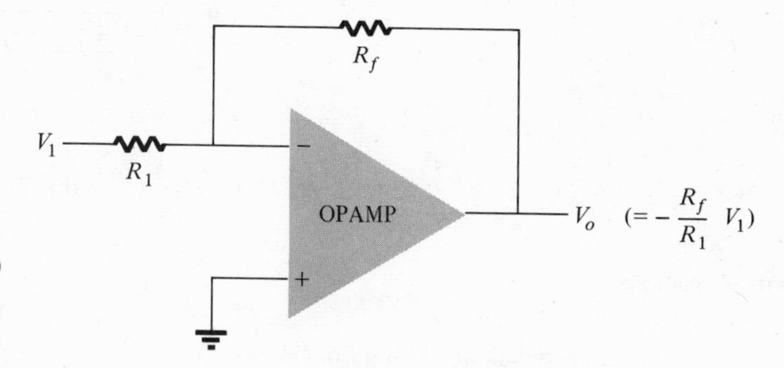

FIG. 11.30
Inverting constant-gain multiplier.

EXAMPLE 11.3 The circuit of Fig. 11.30 has $R_1 = 100$ K and $R_f = 500$ K. What is the output voltage for an input of $V_1 = -2$ V?

Solution: Using Eq. (11.23)

$$V_0 = -\frac{R_f}{R_1}V_1 = -\frac{500 \text{ K}}{100 \text{ K}}(-2) = +10 \text{ V}$$

Noninverting Amplifier

The connection of Fig. 11.31 shows an OPAMP circuit that works as a noninverted constant-gain multiplier. To determine the voltage gain of the circuit we can use the equivalent virtual ground representation in Fig. 11.31b. Note that the voltage across R_1 is V_1, since $V_i \cong 0$ V. This must be equal to the voltage due to the output, V_o, through a voltage divider of R_1 and R_f so that

$$V_1 = \frac{R_1}{R_1 + R_f} V_o$$

and

$$\frac{V_o}{V_1} = \frac{R_1 + R_f}{R_1} = 1 + \frac{R_f}{R_1} \tag{11.34}$$

EXAMPLE 11.4 Calculate the output voltage of a noninverting constant-gain multiplier (as in Fig. 11.31) for values of $V_1 = 2$ V, $R_f = 500$ K, and $R_1 = 100$ K.

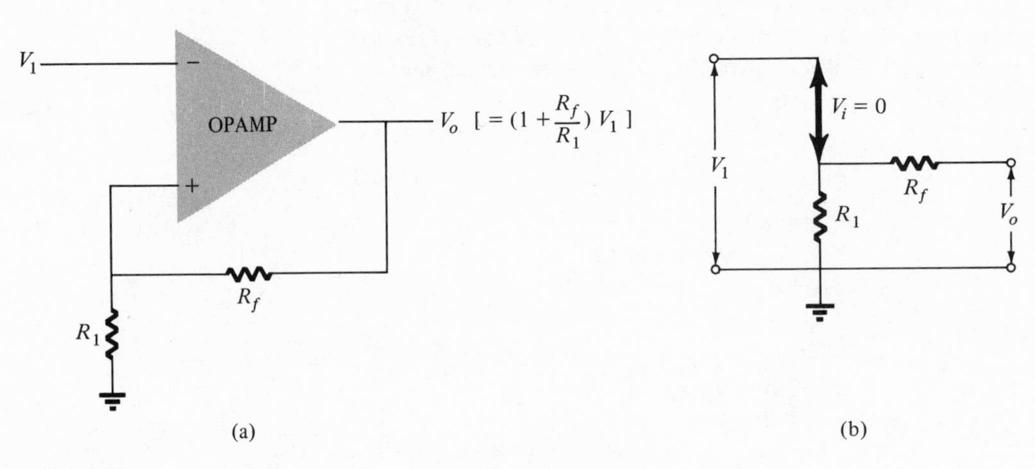

FIG. 11.31
Noninverting constant-gain multiplier.

Solution: Using Eq. (11.34)

$$V_0 = \left(1 + \frac{R_f}{R_1}\right)V_1 = \left(1 + \frac{500 \text{ K}}{100 \text{ K}}\right)(2 \text{ V}) = 6(2) = +12 \text{ V}$$

Unity Follower

The unity follower, as in Fig. 11.32 provides a gain of 1 with no phase reversal. From the equivalent circuit with virtual ground it is clear that

$$V_o = V_1 \tag{11.35}$$

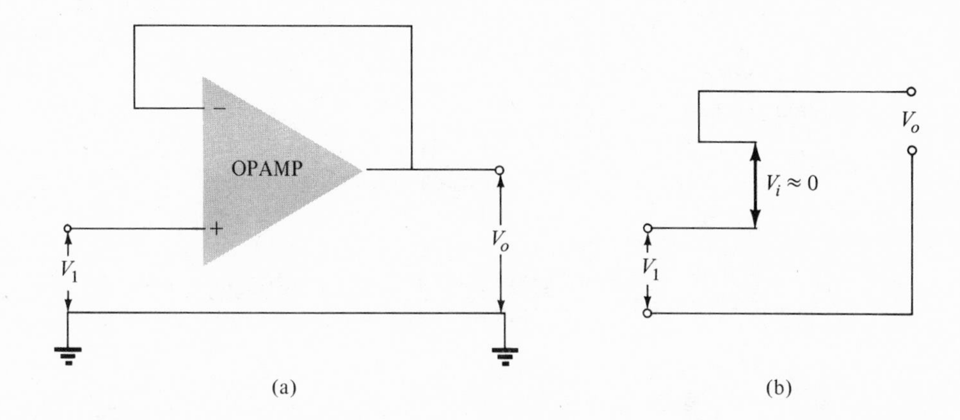

FIG. 11.32
(a) Unity follower; (b) virtual ground equivalent circuit.

and that the output is the same polarity and magnitude as the input. The circuit acts very much like an emitter follower except that the gain is very much closer to being exactly unity.

Summer

Probably the most useful of the OPAMP circuits used in analog computers is the summer circuit. Figure 11.33 shows a three-input summing circuit, which provides a means of algebraically summing (adding) three-input voltages, each multiplied by a constant-gain factor.

Using the virtual equivalent circuit, the output voltage can be expressed in terms of inputs as

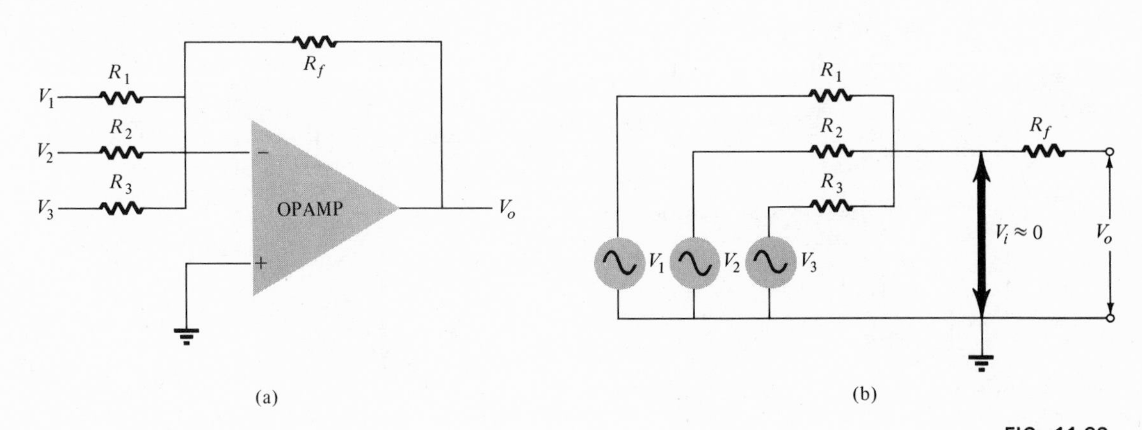

FIG. 11.33
(a) Summer; (b) virtual ground equivalent circuit.

$$V_o = -\left(\frac{R_f}{R_1}V_1 + \frac{R_f}{R_2}V_2 + \frac{R_f}{R_3}V_3\right) \tag{11.36}$$

In other words, each input adds a voltage to the output as obtained for an inverting constant-gain circuit. If more inputs are used they add additional components to the output.

EXAMPLE 11.5 What is the output voltage of an OPAMP summer for the following sets of input voltages and resistors ($R_f = 1$ M in all cases)?

(a) $V_1 = +1$ V, $V_2 = +2$ V, $V_3 = +3$ V,
$R_1 = 500$ K, $R_2 = 1$ M, $R_3 = 1$ M

(b) $V_1 = -2$ V, $V_2 = +3$ V, $V_3 = +1$ V,
$R_1 = 200$ K, $R_2 = 500$ K, $R_3 = 1$ M

Solution: Using Eq. (11.36)

(a)
$$V_o = -\left[\frac{1000\text{ K}}{500\text{ K}}(+1) + \frac{1000\text{ K}}{1000\text{ K}}(+2) + \frac{1000\text{ K}}{1000\text{ K}}(+3)\right]$$
$$= -[2(1) + 1(2) + 1(3)] = \mathbf{-7\text{ V}}$$

(b)
$$V_o = -\left[\frac{1000\text{ K}}{200\text{ K}}(-2) + \frac{1000\text{ K}}{500\text{ K}}(+3) + \frac{1\text{ M}}{1\text{ M}}(+1)\right]$$
$$= -[5(-2) + 2(+3) + 1(1)] = -[-10 + 6 + 1]$$
$$= \mathbf{+3\text{ V}}$$

Integrator

So far the input and feedback components have been resistors. If the feedback component used is a capacitor, as in Fig. 11.34, the resulting circuit is an integrator.

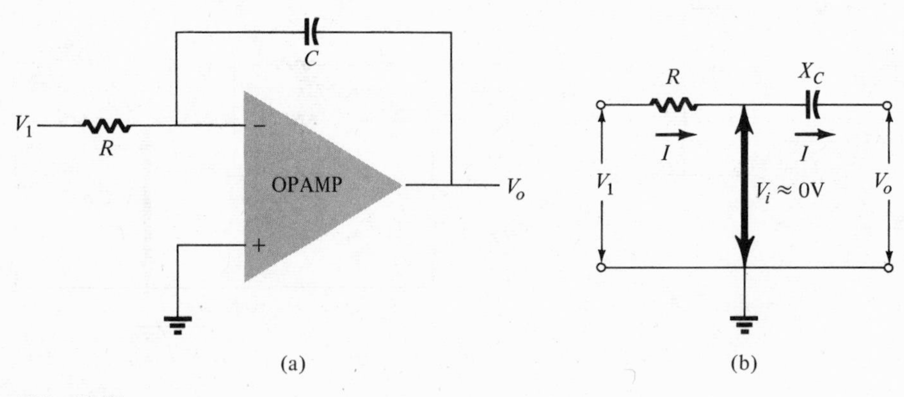

FIG. 11.34
Integrator.

The virtual ground equivalent circuit shows that an expression between input and output voltages can be derived from the current I, which flows from input to output. Recall that virtual ground means that we can consider the voltage at the junction point of R and X_C to be ground (since $V_i \cong 0$ V) but that no current flows to ground at that point. The capacitive impedance can be expressed as

$$X_C = \frac{1}{j\omega C} = \frac{1}{sC}$$

where $s = j\omega$ is the Laplace notation.
Solving for V_o/V_1

$$I = \frac{V_1}{R} = -\frac{V_o}{X_C} = \frac{V_o}{1/sC} = -sCV_o$$

$$\boxed{\frac{V_o}{V_1} = \frac{-1}{sCR}}$$

The last expression can be rewritten in the time domain as

$$V_o(t) = -\frac{1}{RC}\int V_1(t)\,dt \tag{11.37}$$

Equation (11.37) shows that the output is the integral of the input, with an inversion and scale multiplier of $1/RC$. The ability to integrate a given signal provides the analog computer with the ability to solve differential equations and therefore allows setup of a wide variety of electrical circuit analogs of physical system operations.

As an example, consider an input step voltage shown in Fig. 11.35a.

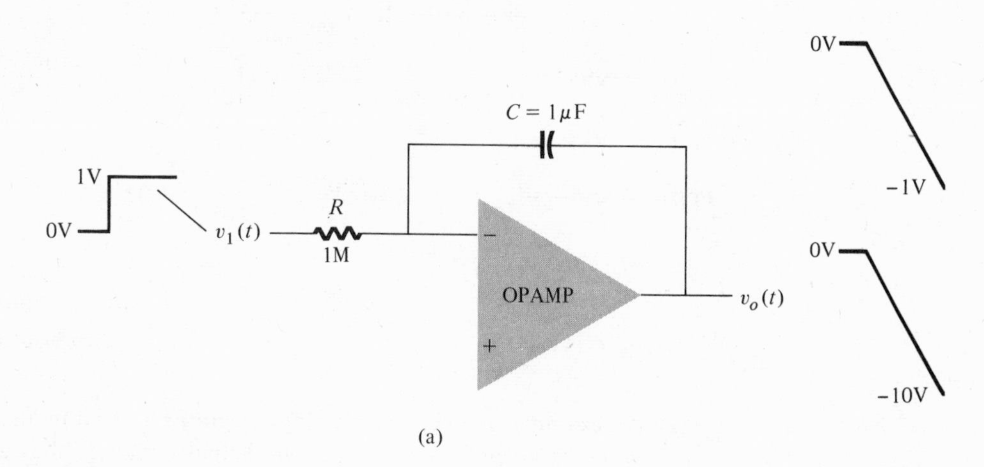

FIG. 11.35
Operation of integrator with step input.

The integral of the step voltage is a ramp or linearly changing voltage. The circuit scale factor of $-1/RC$ is

$$-\frac{1}{RC} = -\frac{1}{10^6 \times 10^{-6}} = -1$$

so that

$$V_o(t) = -\int V_i(t)\,dt$$

and the output is a negative ramp as shown in Fig. 11.35b.

If the scale factor is changed by making $R = 100$ K, for example, then

$$-\frac{1}{RC} = -\frac{1}{10^5 \times 10^{-6}} = -10$$

and the output is

$$V_o(t) = -10\int V_i(t)\,dt$$

which is shown in Fig. 11.35c.

More than one input may be applied to an integrator as shown in Fig. 11.36 with the resulting operation given by

$$V_o(t) = -\left[\frac{1}{R_1C_1}\int V(_1t)\,dt + \frac{1}{R_2C_2}\int V_2(t)\,dt + \frac{1}{R_3C_3}\int V_3(t)\,dt\right] \tag{11.38}$$

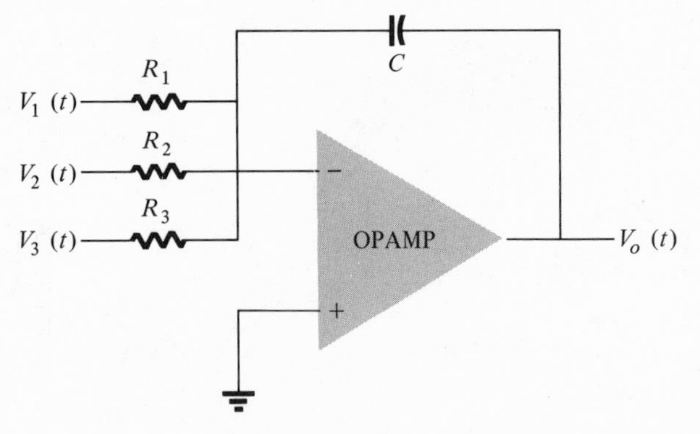

FIG. 11.36
Summing-integrator circuit.

An example, showing a summing integrator as used in an analog computer, is given in Fig. 11.37. The actual circuit is shown with input resistors and feedback capacitor, whereas the analog computer representation only indicates the scale factor for each input.

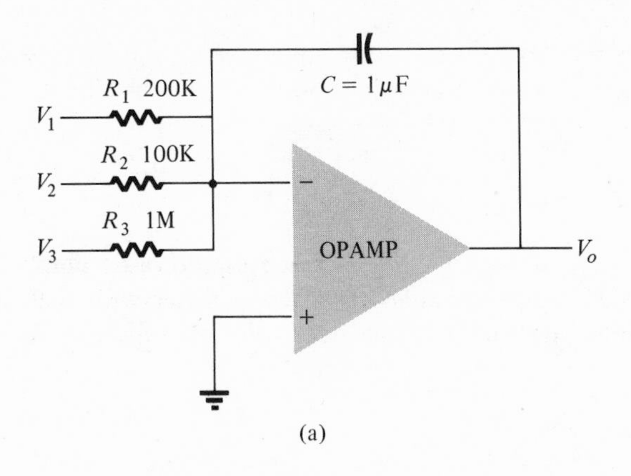

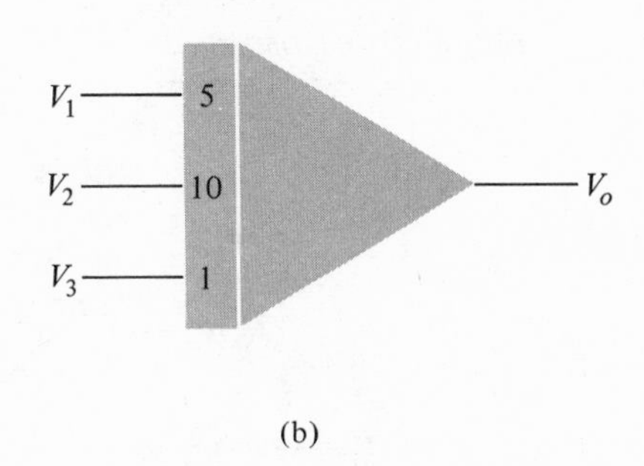

FIG. 11.37
(a) OPAMP and (b) analog computer integrator circuit representation.

Differentiator

The differentiator circuit of Fig. 11.38 is not as useful a computer circuit as the integrator because of practical problems with noise. The resulting relation for the circuit is

$$V_o(t) = -RC\frac{dV_1(t)}{dt}$$

where the scale factor is $-RC$. Reference to any text on analog computers will show how differential equations are set up for solution using mainly summing and integrator circuits.

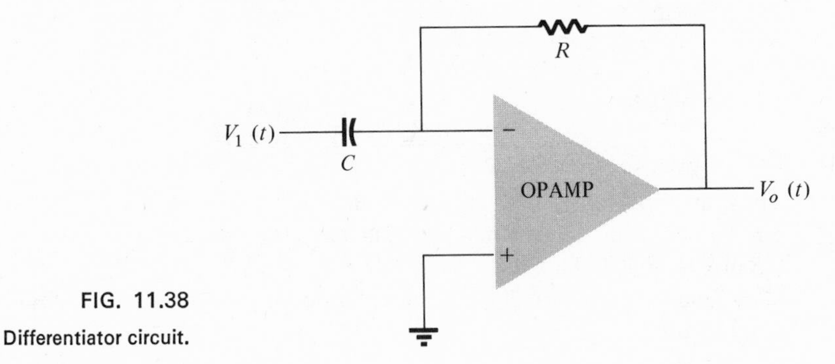

FIG. 11.38
Differentiator circuit.

11.7 OPAMP APPLICATIONS

To provide some indication of how useful OPAMPS can be in other than just analog computer circuits (which is a very large OPAMP application area) a few miscellaneous applications will be considered here, with additional oscillator applications in Chapter 12.

dc Millivoltmeter

Figure 11.39 shows an OPAMP used as the basic amplifier in a dc millivoltmeter. The amplifier provides a meter with high input impedance and scale factors dependent only on resistor value and

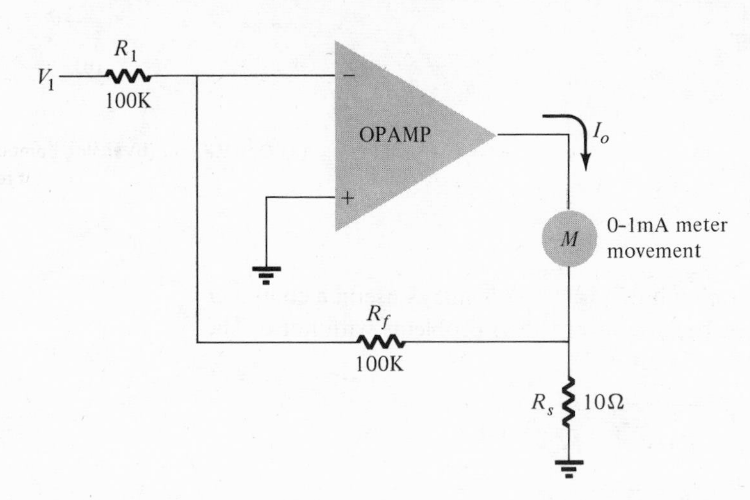

FIG. 11.39
OPAMP dc millivoltmeter.

accuracy. Notice that the meter reads millivolts of signal at the circuit input. An analysis of the OPAMP circuit yields the circuit transfer function

$$\frac{I_o}{V_1} = -\frac{R_f}{R_i}\left(\frac{1}{R_S}\right)$$
$$= -\frac{100\text{ K}}{100\text{ K}} \times \frac{1}{10} = \frac{1\text{ mA}}{10\text{ mV}}$$

Thus, an input of 10 mV will result in a current through the meter of 1 mA. If the input is 5 mV, the current through the meter will be 0.5 mA, which is half-scale deflection.

Changing R_f to 200 K, for example, would result in a circuit scale factor of

$$\frac{I_o}{V_1} = -\frac{200\text{ K}}{100\text{ K}} \times \frac{1}{10} = \frac{1\text{ mA}}{5\text{ mV}}$$

showing that the meter now reads 5 mV, full scale. It should be kept in mind that building such a millivoltmeter requires purchasing an OPAMP circuit, a few resistors, and a meter movement. The ability to obtain a completely operating, tested OPAMP unit makes the overall meter unit easy to set up.

ac Millivoltmeter

As another example, an ac millivoltmeter circuit is shown in Fig. 11.40. The resulting circuit transfer function is

$$\frac{I_o}{V_1} = -\frac{R_f}{R_1}\left(\frac{1}{R_s}\right) = \frac{100\text{ K}}{100\text{ K}} \times \frac{1}{10} = \frac{1\text{ mA}}{5\text{ mV}}$$

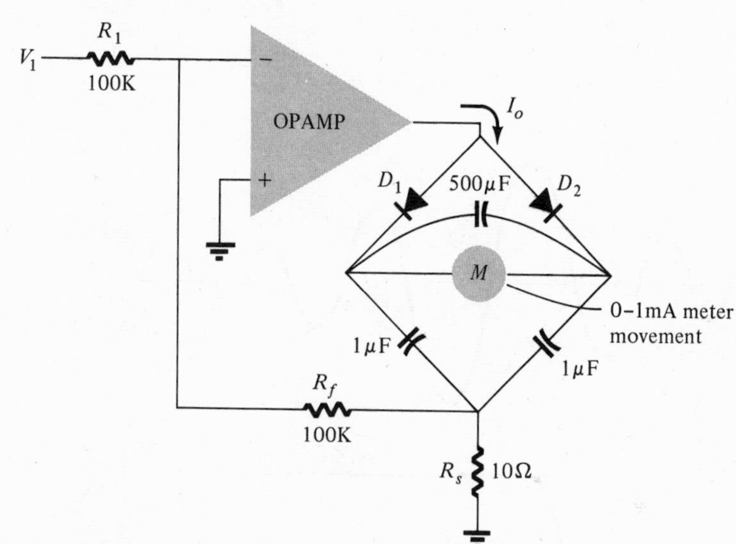

FIG. 11.40
ac millivoltmeter using OPAMP.

which appears the same as the dc millivoltmeter, except that in this case it is for ac signals. The meter indication provides a full-scale deflection for an ac input voltage of 10 mV. An ac input signal of 10 mV will result in full-scale deflection, while an ac input of 5 mV will result in half-scale deflection, and the meter reading can be interpreted in millivolt units.

Peak Follower

A peak follower circuit accepts an ac input voltage and provides as output a voltage whose magnitude follows only the peak of the input signal. It remains at the level of the highest peak of the ac input signal and changes only if the input signal goes higher than the previous peak level to a new higher peak level.

Figure 11.41a shows an OPAMP peak follower circuit. The diode-capacitor circuit acts as a simple capacitor filter, charging the capacitor

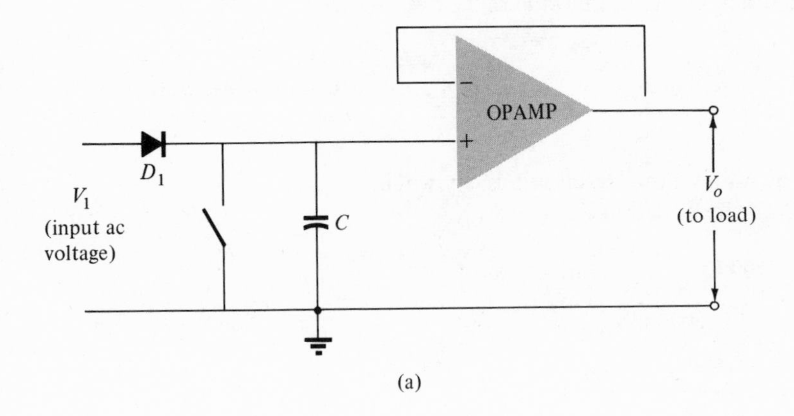

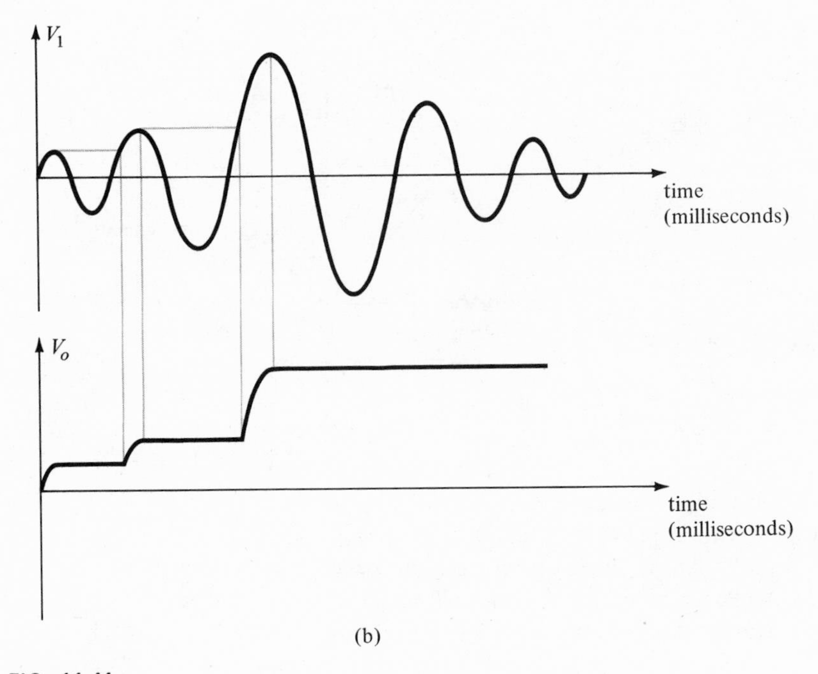

FIG. 11.41
(a) Peak follower circuit; (b) typical operation.

up to the peak of the input voltage. If the input voltage then decreases below this level, the diode becomes reverse biased and the capacitor remains charged. The OPAMP provides the output voltage to a load without "loading down" the capacitor. In effect the OPAMP isolates the capacitor from the load to allow the capacitor to remain charged. To restart a new peak follower operating cycle, the switch is closed and the capacitor discharged. Figure 11.41b shows an input ac voltage waveform and the peak follower output voltage.

A constant-current generator circuit provides a fixed constant current to a load, regardless of the load value. A constant-current source is shown as a possible part of the differential amplifier. An OPAMP version of a constant-current circuit is shown in Fig. 11.42.

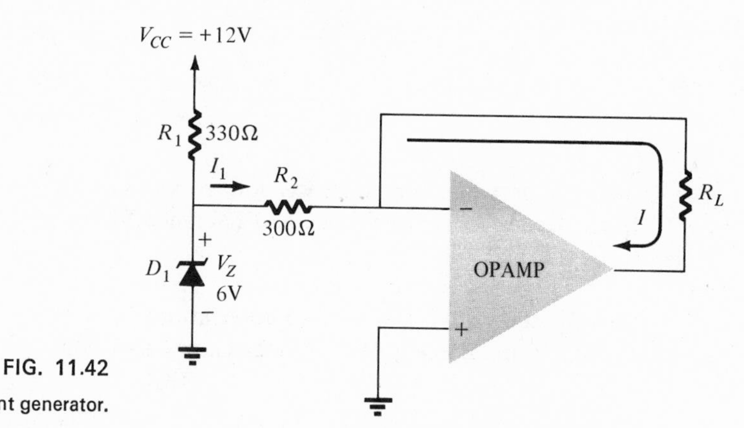

FIG. 11.42
Constant-current generator.

The resistor-diode connection of R_1 and D_1 operates the Zener diode in its Zener region with the input voltage held fixed at $V_Z = 6$ V, in this case. The voltage at the minus input terminal is about 0 V so that the current I_1 is

$$I_1 = \frac{V_Z}{R_2} = \frac{6\text{ V}}{300\ \Omega} = 20\text{ mA}$$

We know from OPAMP theory that

$$I = I_1 = 20\text{ mA}$$

so that the Zener and resistor R_2 fix the current through the output load resistor, R_L, for a wide range of values of R_L.

Voltage-Level Indicator

The circuit of Fig. 11.43 provides a bistable output, which is either at 0 V or $+V_{CC}$. Using a reference voltage at the input the circuit provides the following operation:

$$\text{if } V_1 < 1.5\text{ V},\ V_o = +V_{CC}$$
$$\text{if } V_1 > 1.5\text{ V},\ V_o = 0\text{ V}$$

The circuit thus operates as a comparator that provides indication of whether an input voltage exceeds a preset voltage level. Changing

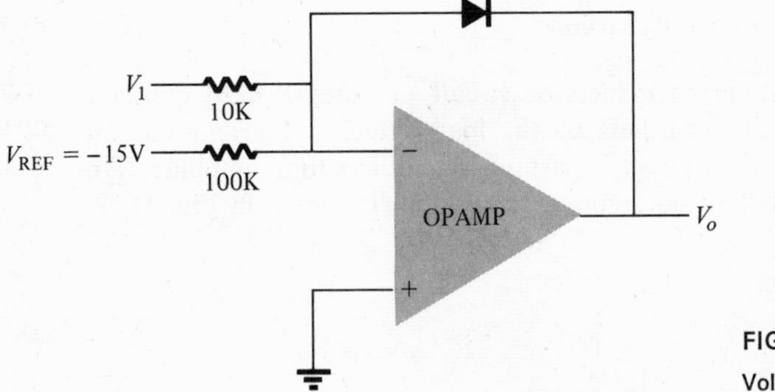

FIG. 11.43

Voltage-level indicator.

the resistor values or reference voltage allows changing the voltage level. Also reversing the diode and reference supply allows the output to be 0 V and $-V_{CC}$.

These few examples just provide some small indication of how versatile the OPAMP can be in applications. A number of additional oscillator circuit examples will be given in Chapter 12.

PROBLEMS

§ 11.1

1. Draw the output waveforms for the input signal and differential amplifier of Fig. 11.44.

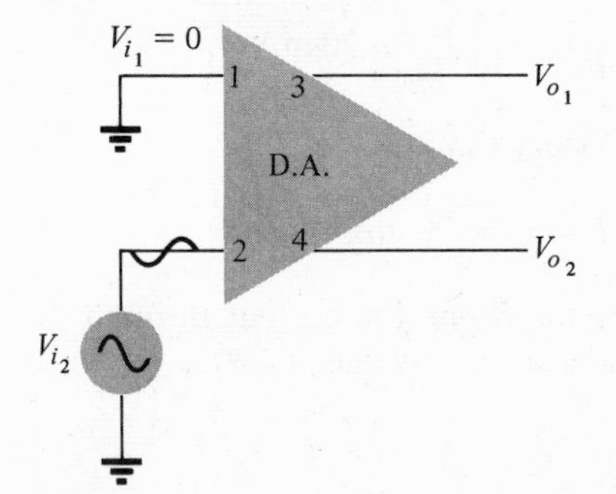

FIG. 11.44

Differential amplifier and input waveform for Problem 11.1.

2. Draw the output waveform for the differential amplifier and difference input signal of Fig. 11.45.

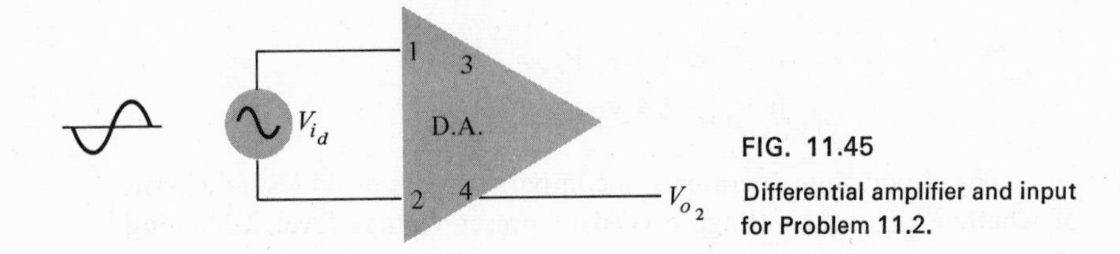

FIG. 11.45

Differential amplifier and input for Problem 11.2.

§ 11.2

3. Determine the dc voltages and currents in the circuit of Fig. 11.46.

4. Calculate the ac voltage gain of the difference amplifier in Fig. 11.46.

5. Calculate the input and output resistance of the circuit in Fig. 11.46.

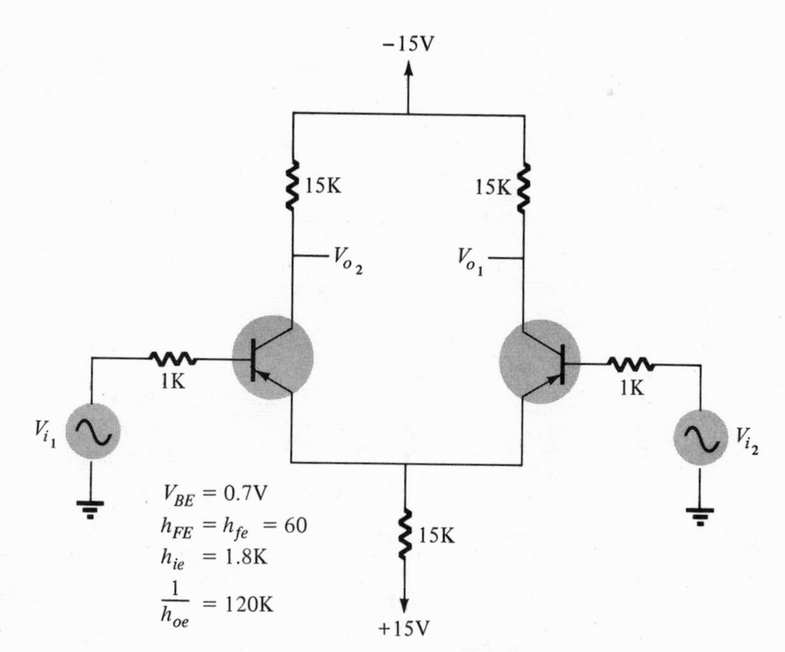

FIG. 11.46

Circuit for Problem 11.3.

6. Calculate the value of the constant current for the circuit of Fig. 11.47.

7. Calculate the dc voltages and currents in the circuit of Fig. 11.47.

§ 11.3

8. Calculate the output voltage of a difference amplifier for inputs of $V_{i_1} = 0.5$ mV, $V_{i_2} = 0.45$ mV, $A_d = 4500$, and CMRR $= 10^4$.

§ 11.4

9. The input impedance of a difference amplifier is measured using a 25-K resistor in series with an input voltage of 5 V. What is the value of R_i?

10. A difference amplifier has single-ended gain of $A_1 = 120$. When determining A_c the circuit measurements are $V_i = 2$ V and $V_0 = 20$ mV. Calculate CMRR in decibels.

§ 11.5

11. What is meant by "virtual ground"?

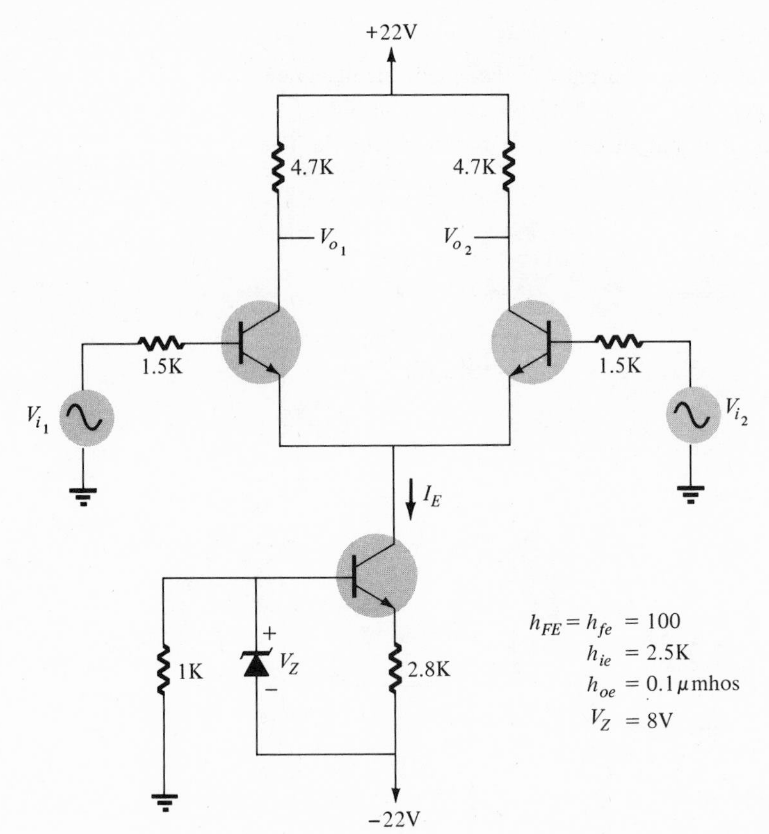

FIG. 11.47
Circuit for Problem 11.6.

§ 11.6

12. Calculate the output voltage of a noninverting OPAMP circuit (as in Fig. 11.31) for values of $V_1 = 4$ V, $R_f = 250$ K, and $R_1 = 50$ K.
13. Calculate the output voltage of a three-input summer (as in Fig. 11.33) for the values: $R_1 = 200$ K, $R_2 = 250$ K, $R_3 = 500$ K, $R_f = 1$ M, $V_1 = -2$ V, $V_2 = +2$ V, and $V_3 = 1$ V.
14. Determine the output voltage of the circuits of Fig. 11.48.

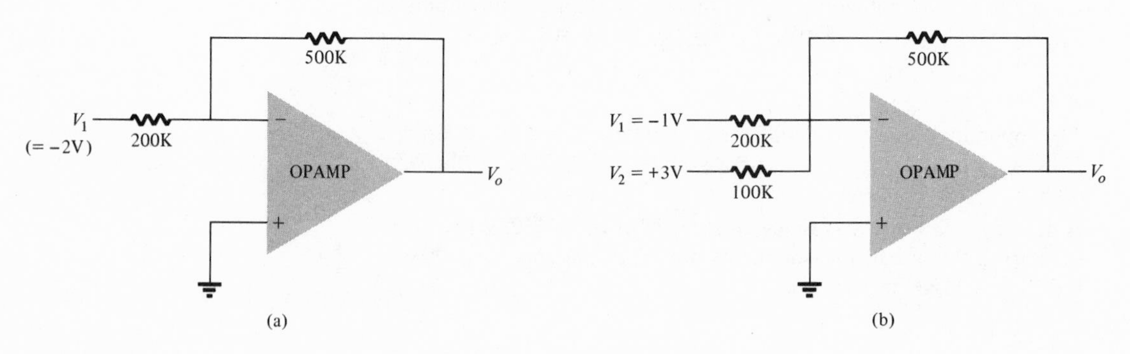

FIG. 11.48
Circuits for Problem 11.14.

15. Repeat Problem 11.14 for the circuit of Fig. 11.49.

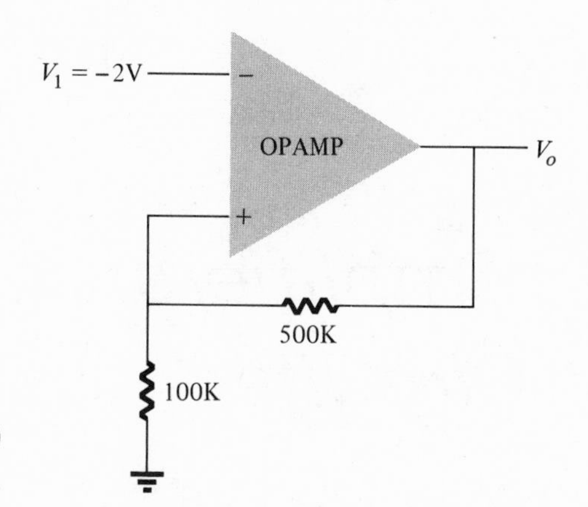

FIG. 11.49
Circuit for Problem 11.15.

16. Draw the output waveform for the circuit and inputs of Fig. 11.50.

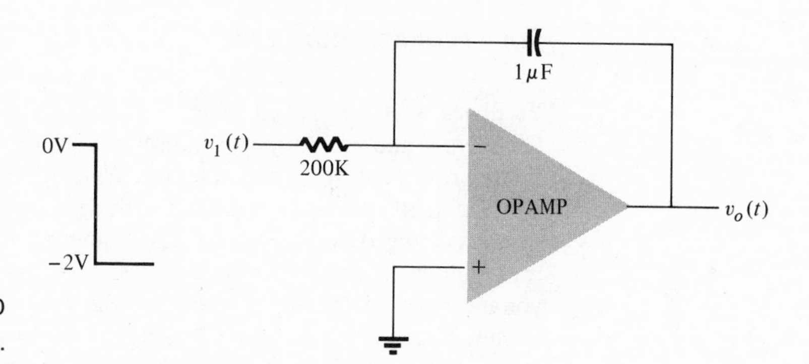

FIG. 11.50
Circuit for Problem 11.16.

17. Repeat Problem 11.16 for Fig. 11.51.

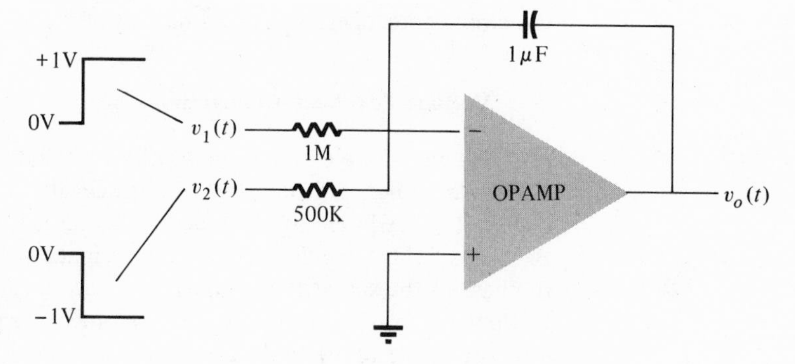

FIG. 11.51
Circuit for Problem 11.17.

12

Feedback Amplifiers and Oscillator Circuits

12.1 FEEDBACK CONCEPTS

Feedback was mentioned when considering dc bias stabilization in Chapters 4 and 11. Amplifier gain was sacrificed in the circuit design for improvement in dc bias stability. We might say that a trade-off of gain for stability was made in the circuit design. Such trade-off is typical of engineering design compromises. Using negative voltage feedback, for example, a circuit can be designed to couple some of the output voltage back to the input, reducing the overall voltage gain of the circuit. For this loss of gain, however, it is possible to obtain higher input impedance, lower output impedance, more stable amplifier gain, or higher cutoff frequency operation.

If the feedback signal is connected so as to aid or add to the input signal applied, however, *positive* feedback occurs, which could drive the circuit into operation as an oscillator.

Voltage Feedback Connection

As an example of voltage feedback the circuit of Fig. 12.1 shows a FET amplifier with negative voltage feedback. Resistor R_f and capacitor C_f (used here to block dc bias voltage) form the feedback path. Because of the amplifier inversion any signal at the output is opposite in phase to the signal at the input. The output signal fed back will then be opposite in phase or polarity and is thus a negative feedback signal. The net result of the feedback action will be to decrease the overall voltage gain to a lower amount dependent on the original gain of the amplifier without feedback and on the amount of the feedback.

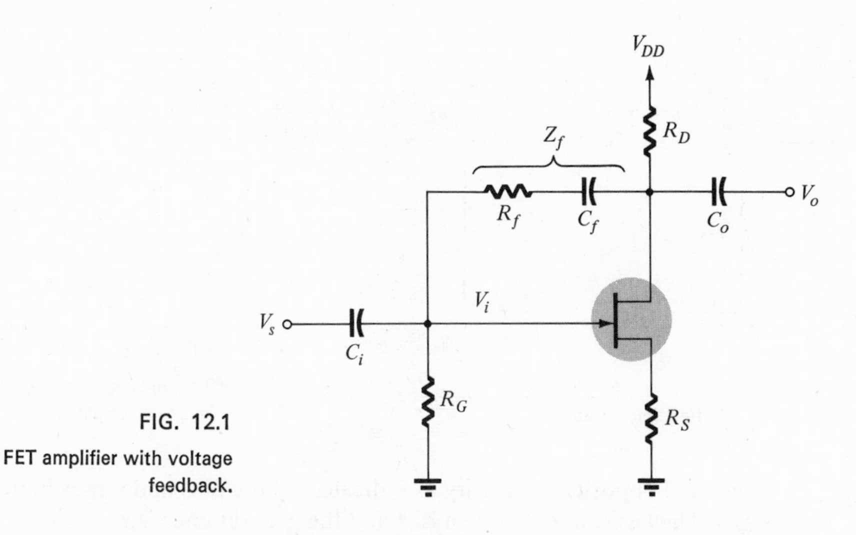

FIG. 12.1
FET amplifier with voltage feedback.

A general block diagram of a feedback circuit is shown in Fig. 12.2. The input signal (V_s) and feedback signal (V_f) are *mixed* or combined to form the single signal (V_i) which is then amplified by the amplifier section of the circuit. The amplifier output then goes into a sampler circuit which feeds part of the amplified signal to the load and part of the signal to the feedback network.

In the circuit of Fig. 12.1 the components C_i and R_g form the mixer network combining input and feedback signals. No sampling network is used in this circuit as the output and signal to the feedback network are the same. The feedback network is composed of resistor R_f and capacitor C_f.

A simpler version of the feedback amplifier of Figs. 12.1 and 12.2 is that of Fig. 12.3. The mixer is shown as a circle with two inputs

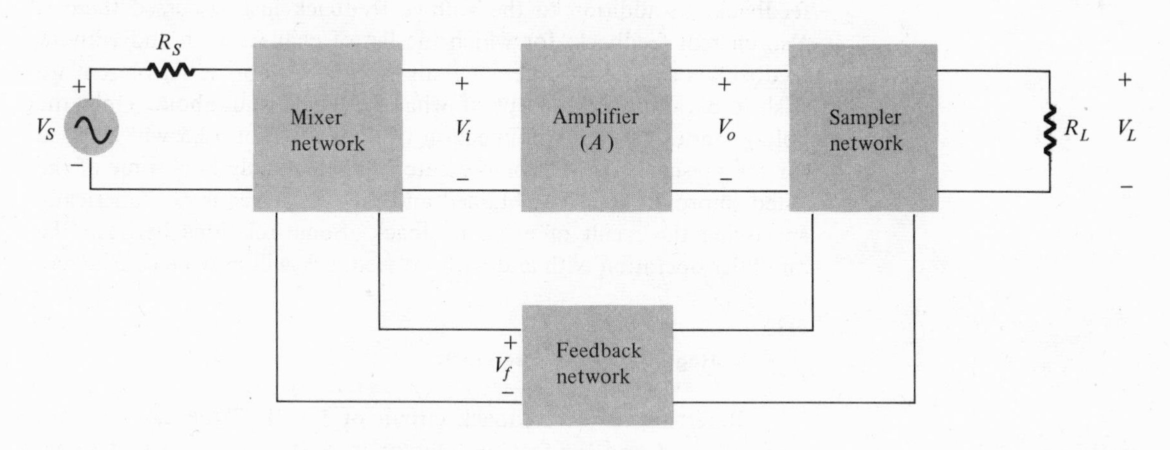

FIG. 12.2
Feedback amplifier, block diagram.

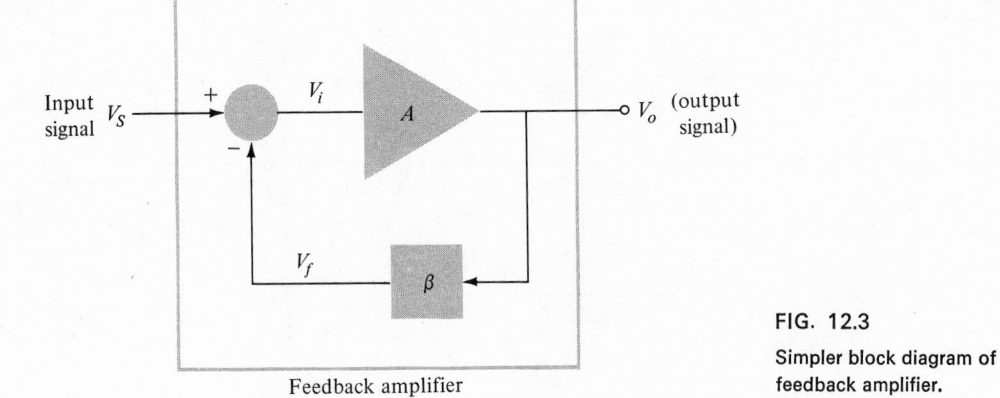

FIG. 12.3
Simpler block diagram of feedback amplifier.

which are opposite in polarity as indicated by the plus and minus input signs. The basic amplifier gain is A and the gain (attenuation, normally) of the feedback network is given as β (beta). Usually, a part of the output signal is coupled back to the input so as to oppose the applied input signal V_s. In return for this gain reduction a number of improvements can be obtained, such as the following:

1. Higher input impedance.
2. Better stabilized voltage gain.
3. Improved frequency response.
4. More linear operation.
5. Lower output impedance.
6. Reduction of noise.

These improvements all occur with a voltage-series type of feedback. In addition to the voltage feedback just discussed there is also current feedback, for which the list of changes with and without feedback is somewhat different than those listed above. At present we wish to obtain some concept of what feedback is all about. Only the voltage-series feedback connection of Figs. 12.2 or 12.3 will be used for the present. We will demonstrate mathematically how some of the listed improvements are obtained and provide a means of numerically specifying the result of using feedback. Some relations between the amplifier operation with and without feedback will now be considered.

Voltage Gain with Feedback

Referring to the feedback circuit of Fig. 12.3 the gain without feedback is A and the feedback factor is β. It is assumed that signal transmission goes only from input (V_i) to output (V_o) for the amplifier stage and only from output (V_o) to feedback input (V_f) for the feedback network.

The input voltage to the basic amplifier is the difference between signal and feedback voltage

$$V_i = V_s - V_f \tag{12.1}$$

where the feedback voltage is a portion of the output voltage

$$V_f = \beta V_o \tag{12.2}$$

the proportionality factor being β. The gain of the basic amplifier is simply

$$A = \frac{V_o}{V_i}$$

so that the output voltage is given by

$$V_o = AV_i \tag{12.3}$$

With feedback employed, the overall gain of the circuitry represented by Fig. 12.3 is

$$A_f = \frac{V_o}{V_s} \tag{12.4}$$

We can solve for this factor using Eqs. (12.1), (12.2), and (12.3) as follows:

$$\frac{V_o}{A} = V_i = V_s - V_f = V_s - \beta V_o$$

$$V_o + \beta A V_o = A V_s$$

$$\boxed{A_f = \frac{V_o}{V_s} = \frac{A}{1 + \beta A}} \tag{12.5}$$

Thus, Eq. (12.5) shows that the gain with feedback depends on the basic amplifier gain and the amount of the feedback factor.

If, for example, the quantities β and A are $\beta = 1/10$, $A = 90$, then the gain *without* feedback is 90 and the gain *with* feedback is

$$A_f = \frac{A}{1 + \beta A} = \frac{90}{1 + (1/10)(90)} = \frac{90}{10} = 9$$

the gain being reduced by a factor of 10. This is *negative feedback*. We can show that the gain *with* feedback is more stable than that without feedback. A change in amplifier gain from 90 to 100 due to component value changes with temperature represents a change of 11.1%. With feedback the resulting gain is

$$A_f = \frac{100}{1 + (1/10)(100)} = 9.1$$

which represents a change of only 1.11%, an improvement by a factor of nearly 10.

If the amplifier gain and feedback values are $A = 90$, $\beta = -1/100$, the gain with feedback is then

$$A_f = \frac{90}{1 + (-1/100)(90)} = \frac{90}{1 - 0.9} = \frac{90}{0.1} = 900$$

The gain has been increased by a factor of 10. This is *positive feedback* and is the principle by which a feedback amplifier can be made into an oscillator circuit. Detailed discussion of the oscillator is deferred to Section 12.7.

Thus, as a general statement: if $|A_f| < |A|$, feedback is negative; if $|A_f| > |A|$, feedback is positive.

GAIN STABILITY OF NEGATIVE FEEDBACK AMPLIFIERS

For a negative feedback amplifier we see that $|1 + \beta A| > 1$. Typically the value of $|\beta A| \gg 1$ so that

$$\boxed{A_f = \frac{A}{1 + \beta A} \simeq \frac{A}{\beta A} = \frac{1}{\beta}} \tag{12.6}$$

In other words, the feedback gain is dependant mainly on the factor β for the case of negative feedback with $|\beta A| \gg 1$. Whereas the amplifier gain A is quite dependant on temperature, device parameters, and so on, and may vary considerably, the reduced gain with negative feedback can be quite stable, typically depending on a resistor feedback network. Since resistors can be selected precisely and with small change in resistive value due to temperature, highly precise and stable gain with negative feedback is possible.

EXAMPLE 12.1 Calculate the gain of a negative feedback amplifier circuit having $A = 1000$, and $\beta = 1/10$.

Solution: Since $\beta A = 1/10(1000) = 100 \gg 1$ the gain with feedback is

$$A_f \cong \frac{1}{\beta} = \frac{1}{0.1} = 10$$

In addition to the β factor setting a precise gain value we are also interested in how stable the feedback amplifier is compared to an amplifier without feedback. Differentiating Eq. (12.5) leads to

$$\boxed{\frac{dA_f}{A_f} = \frac{1}{|1 + \beta A|} \frac{dA}{A}} \tag{12.7a}$$

$$\boxed{\frac{dA_f}{A_f} \cong \frac{1}{\beta A} \frac{dA}{A}, \qquad \text{for } \beta A \gg 1} \tag{12.7b}$$

This shows that the change in gain (dA) is reduced by the factor βA when feedback is employed.

EXAMPLE 12.2 If the amplifier in Example 12.1 has a gain change of 20% due to temperature, calculate the change in gain of the feedback amplifier.

Solution: Using Eq. (12.7b),

$$\frac{dA_f}{A_f} \cong \frac{1}{\beta A} \frac{dA}{A} = \frac{1}{0.1(1000)}(20\%) = 0.2\%$$

The improvement is 100 times. Thus, while the amplifier gain changes from $A = 1000$ by 20%, the feedback gain changes from $A_f = 100$ by only 0.2%.

12.2 FEEDBACK CONNECTION TYPES

There are four basic ways of connecting the feedback signal. Both *voltage* and *current* can be fed back to the input either in *series* or *parallel*. Specifically, there can be

1. Voltage-series feedback (Fig. 12.4a).
2. Voltage-shunt feedback (Fig. 12.4b).
3. Current-series feedback (Fig. 12.4c).
4. Current-shunt feedback (Fig. 12.4d).

In the above listing *voltage* refers to connecting the output voltage as input to the feedback network whereas *current* refers to

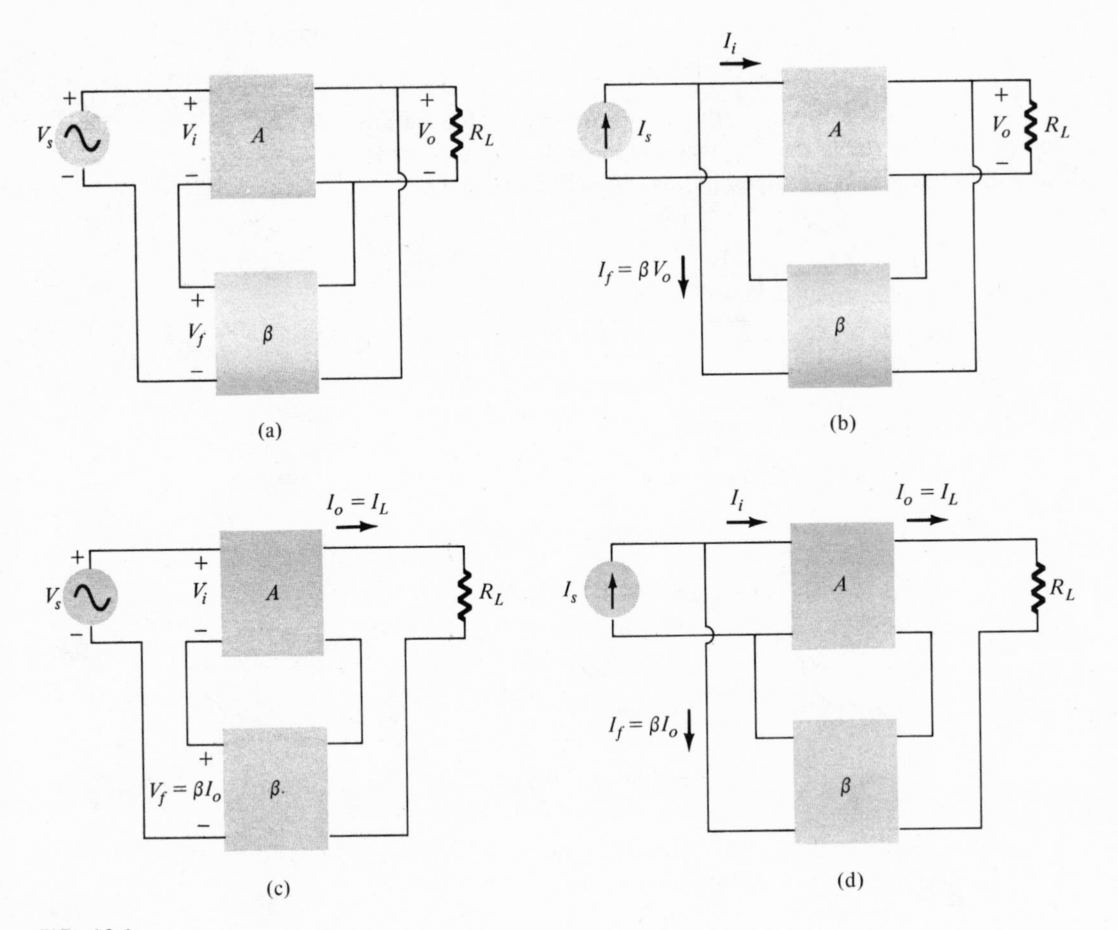

FIG. 12.4

Feedback amplifier connection types: (a) voltage-series feedback; (b) voltage-shunt feedback; (c) current-series feedback; (d) current-shunt feedback.

tapping off some output current through the feedback network. *Series* refers to connecting the feedback signal in series with the input signal voltage whereas *shunt* refers to connecting the feedback signal in shunt (parallel) with an input current source.

Series feedback connections tend to *increase* the input resistance while shunt feedback connections tend to *decrease* the input resistance. Voltage feedback will tend to *decrease* the output impedance while current feedback tends to *increase* the output impedance. Typically, higher input and lower output impedances are desired for most cascade amplifiers. Both of these are provided using the voltage-series feedback connection. We shall therefore concentrate first on this amplifier connection for practical feedback circuits.

A block diagram of a voltage-series feedback amplifier is shown in Fig. 12.5. A practical voltage source having voltage V_S and source resistance R_S is shown in *series* with the feedback signal, V_f, the resulting input signal to the amplifier stage being V_i. The output voltage of the amplifier stage (with gain A_v) is V_o, which is developed directly across load resistor R_L. A parallel *voltage* pickup of the output voltage is connected to the feedback network.

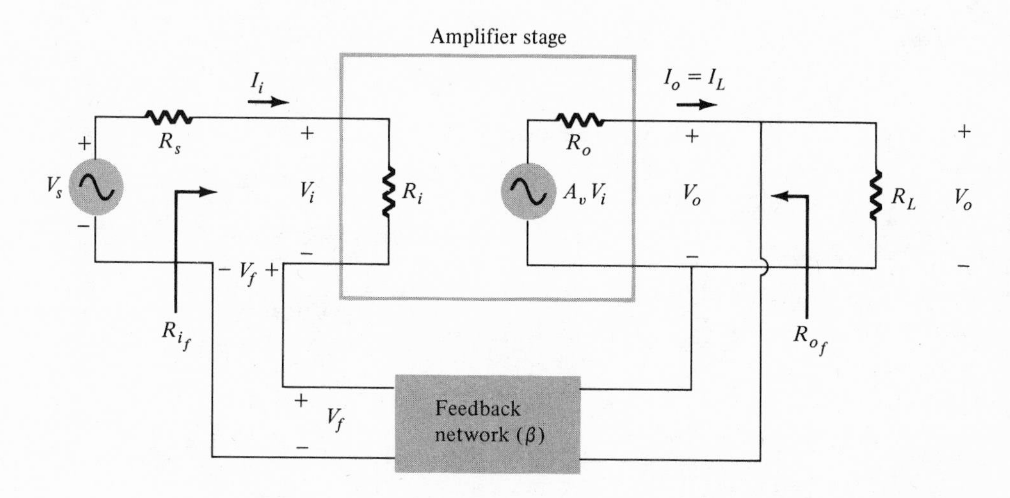

FIG. 12.5
Practical voltage-series feedback amplifier.

If $V_f = 0$ (no feedback) the overall voltage gain A, including loading effects of R_s and R_i, is defined as

$$A_v = \frac{V_o}{V_s}\bigg|_{R_L=\infty \text{ (output open circuit)}}$$

where

$$V_o = A_v V_i - I_o R_o$$

$$= \underbrace{A_v \frac{R_i}{R_i + R_s}}_{A} V_s - I_o R_o$$

$$V_o = AV_s - I_o R_o \qquad \text{(without feedback)}$$

(if $R_s = 0$ then $A = A_v$).

Including feedback connection ($V_f \neq 0$) the voltage gain is

$$A_f = \frac{V_o}{V_s}\bigg|_{R_L=\infty}$$

where

$$V_o = AV_i - I_oR_o$$
$$= A(V_s - V_f) - I_oR_o$$

Now,

$$A(V_s - V_f) = AV_s - AV_f = AV_s - A(\beta V_o)$$
$$= A(V_s - \beta V_o)$$

We have then,

$$V_o = AV_s - A\beta V_o - I_oR_o$$
$$V_o(1 + A\beta) = AV_s - I_oR_o$$
$$V_o = \left(\frac{A}{1 + \beta A}\right)V_s - I_o\frac{R_o}{1 + \beta A}$$
$$V_o = A_fV_s - I_oR_{of}$$

with

$$\boxed{A_f = \frac{A}{1 + \beta A}} \tag{12.8}$$

$$\boxed{R_{of} = \frac{R_o}{1 + \beta A}} \tag{12.9}$$

The last two equations show that the gain without feedback is reduced by the factor $(1 + \beta A)$ with feedback connected. In addition the output resistance is seen to be reduced from R_o (without feedback) by the factor $(1 + \beta A)$ with feedback. The larger the factor $(1 + \beta A)$, the lower the output resistance.

The input resistance with feedback is defined by

$$R_{if} \equiv \frac{V_s}{I_i} - R_s$$

which can be shown to be

$$\boxed{R_{if} = R_i(1 + \beta A)} \tag{12.10}$$

This time the factor $(1 + \beta A)$ makes the feedback input resistance larger than for the amplifier without feedback. In summary, then, a voltage-series feedback amplifier circuit can improve the operation of a nonfeedback amplifier circuit (having gain A, input resistance R_i, and output resistance R_o) as follows:

1. Stabilized voltage gain $A_f = A/(1 + \beta A)$.
2. Higher input resistance $R_{if} = R_i(1 + \beta A)$.
3. Lower output impedance $R_{of} = R_o/(1 + \beta A)$.

EXAMPLE 12.3 Calculate the gain and input and output impedance of a voltage-series feedback amplifier if the amplifier without feedback has $A = 100$, $R_i = 2$ K, $R_o = 40$ K and the amount of feedback is $\beta = 1/10$.

Solution: For a feedback amplifier using voltage-series feedback we can use Eqs. (12.8)–(12.10)

$$A_f = \frac{A}{1 + \beta A} = \frac{100}{1 + (1/10)(100)} = \frac{100}{11} = \mathbf{9.1}$$

$$R_{if} = R_i(1 + \beta A) = 2\text{ K}(11) = \mathbf{22\text{ K}}$$

$$R_{of} = \frac{R_o}{1 + \beta A} = \frac{40\text{ K}}{11} = \mathbf{3.63\text{ K}}$$

REDUCTION IN FREQUENCY DISTORTION

Recall that Eq. (12.6) shows that for a negative feedback amplifier having $\beta A \gg 1$ the gain with feedback is $A_f \cong 1/\beta$. It follows from this that if the feedback network is purely resistive the gain with feedback is not dependent on frequency even though the basic amplifier gain is frequency dependent. Practically, the frequency distortion arising because of varying amplifier gain with frequency is considerably reduced in a negative-voltage feedback amplifier circuit.

REDUCTION IN NOISE AND NONLINEAR DISTORTION

Signal feedback connected to oppose the input signal as in a negative feedback amplifier tends to hold down the amount of noise signal (such as power supply hum) and nonlinear distortion. The factor $(1 + \beta A)$ reduces both input noise and resulting nonlinear distortion for considerable improvement. However, it should be noted that there is a reduction in overall gain (the price required for the improvement in circuit performance). If additional stages are used to bring the overall gain up to the level without feedback it should be noted that the extra stage(s) might introduce as much noise back into the system as that reduced by the feedback amplifier. This problem can be somewhat alleviated by readjusting the gain of the feedback amplifier circuit to obtain higher gain while also providing reduced noise signal.

EFFECT OF NEGATIVE FEEDBACK ON GAIN AND BANDWIDTH

In Eq. (12.6) the overall gain with negative feedback is shown to be

$$A_f = \frac{A}{(1+\beta A)} \cong \frac{A}{\beta A} = \frac{1}{\beta} \qquad \text{for } \beta A \gg 1$$

As long as $\beta A \gg 1$ the overall gain is approximately $1/\beta$. We should realize that for a practical amplifier the open-loop gain drops off at high frequencies due to the active device and circuit capacitances. Gain may also drop off at low frequencies for capacitively coupled amplifier stages. Once the open-loop gain A drops low enough and the factor βA is no longer much larger than 1, the conclusion of Eq. (12.6) that $A_f \cong 1/\beta$ no longer holds true.

Figure 12.6 shows that the amplifier with negative feedback has more bandwidth (BW) than the amplifier without feedback. The feedback amplifier has a higher upper 3 dB frequency and smaller lower 3 dB frequency.

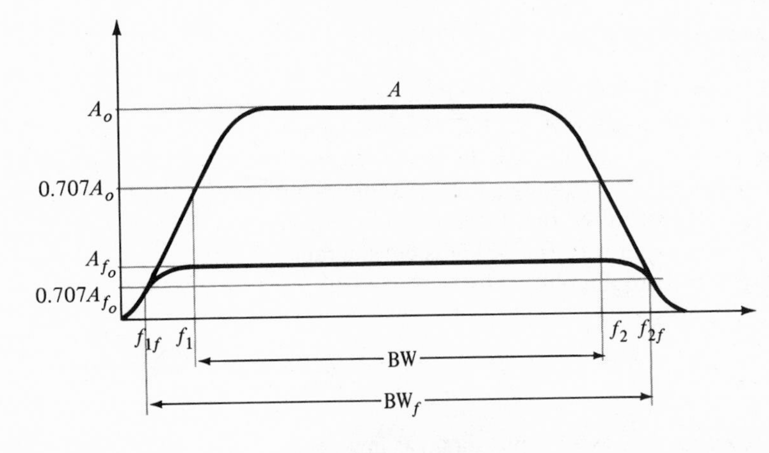

FIG. 12.6
Effect of negative feedback on gain and bandwidth.

It is interesting to note that the use of feedback, while resulting in a lowering of voltage gain, has provided an increase in BW and in the upper 3 dB frequency, particularly. In fact the product of gain and frequency remains the same so that the gain-bandwidth product of the basic amplifier is the same value for the feedback amplifier. However, since the feedback amplifier has lower gain, the net operation was to *trade* gain for bandwidth (we use bandwidth for the upper 3 dB frequency since typically $f_2 \gg f_1$).

12.3 PRACTICAL VOLTAGE-SERIES NEGATIVE FEEDBACK AMPLIFIER CIRCUITS

Transistor Stage

Figure 12.7a shows a transistor amplifier circuit with the output taken from the emitter terminal. Figure 12.7b shows an approximate small-signal equivalent circuit. The feedback signal is shown connected to the input in series with input voltage (V_s).

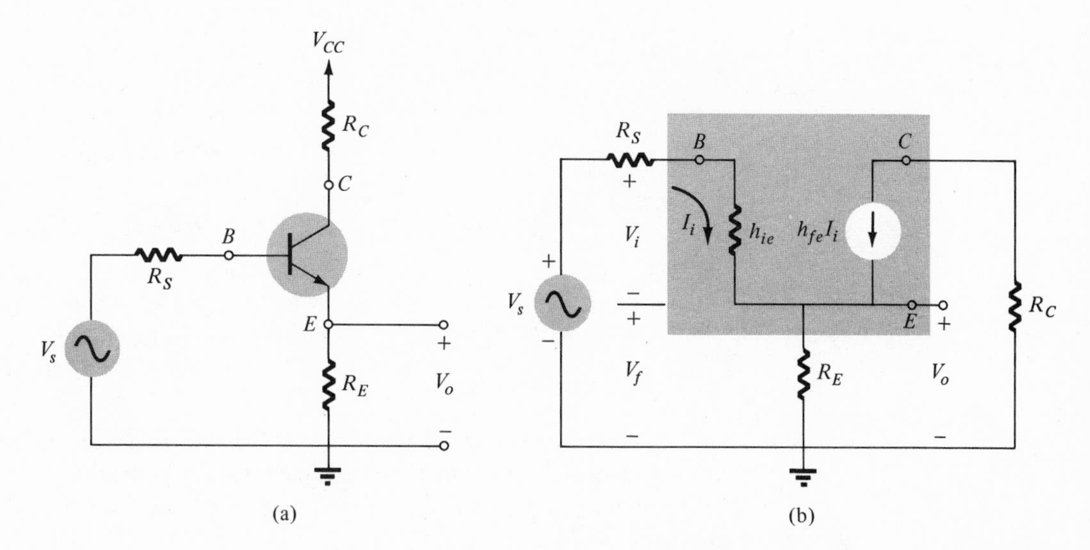

FIG. 12.7
Transistor amplifier with voltage-series feedback: (a) amplifier circuit; (b) equivalent ac circuit.

VOLTAGE GAIN

At first glance it might not appear that any feedback exists in this simple one-stage amplifier circuit. However, the ac equivalent circuit of Fig. 12.7b shows that the input voltage (V_i) is the difference of the input voltage (V_S) and feedback voltage (V_f)

$$V_i = V_s - V_f$$

The feedback voltage is equal in magnitude to the output voltage so that in this case we have $\beta = +1$.

The gain of a basic amplifier with load resistance R_C is

$$\boxed{A = \frac{h_{fe}R_E}{R_S + h_{ie}}} \qquad (12.11)$$

Using Eq. (12.6) we calculate the gain with feedback to be

$$A_f = \frac{A}{1+\beta A} = \frac{h_{fe}R_E/(R_S+h_{ie})}{1+(1)[h_{fe}R_E/(R_S+h_{ie})]}$$

$$\boxed{A_f = \frac{h_{fe}R_E}{R_S+h_{ie}+h_{fe}R_E}} \tag{12.12}$$

INPUT RESISTANCE

Without feedback we note that

$$R_i = h_{ie}$$

With feedback and neglecting R_S ($R_S = 0$) we find, using Eq. (12.10),

$$R_{if} = R_i(1+\beta A) = h_{ie}\left(1+\frac{h_{fe}R_E}{h_{ie}}\right)$$

$$\boxed{R_{if} = h_{ie} + h_{fe}R_E} \tag{12.13}$$

OUTPUT RESISTANCE

Without feedback the equivalent circuit used has an output resistance of infinity. Had $1/h_{oe}$ been included in the transistor equivalent circuit a practical value of around 50 K would be present. Including feedback and using Eq. (12.9),

$$\boxed{R_{of} = \frac{R_o}{1+\beta A} \cong \frac{R_S+h_{ie}}{h_{fe}}} \tag{12.14}$$

A numerical example using the above equations should show that negative feedback in a voltage-series connection reduces the gain while increasing input resistance and lowering output resistance.

EXAMPLE 12.4 For the circuit of Fig. 12.7 and the following circuit values, calculate the voltage gain and input and output resistances, with and without feedback: $R_E = 1.5$ K, $R_S = 1$ K, $h_{ie} = 2$ K, $h_{fe} = 50$.

Solution: Without feedback

$$A = \frac{h_{fe}R_E}{R_S+h_{ie}} = \frac{50(1.5\text{ K})}{1\text{ K}+2\text{ K}} = \mathbf{25}$$

$$R_i = h_{ie} = \mathbf{2\text{ K}}$$

$$R_o = \mathbf{\infty} \qquad \text{(or } l/h_{oe}\text{, if included in equivalent circuit)}$$

With feedback

$$A_f = \frac{h_{fe}R_E}{R_S + h_{ie} + h_{fe}R_E} = \frac{50(1.5\text{ K})}{1\text{ K} + 2\text{ K} + 50(1.5\text{ K})} = \frac{75}{78} = \mathbf{0.96}$$

$$R_{if} = h_{ie} + h_{fe}R_E = 2\text{ K} + 50(1.5\text{ K}) = \mathbf{77\text{ K}}$$

$$R_{of} = \frac{R_S + h_{ie}}{h_{fe}} = \frac{1\text{ K} + 2\text{ K}}{50} = \mathbf{60\ \Omega}$$

FET Stage

Fig. 12.8 shows a single-stage RC-coupled FET amplifier with negative feedback. A part of the output signal (V_o) is picked off by a feedback network made up of resistors R_1 and R_2. The feedback voltage V_f is connected in series with the source signal V_s, and their difference is the input signal V_i (measured from gate to drain).

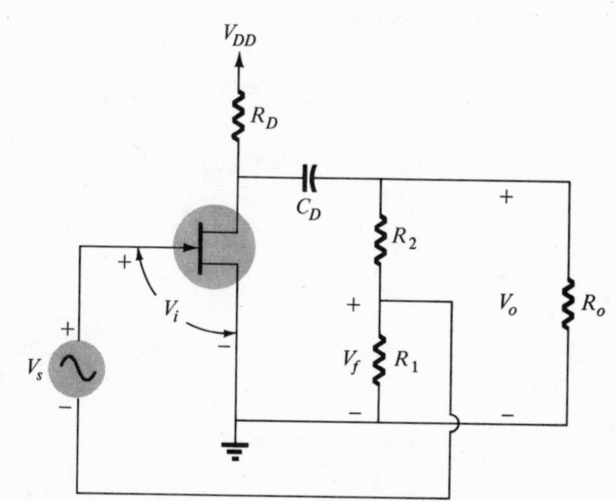

FIG. 12.8
FET amplifier stage with voltage-series feedback.

Without feedback the amplifier gain is

$$\boxed{A = g_m R_L} \tag{12.15}$$

where R_L is the parallel combination of resistors R_D, R_o, and a series equivalent of R_1 and R_2.

The feedback network provides a feedback factor of

$$\beta = \frac{R_1}{R_1 + R_2} \tag{12.16}$$

Using the above values of A and β in Eq. (12.6), the gain with negative feedback is found to be

$$A_f = \frac{A}{1 + \beta A} = \frac{g_m R_L}{1 + [R_1 R_L/(R_1 + R_2)]g_m} \tag{12.17a}$$

If $\beta A \gg 1$ we have

$$\boxed{A_f \cong \frac{1}{\beta} = \frac{R_1 + R_2}{R_1}} \tag{12.17b}$$

EXAMPLE 12.5 Calculate the gain without and with feedback for the FET amplifier circuit of Fig. 12.8 and the following circuit values: $R_1 = 20$ K, $R_2 = 80$ K, $R_o = 10$ K, $R_D = 10$ K, and $g_m = 4000$ μmhos.

Solution

$$R_L \cong \frac{R_o R_D}{R_o + R_D} = \frac{10(10)}{10 + 10} = 5 \text{ K}$$

(neglecting 100 K resistance of R_1 and R_2 in series)

$$A = g_m R_L = (4000 \times 10^{-6})(5 \text{ K}) = \mathbf{20}$$

The feedback factor is

$$\beta = \frac{R_1}{R_1 + R_2} = \frac{20}{20 + 80} = 0.2$$

The gain with feedback is

$$A_f = \frac{A}{1 + \beta A} = \frac{20}{1 + 0.2(20)} = \frac{20}{5} = \mathbf{4}$$

Vacuum-Tube Feedback Circuit

Figure 12.9 shows a voltage-series feedback amplifier using a vacuum-tube circuit. As in the FET amplifier stage a portion of the output voltage is taken off by a resistor feedback network. The feedback signal is connected in series with the input source signal with amplifier input voltage V_i (measured from grid to cathode). A bypassed bias network of resistor R_K and capacitor C_K serve only for dc bias. The combination of resistors R_1 and R_2 are also seen to act as an effective grid bias resistor (typical value 500 K to 1 M).

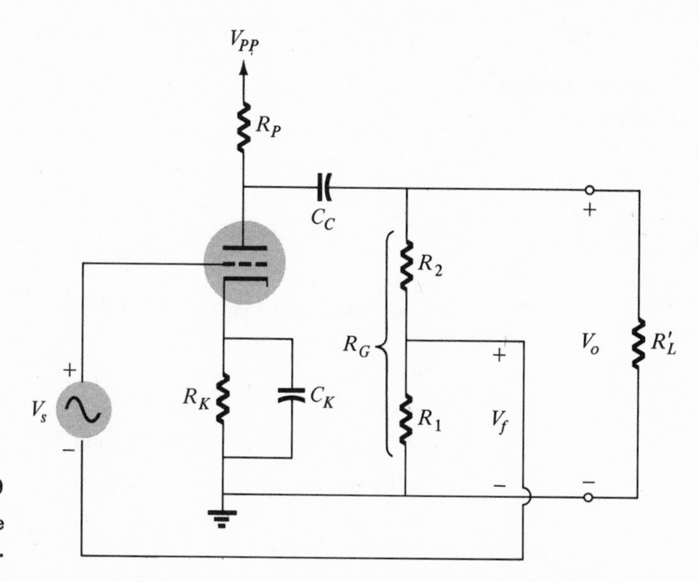

FIG. 12.9
Vacuum-tube amplifier stage with voltage-series feedback.

Neglecting the series combination of resistors R_1 and R_2, which is typically much larger than R'_L or R_P in value, the effective ac load resistance is

$$R_L = \frac{R_P R'_L}{R_P + R'_L} \tag{12.18}$$

and the gain of the stage without feedback is

$$\boxed{A = \frac{\mu R_L}{r_p + R_L}} \tag{12.19}$$

where μ and r_p are the tube gain and output resistance.

The output resistance of the stage without feedback as seen by the load is the parallel combination of R_P and r_p

$$R_o = \frac{R_P r_p}{R_P + r_p} \tag{12.20}$$

The feedback factor is the same as that given in Eq. (12.16) and the stage gain with feedback is given by Eq. (12.8). Similarly the output resistance with feedback is given by Eq. (12.9) using Eq. (12.20) for the output resistance without feedback.

EXAMPLE 12.6 Calculate the gain and output impedance without and with feedback for the circuit of Fig. 12.9 for the following circuit values: $R_P = 40$ K, $R'_L = 5$ K, $R_1 = 100$ K, $R_2 = 900$ K, $\mu = 100$, and $r_P = 20$ K.

Solution: Using Eq. (12.18),

$$R_L = \frac{R_P R'_L}{R_P + R'_L} = \frac{40(5)}{40 + 5} = 4.44 \text{ K}$$

The gain without feedback is then [Eq. (12.19)]

$$A = \frac{\mu R_L}{r_p + R_L} = \frac{100(4.44)}{20 + 4.44} = \mathbf{18.1}$$

The feedback factor is [Eq. (12.16)]

$$\beta = \frac{R_1}{R_1 + R_2} = \frac{100}{100 + 900} = 0.1$$

and the gain with feedback is [Eq. (12.8)]

$$A_f = \frac{A}{1 + \beta A} = \frac{18.1}{1 + 0.1(18.1)} = \mathbf{6.45}$$

Without feedback output resistance is [Eq. (12.20)]

$$R_o = \frac{R_P r_p}{R_P + r_p} = \frac{40(20)}{40 + 20} = \mathbf{13.3\ K}$$

With feedback the output resistance is [Eq. (12.9)]

$$R_{of} = \frac{R_o}{1 + \beta A} = \frac{13.3}{1 + (0.1)(18.1)} = \mathbf{4.74\ K}$$

12.4 OTHER PRACTICAL FEEDBACK CIRCUIT CONNECTIONS

Two-Stage Voltage-Series Feedback

A popular means of incorporating negative feedback to stabilize the gain of an amplifier is signal feedback in a multistage circuit. As an example, Fig. 12.10 shows two cascaded stages with voltage-series feedback. Cascaded amplifier stages with transistors Q_1 and Q_2 provide an overall gain A. A feedback network of resistors R_1 and R_2 is coupled by a capacitor (to block dc) between output and input while allowing feedback of the ac output signal. The feedback signal is taken from the collector of the second stage and is connected to the emitter (neglecting the bypassed 3.6-K resistor) of the first amplifier stage. Negative feedback results from this connection since the in-phase output signal of the second stage collector connected through the feedback network to the emitter opposes the input signal between the base-emitter of the first stage.

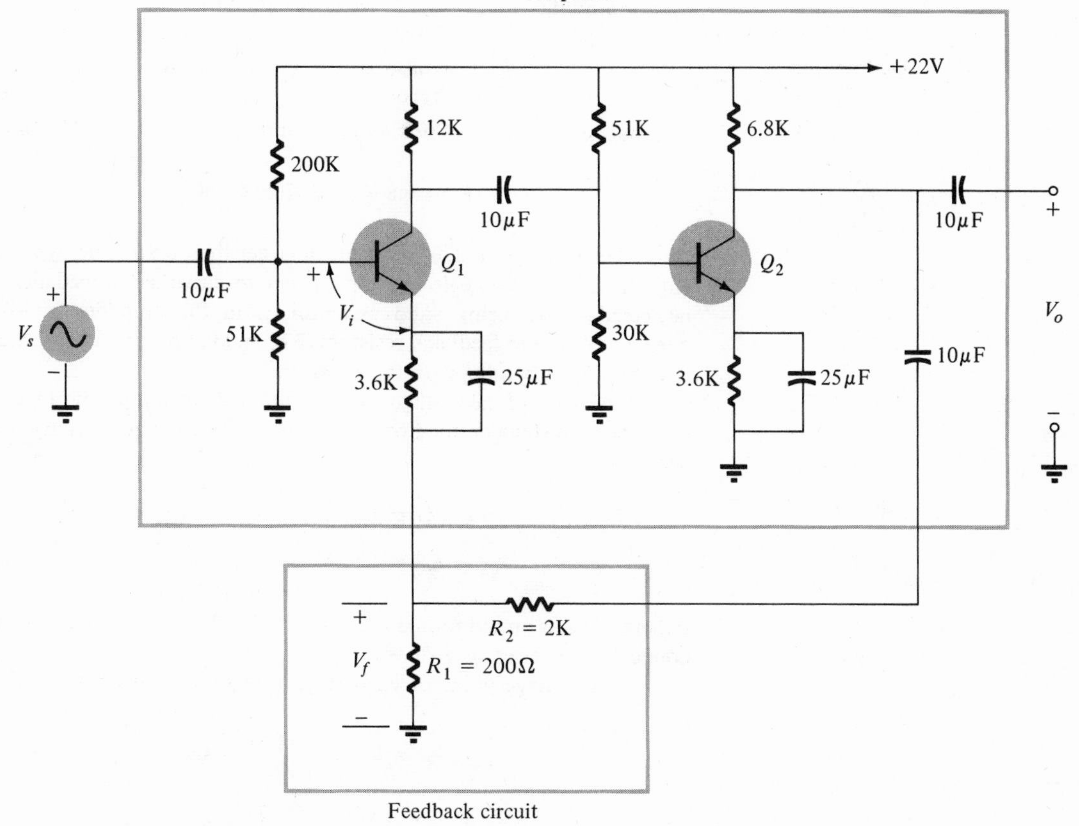

FIG. 12.10
Cascaded voltage-series feedback amplifier.

Calculations of gain, input, and output resistance with and without feedback require no new theory or equations. The techniques for cascaded stages developed in Chapter 7 and the basic equations [Eqs. (12.8)–(12.10)] for voltage-series feedback are used in the following example to show how a circuit such as in Fig. 12.10 can be analyzed.

EXAMPLE 12.7 Calculate A, R_i, and R_o for the cascaded amplifier of Fig. 12.10 omitting feedback, and then A_f, R_{if}, and R_{of} with the feedback connection considered. Use transistor parameters $h_{fe} = 65$, $h_{ie} = 1.8$ K, and $1/h_{oe} = \infty$.

Solution: Without the feedback network: Looking into the base of transistor Q_1 the resistance seen is $R_{i_1} = h_{ie} = 1.8$ K. The emitter 3.6-K resistor is neglected here for ac calculations due to the bypass capacitor. Since the parallel combination of the 51-K and 200-K bias resistors is in parallel with the input (as seen by the source signal) the overall amplifier input impedance is calculated to be

$$R_i = 1.8\text{ K} \,||\, 51\text{ K} \,||\, 200\text{ K} \cong \mathbf{1.7\text{ K}}$$

The output impedance looking back into stage 2 is approximately

$$R_o = 6.8\text{ K} \,||\, 2.2\text{ K} \cong \mathbf{1.7\text{ K}}$$

where $1/h_{oe}$ is neglected as being much larger than 6.8 K, the bypassed emitter resistor is neglected, and the capacitive ac impedance is neglected—these being valid assumptions in the amplifier mid-frequency range, and feedback resistors (2 K and 0.2 K) provide an effective resistor in parallel with the output.

The gain of each stage can be obtained (including the loading of the second stage on the first) as follows: effective load resistances of each stage are

$$R_{L_1} = 12\text{ K} \,||\, 51\text{ K} \,||\, 30\text{ K} \,||\, 1.8\text{ K} \cong 1.5\text{ K}$$

$$R_{L_2} = 6.8\text{ K} \,||\, 2.2\text{ K} \cong 1.7\text{ K}$$

(where 2.2 K is an output load resistance of the 2-K and 0.2-K resistors connected in series to ground).

The voltage gains of each stage (magnitude only) are then

$$A_{v_1} \cong \frac{h_{fe}R_{L_1}}{h_{ie}} = \frac{65(1.5)}{1.8} = 54$$

$$A_{v_2} \cong \frac{h_{fe}R_{L_2}}{h_{ie}} = \frac{65(1.7)}{1.8} = 61$$

The overall gain of the cascaded amplifier, neglecting feedback, is then

$$A = A_{v_1}A_{v_2} = 54(61) = \mathbf{3300}$$

We can calculate the feedback factor to be

$$\beta = \frac{R_1}{R_1 + R_2} = \frac{0.2}{0.2 + 2} = \frac{1}{11} = 0.091$$

With feedback: Using Eqs. (12.8)–(12.10)

$$R_{of} = \frac{R_o}{1 + \beta A} = \frac{1.7\text{ K}}{1 + 0.091(3300)} = \frac{1.7\text{ K}}{301} = \mathbf{5.67\ \Omega}$$

$$R_{if} = R_i(1 + \beta A) = 1.7\text{ K}(301) = \mathbf{510\text{ K}}$$

$$A_f = \frac{A}{1 + \beta A} = \frac{3300}{301} \cong \mathbf{11}$$

So far we have considered only a feedback connection which samples the output voltage and feeds a portion of that voltage back to the input in series opposition with the source signal. Another feedback technique is to sample the output current (I_o) and return a proportional voltage in series with the input. While stabilizing the amplifier gain, the current-series feedback connection increases *both* input and output resistance.

Figure 12.11 shows a simple version of an amplifier with current-series negative feedback. If the input current is negligible as in a tube or FET amplifier then the current through resistor R is the output current I_o. The voltage developed across R is a feedback voltage connected in series with the source signal. We can consider the amplifier as providing an output current dependent on the input source voltage—the amplifier then acting as a transconductance amplifier. Feedback acts to stabilize the transconductance, so that the load current depends on the signal voltage and resistance R only, and is stabilized in regard to any other circuit changes.

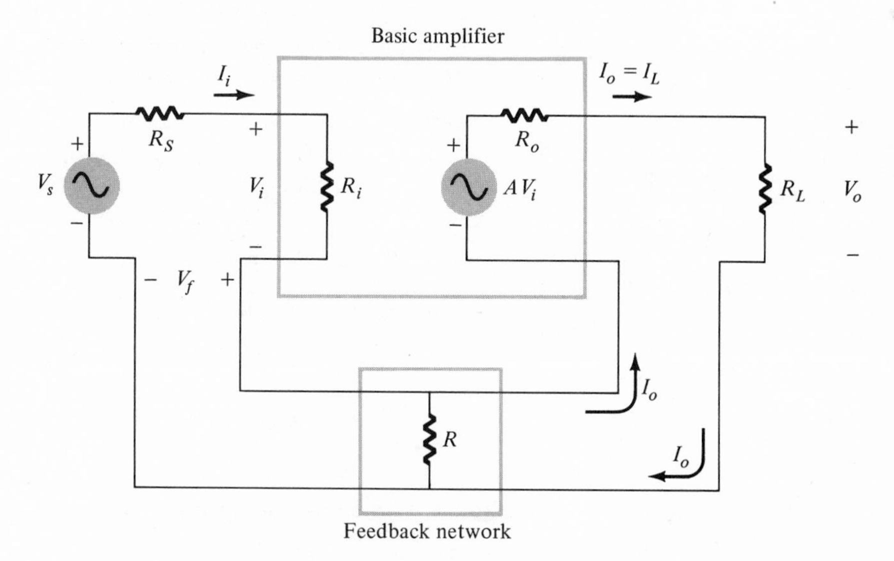

FIG. 12.11
Amplifier with current-series negative feedback connection.

Figure 12.12 shows a single transistor amplifier stage. Since the emitter of this stage has an unbypassed emitter it effectively has current-series feedback. The current through resistor R_E results in a feedback

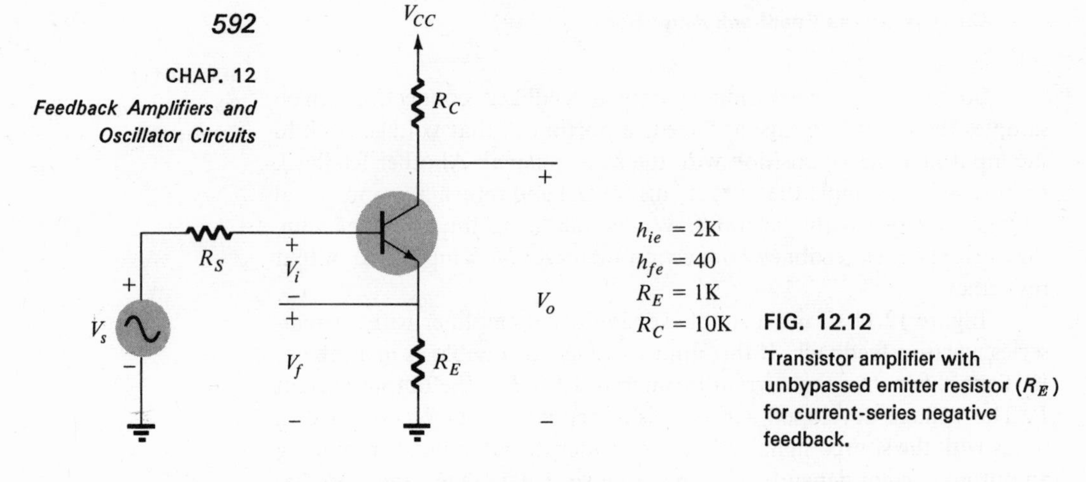

FIG. 12.12 Transistor amplifier with unbypassed emitter resistor (R_E) for current-series negative feedback.

voltage which opposes the source signal applied so that the output voltage V_o is reduced. To remove the current-series feedback the emitter resistor must be either removed or bypassed by a capacitor (as is usually done). An example will show how the amplifier operation is affected by current-series feedback due to resistor R_E.

EXAMPLE 12.8 For the circuit of Fig. 12.12 calculate the gain, input resistance, and output resistance without feedback (bypassed emitter resistor, $R_E = 0$) and with feedback (R_E present), and $1/h_{oe} = \infty$.

Solution: With R_E removed (bypassed)

$$A = \frac{h_{fe}R_C}{h_{ie}} = \frac{40(10\text{ K})}{2\text{ K}} = 200$$

$$R_i = h_{ie} = 2\text{ K}$$

$$R_o = R_C = 10\text{ K}$$

With R_E in the circuit

$$A_f = \frac{h_{fe}R_C}{h_{ie} + h_{fe}R_C} = \frac{40(10\text{ K})}{2\text{ K} + 40(1\text{ K})} = \frac{400}{42} = 9.5$$

$$R_{if} = h_{fe}R_C + h_{ie} = 40(1\text{ K}) + 2\text{ K} = 42\text{ K}$$

$$R_{of} \cong R_C = 10\text{ K}$$

Example 12.8 shows that current-series feedback

1. Reduces amplifier gain.
2. Increases input resistance.

Voltage-Shunt Feedback

Negative feedback can be obtained by coupling a portion of the output voltage in parallel (shunt) with the input signal. Figure 12.13 shows a typical voltage-shunt feedback connection. A portion of the

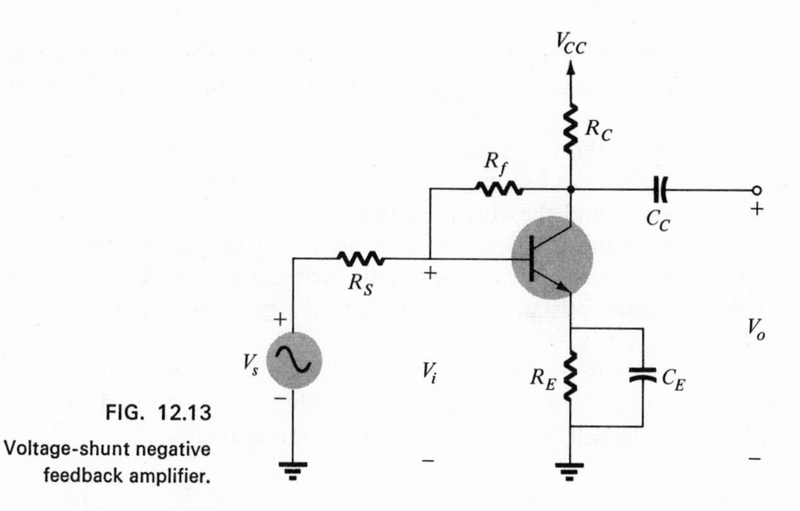

FIG. 12.13
Voltage-shunt negative feedback amplifier.

output voltage (which is opposite in polarity to the input voltage) is connected to the base through resistor R_f. A voltage-shunt connection stabilizes amplifier overall gain while decreasing both input and output resistances.

Current-Shunt Feedback

A fourth feedback connection samples the output current and develops a feedback voltage in shunt with the input signal. A practical circuit version is the two-stage amplifier of Fig. 12.14. The unbypassed

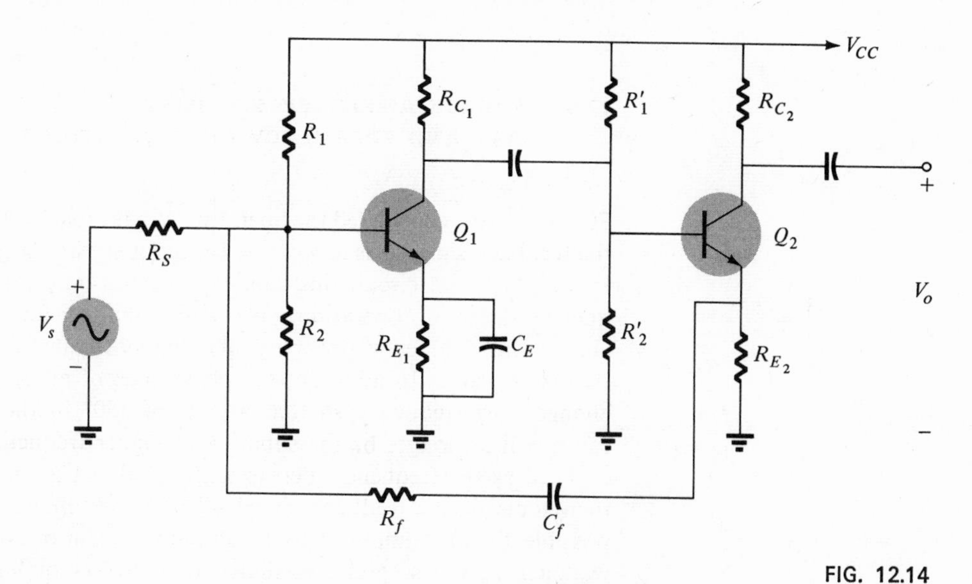

FIG. 12.14
Amplifier with current-shunt negative feedback connection.

emitter resistor of stage 2 provides current sensing. The feedback signal is connected in shunt with the first stage input through a feedback network.

Checking the feedback signal polarity for an input to the base of stage 1, the output of stage 1 is opposite in phase. The input to the base of stage 2 and the voltage across emitter R_{E_1} is then opposite in phase to the input to stage 1 so that negative feedback is achieved. A current-shunt feedback circuit typically increases output resistance and decreases input resistance while holding the gain with feedback constant.

The operation of the four types of feedback connections is summarized in Table 12.1. All types provide stabilized but reduced gain, increased bandwidth, and decreased nonlinear distortion.

TABLE 12.1

Effect of Feedback Connection Type on Input and Output Resistance

	VOLTAGE-SERIES	CURRENT-SERIES	VOLTAGE-SHUNT	CURRENT-SHUNT
R_{if}	increased	increased	decreased	decreased
R_{of}	decreased	increased	decreased	increased

12.5 FEEDBACK AMPLIFIER STABILITY—PHASE AND FREQUENCY CONSIDERATIONS

So far we have considered the operation of a feedback amplifier in which the feedback signal was *opposite* to the input signal—negative feedback. In any practical circuit this condition occurs only for some mid-frequency range of operation. We know that an amplifier gain will change with frequency, dropping off at higher frequencies from the mid-frequency value. In addition, the phase shift of an amplifier will also change with frequency so that a shift of 180° in the mid-frequency range will no longer be the situation at higher frequencies.

If, as the frequency increases, the phase shift changes from 180° then some of the feedback signal *adds* to the input signal. It is then possible for the amplifier to break into oscillations due to positive feedback. If the amplifier oscillates at some low or high frequency it is no longer useful as an amplifier. Proper feedback-amplifier design requires that the circuit be stable at *all* frequencies, not merely those in the range of interest. Otherwise a transient disturbance could cause a seemingly stable amplifier to suddenly start oscillating.

Nyquist Criterion

In judging the stability of a feedback amplifier, as a function of frequency, the factors of loop gain A_f, amplifier gain A, and feedback attenuation β as functions of frequency, can be used. One of the most popular techniques used to investigate stability is the Nyquist method. A Nyquist diagram is used to plot gain and phase shift as a function of frequency on a complex plane. The Nyquist plot, in effect, combines the two Bode plots of gain vs. frequency and phase-shift vs. frequency on a single plot. A Nyquist plot is used to quickly show whether an amplifier is stable for all frequencies and how stable the amplifier is relative to some gain or phase-shift criteria.

As a start, consider the *complex plane* shown in Fig. 12.15. A few points of various gain (βA) values are shown at a few different

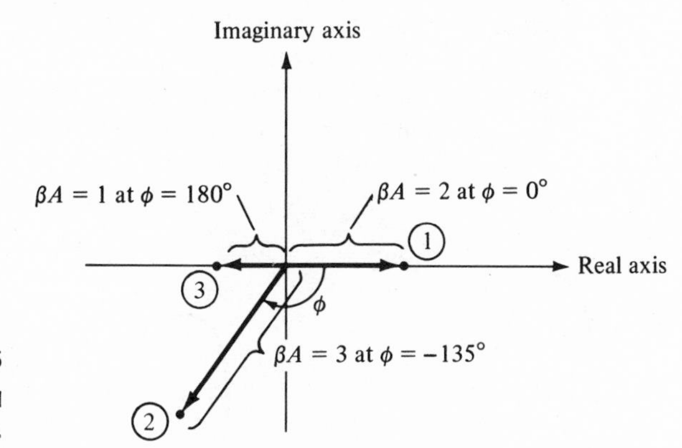

FIG. 12.15
Complex plane showing typical gain-phase points.

phase-shift angles. Using the positive real axis as reference (0°) a magnitude of $\beta A = 2$ is shown at a phase shift of 0° at point 1. Additionally, a magnitude of $\beta A = 3$ at a phase shift of $-135°$ is shown at point 2 and a magnitude/phase of $\beta A = 1$ at 180° is shown at point 3. Thus, points on this plot can represent *both* gain magnitude of βA and phase shift. If the points representing gain and phase shift for an amplifier circuit are plotted at increasing frequency then a Nyquist plot is obtained as shown by the plot in Fig. 12.16. At the origin the gain is 0 at a frequency of 0 (for RC-type coupling). At increasing frequency points f_1, f_2, and f_3 the phase shift increased as did the magnitude of βA. At a representative frequency f_4 the value of A is the vector length from the origin to point f_4 and the phase shift is the angle ϕ. At a frequency f_5 the phase shift is 180°. At higher frequencies the gain is shown to decrease back to 0.

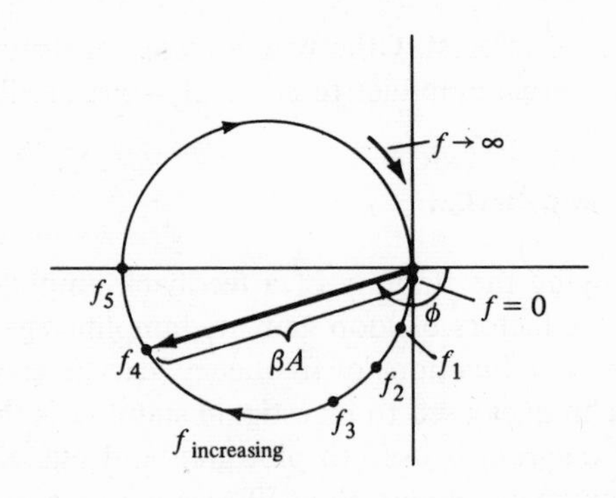

FIG. 12.16
Nyquist plot.

The Nyquist criteria for stability can be stated as follows:

The amplifier is unstable if the Nyquist curve plotted encloses (encircles) the -1 *point, and stable otherwise.*

An example of the Nyquist criteria is demonstrated by the curves in Fig. 12.17. The Nyquist plot in Fig. 12.17a is stable since it does not encircle the -1 point, whereas that shown in Fig. 12.17b is unstable since the curve does encircle the -1 point. Keep in mind that encircling the -1 point means that at a phase shift of 180° the loop gain (βA) is greater than 1, so the feedback signal is in phase with the input and large enough to result in a larger input signal than that applied, with the result that oscillation occurs.

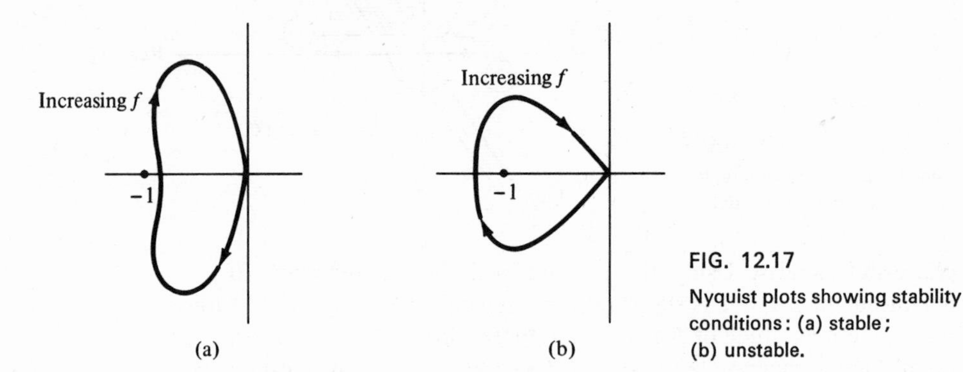

FIG. 12.17
Nyquist plots showing stability conditions: (a) stable; (b) unstable.

Gain and Phase Margins

From the Nyquist criterion we know that a feedback amplifier is stable if the loop gain (βA) is less than unity (0 dB) when its phase angle is 180°. We can additionally determine some margins of stability to indicate how close to instability the unit is. That is, if the gain (βA) is less than unity but, say, 0.95 in value, this would not be as relatively stable as another amplifier having, say, $(\beta A) = 0.7$ (both measured at 180°). Of course, amplifiers with loop gains 0.95 and 0.7 are both stable, but one is closer to instability, if the loop gain increases, than the other. We can define the following terms:

Gain margin (GM) is defined as the value of βA in decibels at the frequency at which the phase angle is 180°. Thus, 0 dB, equal to a value of $\beta A = 1$, is on the border of stability and any negative decibel value is stable. The more negative the decibel gain value the more stable the feedback circuit. The GM may be evaluated in decibels from the curve of Fig. 12.18.

Phase margin (PM) is defined as the angle of 180° minus the magnitude of the angle at which the value βA is unity, 0 dB. The PM may also be evaluated directly from the curve of Fig. 12.18.

An example of these two amplifier factors is shown on the Bode plots of Fig. 12.18. Instability occurs, therefore, with a positive GM and PM greater than 180°.

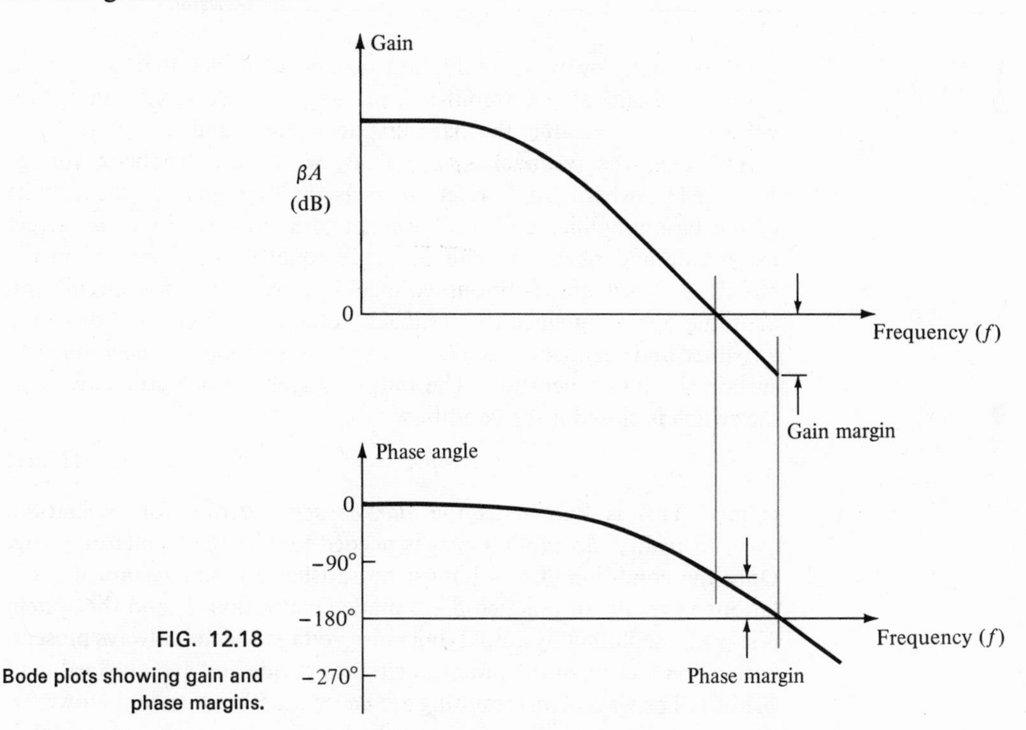

FIG. 12.18
Bode plots showing gain and phase margins.

12.6 OPERATION OF FEEDBACK CIRCUIT AS AN OSCILLATOR

The use of positive feedback results in a feedback amplifier having closed-loop gain A_f greater than open-loop gain A, which results in instability and operation as an oscillator circuit. An oscillator circuit provides a constantly varying amplified output signal. If the output signal varies sinusoidally then the circuit is referred to as a *sinusoidal oscillator*. If the output voltage rises quickly to one voltage level and later drops quickly to another voltage level then the circuit is generally referred to as a *pulse* or *squarewave oscillator*.

To understand how a feedback circuit performs as an oscillator consider the feedback circuit of Fig. 12.19. With the switch at the

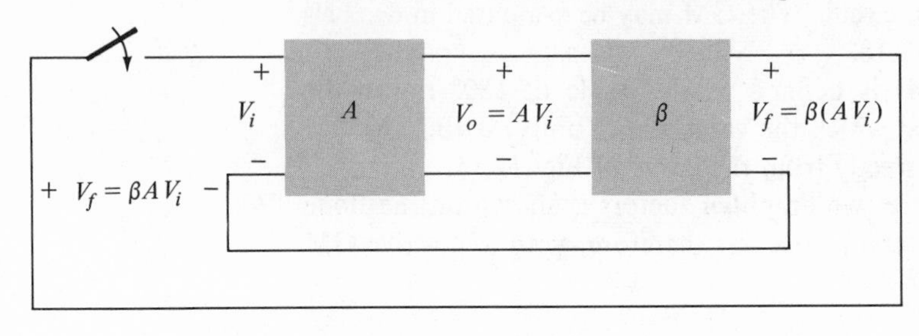

FIG. 12.19

Feedback circuit used as an oscillator.

amplifier input open no oscillation occurs. Consider that we have a *fictitious* voltage at the amplifier input (V_i). This results in an output voltage $V_o = AV_i$ after the base amplifier stage, and a voltage $V_f = \beta(AV_i)$ after the feedback stage. Thus, we have a feedback voltage $V_f = \beta AV_i$, where βA is referred to as the *loop gain.* If the circuits of the base amplifier and feedback network provide βA of a correct magnitude and phase, V_f can be made equal to V_i. Then, when the switch is closed and fictitious voltage V_i is removed, the circuit will continue operating since the feedback voltage is sufficient to drive the amplifier and feedback circuits resulting in a proper input voltage to sustain the loop operation. The output waveform will still exist after the switch is closed if the condition

$$\beta A = 1 \tag{12.21}$$

is met. This is known as the *Barkhausen criterion* for oscillation.

In reality, no input signal is needed to start the oscillator going. Only the condition $\beta A = 1$ must be satisfied for self-sustained oscillations to result. In practice βA is made greater than 1, and the system is started oscillating by amplifying noise voltage which is always present. Saturation factors in the practical circuit provide an "average" value of βA of 1. The waveforms resulting are never exactly sinusoidal However, the closer the value βA is to exactly 1 the more nearly sinusoidal is the waveform. Figure 12.20 shows how the noise signal results in a build-up of a steady-state oscillation condition.

Another way of seeing how the feedback circuit provides operation as an oscillator is obtained by noting the denominator in the basic feedbacke quation, (12.6) $A_f = A/(1 + \beta A)$. When $\beta A = -1$ or magnitude 1 at a phase angle of 180°, the denominator becomes 0 and the gain with feedback, A_f, becomes infinite. Thus, an infinitesimal signal (noise voltage) can provide a measureable output voltage, and the circuit acts as an oscillator even without an input signal.

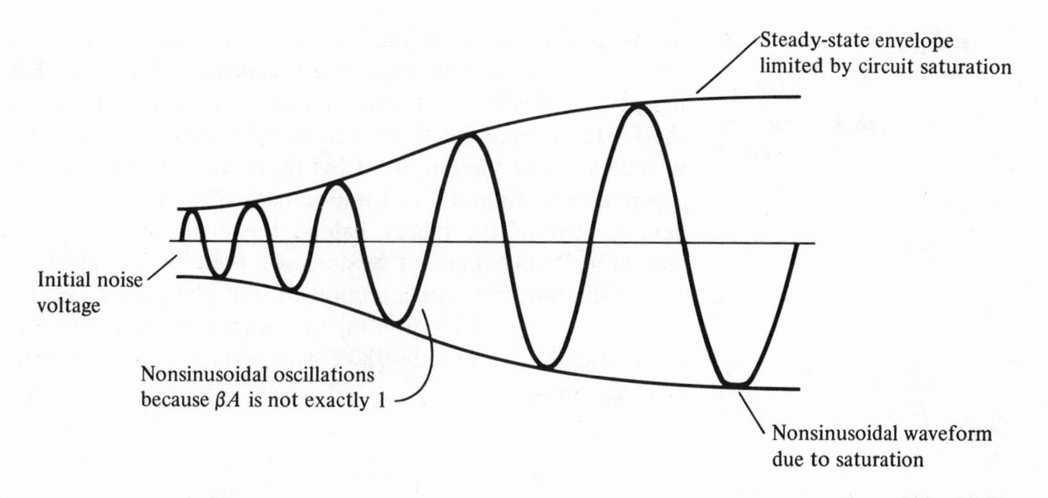

FIG. 12.20

Build-up of steady-state oscillations.

The remainder of this chapter is devoted to various oscillator circuits, using a variety of components. Practical considerations are included so that workable circuits in each of the various cases are discussed.

12.7 PHASE-SHIFT OSCILLATOR

An example of an oscillator circuit which follows the basic development of a feedback circuit is the *phase-shift oscillator*. An idealized version of this circuit is shown in Fig. 12.21. Recall that the require-

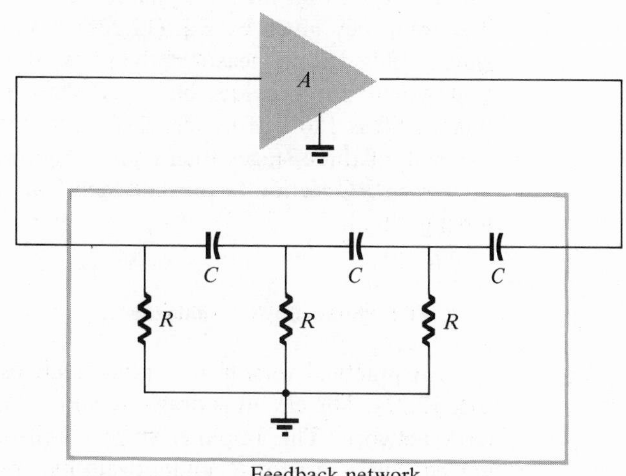

FIG. 12.21

Idealized phase-shift oscillator.

ments for oscillation are that the loop gain, βA is greater than unity *and* that the phase shift around the feedback network is 180° (providing positive feedback). In the present idealization we are considering the feedback network to be driven by a perfect source (zero source impedance) and the output of the feedback network is connected into a perfect load (infinite load impedance). The idealized case will allow development of the theory behind the operation of the phase-shift oscillator. Practical circuit versions will then be considered.

Concentrating our attention on the phase-shift network we are interested in the attenuation of the network at the frequency at which the phase shift is exactly 180°. The results, using classical network analysis, are

$$f = \frac{1}{2\pi RC\sqrt{6}} \tag{12.22a}$$

$$\beta = \frac{1}{29} \tag{12.22b}$$

and the phase shift is 180°.

For the loop gain βA to be greater than unity the gain of the amplifier stage must be greater than $1/\beta$ or 29

$$A > 29 \tag{12.22c}$$

When considering the operation of the feedback network one might naively select the values of R and C to provide (at a specific frequency) 60° phase shift per section for three sections, resulting in 180° phase shift as desired. This, however, is not the case, since each section of the RC in the feedback network loads down the previous one. The net result that the *total* phase shift be 180° is all that is important. The frequency given by Eq. (12.22a) is that at which the *total* phase shift is 180°. If one measured the phase shift per RC section, each section would not provide the same phase shift (although the overall phase shift is 180°). If it were desired to obtain exactly 60° phase shift for each of three stages then emitter follower stages would be needed after each RC section to prevent each from being loaded from the following circuit.

FET Phase-Shift Oscillator

A practical version of a phase-shift oscillator circuit is shown in Fig. 12.22a. The circuit is drawn to show clearly the amplifier and feedback network. The amplifier stage is self-biased with a capacitor bypassed source resistor R_S and a drain bias resistor R_D. The FET device

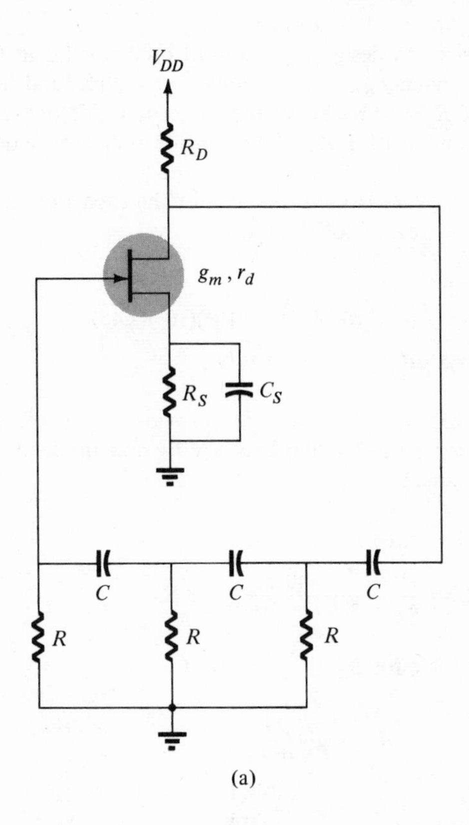

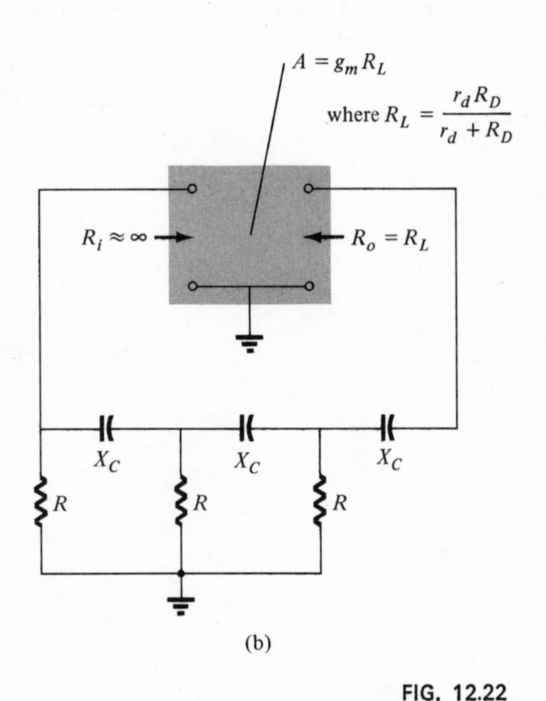

FIG. 12.22
FET phase-shift oscillator circuit.

parameters of interest are g_m and r_d. From FET amplifier theory the amplifier gain is calculated from

$$A = g_m R_L \tag{12.23}$$

where R_L in this case is the parallel resistance of R_D and r_d

$$R_L = \frac{R_D r_d}{R_D + r_d} \tag{12.24}$$

We will assume as a very good approximation that the input impedance of the FET amplifier stage is infinite (see Fig. 12.22b). This assumption is valid as long as the oscillator operating frequency is low enough so that FET capactive impedances can be neglected. The output impedance of the amplifier stage given by R_L should also be small compared to the impedance seen looking into the feedback network so that no attenuation due to loading occurs. In practice these considerations are not always negligible and the amplifier stage gain is then selected somewhat larger than the needed factor of 29 to assure oscillator action.

EXAMPLE 12.9 It is desired to design a phase-shift oscillator (as in Fig. 12.22a) using an FET having $g_m = 5000\ \mu$mhos, $r_d = 40$ K, and feedback circuit value of $R = 10$ K. Select the value of C for oscillator operation at 1 kHz, and R_D for $A > 29$ to ensure oscillator action.

Solution: Equation (12.22a) is used to solve for the capacitor value. Since $f = 1/2\pi RC\sqrt{6}$ we can solve for C

$$C = \frac{1}{2\pi R f \sqrt{6}} = \frac{1}{(6.28)(10 \times 10^3)(10^3)(2.45)}$$

$$C = \mathbf{0.0065\ \mu F}$$

Using Eq. (12.23) we solve for R_L to provide a gain of, say, $A = 40$ (this allows for some loading between R_L and the feedback network input impedance)

$$A = g_m R_L$$

$$R_L = \frac{A}{g_m} = \frac{40}{5000 \times 10^{-6}} = 8\ \text{K}$$

Using Eq. (12.24) we solve for R_D

$$R_L = \frac{R_D r_d}{R_D + r_d}$$

$$8\ \text{K} = \frac{R_D(40\ \text{K})}{R_D + 40\ \text{K}}$$

$$R_D = \mathbf{10\ K}$$

Transistor Phase-Shift Oscillator

Using a transistor as the active element of the amplifier stage, the output of the feedback network is loaded appreciably by the relatively low input resistance (h_{ie}) of the transistor. Of course, an emitter follower input stage followed by a common-emitter amplifier stage could be used. If a single transistor stage is desired, however, the use of voltage-shunt feedback (as shown in Fig. 12.23a) is more suitable. In this connection the feedback signal is coupled through the feedback resistor R_S in *series* with the amplifier stage input resistance (R_i).

An ac equivalent circuit is shown in Fig. 12.23b. The figure shows that the input resistance R_i in series with feedback resistor R', is the parallel combination of resistors R_1, R_2, and h_{ie}. Also, the effective resistance for the third leg of the feedback network, the series combination of resistors R' and R_i, is made the same value as the resistance of the other two resistors of the feedback network to make calculations simpler. We will assume the transistor output impedance $1/h_{oe}$ is much larger than R_C.

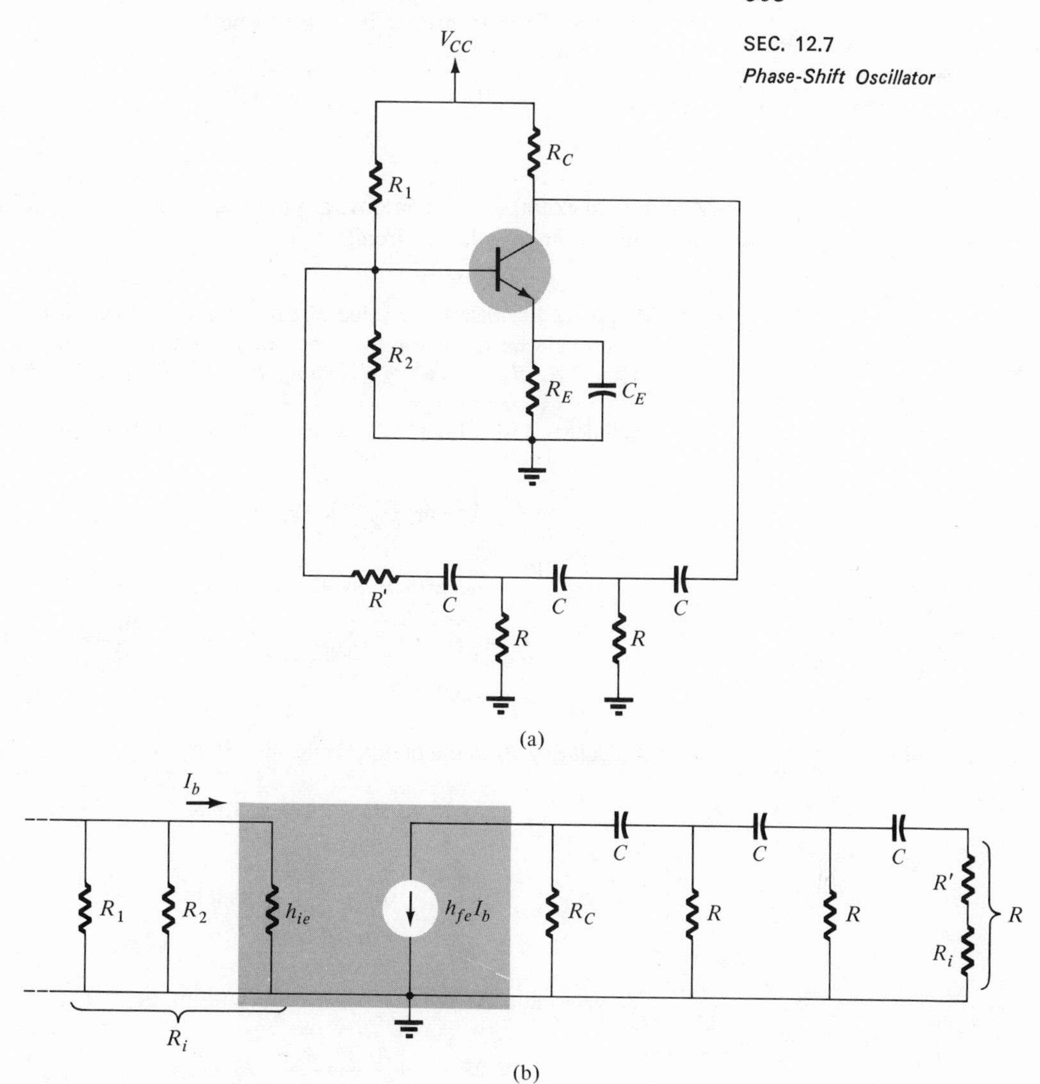

FIG. 12.23
Transistor phase-shift oscillator: (a) transistor circuit; (b) ac equivalent circuit.

Analysis of the ac circuit provides the following equation for the resulting oscillator frequency

$$f = \left(\frac{1}{2\pi RC}\right)\frac{1}{\sqrt{6 + 4(R_C/R)}} \tag{12.25}$$

For the loop gain to be greater than unity, the requirement on the current gain of the transistor is found to be

$$h_{fe} > 23 + \frac{R_C}{29R} + \frac{R}{4R_C} \tag{12.26}$$

A practical example will demonstrate the use of the above information in designing an oscillator circuit.

EXAMPLE 12.10 Select the value of capacitor C and transistor gain h_{fe} to provide an oscillator frequency of $f = 2$ kHz. Circuit values are $h_{ie} = 2$ K, $R_1 = 20$ K, $R_2 = 80$ K, $R_C = 10$ K, and $R = 8$ K.

Solution: Using Eq. (12.25) we can determine the required value of C

$$f = \left(\frac{1}{2\pi RC}\right)\frac{1}{\sqrt{6 + 4R_C/R}}$$

$$2 \times 10^3 = \left[\frac{1}{6.28(8 \times 10^3)C}\right]\frac{1}{\sqrt{6 + 4(1.25)}}$$

$$C = \left[\frac{1}{6.28(8 \times 10^3)(2 \times 10^3)}\right]\frac{1}{3.32} = \frac{10^{-6}}{332} = 3 \times 10^{-9}$$

$$= \mathbf{0.003\ \mu F}$$

Calculating R_i as the parallel resistance of R_1, R_2, and h_{ie} gives

$$R_i = 20\text{ K} \,||\, 80\text{ K} \,||\, 2\text{ K} \cong 1.8\text{ K}$$

For

$$R' + R_i = R = 8\text{ K}$$

$$R' = R - R_i = 8 - 1.8 = 6.2\text{ K}$$

To determine the value of h_{fe} necessary [using Eq. (12.26)]

$$h_{fe} > 23 + 29\frac{R}{R_C} + 4\frac{R_C}{R} = 23 + 29\left(\frac{8}{10}\right) + 4\left(\frac{10}{8}\right)$$

$$= 23 + 23.2 + 5 = \mathbf{51.2}$$

A transistor with $h_{fe} > 51.2$ will provide sufficient loop gain for the circuit to operate as an oscillator. Practically, a transistor with at least $h_{fe} > 60$ would be selected.

Phase-shift oscillators are suited to operating frequencies in the range of a few hertz to a few hundred kilohertz. Other oscillator configurations (typically the tuned circuits) are more suitable to frequency in the megahertz range. To adjust the frequency of a phase-shift oscillator it is necessary to operate the three feedback capacitors as a single ganged component so that all three capacitor values are kept the same. A phase-shift oscillator is operated class-A to keep distortion low while providing a sinusoidal output waveform taken from the output of the amplifier stage. It should be clear that the output signal should not directly drive any low-impedance circuit which will load down the output. This would decrease the output voltage and thereby drop the loop gain below the necessary value for oscillator action. Feeding the output to a high-impedance stage, such as a FET amplifier stage or emitter follower transistor stage, provides negligible loading of this oscillator circuit.

12.8 THE LC-TUNED OSCILLATOR CIRCUIT

An oscillator circuit can be made using a transformer for the feedback network. In addition the inductance of the transformer and a parallel capacitor can be used to tune the circuit to the desired oscillator frequency. Figure 12.24a shows a FET amplifier with positive feedback provided by the transformer and tuning by an *LC* circuit. The circuit is described as a tuned-drain, untuned-gate oscillator.

An ac equivalent circuit is shown in Fig. 12.24b. The input resistance to the gate is assumed very high and is shown as an open circuit. The FET ac equivalent circuit is shown as a current source ($g_m V_{gs}$) in parallel with an output resistance r_d. The transformer can be represented as

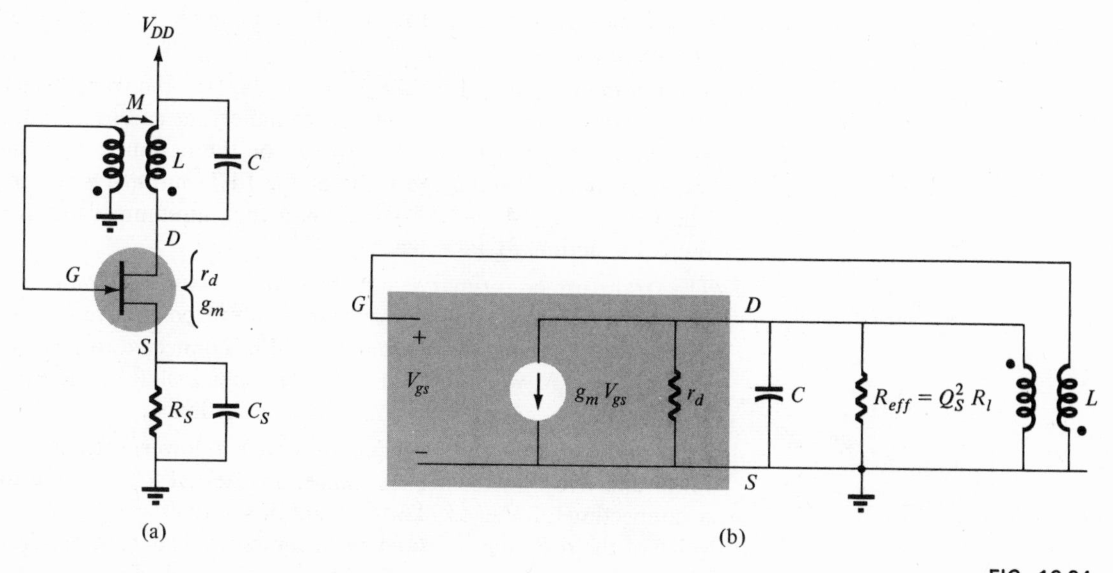

FIG. 12.24
Tuned-drain oscillator circuit.

an inductance L in the primary side and a mutual coupling factor M. The series resistance of the transformer (representing its losses) can be accounted for by an effective resistance in parallel with the primary shown as $R_{eff} = Q_s^2 R_s$ in the circuit. The factor Q_s is the series Q of the transformer, defined as

$$Q_s = \frac{\omega_o L}{R_s} \tag{12.27}$$

Analysis of the ac equivalent circuit provides the following results

$$f_o = \frac{1}{2\pi\sqrt{LC}} \tag{12.28}$$

$$g_m = \frac{1}{R}\frac{L}{M} \tag{12.29}$$

where

$$R = r_d || Q_s^2 R_s \tag{12.30}$$

A summary of the results provided in the above equations follows:

1. The oscillator frequency is determined by the LC-resonant ("tank") circuit. This relationship is sometimes modified slightly due to circuit nonlinearities and unaccounted for resistances, capacitances, and so on. However, it is quite good as a first-order approximation.
2. The minimum required FET g_m is dependant on the transformer effective resistance, the FET output resistance, the coil inductance and mutual coupling—in other words, on the parameters of the transformer chosen and the value of the FET output resistance. The value of g_m should be larger than this minimum value for oscillator action to take place.
3. The transformer primary and secondary windings must be connected in proper polarity sense to result in positive feedback. For this to occur the transformer should be connected to provide 180° phase shift, which, added to the 180° phase shift of the FET amplifier stage, results in overall positive feedback.
4. Loading of the circuit provided by either a lower value of r_d, lower effective transformer resistance, or external loading due to a connection of the oscillator to another circuit results in the value of the resulting resistance R being lower. This then requires a larger value of g_m to provide sufficient loop gain for oscillator action.

Another example of a tuned-circuit oscillator is the tuned-collector circuit shown in Fig. 12.25. The primary side of the transformer forms a tuned-tank circuit to set the oscillator frequency. The transformer is connected to provide positive feedback and the amplifier provides sufficient gain for oscillator action to take place.

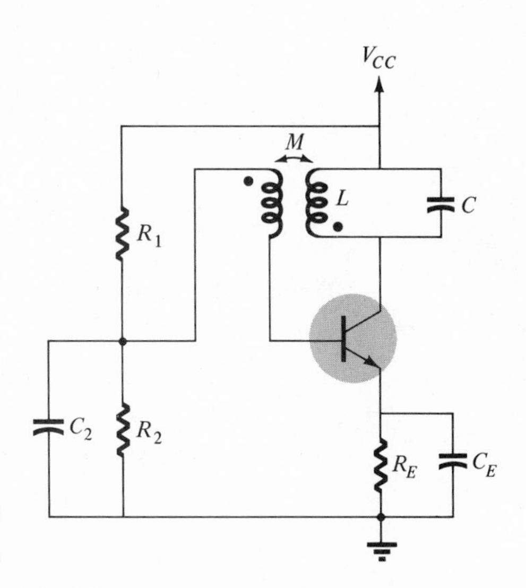

FIG. 12.25
Bipolar transistor-tuned LC oscillator circuit.

Resistors R_1, R_2, and R_E are used to dc bias the transistor. Capacitors C_E and C_2 act to bypass resistors R_E and R_2, respectively, so that they have no effect on the ac operation of the circuit. Notice that although the low-resistance secondary winding of the transformer provides the dc bias voltage set by R_1 and R_2 to be connected to the base, the secondary essentially provides an ac feedback voltage in shunt with the transistor base emitter since the junction point of R_1 and R_2 is at ac ground (due to bypass capacitor C_2).

EXAMPLE 12.11 For the oscillator circuit of Fig. 12.24 and the following circuit values calculate the circuit frequency of oscillation and the minimum gain (g_m) of the FET unit: $r_d = 40$ K, $L = 4$ mH, $M = 0.1$ mH, $R_s = 50\ \Omega$, and $C = 0.001\ \mu$F.

Solution: The resonant frequency of the oscillator circuit is calculated using Eq. (12.28)

$$f_o = \frac{1}{2\pi\sqrt{LC}} = \frac{1}{6.28\sqrt{(4 \times 10^{-3})(0.001 \times 10^{-6})}}$$

$$= \frac{10^6}{6.28(2)} \cong \mathbf{80\ kHz}$$

$$\omega_o = 2\pi f_o = 6.28(80\text{ KHz}) = 500 \times 10^3 \text{ rad/sec}$$

The coil Q is then

$$Q_s = \frac{\omega_o L}{R_s} = \frac{(500 \times 10^3)(4 \times 10^{-3})}{50} = 40$$

The effective coil resistance is

$$R_{eff} = Q_s^2 R_s = (40)^2(50) = 80\ \text{K}$$

Using Eq. (12.30) to calculate R

$$R = \frac{r_d R_{eff}}{r_d + R_{eff}} = \frac{40(80)}{40 + 80} = 26.7\ \text{K}$$

The minimum value of g_m can now be calculated using Eq. (12.29)

$$g_m = \frac{1}{R}\frac{L}{M} = \frac{1}{26.7 \times 10^3}\frac{4 \times 10^{-3}}{0.1 \times 10^{-3}} = 1.5 \times 10^{-3} = \mathbf{1500\ \mu mhos}$$

The FET selected should have a value of g_m greater than 1500 μmhos.

12.9 TUNED-INPUT, TUNED-OUTPUT OSCILLATOR CIRCUITS

A variety of circuits shown in Fig. 12.26 provide tuning in both the input and output sections of the circuit. Analysis of the circuit of Fig. 12.26 reveals that the following types of oscillators are obtained when the reactance elements are as designated.

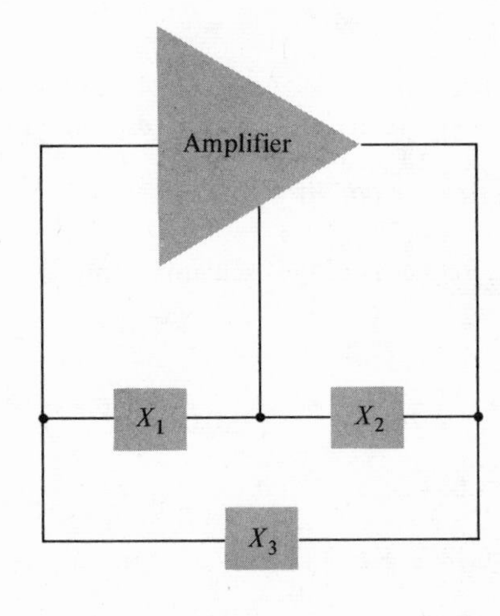

FIG. 12.26
Basic configuration of resonant circuit oscillator.

OSCILLATOR TYPE	REACTANCE ELEMENTS		
	X_1	X_2	X_3
1. Colpitts oscillator	C	C	L
2. Hartley oscillator	L	L	C
3. Tuned input, tuned output	L	L	—

Regardless of the nature of the active amplifier component (tube, FET, or transistor) the configurations shown in Fig. 12.26 will be true of those circuits discussed below.

The circuit form of the oscillator using a FET amplifier is shown in Fig. 12.27a. For operation as an oscillator the Barkhausen criterion is

$$\beta A = 1$$

The amplifier gain A is given simply by

$$A = -g_m Z_L \tag{12.31}$$

where Z_L is a parallel combination of impedances

$$Z_L = r_d \,||\, X_2 \,||\, (X_1 + X_3) \tag{12.32}$$

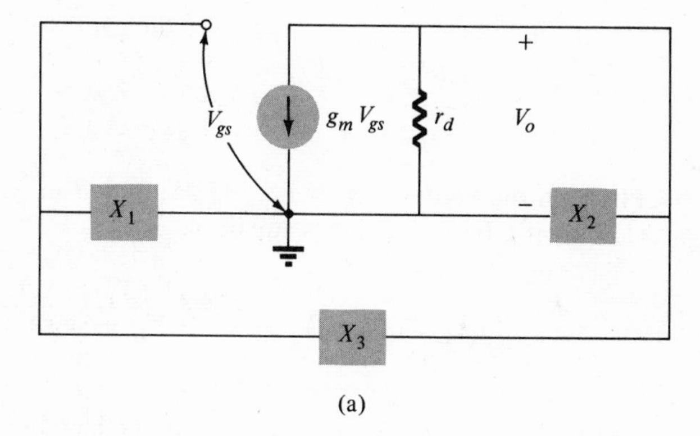

(a)

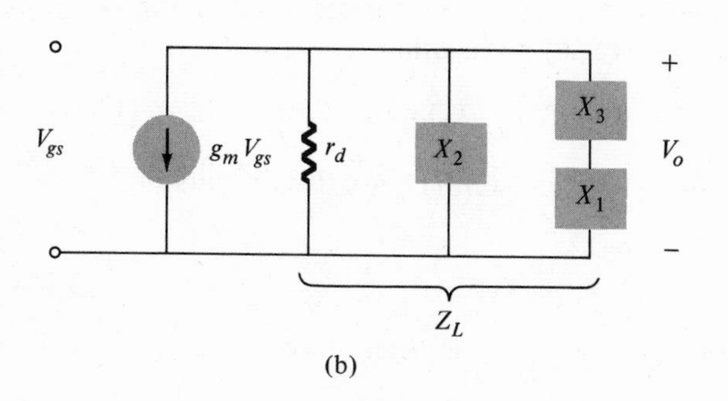

(b)

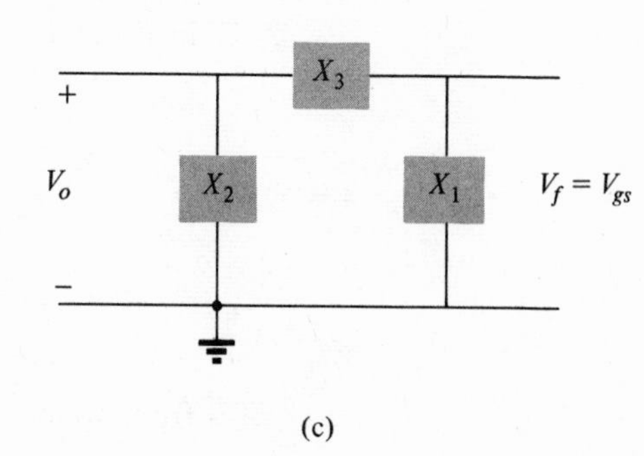

(c)

FIG. 12.27
Basic resonant circuit oscillator using FET amplifier.

The circuit is redrawn in Fig. 12.27b so that the gain network is clearly shown. Figure 12.27b shows the parallel components which form an equivalent ac impedance, Z_L, as given in Eq. (12.32).

Figure 12.27c shows the feedback section of the circuit. Notice that the output voltage (V_o) is developed across X_2 and that the resulting feedback voltage (V_f) across X_1 is the input voltage to the amplifier (V_g). The feedback factor β is given by

$$\boxed{\beta = \frac{X_1}{X_1 + X_3}} \tag{12.33}$$

Plugging Eq. (12.31) for A and Eq. (12.33) for β into the basic equation (12.21) and using Eq. (12.32) for Z_L provides a means of determining the necessary device gain (g_m) and resulting oscillator frequency (f_o).

Expressing Eq. (12.32) for Z_L in more detailed form

$$Z_L = \frac{[r_d X_2/(r_d + X_2)](X_1 + X_3)}{[r_d X_2/(r_d + X_2)] + (X_1 + X_3)}$$

$$= \frac{r_d X_2(X_1 + X_3)}{r_d X_2 + (X_1 + X_3)(r_d + X_2)} \tag{12.34}$$

Using Eq. (12.34) in Eq. (12.31) gives the gain A

$$A = \frac{-g_m r_d X_2(X_1 + X_3)}{r_d(X_1 + X_2 + X_3) + X_2(X_1 + X_3)} \tag{12.35}$$

Plugging the value of A in Eq. (12.35) and β in Eq. (12.33) into Eq. (12.21), the basic oscillator equation, gives

$$\beta A = \frac{X_1}{(X_1 + X_3)} \frac{-g_m r_d X_2(X_1 + X_3)}{r_d(X_1 + X_2 + X_3) + X_2(X_1 + X_3)} = 1 \tag{12.36}$$

The Barkhausen criterion requires that (1) there be no imaginary terms in βA and (2) the value of βA equal 1 (or be greater than 1).

A close look at Eq. (12.36) reveals that

$$r_d(X_1 + X_2 + X_3) = 0 \tag{12.37}$$

for no resulting imaginary parts in Eq. (12.36). This results in

$$\frac{-g_m r_d X_1}{X_1 + X_3} = 1$$

from which the necessary value of g_m is determined to be

$$g_m = \frac{X_1 + X_3}{r_d X_1} \tag{12.38}$$

From Eq. (12.37) we note that since

$$(X_1 + X_2 + X_3) = 0 \tag{12.39}$$

one of the inductive components must be different than the other two. For example, with

$$(X_1 + X_3) = -X_2 \tag{12.40}$$

Eq. (12.37) is satisfied. Using the result of Eq. (12.40) in Eq. (12.38) gives

$$g_m = -\frac{(X_1 + X_3)}{r_d X_1} = -\frac{(-X_2)}{r_d X_1} = \frac{X_2}{r_d X_1}$$

A summary of the above discussion is that for oscillation to take place the FET gain must be at least

$$\boxed{g_m = \frac{X_2}{r_d X_1}} \tag{12.41}$$

The oscillator frequency is then obtained from Eq. (12.39)

$$(X_1 + X_2 + X_3) = 0$$

In addition, Eq. (12.41) shows that X_1 and X_2 must be the same type (both inductors or both capacitors) so that g_m is a real positive value and Eq. (12.42) shows that X_3 must be opposite X_1 or X_2. With X_1, X_2 capacitors and X_3 an inductor the circuit is called a Colpitts oscillator, and with X_1, X_2 inductors and X_3 a capacitor the circuit is called a Hartley oscillator.

12.10 COLPITTS OSCILLATOR

FET Colpitts oscillator

A practical version of a FET Colpitts oscillator is shown in Fig. 12.28a. The circuit is basically the same form as shown in Fig. 12.27 with the addition of the components needed for dc bias of the FET amplifier. Figure 12.28b shows the ac equivalent of the FET Colpitts oscillator. Comparing this with Fig. 12.27 we see that the circuits are the same with $X_1 = X_{C_1}$ $X_3 = X_{C_2}$ and $X_3 = X_L$. From the previous analysis of this circuit form we can obtain the following results. Using Eq. (12.41) with the present circuit inductive components

$$g_m \geq \frac{X_2}{r_d X_1} = \frac{1/j\omega C_2}{r_d(1/j\omega C_1)} = \frac{C_1}{r_d C_2} \tag{12.42}$$

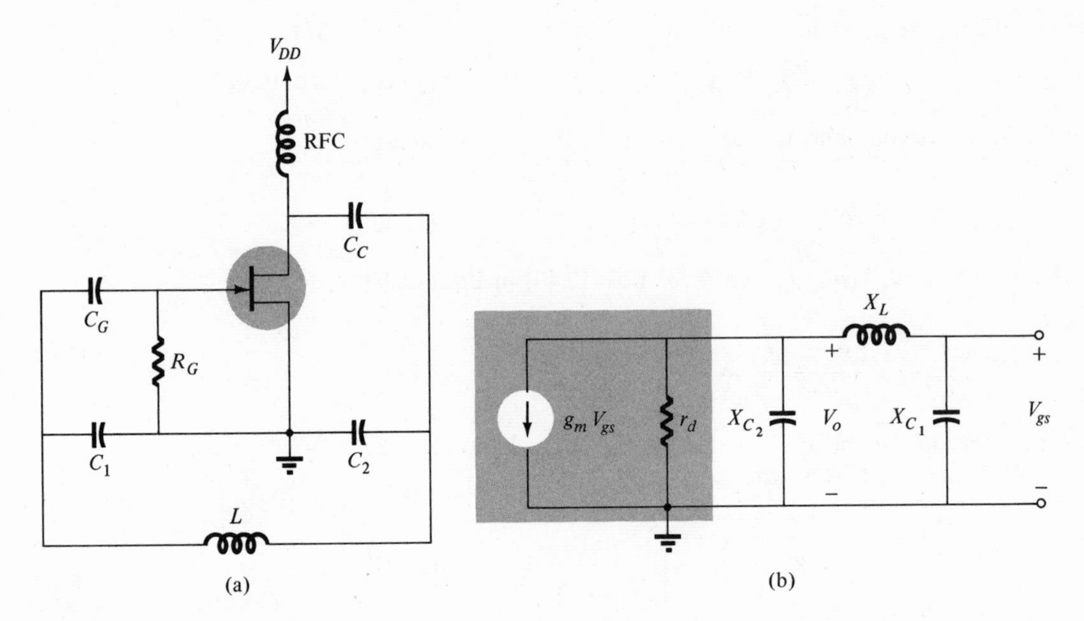

FIG. 12.28
FET Colpitts oscillator.

Equation (12.39) results in

$$\frac{1}{j\omega C_1} + \frac{1}{j\omega C_2} + j\omega L = 0$$

$$\frac{j}{\omega}\left(\frac{C_1 + C_2}{C_1 C_2}\right) = -j\omega L$$

$$\omega^2 = \frac{1}{LC_{eq}}, \qquad \text{where } C_{eq} = \frac{C_1 C_2}{C_1 + C_2} \tag{12.43}$$

$$\omega_0^2 = \frac{1}{\sqrt{LC_{eq}}}$$

$$\boxed{f_o = \frac{1}{2\pi\sqrt{LC_{eq}}}} \tag{12.44}$$

DC CIRCUIT AND COUPLING COMPONENTS

Referring to Fig. 12.28a dc bias is provided by supply voltage V_{DD}, RFC (radio frequency choke), gate-leak bias network C_G and R_G. Capacitor C_C couples the ac output voltage to the feedback network without affecting the circuit dc bias.

The gate-leak bias develops a negative dc voltage from gate to source due to the average rectified sinusoidal signal fed to the amplifier input. Since the FET will conduct when the gate-source goes positive it acts to rectify the input voltage with a dc value developed across capacitor C_G, providing a negative bias voltage. Choosing the time constant of $R_G C_G$ large compared to the time constant of the oscillator $(1/f_o)$ results in the ac oscillator signal at the gate riding on a negative bias voltage. The dc bias voltage is affected by the amplitude of the oscillator signal, providing more negative bias for larger signal amplitude. As a rule of thumb the time constant of the gate-leak bias circuit should be about 10 times greater than the oscillator time constant.

$$R_G C_G \geq \frac{10}{f_o} \tag{12.45}$$

The RFC choke along with the capacitor provide for feeding the ac oscillator signal to the feedback network while blocking any ac from the power supply.

As a general rule we should expect the coil impedance at least 10 times the impedance of the coupling capacitor

$$X_L \geq 10 X_C \tag{12.46}$$

The impedance of the choke is selected to be very high at the oscillator frequency while the impedance of C_C is very low. This results in a decoupling action which passes almost all the oscillator signal to the feedback network while greatly attenuating any ac in the supply lines. In addition the choke provides dc bias to operate the FET and capacitor C_C blocks any dc from the feedback circuit.

An example of a practical circuit design should help emphasize the important points covered above.

EXAMPLE 12.12 For the circuit of Fig. 12.28a and circuit values below, determine the needed value of FET g_m and circuit frequency of oscillation (f_o). Also, check values given for proper circuit operation.

$$r_d = 100\text{ K},\ C_1 = 500\text{ pF},\ C_2 = 2000\text{ pF},\ L = 25\ \mu\text{H},$$

$$R_g = 0.5\text{ M},\ C_g = 500\text{ pF},\ L_{RFC} = 0.5\text{ mH},\ C_C = 1000\text{ pF}$$

Solution: From Eq. (12.42)

$$g_m \geq \frac{C_1}{r_d C_2} = \frac{500\text{ pF}}{(100\text{ K})(2000\text{ pF})} = 2.5 \times 10^{-6} = 2500\text{ mmhos}$$

The equivalent capacitance calculated using Eq. (12.43) is

$$C_{eq} = \frac{C_1 C_2}{C_1 + C_2} = \frac{500(2000)}{500 + 2000} = 400 \text{ pF}$$

and using Eq. (12.44) the circuit resonant frequency is

$$f_o = \frac{1}{2\pi\sqrt{LC_{eq}}} = \frac{1}{6.28\sqrt{(25 \times 10^{-6})(400 \times 10^{-12})}}$$

$$= \frac{1}{6.28\sqrt{10^{-14}}} \cong 1.6 \text{ MHz.}$$

Checking the self-bias circuit time constant [using Eq. (12.45)]

$$R_g C_g \geq \frac{10}{f_o}$$

$$(0.5 \times 10^6)(500 \times 10^{-12}) \geq \frac{10}{1.6 \times 10^6} = 6.3 \times 10^{-6}$$

$$250 \times 10^{-6} \geq 6.3 \times 10^{-6}$$

therefore, the self-bias values are correct.
Checking RFC coil at $f_o = 1.6$ MHz,

$$X_L = 2\pi f_o L_{RFC} = 6.28(1.6 \times 10^6)(0.5 \times 10^{-3})$$
$$= 5 \text{ K}$$

Checking C_C impedance at $f_o = 1.6$ MHz

$$X_C = \frac{1}{\omega_o C_C} = \frac{1}{6.28(1.6 \times 10^6)(100 \times 10^{-12})} = 100\ \Omega$$

Since X_C is much smaller than X_L of the RFC coil, the ac oscillator signal from the FET drain is coupled almost entirely to the feedback network through coupling capacitor C_C. In addition any signal coming from the power supply is attenuated by the RFC choke and coupling capacitor C_C so that very little reaches the osicllator loop.

Vacuum-Tube Colpitts Oscillator

A vacuum-tube version of a Colpitts oscillator is similar to that using a FET as seen in Fig. 12.29. Circuit operation can be obtained using Eqs. (12.42)–(12.46). However, the tube equivalent parameters are usually specified as μ and r_p, which can be related to g_m and r_d in the FET circuit by using r_p instead of r_d and $\mu = g_m r_d$. Thus, Eq. (12.42) is now

$$\mu \geq \frac{C_1}{C_2}$$

for oscillator action to occur (loop gain greater than unity).

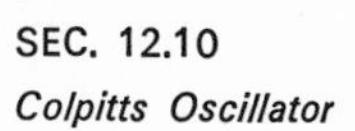

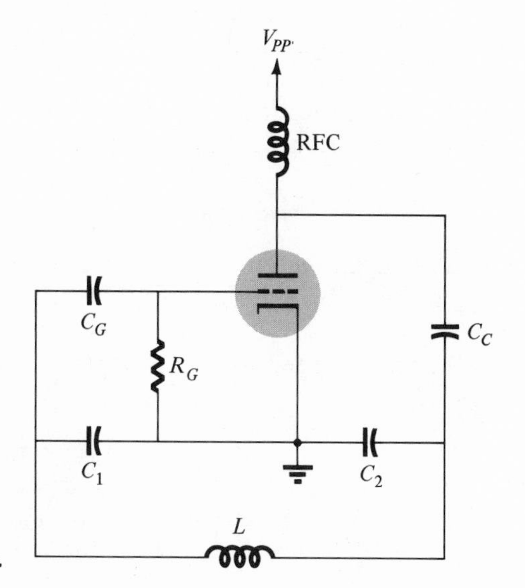

FIG. 12.29
Vacuum-tube Colpitts oscillator.

Transistor Colpitts Oscillator

A transistor Colpitts oscillator circuit can be more complex, as shown in Fig. 12.30. In addition the circuit calculations are more

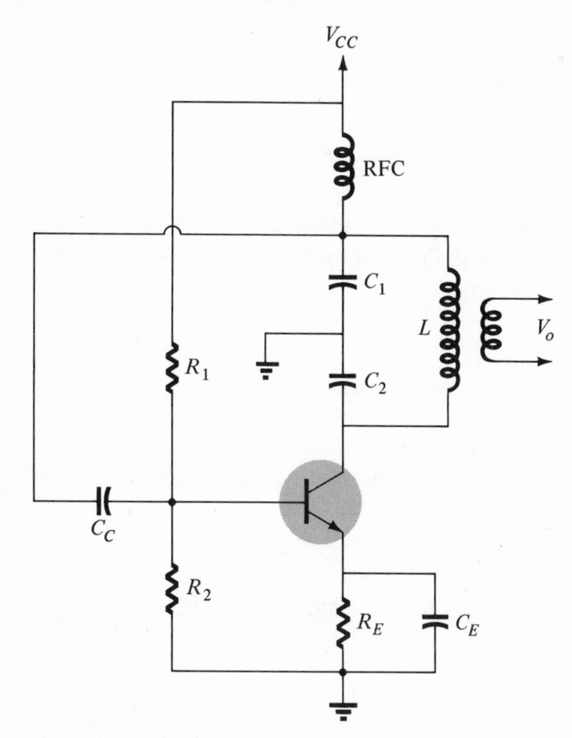

FIG. 12.30
Transistor Colpitts oscillator.

involved because of the low input impedance of the transistor (as compared to an essentially open circuit for a FET or vacuum-tube). In order to consider some details of circuit operation the dc and ac equivalent circuits will be handled separately. Also, simplifying approximations will be made wherever possible—these simplifications being specifically mentioned when used.

AC CIRCUIT OPERATION

A simplified ac circuit is shown in Fig. 12.31a. Only the basic amplifying transistor and feedback network are shown. The RFC coil, coupling capacitor C_C, bypass capacitor C_E, and bias resistors R_1, R_2, and R_E are neglected for the present.

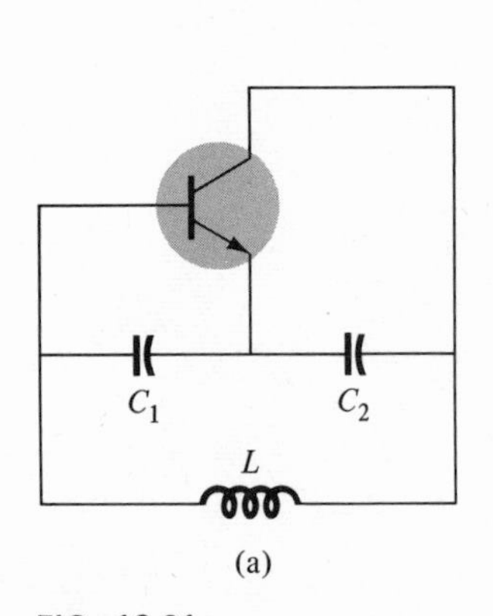

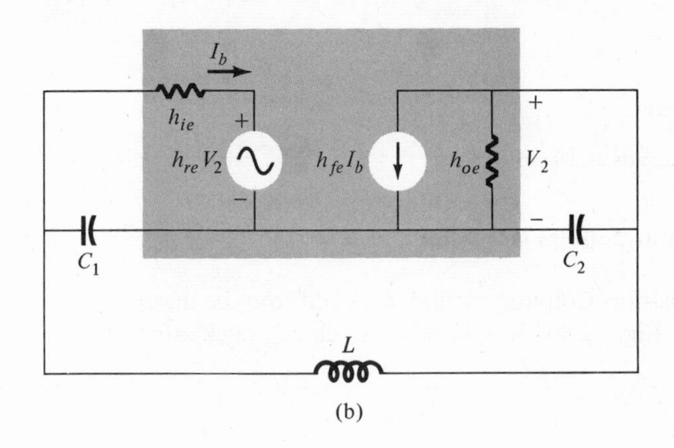

FIG. 12.31
ac Colpitts oscillator circuit: (a) ac part of circuit; (b) ac equivalent circuit.

Using a hybrid equivalent transistor circuit, as in Fig. 12.31b the condition of loop gain being greater than unity for oscillator action results in*

$$h_{fe} \geq \Delta^{he} \frac{C_2}{C_1} \tag{12.47a}$$

$$\Delta^{he} = h_{ie}h_{oe} - h_{re}h_{fe} \tag{12.47b}$$

Determination of the circuit resonant frequency results in

$$\omega_o^2 = \frac{h_{oe}}{C_1 C_2 h_{ie}} + \frac{1}{L}\left(\frac{1}{C_1} + \frac{1}{C_2}\right)$$

A good approximation neglecting the first term is

* See Eq. (4-64) in Robert Sentz and Robert Bartkowiak, *Feedback Amplifiers and Oscillators*, New York: Holt, Reinhart & Winston, 1968.

$$\omega_o = \frac{1}{\sqrt{LC_{eq}}} \qquad (12.48)$$

where

$$C_{eq} = \frac{C_1 C_2}{C_1 + C_2}$$

(It is interesting to note that the resonant frequency is basically that due to the tank circuit made of inductor L and the series equivalent of capacitors C_1 and C_2).

One ac circuit approximation to be considered is the ac bypass of resistor R_E by capacitor C_E for which we can state

$$X_{C_E} \leq \frac{1}{10} R_E \qquad (12.49)$$

for satisfactory operation (at the circuit resonant frequency).

A second approximation to be checked is that the RFC coil impedance is much larger than the coupling capacitor impedance:

$$X_L \geq 10 X_C \qquad (12.50)$$

at the circuit resonant frequency.

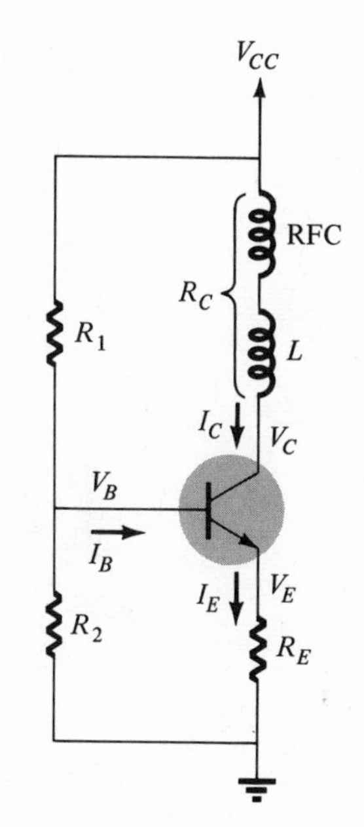

FIG. 12.32
dc bias part of transistor-Colpitts oscillator circuit.

DC CIRCUIT OPERATION

The circuit components affecting dc bias are shown in Fig. 12.32. The circuit is essentially a voltage-divider stabilized dc bias circuit, as discussed in Chapter 4. One main difference is that the effective collector dc resistance is that of the coil dc resistances of the RFC choke and feedback network inductor. Selection of a bias condition operating the circuit initially in class-A is usual.

EXAMPLE 12.13 For the circuit of Fig. 12.30 and circuit values listed below calculate the value of C_1, C_2, and L needed for operation at $f_o = 2$ mHz.

$$h_{ie} = 2\text{ K},\ h_{oe} = 10\ \mu\text{mhos},\ h_{fe} = 50,\ h_{re} = 0.5 \times 10^{-4}$$

Solution: Calculating Δ^{h^e} [Eq. (12.47b)]

$$\Delta^{h^e} = h_{ie}h_{oe} - h_{fe}h_{re} = (2 \times 10^3)(10 \times 10^{-6}) - (50)(0.5 \times 10^{-4})$$
$$= 20 \times 10^{-3} - 2.5 \times 10^{-3} = 17.5 \times 10^{-3}$$

From Eq. (12.47a) we get

$$\frac{h_{fe}}{\Delta^{h^e}} \geq \frac{C_2}{C_1} = \frac{50}{17.5 \times 10^{-3}} = 2.86 \times 10^3$$

Choosing C_1 of 20 pF as a reasonable small value we calculate C_2

$$C_2 = (2.86 \times 10^3)(20 \text{ pF})$$
$$= 0.0572\ \mu\text{F} \cong \mathbf{0.06\ \mu F}$$

Using Eq. (12.48)

$$C_{eq} = \frac{C_1 C_2}{C_1 + C_2} = \frac{(20 \text{ pF})(0.06\ \mu\text{F})}{(20 \text{ pF} + 0.06\ \mu\text{F})} \cong 20 \text{ pF}$$

and at $f_o = 2$ mHz,

$$\omega_o = 2\pi f_o = \frac{1}{\sqrt{LC_{eq}}}$$
$$6.28(2 \times 10^6) = \frac{1}{\sqrt{L(20 \times 10^{-12})}}$$

so that

$$L = \frac{1}{(12.56 \times 10^6)^2(20 \times 10^{-12})} = \frac{1}{3160} = \mathbf{0.316\ mH}$$

EXAMPLE 12.14 Using the circuit of Fig. 12.30, circuit values of Example 12.13, and additional circuit values below, check the ac approximations and dc bias of the circuit.

$L_{RFC} = 2$ mH, $R_{RFC} = 200\ \Omega$, $C_C = 0.1\ \mu$F, $C_E = 0.1\ \mu$F,
$R_1 = 10$ K, $R_2 = 5$ K, $R_E = 300\ \Omega$, $V_{CC} = +12$ V, $R_L = 150\Omega$

Solution: Checking the effectiveness of bypass capacitor C_E at $f_o =$ 2 mHz

$$X_{C_E} = \frac{1}{\omega_o C_E} = \frac{1}{6.28(2 \times 10^6)(0.1 \times 10^{-6})}$$
$$= 0.8 \leq \frac{1}{10} R_E = \frac{300}{10} = 30$$

Checking RFC coil and coupling capacitor impedances,

$$X_L \geq 10 Xc$$
$$\omega_o L_{RFC} = 6.28(2 \times 10^6)(2 \times 10^{-3})$$
$$\cong 25 \times 10^3 \geq 10 \frac{1}{\omega_o C} = \frac{10}{6.28(2 \times 10^6)(0.1 \times 10^{-6})} = 8$$

dc bias voltage calculations give

$$V_B \cong \frac{R_2}{R_1 + R_2}(+V_{CC}) = \frac{5}{10 + 5}(12) = 4 \text{ V} \cong V_E$$
$$I_C \cong I_E = \frac{V_E}{R_E} = \frac{4 \text{ V}}{0.3 \text{ K}} = 13.3 \text{ mA}$$
$$V_C = V_{CC} - I_C R_C = 12 - 13.3 \text{ mA } (200 + 150) = 7.35 \text{ V}$$
$$V_{CE} = V_C - V_E = 7.35 - 4 = 3.35 \text{ V}$$

The Colpitts oscillator using FET, vacuum-tube, or transistor amplifiers can be tuned by a variable core inductor or split-stator tuning capacitor. The later tunes C_1 and C_2 simultaneously to provide required circuit operating ratio between C_1 and C_2. It should be kept in mind that stray circuit capacitances must be included in calculations of tank circuit resonant frequency, especially at higher operating frequencies.

12.11 HARTLEY OSCILLATOR

If the elements in the basic resonant circuit of Fig. 12.27 are X_1 and X_2—inductors, X_3—capacitor, then the circuit is a Hartley oscillator.

FET Oscillator

A FET Hartley oscillator circuit is shown in Fig. 12.33. The circuit is drawn so that the feedback network conforms to the form shown in the basic resonant circuit (Fig. 12.27). Note, however, that inductors L_1 and L_2 have a mutual coupling, M, which must be taken into account in determining the equivalent inductance for the resonant tank circuit.

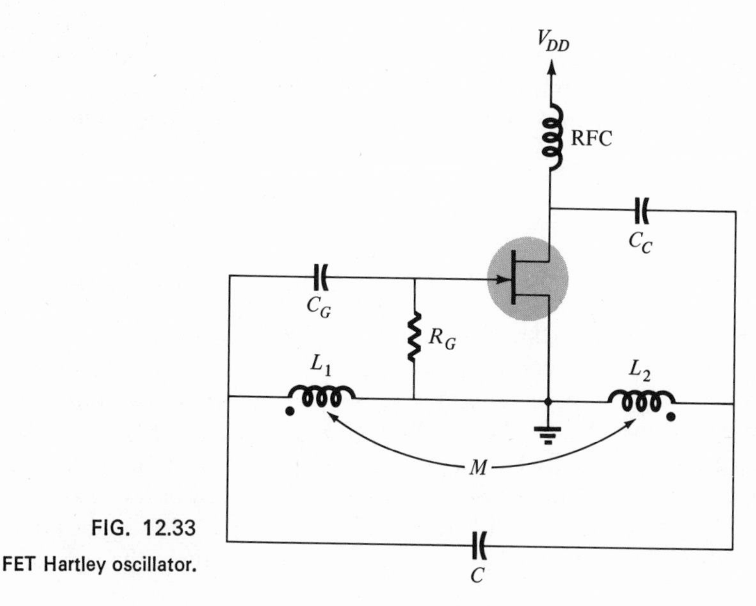

FIG. 12.33
FET Hartley oscillator.

We can obtain an ac equivalent circuit for Fig. 12.33 by replacing capacitors C_C and C_G by shorts, RFC coil by an open circuit, and the FET by its equivalent circuit as shown in Fig. 12.34. The tank circuit of mutually linked inductors L_1 and L_2 and capacitor C can be shown

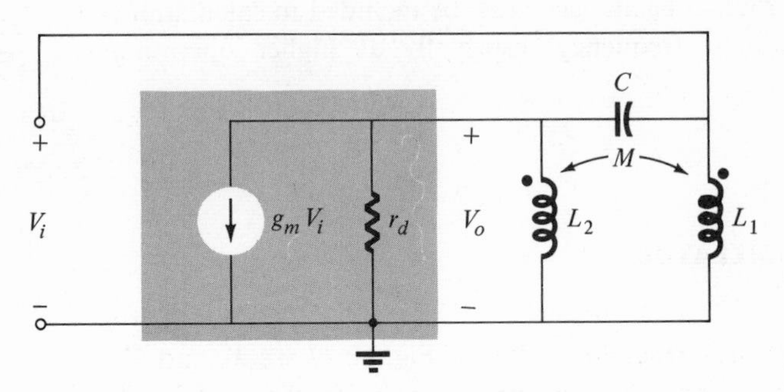

FIG. 12.34
FET Hartley oscillator ac equivalent circuit.

equivalent to a tank circuit of capacitor C in parallel with an equivalent inductance

$$L_{eq} = L_1 + L_2 + 2\,\mathrm{M} \tag{12.51}$$

The circuit frequency of oscillation is then given approximately by

$$f_o = \frac{1}{2\pi\sqrt{L_{eq}C}} \tag{12.52}$$

with L_{eq} given in Eq. (12.51).

The needed transistor gain to provide loop gain greater than unity can be obtained using Eq. (12.41) with X_1 and X_2 as inductors, that is,

$$g_m \geq \frac{L_2}{r_d L_1} \tag{12.53}$$

It is only necessary that the loop gain be slightly greater than unity for oscillation to occur. In practice a loop gain of about 1.5 or 2 is desired to insure oscillation. The closer to unity, however, the more nearly sinusoidal is the output waveform. Since the FET provides gain greater than unity typical practice is to choose L_1 and L_2 to provide attenuation. From Eq. (12.53) the value of L_1 should be less than that L_2. If a single tapped inductor is used to provide L_1 and L_2 the tap is placed so that L_1 is smaller—typically at about 0.1 times the L_2 value.

EXAMPLE 12.15 For the FET Hartley oscillator of Fig. 12.33 and circuit values below, calculate the circuit frequency of oscillation and necessary FET gain for oscillation to take place $C = 100$ pF, $L_1 = 100\ \mu$H, $L_2 = 1$ mH, M $= 20\ \mu$H, $r_d = 20$ K.

Solution: Using Eq. (12.51) to calculate L_{eq}

$$L_{eq} = L_1 + L_2 + 2M = 100 + 1000 + 2(20)$$
$$= 1140\ \mu\text{H} = \mathbf{1.14\ mH}$$

The resonant frequency, using Eq. (12.52), is then

$$f_o = \frac{1}{2\pi\sqrt{L_{eq}C}} = \frac{1}{6.28\sqrt{(1.14 \times 10^{-3})(100 \times 10^{-12})}} = \mathbf{470\ kHz}$$

From Eq. (12.53)

$$g_m \geq \frac{L_2}{r_d L_1} = \frac{10^{-3}}{(20 \times 10^3)(100 \times 10^{-6})} = \frac{1}{2000} = \mathbf{500\ \mu mhos}$$

EXAMPLE 12.16 Select values for L_1, L_2, C, FET g_m, and r_d for a Hartley oscillator for operation at 10 MHz.

Solution: We desire operation at $f_o = 10$ mHz or

$$\omega_o = 2\pi f_o = 62.8 \times 10^6 \text{ rad/sec}$$

We note from previous discussion that L_{eq} is usually most dependent on the value of L_2. If we choose L_2 10 times greater than L_1, then an L_2 value of, say, 100 μH would result in $L_1 = 1/10(L_2) = 10\ \mu$H. Using Eq. (12.52) with $L_{eq} \cong L_2$ as a first-order calculation, we determine that C should be

$$C \cong \frac{1}{\omega_0^2 L_2} = \frac{1}{(62.8 \times 10^6)^2(100 \times 10^{-6})} = \mathbf{2.54\ pF}$$

This is quite a small capacitance value, so we now choose C of, say, 10 pF and calculate the value of L_{eq} required.

$$L_{eq} = \frac{1}{\omega_0^2 C} = \frac{1}{(62.8 \times 10^6)^2(10 \times 10^{-12})} = 25.4\ \mu\text{H}$$

Neglecting mutual coupling we choose $L_2 = 10L_1$ so that

$$L_2 \cong \mathbf{20\ \mu H}$$
$$L_1 \cong \mathbf{2\ \mu H}$$

If the mutual coupling is then, say, only 1.5 μH we have (as a check)

$$L_{eq} = L_1 + L_2 + 2\text{ M} = 2 + 20 + 2(1.5) = 25\ \mu\text{H}$$

For the necessary FET parameters, using a value of $r_d = 40$ K we determine the value of transistor g_m to be

$$g_m \geq \frac{L_2}{r_d L_1} = \frac{20 \times 10^{-6}}{(40 \times 10^3)(2 \times 10^{-6})} = \mathbf{250\ \mu mhos}$$

Thus, we choose a FET having $r_d = 40$ K and $g_m = 500\ \mu$mhos (twice the value needed just to obtain oscillation). In summary, component values are

$$C = 10\ \text{pF},\ L_1 = 2\ \mu\text{H},\ L_2 = 20\ \mu\text{H}$$

CIRCUIT APPROXIMATIONS

For the remainder of the Hartley oscillator circuit of Fig. 12.33 circuit approximations are that the time constant of R_G and C_G be at least 10 times larger than that of f_o [Eq. (12.45)] and that the impedance of the RFC coil be at least 10 times larger than that of coupling capacitor C_C [Eq. (12.50)].

Vacuum-Tube Hartley Oscillator

A vacuum-tube version of a Hartley oscillator is shown in Fig. 12.35a. It is the same circuit as in Fig. 12.33 with the FET replaced by

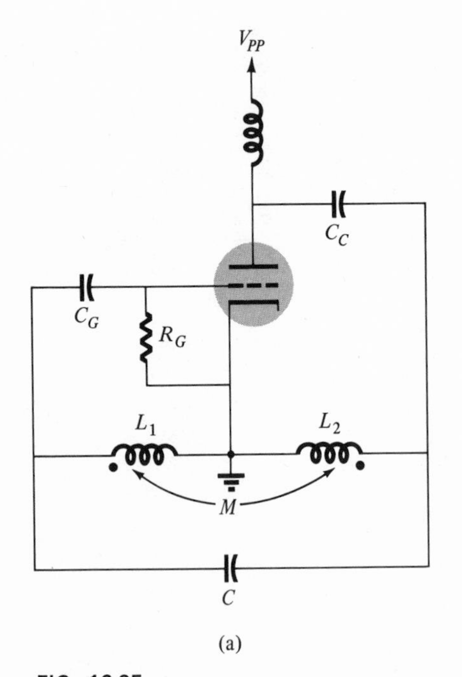

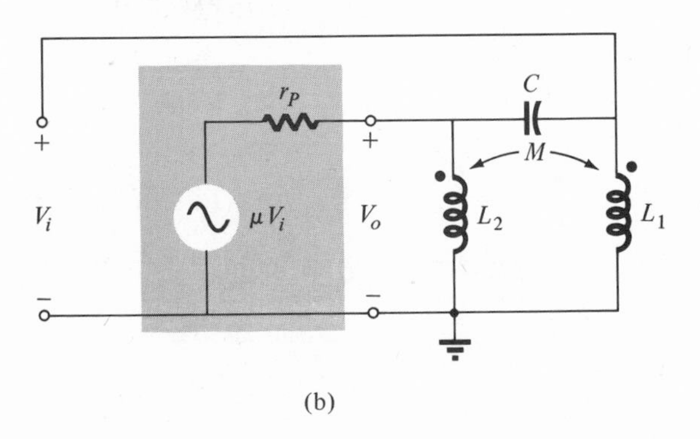

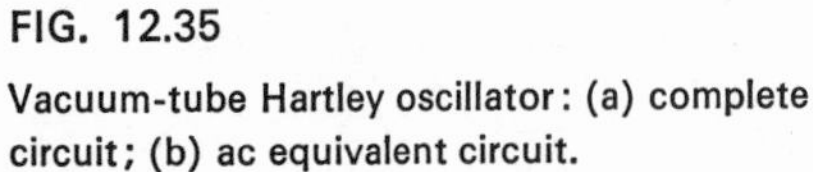

FIG. 12.35
Vacuum-tube Hartley oscillator: (a) complete circuit; (b) ac equivalent circuit.

the vacuum tube. The analysis of this circuit is the same as for the FET circuit version. Since the vacuum tube is often characterised by its plate resistance (r_p) and voltage gain (μ) the ac equivalent circuit shown in Fig. 12.35b shows these tube parameters in the equivalent circuit of the tube. The gain of the tube required for oscillation is then specified as

$$\boxed{\mu \geq \frac{L_2}{L_1}} \tag{12.54}$$

Thus, if L_2 is chosen, say, 10 times greater than L_1 and the vacuum tube has a gain greater than 10, the circuit will oscillate.

The oscillator frequency will be calculated using Eqs. (12.51) and (12.52).

EXAMPLE 12.17 Design a vacuum-tube Hartley oscillator to operate at 2 MHz using a vacuum tube with $\mu = 40$ and $r_p = 40$ K.

Solution: From the gain equation (12.54) the ratio of L_1 to L_2 should be

$$\frac{L_2}{L_1} \leq \mu = 40$$

Choosing $L_2/L_1 = 20$ we note that (neglecting mutual coupling)

$$L_{eq} \cong L_2$$

So that, using Eq. (12.52)

$$L_{eq}C = \frac{1}{\omega_0^2} = \frac{1}{(2\pi \times 2 \times 10^6)^2} = 0.6 \times 10^{-14}$$

If we select $C = 20$ pF then

$$L_{eq} = \frac{0.6 \times 10^{-14}}{20 \times 10^{-12}} = 0.3 \text{ mH}$$

Thus,

$$L_1 \cong \mathbf{15\ \mu H} \qquad L_2 \cong \mathbf{285\ \mu H}$$

Transistor Hartley Oscillator

Figure 12.36a shows a transistor Hartley oscillator circuit. It seems more complicated because of the dc bias components. Since the dc bias considerations are so different than the FET and vacuum-tube circuits we will consider it first.

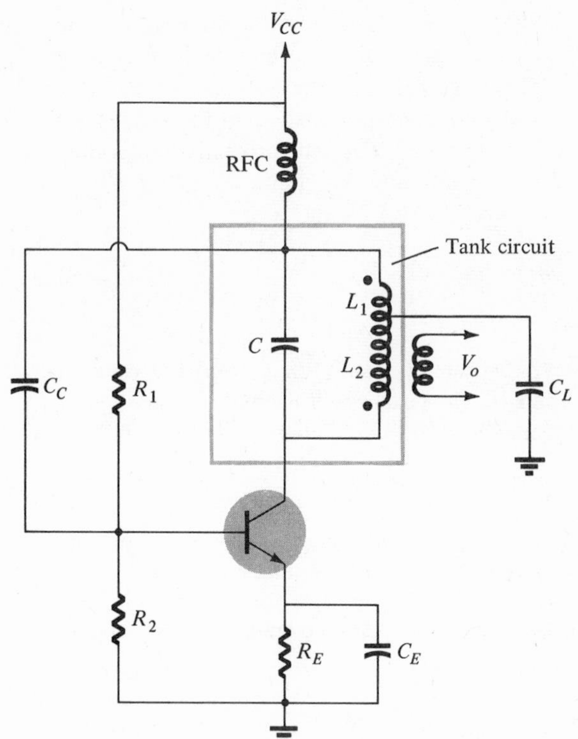

FIG. 12.36
Transistor Hartley oscillator circuit.

DC BIAS AND APPROXIMATIONS

DC bias is the voltage-divider stabilized type covered previously. Collector current flows through the RFC and L_1 and L_2 inductors. The dc resistance of these coils provides a collector resistance of equivalent value, R_C. Capacitor C_E should be selected so that its impedance is small compared to that of R_E. Capacitor C_C should have low impedance compared to the circuit input resistance. Also, the impedance of the RFC coil should be large compared to the impedance of C_C. These criteria had been covered in the discussion of a transistor Colpitts oscillator.

In addition, capacitor C_L provides an ac ground point at the junction of L_1 and L_2. Direct dc ground would prevent the bias current from the positive supply terminal from flowing through the collector. The use of capacitor C_L then allows dc bias current to pass while providing ac ground. In this regard the ac impedance of C_L should be small compared to that of L_1 and L_2. Since L_1 will be the smaller inductor we need only check that X_{C_L} be about 10 times smaller than X_{C_L}:

$$X_{C_L} \leq 0.1 \; X_{L_1} \tag{12.55}$$

From the basic resonant circuit calculations and those of the Colpitts transistor oscillator we can obtain the requirement of transistor gain necessary for oscillation:

$$\boxed{h_{fe} \geq \frac{\Delta^{h^e} L_2}{L_1}} \tag{12.56}$$

The frequency of oscillation is given by Eqs. (12.51) and (12.52).

EXAMPLE 12.18 Determine the operating frequency of a transistor Hartley oscillator as in Fig. 12.36 for the circuit values listed below. Calculate the necessary transistor gain (h_{fe}), select suitable values for coupling and bypass components C_C, C_L, and C_E.

$R_1 = 12$ K, $R_2 = 8$ K, $R_E = 750\ \Omega$, $L_{RFC} = 1$ mH, $L_1 = 100\ \mu$H,
$L_2 = 1$ mH, M $= 20\ \mu$H, $C = 20$ pF, $h_{oe} = 50\ \mu$mhos,
$h_{ie} = 2$ K, $h_{re} \cong 0$.

Solution: Using Eqs. (12.51) and (12.52),

$$L_{eq} = L_1 + L_2 + 2\text{ M} = 100 + 1000 + 2(20) = 1140\ \mu\text{H}$$

For $C = 20$ pF,

$$f_o = \frac{1}{2\pi\sqrt{L_{eq}C}} = \frac{1}{6.28\sqrt{(1140 \times 10^{-6})(20 \times 10^{-12})}} = \mathbf{1.05\ MHz}$$

Using Eq. (12.56)

$$h_{fe} \geq \frac{\Delta^{h^e} L_2}{L_1} = \frac{(h_{ie}h_{oe} - h_{re}h_{fe})L_2}{L_1}$$
$$= \frac{(2 \times 10^3)(50 \times 10^{-6})(10^{-3})}{(100 \times 10^{-6})} = 1$$

For C_E

$$X_{C_E} \leq 0.1 R_E = 0.1(750) = 75\ \Omega$$

At $f_o = 1.05$ MHz ($\omega_o = 6.6 \times 10^6$)

$$X_{C_E} = \frac{1}{\omega_o C_E}$$

$$C_E \geq \frac{1}{\omega_o X_{C_E}} = \frac{1}{(6.6 \times 10^6)(75)} = 2000\text{ pF}$$

(choose $C_E = 2500$ pF)

For C_C and X_{RFC}

$$X_{RFC} \geq 10X_{C_C}$$

$$\omega_o L_{RFC} \geq \frac{10}{\omega_o C_C}$$

$$C_C \geq \frac{10}{\omega_o^2 L_{RFC}} = \frac{10}{(6.6 \times 10^6)^2(10^{-3})} = \mathbf{230\ pF}$$

For C_L

$$X_{C_L} \leq 0.1X_{L_1}$$

$$\frac{1}{\omega_o C_L} \leq \frac{\omega_o L_1}{10}$$

$$C_L \geq \frac{10}{\omega_0^2 L_1} = \frac{10}{(6.6 \times 10^6)^2(100 \times 10^{-6})} = 2300\ \text{pF}$$

(choose $C_L = \mathbf{2500\ pF}$)

12.12 CRYSTAL OSCILLATOR

A crystal oscillator is basically a tuned-circuit oscillator using a piezoelectric crystal as a resonant tank circuit. The crystal (usually quartz) has a greater stability in holding constant at whatever frequency the crystal is originally cut to operate. Crystal oscillators are used, then, where great stability is required, as in communication transmitters and receivers.

Characteristics of a Quartz Crystal

A quartz crystal (one of a number of crystal types) exhibits the property that when mechanical stress is applied across the faces of the crystal a difference of potential develops across opposite faces of the crystal. This property of a crystal is called the *piezoelectric effect*. Similarly, a voltage applied across one set of faces of the crystal causes mechanical distortion in the crystal shape.

When alternating voltage is applied to a crystal, mechanical vibrations are set up—these vibrations having a natural resonant frequency dependent on the crystal. Although the crystal has electromechanical resonance we can represent the crystal action by an equivalent electrical resonant circuit as shown in Fig. 12.37. The inductor L and capacitor C represent electrical equivalents of crystal mass and compliance while resistance R is an electrical equivalent of the crystal structure's internal friction. The shunt capacitance C_M represents the capacitance due to mechanical mounting of the crystal. Because the crystal losses, represented by R, are quite small, the equivalent crystal

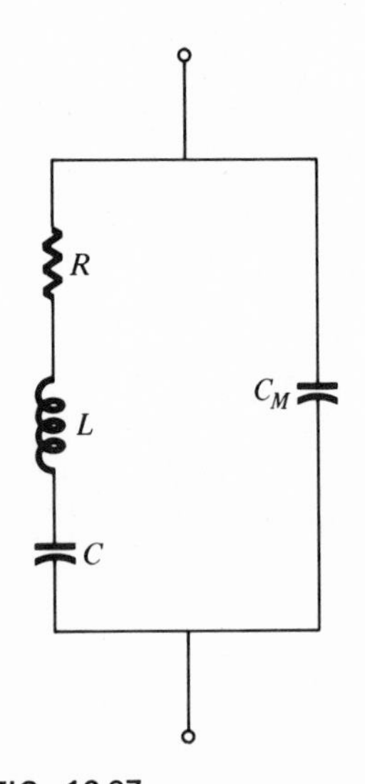

FIG. 12.37
Electrical equivalent circuit of a crystal.

Q (quality factor) is quite high—typically 20,000. Values of Q up to almost 10^6 can be achieved using crystals.

The crystal as represented by the equivalent electrical circuit of Fig. 12.37 can have two resonant frequencies. One resonant condition occurs when the reactances of the series RLC leg are equal (and opposite). For this condition the *series-resonant* impedance is very low (equal to R). The other resonant condition occurs at a higher frequency when the reactance of the series resonant leg equals the reactance of capacitor C_M. This is a parallel resonance or antiresonance condition of the crystal. At this frequency the crystal offers a very high impedance to the external circuit. The impedance versus frequency of the crystal is shown in Fig. 12.38. In order to use the crystal properly it must be

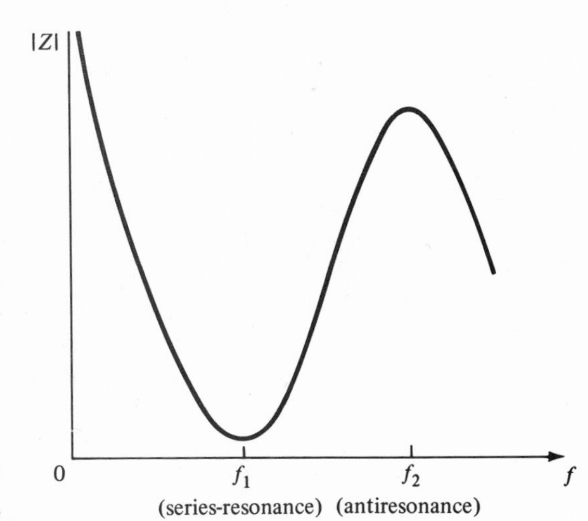

FIG. 12.38
Crystal impedance vs. frequency.

connected in a circuit so that its low impedance in the series-resonant operating mode or high impedance in the antiresonant operating mode is selected.

Series-Resonant Circuits

To excite a crystal for operation in the series-resonant mode it may be connected as a series element in a feedback path. At the series-resonant frequency of the crystal its impedance is smallest and the amount of (positive) feedback is largest. A typical transistor circuit is shown in Fig. 12.39. Resistors R_1, R_2, and R_E provide a voltage-divider stabilized dc bias circuit. Capacitor C_E provides ac bypass of the emitter resistor and the RFC coil provides for dc bias while decoupling any ac signal on the power lines from affecting the output signal. The voltage feedback from collector to base is a maximum when the crystal impedance is minimum (in series-resonant mode). The coupling ca-

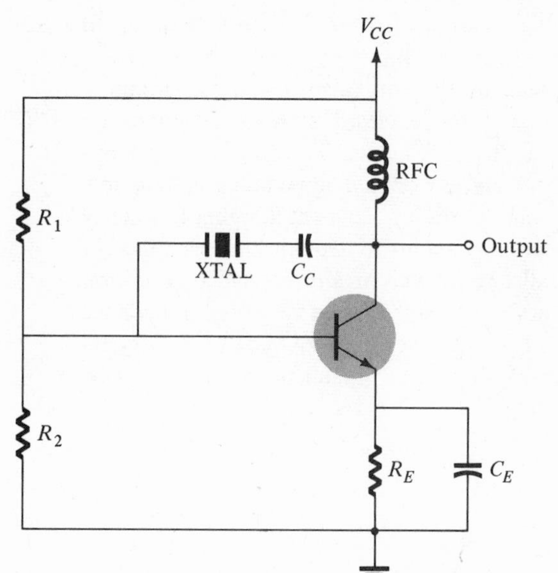

FIG. 12.39

Crystal controlled oscillator using crystal in series-feedback path.

pacitor C_C has neglibible impedance at the circuit operating frequency but blocks any dc between collector and base.

The resulting circuit frequency of oscillation is set, then, by the series-resonant frequency of the crystal and changes in supply voltage, transistor device parameters, and so on, have no effect on the circuit

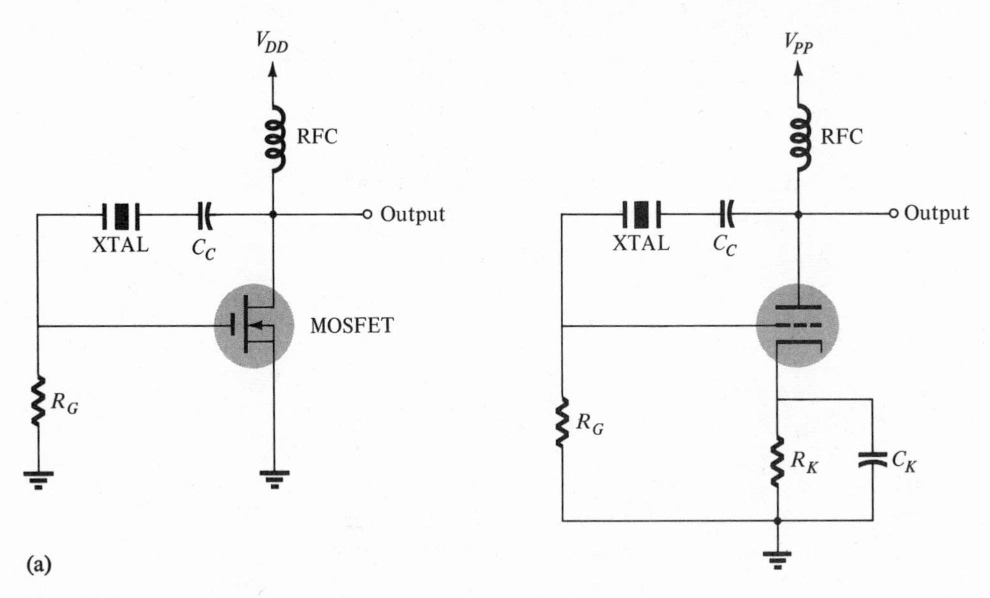

FIG. 12.40

FET and vacuum-tube Pierce crystal controlled oscillator circuits: (a) FET circuit; (b) vacuum-tube circuit.

operating frequency which is held stabilized by the crystal. The circuit frequency stability is set by the crystal frequency stability—which is quite good.

The circuit shown in Fig. 12.39 is generally called a Pierce crystal-controlled oscillator. Other versions of the circuit using FETs and vacuum tubes are shown in Fig. 12.40.

Another transistor circuit is shown in Fig. 12.41. The circuit provides tuning by an *LC* tank circuit in the collector and tuning by a series-resonant excited crystal connected as feedback from a capacitive voltage divider. The *LC* circuit is adjusted near the desired operating crystal frequency, but the exact circuit frequency is set by the crystal and stabilized by the crystal.

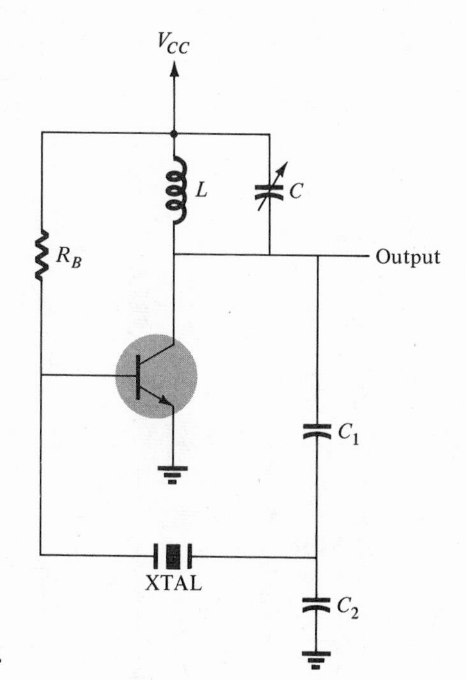

FIG. 12.41
Transistor crystal oscillator.

Parallel-Resonant Circuits

Since the parallel-resonant impedance of a crystal is a maximum value it is connected in shunt. At the parallel-resonant operating frequency a crystal appears as an inductive reactance of largest value. Figure 12.42 shows a crystal connected as the inductor element in a modified Colpitts circuit. The basic dc bias circuit should be evident. Maximum voltage is developed across the crystal at its parallel-resonant frequency. The voltage is coupled to the emitter by a capacitor voltage divider—capacitors C_1 and C_2.

A *Miller* crystal controlled oscillator circuit is shown in Fig. 12.43. A tuned *LC* circuit in the drain section is adjusted near the

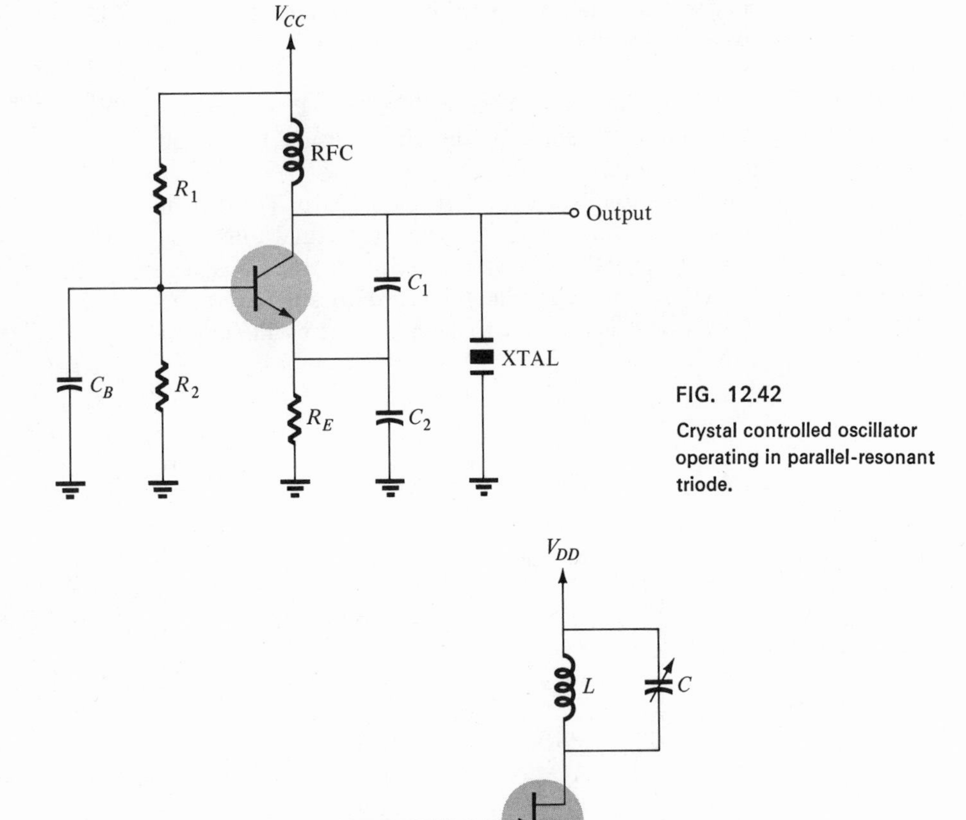

FIG. 12.42
Crystal controlled oscillator operating in parallel-resonant triode.

FIG. 12.43
Miller crystal controlled oscillator.

crystal parallel-resonant frequency. The maximum gate-source signal occurs at the crystal antiresonant frequency controlling the circuit operating frequency.

12.13 OPAMP OSCILLATOR CIRCUITS

Phase-Shift Oscillator

As IC circuits have become more popular they have been adapted to operate in oscillator circuits. One need buy only an OPAMP to obtain an amplifier circuit of stabilized gain setting and incorporate some means of signal feedback to produce an oscillator circuit. For example, a phase-shift oscillator is shown in Fig. 12.44. The output of the OPAMP is fed to a three-stage *RC* network which provides the needed 180° of phase shift (at an attenuation factor of 1/29). If the OPAMP provides gain (set by resistors R_i and R_f) of greater than 29 a loop gain greater than unity results and the circuit acts as an oscillator [Oscillator frequency is given by Eq. (12.22a)].

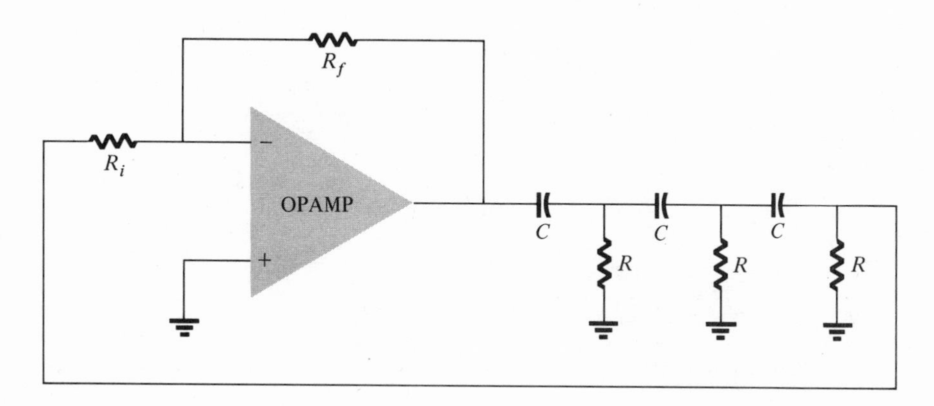

FIG. 12.44
Phase-shift oscillator using OPAMP.

Colpitts Oscillator

An OPAMP Colpitts oscillator circuit is shown in Fig. 12.45. Again the OPAMP provides the basic amplification needed while the

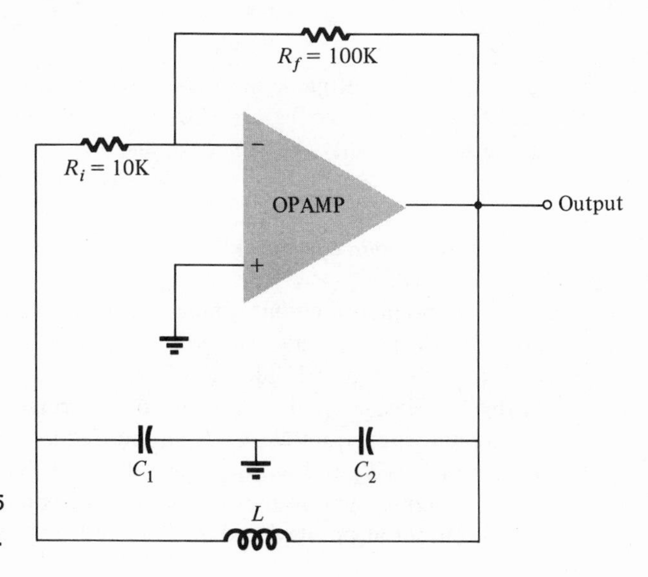

FIG. 12.45
OPAMP Colpitts oscillator.

oscillator frequency is set by an LC feedback network of a Colpitt configuration. The oscillator frequency is given by Eq. (12.44).

Crystal Oscillator

An OPAMP can be used in a crystal oscillator as shown in Fig. 12.46. The crystal is connected in the series-resonant path and operates at the crystal series-resonant frequency. The present circuit has a high

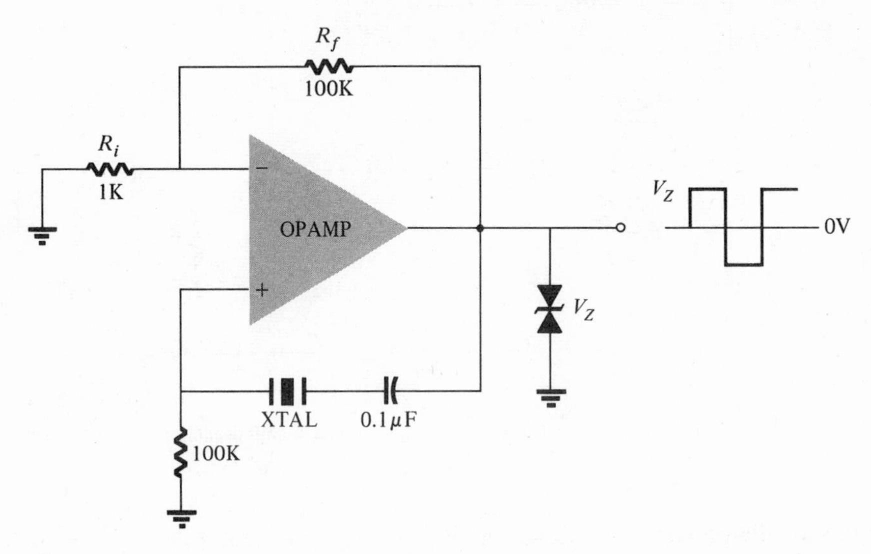

FIG. 12.46
Crystal oscillator using OPAMP.

gain so that an output squarewave signal results as shown in the figure. A pair of Zener diodes are shown at the output to provide output amplitude at exactly the Zener voltage (V_Z).

Wien Bridge Oscillator

An oscillator circuit which is quite practical uses an OPAMP and RC bridge circuit, with the oscillator frequency set by the R and C components. Figure 12.47 shows a basic version of a Wien bridge oscillator circuit. Note the basic bridge connection. Resistors R_1, R_2 and capacitors C_1, C_2 form the frequency adjustment elements, while resistors R_3 and R_4 form part of the feedback path. The OPAMP output is connected as the bridge input at points a and c. The bridge circuit output at points b and d is the input to the OPAMP.

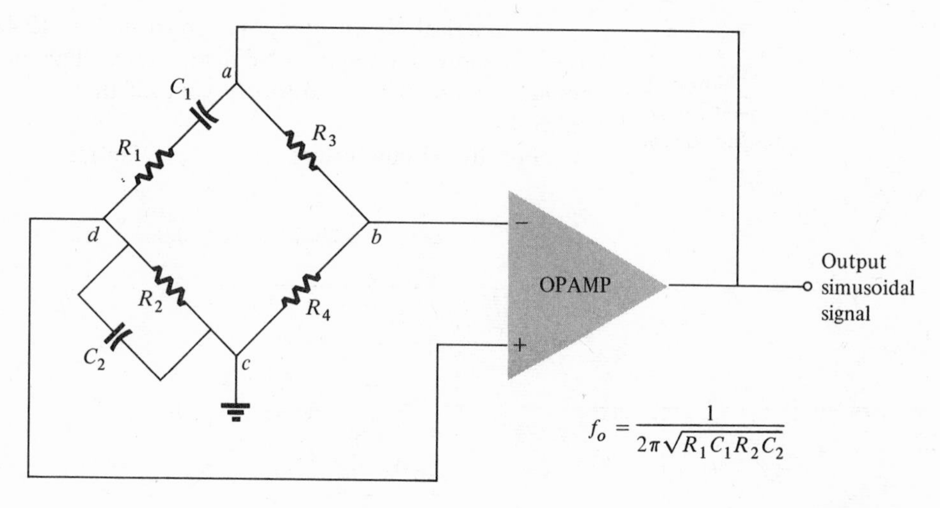

FIG. 12.47
Wien bridge oscillator circuit using OPAMP amplifier.

Neglecting loading effects of the OPAMP input and output impedances, the analysis of the bridge circuit results in

$$\boxed{\frac{R_3}{R_4} = \frac{R_1}{R_2} + \frac{C_2}{C_1}} \tag{12.57a}$$

and

$$\boxed{\omega_o = \frac{1}{\sqrt{R_1C_1R_2C_2}}} \tag{12.57b}$$

If, in particular, the values are $R_1 = R_2 = R$ and $C_1 = C_2 = C$, the resulting oscillator frequency is

$$\omega_o = \frac{1}{RC} \tag{12.58a}$$

and

$$\frac{R_3}{R_4} = 2 \tag{12.58b}$$

Thus, a ratio of R_3 to R_4 greater than 2 will provide sufficient loop gain for the circuit to oscillate at the frequency calculated using Eq. (12.58a).

A practical circuit design is shown in Fig. 12.48. Although the circuit is shown in somewhat different form than in Fig. 12.47 you should compare the two to satisfy yourself that they are indeed identical in form.

For the circuit values given we calculate

$$f_o = \frac{1}{2\pi RC} = \frac{1}{6.28(50 \times 10^3)(0.001 \times 10^{-6})} = 3.18 \text{ kHz}$$

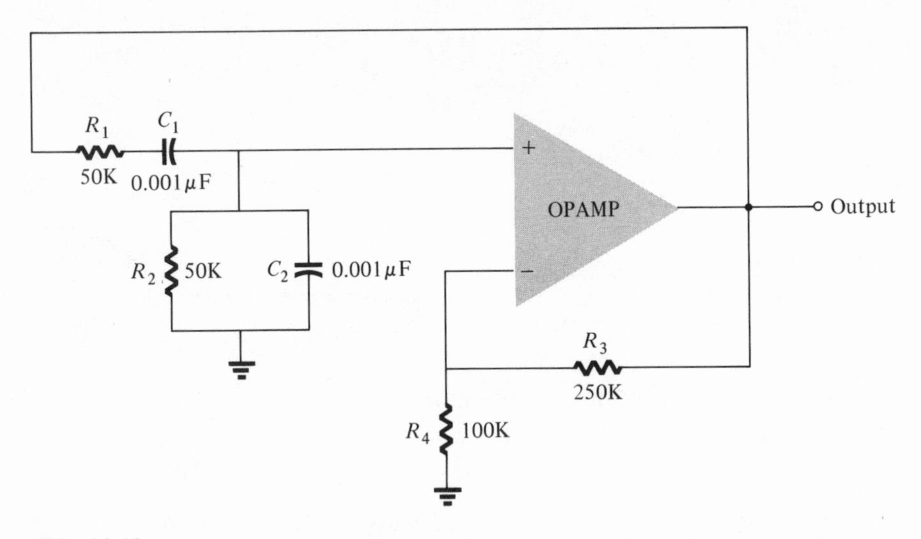

FIG. 12.48
Practical Wien bridge oscillator circuit.

EXAMPLE 12.19 Design the *RC* elements of a Wien bridge oscillator as in Fig. 12.48 for operation at $f_o = 10$ kHz.

Solution: Using equal values of *R* and *C* we can select $R = 100$ K and calculate the required value of *C* using Eq. (12.58a):

$$f_o = \frac{1}{2\pi RC}$$

$$C = \frac{1}{2\pi f_o R} = \frac{1}{(6.28(10 \times 10^3)(100 \times 10^3)} = \frac{10^9}{6.28} = \mathbf{159\ pF}$$

We can use $R_3 = 250$ K, $R_4 = 100$ K to provide a ratio R_3/R_4 greater than 2 for oscillation to take place.

12.14 UNIJUNCTION OSCILLATOR

A particular device, the unijunction transistor (discussed in Chapter 9) can be used in a single-stage oscillator circuit; providing a pulse signal suitable for digital circuit applications. The unijunction transistor can be used in what is called a relaxation oscillator as shown by the basic circuit of Fig. 12.49. Resistor R_T and capacitor C_T are the timing components which set the circuit oscillating rate. The oscillating frequency may be calculated using Eq. (12.59) which includes the unijunction

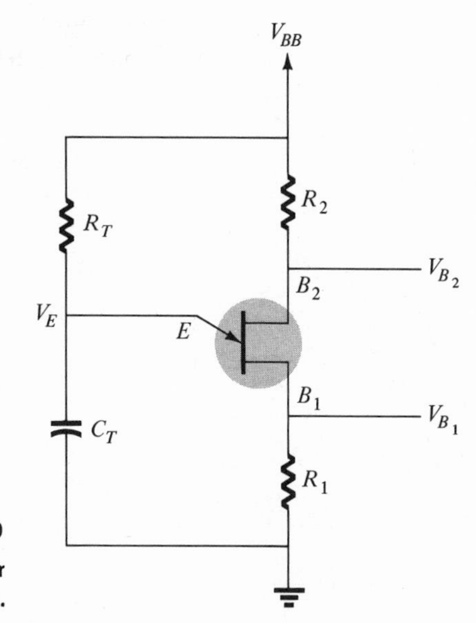

FIG. 12.49
Basic unijunction oscillator circuit.

transistor *intrinsic standoff ratio* η, as a factor (in addition to R_T and C_T) in the oscillator operating frequency.

$$f_o \cong \frac{1}{R_T C_T \ln [1/(1-\eta)]} \tag{12.59}$$

Typically, a unijunction transistor has a standoff ratio from 0.4 to 0.6. Using a value of $\eta = 0.5$,

$$f_o \cong \frac{1}{R_T C_T \ln [1/(1-0.5)]} = \frac{1}{R_T C_T \ln 2} = \frac{1.44}{R_T C_T}$$

$$\cong \frac{1.5}{R_T C_T} \tag{12.60}$$

Capacitor C_T is charged through resistor R_T toward supply voltage V_{BB}. As long as the capacitor voltage V_E is below a standoff voltage (V_P) set by the voltage across $B_1 - B_2$, and the transistor standoff ratio η is given by Eq. (12.61),

$$V_P = \eta V_{B_1} V_{B_2} - V_D \tag{12.61}$$

and the unijunction emitter lead appears as an open circuit. When the emitter voltage across capacitor C_T exceeds this value (V_P) the unijunction circuit fires, discharging the capacitor, after which a new charge cycle begins. When the unijunction fires a voltage rise is developed across R_1 and a voltage drop across R_2 as shown in Fig. 12.50. The signal at the emitter is a sawtooth voltage waveform, that at base 1 a positive-going pulse, and that at base 2 a negative-going pulse.

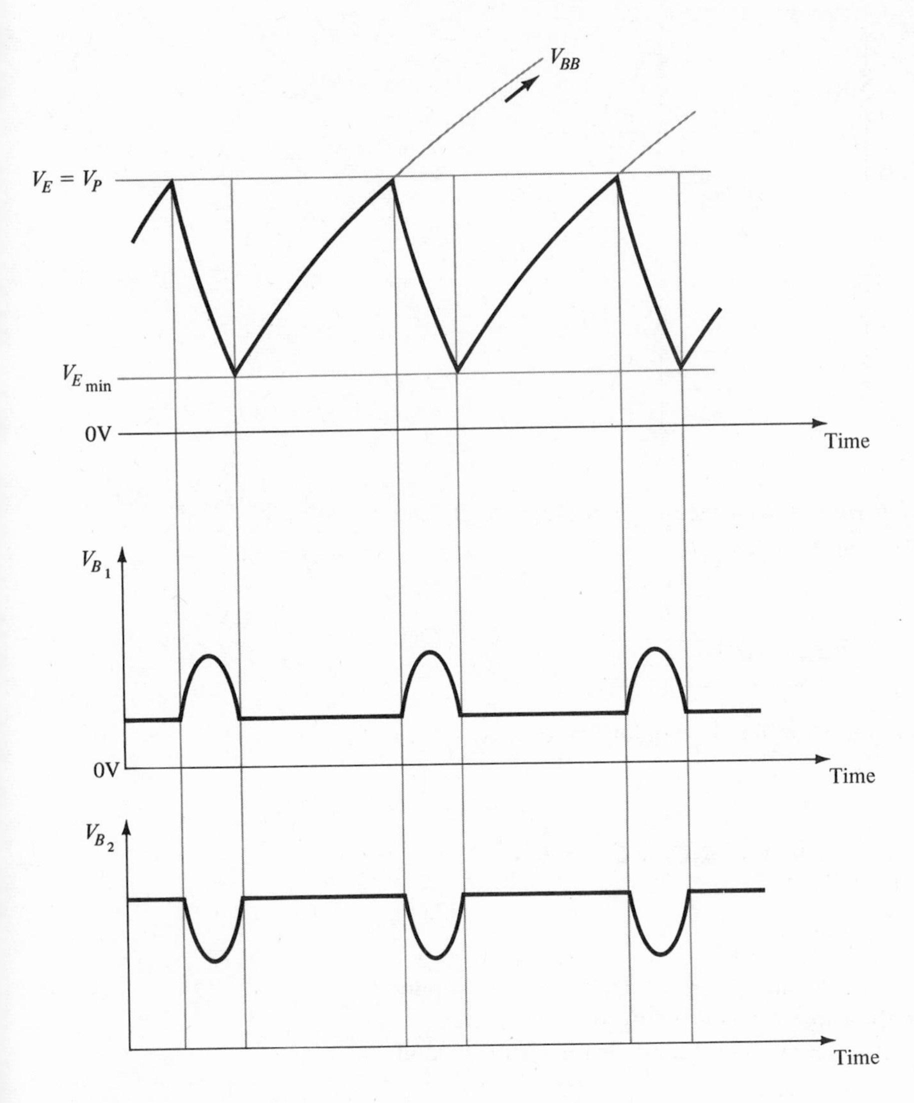

FIG. 12.50
Unijunction oscillator waveforms.

For the basic circuit of Fig. 12.49 the relation between R_T, C_T, and the oscillator frequency (at a fixed value of η) can be given as a nomograph (see Fig. 12.51) instead of as an equation [Eqs. (12.59) and (12.60)]. The line shown on the nomograph shows how to obtain

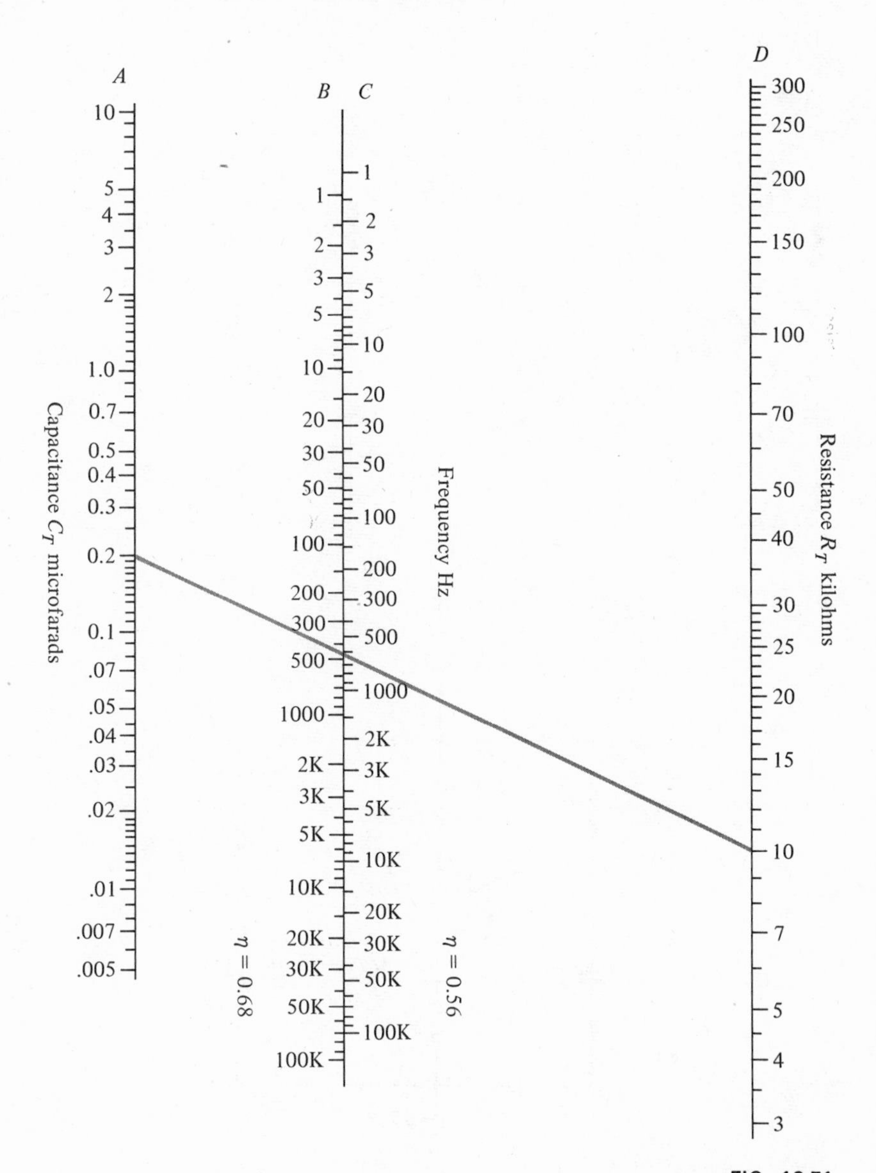

FIG. 12.51
Nomograph to calculate unijunction oscillator frequency.

the operating frequency for values of $R_T = 10$ K and $C_T = 0.2\ \mu$F. As shown the straight line connecting these points intersects the frequency axis at about $f_o = 650$ Hz for $\eta = 0.56$, or $f_o = 480$ for $\eta = 0.68$. A transistor with $\eta = 0.6$, for example, would operate at a frequency between 480 and 650 Hz.

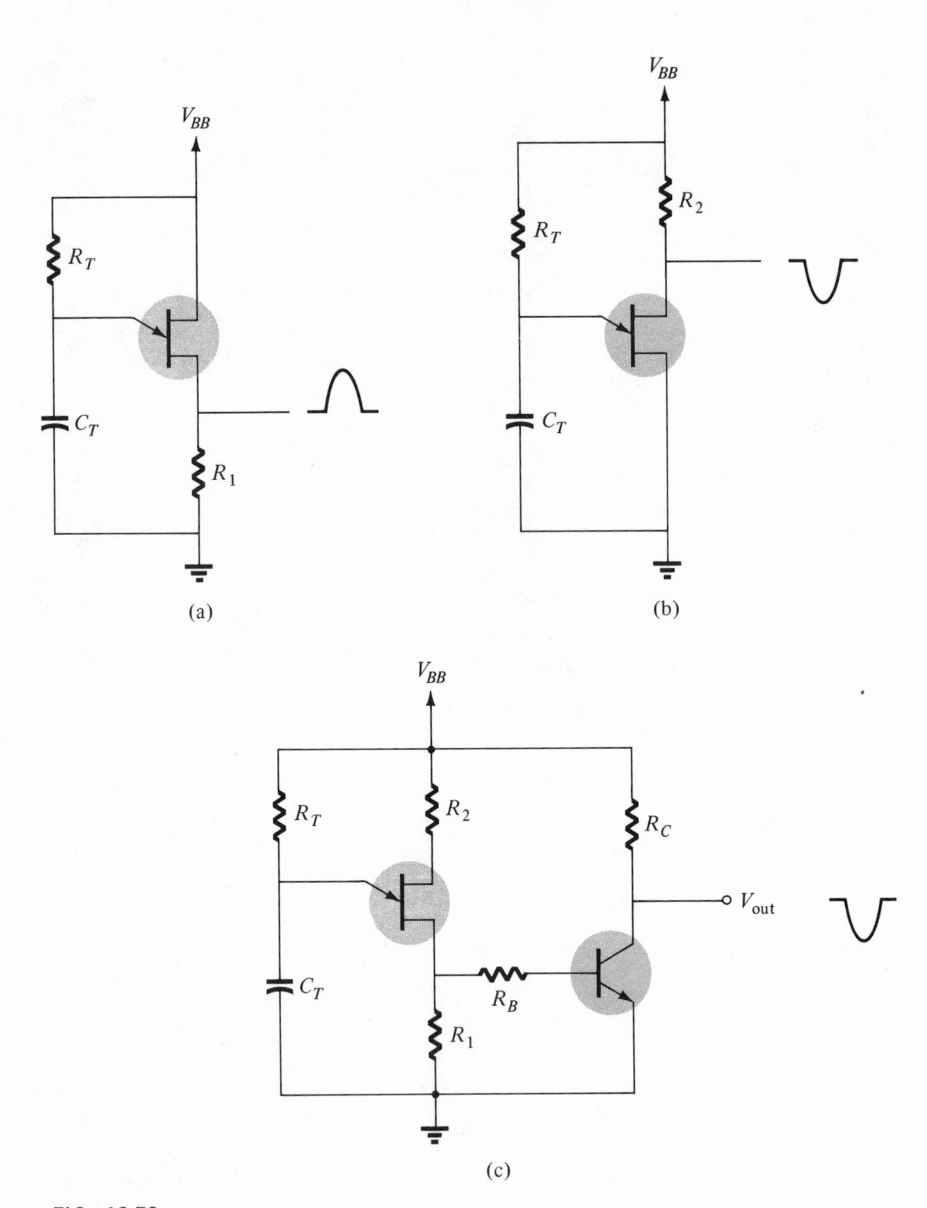

FIG. 12.52

Some unijunction oscillator circuit configurations.

EXAMPLE 12.20 Using the nomograph determine the value of R_T for $C_T = 0.01\ \mu F$ and $\eta = 0.56$ for operation at $f_o = 10$ kHz.

Solution: Connecting a straight line between $C_T = 0.01\ \mu F$ and $f_o = 10$ kHz (on $\eta = 0.56$ scale) points, the intersection with the R_T axis is read as

$$R_T = 12\ K$$

[As an exercise, check the result using Eq. (12.59).]

A few circuit variations of the unijunction oscillator are provided in Fig. 12.52.

PROBLEMS

§ 12.1

1. Calculate the gain of a negative feedback amplifier having $A = 2000$, $\beta = 1/10$.
2. If the gain of an amplifier changes from a value of 1000 by 10% calculate the gain change if the amplifier is used in a feedback circuit having $\beta = 1/20$.

§ 12.2

3. Calculate the gain, input, and output impedances of a voltage-series feedback amplifier having $A = 300$, $R_i = 1.5$ K, $R_o = 50$ K, and $\beta = 1/15$.

§ 12.3

4. Calculate the voltage gain, input, and output impedance with and without feedback for a circuit as in Fig. 12.7 for circuit values $R_e =$ 2 K, $R_s = 600\ \Omega$, $h_{ie} = 1.5$ K, and $h_{fe} = 100$.
5. Calculate the gain with and without feedback for a FET amplifier as in Fig. 12.8 for circuit values $R_1 = 200$ K, $R_2 = 800\ \Omega$, $R_o = 40$ K, $R_D = 8$ K, and $g_m = 5000\ \mu$mhos.
6. Calculate the gain and output impedance with and without feedback for a vacuum-tube circuit as in Fig. 12.9 for circuit values $R_p = 50$ K, $R'_L = 4$ K, $R_1 = 250$ K, $R_2 = 750$ K, $\mu = 80$, $r_p = 25$ K.

§ 12.4

7. Calculate A, R_i, and R_o with and without feedback for a circuit as in Fig. 12.10 for transistor parameters $h_{fe} = 80$, $h_{ie} = 2.2$ K.
8. For a circuit as in Fig. 12.12 and the following circuit values, calculate the circuit gain, and input and output impedances with and without feedback: $R_E = 1.2$ K, $R_C = 12$ K, $h_{ie} = 2$ K, and $h_{fe} = 75$.

§ 12.7

9. A FET phase-shift oscillator having $g_m = 6000\ \mu$mhos, $r_d = 36$ K, and feedback resistor $R = 12$ K is to operate at 2.5 kHz. Select R_D and C for specified oscillator operation.
10. Select values of capacitor C and transistor gain h_{fe} to provide operation of a transistor phase-shift oscillator at 5 kHz for circuit values $R_1 =$ 24 K, $R_2 = 75$ K, $R_C = 18$ K, $R = 6$ K, and $h_{ie} = 2$ K.

§ 12.8

11. Calculate the minimum gain (g_m) of the FET in the oscillator circuit of Fig. 12.24 and the circuit frequency of oscillation for circuit values $r_d = 50$ K, $L = 5$ mH, $M = 0.2$ mH, $R_s = 60\ \Omega$, and $C = 0.002\ \mu$F.
12. For a FET Colpitts oscillator as in Fig. 12.28a and circuit values below, determine
    (a) minimum value of FET g_m for oscillator action;
    (b) circuit frequency of oscillation.

$$r_d = 60 \text{ K},\ C_1 = 750 \text{ pF},\ C_2 = 2500 \text{ pF},$$
$$L = 40\ \mu\text{H},\ R_g = 750 \text{ K},\ C_g = 470 \text{ pF},$$
$$L_{RFC} = 0.2 \text{ mH},\ C_C = 2000 \text{ pF}$$

13. For the circuit of Fig. 12.28a and component values in Problem 12.12, check the following for proper circuit operation.
    (a) Self-bias circuit time constant.
    (b) RFC coil impedance compared to coupling capacitor impedance.
14. For the transistor Colpitts oscillator of Fig. 12.30 and circuit values below, calculate values of C_1, C_2, and L needed for operation at $f_o = 1.5$ mHz. $h_{ie} = 1.5$ K, $h_{oe} = 12.5 \times 10^{-6}$ mhos, $h_{fe} = 75$, $h_{re} = 0.3 \times 10^{-4}$.
15. Calculate the dc bias voltages in the circuit of Problem 12.14. Additional circuit values are $R_1 = 12$ K, $R_2 = 4.7$ K, $R_E = 470\ \Omega$, $V_{cc} = +15$ V, $R_L = 240\ \Omega$, $R_{RFC} = 150\ \Omega$.
16. Check circuit of Fig. 12.30 using circuit values of Problem 12.14 and additional values below for proper operation of
    (a) bypass capacitor C_E;
    (b) RFC coil.

$$L_{RFC} = 1.5 \text{ mH},\ C_c = 0.25\ \mu\text{F},\ C_E = 0.2\ \mu\text{F}$$

§ 12.11

17. Calculate the required FET gain (g_m) and oscillator frequency for a FET Hartley oscillator as in Fig. 12.33 for the following circuit values $C = 180$ pF, $L_1 = 150\ \mu$H, $L_2 = 1.5$ mH, M $= 25\ \mu$H, $r_d = 30$ K.
18. Design a vacuum-tube Hartley oscillator for operation at 2.5 mHz using a tube with parameters $\mu = 80$, $r_p = 25$ K.
19. Calculate the required transistor gain for oscillator action in the circuit of Fig. 12.36 and the following circuit values $R_1 = 15$ K, $R_2 =$ 7.5 K, $R_E = 680\ \Omega$, $L_{RFC} = 0.5$ mH, $L_1 = 75\ \mu$H, $L_2 = 750\ \mu$H, $M = 15\ \mu$H, $C = 35$ pF, $h_{oe} = 1/40$ K, $h_{ie} = 2$ K, $h_{re} \cong 0$.

20. (a) Determine the oscillator frequency for the circuit and values of Problem 12.19.
(b) Select suitable values of C_C, C_L, and C_E.

§ 12.12

21. Draw circuit diagrams of
(a) series-operated crystal oscillator;
(b) shunt-excited crystal oscillator.

§ 12.13

22. Design the RC elements of a Wien bridge oscillator circuit (as in Fig. 12.48) for operation at $f_o = 2$ kHz.

§ 12.14

23. Design a unijunction oscillator circuit for operation at
(a) 1 kHz;
(b) 150 kHz.

13

Pulse and Digital Circuits

13.1 GENERAL

An area of electronics that has become extremely popular over the past decade is the use of pulse and digital circuits in digital computers. Digital circuits are built using transistors, FETs, or integrated circuits (ICs). Basically, digital circuits are simple since they are operated either fully saturated (ON) or in cutoff (OFF). Such operation is simpler than operation designed for linear gain of a specific amount, as in the case of amplifier circuits.

Our concern, then, is with the two output states of a digital circuit. These can be at either one of two preselected voltage levels—$+10$ and 0 V, -10 and 0 V, $+6$ and -6 V, and so on. Logically, these levels may be related to the binary conditions of 1 and 0. For example, $+10\text{ V} = 1$ and $0\text{ V} = 0$, or $-10\text{ V} = 1$ and $0\text{ V} = 0$ are possible relations that may be assigned to voltage levels and logic conditions. Digital circuits provide manipulation of the logical conditions representing the two different voltage levels of the circuit. A number of logic gates or circuits are covered in the present chapter. A logical AND gate, for example, provides an output only when *all* inputs are present, whereas a logical OR gate provides an output if *any* one input is present. An inverter circuit provides logical inversion—a 1 output for 0 input, or vice versa. The most popular configurations for a logic gate include an AND or OR gate, followed by an inverter, an AND-gate-inverter combination called a not-AND or NAND gate, and an OR inverter called a not-OR or NOR gate.

Also quite important in digital circuits is a class of multivibrator circuits. These circuits have two opposite output terminals (if one output is logical-1 the other is logical-0, or vice versa). The most useful of these is the bistable multivibrator or *flip-flop*, which can remain in either stable condition of output. The flip-flop can be used as a memory device—holding the state it was placed in after the initial pulse operating the circuit has passed. It can also be used to build a binary counter or a shift-register for use in digital computers.

A second circuit, the monostable multivibrator, provides opposite voltage outputs but, as the name implies, it can remain stable in only one state. If the circuit is triggered to operate it can go into the opposite state of output voltages but can remain in this state only for a fixed time interval. The monostable multivibrator, or ONE-SHOT, is used for pulse shaping, time delay, and other timing actions.

An astable multivibrator or CLOCK circuit has no stable operating state and provides a constantly changing output voltage (pulse train). The circuit is therefore a square-wave oscillator providing pulses to activate various computer operations.

A circuit that appears similar to the multivibrator class is the Schmitt trigger. This circuit operates from a slowly varying input signal and switches output voltage state when the input voltage goes above a preset voltage level or below a second voltage level.

13.2 DIODE LOGIC GATES—AND, OR

A logic gate provides an output signal for a desired logical combination of input signals. An AND gate, for example, provides an output of "logical-1" only if all the inputs are present, that is, if each is at 1 (logical-1). This type of circuit can be compared to one using switches connected in series. Only if, as in Fig. 13.1, all switches are closed (at 1) does an output voltage appear (and drive the light indicator ON). Thus, a light ON (logical-1) appears if switch *A*, AND switch *B*, AND switch *C*, are closed. This can also be written as the Boolean or logical expression

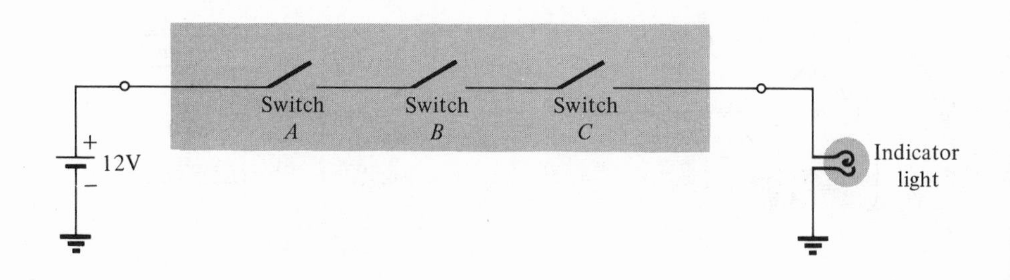

$$\text{Light ON} = A \cdot B \cdot C$$

(The dot indicates AND; read as A *AND* B *AND* C.)

Diode AND Gate

A more practical form of AND gate using diodes is shown in Fig. 13.2. Using positive-voltage levels of $+V$ for 1 and 0 V as 0, the circuit shown provides a 1 output of $+V$ only if *all* inputs are $+V$ or 1. More specifically, any input at 0 V will hold the output at about 0 V, as shown below.

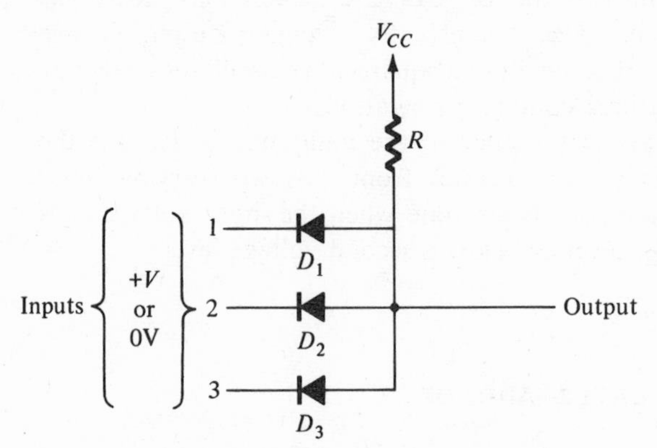

FIG. 13.2
Diode AND-gate circuit.

1. A zero, ground-level input at any (one or more) input terminal(s) will cause that diode to short the output to ground. In logic form this means that 0 at any input produces 0 output.
2. A 1 input at inputs 1 and 2, but a 0 input at 3, will produce a 0 output because diode 3 shorts the output to ground.

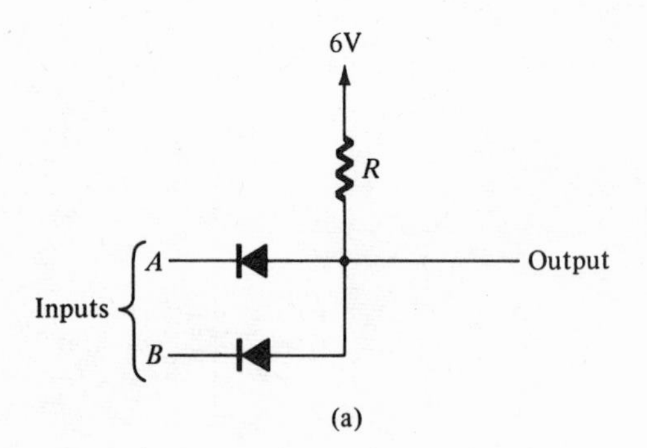

A	*B*	Output
0V	0V	0V
0V	+6V	0V
+6V	0V	0V
+6V	+6V	+6V

(b)

FIG. 13.3
(a) Diode logic circuit; (b) voltage truth table.

3. Only when a 1 input is provided at inputs 1 AND 2 AND 3, none of the diodes is conducting and the output is 1.

Figure 13.3 shows a typical diode gate circuit and voltage truth table. Figure 13.4 shows the logic symbol and logic truth table for the circuit of Fig. 13.3 for positive logic operation (defined below).

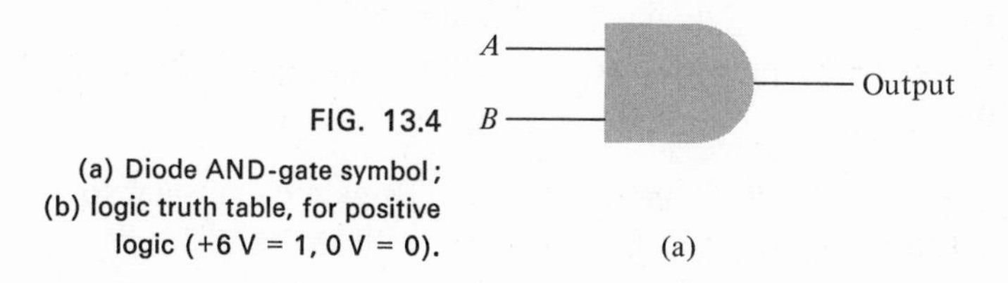

A	B	Output
0	0	0
0	1	0
1	0	0
1	1	1

(b)

FIG. 13.4
(a) Diode AND-gate symbol; (b) logic truth table, for positive logic (+6 V = 1, 0 V = 0).

Diode OR Gate

The circuit of Fig. 13.5 shows an OR gate circuit connection using three switches connected in parallel. If either switch A OR B OR C (or any combination of these) is closed (logical-1) the indicator light will be turned ON. A practical version of such a circuit uses diodes as shown in Fig. 13.6. The circuit is a positive-logic diode OR

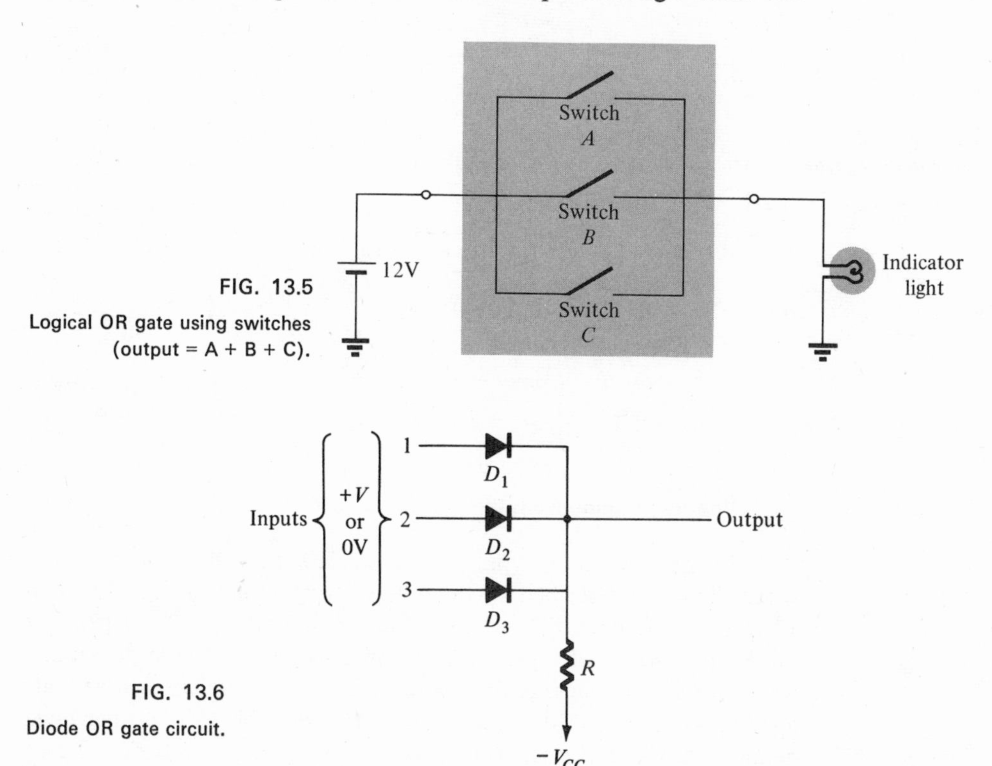

FIG. 13.5
Logical OR gate using switches (output = A + B + C).

FIG. 13.6
Diode OR gate circuit.

gate, and for the present example uses the same $+V$ and 0-V levels as the AND-gate circuit previously considered.

The operation of the circuit of Fig. 13.6 is the following:

1. A 1 ($+V$) input at any (one or more) input terminal(s) will cause that diode to conduct, placing the output at the 1 level.
2. A 0 input at inputs 1 and 2, but a 1 input at 3, will produce a 1 output because diode 3 conducts, placing the output at $+V$, and thereby holding diodes 1 and 2 in cutoff.
3. Only when a 0 input is provided at all three inputs will the output be 0.

Figures 13.7 and 13.8 show a typical OR-gate circuit and logic symbol and the respective voltage and logic truth tables.

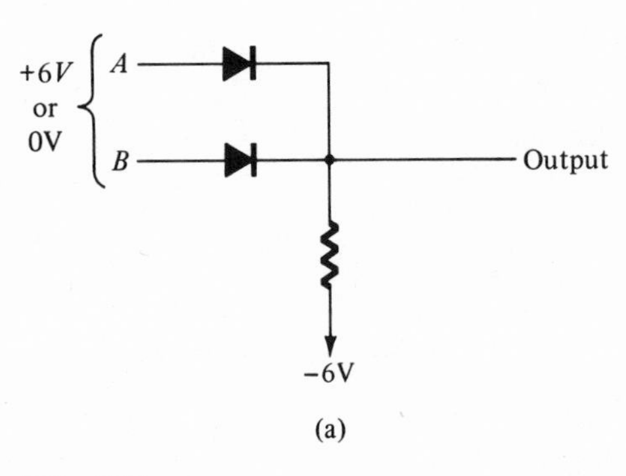

A	B	Output
0V	0V	0V
0V	+6V	+6V
+6V	0V	+6V
+6V	+6V	+6V

(b)

FIG. 13.7
(a) OR-gate circuit; and (b) voltage truth table.

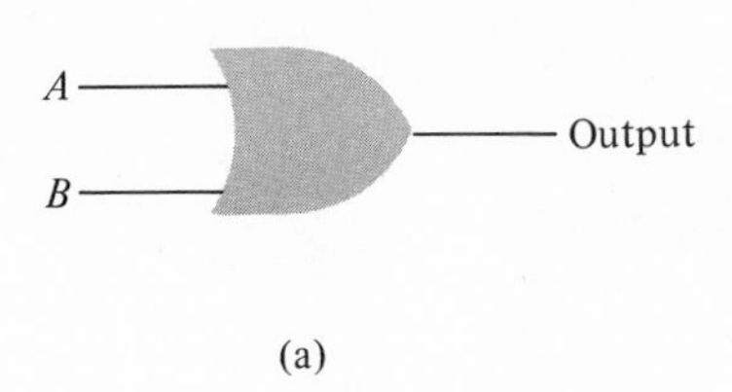

A	B	Output
0	0	0
0	1	1
1	0	1
1	1	1

(b)

FIG. 13.8
(a) OR-gate logic symbol; (b) logic truth table.

Positive Logic—Negative Logic

In the two circuit types just considered the voltage operation of these circuits is described by the voltage truth tables of Figs. 13.3 and 13.7. The logic operation of these two circuits, however, is dependent on the definitions of logical-1 and logical-0. Positive-logic definitions were used so far, where **positive logic** meant that the *more positive* voltage was assigned as the logical-1 state. It is possible to use other logic definitions. Using the same two circuits and voltage levels of $-V$ and

0 V provides **negative-logic** operation—with the definition of the *more negative* voltage ($-V$) as the logical-1 level (and 0 V as logical-0).

A summary of positive- and negative-logic gates is given in Fig. 13.9 and Fig. 13.10.

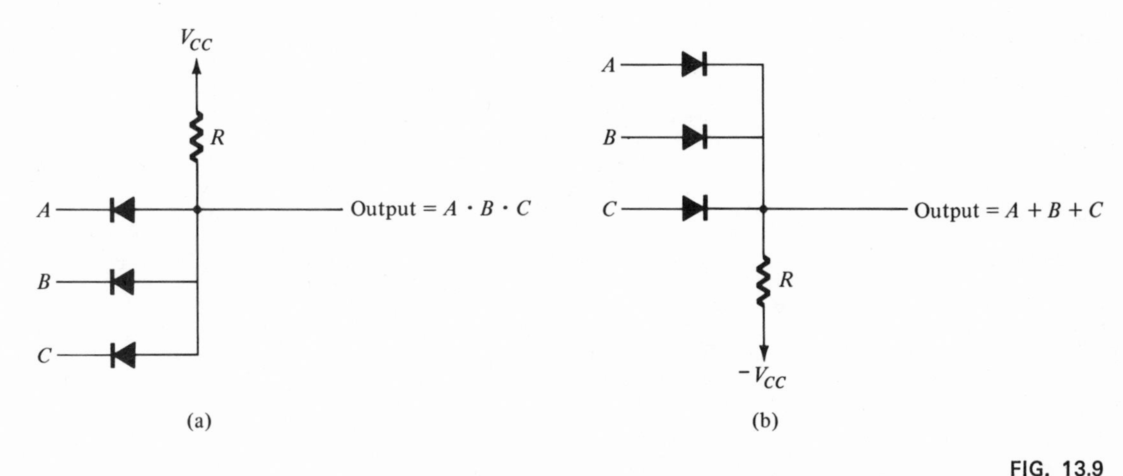

FIG. 13.9
Positive-logic circuits: (a) AND gate; (b) OR gate.

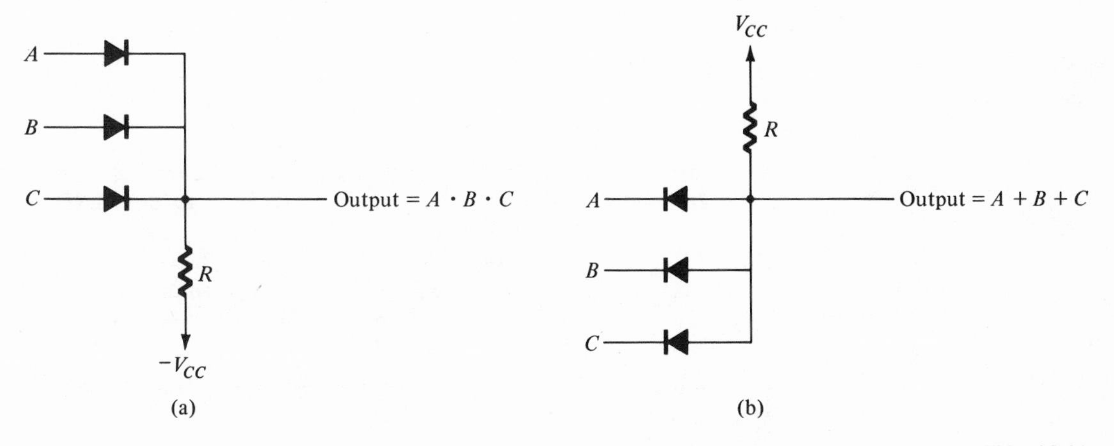

FIG. 13.10
Negative-logic circuits: (a) AND gate; (b) OR gate.

EXAMPLE 13.1 Determine the output voltage for each of the three logic circuits shown in Fig. 13.11. Indicate what type of logic circuit each is for positive-logic operation.

Solution: (a) (Fig. 13.11a) More positive output voltage at output would be +10 V with diode D_1 conducting and diode D_2 reverse-biasing to cutoff. The circuit is a positive-logic OR gate.
(b) (Fig. 13.11b) The output is more positive voltage: 0 V since D_2 conducts. The circuit is a positive-logic OR gate.

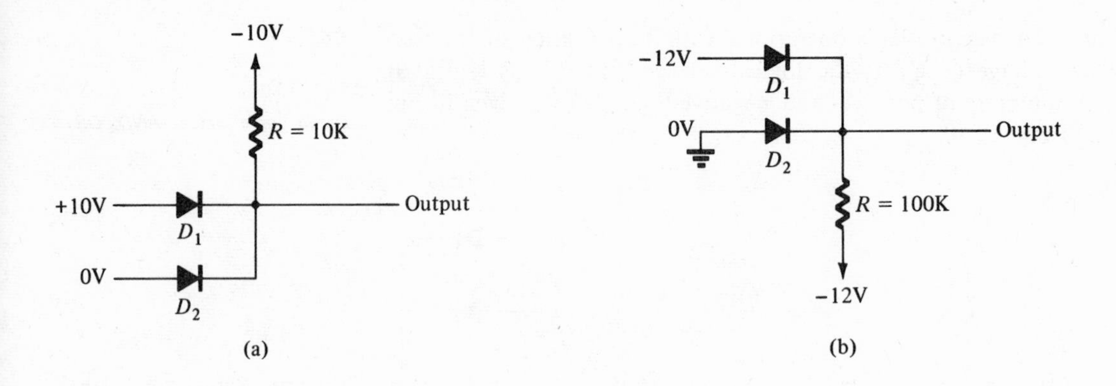

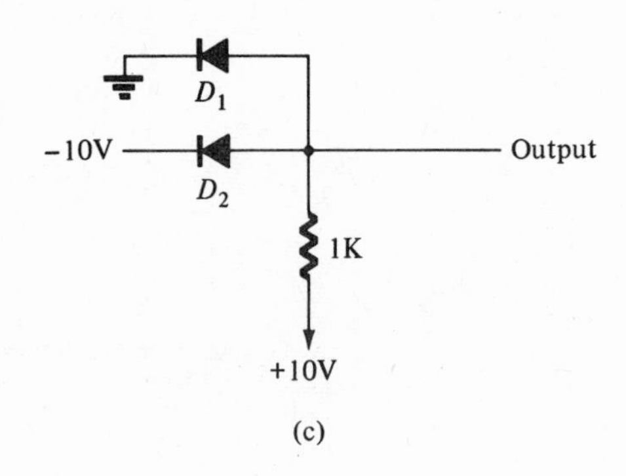

FIG. 13.11
Circuits for Example 13.1.

(c) (Fig. 13.11c) The less positive voltage at the output is -10 V as D_2 conducts, cutting off D_1. The circuit is a positive-logic AND gate.

EXAMPLE 13.2 Draw the circuit diagram of a three-input negative-logic OR gate, and prepare voltage and logic truth tables for this circuit. (Hint: for three inputs there are 2^3 or 8 possible combinations.) Use supply and voltage levels of $+10$ and 0 V.

Solution: Figure 13.12 shows the circuit diagram. Voltage and logic truth tables are given on page 649.

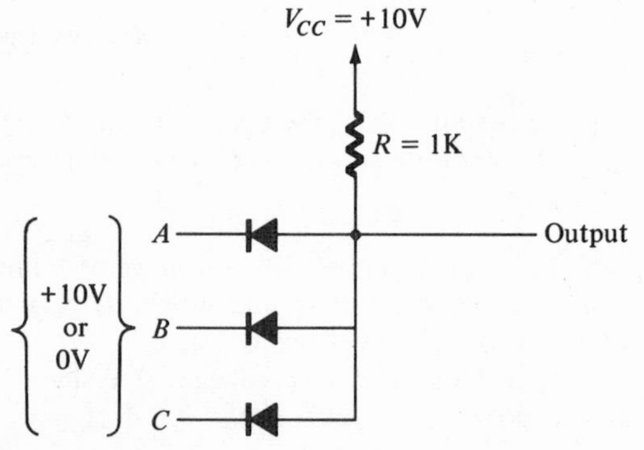

FIG. 13.12
Solution to Example 13.2.

VOLTAGE TRUTH TABLE			
A	*B*	*C*	*Output*
0 V	0 V	0 V	0 V
0 V	0 V	+10 V	0 V
0 V	+10 V	0 V	0 V
0 V	+10 V	+10 V	0 V
+10 V	0 V	0 V	0 V
+10 V	0 V	+10 V	0 V
+10 V	+10 V	0 V	0 V
+10 V	+10 V	+10 V	+10 V

LOGIC TRUTH TABLE (OUTPUT = $A \cdot B \cdot C$)			
A	*B*	*C*	*Output*
0	0	0	0
0	0	1	0
0	1	0	0
0	1	1	0
1	0	0	0
1	0	1	0
1	1	0	0
1	1	1	1

13.3 TRANSISTOR INVERTER

A simple but quite important digital circuit is the inverter. Using voltage levels of 0 and +10 V, as an example, the inverter circuit will provide an output of 0 V for an input of +10 V and an output of +10 V for an input of 0 V. In logical terms the inverter, having single-input and single-output terminals, provides the opposite output—a 1 output for 0 input or vice versa.

Figure 13.13 shows the circuit diagram of an inverter. The transistor in this type of circuit is operated either in saturation (ON) or in cutoff (OFF). To operate the transistor in saturation there must be sufficient base current drive corresponding to the current drawn by the

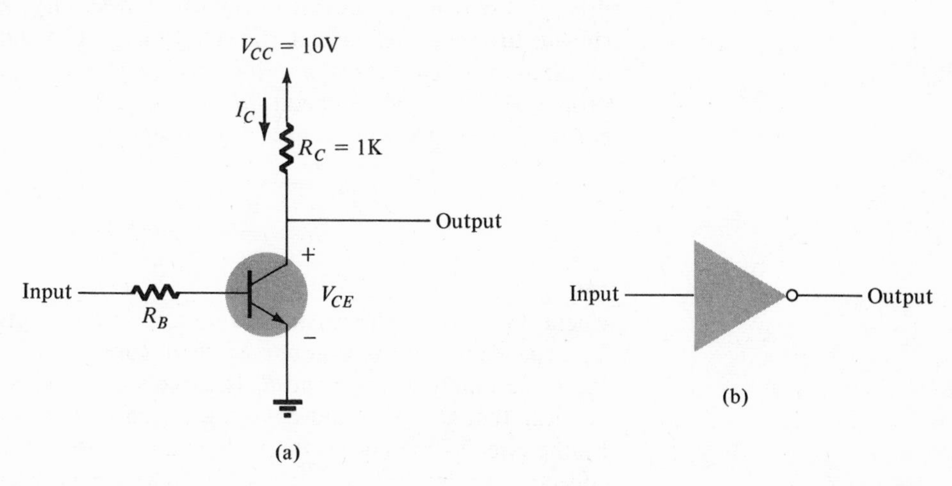

FIG. 13.13

Inverter circuit and characteristics: (a) transistor inverter circuit; (b) inverter logic symbol.

collector. This requires that for a particular amount of collector current the base current must be larger than the amount specified by the transistor current gain (h_{fe} or β). Referring to Fig. 13.14 the following relation is obtained for the base current

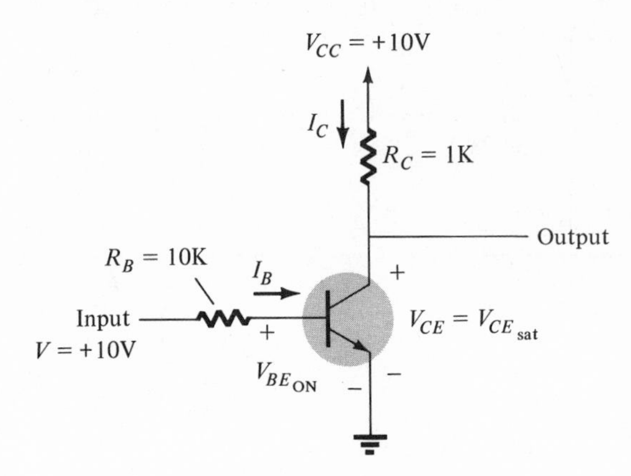

FIG. 13.14
Transistor considerations in saturation.

$$I_B = \frac{V - V_{BE\ ON}}{R_B} \tag{13.1}$$

where $V_{BE_{ON}}$ is the voltage drop across a forward-biased base-emitter junction—0.7 V for a silicon and 0.3 V for a germanium transistor, typical—and V is the input logic voltage.

Since the value of $V_{BE_{ON}}$ is relatively fixed, as are the values of the input voltage V and base resistor R_B, the amount of base current drive is fixed in the present circuit. It is necessary, as will soon be shown, to make this current sufficiently large to handle all extremes of variation in base resistance, input voltage, and base-emitter voltage drop with load and temperature changes. The amount of collector current is also generally a fixed (or maximum) amount

$$I_C = \frac{V_{CC} - V_{CE_{sat}}}{R_C} \tag{13.2}$$

where $V_{CE_{sat}}$ is the transistor saturation voltage, typically 0.05–0.5 V, dependent on the amount of load current (I_C) being handled; V_{CC} is the supply voltage; and R_C is the collector load resistor. It should be clear that the base and collector currents are fixed independently by the circuit—mainly by the voltage and resistor values. A ratio of collector current to base current using the circuit-derived values only specifies the amount of current gain required for the circuit to operate as expected—namely, with the transistor in saturation. What is, therefore, necessary is that the transistor current gain itself be larger in value than this required (forced) current gain of the circuit. Stated mathematically, the transistor will be in saturation if

$$\boxed{h_{FE} > \beta_C = \frac{I_C}{I_B}} \tag{13.3}$$

where h_{FE} is the transistor current gain (specified by the manufacturer), and β_C is the forced beta of the circuit (or circuit current gain) dependent on the values of I_C and I_B, given in Eq. (13.1) and (13.2). An example will help clarify how these different relations are properly used.

EXAMPLE 13.3 For the inverter circuit and values of Fig. 13.14, with logic inputs of 10 and 0 V, determine whether the transistor is properly in saturation, if the transistor current gain is $h_{FE} = 20$. Use $V_{BE_{ON}} = 0.7$ V and $V_{CE_{sat}} = 0.2$ V.

Solution: Calculating the base and collector currents using Eqs. (13.1) and (13.2) gives

$$I_B = \frac{V - V_{BE_{ON}}}{R_B} = \frac{10 - 0.7\text{ V}}{10\text{ K}} = \frac{9.3\text{ V}}{10\text{ K}} = 0.93\text{ mA}$$

$$I_C = \frac{V_{CC} - V_{CE_{sat}}}{R_C} = \frac{10 - 0.2\text{ V}}{1\text{ K}} = \frac{9.8\text{ V}}{1\text{ K}} = 9.8\text{ mA}$$

Using Eq. (13.3) to check

$$h_{FE} = 20 > \beta_C = \frac{I_C}{I_B} = \frac{9.8}{0.93} \cong 10.5$$

Since the transistor gain is larger than the required forced beta of the circuit the transistor is driven into saturation.

If, in Example 13.3, the transistor gain were only 10, the transistor would not be driven fully into saturation as desired. However, as presently designed the transistor has a current gain twice the circuit forced current gain providing a safety margin to account for changes in V_{CC}, V, and $V_{BE_{ON}}$.

13.4 LOGIC NAND AND NOR GATES

It was previously pointed out that the combination of logic gate followed by inverter is extremely popular. A loading problem exists when connecting AND or OR gates one after the other as is required in logic systems. An inverter circuit provides a buffer, minimizing loading effects. If a basic logic circuit is built using an inverter after each logic AND or OR circuit then the overall circuit has much better characteristic in terms of loading and in terms of how fast the gate operates (passes signals). Using mass production to build these logic circuits makes it more economical in the long run to build only NAND gates or only NOR gates for a particular logic system, either type of logic gate being capable of performing all required logical operations. Having considered some rationale for the existence of NAND or NOR gates, let us examine the construction and operation of these gates.

Diode-Coupled Transistor Logic (DCTL Gate)

One of the popular versions of a logic-inverter gate uses the type of diode logic gate previously considered followed by an inverter. A complete form of this combination circuit using an NPN transistor is shown in Fig. 13.15. Considering positive-logic operation, with logic levels of +5 V as logical-1 and 0 V as logical-0, the gate shown is a NAND gate.

TRANSISTOR IN SATURATION

The NPN transistor in the circuit of Fig. 13.15 will be in saturation when the inputs are all $+V_{CC}$ (logical-1). The circuit requirements to

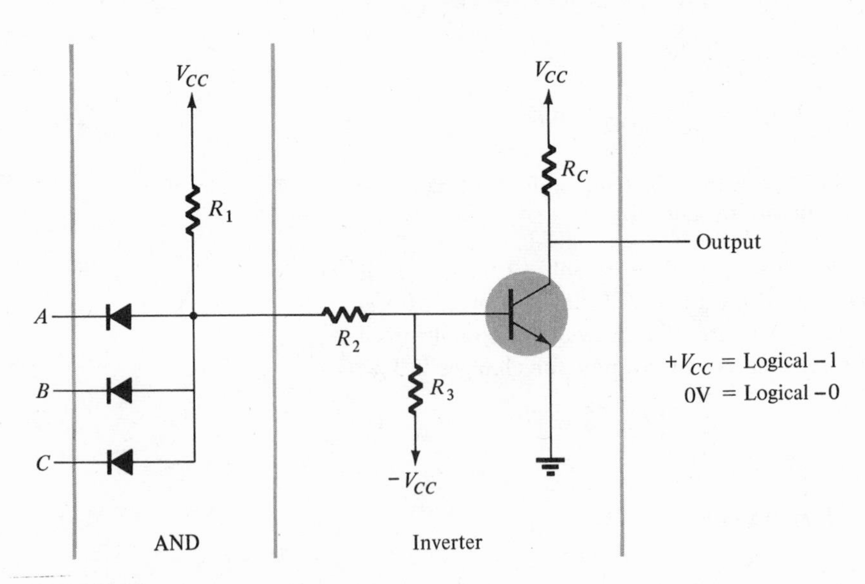

FIG. 13.15
DCTL NAND-gate circuit (positive-logic).

ensure transistor saturation are given by the partial circuit diagram of Fig. 13.16a. For this condition of operation the requirement is much the same as for the inverter circuit considered previously. With the transistor ON the current relations and saturation requirements are as follows:

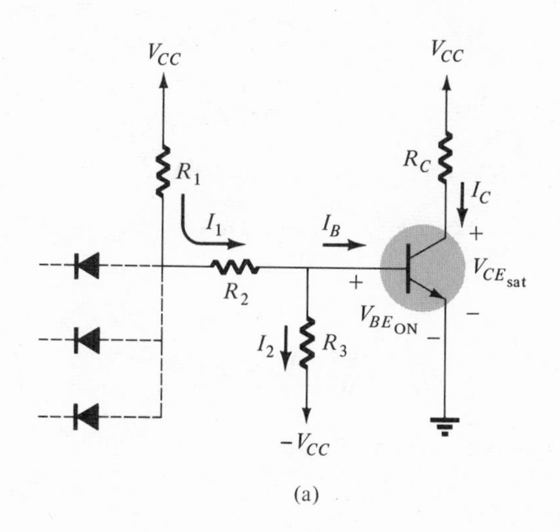

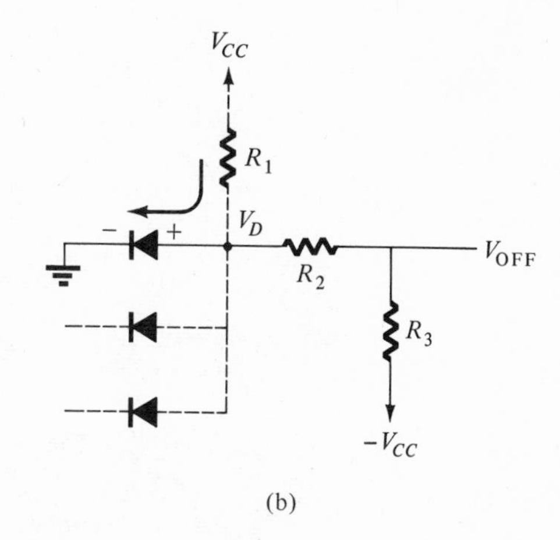

FIG. 13.16
Partial circuit operation of NPN NAND-NOR gate: (a) transistor in saturation; (b) transistor in cutoff.

$$I_1 = \frac{V_{CC} - V_{BE_{ON}}}{R_1 + R_2} \tag{13.4}$$

$$I_2 = \frac{V_{CC} + V_{BE_{ON}}}{R_3} \tag{13.5}$$

$$I_B = I_1 - I_2 \tag{13.6}$$

$$I_C = \frac{V_{CC} - V_{CE_{sat}}}{R_C} \tag{13.7}$$

For saturation

$$h_{FE} > \beta_C = \frac{I_C}{I_B} \tag{13.8}$$

With the base and collector currents fixed by the circuit component values and voltages, the circuit forced beta is obtained. As long as the transistor has more current gain than that required by the circuit the transistor will be driven into saturation.

EXAMPLE 13.4 For the NAND circuit of Fig. 13.15 determine whether the transistor is in saturation for the following circuit and transistor values: $R_1 = 1.5$ K, $R_2 = 5.6$ K, $R_3 = 10$ K, $V_{CC} = 6$ V, $R_C = 1$ K, $h_{FE} = 15$, $V_{BE_{ON}} = 0.7$ V, and $V_{CE_{sat}} = 0.3$ V.

Solution: Using Eqs. (13.4)–(13.8)

$$I_1 = \frac{V_{CC} - V_{BE_{ON}}}{R_1 + R_2} = \frac{6 - 0.7}{1.5 + 5.6} = \frac{5.3 \text{ V}}{7.1 \text{ K}} = 0.747 \text{ mA}$$

$$I_2 = \frac{V_{CC} + V_{BE_{ON}}}{R_3} = \frac{6 + 0.7}{10 \text{ K}} = 0.67 \text{ mA}$$

$$I_B = I_1 - I_2 = 0.747 - 0.670 = 0.077 \text{ mA}$$

$$I_C = \frac{V_{CC} - V_{CE_{sat}}}{R_C} = \frac{6 - 0.3}{1 \text{ K}} = 5.7 \text{ mA}$$

is

$$h_{FE} = 15 > \beta_C = \frac{I_C}{I_B} = \frac{5.7 \text{ mA}}{0.077 \text{ mA}} = 74$$

Therefore, the transistor will not be saturated.

TRANSISTOR IN CUTOFF

If any one of the diode inputs is 0 V the transistor should be in cutoff. A partial circuit showing the details of this operating condition is that of Fig. 13.16b. The voltage at the junction point of the diode anodes and resistors R_1 and R_2 is $+V_D$, the voltage drop across a forward-biased diode. The OFF voltage at the transistor base is determined by the voltage divider made up of resistors R_2 and R_3 with the voltages $+V_D$ and $-V_{CC}$. The OFF voltage and holdoff (noise) margin (based on a base off voltage of 0 V) are determined as follows:

$$\boxed{V_{\text{OFF}} = \frac{R_3}{R_2 + R_3}(V_D + V_{CC}) - V_{CC}} \tag{13.9}$$

for $V_{\text{OFF}} = 0$ V

$$0 = \frac{R_3}{R_2 + R_3}(V_i + V_{cc}) - V_{CC}$$

$$\boxed{\text{holdoff (noise) margin} = V_i = \frac{R_2}{R_3} V_{CC}} \tag{13.10}$$

EXAMPLE 13.5 Using the circuit values of Example 13.4 determine V_{OFF} and the holdoff margin (use $V_D = 0.7$ V).

Solution

$$V_{OFF} = \frac{R_3}{R_2 + R_3}(V_D + V_{CC}) - V_{CC}$$
$$= \frac{10}{5.6 + 10}(0.7 + 6) - 6 = \frac{10}{15.6}(6.7) - 6$$
$$= 4.3 - 6 = -1.7\text{ V}$$
$$\text{holdoff (noise) margin} = \frac{R_2}{R_3}(V_{CC}) = \frac{5.6}{10}(6) = 3.36\text{ V}$$

(It would take more than 3.36 V at the junction of diode anodes to result in 0 V at the transistor base; otherwise, V_{OFF} is negative and the transistor is held in cutoff.)

13.5 INTEGRATED-CIRCUIT (IC) LOGIC DEVICES

Digital integrated circuits (ICs) of various types find widespread use. It is economical and attractive to purchase a complete circuit of small size so that most users of digital logic circuits depend on the IC units provided by the numerous manufacturers. At present there are a number of different types of logic circuits popular, each having some advantages and disadvantages. No one circuit type has been universally accepted so it seems reasonable to consider some of the more popular circuit types to understand their operation and their relative advantages and disadvantages. It should be clear that each type provides the same basic logical function and that other more practical factors about the overall system generally dictate which particular logic circuit type is chosen.

As a partial summary of the important factors used in selecting a circuit type we have (not necessarily in order of importance)

1. Cost.
2. Power dissipation.
3. Speed of operation.
4. Noise immunity.

One of the circuit types has already been discussed in Section 13.4. This is the diode-transistor-coupled logic (DCTL) circuit. A modified version of the circuit considered includes a diode-coupling element as shown in Fig. 13.17. This circuit version is referred to as a DTL gate.

DTL Circuit

The DCTL circuit previously considered allowed selecting resistor values to obtain a wide range of adjustment on the holdoff (noise) margin. The DTL circuit of Fig. 13.17 provides a noise margin based mainly on the voltage drop across diode D_4.

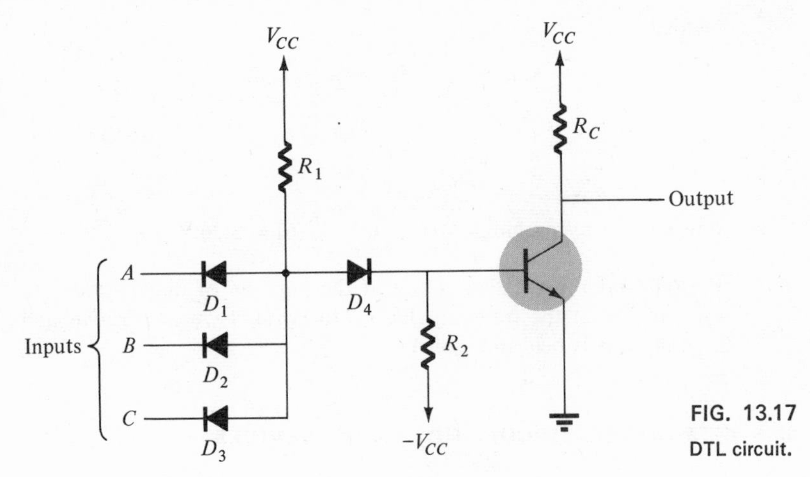

FIG. 13.17
DTL circuit.

Resistor-Transistor Logic (RTL) Circuit

Another popular IC circuit is the resistor-transistor logic circuit (called RTL). Figure 13.18a shows a three-input NOR gate. Notice that each input requires a resistor and transistor. Recalling that the circuit is manufactured as an integrated circuit the use of a transistor for each input is not unreasonable. In fact, each resistor causes more problem in manufacture than each transistor. An IC resistor can be more expensive to make, is not easily obtained in close tolerance or nominal value, and requires more space than a transistor.

The NOR part of the circuit operation is simply seen. With no inputs (or inputs of 0 V or less) all transistors are held in cutoff and the output voltage is the supply voltage, $+V_{CC}$. If any one input is $+V_{CC}$, that transistor will be driven ON and the output terminal will then be at near 0 V, $V_{CE_{sat}}$ to be exact. This operation is that of a NOR gate for positive logic. Voltage and logic truth tables in Fig. 13.18 show this in more detail. The NOR operation provides that the output be logical-0 if inputs *A OR B OR C* (or any combination of these) are logical-1.

Some advantages of the RTL circuit are the relative simplicity for manufacture as integrated circuit and the requirement of only a single supply voltage. Some disadvantages are the low-noise immunity (low-noise margin) of the circuit because of the single supply voltage, and the slower speed of this circuit compared with, say, DTL or DCTL circuits.

Transistor-Transistor Logic (TTL) Circuit

Transistor-transistor logic is one circuit form of logic gate that, although possible as discrete components, is quite appropriate for manufacture in integrated form. Figure 13.19a shows a discrete form of the logic circuit. Notice that each input is made using a transistor.

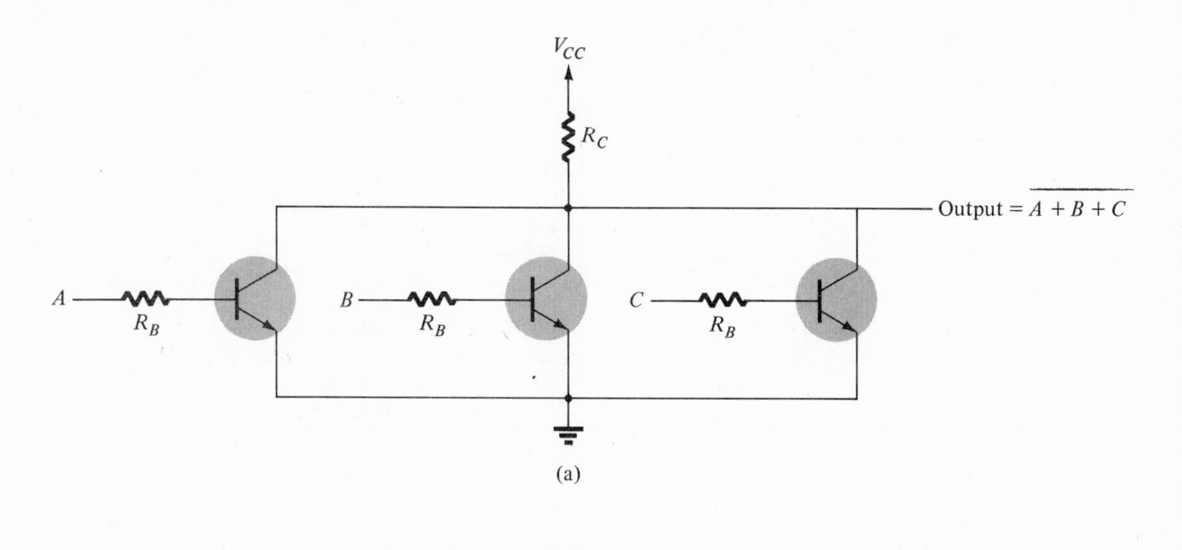

$$\text{Positive logic} \begin{cases} +V_{CC} = +V \equiv \text{logical} - 1 \\ \approx 0\text{V} \equiv \text{logical} - 0 \end{cases}$$

$$\text{Output} = \overline{A + B + C}$$

A	B	C	Output
0V	0V	0V	+V
0V	0V	+V	$V_{CE_{sat}} \approx$ 0V
0V	+V	0V	$\approx$ 0V
0V	+V	+V	$\approx$ 0V
+V	0V	0V	$\approx$ 0V
+V	0V	+V	$\approx$ 0V
+V	+V	0V	$\approx$ 0V
+V	+V	+V	$\approx$ 0V

A	B	C	Output
0	0	0	1
0	0	1	0
0	1	0	0
0	1	1	0
1	0	0	0
1	0	1	0
1	1	0	0
1	1	1	0

FIG. 13.18

(a) RTL NOR logic gate: (b) symbol; (c) voltage and logic truth table.

Only one resistor is required for all inputs, in addition to the load resistor (R_C). It is this savings in number of resistors, or any components other than transistors, that makes the circuit so suitable for IC production.

Using input voltage levels of $+V_{CC}$ and 0 V, and positive-logic definitions, the circuit of Fig. 13.18a operates as a NAND gate. If all

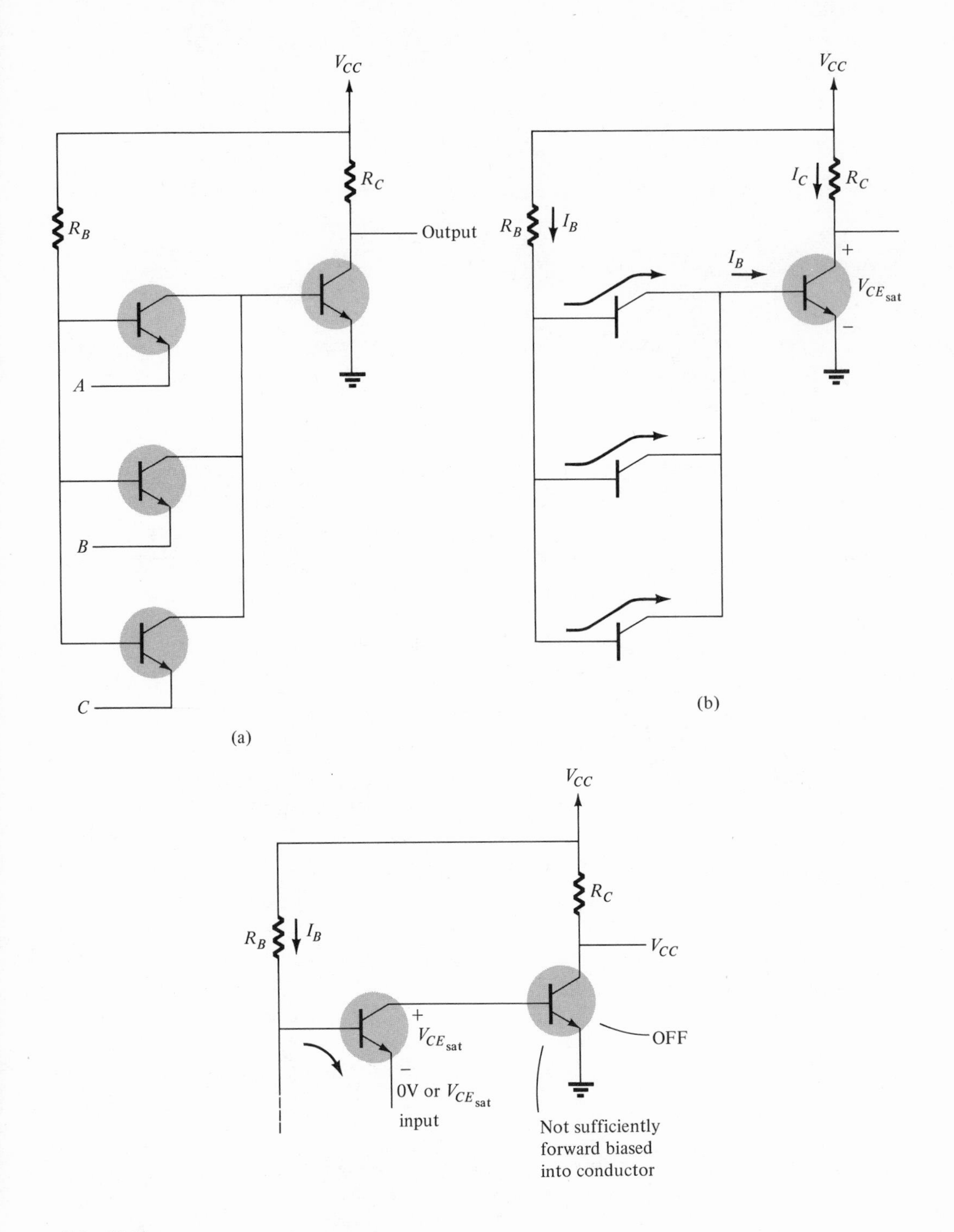

FIG. 13.19

(a) Discrete form of TTL logic circuit: (b) output transistor ON; (c) output transistor OFF.

inputs are left open (or are $+V_{CC}$) then the output transistor will be driven ON with base current flowing through biasing resistor R_B, and the forward-biased diode junction from base to collector of each input transistor (see Fig. 13.19b). This is not quite obvious since the transistor marking usually shows only the base-emitter diode polarity. However, for an NPN transistor, as shown, the base-collector is also a p-n junction, which, with the base-emitter open or off, can be used as a simple diode p-n junction. With the output transistor driven ON, the output voltage is the transistor saturation value, $V_{CE_{sat}}$, which is approximately the 0 V (logical-0) level.

If any one input is 0 V (or approximately 0 V) then that transistor will be driven ON with base current through resistor R_1 now flowing through the ON transistor base-emitter junction (Fig. 13.19c). Thus, we can see that the switching action of the circuit transfers the base drive current from flowing through a base-collector junction, to the output transistor to a base-emitter junction, and to ground (or through the saturated junction of the transistor providing the logical-0 input), With an input transistor driven ON (saturated) the voltage at the output transistor base is $V_{CE_{sat}}$, which will be low enough compared to that required to drive a transistor ON so that the output transistor will be held OFF and the output voltage will then be $+V_{CC}$.

A summary of this operation is provided in the voltage and logic truth tables of Fig. 13.20. The circuit is shown, by the truth tables,

Positive logic $\begin{cases} +V_{CC} \equiv \text{logical} - 1 \\ 0\text{V} = \text{logical} - 0 \end{cases}$

Output $= \overline{A \cdot B \cdot C}$ (NAND gate)

A	B	C	Output
0V	0V	0V	$+V_{CC}$
0V	0V	$+V_{CC}$	$+V_{CC}$
0V	$+V_{CC}$	0V	$+V_{CC}$
0V	$+V_{CC}$	$+V_{CC}$	$+V_{CC}$
$+V_{CC}$	0V	0V	$+V_{CC}$
$+V_{CC}$	0V	$+V_{CC}$	$+V_{CC}$
$+V_{CC}$	$+V_{CC}$	0V	$+V_{CC}$
$+V_{CC}$	$+V_{CC}$	$+V_{CC}$	0V ($V_{CE_{sat}}$)

(a)

A	B	C	Output
0	0	0	1
0	0	1	1
0	1	0	1
0	1	1	1
1	0	0	1
1	0	1	1
1	1	0	1
1	1	1	0

(b)

FIG. 13.20

TTL NAND-gate truth tables: (a) voltage truth table; (b) logic truth table.

to be a positive-logic NAND gate. Exactly the same circuit can be made as an integrated circuit, as shown in Fig. 13.21. Comparing this to Fig. 13.19a, we see that input transistor bases are connected in common, as well as connection of the transistors, these being made as one piece in the IC manufacture. The input then appears as a single transistor with multiple emitter leads, one for each input of the logic gate.

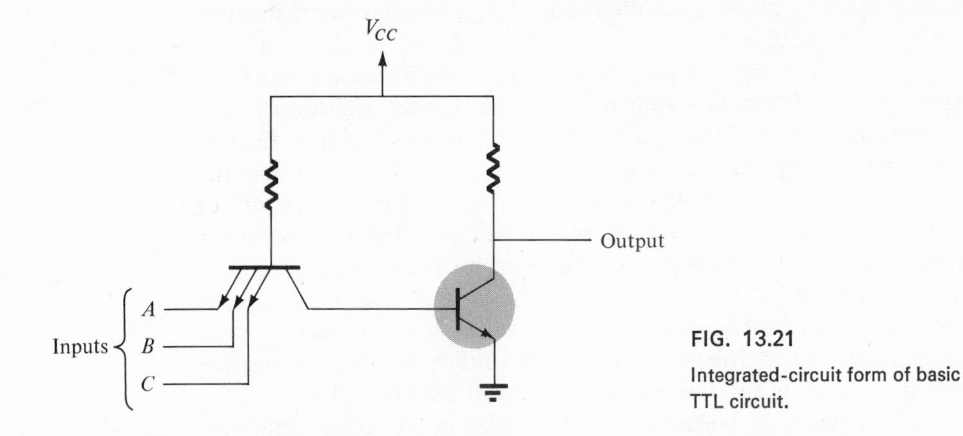

FIG. 13.21
Integrated-circuit form of basic TTL circuit.

The obvious advantage of this circuit form is the fact that it can be manufactured easily as an integrated circuit, using mostly transistors. The circuit, in the form shown, also uses only a single supply voltage, which is another advantage. A third positive factor is the switching action of the circuit. The output transistor is turned off through a saturated input transistor, which provides a low-impedance path for the base current of the output transistor (the base current that must be removed before the transistor turns OFF).

Emitter-Coupled Logic (ECL) Circuit

The technique of emitter-coupled logic (ECL) or current-mode logic (CML) circuits differs from those covered previously in one main respect. All previous circuits allowed the transistors to saturate. This means that an amount of charge is stored in the transistor base and collector regions resulting in a time delay in turning off the transistor. Current-mode logic operates the transistor in a nonsaturated condition, thereby providing shorter propagation delays through the circuits.

An emitter-coupled logic circuit, such as that of Fig. 13.22, uses a transistor for each input of the circuit, which is desirable for IC manufacture. Whereas a direct-coupled logic circuit would connect the transistor emitters to ground (thereby allowing the base drive to be

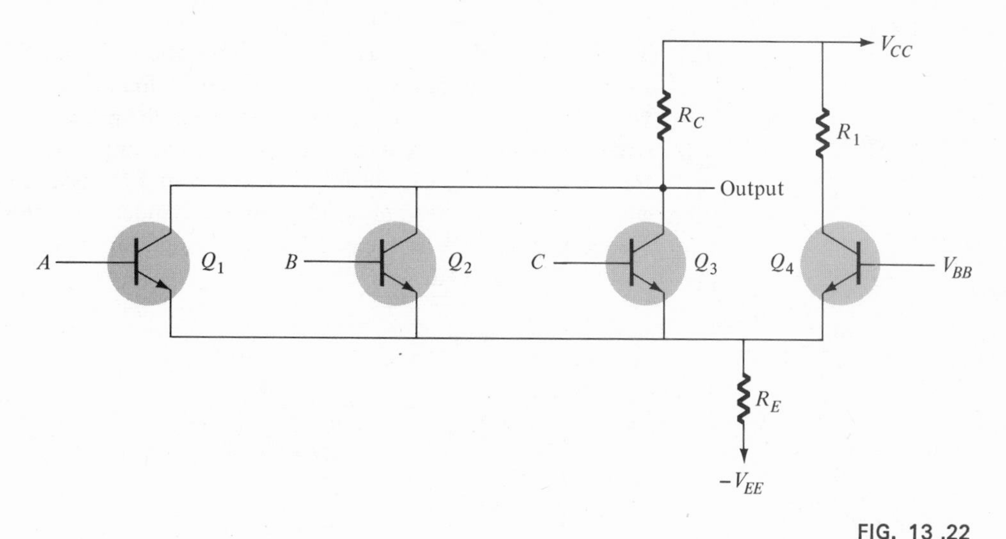

FIG. 13.22
Emitter-coupled logic (ECL) circuit.

dependent on the input voltage), the present circuit uses an additional transistor stage providing a reference point for the common-emitter terminals. If more than the needed turn-on drive is applied this will cause the emitter point to rise in voltage, maintaining the logic transistor in a nonsaturated mode of operation.

In summary, then, the nonsaturated operation of an emitter-coupled logic (ECL) circuit provides very fast switching speeds, simultaneous OR and NOR outputs from each circuit, high fan-in and fan-out capacity, and constant noise immunity of the power supply due to the relative constant-current demand of the logic circuit whether OFF or ON. Disadvantages are the need for three voltage supplies (which is reduced to only one using ground as one), the need for a bias driver to provide the reference supply voltage, and the higher cost of ECL circuits.

Field Effect Transistor (FET) Logic Circuits

Up to now the logic circuits considered used bipolar transistors. Figure 13.23 shows a NOR-gate logic circuit using unipolar field effect transistors. Some advantages of such a circuit are low-power operation, small physical size, and the requirement of only FET devices to make up the complete logic circuit.

The FETs used in Fig. 13.23 are all insulated-gate, *p*-channel units. Input logic voltages are applied to the gate of each input FET transistor similar in manner to the operation of a DCTL logic circuit.

Typically, one would expect the common point of the FET transistors used as the output point to connect through a load resistor to the supply voltage. Since a resistor requires considerable space on an integrated circuit, the present circuit provides a resistor component in the form of the FET unit. The gate terminal of the load FET is connected to the supply voltage with drain-current value dependent on the source-drain voltage (increasing as the source-drain voltage is made larger by an input FET being driven ON).

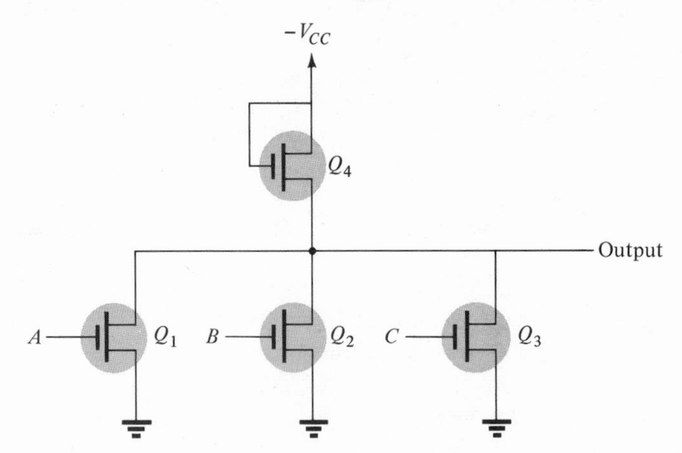

FIG. 13.23
Field effect transistor NOR gate.

The advantages of smaller size and low-power dissipation are offset by considerably lower speed of operation, although it can be anticipated that FET switching speeds will improve with manufacturing advances. Where speed is not critical the advantages of size and power for a FET circuit are quite attractive. As an example, an insulated-gate FET logic circuit is presently about 5 times smaller than its bipolar counterpart. Additionally, the manufacturing process for FET construction requires significantly fewer processing steps, thereby increasing production yield.

Compatibility of Integrated-Logic Circuits

Having shown that a number of different types of logic circuits are presently available and utilized, it would be helpful at this time to indicate some general factors about the various types of logic techniques. In more general terms there are three types of logic systems—current-sinking, current-sourcing, and current-mode. These types are basically incompatible. Different logic circuits of the same type can be used in the same operating system with the possibility of driving one directly from another.

As indicated in Fig. 13.24a current-sink circuits draw current from the input of the following stage into the driving stage. Examples of current-sink operating logic are DTL and TTL. Assuming that speed and voltage levels are set compatibly these units could be used in the same system to drive each other.

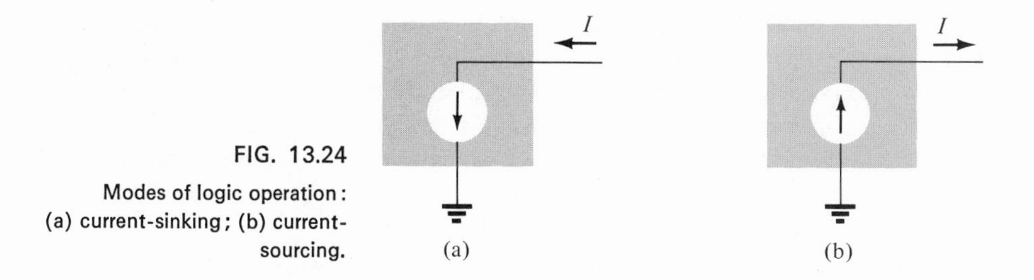

FIG. 13.24
Modes of logic operation: (a) current-sinking; (b) current-sourcing.

CURRENT-SOURCING LOGIC

Figure 13.24b shows that a current-sourcing circuit provides current to the input being driven by the logic circuit. Both DCTL and RTL operate in this manner. It should be obvious that the opposite current-flow directions of current-sink and -source circuits make these two types completely incompatible.

CURRENT-MODE LOGIC

Although current-mode logic circuits, such as an ECL, appear as current-sourcing operating circuits, the logic levels, switching speeds, and power supply requirements make this type incompatible for use with the other two types.

13.6 BISTABLE MULTIVIBRATOR CIRCUITS

Of equal importance to logic circuits in digital circuitry is the class of multivibrator circuits. There are three basic forms of the multivibrator —bistable, monostable, and astable; the most important of these, by far, is the bistable multivibrator or flip-flop. As an indication of the applications of the multivibrator circuits, consider the following:

Bistable (flip-flop)—storage stage, counter, shift-register
Monostable (ONE-SHOT)—delay circuit, waveshaping, timing circuit
Astable (CLOCK)—timing oscillator (square-wave)

In a logic system there will typically be a large number of flip-flop stages used as counters, storage registers, shift-registers, a few one-shot circuits in special timing or pulse-shaping uses, and a limited number of CLOCK circuits (typically only one).

Characteristic of all three circuits is the availability of two outputs, where the outputs are logically inverse signals. One output is selected as the reference, this designation being indicated in a number of ways. The two outputs are sometimes marked as 0 and 1, FALSE and TRUE, or $\bar{A}$ and A, etc. The main point of the designation is to indicate that the outputs are logically opposite and to mark the output chosen as the reference output. Another means of indicating the state of the multivibrator circuit is the use of the designation of SET and RESET. When referring to the state of the circuit the definitions of SET and RESET are the following:

SET:	TRUE output is logical-1
	FALSE output is logical-0
RESET:	TRUE output is logical-0
	FALSE output is logical-1

Bistable Multivibrator (Flip-Flop)

The flip-flop circuit, the most important of the multivibrator circuits, will be covered first. To provide some basic consideration of this circuit's operation a simple form of bistable circuit using two inverters is shown in Fig. 13.25a. The inverters are essentially connected in series, with two output points indicated. The two outputs are labelled 1 and 0 or A and $\bar{A}$, respectively. If the A output is a

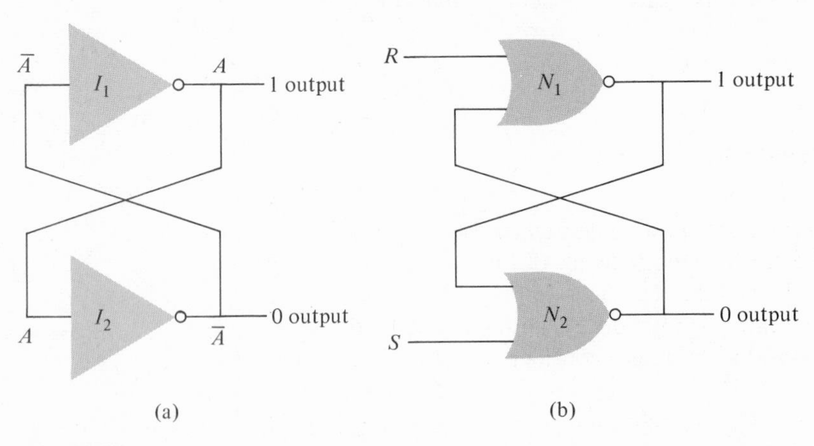

FIG. 13.25
(a) Two-inverter flip-flop circuit; (b) RESET and SET operation of flip-flop.

logical-1 then inverter I_2 will provide $\bar{A}$ as a logical-0. Since $\bar{A}$ is connected as input to inverter I_1, it will cause the output of that stage to be a logical-1, as assumed. Thus, the state of logical conditions, or the voltage they represent, forms a stable situation with A output logical-1 and $\bar{A}$ output logical-0. If some external means is used to cause the A signal to change to logical-0, then, through inverter I_2, $\bar{A}$ would change to logical-1. The $\bar{A}$ input of logical-1 would then result in A being logical-0 as initially proposed. Thus, the circuit will also remain in a stable condition if the A output is logical-0 and the $\bar{A}$ output is logical-1. In effect, then, the circuit has two stable operating states, acting as a memory of the last state it was placed into. Some external means is necessary, however, to cause the circuit to change state.

Figure 13.25b shows the use of NOR gates connected in series with additional inputs providing signals to cause the circuit to change state. The inputs are marked R for RESET and S for SET. Recall that a NOR gate provides logical-0 output if any of its inputs is logical-1. A logical-1 input to the S terminal will cause the output of N_2 to be logical-0. Assuming that no input is connected to the R terminal at this time, the inputs to N_1 are both logical-0 with output of logical-1. The result of a SET input signal then is to cause the circuit to become SET, where the SET state was previously defined as 1 output = logical-1 and 0 output = logical-0. Similarly, the application of only a logical-1 to the R input will cause the 1 output to become logical-0 and 0 output logical-1, which is the RESET state of the circuit. It should be obvious that simultaneous application of logical-1 signals to both S and R inputs is ambiguous, forcing both outputs to the logical-0 condition. This would not be an accepted operation of this circuit, in which the two output signals should be always logically opposite. If the R and S inputs are both logical-0 then the circuit remains in whatever state it was last placed into.

RS Flip-Flop

A basic version of the flip-flop circuit is the RS or RESET-SET flip-flop. A block version of this circuit is that of Fig. 13.26a showing R and S input terminals and 1 and 0 output terminals. As previously considered the application of a SET signal to the S input terminal will cause the circuit to become SET (1 output = logical-1 and 0 output = logical-0). A detailed circuit diagram is shown in Fig. 13.26b with the inputs and outputs indicated. Positive-logic voltage definitions of $+V$ = logical-1 and 0 V = logical-0 are used. If a 1 input is applied to the R terminal, for example, the $+V$ will provide a turn-on voltage through resistor R_4 to the base of transistor Q_2, causing that transistor to turn ON. With Q_2 ON the 1 output voltage is near 0 V (actually $V_{CE_{sat}}$, for transistor Q_2), and transistor Q_1 is held OFF through the

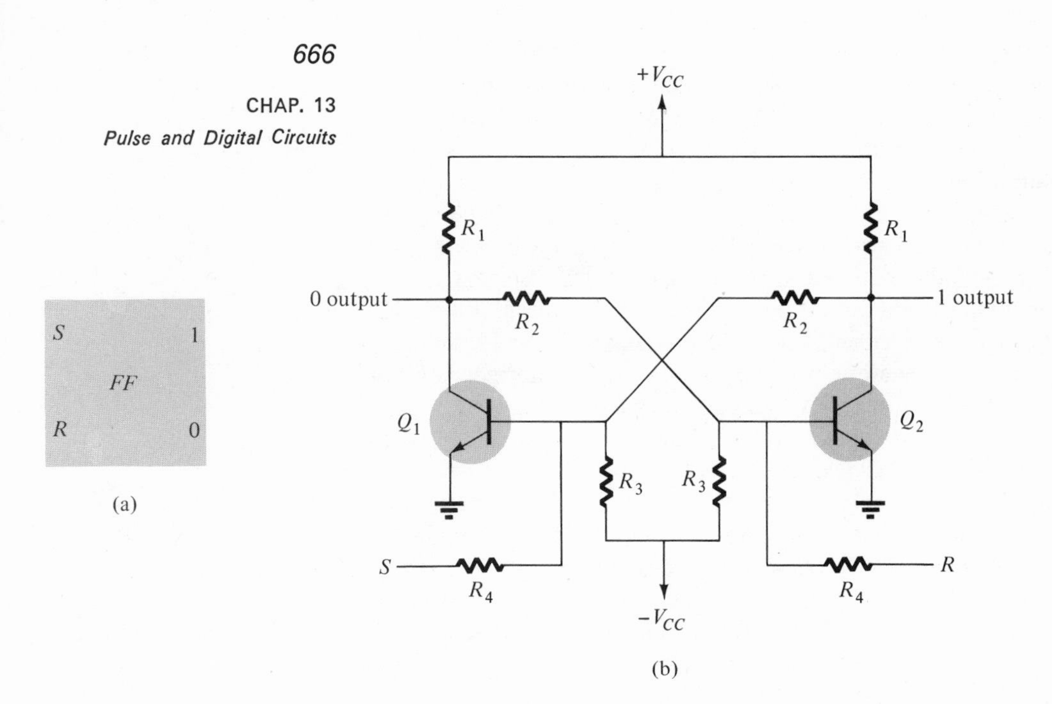

FIG. 13.26
RS flip-flop: (a) block symbol; (b) circuit diagram.

cross-coupling connection of resistor R_2 from the collector of Q_1 to the base of Q_2. With Q_1 OFF and Q_2 ON the circuit is in the RESET state, as desired.

Similarly, a logical-1 applied to the S terminal will drive transistor Q_1 ON thereby forcing Q_2 OFF, placing the circuit in the SET state. Note that the signal to switch the circuit need only be applied momentarily—long enough for the cross-coupling voltage to turn the opposite transistor OFF. Once this condition is reached removal of the original input signal will leave the circuit in its present state. On the other hand it should be clear that as long as a logical-1 signal is maintained at the input the circuit cannot change state since this input signal holds the circuit locked. This is important when the circuit has additional means of input triggering as will be discussed.

T-Type Flip-Flop

One of the more common types of flip-flop circuit is the *T*-type or triggered flip-flop. This circuit is also called a complementing flip-flop, or toggle flip-flop, since its action is to change state every time an input pulse is applied to the single *T* input terminal. Figure 13.27a shows a

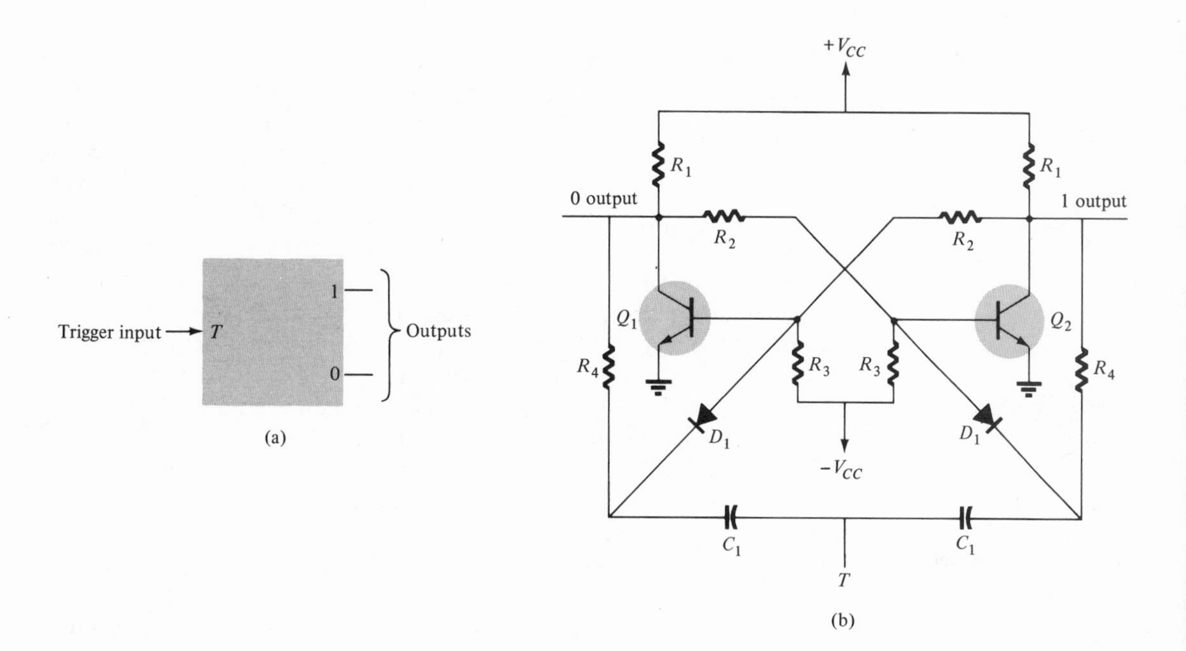

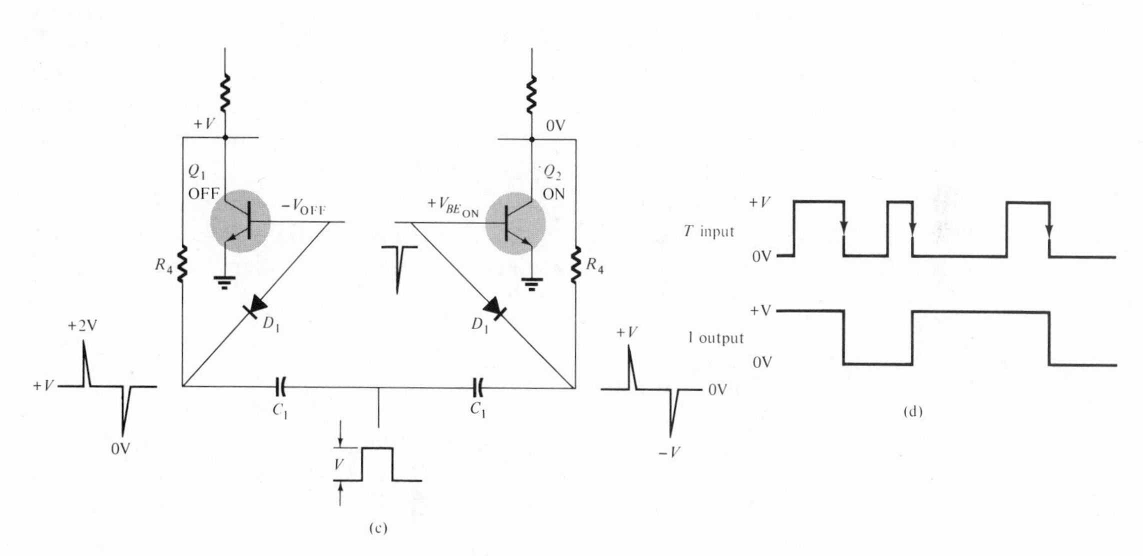

FIG. 13.27

T flip-flop; (a) block symbol; (b) circuit diagram; (c) partial circuit showing action of steering resistors; (d) typical pulse pattern for trigger (*T*) input and 1 output terminals.

block symbol of the *T* flip-flop and Fig. 13.27b shows the circuit details using NPN transistors.

A pulse applied to the *T* input terminal is coupled to both input circuit capacitors. The input pulse is differentiated into positive- and negative-voltage spikes, which appear at the two junction points of resistor R_4, diode D_1, and capacitor C_1. The connection of the R_4 resistors to appropriate collector terminal points provides a steering action at each of these junction points. Assume, as in Fig. 13.27c, that transistor Q_1 is OFF and Q_2 is ON. The collector voltage of transistor Q_1 provides a $+V$ voltage level around which the positive and negative spikes vary as shown in the figure. For ON transistor Q_2 the collector voltage level is near 0 V with the positive- and negative-going voltage spikes varying around this 0 V level. Note now that the cathode voltage of the diode connected to the base of transistor Q_1 never goes below 0 V, and with the diode-anode voltage at some negative voltage holding transistor Q_1 OFF, the negative spike does not pass through the diode. On the other side of the circuit, the negative-going voltage spike will drive that diode ON providing a negative-going voltage pulse at the base of transistor Q_1 as shown in Fig. 13.27c. Thus, the action of the steering resistors R_4 is to pass a negative spike to the base of the conducting transistor only (that having a collector voltage of near-0 V). By this steering voltage action the ON transistor will receive the negative-voltage spike turning the ON transistor OFF.

On the next negative-going voltage change of the input trigger signal the ON transistor (Q_1 at that time) will be turned OFF and Q_2 will again turn ON, the circuit again changing stage. In summary, then, the triggered flip-flop will change state whenever a specified voltage polarity change occurs—typically negative-going for NPN transistors and positive-going for PNP transistors.

Figure 13.27d shows an input pulse train waveform and the resulting output voltage from the 1 output side of the flip-flop circuit. Note that the circuit changes state on every negative-going voltage change of the pulse input signal.

RST Flip-Flop

A versatile multivibrator circuit combines the SET, RESET and toggle features in a single unit called an RST flip-flop. The *R* input (see Fig. 13.28) provides direct RESET operation, the *S* input provides direct SET operation, and the *T* input provides the toggle or complementing operation.

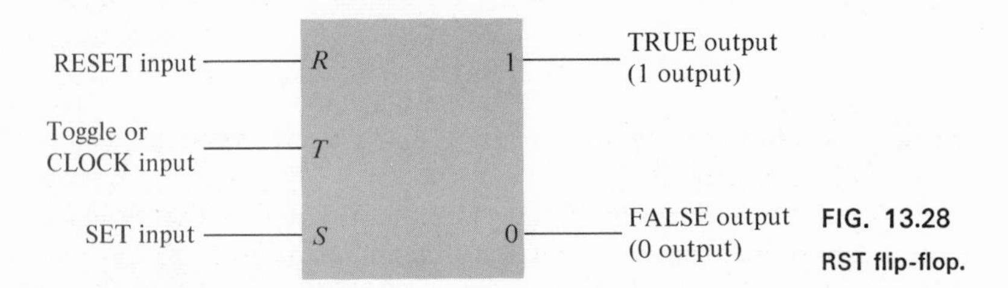

FIG. 13.28
RST flip-flop.

When a logical-1 input is applied to the *R* input the circuit will be RESET. If the logical-1 dc voltage is maintained at the *R* terminal the circuit will be held "locked" or forced in the RESET state. Even if a CLOCK signal is applied to the *T* input, the circuit will remain in RESET. Similarly, a logical-1 input only to the *S* terminal will hold the circuit locked in the SET state. Only if both *R* and *S* inputs are logical-0 can the *T* input signal operate the flip-flop. With *R* and *S* inputs logical-0, a CLOCK input to the *T* terminal will complement the stage output for every CLOCK pulse received. If, however, both *R* and *S* inputs are logical-1, the output state is not defined (usually, the two outputs will be logical-0—not a proper condition).

JK Flip-Flop

The *J* and *K* input terminals (Fig. 13.29a) are used to provide information or data inputs. When a trigger pulse is then applied the circuit changes state corresponding to the inputs to the *J* and *K* terminals. A *JK* circuit can be built in integrated circuit form using DTL, TTL, or even ECL logic manufacture. All these circuits have the same

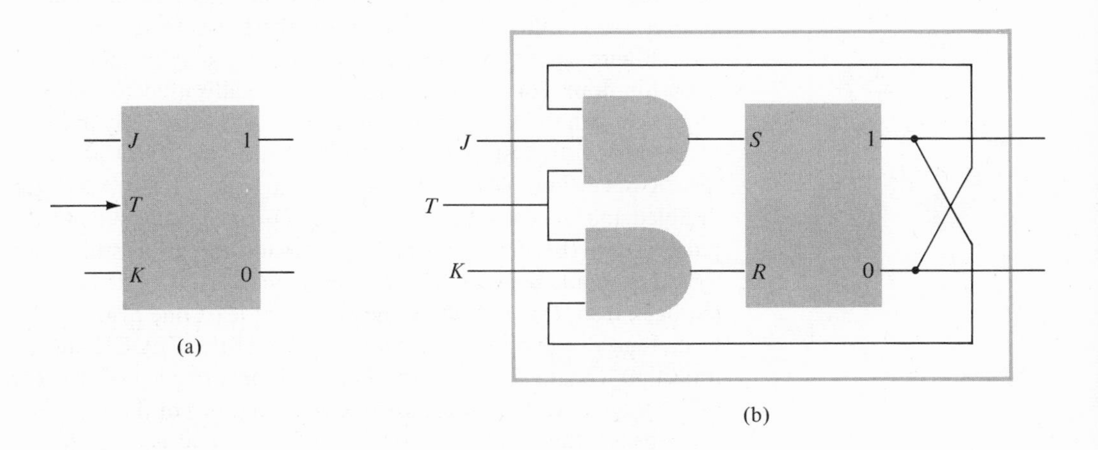

J	*K*	Circuit Action
0	0	Remains in same state
0	1	RESETS
1	0	SETS
1	1	Circuit toggles

(c)

FIG. 13.29

Basics of *JK* flip-flop: (a) logic symbol; (b) *JK* flip-flop made using *RS* flip-flop, plus gating; (c) truth table.

basic operating features. It is not at all important here to consider the circuit details and in fact one only purchases a complete unit in IC form and has little to do with the details of circuit operation. It is important to be aware of the differences in using these different logic types, in knowing whether they use current-sourcing, sinking, or emitter-coupled logic, in details of speed of operation, noise margin, power supply voltages, etc. These factors, however, are descriptive of the overall circuit, and the details of what actually goes into building the actual circuit are often of little importance. For the *JK* flip-flop, then, we shall discuss its operation from the logic or block diagram point of view and consider some applications using the circuit. The *JK* circuit is quite versatile and is presently the most popular version of the flip-flop circuit. A logic symbol of a *JK* flip-flop is shown in Fig. 13.29a. The *J* and *K* terminals shown are the data input terminals receiving the information of logical-1 or logical-0. These information inputs do not, however, change the state of the flip-flop circuit, which will remain in its present state until a CLOCK or trigger pulse is applied. Thus, for example, if the circuit were presently RESET and the input data were such as to result in the SET condition, the circuit would still maintain the RESET condition, even with the *J* and *K* input data signals applied. Only when the trigger pulse occurs are the input data used to determine the new state of the circuit—the SET state for the present example.

Figure 13.29b shows how an *RS* flip-flop may be modified to form a *JK* flip-flop. This connection is not typically used to make *JK* flip-flops although the practical circuit version acts essentially in the manner to be described. Using an *RS* flip-flop and two AND gates provides the *JK* operation. With the trigger input logical-0 the AND gates are disabled and the circuit will maintain its present state. When the trigger pulse occurs (becomes logical-1) the circuit operation still depends on the other inputs to the AND gates. Since each AND gate had one input fed back from the outputs of the circuit, at least one of the AND gates has a logical-1 in addition to the logical-1 of the CLOCK signal (when it occurs). Finally, we have the *J* and *K* inputs to each of the respective AND gates. There are four possible combinations of the *J* and *K* inputs. To consider the complete operation of the circuit each of the possible conditions is listed in the truth table of Fig. 13.29. If both *J* and *K* inputs are logical-0 then both AND gates are disabled and the circuit remains in the same state (no change takes place). If the *J* input is logical-1 and *K* input logical-1, and, further, if the 1 output is logical-1, then the CLOCK pulse going to the logical-1 level will provide all logical-1 inputs to the AND gate connected to the *R* input so that the circuit ends up in the RESET state. If the *K* input is logical-0 and the *J* input logical-1, and, further, if the 0 output is logical-1, then the occurrence of the CLOCK pulse will cause the circuit to be SET. Finally, if both *J* and *K* inputs are logical-1 the action of the CLOCK pulse becoming logical-1 is to toggle or complement the circuit. If the 0 output were logical-1 (circuit RESET) the CLOCK pulse occurring would result in the circuit's being SET, thereby causing it to change state. Similarly, with the 1 output logical-1, the occurrence of a CLOCK pulse would result in a RESET signal, thereby changing the circuit state. Thus, the circuit would toggle (change state) from whichever condition it happened to be in on application of the CLOCK pulse. In this last case, with J and K inputs both logical-1, the circuit would operate as a T flip-flop and could be used as such. When opposite data input signals are applied as J and K inputs the CLOCK pulse will shift the data into the present flip-flop state, the stage then acting as a shift-register stage. Thus, the JK flip-flop can be used as a shift-register stage, a toggle stage for counting operations, or generally as a control logic stage.

Using positive logic (0 V = logical-0 and $+V$ as logical-1, for example) the circuit triggers when the CLOCK goes from the logical-0 to logical-1 condition—this being referred to as positive-edge triggering since the circuit is triggered at the time the voltage goes positive (from 0 to $+V$). As opposed to this the circuit used in the T flip-flop discussion (Fig. 13.29b) triggered when the voltage went from $+V$ to 0 V, this being trailing-edge logic triggering. The manufacturer's information sheets should indicate the type of triggering required to operate the particular circuit so that it may be properly used.

13.7 MONOSTABLE AND ASTABLE MULTIVIBRATOR CIRCUITS AND SCHMITT TRIGGER CIRCUIT

Monostable Multivibrator (ONE-SHOT)

As a characteristic property of a multivibrator circuit the monostable provides two opposite-state output signals. As the name implies the outputs are stable in only one of the two possible states (SET and RESET). Figure 13.30a shows a logic block symbol of a ONE-SHOT circuit in the stable RESET state (1 output = logical-0, and 0 output = logical-1). The input trigger signal is a pulse that operates the circuit in an edge-triggered manner. Figure 13.30b shows a typical input trigger pulse and corresponding output waveform (assuming triggering on the trailing edge of the trigger pulse). The 1 output is normally LO, (RESET state). When a negative-going voltage change triggers the circuit the 1 output goes HI (SET state), which is the unstable circuit state. It will remain HI only for a fixed time interval, T, which is determined basically by a timing capacitor whose value may be selected externally. Thus, the output state remains in the SET state

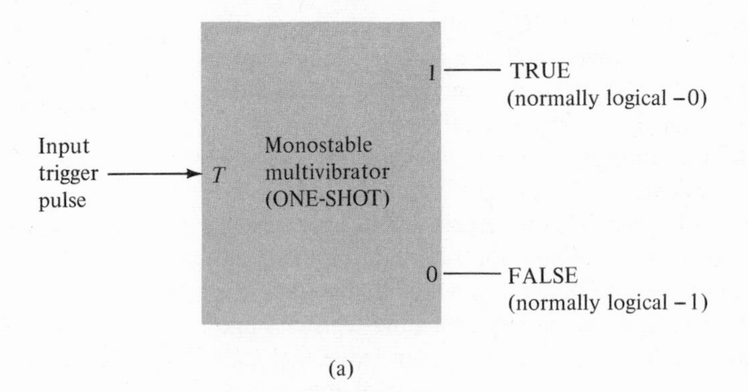

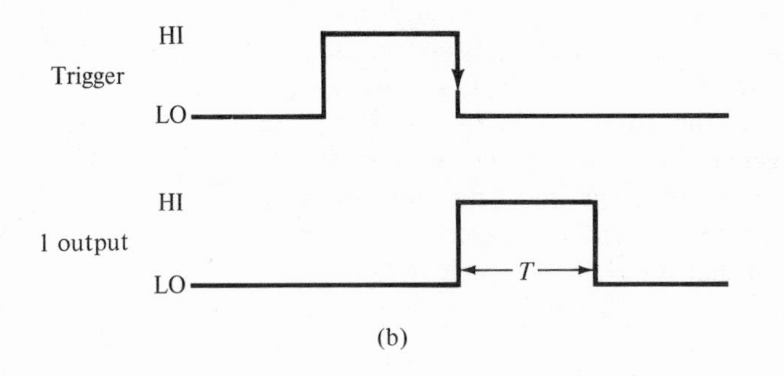

FIG. 13.30
Operation of monostable multivibrator: (a) ONE-SHOT logic symbol; (b) input trigger and 1 output waveforms.

only for a preselected time T, after which the output returns to the RESET state, where it remains until another trigger pulse is applied.

Referring to the waveform of Fig. 13.30b, the output of the ONE-SHOT can be viewed as a delayed pulse whose negative-going edge occurs at some set time T after the trigger pulse. Figure 13.31 shows a number of additional actions possible using the ONE-SHOT circuit. For example, it is possible to accept a train of narrow pulses as trigger signal and provide as output a corresponding train of wider pulses as shown in Fig. 13.31a. If the pulses received are quite narrow, then they may be widened to a pulse interval T, where T is, of course, less than the interval between trigger pulses. Any input pulse received during the timing interval T will be ignored by the usual ONE-SHOT circuit.

Figure 13.31b shows a series of wide input pulses and corresponding narrow output pulses from the ONE-SHOT circuit. Note that for the present example the narrow pulse is initiated by the negative-going edge (HI to LO) of the input trigger signal. Figure 13.31c shows how the ONE-SHOT may be used to provide only a single pulse when the trigger signal is a number of pulses. This is quite useful, for example, if the input pulses are obtained from a mechanical switch. When the switch is moved to change to signal from HI to LO the contact bounce

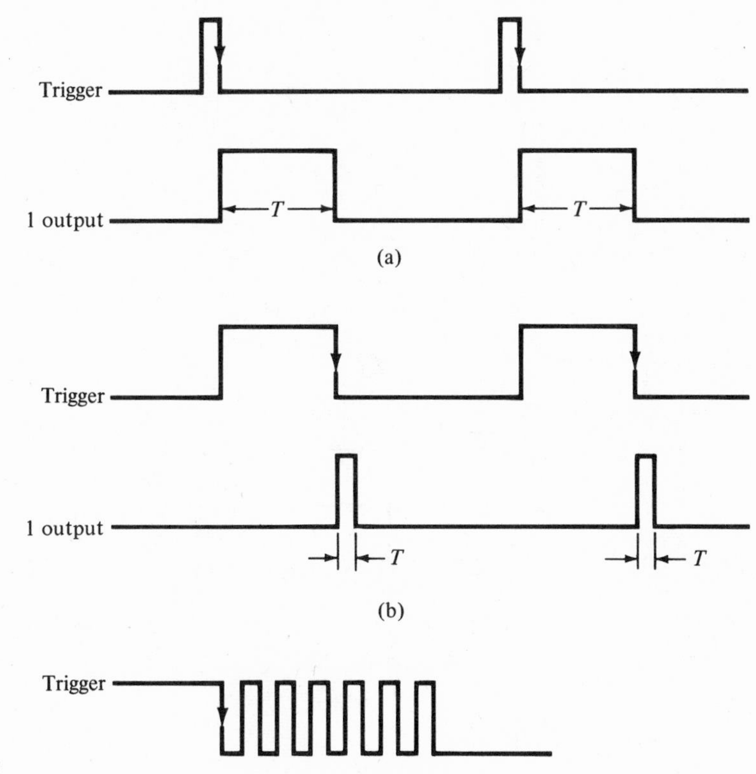

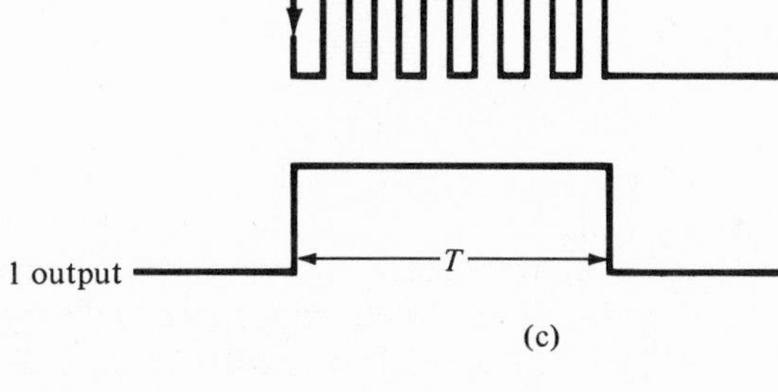

FIG. 13.31

Some pulse-shaping actions of a ONE-SHOT circuit: (a) Pulse shaping—widening a narrow pulse; (b) Pulse shaping—narrowing a wide pulse; (c) blocking multiple pulse—providing a single output pulse.

will usually result in a number of transitions between LO and HI states. If the switch signal is used directly, one throw of the switch may result in one, two, three, or more HI-to-LO transitions—which would result in erroneous system action. Using the ONE-SHOT, only a single pulse is provided—even with the multiple input pulses shown—as long as the interval of contact bounce is less than the pulse interval, T, of the ONE-SHOT.

A ONE-SHOT circuit diagram is shown in Fig. 13.32. A basic difference between this ONE-SHOT circuit and that of the bistable multivibrator of Fig. 13.26 is the cross-coupling component, capacitor C_T. Normally, transistor Q_2 is held ON and transistor Q_1 is held OFF through cross-coupling resistor R_2. A trigger pulse applied through a differentiating and diode clipping circuit provides a negative spike to turn transistor Q_2 OFF. By means of the direct cross coupling of re-

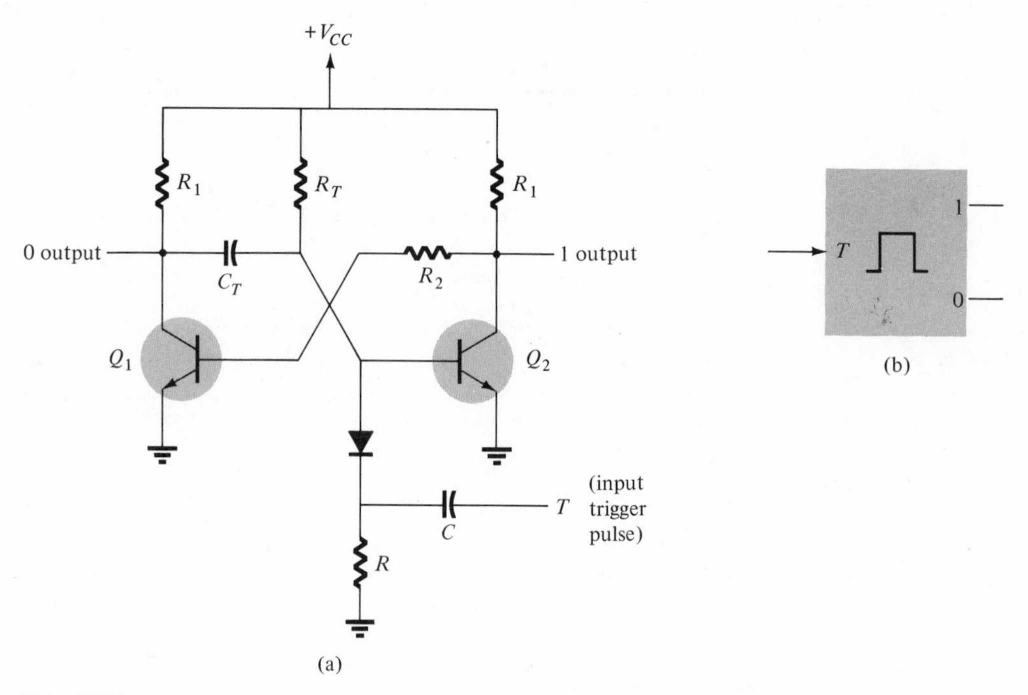

FIG. 13.32
(a) ONE-SHOT circuit schematic diagram;
(b) logic symbol.

sistor R_2, transistor Q_1 is turned ON. Transistor Q_2 is held OFF during a timing interval determined essentially by components R_T and C_T as will be shown. After a fixed time interval T, transistor Q_2 turns ON thereby causing transistor Q_1 to turn OFF, and the circuit is back in a stable operating state.

Both 1 and 0 outputs are available—these outputs being in the RESET condition until the circuit is pulsed, the outputs going to the SET condition for a period of time, T dependent on the circuit timing components C_T and R_T.

Analysis of the charging operation provides a time interval that is approximately

$$T \cong 0.7 R_T C_T \tag{13.11}$$

Typically, R_T is set at a fixed value since it also provides base drive current to Q_2, when ON. The connections for C_T are then provided so that an external capacitor may be connected to adjust the circuit time valve, T. Figure 13.33 shows a few of the circuit waveforms.

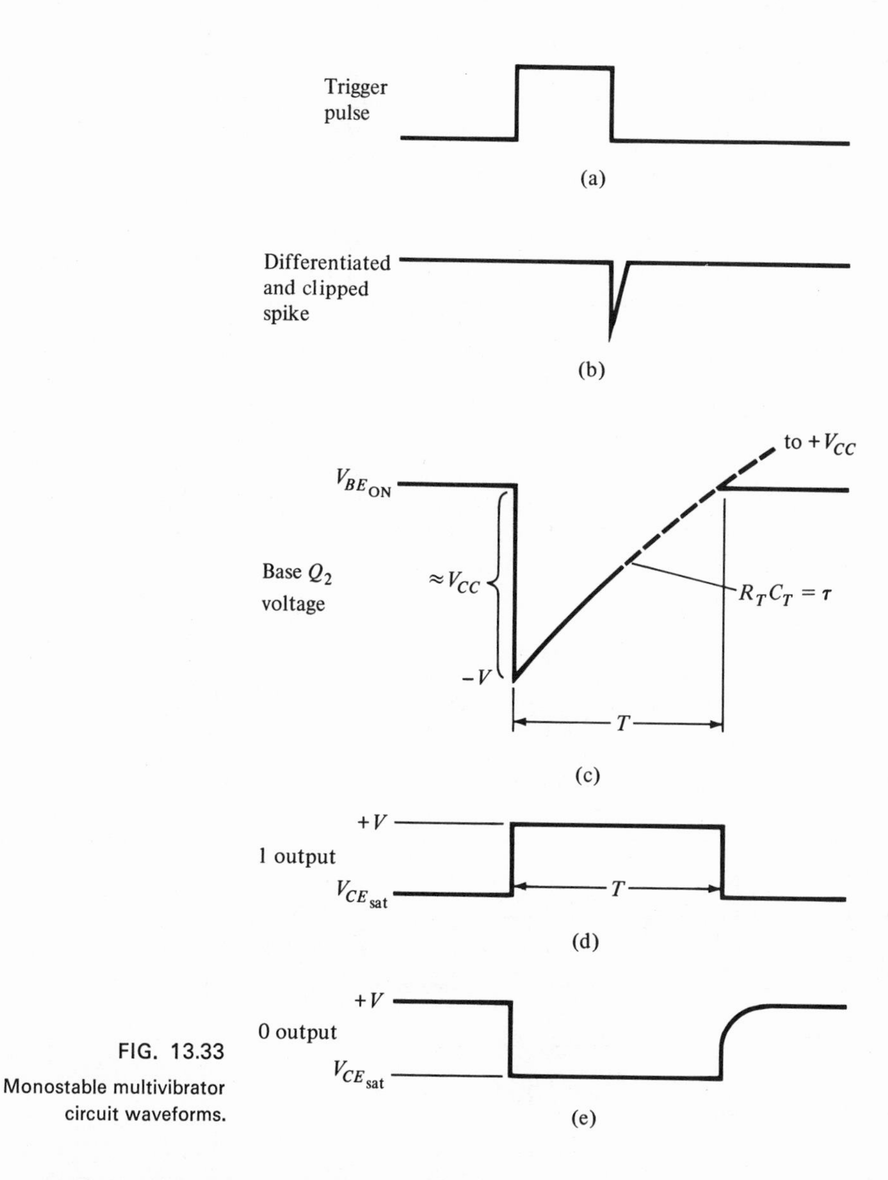

FIG. 13.33
Monostable multivibrator circuit waveforms.

EXAMPLE 13.6 Draw the pulse waveform of the 1 output from a ONE-SHOT circuit as in Fig. 13.32 for a 100-kHZ square wave used as trigger input. Circuit timing components are $R_T = 10\text{ K}$ and $C_T = 100\text{ pF}$.

Solution

$$T = 0.7R_T C_T = 0.7(10 \times 10^3)(100 \times 10^{-12}) = \mathbf{0.7\ \mu sec}$$

$$\frac{1}{f} = \frac{1}{100 \times 10^3} = \mathbf{10\ \mu sec} \qquad \text{(CLOCK period)}$$

Figure 13.34 shows the input and output waveforms.

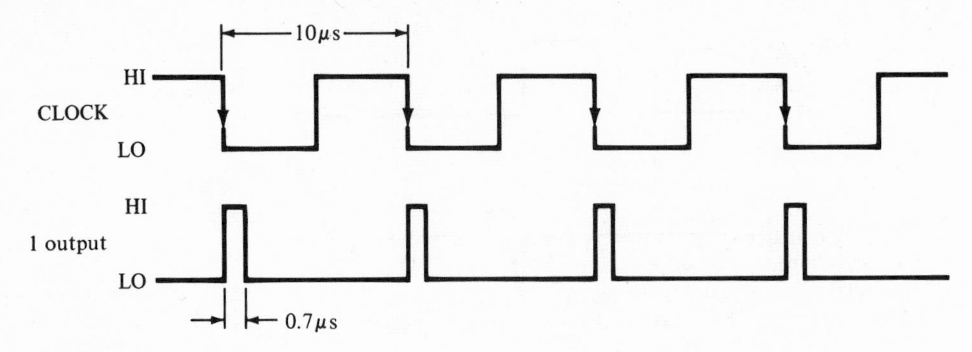

FIG. 13.34
Solution of Example 13.6.

EXAMPLE 13.7 A trigger signal of 12.5 kHz is used to operate a PNP transistor monostable circuit having $R_T = 7.5$ K and $C_T = 0.001\ \mu$F. Draw the 1 output waveform.

Solution

$$T = 0.7R_TC_T = 0.7(7.5 \times 10^3)(0.001 \times 10^{-6}) = \mathbf{5.25\ \mu sec}$$

$$\frac{1}{f} = \frac{1}{12.5 \times 10^3} = \mathbf{80\ \mu sec} \qquad \text{(CLOCK period)}$$

Figure 13.35 shows the input and output waveforms.

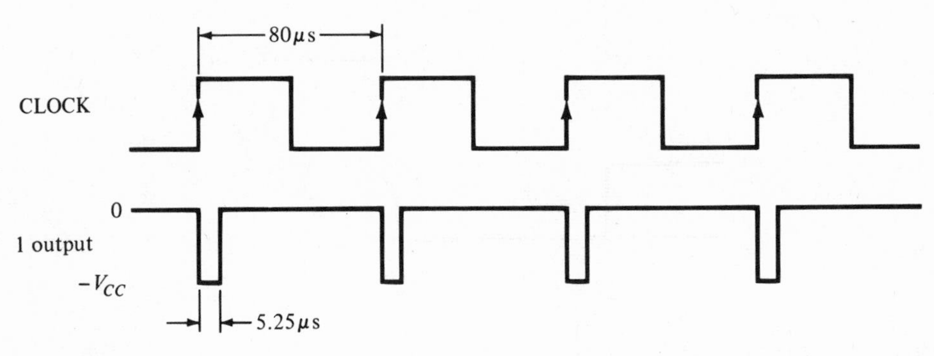

FIG. 13.35
Solution of Example 13.7.

Astable Multivibrator (CLOCK)

A third version of multivibrator has no stable operating state—it oscillates back and forth between RESET and SET states. The circuit provides a CLOCK signal for use as a timing train of pulses to operate digital circuits. Figure 13.36 shows a circuit diagram of an as-

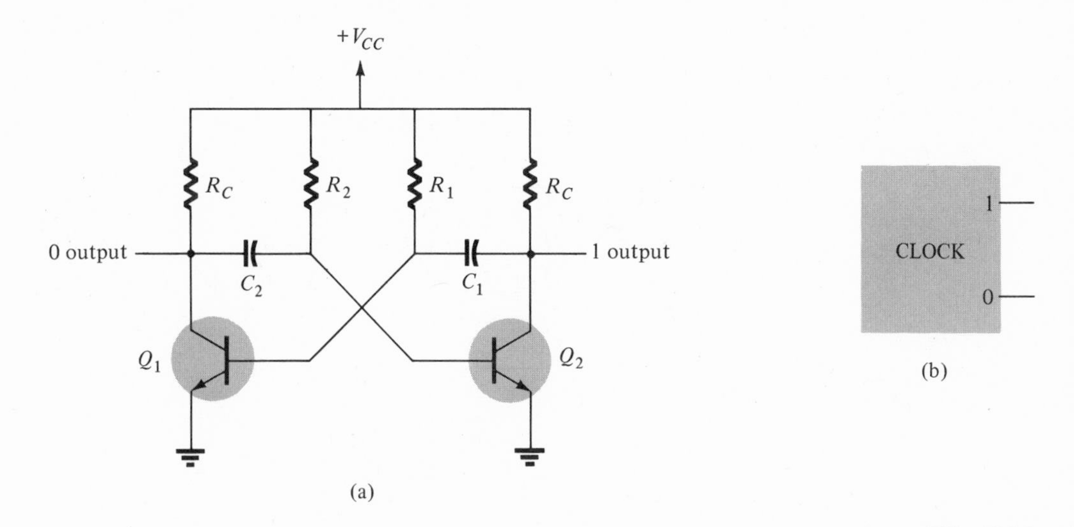

FIG. 13.36
(a) Astable multivibrator circuit; (b) logic symbol.

table multivibrator. Notice that both cross-coupling components are capacitors, thereby allowing no stable operating state.

To study the circuit operation consider transistor Q_1 just turning ON and Q_2 OFF. The base voltage of transistor Q_2 will then rise exponentially from a negative value near $-V_{CC}$ toward $+V_{CC}$. When the base voltage reaches $+V_{BE_{ON}}$ transistor Q_2 then turns ON. The interval that Q_2 is OFF is determined mainly by components R_2 and C_2. When Q_2 turns ON with collector Q_2 voltage dropping from $+V_{CC}$ to near-0 V the base Q_1 voltage drops from V_{BE} by about V_{CC} so that Q_1 is turned OFF. The interval that Q_1 is OFF is determined mainly by components R_1 and C_1. After the base Q_1 voltage rises to $+V_{BE_{ON}}$ transistor Q_1 turns ON, Q_2 then is turned OFF, and the operation will continue to repeat indefinitely (as long as supply voltage $+V_{CC}$ is present). Figure 13.37 shows the base and collector waveforms example of the astable operation.

The frequency of the astable circuit can be determined as follows:

$$f = \frac{1}{T_1 + T_2} = \frac{1}{0.7R_1C_1 + 0.7R_2C_2} = \frac{1.4}{R_1C_1 + R_2C_2} \quad (13.12a)$$

If the resistors and capacitors used are of equal value the frequency of the CLOCK is

$$\boxed{f = \frac{1}{2T} = \frac{1}{2(0.7)RC} = \frac{1}{1.4RC} = \frac{0.7}{RC}} \quad (13.12b)$$

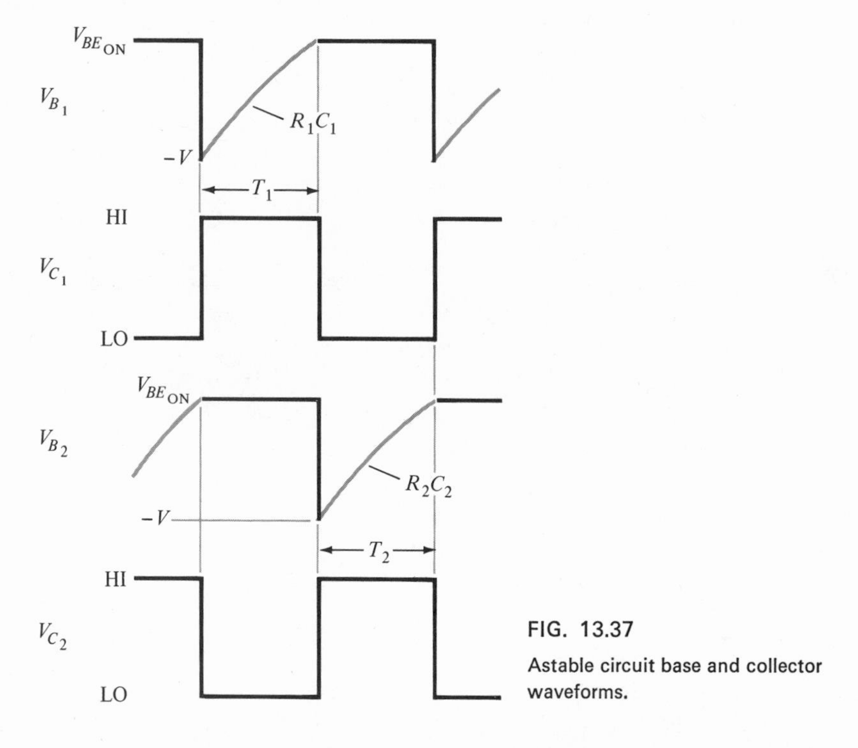

FIG. 13.37
Astable circuit base and collector waveforms.

EXAMPLE 13.8 An astable multivibrator has component values $R_1 = R_2 = 20$ K, and $C_1 = C_2 = 120$ pF. Calculate the oscillator CLOCK frequency.

Solution

$$f = \frac{0.7}{RC} = \frac{0.7}{(20 \times 10^3)(120 \times 10^{-12})} = 0.292 \times 10^6 = \mathbf{292\ kHz}$$

Schmitt Trigger

A circuit that is somewhat like the multivibrator circuits considered is the Schmitt trigger circuit shown in Fig. 13.38. Somewhat analogous to the ONE-SHOT, the Schmitt trigger is used for waveshaping purposes. Basically, the circuit has two opposite operating states as do all the multivibrator circuits. The trigger signal, however, is not typically a pulse waveform but a slowly varying ac voltage. The Schmitt trigger is level sensitive and switches the output state at two distinct triggering levels, one called a lower-trigger level (LTL) and the other an upper-trigger level (UTL). The circuit generally operates from a slowly varying input signal, such as a sinusoidal waveform, and provides a digital output—either the logical-0 or logical-1 voltage level.

The typical waveform of Fig. 13.39 shows a sinusoidal waveform input and squared waveform output. Note that the output signal frequency is exactly that of the input signal, except that the output has a sharply shaped slope and remains at the LO or HI voltage level until it

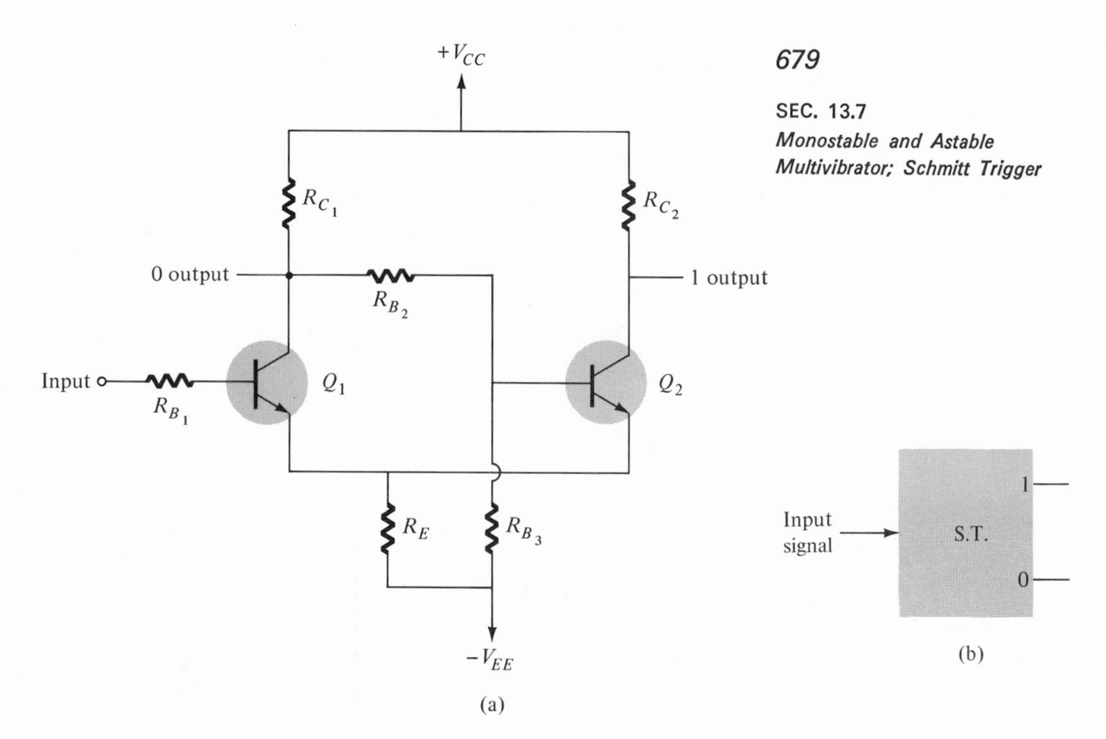

FIG. 13.38
(a) Schmitt trigger circuit; (b) block symbol.

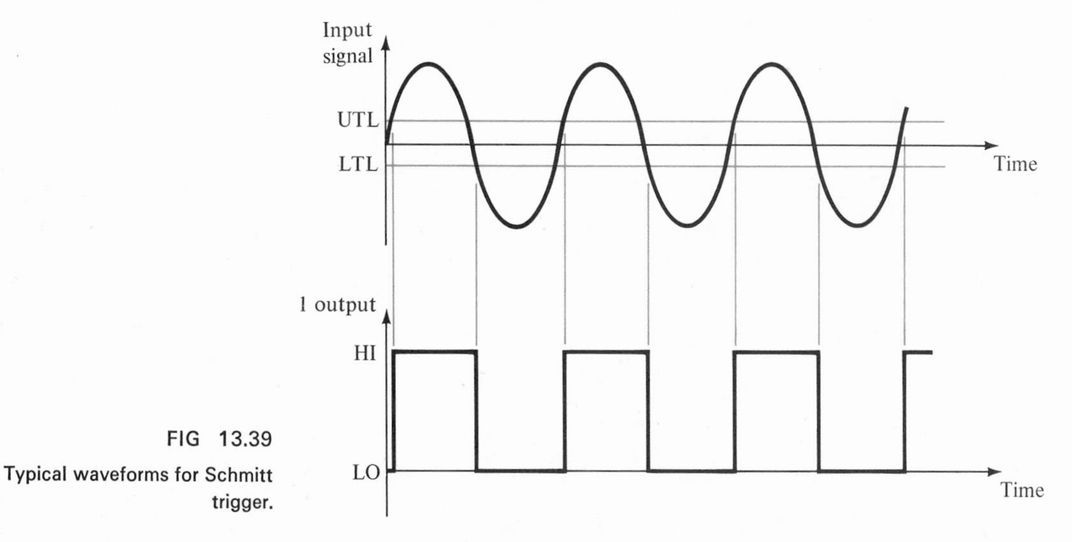

FIG 13.39
Typical waveforms for Schmitt trigger.

switches. One example of a Schmitt trigger application is converting a sinusoidal signal to one that is useful with digital circuits. Signals such as a 60-Hz line voltage, or a slowly varying voltage obtained from a magnetic pickup, are squared up for digital use. Another possibility is using the Schmitt trigger to provide a logic signal that indicates whenever the input goes above a threshold level (UTL).

§ 13.2

1. A positive-logic diode AND gate has voltage levels of 0 and +5 V. Draw a circuit diagram and prepare voltage and logic truth tables for a two-input gate. (Assume ideal diodes.)
2. Draw the circuit diagram of a negative-logic AND gate for voltage levels of −8 V and 0 V. Assume ideal diodes and prepare voltage and logic truth tables for a two-input gate.
3. Determine the output voltage of circuits of Fig. 13.40 (assume ideal diodes).

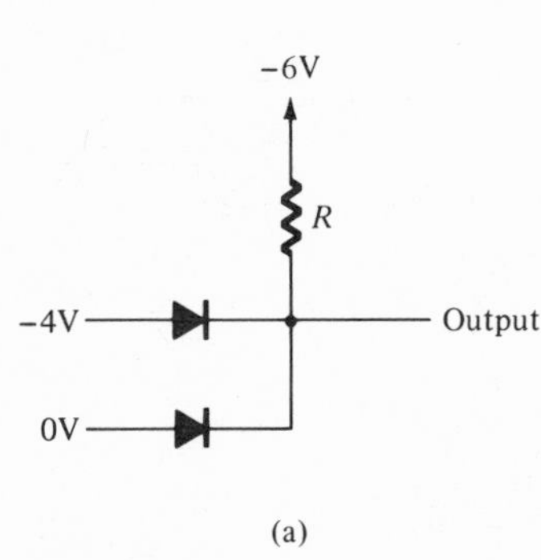

+2V
+8V
R
+12V
Output
(b)

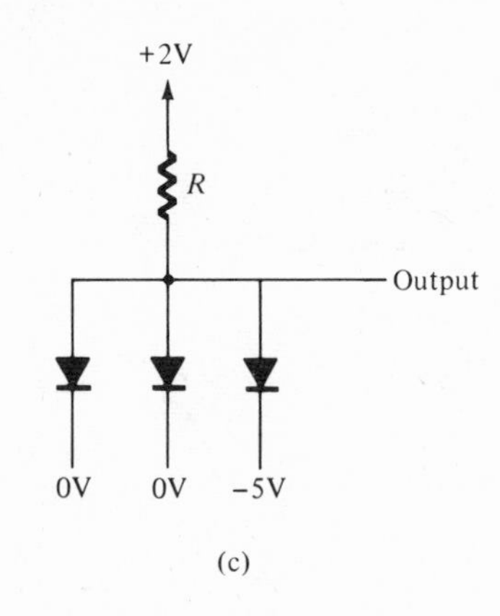

FIG. 13.40

Diode circuits for Problem 13.3.

§ 13.3

4. Determine whether the inverter circuit of Fig. 13.13a is driven into saturation for the following circuit values: $R_B = 18$ K, $R_C = 1.8$ K, $V_{CC} = 6$ V, $V_{BE_{ON}} = 0.7$ V, $h_{FE} = 20$, and $V_{CE_{sat}} = 0.1$ V, logic voltage levels are +6 and 0 V.
5. Determine the base OFF voltage of the circuit of Fig. 13.41.

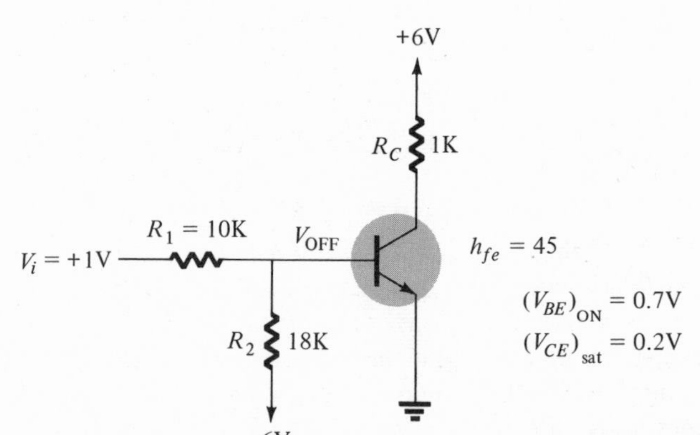

FIG. 13.41

Circuit for Problem 13.5.

6. For the circuit of Fig. 13.41 determine the ratio of transistor gain to circuit (forced) beta ($V_i = 6$ V).

§ 13.4

7. Determine the circuit (forced) beta for the circuit of Fig. 13.15 for the following circuit values: $R_1 = 2.7$ K, $R_2 = 3.6$ K, $R_3 = 18$ K, $R_C = 2.4$ K, $V_{BE_{ON}} = 0.7$ V, $V_{CE_{sat}} = 0.25$ V, $\pm V_{CC} = 8$ V, and $h_{FE} = 25$.
8. For a DTL circuit determine the circuit beta (β_C) for the circuit values: $R_1 = 1.5$ K, $R_2 = 7.5$ K, $V_{CC} = \pm 4$ V, $V_D = 0.7$ V, $R_C = 1$ K, $V_{BE_{ON}} = 0.6$ V, $V_{CE_{sat}} = 0.1$ V, and $h_{FE} = 20$.
9. For the circuit of Fig. 13.42 determine the circuit noise margin using $V_{OFF} = 0.4$ V.

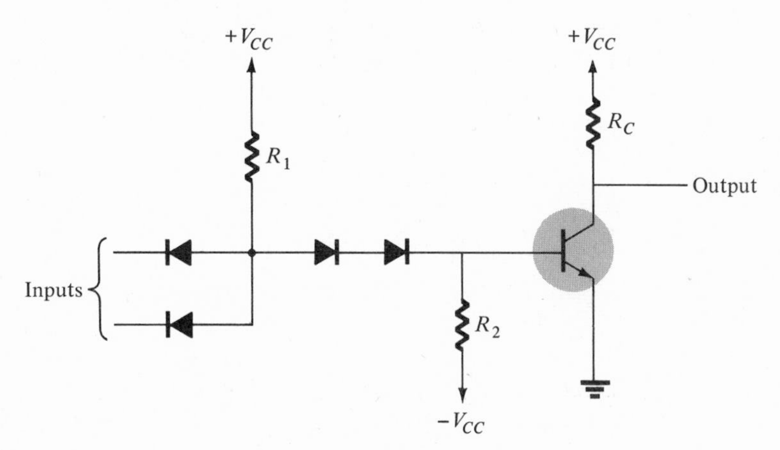

FIG. 13.42
Circuit for Problem 13.9.

10. Draw the circuit diagram of a three-input RTL NOR gate. Define voltage level and logic relations.
11. Draw the circuit diagram of a four-input TTL NAND gate indicating voltage levels and logic definitions.
12. State two differences between RTL and TTL operation.
13. Draw the circuit diagram of a three-input ECL circuit.
14. How does an ECL-type gate differ from RTL?
15. Draw the circuit diagram of a two-input FET logic gate. Indicate voltage and logic levels.

§ 13.6

16. Draw the circuit diagram of an *RS* flip-flop using PNP transistors.
17. Draw the circuit diagram of a *T* flip-flop using PNP transistors, indicating which polarity voltage change of the trigger signal operates the circuit.
18. Describe the operation of a *JK* flip-flop.

§ 13.7

19. Draw the circuit diagram of a ONE-SHOT circuit using PNP transistors. Define output logic levels and show which polarity voltage change triggers the circuit.

20. Draw the output wave form from the TRUE (1) side of a ONE-SHOT circuit using PNP transistors for an input CLOCK signal of 10 kHz. The timing component values are $R_T = 10$ K and $C_T = 1000$ pF.
21. What is the frequency of an astable multivibrator circuit having timing component values of $R_T = 2.7$ K and $C_T = 750$ pF?
22. Draw the TRUE output voltage signal of an NPN Schmitt trigger circuit for the input trigger signal of 5 V (rms) at 60 Hz. Circuit trigger levels are UTL = +5 V, and LTL = 0 V.

14

Regulators and Miscellaneous Circuit Applications

14.1 INTRODUCTION

This chapter examines both the voltage and current regulators and introduces a few circuits of general interest. There is, throughout the chapter, an application of the usually appreciated, approximate technique of semiconductor circuit analysis. In this manner, the "total" circuit behavior can more easily be understood. This chapter will also serve, in an indirect manner, as a test of the reader's comprehension of some of the important fundamental concepts introduced in earlier chapters. If the circuits described are, in general, understood with little necessity to review past discussions, the reader has every right to feel he has reached a first, important plateau of sophistication in semiconductor circuit analysis.

REGULATORS

14.2 REGULATION DEFINED

The wide variety of circuit configurations capable of performing voltage or current regulation necessitates that only a few of the more commonly applied be considered in this text. Voltage and current regulation can best be defined through the use of the circuits of Figs. 14.1 and 14.3, respectively.

In Fig. 14.1a, the no-load (open-circuit) terminal voltage of the supply is designated by V_{NL}. The corresponding load current is $I_{NL} = 0$.

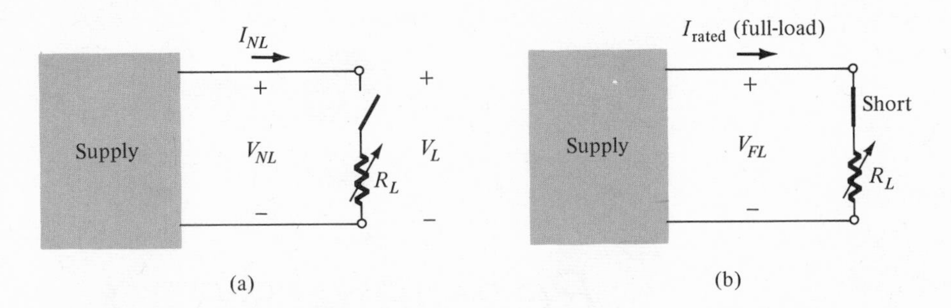

FIG. 14.1
Voltage regulation: (a) no-load (NL) state; (b) full-load (FL) state.

The full-load conditions are shown in Fig. 14.1b. The ideal situation would require that $V_L = V_{FL} = V_{NL}$ for every value of R_L between no-load and full-load conditions. In other words, the terminal voltage V_L would be unaffected by variations in R_L. Unfortunately, there is no supply available today, whether it be tube, semiconductor, or electromechanical (generator), that can provide a terminal voltage completely independent of the load applied (even for a specified range). However, for most applications, the level of sophistication has reached the point where the ideal response curve can be assumed. An ideal curve has been superimposed on the cumulative (flat) compound dc generator characteristics of Fig. 14.2.

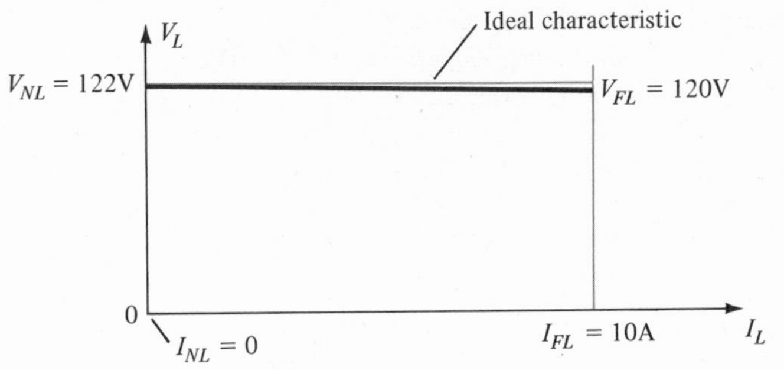

FIG. 14.2
Terminal characteristics of a cumulative compound dc generator.

Note that the load voltage V_L drops 2 V from its no-load to full-load value. The drop in terminal voltage is due to a number of changes in potential levels internal to the generator as a result of increased load.

Voltage regulation of any supply, in per cent, is defined by

$$\text{Voltage regulation (VR) (\%)} = \frac{V_{NL} - V_{FL}}{V_{FL}} \times 100\% \tag{14.1}$$

Substituting the ideal values of $V_L = V_{FL} = V_{NL}$ into Eq. (14.1), VR = 0%. Obviously, a *low* voltage regulation is desirable and 0% is ideal. Semiconductor and tube supplies with voltage regulations of 0.01% or lower are quite common today. For the electromechanical generator of Fig. 14.2

$$\text{VR\%} = \frac{122 - 120}{120} \times 100\% \cong 1.7\%$$

The circuit configurations associated with the no-load and full-load conditions for *current regulation* are provided in Fig. 14.3. The defining terminal characteristics are provided in Fig. 14.4.

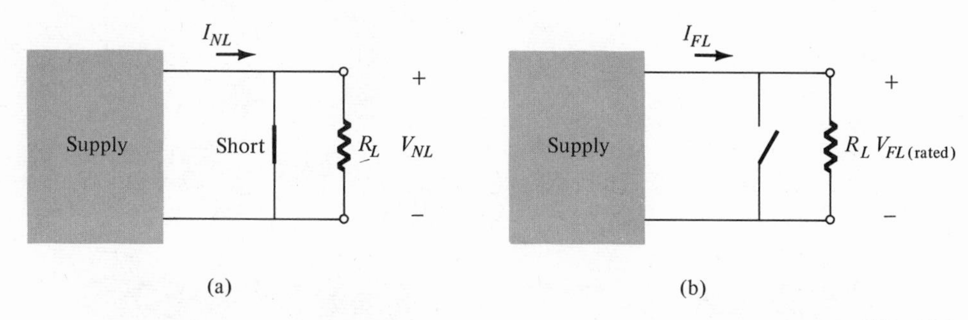

FIG. 14.3
Current regulation: (a) no-load (*NL*) state; (b) full-load (*FL*) state.

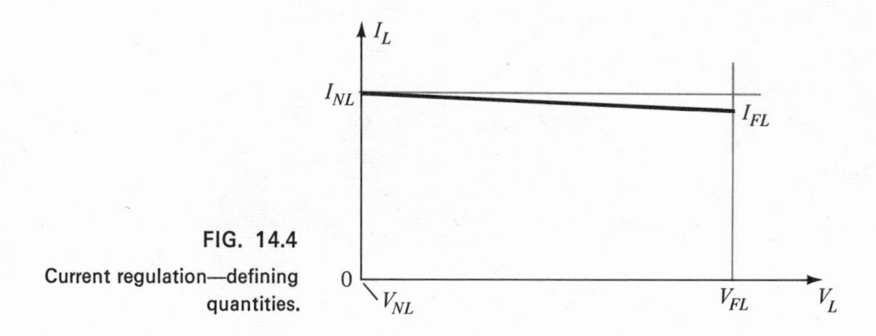

FIG. 14.4
Current regulation—defining quantities.

Note for this situation that a constant I_L is desired for a variable terminal voltage. Current regulation is defined by

$$\text{Current regulation (IR) (\%)} = \frac{I_{NL} - I_{FL}}{I_{FL}} \times 100\% \qquad (14.2)$$

where I_{NL} and I_{FL} and the no-load and full-load currents as defined by Fig. 14.3.

Three voltage- and one current-regulated supply appear in Fig. 14.5. The necessity for low voltage and current regulation characteristics should be quite obvious. Without them, the bias conditions of electronic systems, the speed of dc and ac motors, and the logic of computer circuits would all be severely affected.

(Courtesy Kepco, Inc.)

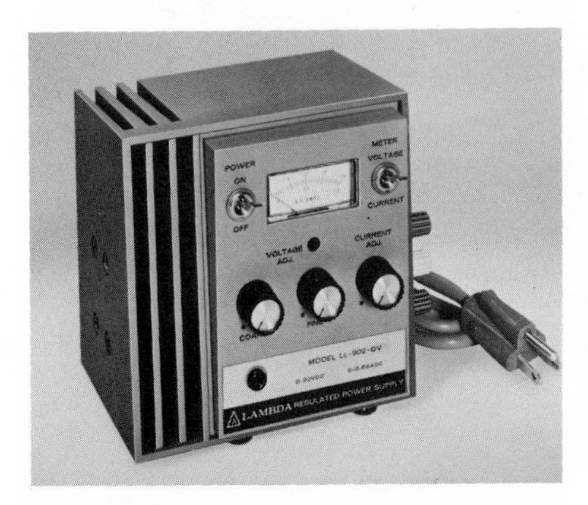

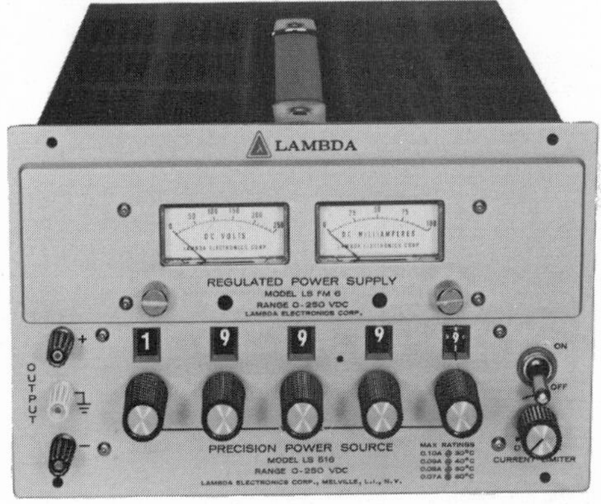

(Courtesy Lambda Electronics Corp.)

(a)

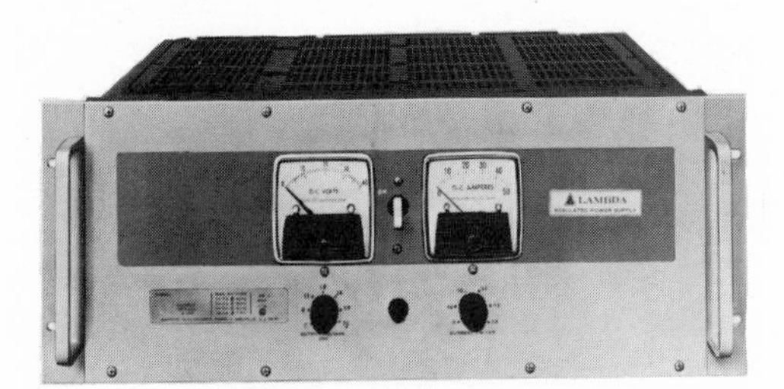

(Courtesy Lambda Electronics Corp.)

(b)

FIG. 14.5

Regulated supplies: (a) voltage regulated; (b) current regulated.

14.3 ZENER AND THERMISTOR VOLTAGE REGULATORS

There are, fundamentally, two basic configurarions for establishing voltage or current regulation. Each is shown in Fig. 14.6. The choice of terminology for each as indicated in the figure is quite obviously derived. You will find as you progress through some of the more typical circuit configurations that the more sophisticated regulators will apply the

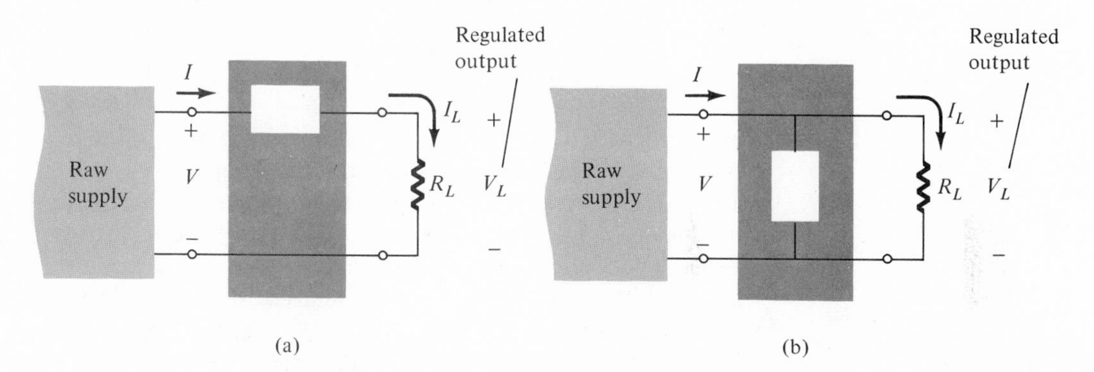

FIG. 14.6
Regulators: (a) series; (b) parallel (shunt).

benefits to be derived from both series and shunt regulation in the same system. Before continuing, let us examine the unregulated supply of Fig. 14.7. It clearly demonstrates the need for voltage and/or current regulators. Recall from your experimental work that when your dc supply indicates 20 V, you want it to be that value for any load you apply to its terminals. At any except infinite ohms (open circuit) in Fig. 14.7, would this be the case? Obviously not. As R_L increases the voltage across R_L increases also, and V_L does not remain fixed. The function of a voltage regulator is to maintain V_L at 20 V for any R_L from 0 to 1 K. The voltage regulator appears, therefore, between the unregulated supply and the load as shown in Fig. 14.8.

Our interest for the moment is in the voltage-regulating device. A simple shunt voltage regulating system is shown in Fig. 14.8. As

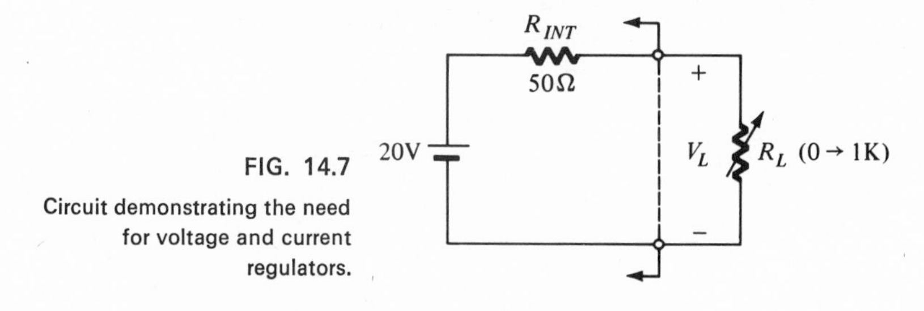

FIG. 14.7
Circuit demonstrating the need for voltage and current regulators.

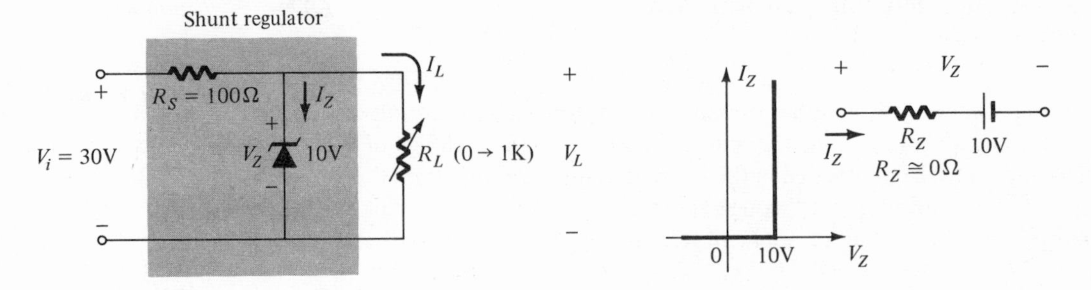

FIG. 14.8
Zener diode shunt regulator.

indicated, it consists simply of a Zener diode and series resistance, R_s. Proper operation requires that the Zener diode be in the *on* state. The first requirement, therefore, is to find the minimum R_L (and corresponding I_L) to ensure that this condition is established. Before Zener conduction, the Zener diode is fundamentally an open circuit and the circuit of Fig. 14.8 can be replaced by that indicated in Fig. 14.9. As indicated in the figure, the load voltage will be determined by the voltage divider rule (similar to the system of Fig. 14.7). At Zener conduction, $V_L = V_Z = 10$ V. Using this data, it is possible to find the minimum value of R_L that the Zener will tolerate in maintaining

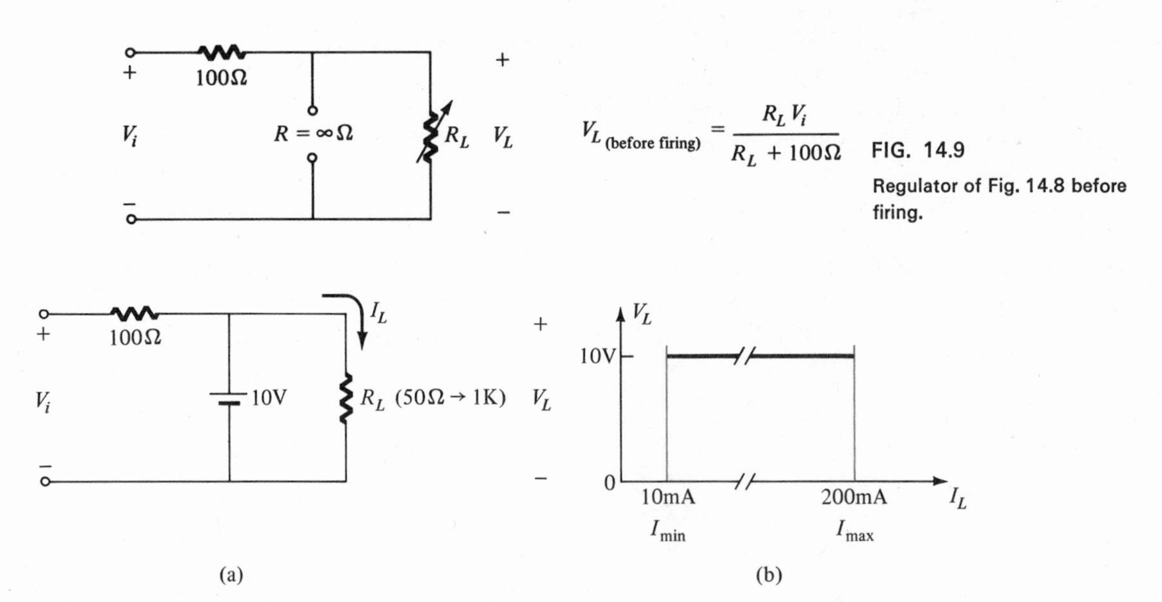

FIG. 14.9
Regulator of Fig. 14.8 before firing.

FIG. 14.10
Zener diode shunt regulator: (a) after firing; (b) regulated output.

constant V_L. Applying the voltage divider rule to the circuit of Fig. 14.10a

$$V_L = \frac{R_L V_i}{R_L + R_s}$$

Substituting values

$$10 = \frac{R_L \times 30}{R_L + 0.1\text{ K}}$$

$$10R_L + 1\text{ K} = 30R_L$$

$$20R_L = 1\text{ K}$$

and

$$R_L = 50\ \Omega$$

The *minimum* load R_L for this supply is therefore 50 Ω, corresponding to a *maximum* load current of

$$I_{\text{max}} = \frac{10\text{ V}}{0.05\text{ K}} = 200\text{ mA}$$

The maximum current is indicated on the plot of Fig. 14.10b. For any value of R_L from 50 Ω to 1 K the Zener diode will be in the *on* state. An application of the voltage divider rule to the circuit of Fig. 14.9 for any value of R_L less than 50 Ω will result in $V_L < V_Z$ and the diode will remain in the *off* state. At $R_L = 1$ K, $I_L = (10/1\text{ K}) = 10$ mA (indicated in Fig. 14.10b as I_{min})

and

$$I_{R_s} = \frac{30 - 10}{0.1\text{ K}} = 200\text{ mA}$$

with

$$I_Z = I_{R_s} - I_L = 190\text{ mA}$$

Our approximation, $R_Z \cong 0\ \Omega$, has resulted in the ideal characteristics of Fig. 14.10b between 10 and 200 mA. For this regulator the per cent regulation would be determined between these two points of normal operation. The effect of R_Z on the regulation can be determined quite easily if we find the Thévenin equivalent circuit for the portion of the network of Fig. 14.11 at the left of points 1 and 2.

$$V_{\text{Th}} = 10 + \frac{2(30 - 10)}{102} = 10 + \frac{40}{102} \cong 10.4\text{ V}$$

$$R_{\text{Th}} = 100\,||\,2\ \Omega \cong 2\ \Omega$$

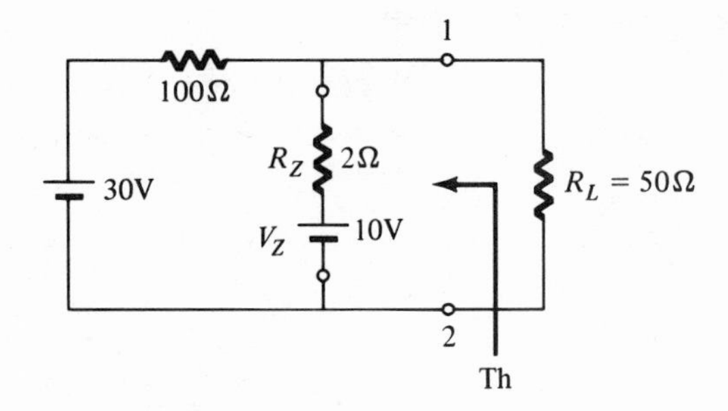

FIG. 14.11
Determining the effect of R_z on the output of the Zener shunt regulator.

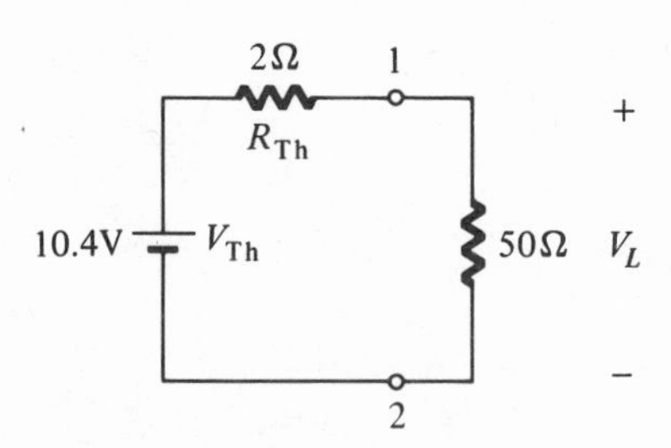

FIG. 14.12
Thévenin equivalent circuit for the circuit of Fig. 14.11.

Substitute the Thévenin equivalent circuit (Fig. 14.12) and

$$V_L = \frac{50(10.4)}{52} = 10\text{ V}$$

R_Z, therefore, has a negligible effect at minimum R_L (maximum I_L). For $R_L = 1$ K, (minimum I_L).

$$V_L = \frac{1\text{ K}(10.4)}{1\text{ K} + 2\,\Omega} \cong 10.4\text{ V} > 10\text{ V (obtained above for } R_L \cong 0\,\Omega)$$

and

$$\text{VR\%} = \frac{V_{1\text{K}} - V_{50\Omega}}{V_{50\Omega}} \times 100\% = \frac{10.4 - 10.0}{10.0} \times 100\%$$

$$= \frac{0.4}{10.4} \times 100\% = 4\%$$

More than one regulated output can be supplied at the same time from the same unregulated input using the circuit configuration of Fig. 14.13.

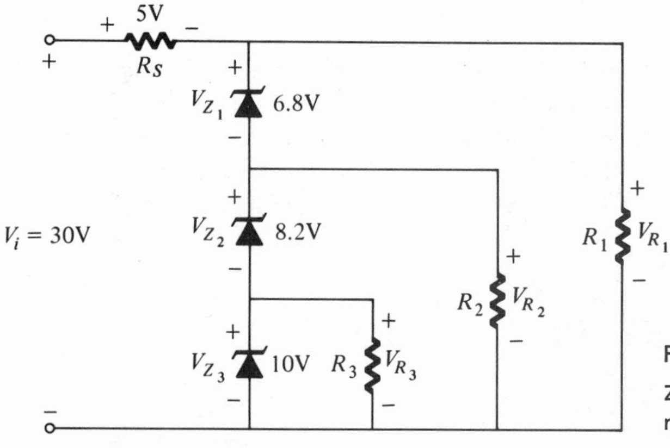

FIG. 14.13
Zener regulator having three reference potentials.

It should be quite clear that

$$V_{R_1} = V_{Z_1} + V_{Z_2} + V_{Z_3} = 25\text{ V}$$
$$V_{R_2} = V_{Z_2} + V_{Z_3} = 18.2\text{ V}$$

and

$$V_{R_3} = V_{Z_3} = 10\text{ V}$$

A shunt regulator employing a thermistor appears in Fig. 14.14. Any tendency for V_L to decrease due to a change in load will result in a decrease in current through the thermistor. The temperature of the

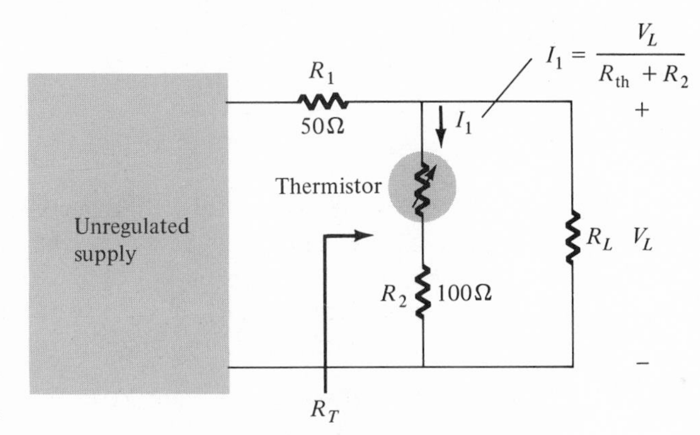

FIG. 14.14
Thermistor shunt regulator.

thermistor element will thereby decrease, resulting in an increase in its resistance. The resulting resistance $R_T = R_L || (R_{\text{Th}} + 100\ \Omega)$ will increase somewhat and the load voltage $V_L = (R_T V_i / R_T + R_1)$ will tend to increase, offsetting the initial drop in V_L. Assigning the symbols ↑ to an increasing quantity and ↓ to a decreasing quantity, the following summary of the voltage-regulating action of this system will result (read from left to right):

$$V_L\downarrow,\ I_{\text{Th}}\downarrow,\ R_{\text{Th}}\uparrow,\ R_T\uparrow,\ V_L\uparrow$$
balance

An increasing V_L will have the opposite effect on each element and quantity of the above summary.

14.4 TRANSISTOR VOLTAGE REGULATORS

The characteristics of a voltage regulator can be markedly improved through the use of active devices such as the transistor. The simplest of the transistor-*series*-type voltage regulator appears in Fig. 14.15a. In this configuration, the transistor behaves like a simple variable resistor,

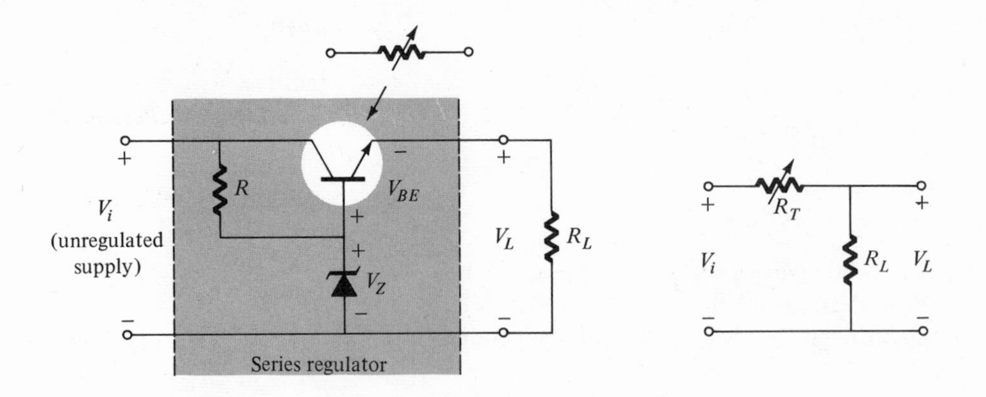

FIG. 14.15
Transistor series voltage regulator.

whose resistance is determined by the operating conditions. The basic operation of the regulator is best described using the circuit of Fig. 14.15b, in which the transistor has been replaced by a variable resistor, R_T. For variations in R_L, if V_L is to remain constant the ratio of R_L to R_T must remain fixed. Applying the voltage divider rule

$$V_L = \overbrace{\frac{R_L}{R_L + R_T}}^{\text{fixed for constant } V_L (V_i = \text{constant})} V_i$$

For

$$\frac{R_L}{R_T} = k_1 \qquad \text{or} \qquad R_L = k_1 R_T$$

$$\frac{R_L}{R_L + R_T} = \frac{k_1 R_T}{k_1 R_T + R_T} = \frac{k_1}{k_1 + 1} = k \qquad \text{(constant as required)}$$

In summary, for a decreasing or increasing load (R_L), R_T must change in the same manner, at the same rate, to maintain the same voltage division.

You will recall, from Section 14.2, that the voltage regulation is determined by noting the variations in terminal voltage versus the load current demand. For this circuit, an increasing current demand associated with a *decreasing* R_L will result in a tendency on the part of V_L to decrease in magnitude also. However, if we apply Kirchhoff's voltage law around the output loop

$$V_{BE} = \overbrace{V_Z}^{\text{(fixed)}} - V_L$$

A decrease in V_L (since V_Z is fixed in magnitude) will result in an increase in V_{BE}. This effect will, in turn, increase the level of conduction of the transistor, resulting in *decrease* in its terminal (collector-to-emitter) resistance. This is, as described in the previous paragraphs of this section, the effect desired to maintain V_L at a fixed level.

A voltage regulator employing a transistor in the shunt configuration is provided in Fig. 14.16. Any tendency on the part of V_L to in-

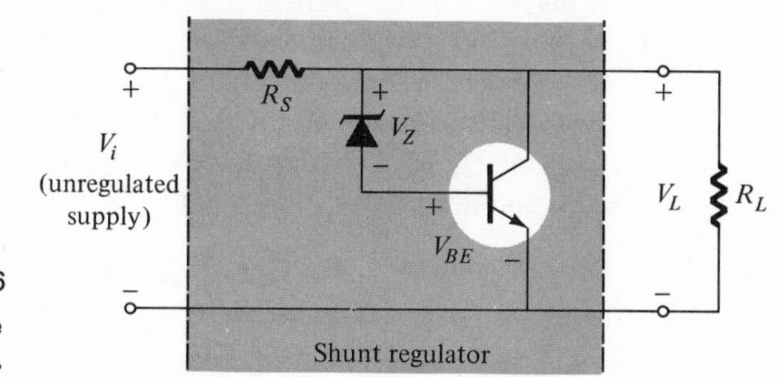

FIG. 14.16
Transistor shunt voltage regulator.

crease or decrease in magnitude will have the corresponding effect on V_{BE} since

$$V_{BE} = V_L - \overbrace{V_Z}^{\text{(fixed)}}$$

For decreasing V_L, the result is a decrease in the current through the resistor R_S since the conduction level of the transistor has dropped ($V_{BE\downarrow}$). The reduced drop in potential across R_S will offset any tendency on the part of V_L to decrease in magnitude. In sequential logic

$$V_L\downarrow,\ V_{BE}\downarrow,\ I_B\downarrow,\ I_C\downarrow, I_{R_s}\downarrow,\ V_{R_s}\downarrow, V_L\uparrow \quad \text{(balance)}$$

A similar discussion can be applied to increasing values of V_L.

A series voltage regulator employing a second transistor for control purposes can be found in Fig. 14.17. The base-to-emitter

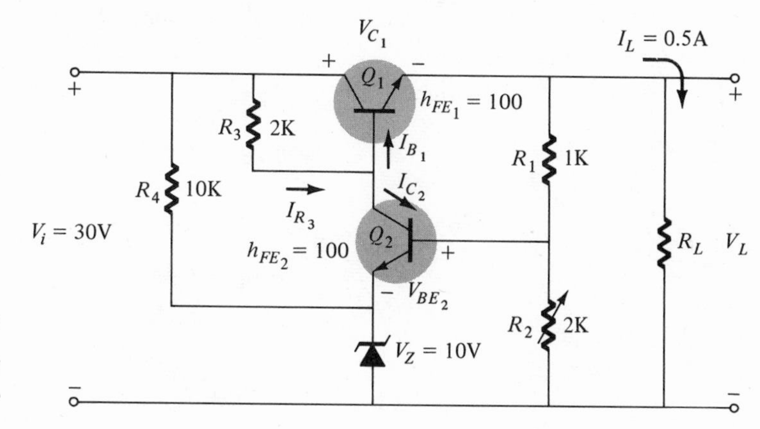

FIG. 14.17
A series voltage regulator employing two transistors.

potential (V_{BE_2}) of the control transistor Q_2 is determined by the difference between V_1 and the reference voltage V_Z. The voltage level V_1 is quite sensitive to changes in the terminal voltage V_L. Any tendency on the part of V_L to increase will result in an increase in V_1 and therefore in V_{BE_2} since $V_{BE_2} = V_1 - V_Z$. The difference in potential is amplified by the control transistor and carried to the variable series resistive element Q_1. An increase in V_{BE_2}, corresponding to an increase in I_{B_2} and I_{C_2} will result in a decreasing I_{B_1} (assuming I_{R_3} to be relatively constant or decreasing only slightly). The net result is a decrease in the conductivity of Q_1 corresponding to an increase in its terminal resistance and a stabilization of V_L. In sequential logic

$$V_L\uparrow,\ V_1\uparrow,\ V_{BE_2}\uparrow,\ I_{C_2}\uparrow,\ I_{B_1}\downarrow,\ R_{(Q_1)}\uparrow,\ V_L\downarrow$$

balance

Again, a similar discussion can be applied to decreasing values of V_L.

EXAMPLE 14.1 We shall now calculate various currents and voltages of the circuit of Fig. 14.17 for the input shown. Approximations, as introduced in previous chapters, will be the working tools of this analysis. Those of primary importance include $I_C \cong h_{FE}I_B$, $V_{BE} \cong 0$ V, and $I_C \cong I_E$.

Solution

$$V_{R_4} = V_i - V_Z = 30 - 10 = \mathbf{20\ V}$$

and

$$I_{R_4} = \frac{20}{10\text{ K}} = \mathbf{2\ mA}$$

$$V_{R_2} \cong V_Z = \mathbf{10\ V} \qquad \text{since } V_{BE_2} \cong 0\text{ V}$$

and

$$I_{R_2} = \frac{10}{2\text{ K}} = \mathbf{5\ mA}$$

Assuming

$$I_{B_2} \ll I_{R_1},\quad I_{R_2}$$

then

$$I_{R_1} = I_{R_2} = \mathbf{5\ mA}$$

and

$$V_L = 5\text{ mA} \times 3\text{ K} = \mathbf{15\ V}$$

with

$$V_{R_3} = V_i - V_L (V_{BE_1} \cong 0\,\text{V}$$
$$= 30 - 15 = \mathbf{15\ V}$$

and

$$I_{R_3} = \frac{15}{2\,\text{K}} = \mathbf{7.5\ mA}$$

Similarly,

$$V_{C_1} = V_i - V_L = 30 - 15 = \mathbf{15\ V}$$
$$I_{E_1} \cong h_{FE} I_{B_1} = 100 I_{B_1}$$

and

$$I_{B_1} = \frac{I_{E_1}}{100} = \frac{(500 + 5)}{100} = \mathbf{5.05\ mA}$$
$$I_{C_2} = I_{R_3} - I_{B_1}$$
$$= 7.5 - 5.05$$
$$= \mathbf{2.45\ mA}$$
$$I_{B_2} \cong \frac{I_{C_2}}{100} = \frac{2.45}{100} = \mathbf{2.45\ \mu A}$$

(Certainly, $I_{B_2} \ll I_{R_2}, I_{R_2}$ as employed above is an excellent approximation,) and, finally,

$$I_Z = I_{R_4} + I_{C_2} = 2 + 2.45 = \mathbf{4.45\ mA}$$

The fact that $I_{B_2} \ll I_{R_1}, I_{R_2}$ permits the use of the circuit of Fig. 14.18 to derive a rather useful equation for the circuit of Fig. 14.17. Applying the voltage divider rule

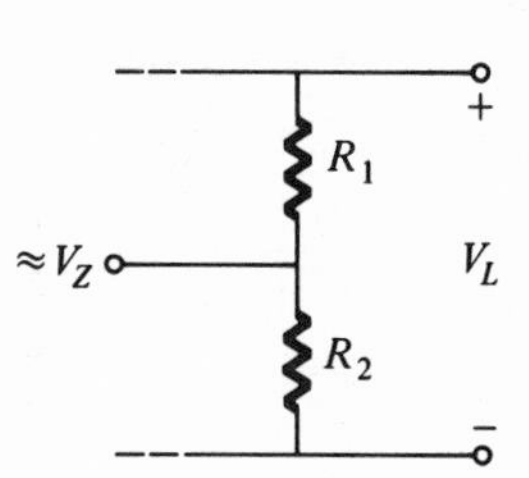

FIG. 14.18
Circuit employed in the derivation of Eq. (14.1).

$$V_Z = \frac{R_2 V_L}{R_1 + R_2}$$

or since V_Z is fixed,

$$\boxed{V_L = V_Z\left(1 + \frac{R_1}{R_2}\right)} \qquad (14.3)$$

For the above case

$$V_L = 10(1 + \tfrac{1}{2}) = 15\ \text{V}$$

You will find in Fig. 14.17 that R_2 is a variable resistor. Variations in this resistance will control V_L as determined by Eq. (14.1). The maximum voltage available is quite obviously 30 V (for $V_i = 30$ V) since at this point $V_{C_1} = 0$ V (saturation). The minimum is 10 V attainable with either $R_1 = 0$ or $R_2 = \infty$.

14.5 COMPLETE POWER SUPPLY (VOLTAGE REGULATED)

A power supply employing a voltage regulator similar to the one described in Fig. 14.17 appears in Fig. 14.19. A *Darlington* circuit has replaced the single series transistor of Fig. 14.7 in an effort to increase the sensitivity of the regulator to changes in V_L. In the circuit of Fig. 14.17, changes in I_{C_2} will be reflected in changes in I_{R_3} causing a reduc-

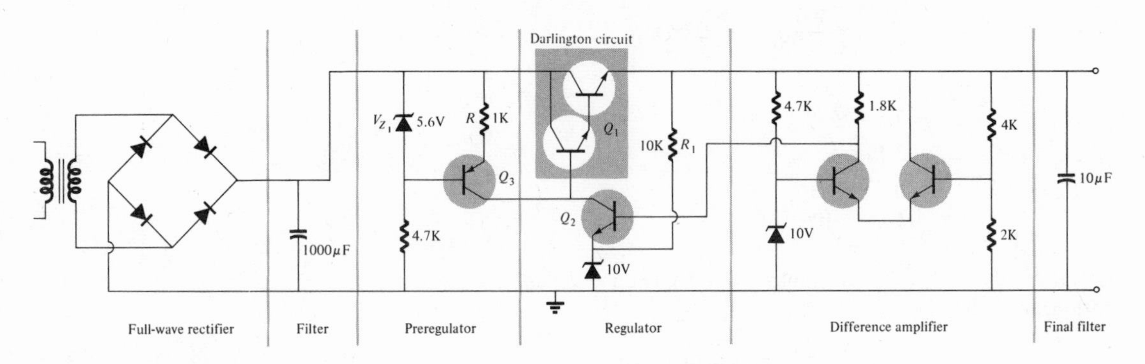

FIG. 14.19
Complete voltage-regulated power supply.

tion in the sensitivity of I_{B_1} to changes in V_L. To minimize this undesirable effect R_3 should be as large as possible while still permitting the necessary I_{R_3} for proper circuit behavior. This is most efficiently achieved by employing a current source in place of R_3. The current source has, ideally, infinite terminal resistance along with the capability to supply the necessary current. This portion of a power supply as indicated in Fig. 14.19 is sometimes referred to as a *preregulator*. For the circuit of Fig. 14.19

$$I_{\text{current source}} = I_{C_3} \cong \frac{V_{Z_1}}{R}$$

To improve further the sensitivity of the regulator to changes in V_L a *difference* amplifier has been introduced, the output of which is fed to the control transistor. The unregulated input is a full-wave rectified signal to be passed through a RC filter. The 10-μF capacitor at the output is to reduce the possibility of oscillations and further filter the supply voltage. The supply voltage V_L can be varied by changing R_1 while still maintaining regulation.

14.6 CURRENT REGULATOR

The analysis of current regulators will be limited to a brief discussion of the circuit of Fig. 14.20. Recall from the introductory discussion that a current regulator is designed to maintain a fixed current through a load for variations in terminal voltage. There is a tendency for I_L

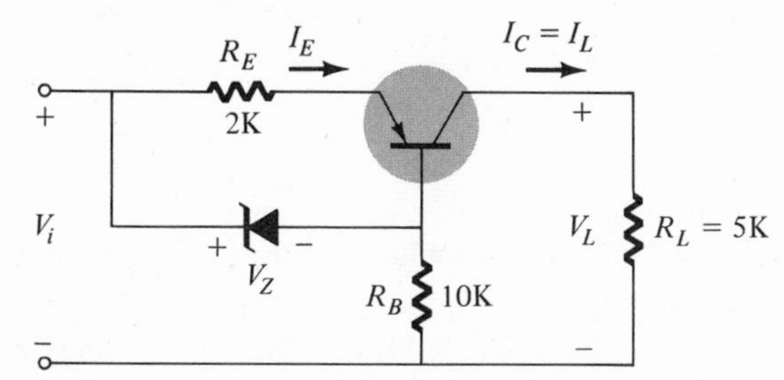

FIG. 14.20
Series current regulator.

to decrease with decrease in R_L. However, a decrease in $I_L = I_C$ would result in a decrease in $I_E \cong I_C$ and, in turn, a drop in V_{R_E}. The base-to-emitter potential is

$$V_{BE} = \overbrace{V_Z}^{\text{(fixed)}} - V_{R_E}$$

A decrease in V_{R_E} will result in an increase in V_{B_E} and the conductivity of the transistor, balancing the effect of the decreasing load to maintain I_L.

MISCELLANEOUS CIRCUIT APPLICATIONS

14.7 CAPACITIVE-DISCHARGE IGNITION SYSTEM

The capacitive-discharge (CD) ignition system has become increasingly popular in recent years due to the resulting improvement in engine performance. Tests have indicated that there is not only an increase in available power, but also better mileage and extended spark plug life. This system will rechannel the heavy firing currents that otherwise would pass through, and thereby cause the rapid wear of, the distributor points. In the CD system the heavy currents are carried by a semicon-

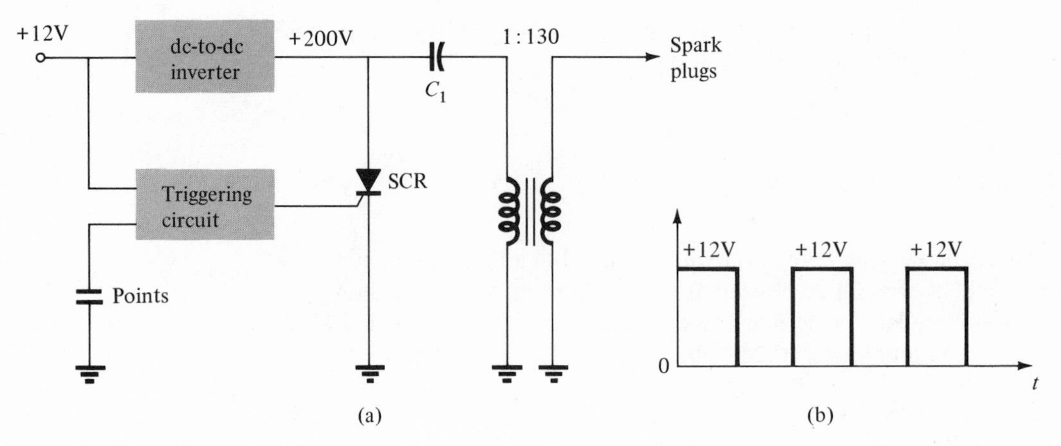

FIG. 14.21
Capacitive-discharge ignition system.

ductor device (SCR) whose operation is simply controlled by the distributor points.

The essential components of a capacitive-discharge ignition system appear in Fig. 14.21a. The dc-to-dc inverter uses an oscillator chopper to convert the 12-V dc to an ac pulse (Fig. 14.21b) so that transformer action can be employed to step up the voltage. Recall that only time-varying primary inputs can be transformed at a higher or lower level to the secondary of a transformer. The transformer output is then filtered to produce the indicated 200-V dc level. When the distributor points are closed, the SCR will be in the open-circuit state and the capacitor (C_1) will charge to 200 V at a rapid rate since the charging time constant ($\cong R_p \times C$) is very small. R_p is simply the dc resistance of the primary coil of the transformer. At the instant the points are open, the trigger circuit will turn the SCR on (short-circuit state) and the capacitor will discharge across the primary coil. The indicated turns ratio will result in a high secondary voltage, which will appear directly across the spark plugs. For a change (ΔV_p) in primary voltage of 200 V, the change in secondary voltage (ΔV_s) as determined by the turns ratio will be

$$\frac{\Delta V_p}{\Delta V_s} = \frac{N_p}{N_s} \Rightarrow \Delta V_s = \frac{N_s}{N_p} \Delta V_p = 130(200) = 26 \text{ Kv}$$

14.8 COLOR ORGAN

The present interest in the unique has resulted in a rather interesting device called the color organ. It is an electronic package that will produce a visual display of colored lights, at various intensities, that will correspond directly with the intensity of the various tones of the audio signal. One workable design appears in Fig. 14.22. Fundamentally, the circuit package has an active filter associated with each color, with the result that the circuit to the right of the dashed line must be repeated for each color. Each filter is designed to pass only those frequencies associated with a particular color. The greater the intensity of a particular frequency, the brighter will be that particular color. The notation included in Fig. 14.22 clearly describes the function of each section of the system. For this system the circuit at the right of the dashed line is repeated 4 times with the indicated value of C for each color.

The power supply is nothing more than a full-wave bridge-rectifier with a RC filter and output reference Zener diode of 18 V. The color control resistor R_1 is simply controlling the strength of the signal appearing at the base of the active filter. A signal of the proper strength and frequency will activate the trigger circuit (simple transistor amplifier) to turn on the SCR. Once the SCR is in its short-circuit state, the colored lamp will light. The brightness of the bulb is controlled by the strength of the "on" state of the SCR.

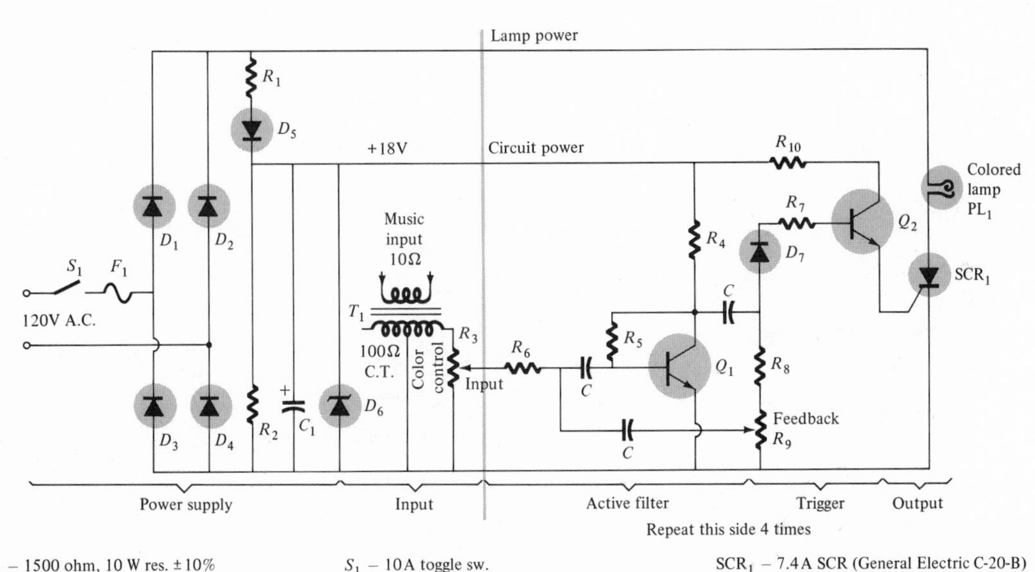

R_1 – 1500 ohm, 10 W res. ±10%
R_2 – 820 ohm, 1 W res. ±10%
R_3 – 50 ohm pot "Color Control"
C_1 – 15 μF, 35 V elec. capacitor
D_1, D_2 – 1N3495 rectifier (Motorola)
D_3, D_4 – 1N3495 R rectifier (Motorola)
D_5, D_7 – 1N40001 diode
D_6 – 18 V zener diode (Motorola 1N4746)
T_1 – Interstage trans. 100 ohms c.t./10 ohms c.t. (Stancor TA-2 or equiv.)

S_1 – 10 A toggle sw.
F_1 – 10 A, 120 V fuse
The following parts are for a single channel. Four channels are required.
R_4 – 3300 ohm, ½ W res. ±10%
R_5 – 1 megohm, ½ W res. ±10%
R_6 – 4700 ohm, ½ W res. ±10%
R_7 – 10,000 ohm, ½ W res. ±10%
R_8 – 2700 ohm, ½ W res. ±10%
R_9 – 2000 ohm pot (Mallory "Trim-Pot" MTC-1)
R_{10} – 560 ohm, ½ W res. ±10%

SCR_1 – 7.4 A SCR (General Electric C-20-B)
PL_1 – 120 V incandescent bulb–in color (20 to 450 W total per channel)
Q_1, Q_2 – 2N3391 transistors
C – 0.1 μF, 50 V capacitor (for l.f. green channel)
C – 0.047 μF, 50 V capacitor (for medium-l.f. blue channel)
C – 0.022 μF, 50 V capacitor (for medium-h.f. red channel)
C – 0.01 μF, 50 V capacitor (for h.f. yellow channel)

FIG. 14.22

Color organ.
(Courtesy Electronics World)

The circuit of Fig. 14.23 can be employed as a light dimmer for lamps up to 400 W or as a light motor speed control (up to 2 A). The bridge rectifier will establish the signal appearing in Fig. 14.24a at point *a*

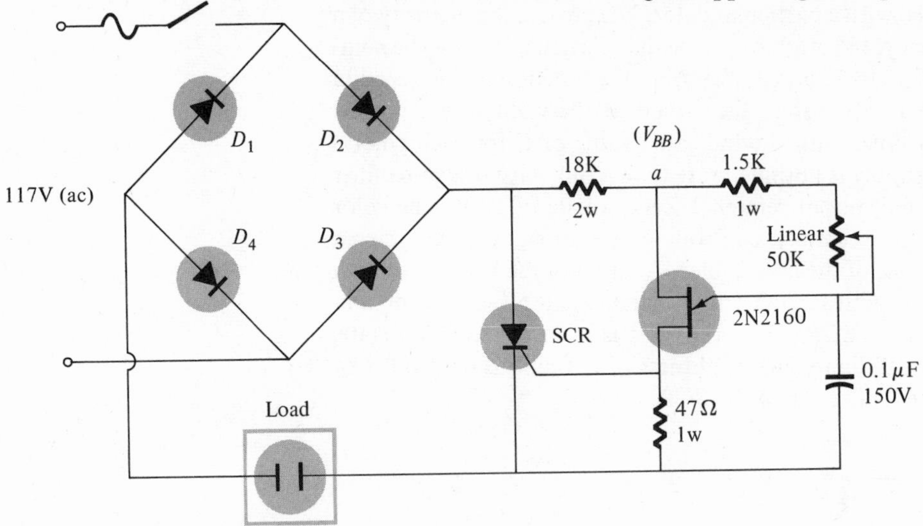

FIG. 14.23

Light dimmer (motor speed control).
(Courtesy Electronics World)

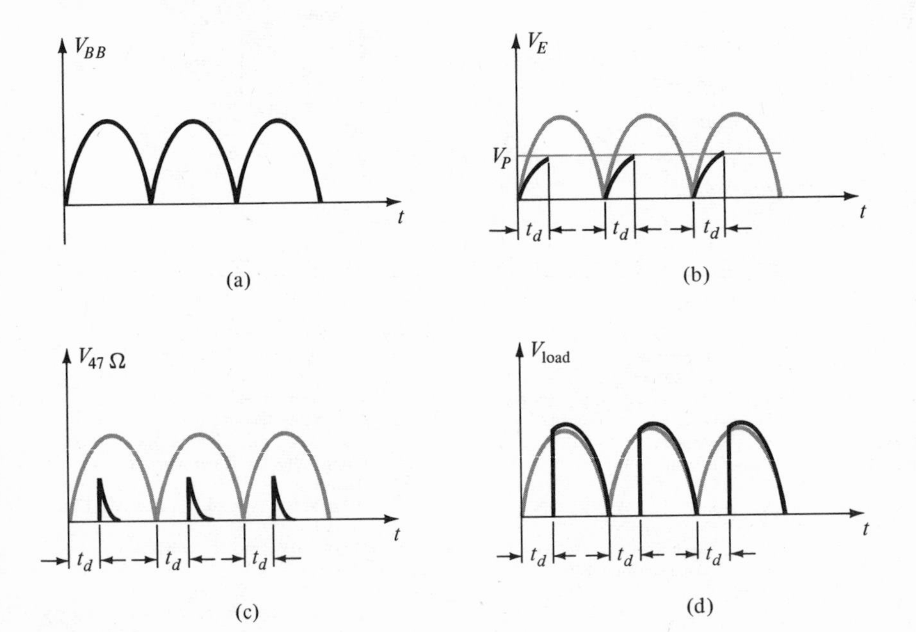

FIG. 14.24

Light dimmer control and load waveforms.
(Courtesy General Electric Co.)

of the circuit of Fig. 14.23. This is, as indicated, V_{BB} for the unijunction transistor as discussed in Chapter 9. It differs to the extent that V_{BB} is a variable quantity here, whereas it was a fixed dc level in Chapter 9. The voltage V_E at the emitter of the unijunction transistor will increase toward the instantaneous value of V_{BB} (as shown in Fig. 14.24b) until the firing potential (V_P) is reached. At this point, the unijunction will enter the conduction state and the discharge of C through R_2 (47 Ω) will result in the waveform of Fig. 14.24c. The pulse appearing across R_2 will trigger the SCR into the conduction state and the output across the load will appear as shown in Fig. 14.24d. Note that the peak value of the load voltage is greater than the peak value of V_{BB} due to the elimination of the voltage divider action of the 18-K resistor. The delay time, and consequently the voltage appearing across the load, can be varied by the linear 50-K potentiometer. An increase in t_d will, of course, result in a reduction in the effective voltage appearing across the load, reducing the brightness of the bulb, or the speed of a motor if employed as a control device.

14.10 UNIJUNCTION CODE PRACTICE OSCILLATOR

A code practice oscillator of relatively simple construction and operation appears in Fig. 14.25. The 50-K volume control is affecting nothing more than the V_{BB} potential of the unijunction transistor. The higher its level, the greater the strength of the resulting output signal. The 25-K tone control and capacitor C_1 are controlling the frequency and thereby the tone of the output signal. You will recall from Chapter 9 that the frequency of the output signal of a unijunction oscillator is directly

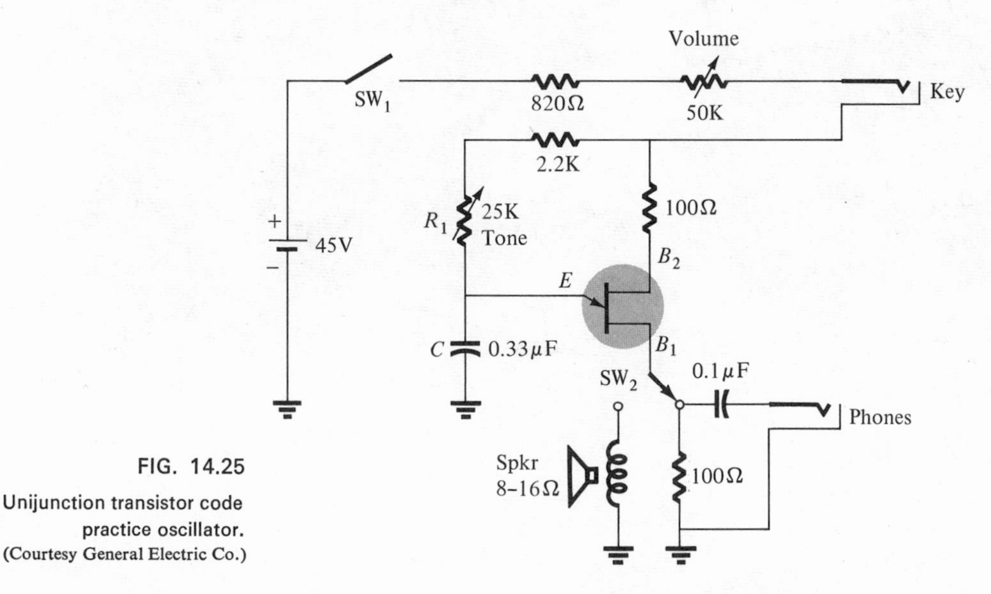

FIG. 14.25
Unijunction transistor code practice oscillator.
(Courtesy General Electric Co.)

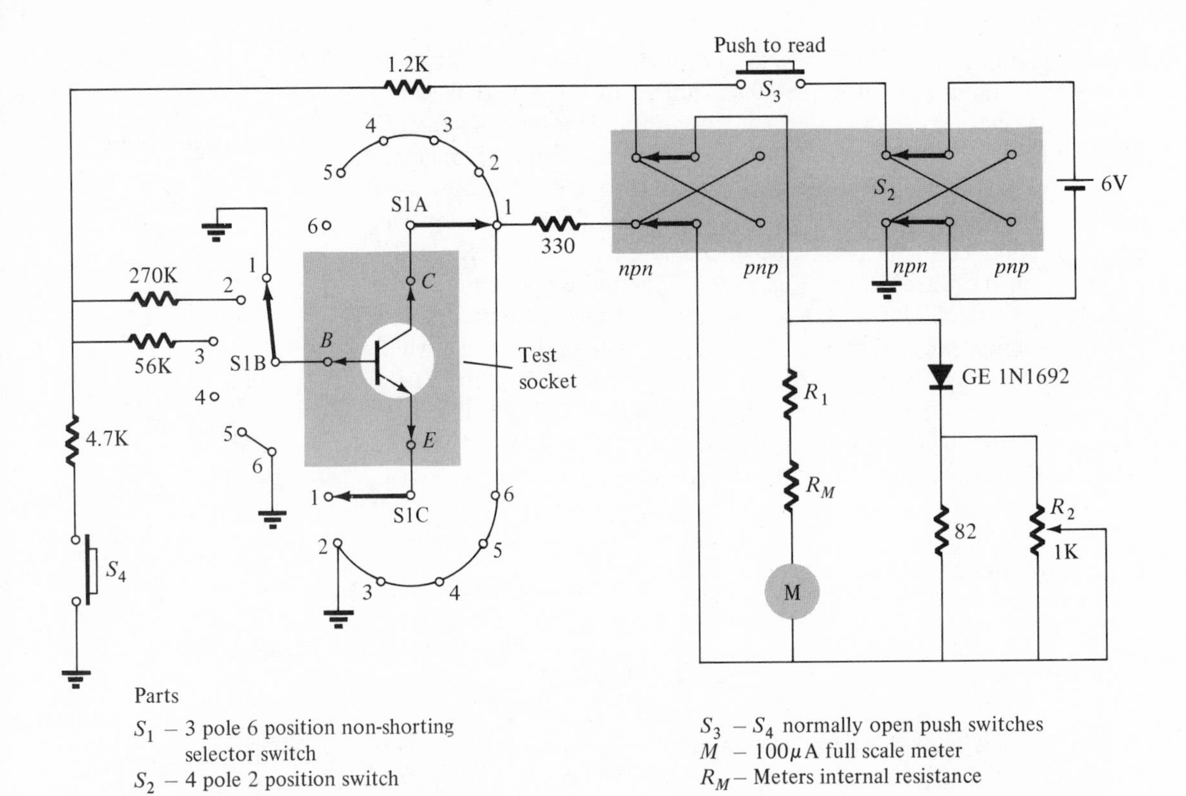

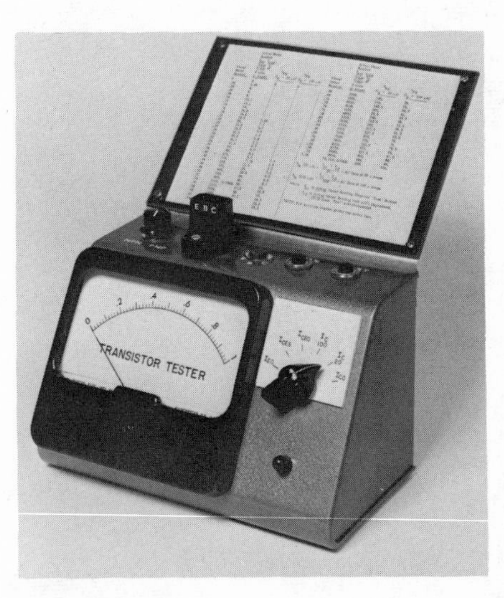

FIG. 14.26 (a)

Transistor tester.
(Courtesy General Electric Co.)

dependent on the magnitude of R_1, R_2, and C. So long as the key is depressed, a pulse having a definite frequency and strength will appear across the phone or speaker with a tone determined by the component values.

14.11 TRANSISTOR TESTER

The circuit for, and actual photographs of a transistor tester, appear in Fig. 14.26a. The basic operation of the tester can best be described by examining the resultant circuit configuration for determining one of the parameters, such as h_{FE}, the static value of the short-circuit forward

To test	When	Adjust Selector switch S_1 to position	Result	
I_{CO}	$V_{CE} = 6V$	1	Read meter direct	
I_C	$I_B = 20\mu A$	2	Read meter direct	
I_C	$I_B = 100\mu A$	3	Read meter direct	
I_{CEO}	$V_{CE} = 6V$	4	Read meter direct	
I_{CES}	$V_{CE} = 6V$	5	Read meter direct	
I_{EO}	$V_{EO} = 6V$	6	Read meter direct	
h_{FE}	$I_B = 20\mu A$	2	Calculate: $h_{FE} = \frac{I_C}{I_B} = \frac{\text{meter reading}}{20\mu A}$	
h_{FE}	$I_B = 100\mu A$	3	Calculate: $h_{FE} = \frac{I_C}{I_B} = \frac{\text{meter reading}}{100\mu A}$	
h_{fe}	$I_B = 20\mu A$	2	Calculate: $h_{fe} = \frac{I_{C_1} - I_{C_2}}{4 \times 10^{-6}}$	Where: I_{C_1} = meter reading; I_{C_2} = meter reading with S_4 closed
h_{fe}	$I_B = 100\mu A$	3	Calculate: $h_{fe} = \frac{I_{C_1} - I_{C_2}}{20 \times 10^{-6}}$	
6V battery	——	4	With 150Ω resistor connected to C–E of test socket, full-scale meter deflection will result when S_3 is pressed.	

FIG. 14.26 (b)
(Courtesy General Electric Co.)

current gain of a transistor. The table of Fig. 14.26b directs that switch S_1 be in position 2 for this parameter. This means that the contacts *SIA*, *SIB*, and *SIC* are all at position 2. We shall assume the transistor is NPN so the switch S_2 will be in the position indicated in Fig. 14.26a. The resulting circuit appears in Fig. 14.27a. It is redrawn in Fig. 14.27b for ease of further analysis. The enclosed area of Fig. 14.27b is the current-limiting, calibrating, fine adjustment for the meter movement. It is quite obvious from Fig. 14.27b that the movement will indicate the magnitude of I_C.

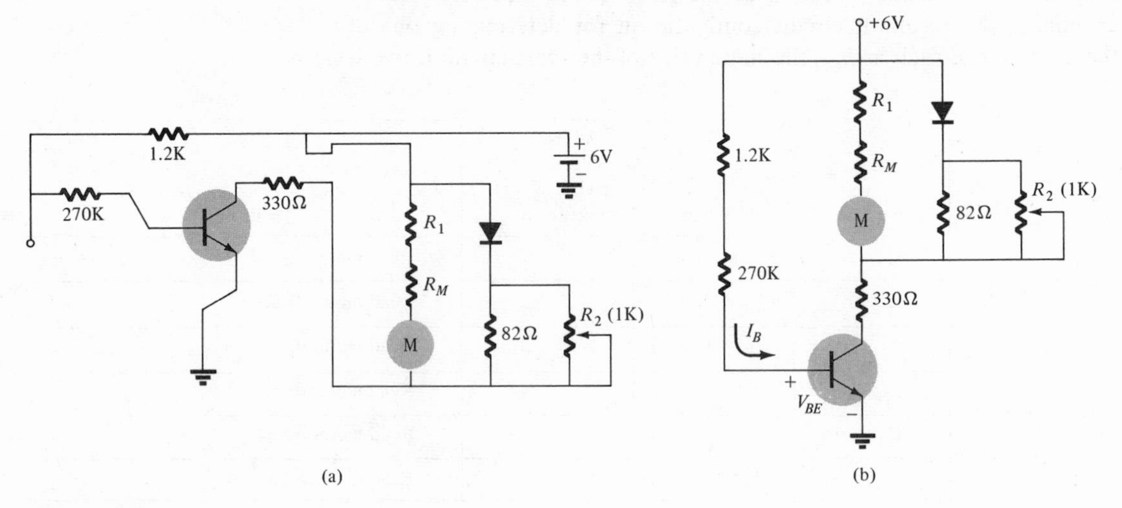

FIG. 14.27
(a) transistor tester circuit for determining h_{be};
(b) circuit redrawn.

If we assume $V_{BE} = 0.6$ V, then

$$I_B = \frac{6 - 0.6}{271.2\text{ K}} = \frac{5.4}{271.2\text{ K}} \cong 20\ \mu\text{A}$$

as indicated on the chart. It should prove somewhat interesting to trace through the resulting circuits for the remaining measurements using this particular transistor tester. In addition it would be time well spent to build the circuit and check its operation. A transistor tester is certainly a useful instrument to have available.

14.12 HIGH-IMPEDANCE FET VOLTMETER

The schematic for a high-impedance FET voltmeter appears in Fig. 14.28. The ultimate in voltmeter design requires that the input impedance be infinite for each scale so that the response of a circuit is not altered when the meter is introduced. The high input impedance of the FET amplifier is employed toward this end in the FET voltmeter. The wiper arm denoted *a* in Fig. 14.28 is set on the appropriate voltage scale. The voltage appearing at the gate of the FET is then determined by a simple voltage divider relationship if we assume the input impedance of the FET to be essentially infinite (open-circuit) ohms; that is, if we define $R_{A-B} = 2\text{ M} + 10\text{ M} + 8\text{ M} + \text{M} + 800\text{ K} + 100\text{ K} + 80\text{ K} + 10\text{ K} + 10\text{ K} = 22\text{ M}$. Then for the 1 V scale

$$V_{\text{gate}} = \frac{(R_{A-B} - 12\text{ M})V_{A-B}}{R_{A-B}} = \frac{10\text{ M }V_{A-B}}{22\text{ M}} \cong 0.45\ V_{A-B}$$

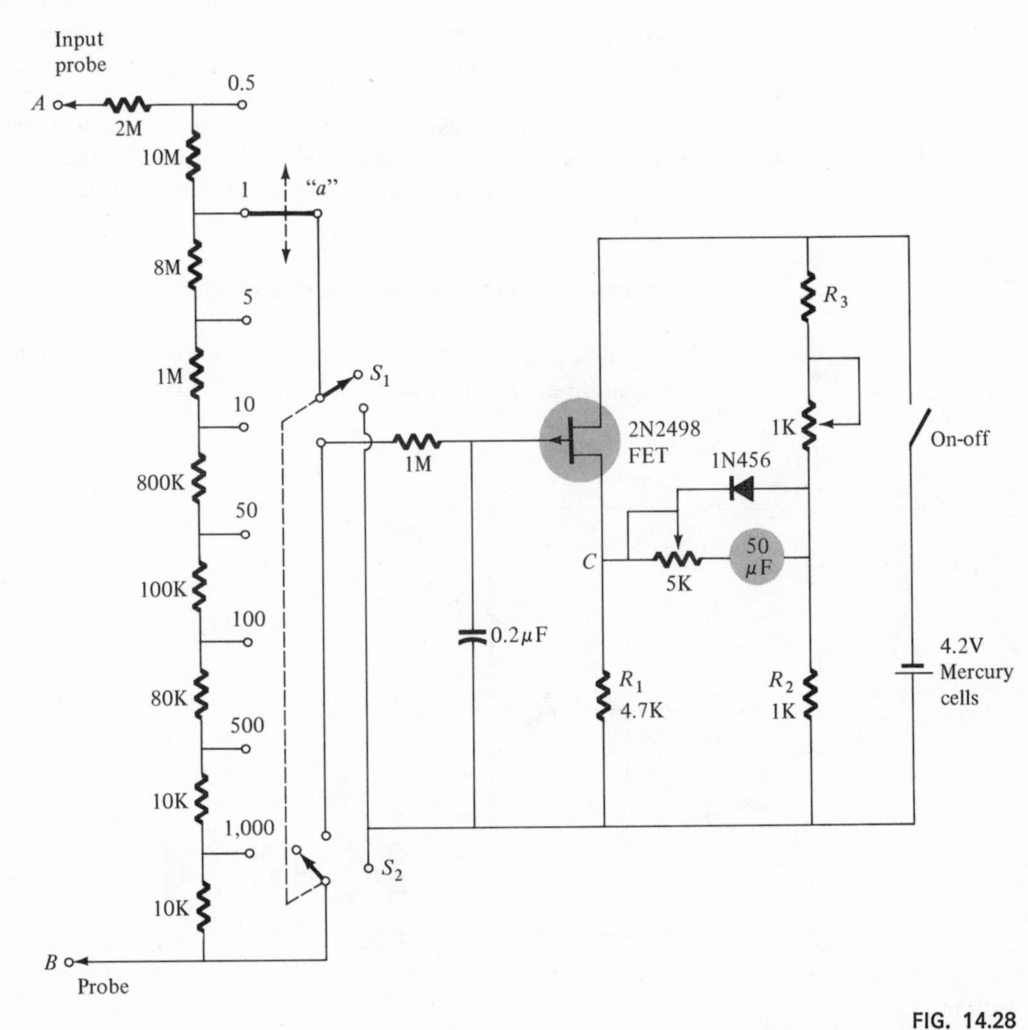

FIG. 14.28

High-impedance FET voltmeter.
(Courtesy Texas Instruments Inc.)

In addition, for the above approximation $Z_{\text{input(FET)}} = \infty\ \Omega$, the input impedance of the voltmeter is simply determined by the series resistance between *A* and *B* independent of scale employed. Therefore,

$$Z_{i(\text{meter})} = 22\ \text{M}$$

A moment's investigation should reveal that the FET and movement are part of a bridge circuit. This in itself should give some hint as to its mode of operation. With no difference in potential across *A–B*, the various calibrating resistors are adjusted to indicate zero deflection. Subsequently, when *A* and *B* are placed across a difference in potential levels, the resulting drain-to-source voltage of the FET will result in an unbalanced condition and a deflection of the movement. If properly calibrated, the movement will indicate the magnitude, in volts, of the voltage across points *A–B*. The 1-M resistor and 0.02-μF capacitor will filter out any stray ac voltages that appear at the input to the voltmeter. Switches S_1 and S_2 will allow the meter to read upscale even if the polarity of the potential level at *A* with respect to *B* is reversed. The vertical dashed lines indicate that they are ganged together so they will both move to the same relative positions.

14.13 UNIJUNCTION HOME SIGNAL SYSTEM

The system of Fig. 14.29 will reveal, by the tone of the audio signal, the location at which entrance is desired. The capacitor *C* will charge

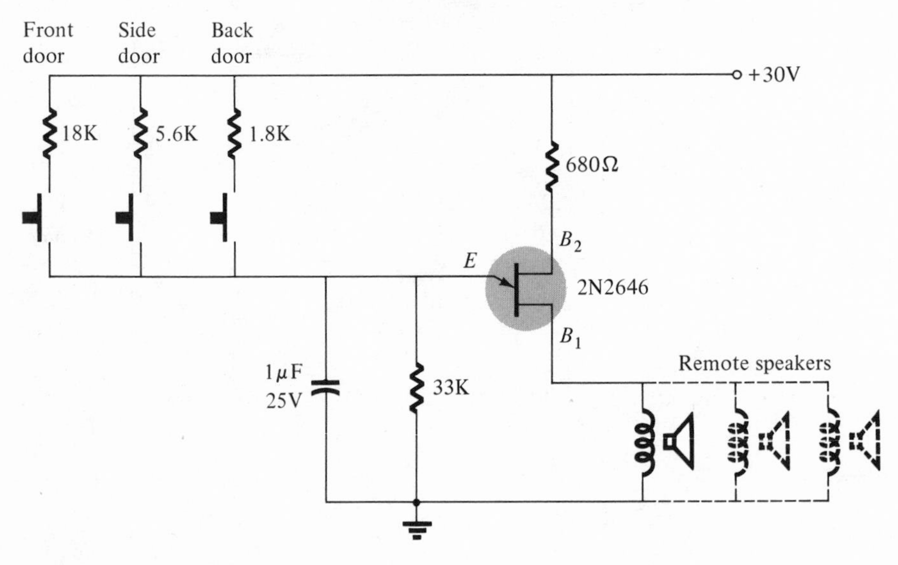

FIG. 14.29

Unijunction home signal system. (Courtesy General Electric Co.)

toward a potential determined by the voltage divider rule; that is,

$$V = \frac{33\ \text{K}(30)}{33\ \text{K} + R\ \text{(determined by button depressed)}}$$

The charging time constant ($\tau = RC$) will be determined by the botton depressed. The smaller the time constant, the sooner the unijunction transistor will enter the conduction state and permit a pulse to reach the remote speakers. Quite naturally, therefore, the smaller the time constant the higher the pulse frequency and the higher the pitch of the resulting signal.

PROBLEMS

§ 14.2

14.1. Calculate the percent voltage regulation of a voltage supply providing 100 V unloaded and 95 V at full-load.

§ 14.3

14.2. Calculate the circuit currents for the circuit of Fig. 14.30 at $R_L = 2$ K (no-load) and $R_L = 200\ \Omega$ (full-load).

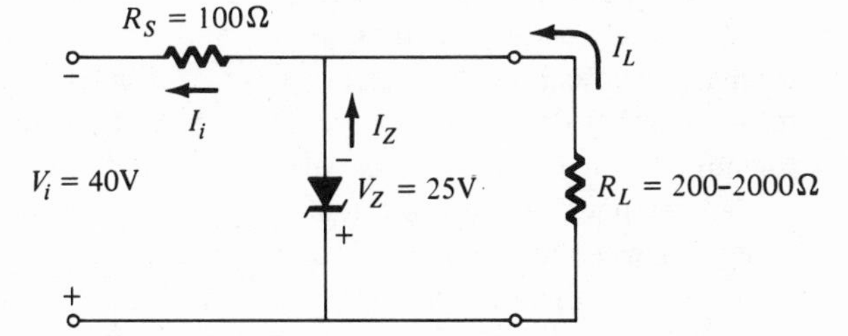

FIG. 14.30
Regulator circuit for Problem 14.2.

§ 14.4

14.3. Calculate the value of dc bias voltages and currents in the circuit of Fig. 14.17 for $V_i = 50$ V, $I_L = 1$ A, $V_Z = 20$ V, $R_L = 30\ \Omega$, and $R_3 = 1$ K (all other circuit values are the same).

14.4. Design a voltage regulator as in Fig. 14.17 to take a voltage of 15 V and provide a regulated voltage of 10 V to a 100-Ω load. Select suitable values for all components.

14.5. Draw the circuit diagram of a voltage regulator to operate with PNP transistors. Show all voltage polarities.

14.6. Calculate the circuit currents of a shunt regulator as in Fig. 14.16 for circuit values of $R_S = 500\ \Omega$, $V_i = 50$ V, $V_Z = 19.3$ V, $R_L = 1$ K, and $h_{FE} = 50$.

15

Cathode Ray Oscilloscope

15.1 GENERAL

One of the basic functions of electronic circuits is the generation and manipulation of electronic waveshapes. These electronic signals may represent audio information, computer data, television pictures, timing information (as used in radar work), and so on. The common meters used in electrical work—the dc or ac voltmeter—(VOM or VTVM) measure either dc, peak, or rms. These measurements are correct only for nondistorted sinusoidal signals or they measure true rms for a particular signal with no indication of how the signal varies with time. Obviously, when signal processing is being done, these overall measurements are essentially meaningless. What is necessary is to "see" what is going on in the circuit, hopefully in the small fractions of time it takes for the signal waveshape to change. The cathode ray oscilloscope (CRO) provides just this type of operation—visual presentation of the signal waveshape, allowing the technician or engineer to look at different points in the electronic circuit to see those changes taking place. In addition, the CRO may be calibrated and used to measure both voltage and time variations so that information is available on how much voltage is present, how much voltage changed, and how long it took to make the change (or a portion of the change).

Consider a radar circuit in which a pulse orginates in an electronic circuit and is radiated by an antenna toward a distant object. At some later time a reflected pulse is received. The time it took to reach the object and return is measured on a CRO to provide an indication of distance. It is necessary when developing and setting up such circuits, to be able to display the pulse sent out—its amplitude, the sharpness of the pulse, etc., and then visually compare it with the return signal after reception and processing of the signal. The return signal can be checked on the CRO for sharpness and voltage amplitude. The time difference between the pulse transmitted and that received may be read on the calibrated time scale of the CRO.

A second example is the use of a CRO to view signals throughout the circuit of a television (TV) receiver (or transmitter). The video signals (information signals for TV) contain complex waveforms, which provide black-white intensity, synchronization information for the picture, and audio information. To properly test and adjust such circuits requires comparing their operating waveforms with expected waveforms. Such adjustments could not be made without the use of the CRO. (It is interesting to note that due to the cathode ray tube (CRT), the TV can be its own test scope, showing where the circuit is operating well or poorly).

Without the scope, electronic work would virtually be impossible—it would be like groping in a dark room. Unless one is able to "see" the circuit waveforms there is no way to correct errors, understand mistakes in design, or make adjustments. As the voltmeter, ammeter, and power

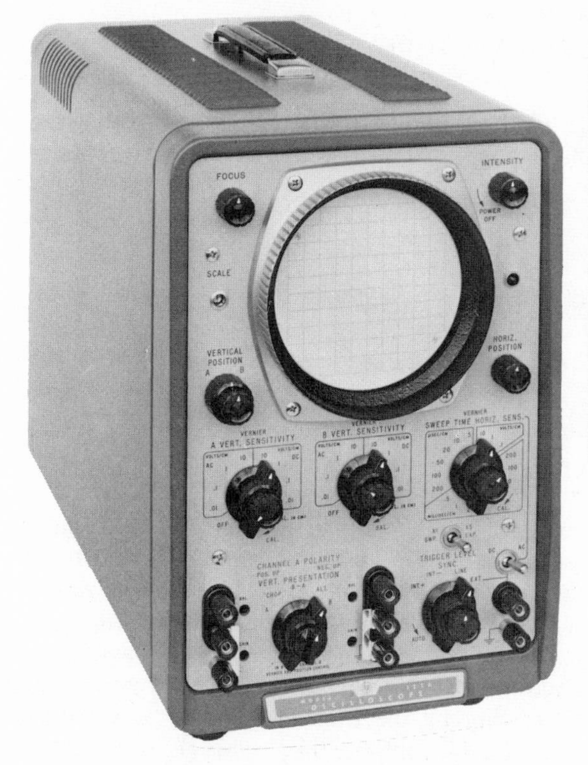

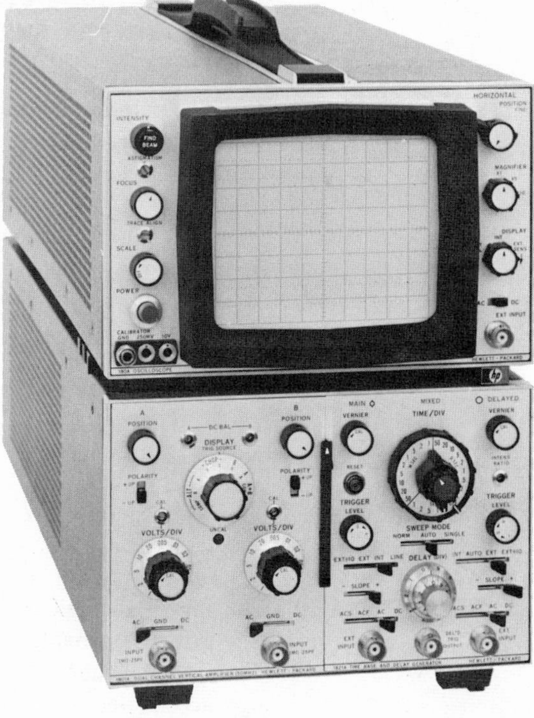

FIG. 15.1
Typical cathode ray oscilloscopes (CROs): HP 122A and HP 180A scopes.

meter are basic tools of the power engineer or electrician, the CRO is the basic tool of the electronic engineer and technician.

There is a wide range of CROs available, some suited for general work in many areas of electronics, others for work only in a specific area. A CRO may be designed to operate from 100 Hz up to 500 kHz, or from dc up to 50 MHz; it may allow viewing signals to within a time span of 1 microsecond, or down to a few nanoseconds (10^{-9} sec); it may provide one waveform or a number of waveforms simultaneously on the face of the CRO. When a number of waveforms are shown simultaneously voltages at different places in the circuit can be viewed at the same time. All these CRO features provide flexibility and enable one to use a CRO that is well suited for the job at hand. Another CRO feature is ability to hold the display for either a very short duration of time or for a long duration. In fact newer CROs (called storage scopes) provide storage of a display for many hours so that an original signal (which appeared long before) may still be analyzed or compared with another signal at a later time. Figure 15.1 shows two representative scopes.

15.2 CATHODE RAY TUBE— THEORY AND CONSTRUCTION

The cathode ray tube (CRT) is the "heart" of the CRO, providing the visual display that makes the instrument so useful. The tube contains four basic parts

1. an *electron gun* to produce a stream of electrons;
2. *focusing and accelerating* elements to produce a well-defined *beam* of electrons;
3. *horizontal and vertical deflecting plates* to control the path of the beam; and
4. an *evacuated glass envelope* with a *phosphorescent screen*, which glows visibly when struck by the electron beam.

Figure 15.2 shows an overall view of a complete electrostatic deflection CRT. We shall first briefly consider its operation. A *cathode* (K) containing an oxide coating is heated indirectly by a filament resulting in the release of electrons from the cathode surface. A control grid (G) provides for control of the number of electrons passing on into the tube. The electrons are then focused into a tight beam and accelerated to higher velocity by the the focusing and accelerating anodes. The parts discussed so far comprise the *electron gun* of the CRT.

The high-velocity, well-defined electron beam then passes through two sets of deflection plates. The first set of plates is oriented to deflect

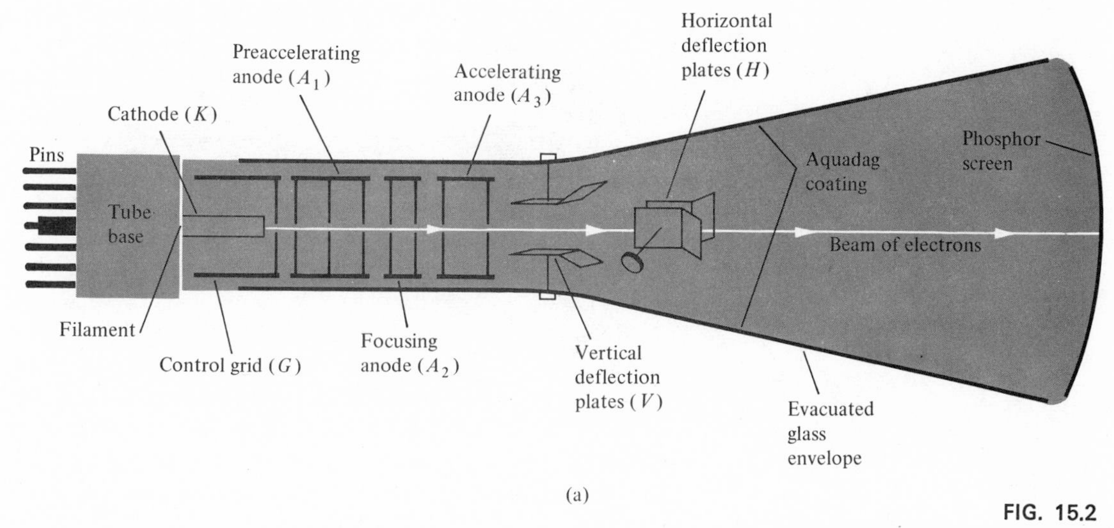

FIG. 15.2
Cathode ray tube: basic construction.

the electron beam *vertically*, up or down. The *direction* of the vertical deflection is determined by the voltage *polarity* applied to the deflecting plates. The amount of deflection is set by the applied voltage *magnitude*. The beam is also deflected horizontally (to the left or right) by a voltage applied to the horizontal deflecting plates. The deflected beam is further accelerated by very high voltages applied to the tube and it finally strikes a *phosphorescent* material on the inside face of the tube. The phosphor glows when struck by the energetic electrons—the visible glow seen at the front of the tube by the person using the scope.

The CRT is a self-contained unit with leads brought out through a base to pins, as in any electron tube. Various types of CRTs are manufactured for a variety of applications. The CRT of Fig. 15.2 is quite basic and allows discussion of the essential elements of the device. We can now consider each part of the tube in more detail and then the external CRO controls, which provide the useful operation of the tube.

Electron Gun

The cathode (Fig. 15.3a) is cylindrical, made of nickel, and capped on the end with an oxide coating. The oxide coating, typically made of oxides of barium and strontium, is applied to the cap of the cathode facing the screen direction. A filament, made of a tungsten or tungsten alloy, indirectly heats the cathode when a current is caused to flow through the filament. The filament acts like the heating elements of a device such as a toaster or rotisserie. Electrons will be liberated from

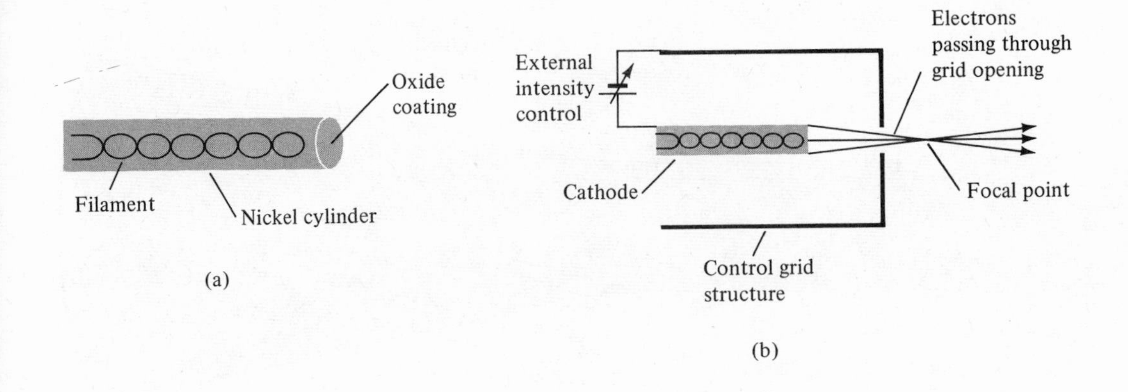

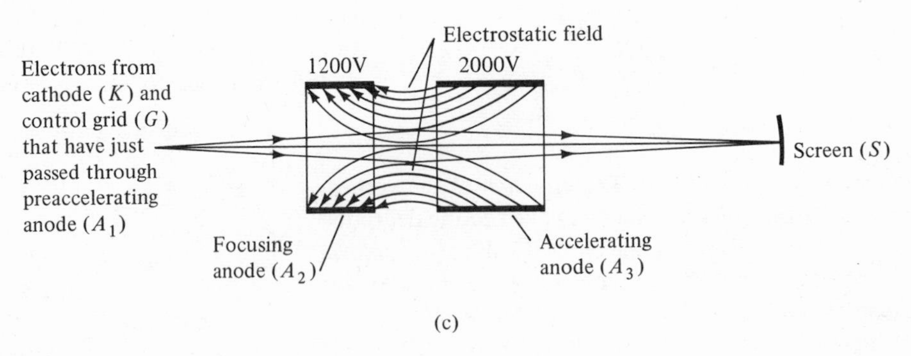

FIG. 15.3

Components of the CRT electron gun: (a) cathode (K) and filament; (b) control grid (G); (c) focusing (A_1) and accelerating anodes (A_2).

the oxide coating on the cathode surface by the heating effect of the filament. These liberated electrons will travel in the general direction of the screen but at various angles and with various velocities.

Focusing and Accelerating Elements

To provide some focusing of the electrons a control grid (Fig. 15.3b) with a small opening in the direction of the screen is placed after the cathode. In addition a biasing voltage is applied to the control grid to control the flow of electrons through the small opening of the grid structure. If the grid voltage is made negative (wtih respect to the cathode) there will be a reduction in the number of negatively charged electrons passing through the grid opening. With a large enough negative voltage all the electrons leaving the cathode due to the heating effect of the filament will be prevented from passing through the grid aperture. The grid therefore permits adjustment of the number of electrons generated by the electron gun with the ability to completely cut off the electron flow to the screen, if desired.

Those electrons that pass through the control grid structure are given an initial acceleration toward the screen of the tube by a pre-accelerator anode (A_1) having a positive potential (100–200 V) with respect to the cathode. The focusing anode (A_2) and accelerating anode (A_3) operate to focus the electrons into a tight beam and also to further accelerate or speed up the electrons coming from the cathode. Fig. 15.3c shows the electric field set up by typical voltage applied to anodes A_2 and A_3 and their effect on the electron flow. The force on the electrons due to the electric field set up by these anodes will result in a well-shaped beam having a focal point at the screen of the tube. The person using the scope can control the focusing of the beam through an external FOCUS control.

The difference in potential between anodes A_2 and A_3 will set up an electrostatic field as shown in Fig. 15.3c. The effect of these electric field force lines will be to focus the electron beam as shown. The deflection will be such as to push those electrons diverging from the center back in toward the center of the beam. Once past the focusing anodes the electrons move past the deflecting plates to the screen in a tight beam.

Deflecting Plates (Vertical and Horizontal)

Figure 15.4 shows the electron beam passing through a pair of plates. If the voltage of the upper plate is more positive than the voltage of the lower plate; the electron beam will be attracted upward. Reversing the plate polarity would cause the beam to be deflected downward. The voltage applied externally (of the CRT) to the deflection plates shown, results in the signal being deflected in a vertical manner; hence, the designation *vertical deflection plates*. This voltage may be a steady (dc) voltage or a varying one. As a point of interest the home TV tube is similar to that discussed so far except for a magnetic deflection system.

An expression relating the factors in calculating the amount of deflection is (see Fig. 15.4)

$$\boxed{D = \frac{lLV_d}{2dV_a}} \tag{15.1}$$

where the distance terms are shown in Fig. 15.4 and the voltages are V_d, the voltage applied to the deflection plates and V_a, the accelerating voltage of the tube (typically thousands of volts). Since all the values

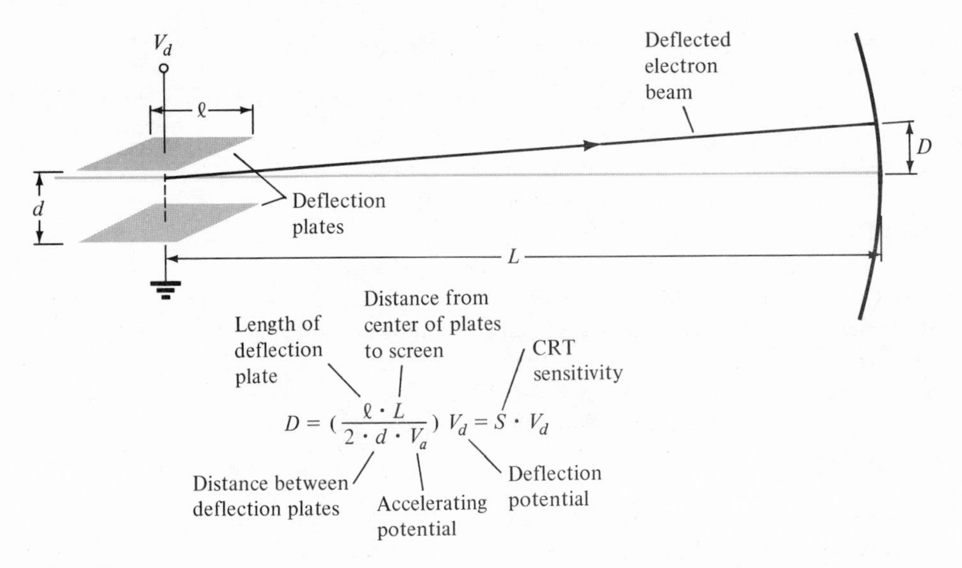

FIG. 15.4
Tube factors affecting beam deflection.

except the deflecting voltage are usually fixed for a tube they may be lumped together as one term defined as the electrostatic deflection sensitivity

$$S \equiv \frac{D}{V_d} = \frac{lL}{2dV_a} \tag{15.2}$$

The sensitivity has units of meters per volt, centimeters per volt, or millimeters per volt, depending on the unit of measurement used. It indicates the amount of deflection of the electron beam at the screen per volt of signal applied to the deflection plates.

The *deflection factor* (G) of a CRT is

$$G \equiv \frac{V_d}{D} = \frac{1}{S} \tag{15.3}$$

Note that G is the reciprocal of the sensitivity and indicates how many volts of deflecting voltage must be applied for each meter of deflection at the screen. Either the deflection sensitivity or deflection factor is provided by the manufacturer for each type of CRT.

EXAMPLE 15.1 A manufacturer of a CRT rates his tube as having a deflection factor of 40 V/in. Calculate the amount of deflection seen on the screen for deflection voltages of 20 and 120 V.

Solution: Using the relation for the amount of deflection in Eqs. (15.2) and (15.3) the solution is

(a) $D = SV_d = (V_d/G) = [20\ \text{V}/(40\ \text{V/in})] = \frac{1}{2}$ in.

(b) $D = [120\ \text{V}/(40\ \text{V/in})] = 3$ in.

EXAMPLE 15.2 If a deflection of 2.5 in is obtained from a deflection voltage of 50 V on the plates of the CRT calculate the deflection factor of the tube. How much will the beam be deflected for a voltage of 75 V for the same tube?

Solution: The deflection factor of the tube is obtained using Eq. (15.3).

$$G = V_d/D = 50/2.5 = 20\ \text{V/in.}$$

Again using Eq. 15.3,

$$D = V_d/G = 75/20 = 3.75\ \text{in.}$$

EXAMPLE 15.3 The sensitivity of a CRT is given as 0.25 mm/V. What voltage must be applied to the deflecting plates to obtain a deflection of 2 in?

Solution: It is first necessary to convert the deflection amount of 2 in. into millimeters. The conversion factor of 25.4 mm/in. is used to obtain

$$D = 2\ \text{in.} \times \frac{25.4\ \text{mm}}{1\ \text{in.}} = 50.8\ \text{mm}$$

Using Eq. (15.2) we calculate the deflection voltage

$$V_d = \frac{D}{S} = \frac{50.8\ \text{mm}}{0.25\ \text{mm/V}} = 203.2\ \text{V}$$

The deflection plates are placed so that the vertical deflection plates are farther from the screen than the horizontal deflection plates. The reason for this is indicated in the above discussion of deflection sensitivity. The vertical plates are used to display the voltage to be viewed or measured. Since the further the deflection plates are from the screen the greater the sensitivity, the usual practice is to obtain the greater sensitivity for the vertical deflection plates.

Electron Acceleration and Phosphor Screen Action

After the beam is acted upon by the deflection plates it passes down the tube in a straight path (although at some deflection angle due to the deflecting voltages). To ensure sufficiently energetic electrons striking the phosphor screen the electron beam may be further accelerated by intensifier bands as indicated in Fig. 15.5.

The high-energy electron beam strikes the phosphor material, causing it to glow. Because of the high energy of the striking electrons a secondary emission of electrons from the phosphor screen will occur. These electrons would build up a layer of negative charge, which would deteriorate the tube operation. A layer of material called Aquadag is coated along the side of the tube, however, and the electrons emitted by secondary action are picked up by the coating and returned to the cathode.

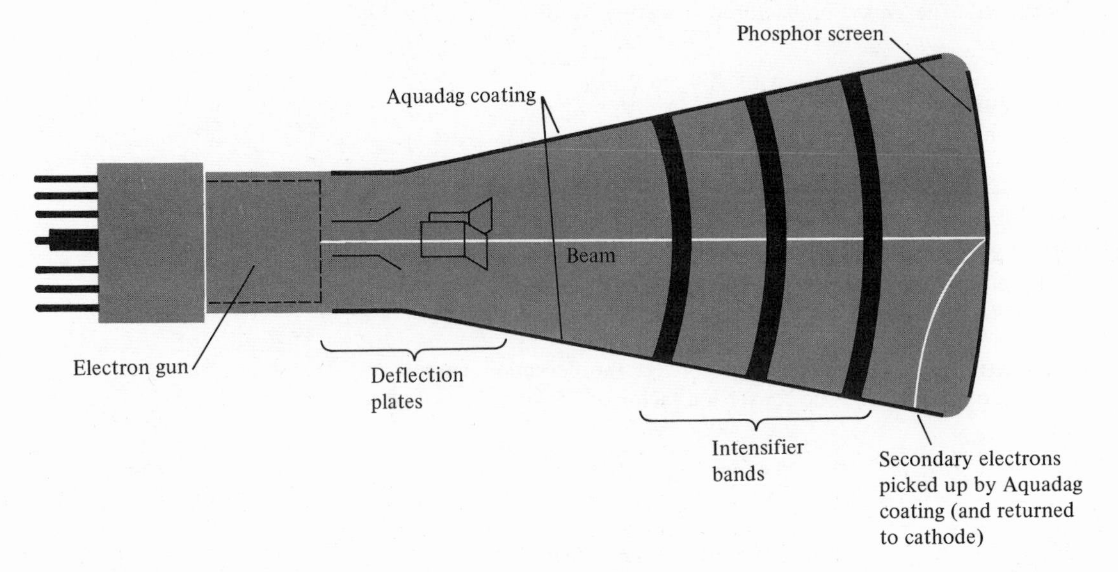

FIG. 15.5
Action of intensifier bands, phosphor screen, and Aquadag coating.

If the glow seen were to be present only as long as the beam strikes the screen, the light given off would be termed fluorescent (as the lights in a home or plant). The phosphor screen, however, will continue to glow even after the beam is turned off and the length of time the glow continues can vary from a few milliseconds to a few seconds depending on the type of phosphor material used. CRO tubes are used to observe very-high-frequency signals have short *persistence* (short amount of glow time after beam is removed). Other scopes used for observing only very-low-frequency signals have persistence times of a few seconds. There is also a special type of CRT used in memory-type scopes that has an effective persistence in the range of hours.

Table 15.1 shows a number of the phosphor types as well as their persistence and color of glow. The color is determined by the phosphor material used. The tube types listed by the manufacturer include the phosphor type. For example, a 5BP4 has a phosphor screen type *P*4. CRO manufacturers will also refer to the phosphor, indicating, for example, that their scope can be purchased with either *P*31, *P*2, *P*7, or *P*11 phosphors, depending on the buyer's choice.

TABLE 15.1

CRT Phosphors

PHOSPHOR NO.	COLOR OF PHOSPHORESCENT GLOW	PERSISTENCE
*P*1	green	medium
*P*2	green	long
*P*4	white	medium
*P*5	blue	very short
*P*7	greenish-yellow	very long
*P*11	blue	short
*P*31	green	medium-short

OPERATION AND CONTROLS AFFECTING CRT

As important as it is to understand how the CRT operates it is equally important to understand how the external controls of the CRO relate to the CRT and how the various control operations are carried out in the scope. For the present we shall concentrate on the operation of the *intensity*, *focusing*, and *positioning* controls of the scope.

Figure 15.6a shows a typical biasing arrangement for a few of the CRO controls. The external scope dials are shown in Fig. 15.6b to provide a clearer picture of the operation of the scope. A voltage divider network containing resistors and potentiometers provides for the positioning, focusing, and intensity control of the CRT. A high-voltage (HV) power supply provides the plus (+)-to-minus (−) voltage indicated in Fig. 15.6a. A potentiometer, allowing adjustment of the difference in voltage between cathode (K) and grid ($G1$) provides the control of beam intensity. The voltage applied to the grid is more negative than that connected to the cathode. By adjusting the grid voltage to be more negative the number of electrons passing through the grid is reduced, providing a less intense electron beam. The adjustment of intensity can be made sufficiently negative (from grid to cathode) to enable complete cutoff of the beam. Thus, the user is provided with the ability to regulate the intensity of the screen glow.

The voltage at anode 1 (A_2) is connected to another potentiometer, allowing adjustment and control of the focusing of the electron beam. Using the adjustment, the focal point of the beam can be set directly at the screen so that a clear spot results. Variation of the intensity

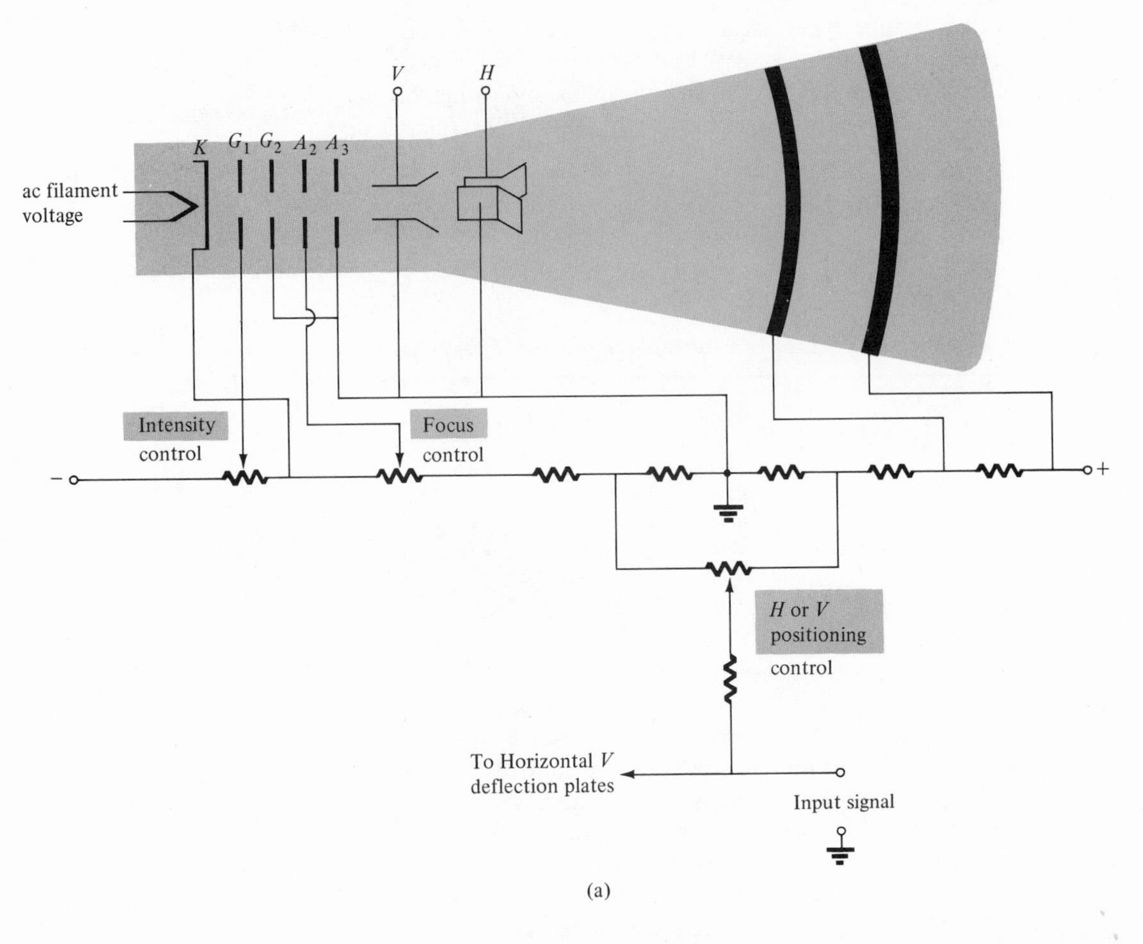

(a)

will have some effect on the focusing so that adjustment of one usually requires adjustment of the other.

The very high voltages at the right side of the voltage divider network are connected to the intensifier bands to provide high acceleration of the electron beam. In general these intensifier band voltages are fixed.

The result of applying voltages to the vertical (or horizontal) plates is to deflect the beam vertically (or horizontally) on the face of the tube. In addition it is usually essential to be able to reposition the picture on the screen; this is provided by external horizontal and vertical positioning controls. This action is shown simply by the addition of an adjustable dc voltage to the input signal (to either the vertical or horizontal plates). The addition of the positioning dc voltage to that applied to the deflection plates moves the electron beam additionally to the right or left (or up or down) so that the overall picture position can be adjusted on the screen. Separate control is provided for the vertical and horizontal sections of the tube.

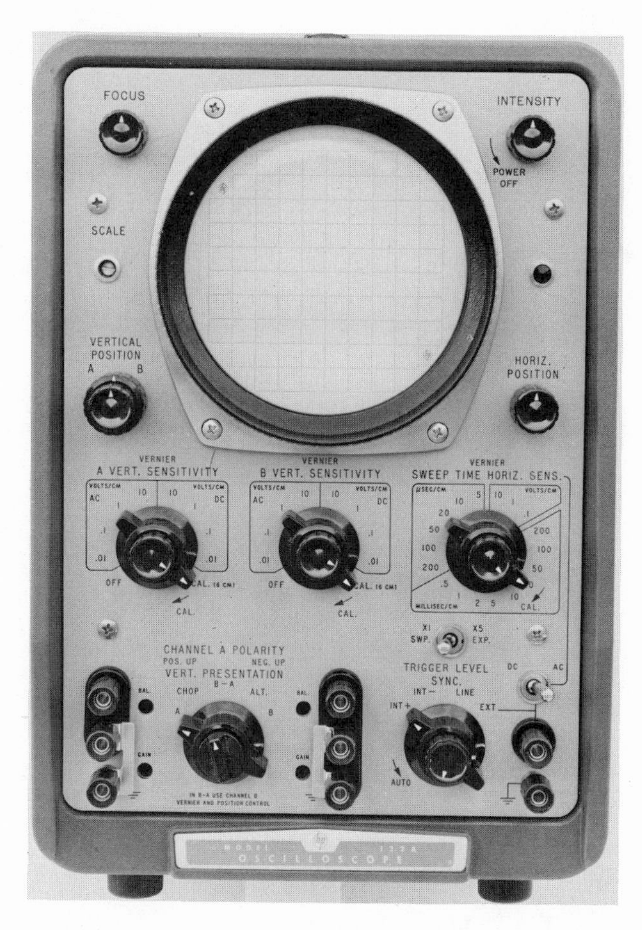

FIG. 15.6

CRO controls in CRT operation.

(b)

15.3 CATHODE RAY OSCILLOSCOPE (CRO) —OPERATION AND CONTROLS

It is necessary that the electron beam be horizontally deflected across the face of the tube to view signal waveforms. In the usual repetitive mode of operation the beam must be returned back across the tube to allow the repeated horizontal sweep of the beam. The beam is deflected horizontally from the left side of the CRT face to the right side during normal operation and then the beam is "blanked" (turned off) during the return sweep of the beam. Additionally, the signal to be viewed must be applied to these electronic circuits, which drive the beam vertically, while it is being swept horizontally across the CRT face.

Typical tube phosphors used have some persistence even to signals of very short time duration. A repetitive tracing of the viewed wave form will then result in a steady continuous display. This requires synchronizing the sweeping of the beam (*sync*) with the signal being viewed. If the signal is properly synced the display will be stationary and appear as if the same signal is being viewed all the time. In the absence of sync the picture will appear to drift or move horizontally across the screen.

The basic parts of a CRO are shown in the block diagram of Fig. 15.7. We shall first consider the operation of the CRO for this simplified general picture.

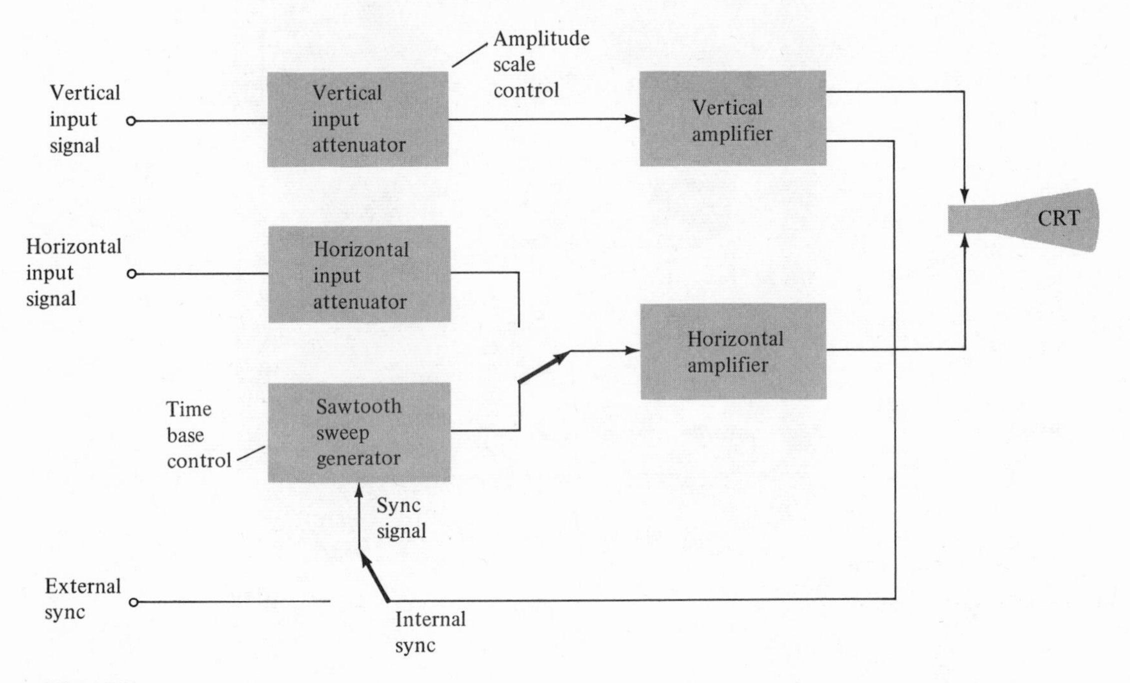

FIG. 15.7

Cathode ray oscilloscope, general block diagram.

To obtain a noticeable beam deflection of a fraction of an inch to a few inches, the usual voltage applied to the deflection plates is on the order of tens to hundreds of volts. Since the signals measured using the CRO may be only a few volts, or even as little as a few microvolts, amplifier circuits are needed to increase the input signal to the voltage levels required to operate the tube. Thus, there are amplifier sections as part of the CRO circuitry for both the vertical and horizontal deflection of the beam. It may also happen that the signal to be viewed is too large (with the amplifier gain) to be viewed within the span of the tube face. When this happens, an attenuator is needed to cut down or reduce the signal amplitude to the amount that results in a visible display within screen dimensional limits.

A synchronizing signal must be available in order to synchronize the horizontal sweep of the electron beam with the signal to be viewed. As shown in Fig. 15.7 the sync signal applied to the horizontal sweep generator can be obtained from either the vertical channel of the CRO as an internal sync signal, or an external sync signal may be connected. Note that the input to the horizontal amplifier is obtained from either the sweep generator or from a horizontal input attenuator. We should be aware that the normal scope operation uses the sweep input to the horizontal channel and that a *different* external horizontal input signal is typically used to obtain a Lissajous pattern for phase or frequency measurement (see Section 15.1).

15.4 CRO—DEFLECTION AND SWEEP OPERATION

We now consider the operation of the CRO for various inputs to the vertical and horizontal deflection plates. Although the actual CRO inputs are not *directly* connected to the deflection plates, but are coupled through attenuator networks and amplifiers, we shall still refer to such signals (applied to the vertical and horizontal inputs or vertical and horizontal channels of the CRO) as the deflection signals. Obviously, the positioning controls and the setting of the attenuators for each channel of the scope will affect the resulting signal actually applied to the deflection plates.

With 0 V connected to the vertical input terminals the electron beam may be positioned to the vertical center of the screen. If 0 V are also applied to the horizontal input, the beam is then at the center of the CRT face and remains a stationary dot on the face of the CRT. Note that the vertical and horizontal positioning controls allow movement of this dot anywhere on the face of the CRT so that the zero input voltages cause the beam to strike the center of the screen *only* if the positioning controls are also zeroed. Unless otherwide noted in the following discussion, assume that the beam is properly centered and that any deflection off center is due to the input signals to the vertical and horizontal channels of the scope.

Zero input signals result in a dot on the screen (Fig. 15.8a). Any dc voltage applied to either input will result in shifting the dot on the screen with the resulting picture still only a dot. Figure 15.8b shows the resulting display due to a negative dc voltage applied at the vertical input and a positive dc voltage at the horizontal input. The negative voltage to the vertical plates deflects the beam downward by an amount proportional to the voltage magnitude and the voltage on the horizontal plates deflects the beam to the right, the resulting dot appearing in the lower right sector (second quadrant) of the screen as shown. The

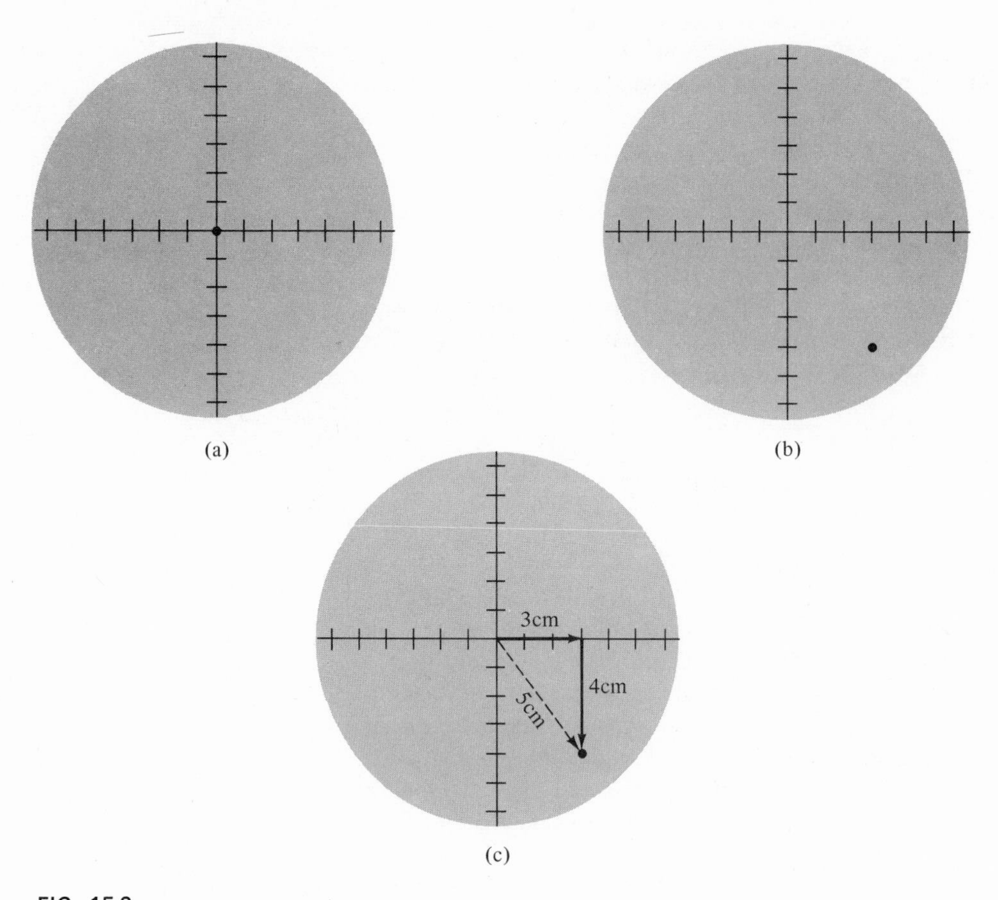

FIG. 15.8

Dot on CRT screen due to stationary electron beam: (a) centered dot due to stationary electron beam; (b) off-center stationary dot; (c) stationary dot showing deflection components.

position of the resultant dot can be considered as the vector sum of the two deflection voltages. A positive horizontal deflection of 3 cm and a negative vertical deflection of 4 cm, for example, would result in the beam spot as shown in Fig. 15.8c, at a distance of 5 cm from the screen center.

To view a signal on the CRT face (note that only a dot was visible even though a steady dc voltage was applied) it is necessary to deflect the beam across the CRT with a horizontal sweep signal so that the variations of the vertical signal can be observed. Figure 15.9 shows the resulting straight line display for the positive dc voltage applied to the vertical input using a linear (sawtooth) sweep signal on the horizontal channel. With the electron beam held at a constant vertical distance, the horizontal voltage, going from negative to zero to positive

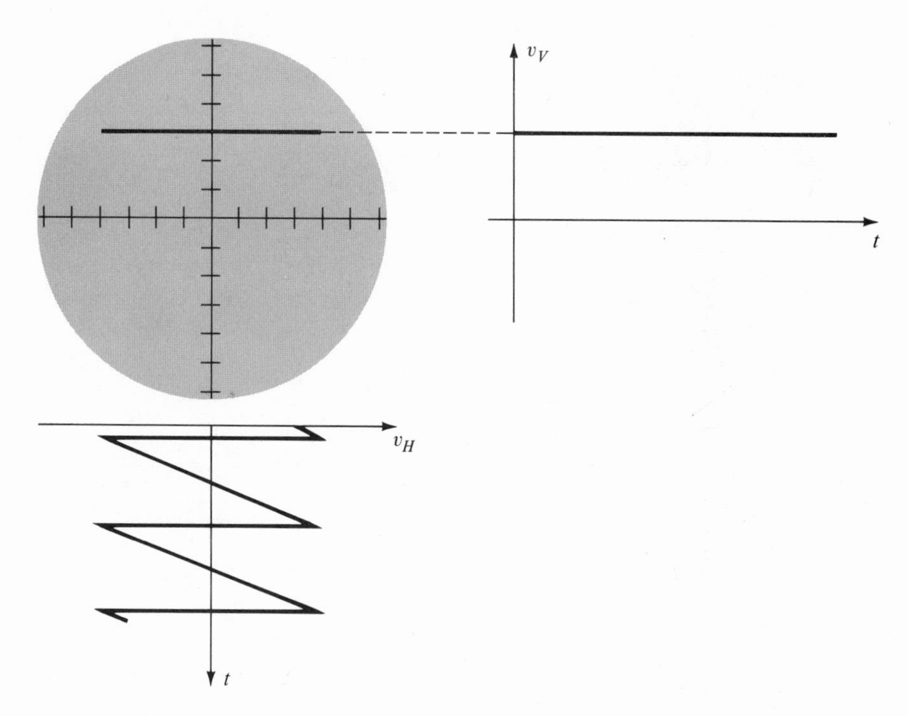

FIG. 15.9
Scope display for dc vertical signal and linear horizontal sweep signal.

voltage, causes the beam to move from the left side of the tube to the center and over to the right side. The resulting display is a straight line above the vertical center and a dc voltage now properly appears as a steady voltage line. The sweep signal is indicated to be a continuous waveform and not just a single sweep. This is a necessary if a long-term display is to be obtained. A single sweep across the tube would quickly fade out. By repeating the sweep, the display is generated over and over again and if enough sweeps are generated per second the display always appears to be present. If the sweep rate is slowed down (the time scale controls of the scope allow this) the actual travel of the beam across the face of the tube can be observed.

Applying only a sinusoidal signal to the vertical input (no horizontal signal) results in a vertical straight line as shown in Fig. 15.10. If the sweep speed (frequency of the sinusoidal signal in this case) is reduced it will be possible to see the electron beam moving up and down along the straight line path.

To view a sinusoidal signal, it is necessary to use a linear sweep signal on the horizontal plates. Then, with linear sweep applied to the horizontal plates the signal viewed on the CRO is that applied to the vertical input. Figure 15.11 shows the resulting CRO display due to a horizontal linear sweep input and a sinusoidal input to the vertical

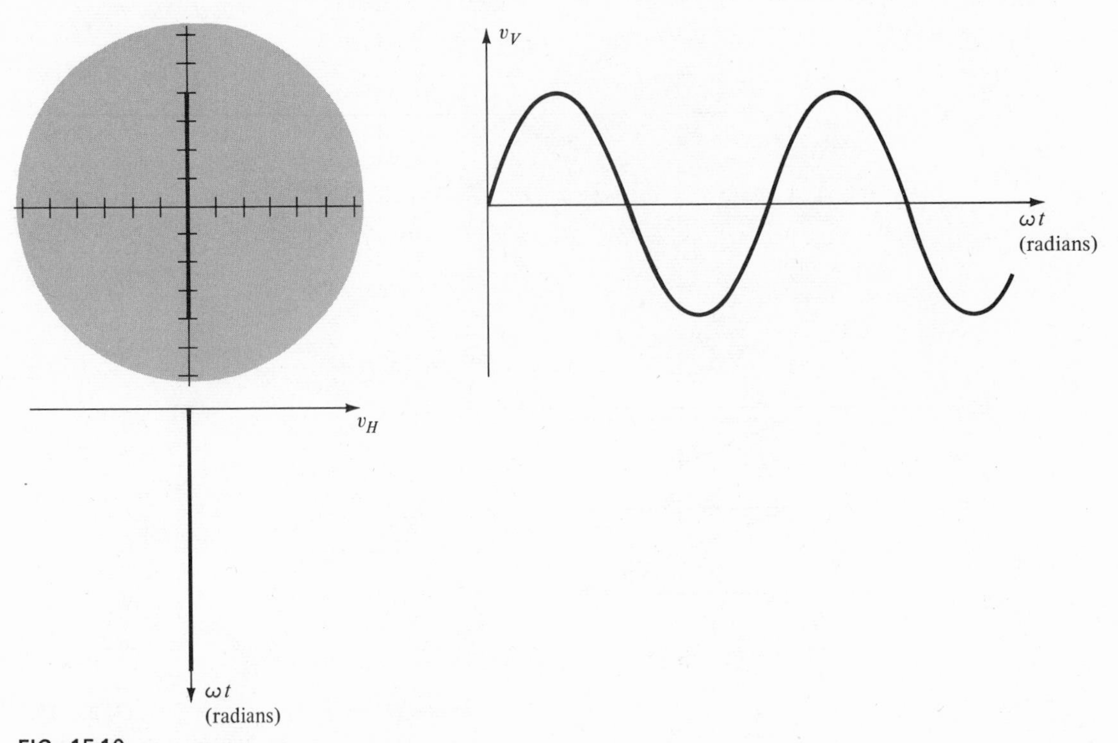

FIG. 15.10
Resulting scope display for sinusoidal vertical input and no horizontal input.

channel. For 1 cycle of the input signal to appear as shown, it is necessary that the signal and linear sweep frequencies be synchronized. If there is any difference the display will appear to move (not be synchronized), unless the sweep frequency is some multiple of the sinusoidal frequency. If, for example, the time for 1 cycle of the sinusoidal signal is 5 msec ($f = 200$ Hz) and that for 1 cycle of the linear sweep is 10 msec ($f = 100$ Hz), the vertical input signal will go through 2 cycles before the sweep has moved the beam across the face of the CRT and the display appears as 2 cycles of the input signal (Fig. 15.11b). Lowering the sweep frequency allows more cycles of the sinusoidal signal to be displayed, whereas increasing the sweep frequency results in less of the sinusoidal vertical input to be displayed, thereby appearing as a *magnification* of a part of the input signal.

Use of Linear Sawtooth Sweep to Display Vertical Input

A signal applied only to the vertical input will cause the electron beam to be deflected only up and down. If the vertical input is a dc voltage the result will be a dot on the screen displaced from the screen

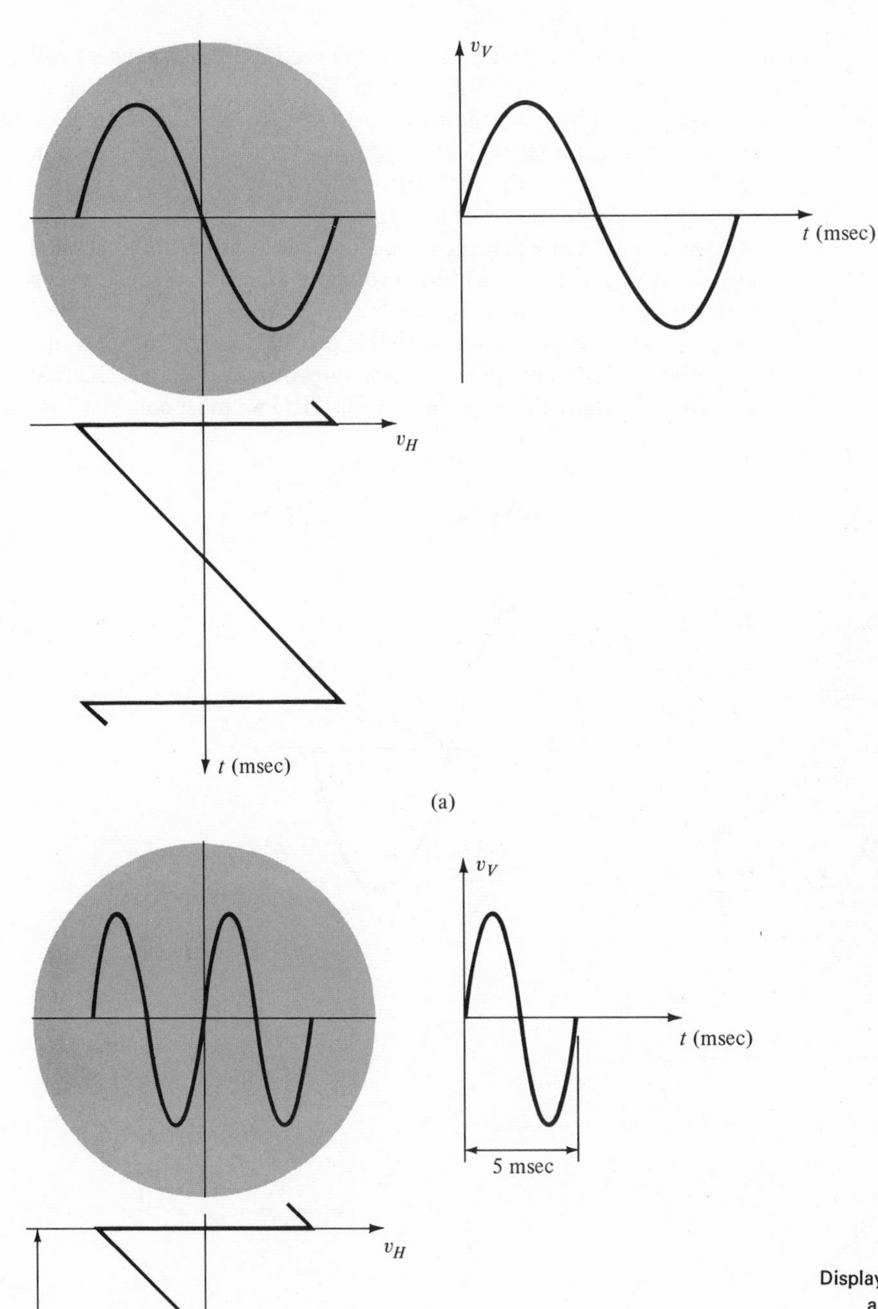

FIG. 15.11
Display of sinusoidal vertical input and horizontal sweep input: (a) display of vertical input signal using linear sweep signal for horizontal deflection; (b) scope display for a sinusoidal vertical input and a horizontal sweep speed equal to one-half that of the vertical signal.

center. If a varying voltage is applied to the vertical input the beam will move up and down resulting in either the dot moving up and down or the appearance of a line, depending on how fast the input signal repeats a cycle and the persistence of the screen phosphor. Similar action results for an input to only the horizontal channel with resulting displacement to the right or left or a horizontal line.

To obtain a display that shows the form of the input signal applied to the vertical channel, it is necessary to apply a linear sawtooth sweep signal to the horizontal channel as well. This is the normal operation of the CRO and the sawtooth sweep signal is provided as part of the scope circuitry with adjustment of the sweep rate provided as an external control. To understand the operation of the CRO we must consider how

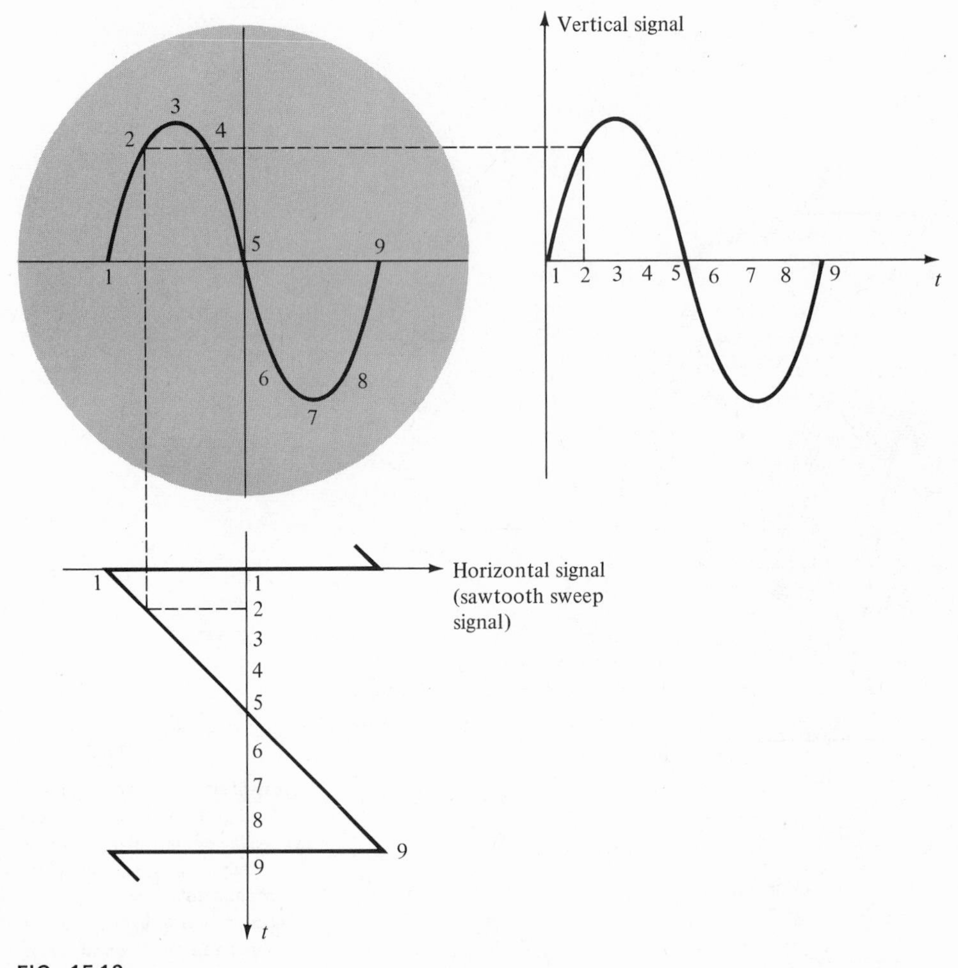

FIG. 15.12
Use of linear sweep signal to view input signal.

the electron beam is deflected as a result of signals applied to both the vertical and horizontal input at the same time.

To understand how the two deflecting voltages result in the CRT display we shall determine the path the beam takes by observing where it is at a few points during a single sweep of the electron beam (see Fig. 15.12). The voltage applied to the horizontal deflection plates is the sawtooth voltage. For convenience the horizontal signal time axis is shown with increasing time in the downward vertical direction. The amount of the sweep voltage is shown horizontally with 0 V at the center of the screen display, positive voltage to the right of center, and negative voltage to the left of center. Considering the beam deflection due to the sweep signal alone, the voltage starting off as negative deflects the beam to the left side of the screen. As the sweep voltage gets less negative the beam will move toward the center. The sweep voltage will pass through zero and go to some positive value at which time the beam will be deflected to the right side of the screen. (Note that the vertical deflection is not being considered for the moment.) When the sweep voltage reaches the largest positive voltage shown (point 9) it very quickly drops back to a negative value and the beam moves back to the starting point on the left of the screen.

The action of the sweep voltage, then, is to move the electron beam across the screen (from left to right) at a constant rate. When we now consider the additional deflection that occurs because of the vertical input signal we shall obtain the actual display shown in Fig. 15.12. It should be clearly understood that the vertical signal applied alone (without horizontal sweep) results in a straight vertical line on the screen. The linear (horizontal) sawtooth sweep voltage must also be applied to cause the beam to move across the CRT so that a display of the input signal is obtained.

Use of a linear sawtooth sweep voltage applied to the horizontal deflection circuit results in the developed display being the same as the input signal applied to the vertical input. Another example of the operation of vertical and horizontal inputs and resulting screen display is the pulse-type signal shown in Fig. 15.13.

The two examples shown in Figs. 15.12 and 15.13 had the same frequencies; that is, the time for a complete cycle of the vertical signal and for a cycle of the sawtooth sweep signal were the same. When this is true the display is a single cycle of the input signal. The horizontal sweep speed, however, is adjustable, and need not be exactly the same as the input signal to be viewed. Figure 15.14 shows the resulting display if the sweep speed is faster than the input signal. In the example shown the time for 1 cycle of the input signal is 4 msec, whereas the time for 1 cycle of the sawtooth signal is only 1 msec. In this case the beam is moved across the tube in 1 msec during which time the vertical input is deflected from zero up to the maximum positive point as shown in the figure. Only a part of the input signal is shown in this case. If the

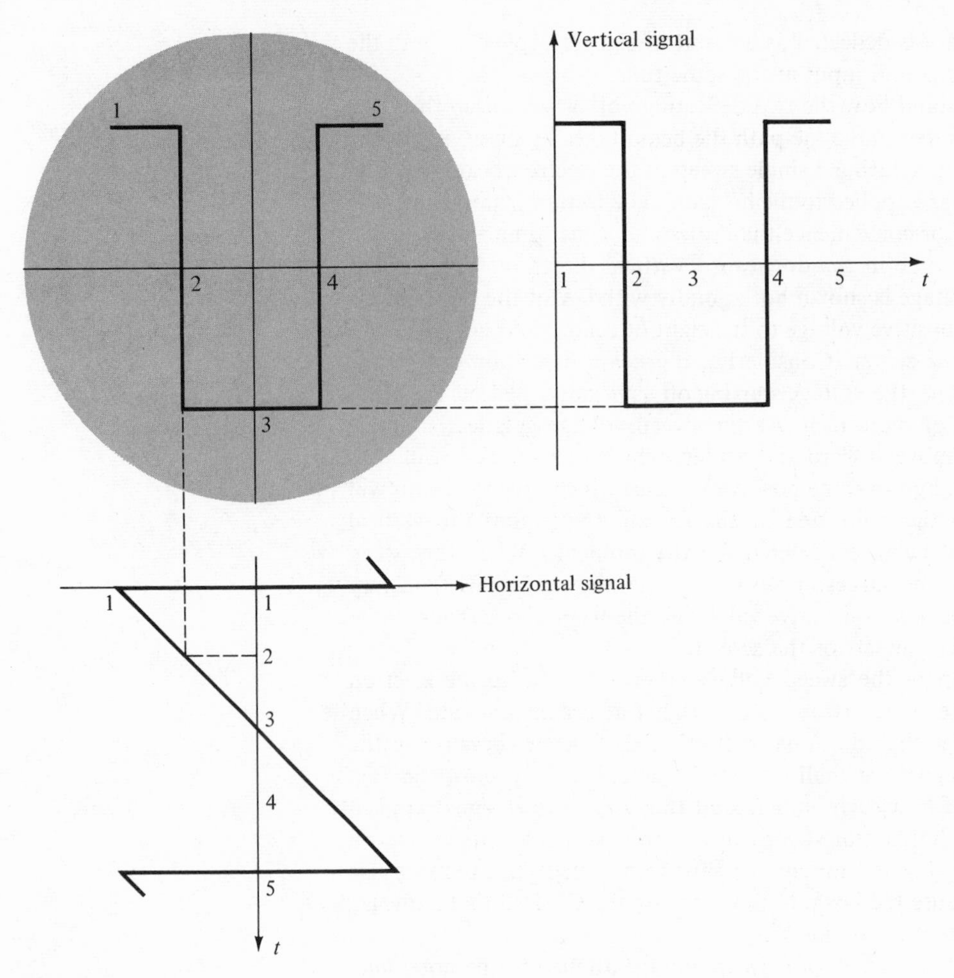

FIG. 15.13

Use of the linear sweep for a pulse-type wave form.

time base of the sawtooth sweep is adjusted even shorter, the amount of the signal displayed will be an even shorter portion. This is an effective *magnification* of the signal, since the full screen width can now display a smaller portion of the input signal. For pulse-type computer signals such magnification is extremely important. On the other hand it may also be necessary to view more than one full cycle of the input signal. In this case the sweep speed is made slower, so that it takes longer for the sweep beam to be deflected once across the screen, thereby allowing a number of cycles of the input signal to be displayed. Figure 15.15 shows the resulting display when the sweep signal takes 16 msec and the input signal only 4 msec for one full cycle. In this case the horizontal sweep speed is 4 times slower and the display shown in 4 cycles of the input signal.

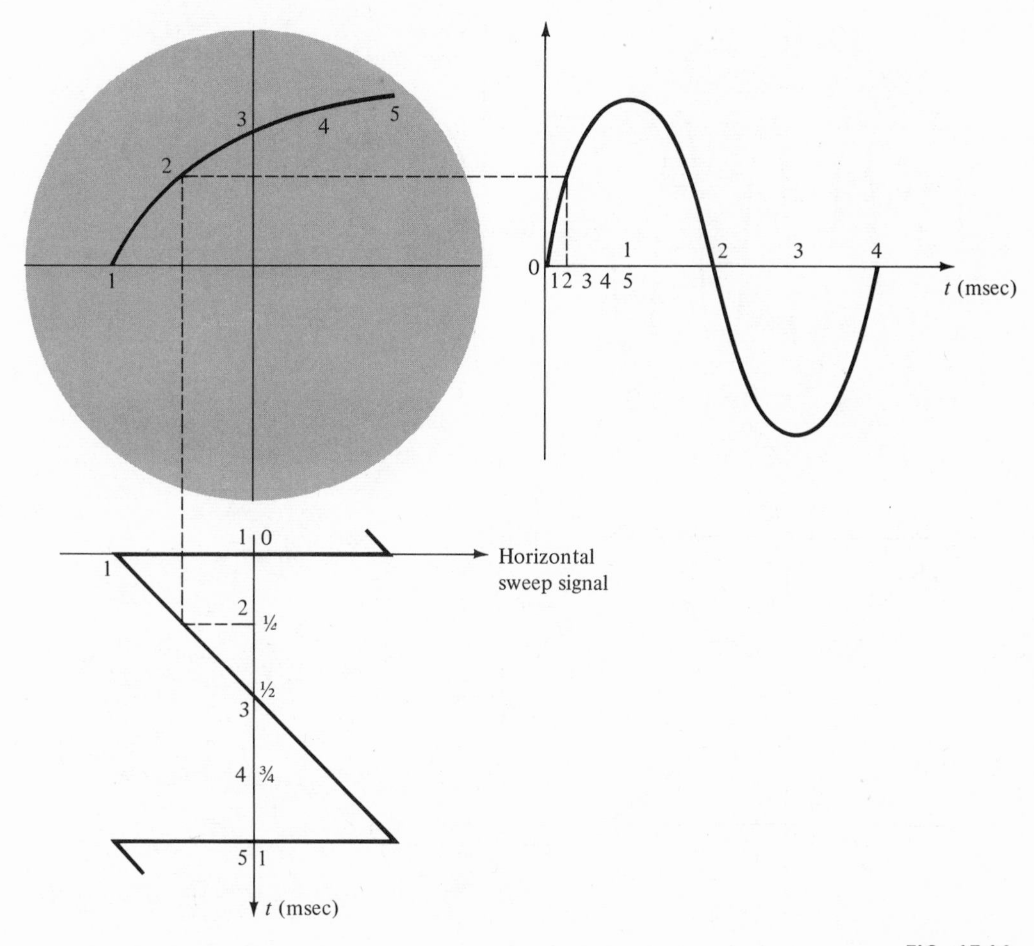

FIG. 15.14
Time base at higher frequency than input signal—magnification of displayed signal.

EXAMPLE 15.4 A sinusoidal waveform at a frequency of 4 kHz is applied to the vertical input of a scope. (a) The horizontal sweep speed is set so that a full cycle takes 0.5 msec. Show the resulting display for one sweep of the beam.

(b) Repeat part (a) for a sweep frequency of 8 kHz.

Solution: (a) At a frequency of 4 kHz the time for one full cycle of the input signal is

$$T = \frac{1}{f} = \frac{1}{4 \times 10^3} = 0.25 \times 10^{-3} \text{ sec} = 0.25 \text{ msec}$$

During the sweep time of 0.50 msec the input signal will go through 2 cycles.

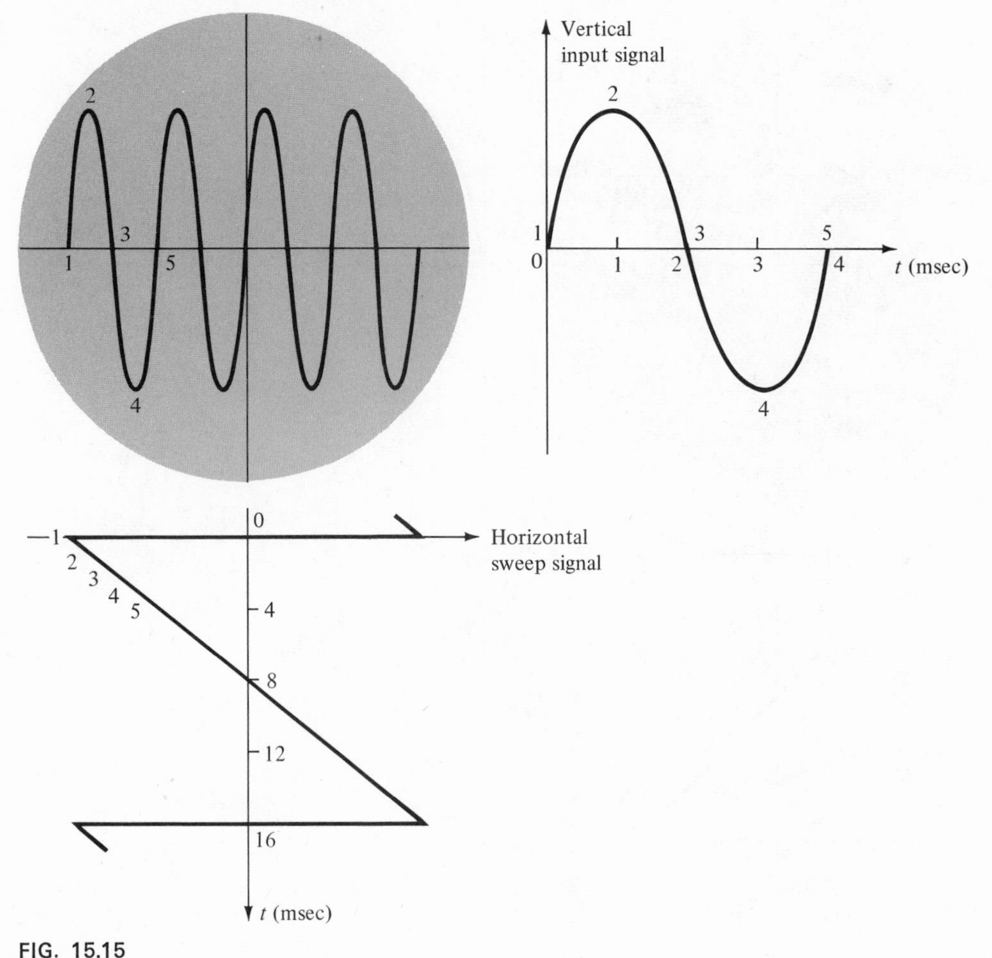

FIG. 15.15

Time base slower than input signal—many cycles viewed in one sweep.

(b) At a sweep speed of 2 kHz the time for 1 cycle of sweep is

$$T = \frac{1}{f} = \frac{1}{8} \times 10^3 = 0.125 \text{ msec}$$

During this amount of time, the input signal will only go through 1 half-cycle so that the display will be only 1 half-cycle of the input signal.

15.5 SYNCHRONIZATION AND TRIGGERING

Synchronization

The CRO display can be adjusted by setting the sweep speed to display either a number of cycles, 1 cycle, or part of a cycle. This is a very valuable feature of the CRO and helps make it the useful instrument it is. However, in discussing the sweep of the beam for a single cycle, we have only considered the case in which the input signal and sweep signal frequencies are related. More generally the horizontal sweep frequency setting is not the same, or even proportional to, the frequency of the input signal. When this occurs the display is not synchronized and either appears to drift or is not recognizable.

Figure 15.16 shows the resulting display for a number of cycles of the sweep signal. Each time the horizontal sawtooth voltage goes through a linear sweep cycle (from maximum negative to zero to maximum positive voltage), the electron beam is caused to move once horizontally across the face of the tube. The sawtooth voltage then drops very quickly back to the negative starting voltage and the electron beam is suddenly caused to move back to the left side of the screen. In most CROs the electron beam is *blanked* during this *retrace* of the beam so that no line is shown on the screen. After the very short retrace time the beam begins another sweep across the tube. If the input voltage is not the same each time a new sweep begins the same display will not be seen each time. To see a steady display it is necessary that the input

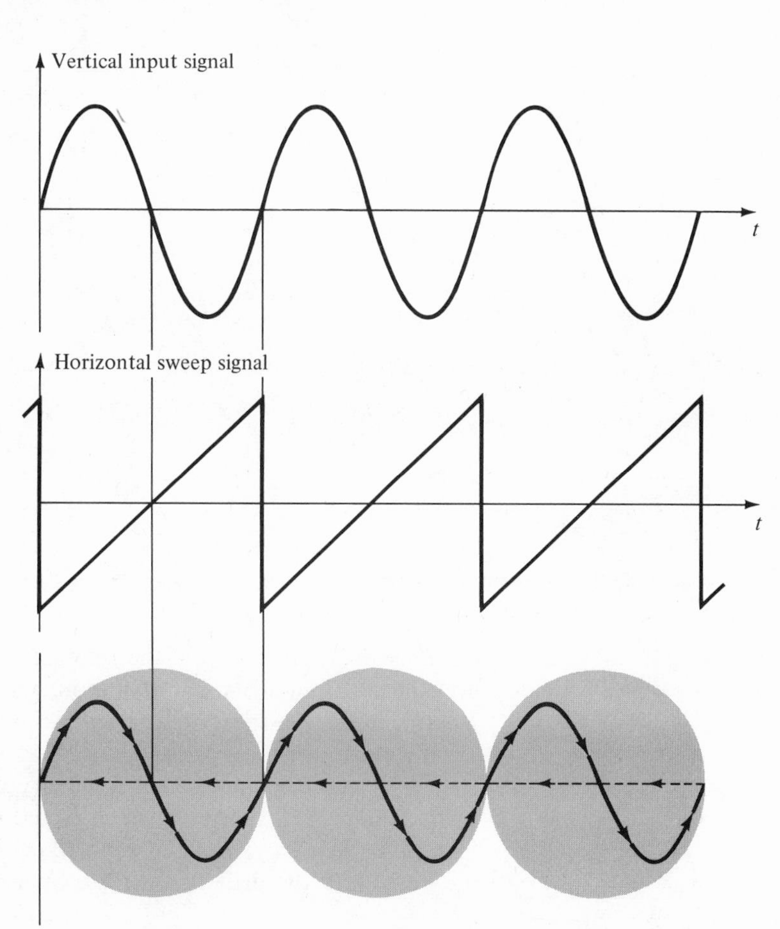

FIG. 15.16

Steady scope display—input and sweep signals synchronized.

signal repeat exactly the same pattern for each sweep of the beam. In Fig. 15.17 the sweep signal frequency is too low and the CRO display will have an apparent "drift" to the left. Actually a different display is formed each sweep of the beam as shown in Fig. 15.17. When viewed on the CROs such a display continuously drifts to the left. Observe

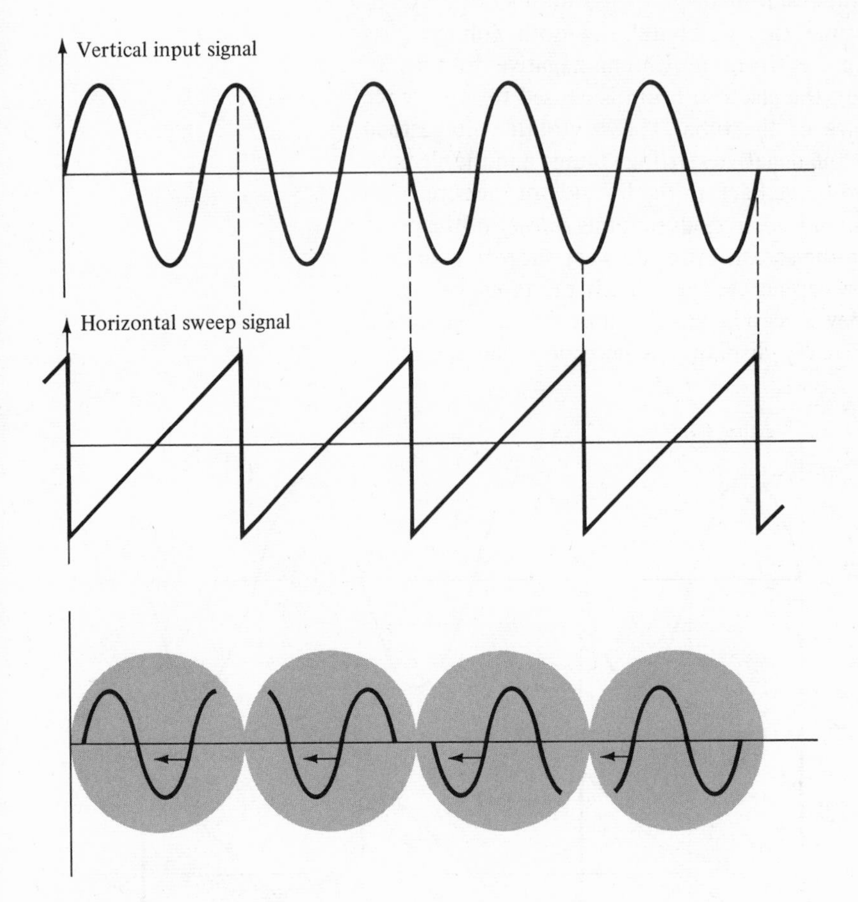

FIG. 15.17
Sweep frequency too *low*—apparent drift to left.

carefully that each sweep of the beam starts at a different point in the cycle of the input signal and that a different display is formed. Adjustment of the sweep speed to a faster sweep time synchronizes or brings the display to a standstill—assuming that neither the sweep generator frequency nor the input signal frequency changes.

Figure 15.18 shows the result of setting the sweep frequency too high. Less than 1 cycle of the input signal is viewed by each sweep of the beam and the drift in this case appears to be toward the right.

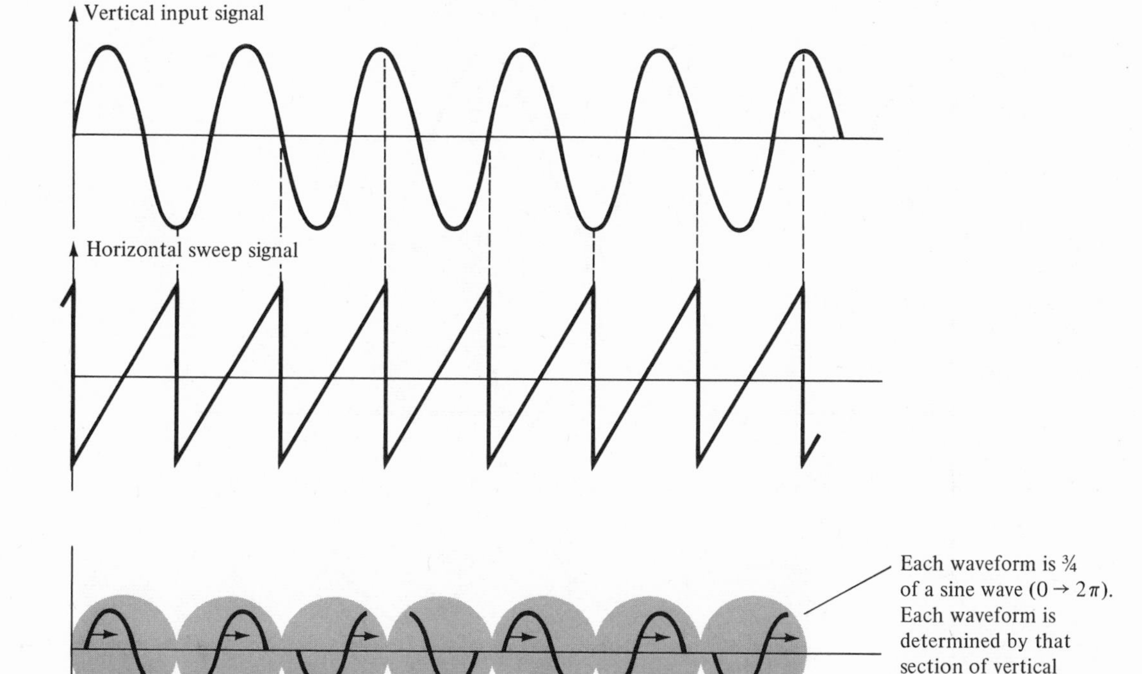

FIG. 15.18
Sweep frequency too *high*—apparent drift to right.

Triggering

The usual method of synchronizing the input signal uses a portion of the input signal to *trigger* the sweep generator so that the rate of the sweep signal is locked or synchronized to the input signal. This is easily done in most CROs using the INTERNAL sync. Figure 15.19 shows that portion of the control panel of a CRO indicating the trigger and sync inputs and controls. We shall refer to these during the following discussion.

FIG. 15.19
Scope sync and trigger controls.

When INTERNAL sync is used, a portion of the vertical input signal is taken from some point in the vertical amplifier circuit and fed as the trigger input signal to the synchronizing circuit section of the horizontal sweep generator. Using triggered sweep, the start of a horizontal linear sweep voltage does not begin immediately after the end of the retrace time (as previously considered) but only when the triggering signal occurs. Thus, the sweep occurs not at a repetitive rate set by the cycle time of the sawtooth signal but by the cycle time of the triggering signal. Using the same triggering signal as that viewed on the vertical

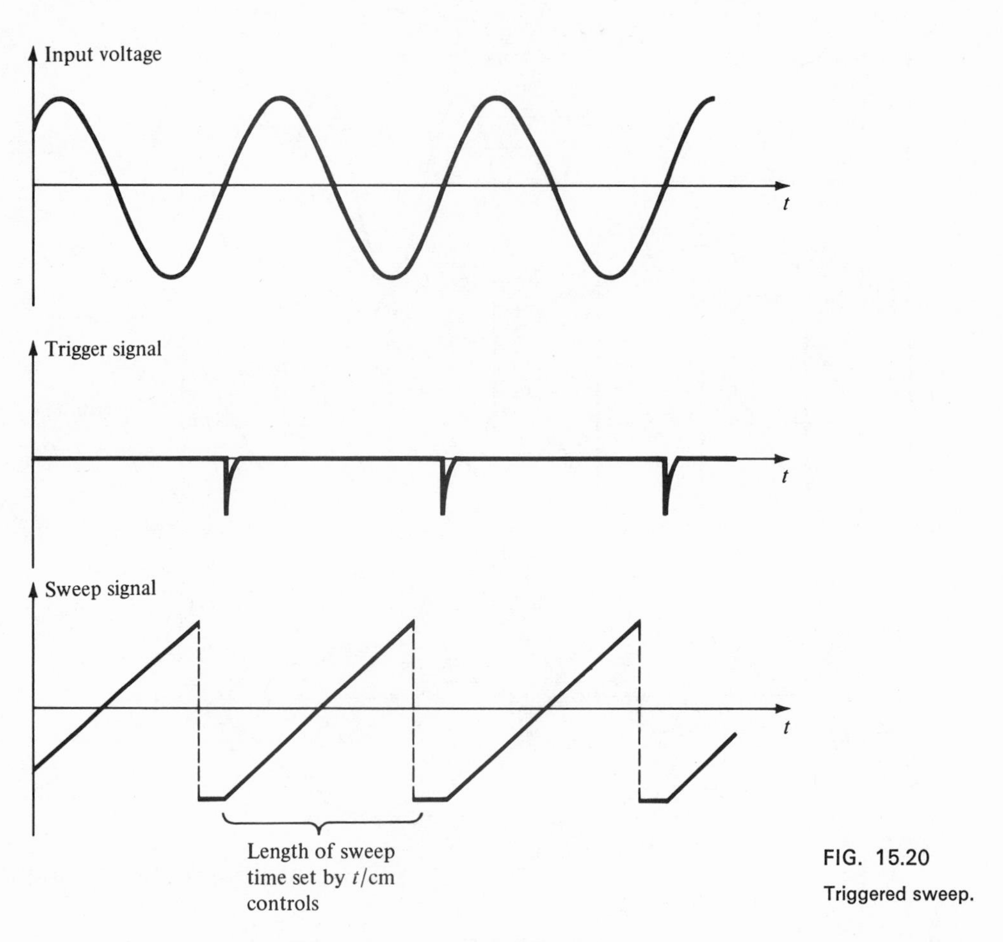

FIG. 15.20
Triggered sweep.

input achieves the desired synchronization without any necessary sweep speed adjustment. Figure 15.20 shows the operation of a few cycles of the sweep signal and the triggering of the sweep generator. Since the trigger signal shown in Fig. 15.20 occurs at the beginning of each sinusoidal cycle, the display starts only when the input voltage at the beginning of a positive-going cycle triggers the sweep generator, and a steady picture is thereby obtained.

The exact triggering point at which the sweep begins can be adjusted using front panel controls (refer back to Fig. 15.19). Setting the trigger control to *INT* +, for example, means that the triggering signal to the sweep circuits will be obtained when the vertical input to the CRO has positive slope (voltage getting more positive with time). Using the position marked *INT* −, the sweep will be started during the time the input signal is going more negative (having negative slope). Figure 15.21 shows the display of the same sinusoidal signal for INT + and INT − trigger settings, respectively. In addition to these two control settings the triggering *level* may also be adjusted. Setting the level to zero results in the display's starting when the input signal level crosses 0 V. This level can be adjusted so that the sweep is triggered to start at *any* point of voltage during either the positive slope or negative slope part of the cycle. Using both trigger level and trigger slope on plus (+) or minus (−) allows a wide range of trigger time adjustment for synchronization of a given waveform.

The controls shown in Fig. 15.19 also allow two other sync modes of operation. The *LINE* sync provides triggering at the line frequency rate (60 Hz) for measurements of signals derived from the main power line. An equally important sync mode is the *EXT* (external) mode of synchronization. A completely separate signal from that applied to the vertical input can be applied to the input terminals marked EXT. This external input signal is then used to trigger the CRO. In many applications a signal taken from a point in the circuit being tested is used as the external sync signal. Thus, any measurements made using the CRO at any other point in the same circuit will be synced by a signal having the same frequency rate. More important, these other signals may be out of phase with the sync signal and this relative phase difference will be both displayed and *measurable* using the CRO. The use of the scope both to display and allow measurement of phase difference (or more generally of time displacement) between signals is quite important and will be covered in detail in Section 15.5.

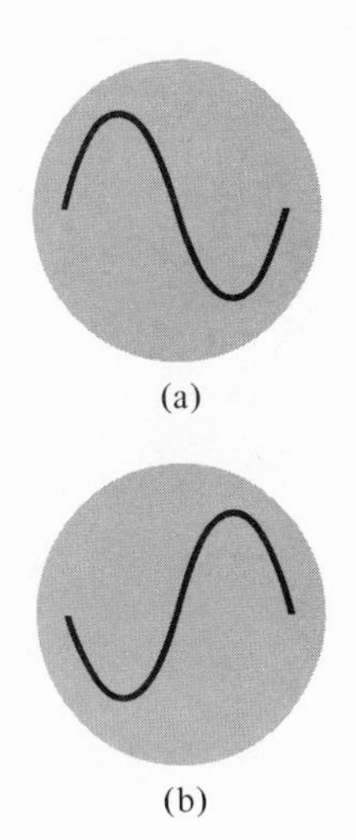

FIG. 15.21
Use of internal sync modes: (a) trigger set to INT+; (b) trigger set to INT−.

BASIC SYNCHRONIZATION AND TRIGGERING CIRCUIT

Synchronization of the CRO deflection is obtained by triggering the start of a new sweep cycle using the signal to be displayed (or some other signal derived from it) as the trigger signal. Figure 15.22 shows a simple block diagram of how the trigger signal and sweep signal are related. Either a part of the vertical input signal to be viewed (INT) or a separate external input signal (EXT) is connected as the trigger input signal. The *coupling* to the trigger circuit (ac or dc), the *slope* of the signal selected (+ or −) and the *level* at which the trigger signal is set to operate are all adjustable using the CRO external controls (see Fig. 15.23). The trigger signal so determined is then used to trigger the gate generator circuit. This causes the gate circuit to begin a timing operation providing an output pulse, which begins when the trigger pulse is received and ends after the amount of time for a single sweep of the beam, which is determined by the external time base settings (e.g., 1 msec, 10 μsec, etc.). A sweep voltage used to drive the horizontal deflection plates is derived from the pulse gate generator signal using an integrator circuit.

Figure 15.24 shows some detail of a simple sweep circuit to provide an indication of how a linear sweep may be obtained. The RC circuit (Fig. 15.24a) provides the integrator operation. As shown, the sweep voltage is obtained using a part of the charge cycle of the capacitor.

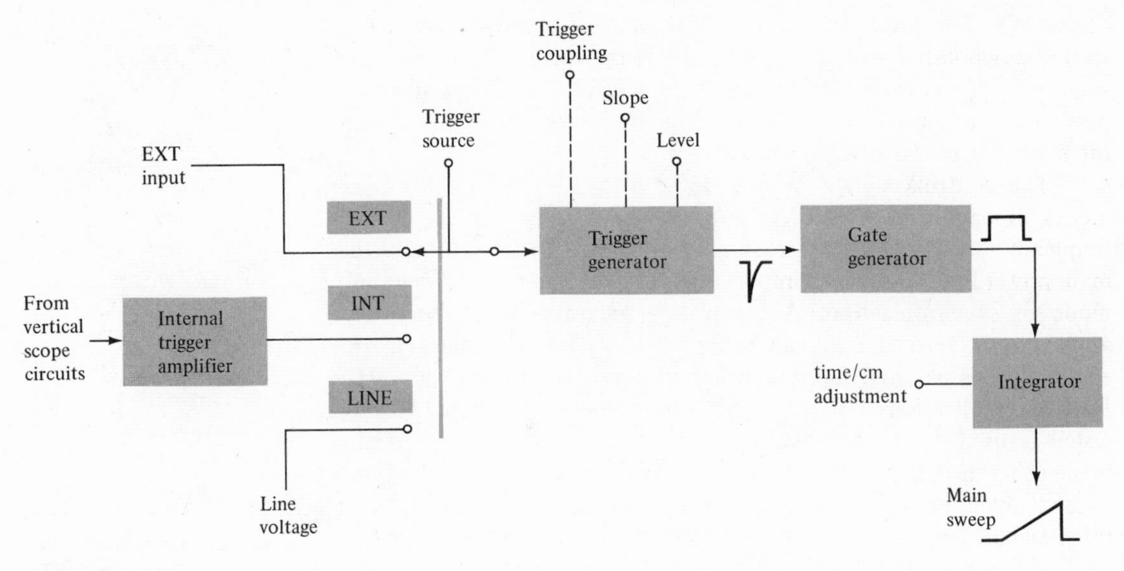

FIG. 15.22

Block diagram showing trigger operation of scope.

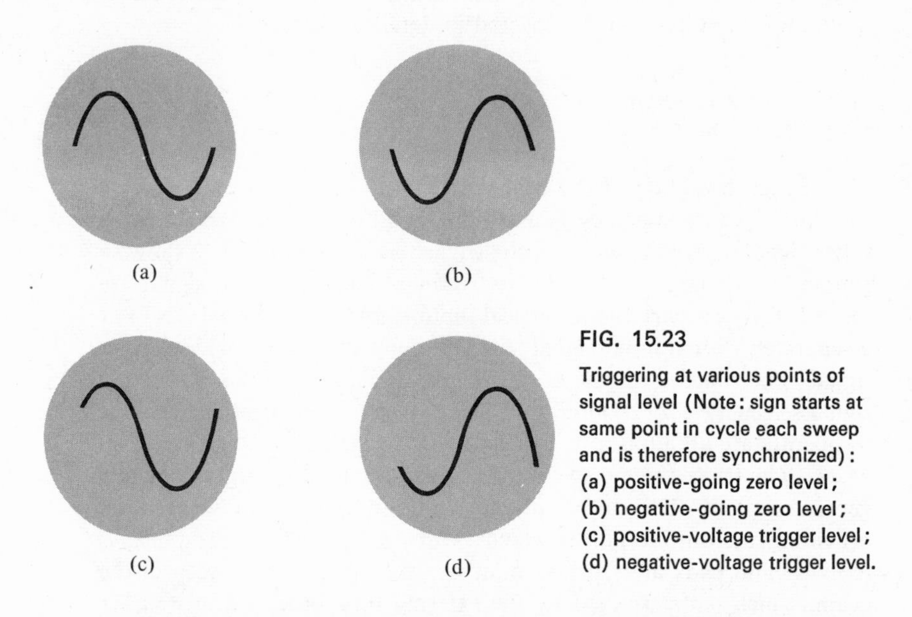

FIG. 15.23

Triggering at various points of signal level (Note: sign starts at same point in cycle each sweep and is therefore synchronized): (a) positive-going zero level; (b) negative-going zero level; (c) positive-voltage trigger level; (d) negative-voltage trigger level.

Although the capacitor charges exponentially at a time rate set by the values of R and C, using only the lower part of the charging voltage waveform provides a fairly linear sweep voltage. Even better linearity is obtained charging a capacitor using a constant-current source, as shown in Fig. 15.24b. Thus, we see that the sweep voltage required to drive the horizontal deflection plates is obtained using some kind of integrator circuit that is triggered by a pulse derived from the input

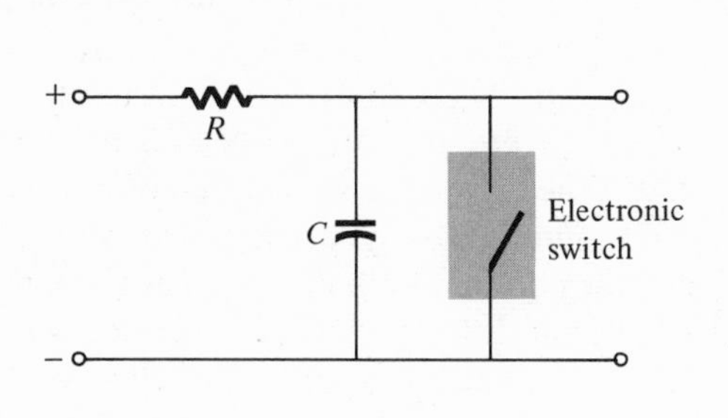

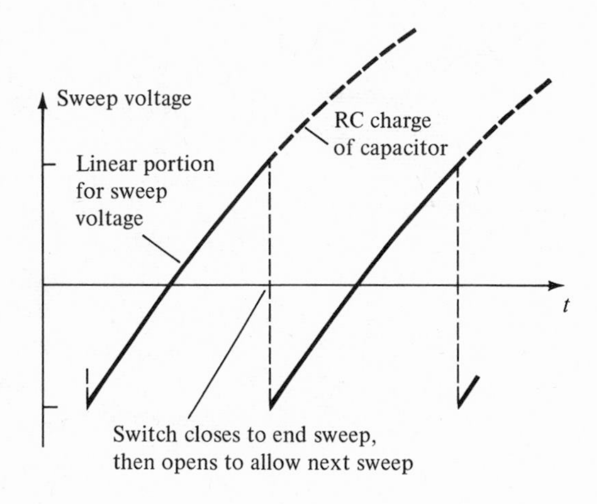

(a)

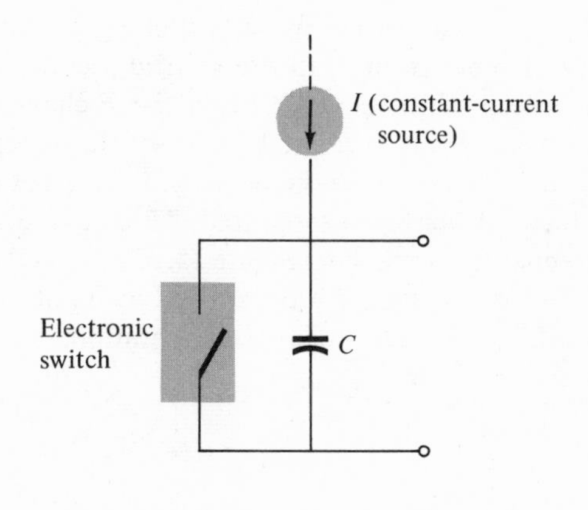

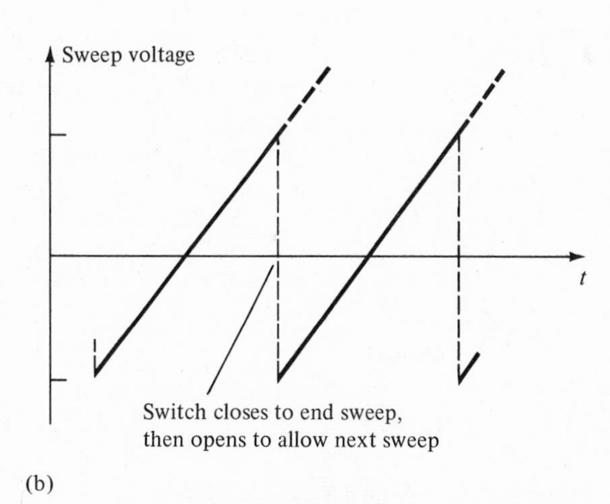

(b)

FIG. 15.24

Linear sweep voltage circuits: (a) part of an RC exponential charge used as a linear sweep; (b) use of constant-current charging of a capacitor to obtain a linear sweep voltage.

vertical signal to be viewed. This provides the necessary synchronization required to obtain a steady display on the CRT screen.

MULTITRACE OPERATION

A useful CRT display is sometimes obtained by using a type of CRO that will provide display of more than one waveform at the same time. This can be done using a multielement CRT gun (more than one electron gun and more than one beam). A CRO using a dual-beam CRT is called a *dual-beam* CRO. It can also be obtained using a single-electron gun and some external electronic switching to obtain the necessary multidisplay on the screen; The CRO is then referred to as *dual-trace*. Two dual-trace features are the ALTERNATE and the CHOPPED modes of display. When using these features of a CRO it is possible to connect and simultaneously display two separate signals via the CRO. Actually, they are not both displayed at the same time since there is only a single beam. It is possible, however, to switch fast enough so that the illusion of two images appearing simultaneously on the screen is produced.

Using the ALTERNATE mode of electronic switching, the two input signals to be viewed, connected to channels A and B of the scope (to differentiate between the two vertical input signals), are connected to the vertical channel of the scope for alternate cycles of the sweep. Thus, on 1 cycle of the sweep the input to channel A is connected to the vertical section of the scope and drives the vertical deflection plates. After 1 cycle of sweep is completed the electronic switching circuits (see Fig. 15.25) connect the channel B input to the vertical section of the scope and on the next sweep of the electron beam the B-channel signal is displayed on the screen. If (as is usually the case) the sweep times are fast enough, the alternate sweep of the beam will trace out a second display *before* the first display has disappeared. Thus, two displays will appear to be present at the same time. Figure 15.26a shows the resulting display for a sine wave and square wave applied as inputs to the CRO when using the ALTERNATE mode of presentation.

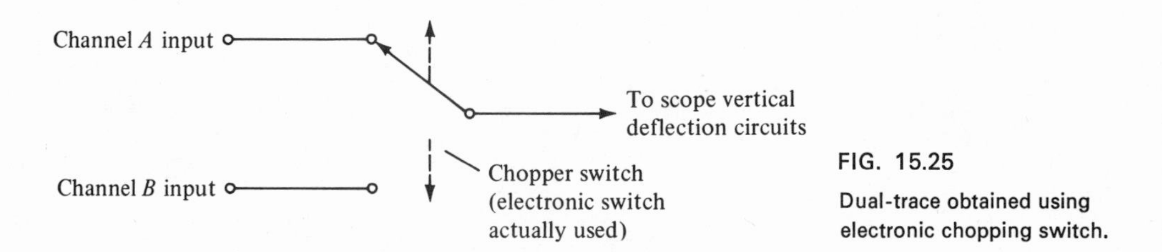

FIG. 15.25
Dual-trace obtained using electronic chopping switch.

When the signals to be viewed are of low frequency a CHOPPED mode of display is used. The CHOPPED mode of electronic switching switches the vertical amplifier input signal from the input connected at channel A to the signal at channel B and back again, repetitively—many times—for a single sweep of the beam. Thus, the display obtained is actually composed of small pieces of each signal with enough of these small pieces to provide the illusion of two steady display signals. Figure 15.26b shows the resulting display of two signals using the CHOPPED mode of electronic switching.

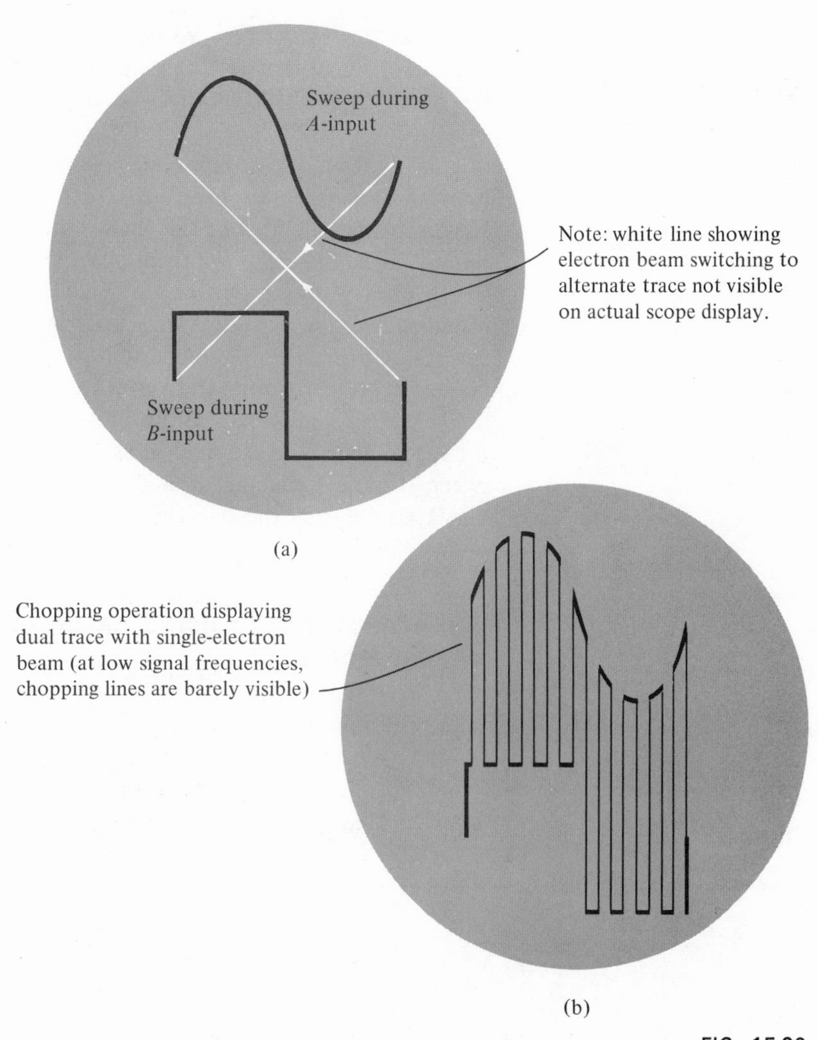

FIG. 15.26

ALTERNATE and CHOPPED mode displays for dual-trace operation: (a) ALTERNATE mode for dual-trace using single electron beam; (b) CHOPPED mode for dual-trace using single electron beam.

An important point to keep in mind in regard to the two modes of display is the triggering operation for each mode. In CHOPPED operation a single trigger signal starts each sweep, which then provides two displays on the screen. Whatever phase displacement exists between the two signals will be displayed *exactly,* since the two inputs are being shown, in time, as they occur. The ALTERNATE mode of display, however, may provide some problem. For example the INTERNAL mode of trigger is used with ALTERNATE sweep and, say, the trigger setting is the zero voltage level with positive slope. When the channel *A* signal crosses 0 V, with positive slope, a trigger signal starts the sweep. The ALTERNATE sweep will be when the channel *B* signal crosses 0 V with positive slope—*whenever* that occurs. The display will thus be that of two in-phase signals and the true phase displacement between the signals is *not* shown. To get around this difficulty, EXTERNAL sync may be used so that each sweep is started by the same input signal and any phase offset from *that* reference signal will be observed. Using one of the inputs as the EXTERNAL sync signal reference provides a display of the two signals in proper phase relation. Some newer multitrace CROs provide additional triggering for ALTERNATE mode display using the same input from, say, channel *B* as the trigger for initiating the sweep of both input signals. The ALTERNATE *B*-trigger mode, for example, uses the *B* input as the trigger for *each* sweep (for both channel *A* and channel *B* display). Any phase shift displayed between the two signals will then be properly shown.

15.6 MEASUREMENTS USING CALIBRATED CRO SCALES

The CRT face has two display axes—vertical and horizontal. In the normal operation of the CRO the input signal to be observed is applied to the vertical input (either single-channel or dual-channel) and the horizontal sweep is obtained using internal sweep circuitry (see Fig.

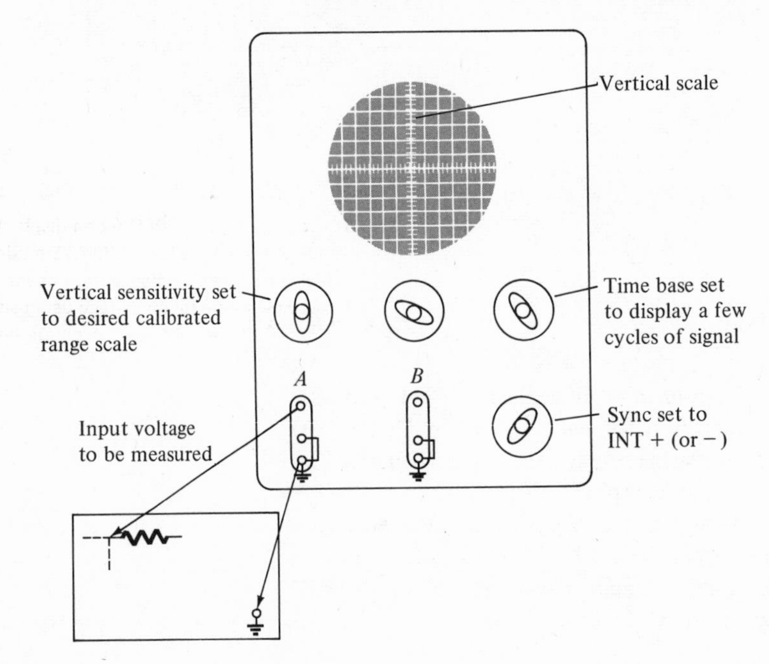

FIG. 15.27
Connections to measure voltage amplitude using the scope.

15.27). If the vertical amplifiers and attenuators are calibrated (or set to the calibrated settings), then the *amplitude* of the vertical (input) signal (or any part of it) can be accurately *measured.*

Using the calibrated horizontal sweep scales of the CRO the amount of time for 1 cycle of the input signal (the *period*) may be measured and used to calculate the signal frequency. In addition, the amount of time between two sinusoidal signals crossing 0 V can be read and used to calculate the phase shift between the two signals. It is also possible to use the horizontal scales to measure the amount of time that separates the two signals being observed. Measurements of this type provide important information in pulse and digital circuitry.

Amplitude Measurements

The vertical scale of the scope is generally calibrated in units of volts per centimeter (V/cm). Figure 15.28a shows the typical CRT screen of a CRO with the vertical scale marked off in centimeters (cm). (The centimeter is a full box as indicated and the scale is sometimes referred to as volts per box.) Each centimeter or box is further subdivided into five parts so that each minor division mark represents 0.2 cm.

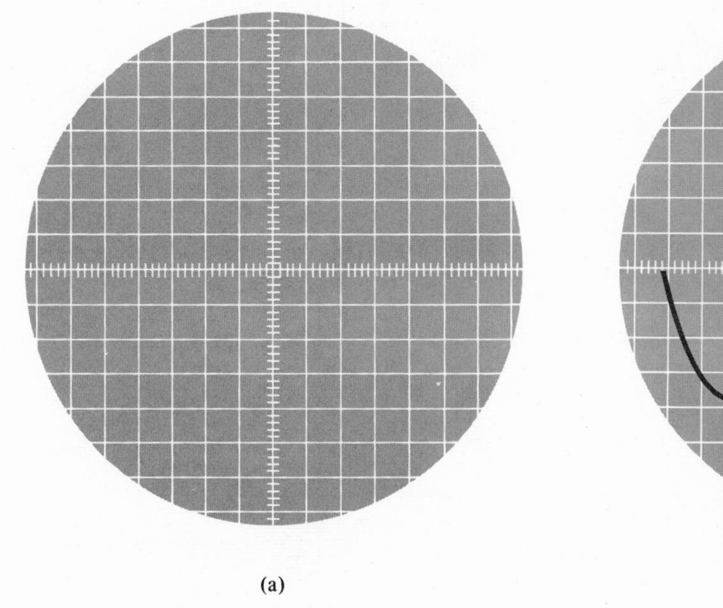
(a)

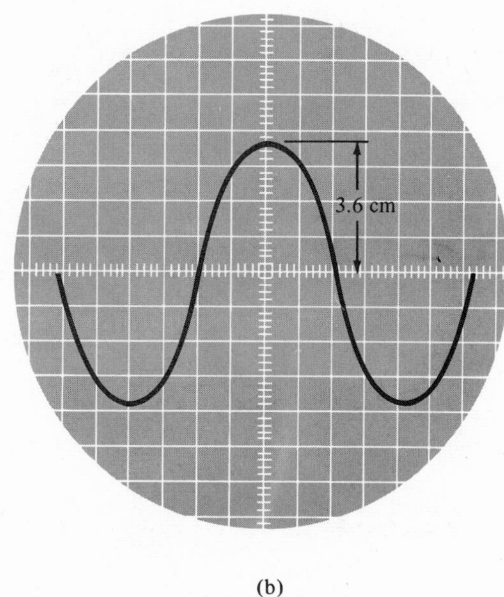

(b)

FIG. 15.28

Scope scale for amplitude measurement: (a) scope scales; (b) measurement of the peak amplitude of a sinusoidal waveform.

As an example of voltage amplitude measurement using the scope calibrated scale, observe the waveform in Fig. 15.28b. The peak amplitude of the sinusoidal waveform shown can be read off the CRT screen as 3.6 cm. If the scale selected by the vertical sensitivity dial setting is 1 V/cm this reading would represent $3.6\ \text{cm} \times 1\ \text{V/cm} = 3.6\ \text{V}$. If the vertical sensitivity dial setting were 0.1 V/cm then the voltage amplitude would be $3.6\ \text{cm} \times 0.1\ \text{V/cm} = 0.36\ \text{V}$.

The vertical and horizontal *position* of the displayed waveform may be adjusted without affecting the amplitude value. *It is important, however, that the vernier part of the sensitivity dial be set to the calibrated* (CAL) *position*. In the CRO shown in Fig. 15.29, the calibrated position of the vertical sensitivity dial is the fully *clockwise* position of the dial. For the measurement considered above, the peak value of the waveform was measured with respect to the center line of the CRT screen. For this to correspond to the center of the waveform, the vertical position of the beam would have had to be centered previously. When centering is *not* desired, the measurement of peak-to-peak amplitude is a more reasonable and accurate choice. The pulse-type waveform shown in

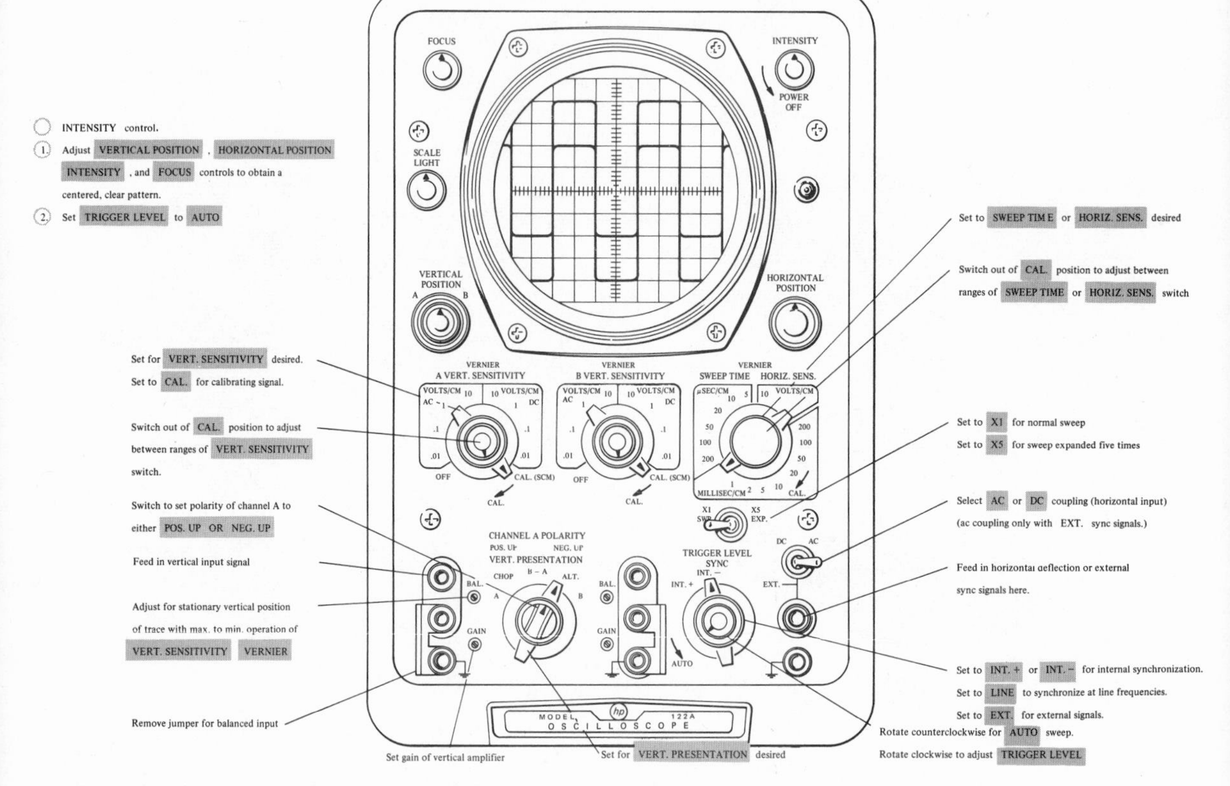

FIG. 15.29
HP 122A scope showing front panel controls.

Fig. 15.30a has a peak-to-peak amplitude of 4.8 cm so that a dial setting of 1 V/cm would indicate a voltage amplitude (peak-to-peak) of 4.8 V/cm × 1 V/cm = 4.8 V.

Figure 15.30b shows another pulse-type waveform. To find the amplitude (voltage) difference between the peaks of the two pulses shown the measurement is

$$\begin{aligned} \text{1st peak} &= 3.5 \text{ cm} \\ \text{2nd peak} &= 2.4 \text{ cm} \\ \text{pulse amplitude difference} &= 3.5 - 2.4 = 1.1 \text{ cm} \end{aligned}$$

If scale setting were 10 V/cm this would correspond to a voltage difference of 1.1 cm × 10 V/cm = 11 V.

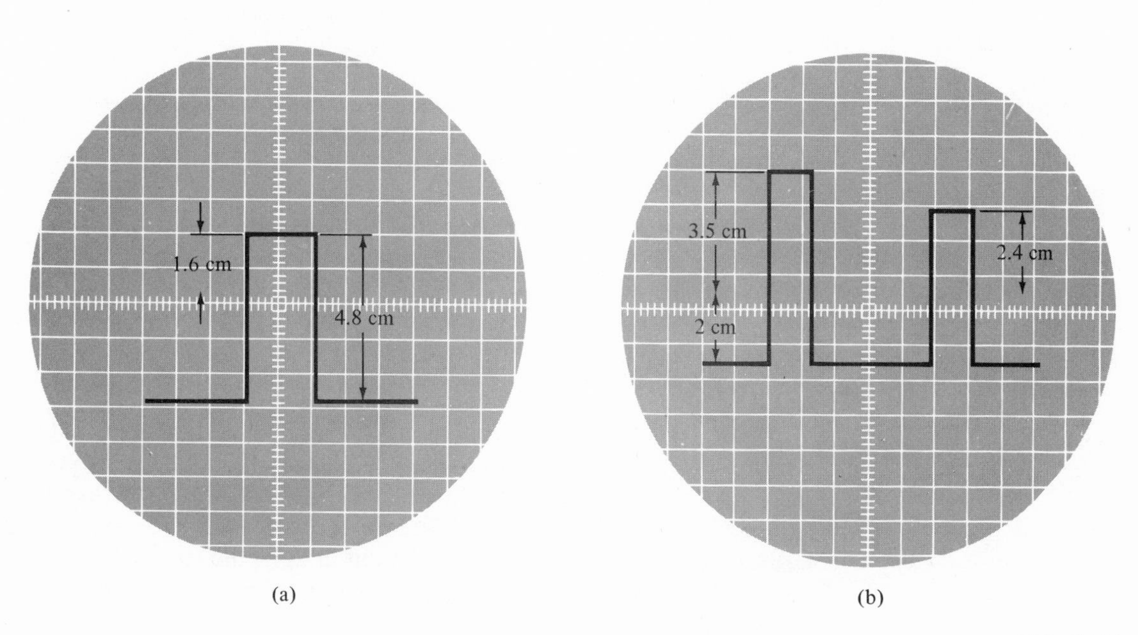

(a) (b)

FIG. 15.30
Measurement of pulse-type waveform amplitudes.

EXAMPLE 15.5 A sinusoidal waveform is observed on a scope as having a peak-to-peak amplitude of 6.4 cm. If the CRO vertical sensitivity setting is 5 V/cm, calculate the peak-to-peak and the rms values of the voltage.

Solution: The peak-to-peak amplitude is

$$6.4 \text{ cm} \times 5 \text{ V/cm} = 32 \text{ V} = V_{p-p}$$

$$V_p = \frac{V_{p-p}}{2} = \frac{32 \text{ V}}{2} = 16 \text{ V}$$

The rms value of voltage is

$$0.707 V_p = 0.707(16) = \mathbf{11.3\ V}$$

EXAMPLE 15.6 How many centimeters (peak-to-peak) of a sinusoidal waveform should be observed corresponding to an rms voltage of 120 V for a scale setting of 50 V/cm?

Solution: Converting the rms voltage to peak-to-peak we get

$$2 \times (120 \times 1.414) = 339.4\ \text{V} \cong 340\ \text{V}, \qquad \text{peak-to-peak}$$

The number of centimeters corresponding to this would be

$$\frac{1\ \text{cm}}{50\ \text{V}} \times 340\ \text{V} = \mathbf{6.8\ cm}, \qquad \text{or 6.8 major divisions (boxes)}$$

EXAMPLE 15.7 A pulse-type waveform is measured as having a peak amplitude of 15 V. If the vertical dial setting was 10 V/cm, how many centimeters of signal amplitude were observed (peak-to-peak)?

Solution: For a peak reading of 15 V, the number of centimeters is

$$15\ \text{V} \times \frac{1\ \text{cm}}{10\ \text{V}} = 1.5\ \text{cm}$$

The peak-to-peak amplitude is 2(1.5 cm) = **3 cm.**

The vertical sensitivity dial shown in Fig. 15.29 indicates separate positions for ac and dc readings. (On many other quality CROs there is one set of sensitivity scales with a separate switch to change from ac to dc operation.) The difference between these two modes of measurement is quite simple but important in using the scope properly. The *dc input* results in the displayed waveform showing the dc level of the signal being measured. If, for example, the signal to be measured is a 2 V, peak-to-peak sinusoidal voltage riding on a 3 V dc level as shown in Fig. 15.31a (assuming the scope position controls were previously centered), the display indicates the presence of dc. We are then able to measure not only the ac variation of the signal, but the exact dc levels at all parts of the signal, as shown in Fig. 15.31a.

When only the ac variation is of interest, then the ac scale setting may be used. Using the ac input setting the same input signal shown in Fig. 15.31a is displayed in Fig. 15.31b. Notice that the dc level has been removed and only the ac variation is shown. Essentially the difference between the two is that the ac input scale position couples the signal through a capacitor to eliminate the dc level of the input signal and provide only the ac variation for measurement.

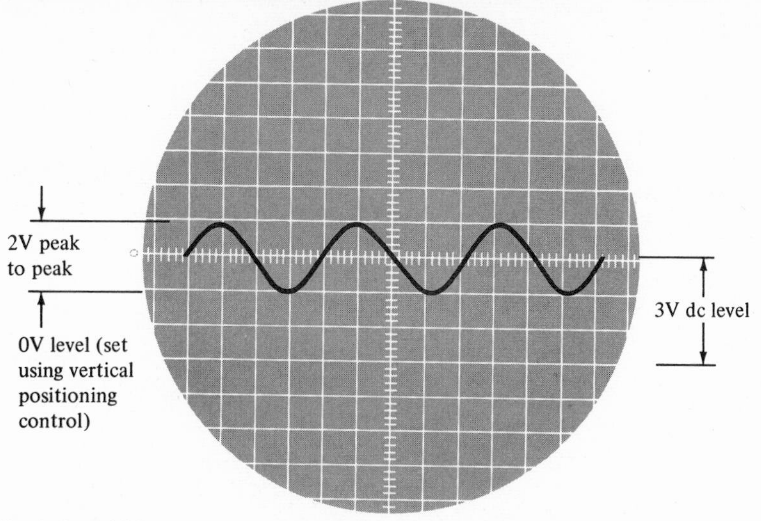

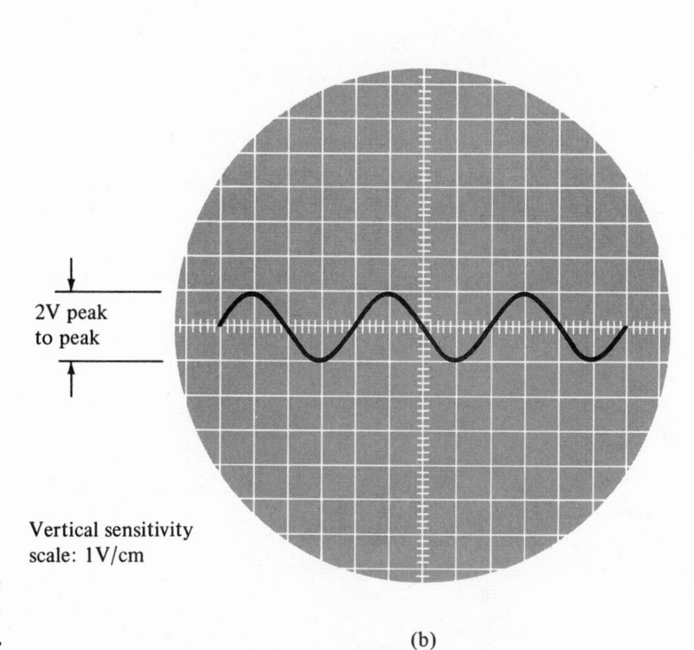

FIG. 15.31
Use of dc and ac input modes: (a) dc input mode; (b) ac input mode.

A practical example of the advantage of the ac over the dc setting is in the measurement of a signal having, say, a dc level of 80 V and an ac variation around this level of only 4 V. Displaying this, using the dc scale setting of 20 V/cm, for example, results in the scope display of Fig. 15.32a. Notice how small the ac variation is compared to the large dc level of 80 V. (If a smaller scale setting, of, say, 2 V/cm were used the ac signal would not be seen at all since a deflection for 80 V would be well off the face of the screen.) Using the ac input scale setting of, say, 1 V/cm provides the display of Fig. 15.32b, where only the ac part of the signal is shown and is expanded to a reasonable viewing and measuring amplitude.

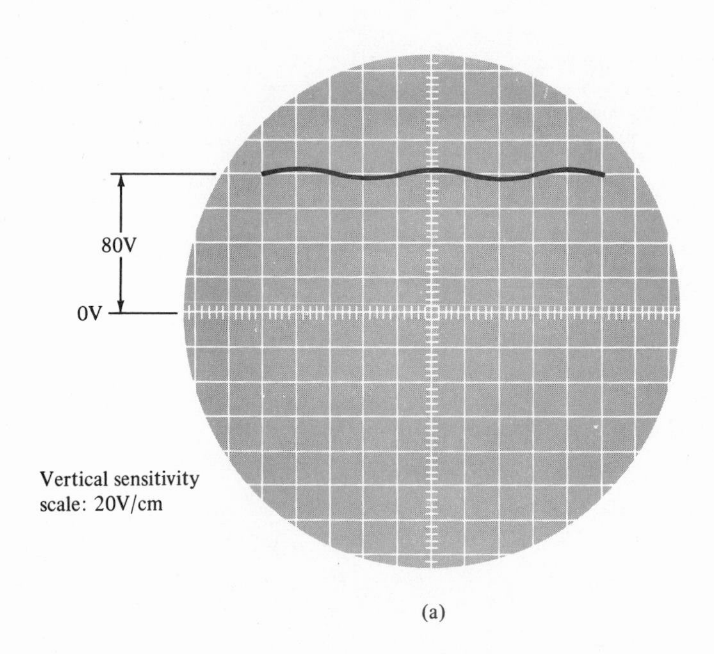

(a)

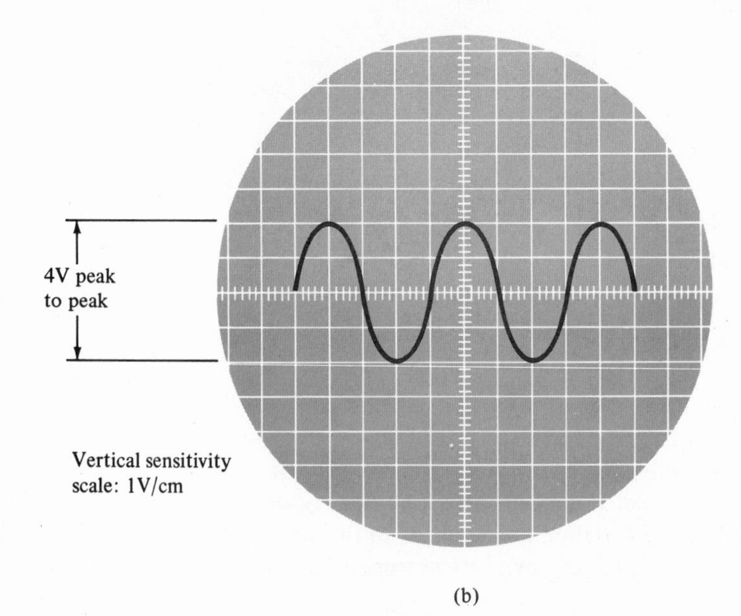

(b)

FIG. 15.32

Using ac mode to view only signal variation: (a) small ac variation around large dc level shown using dc mode for input; (b) observing *only* ac variation of input signal using ac mode for input.

Although the ac scales are good for observing the ac part of a signal having a dc level, the dc scale setting is still important when dc levels must be measured. One additional feature is often found in a good scope—a zero voltage or GROUND position, which connects the input of the vertical amplifier to 0 V without requiring the input signal connection to be removed.

USING CALIBRATED SWEEP FOR TIME MEASUREMENTS

The horizontal sweep signal can be adjusted in calibrated steps from a few seconds to microseconds of time per centimeter. If the sweep time selector were set at 1 msec/cm each box (or centimeter) on the screen would correspond to a time of 1 msec. If a pulse signal, such as the one in Fig. 15.33, were observed to have a pulse width of 2.6 cm at a scale setting of 50 μsec/cm the pulse width could be calculated

$$2.6 \text{ cm} \times 50 \ \mu\text{sec/cm} = 130 \ \mu\text{sec}.$$

Thus, the CRO allows display of waveforms of all shapes and permits measurements of time so that all aspects of the signal observed can be measured accurately.

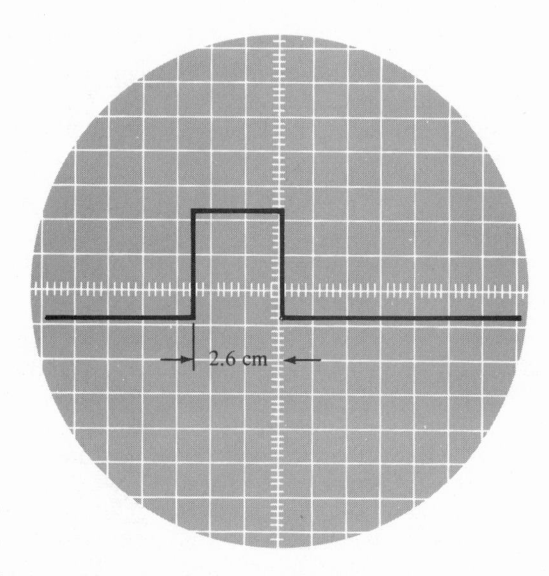

FIG. 15.33
Pulse waveform for time measurement.

With the dual-trace feature of the CRO two different waveforms can be observed simultaneously and any time differences between parts of the respective signals may be observed and measured. For example, the waveforms of Fig. 15.34 show two pulses viewed at once (using, say, external sync to preserve their proper time displacement). Using the calibrated horizontal sweep scales we can measure, for ex-

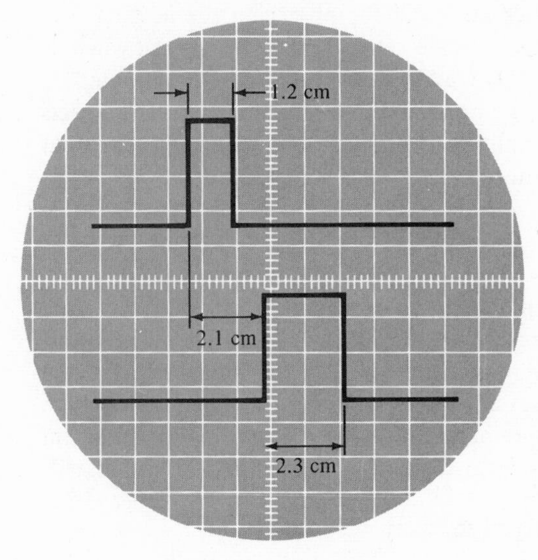

FIG. 15.34
Pulse waveforms showing time measurement using calibrated sweep scales.

ample, the time for each pulse and the time between the two pulses. From the waveforms shown, the pulse widths are calculated as

$$1.2 \text{ cm} \times 20 \ \mu\text{sec/cm} = 24 \ \mu\text{sec}$$

$$2.3 \text{ cm} \times 20 \ \mu\text{sec/cm} = 46 \ \mu\text{sec}$$

The time from the start of the first pulse until the start of the second is calculated as

$$2.1 \text{ cm} \times 20 \ \mu\text{sec/cm} = 42 \ \mu\text{sec}$$

Using the scope in this way provides all information about the two waveforms and their relation to each other.

EXAMPLE 15.8 Two pulse-type signals are observed on a CRO using a sweep scale setting of 5 msec/cm. If the pulse widths are measured as 3.5 and 4.2 cm, respectively, and the distance between the start of each pulse is 2.8 cm, calculate the time measurements for the pulse widths and delay between pulses.

Solution: Pulse width 1 = 3.5 cm × 5 msec/cm = 16.5 msec
Pulse width 2 = 4.2 cm × 5 msec/cm = 21 msec
Delay time = 2.8 cm × 5 msec/cm = 14 msec

EXAMPLE 15.9 Two pulses delayed by 15 μsec are observed on an oscilloscope using a time base setting of 10 μsec/cm. Both pulses have pulse widths of 2 μsec. Calculate the readings in centimeters on the scope for the pulse width and delay time.

Solution: Pulse delay measurement = 15 μsec/10 μsec/cm = 1.5 cm
Pulse width measurement = 2 μsec/10 μsec/cm = 0.2 cm

Frequency Measurements Using Calibrated Scope Scales

It is also possible using the calibrated time scales of the CRO to calculate the *frequency* of the observed signals. This requires using the calibrated horizontal sweep scale to measure the time for 1 cycle of the observed signal and then calculating the signal frequency using the relation

$$f = 1/T$$

where f is the signal frequency and T is the period or the time for one full cycle of the signal.

EXAMPLE 15.10 A squarewave signal is observed on the scope to have 1 cycle measured as 8 cm at a scale setting of 20 μsec/cm. Calculate the signal frequency.

Solution: Calculating, first, the period for 1 cycle of the observed signal:

$$T = 8 \text{ cm} \times 20\ \mu\text{sec/cm} = 160\ \mu\text{sec}$$

The frequency is then calculated to be

$$f = \frac{1}{T} = \frac{1}{160\ \mu\text{sec}} = \frac{1}{160} \times 10^{-6} = \mathbf{6.7\ kHz}$$

EXAMPLE 15.11 A sinusoidal signal is observed on the scope to repeat 1 cycle in 4.8 cm. If the scale setting was 50 μsec/cm calculate the frequency of the sinusoidal signal.

Solution

$$T = 4.8 \text{ cm} \times 50\ \mu\text{sec/cm} = 240\ \mu\text{sec}$$

$$f = \frac{1}{T} = \frac{1}{240\ \mu\text{sec}} = \mathbf{4.16\ kHz.}$$

EXAMPLE 15.12 A 500-kHz signal is observed on the CRO. How many centimeters should be observed for one full cycle of the signal if the sweep setting is 1 μsec/cm?

Solution: Calculating first the time for 1 cycle of the signal

$$T = \frac{1}{f} = \frac{1}{500 \text{ kHz}} = 2\ \mu\text{sec}$$

Calculating the number of centimeters for a full cycle

$$\text{No. of cm} = 2\ \mu\text{sec} \times \frac{1\ \text{cm}}{1\ \mu\text{sec}} = 2\ \text{cm}$$

Phase-Shift Measurements Using Calibrated Scope Scales

The calibrated time scales can also be used to calculate phase shift between two sinusoidal signals (of the same frequency, of course). If a dual-trace or dual-beam CRO is used to display the two sinusoidal signals simultaneously so that one signal is used for the EXT sync input, the two waveforms will appear in proper time perspective and the CRO can be used to measure the amount of time between the start of 1 cycle of each of the waveforms. This amount of time can then be used to calculate the phase angle between the two signals. Figure 15.35 shows two sinusoidal signals having a phase shift of theta (θ) degrees. We can measure distance in centimeters on the CRO scale and use these readings to obtain θ as follows. The value of the phase angle is related to the degrees in one full cycle of the sinusoidal signal. We can set up a simple relation between these values by equating the number of centimeters or boxes for one full cycle to 360° and the number of centimeters or boxes for the phase shift to the desired phase angle in degrees. The relation is

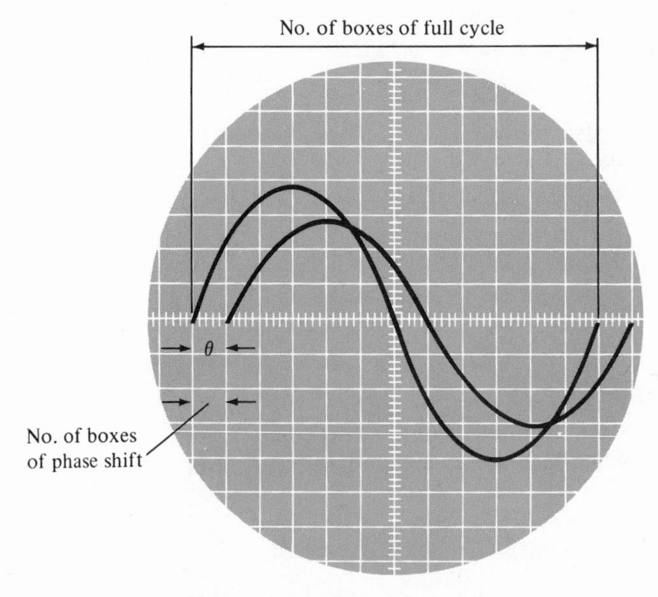

$$\theta = \frac{\text{No. of boxes of phase shift}}{\text{No. of boxes of full cycle}} \times 360^\circ$$

FIG. 15.35

Phase shift measurement using horizontal scope scale.

$$\frac{\text{No. of boxes for one full cycle}}{360^\circ} = \frac{\text{No. of boxes of phase shift}}{\theta} \tag{15.5}$$

where θ is the phase angle (phase shift) in degrees. Using Eq. (15-5) we can calculate the phase angle from

$$\boxed{\theta = \frac{\text{No. of boxes of phase shift}}{\text{No. of boxes of full cycle}} \times 360^\circ} \tag{15.6}$$

Note that the calculation does not involve the actual calibrated time base setting and, in fact, the observed waveform can be varied using the horizontal amplifier vernier adjustment to obtain as many boxes for one full cycle as desired. This will not affect the actual phase shift calculated since the proportionality between the phase shift and one full cycle is preserved at any gain or scale setting used. Adjusting the time base so that, say, 12 boxes (or centimeters) correspond to one full cycle would rescale the measurement so that each box is 30° of phase angle (360°/12). Then, reading 2 boxes of phase shift would quickly convert to 60° phase shift by multiplying the boxes of phase shift by the 30°/box scale factor. If the reading obtained were 1.5 boxes then the phase shift would be 1.5 boxes × 30°/box = 45° phase shift. The relation for calculating phase angle is then

$$\boxed{\theta = \text{scale factor} \times \text{phase distance measured (in boxes or centimeters)}} \tag{15.7}$$

EXAMPLE 15.13 In measuring phase shift between two sinusoidal signals on a CRO, the scale setting is adjusted so that one full cycle is 8 boxes. Calculate the scale factor for this adjustment and the amount of phase shift for a reading of 0.75 boxes.

Solution: Setting one full cycle to 8 boxes results in a scale factor of 360°/8 boxes = 45°/box.

$$\begin{aligned}\theta &= \text{scale factor} \times \text{phase distance measured} \\ &\quad \text{(in boxes or centimeters)} \\ &= \frac{45^\circ}{\text{box}} \times 0.75 \text{ boxes} = \mathbf{33.85^\circ}\end{aligned}$$

EXAMPLE 15.14 One full cycle is set to 9 boxes. The phase displacement is measured as 0.4 boxes. Calculate (a) the scale factor and phase shift in degrees, and (b) the number of boxes to observe for a phase shift of 60°.

Solution: (a) The scale factor set is

$$\text{scale factor} = \frac{360°}{9 \text{ boxes}} = \frac{40°}{\text{box}}$$

$$\text{phase shift} = 0.4 \text{ boxes} \times \frac{40°}{\text{box}} = 16°$$

(b) $$\text{No. of boxes for } 60° \text{ phase shift} = \frac{60°}{40°/\text{box}} = 1.5 \text{ boxes}$$

15.7 USE OF LISSAJOUS FIGURES FOR PHASE AND FREQUENCY MEASUREMENTS

Another method for measuring either phase shift between two signals or the frequency of an unknown signal is the use of Lissajous figures. The technique can be applied to a single-channel CRO and does not require the fine calibration scales previously considered. Basically the two signals under study (to determine the phase shift between the two signals) are connected as vertical and horizontal inputs to the CRO. The usual (internal) horizontal sweep signal is *not* used at this time. A pattern or Lissajous figure is developed on the CRT and is used to determine the amount of the phase shift, or frequency of the unknown signal.

Lissajous figure techniques for measurement are more popular for low-quality CROs and for single-trace scopes (where two inputs cannot be compared at one time). Although not as popular a measurement technique as those considered in Section 15.6, the use of Lissajous figures is still interesting and sometimes helpful.

Use of Lissajous Figures to Calculate Phase Shift

Lissajous figures are obtained on the scope by applying the two inputs to be compared to the vertical and horizontal channels of the oscilloscope. The value of the phase shift is then calculated from measured values taken from the resulting Lissajous pattern.

Figure 15.36 shows a circuit arrangement with two different signals connected to the vertical and horizontal inputs of the CRO, respectively. The resulting pattern on the CRT face is a straight line, a circle, or an ellipse. To understand the relationship between the re-

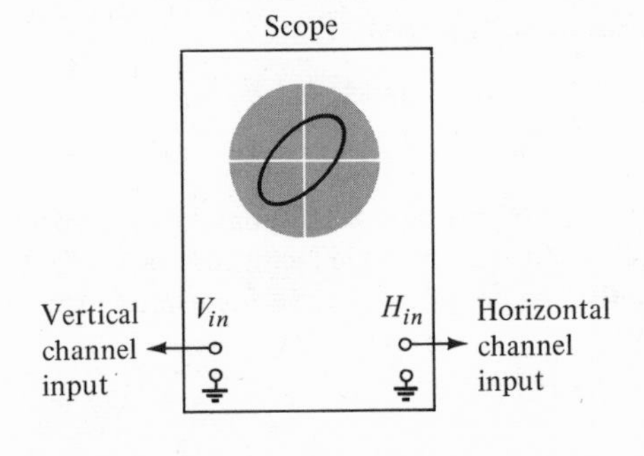

FIG. 15.36
Signal input connection for Lissajous figure measurement of phase angle.

sulting figure and the applied inputs we shall investigate a variety of signals having different phase angles and determine the resulting Lissajous pattern.

The procedure for measuring phase difference between two sinusoidal signals is simply to apply the two signals to the vertical and horizontal inputs and then make two measurements from the resulting display. The usual horizontal linear sweep is not used at all for this procedure. Note that the two signals must be the same frequency (otherwise the parameter phase angle is meaningless). Figure 15.37 shows the resulting display when the *same* signal is applied to *both* channels (0° phase shift). The overall display is that of a straight line. In the example the line is at 45° slope for equal-amplitude signals. If the signals were in phase but not of equal amplitude the line would have slope other than 45°, but the important factor here, the fact that it is a straight line, indicates that the phase shift is zero.

To see more clearly how the straight line is developed on the CRT screen, let us break the sinusoidal cycle into, say, quarter-cycles and follow the resulting beam deflection due to the two sinusoidal signals applied to the vertical and horizontal inputs. Figure 15.37 shows the two input signals and the resulting display. If the signal frequency is very low (a few hertz) the beam movement can be observed

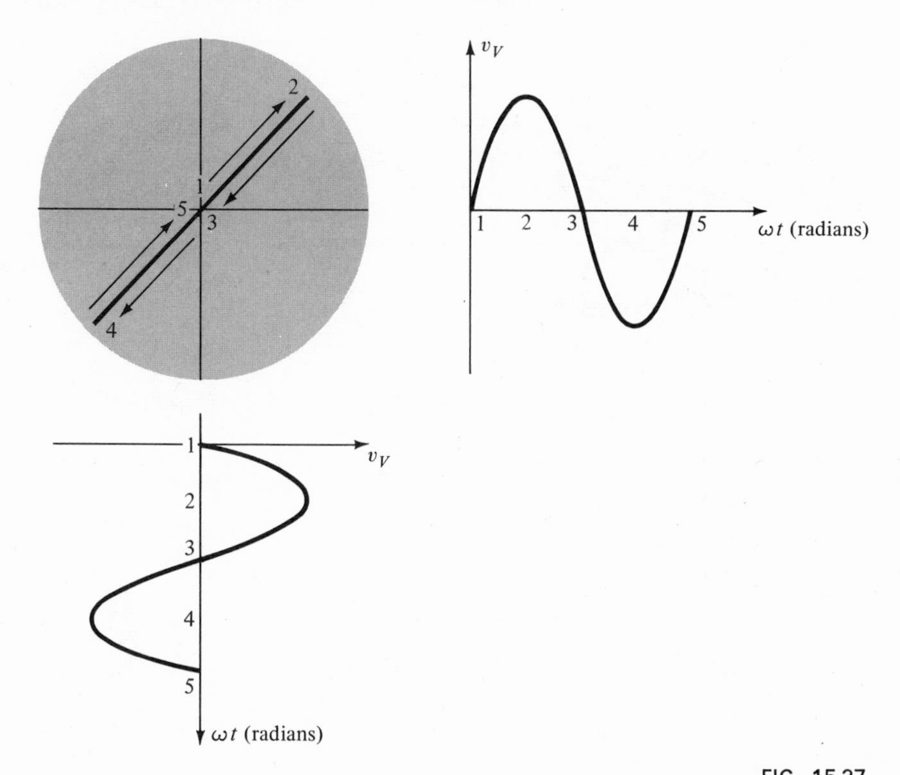

FIG. 15.37
Lissajous figure for 0° phase shift.

on the tube. If the signal frequency is high enough the beam will travel up and back so fast that the display will appear as a solid line. The appearance of a straight line when measuring phase angle between two sinusoidal signals can be directly interpreted as no phase shift (0° phase angle between signals).

Figure 15.38 shows the resulting waveform for two inputs having a phase shift between 0° and 90°. The resulting pattern is an ellipse (at 45° if the two amplitudes are the same). The angle at which the ellipse is generated is of no importance for the phase angle calculation. Noting that the vertical signal amplitude at time 1 is $V = V_m \sin\theta$, we can calculate the angle θ from

$$\theta = \sin^{-1}(V/V_m) \tag{15.8}$$

The values of V and V_m can be easily obtained from the ellipse by measuring the distance (amplitude) of the signal from the center line to where it crosses the center vertical axis, and V_m as the distance from the vertical center line to the top of the ellipse. Using these values in the above relation, we can calculate the phase angle θ.

Since the measurements of V and V_m will be used as a ratio, the actual size or values are not important—only their ratio. This being the case, the actual scale settings of the input signals are unimportant and the CRO adjustments may be used to get an ellipse on the

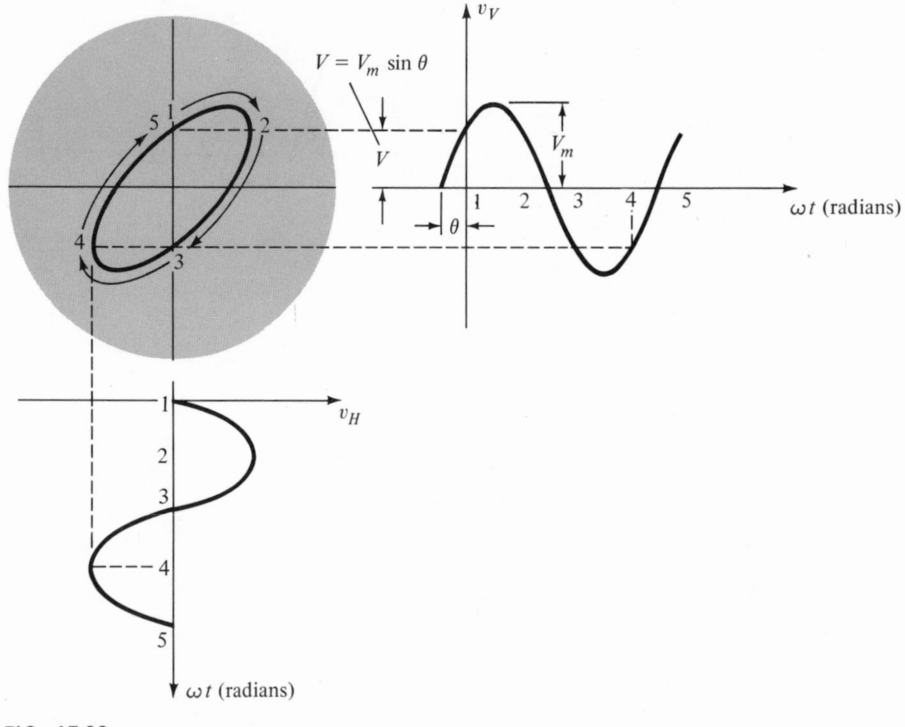

FIG. 15.38

Lissajous figures for θ between 0 and 90° phase shift.

CRT face of about maximum possible size for greatest accuracy. Having properly centered the deflection controls by adjusting the beam to the center of the tube with no input signals, the two measurements marked A and B in Fig. 15.39 are read and the angle θ is calculated from

$$\boxed{\theta = \sin^{-1}\left(\frac{A}{B}\right)} \qquad (15.9)$$

If, for example, the distance B is set to 10 boxes (for whatever scale setting and fine adjustments are necessary) the number of boxes of A can be read and the ratio of A/B obtained. The actual voltages are not important, only their ratio as obtained by the number of scale divisions or boxes on the CRT face.

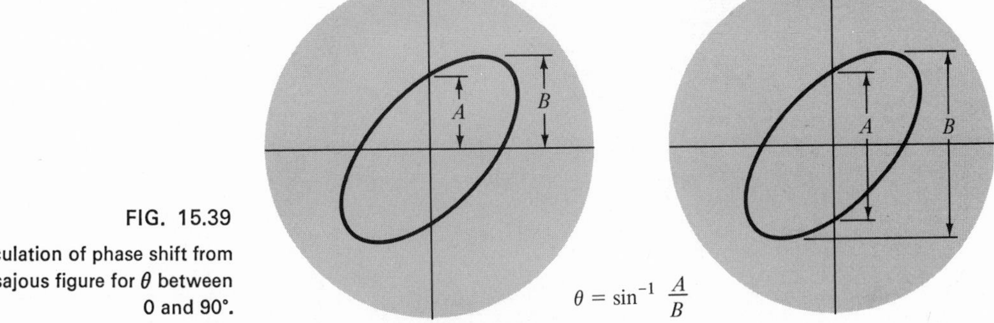

FIG. 15.39
Calculation of phase shift from Lissajous figure for θ between 0 and 90°.

EXAMPLE 15.15 Calculate the phase shift θ for the Lissajous figure in Fig. 15.40.

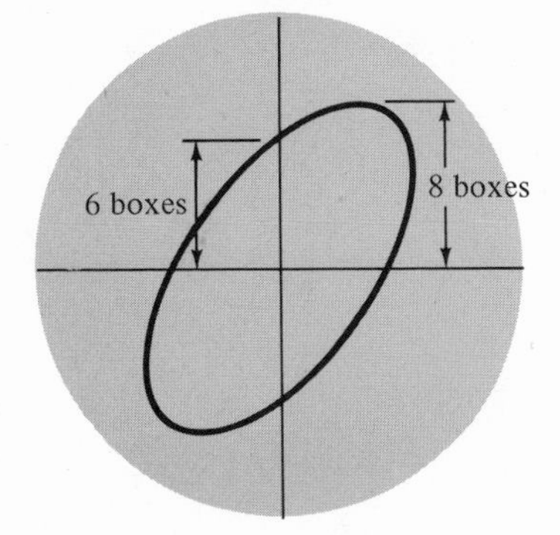

FIG. 15.40
Lissajous figure for Example 15.15.

Solution

$$\theta = \sin^{-1}\frac{A}{B} = \sin^{-1}\frac{6}{8} = \sin^{-1} 0.75 = \mathbf{49°}$$

If the two signals are out of phase by exactly 90°, the resulting waveform is a circle. This result also follows from the above calculation, since for a circle the measured values A and B are equal and the value of θ calculated is $\theta = \sin^{-1}(1) = 90°$. The relation for calculating

θ would also show that for the straight line the measured value of $A = 0$ gives $0/B = 0$, and $\theta = \sin^{-1}(0) = 0°$. Thus, in summary, the values A and B can be measured as indicated above and used to calculate the phase angle within the range of 0 and 90°.

For phase angles of 90–180° the ellipse has a negative slope, as in Fig. 15.41, and the angle calculated by the above method must be subtracted from 180° to obtain the phase shift. Phase angles above

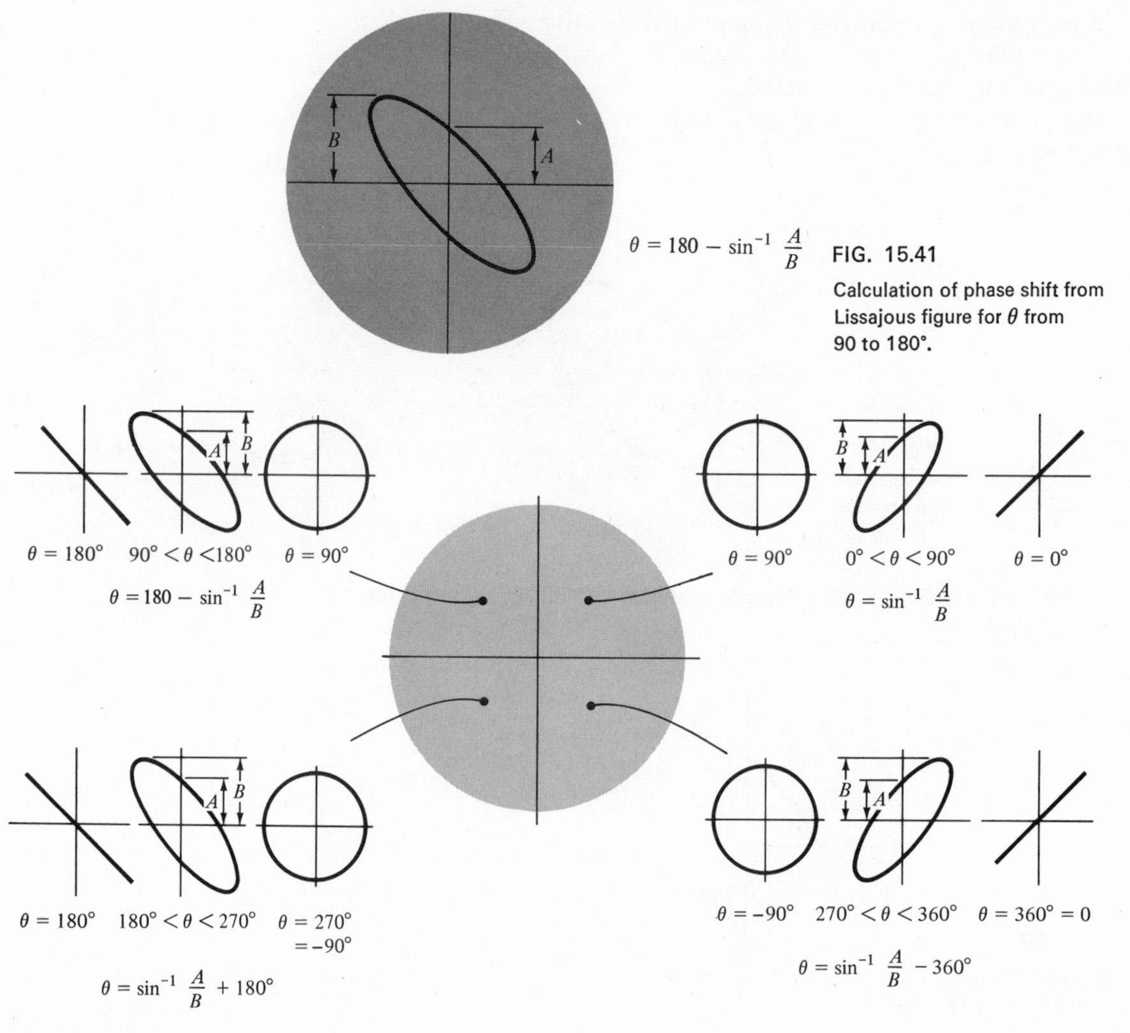

FIG. 15.41

Calculation of phase shift from Lissajous figure for θ from 90 to 180°.

FIG. 15.42

Lissajous phase angle calculation in all quadrants. Note: Additional test adding phase angle shift to signal to be measured is necessary to determine whether observed figure is for upper two quadrants or lower two quadrants. If added phase shift causes positive-sloped ellipse to become larger or negative-sloped ellipse smaller, upper quadrants are indicated; otherwise, lower quadrants.

result in Lissajous figures such as those below 180°, and cannot be directly distinguished. One technique for determining if the measured angle is less or more than 180° is to add an extra (slight) phase shift to the signal being measured. If the phase angle measured increases, the angle was less than 180°. If it decreases, the angle was greater than 180°, and the correct angle is then calculated by adding 180° for the angle computed with negative-sloped ellipse. A comprehensive summary, which clearly shows the required methods to compute the phase angle, is shown in Fig. 15.42.

EXAMPLE 15.16 Calculate the phase angle (within the first two quadrants) for the following Lissajous figures (Fig. 15.43).

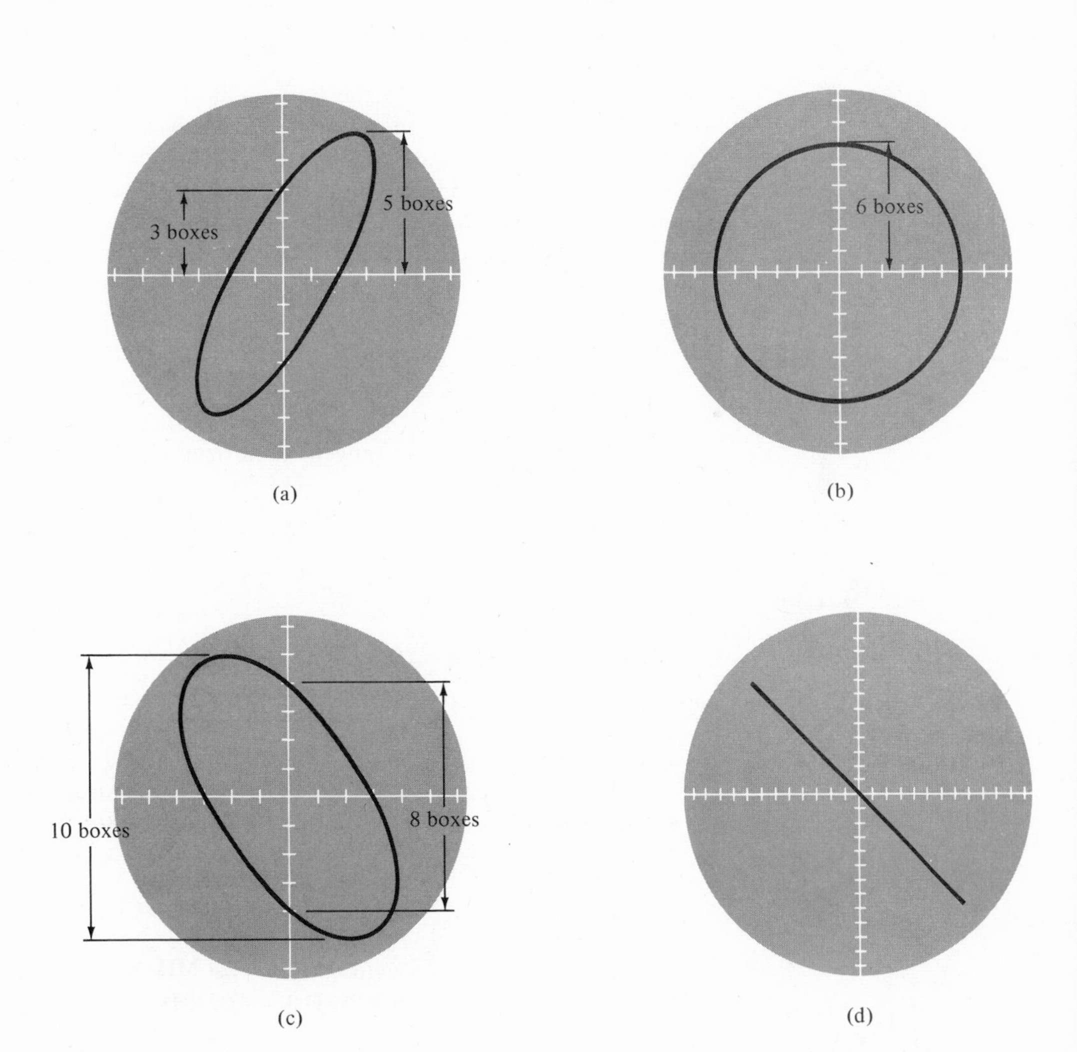

FIG. 15.43
Lissajous figures for Example 15.16.

Solution

(a) $\theta = \sin^{-1}(A/B) = \sin^{-1}(3/5) = \mathbf{37^\circ}$
(b) $\theta = \sin^{-1}(A/B) = \sin^{-1}(6/6) = \mathbf{90^\circ}$ (could have been determined by inspection—circle indicates phase shift of 90°)
(c) $\theta = 180^\circ - \sin^{-1}(A/B) = 180^\circ - \sin^{-1}(4/5) = 180^\circ - 53^\circ = \mathbf{127^\circ}$
(d) by inspection, $\theta = \mathbf{180^\circ}$

Use of Lissajous Figures for Frequency Measurements

If a well-calibrated CRO time base is not available a signal generator can be used to measure the frequency of an unknown sinusoidal signal. Figure 15.44a shows the equipment setup to perform the measurement. The unknown signal is connected to the vertical channel (it could have been fed to the horizontal as well) and the calibrated signal source input is fed to the horizontal channel. The frequency of the signal generator is adjusted until a steady Lissajous pattern is obtained. A sample figure resulting from this procedure appears in Fig. 15.44b. The Lissajous pattern can become quite interesting and involved to analyze. However, for the frequency measurement, all that is needed is the number of tangencies (points at the edge of the arcs) along a vertical and horizontal line at the side of the figure as shown in Fig. 15.44b. The frequency relation between horizontal and vertical inputs is given by

$$\frac{f_H}{f_V} = \frac{\text{No. of tangencies vertical}}{\text{No. of tangencies horizontal}} \tag{15.10}$$

from which we obtain the relation used to calculate the unknown input signal:

$$f_{\text{unknown}} = f_V = \frac{\text{No. of tangencies horizontal}}{\text{No. of tangencies vertical}} \times f_H \tag{15.11}$$

EXAMPLE 15.17 Calculate the frequency of the unknown signal applied to the vertical input for the Lissajous figures shown in Fig. 15.45. The frequency indicated on the figure is that of the known horizontal input.

Solution

(a) $f_u = \frac{2}{1}(1000) = 2000\text{ Hz}$
(b) $f_u = \frac{4}{1}(600\text{ kHz}) = 2.4\text{ MHz}$
(c) $f_u = \frac{5}{3}(50\text{ kHz}) = 83.3\text{ kHz}$

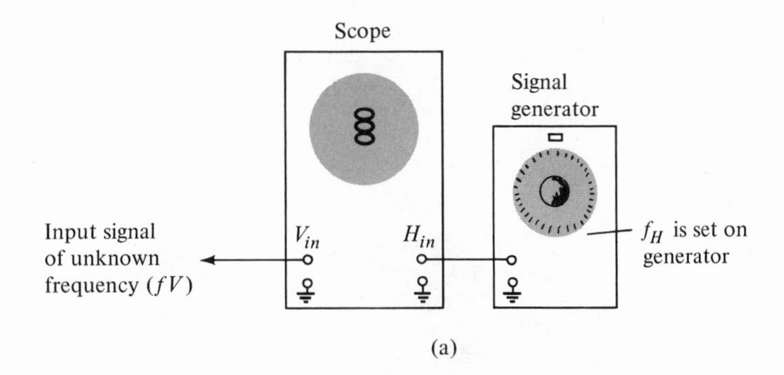

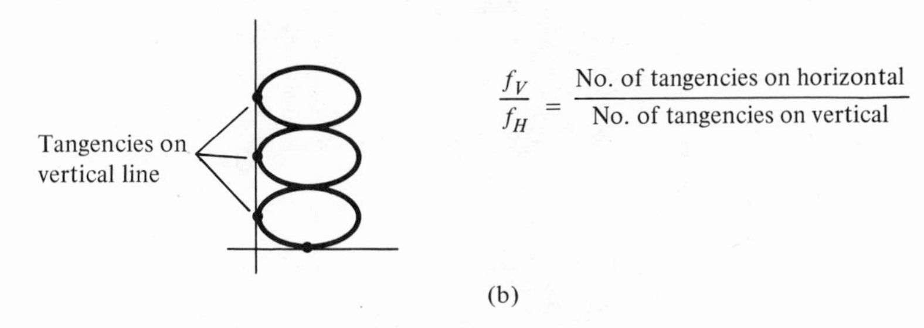

FIG. 15.44

Frequency calculation using Lissajous figure: (a) equipment setup for frequency measurement; (b) Lissajous figure for frequency calculation.

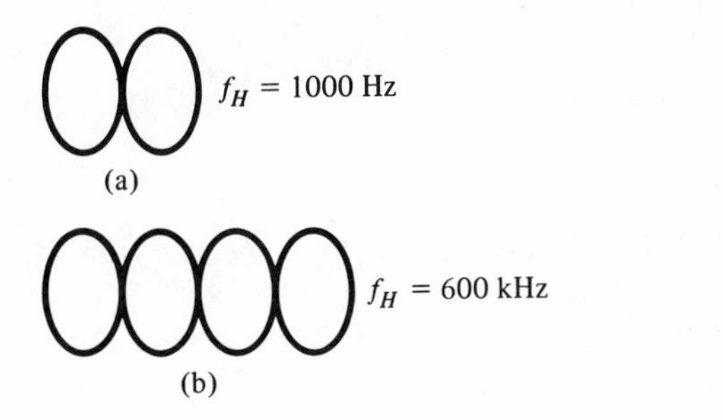

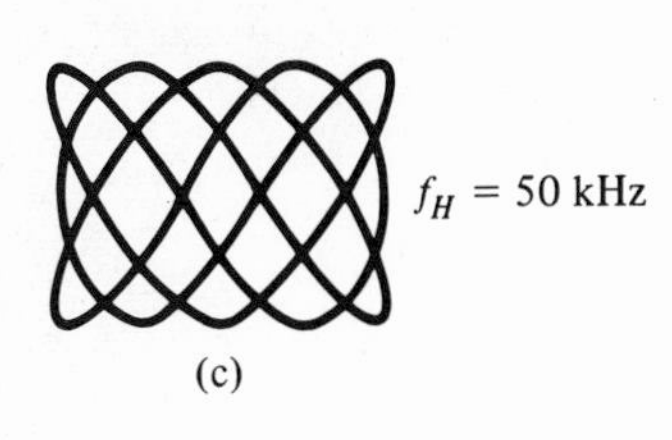

FIG. 15.45

Lissajous figure for Example 15.17.

15.8 SPECIAL CRO FEATURES

The CRO is becoming increasingly more sophisticated and specialized in use. Whereas CROs were originally quite general in range and usage, modern CROs can be geared specifically to the field or area of interest providing those measurements of importance to a particular area of electronics. One important feature of many CROs is the use of *plug-ins*. Rather than manufacture a single integral unit with certain features of interest to a particular area or range of operation, the CRO is manufactured with only the power supply and deflection circuitry as integral to the unit. Figure 15.46 shows a few typical plug-in units

FIG 15.46
Plug-in scope attachments.

and their respective scope main frames (the main frame is the basic CRO body containing power supplies, CRT, and CRT deflection circuitry). Plug-in units are available for both the vertical and horizontal sections of the CRO. These plug-in units may be selected to have features such as the following: single input ranging from 0.005 V/cm to 20 V/cm scale sensitivity; dual-trace capability with two vertical inputs and selection of channel *A* only, channel *B* only, CHOPPED, and ALTERNATE modes of display; differential input for two separate signals; dc–50-MHz frequency range; and dc input from 100 μV to 20 V. Time base plug-ins might provide single time base from 1 μsec to 1 sec/cm selection; two time bases allowing delayed and mixed sweep modes of operation; time base with sampling operation allowing viewing down to a few nanoseconds, etc.

A useful CRO feature uses two time bases to provide a selection of a small part of the signal viewed allowing expanded presentation of only that selected part of the signal. Figure 15.47 shows a digital-type signal containing a number of separate pulses. A delayed type of sweep presentation would be necessary if it were desired to view, say, the third pulse in Fig. 15.47. If the usual single time base were expanded by changing the sweep generator rate so that a shorter period of time were viewed on the screen this would only magnify the signal shown in Fig. 15.47 from the left (start of the sweep) out; that is, the display would be expanded for the part of the signal starting at the beginning of the sweep but the part of the signal over to the right would be out of the screen area for this more detailed sweep setting. The delay sweep feature to be discussed allows selecting a part of the displayed presentation and then changing the display to show only a selected portion at whatever magnification is desired.

The main time base is referred to as the *A* time base and is the horizontal sweep, which provides the picture as shown in Fig. 15.47. An additional time base sweep generator is also provided (called *B* time base) when delayed sweep operation is available. A basic description of this delayed sweep operation is shown in Fig. 15.48. The pulse-type signal shown in Fig. 15.48a indicates a selected part of that signal by the more intensified display. This part of the signal is then shown in Fig. 15.48b in a more detailed (more magnified) presentation. The use of a delayed sweep time base allowed this selection and more detailed presentation and the block diagram of Fig. 15.48c shows how the circuitry is connected to accomplish this operation. Note in Fig. 15.48c that the main and delayed sweep circuitry are two approximately identical units. Their connection in the overall CRO operation is what differentiates them. With the front panel control set to main sweep

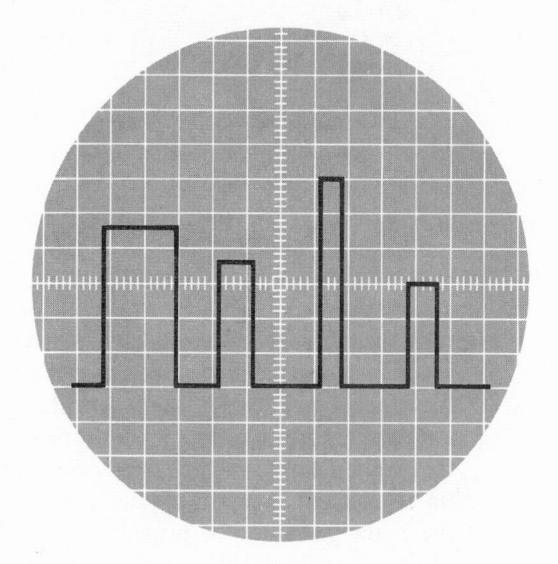

FIG. 15.47
Digital-type pulse viewed on scope.

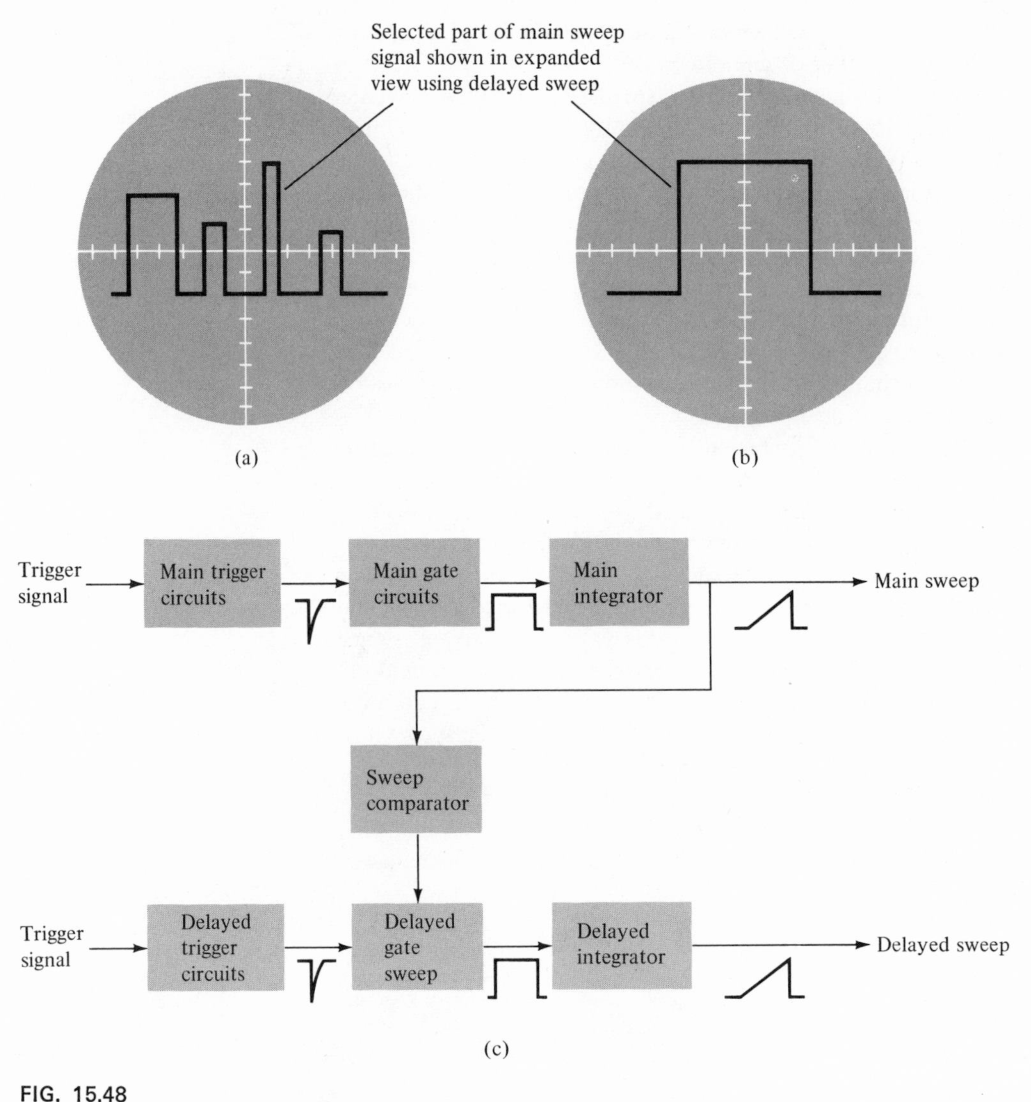

FIG. 15.48

Operation of delayed sweep: (a) main sweep presentation; (b) delayed sweep presentation; (c) delayed sweep operation—block diagram.

only the main sweep generator is activated and the input trigger signal is connected to the main sweep trigger circuit resulting in the main sweep waveform being used as the horizontal deflection signal. When delayed sweep operation is set up, an additional sweep generator is activated—this is the delayed sweep circuit. For operational ease, the CRO is set up so that the sweep of the delayed generator can be added to that of the main generator to provide the intensified display shown in Fig. 15.48a. In effect, then, a potentiometer adjustment is provided on the front panel controls to set the sweep comparator of Fig. 15.48c so that at some selected time in the main sweep the delayed sweep circuitry will be activated (triggered). The amount of time that the delayed sweep is generated is independently set by the delayed sweep time base controls on the scope front panel. This time setting is always less than that of the main sweep. Thus, an intensified presentation is obtained with some adjustment for when the intensified part of the display begins and for how long it lasts.

The CRO mode of operation can now be changed from main time base to delayed time base operation. When this is done the displayed signal is then only the portion previously shown as the intensified section. What has happened is that the main sweep no longer drives the horizontal deflection circuitry—but it still is used to trigger the delayed sweep as before. Thus, the delayed sweep does not start until the delayed time interval previously seen on the screen as the start of the intensified display. When this time in the signal operation occurs the delayed sweep now drives the horizontal deflection circuitry and a display (at the delayed generator sweep rate) is provided. This display, as shown in Fig. 15.48b, shows only the previously intensified part of

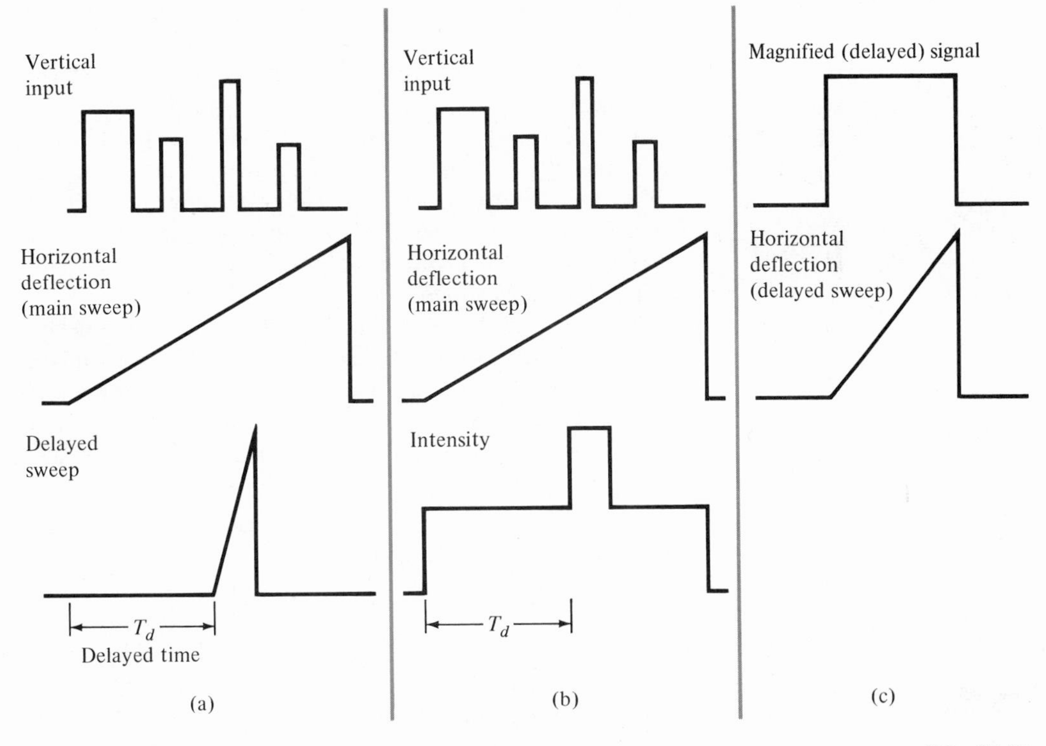

FIG. 15.49

Main and delay sweep displays: (a) main sweep; (b) intensified main sweep; (c) delayed sweep.

the signal, and shows it at the expanded display setting of the delayed sweep generator.

Figure 15.49 shows the sweep waveforms for the main and delayed circuitry to show the resulting operation as described. As shown in Fig. 15.49a the amount of the vertical input signal displayed is set by the main sweep horizontal time base. The delayed sweep time base is adjusted for some faster sweep rate providing less sweep time for a cycle, the start of the delayed sweep being set at some amount of time after the start of the main sweep. With the controls set to main sweep, the 3 pulses shown in Fig. 15.49a are displayed. Figure 15.49b shows the intensified main sweep and the intensity signal, which controls the signal intensity seen on the scope screen. (This intensity signal drives the control grid to vary the number of electrons in the electron beam.) The intensity is kept at a normally lower level and set to a higher level during the delay time interval. The display seen then still shows the 3 pulses selected by the main sweep and additionally the intensified part of the sweep as set by the delayed sweep circuitry.

Finally, when the controls are switched to delayed sweep operation the horizontal sweep is taken from the delayed sweep generator and the previously intensified presentation is now the complete screen display. It must be kept in mind that even when only the delayed sweep is operating the screen display the signal seen is tied to that originally displayed by the main sweep and the delayed time is also based on the start of the main sweep.

As a summary, the use of the delayed sweep additional to the main sweep allows selection of a part of any displayed signal with complete and flexible control in displaying only the selected part of the signal, at whatever shorter time display interval desired. Figure 15.50 shows actual CRO displays of main sweep, main sweep intensified, and, finally, delayed sweep.

Another special feature found in CROs having the delayed sweep operation is *mixed sweep*. This is nothing more than a mixing of the main sweep and delayed sweep signals at one time on the screen. The adjustment of the delayed trigger time sets the point at which the sweep changes from the main sweep rate to the delayed sweep rate as shown in Fig. 15.51.

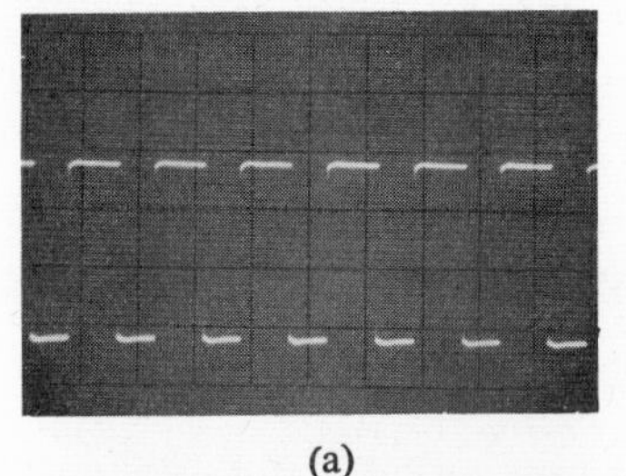

(a)

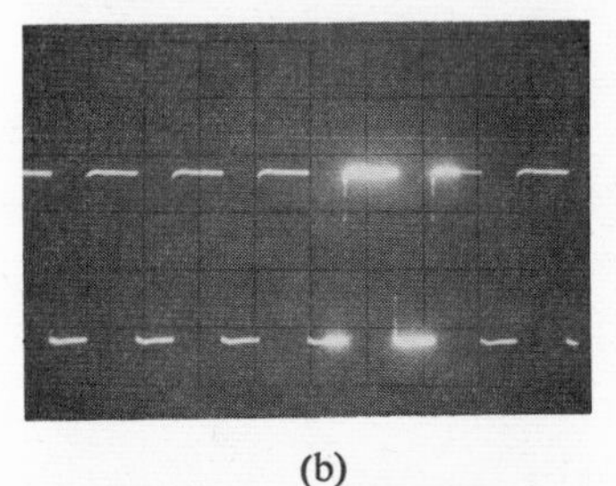

(b)

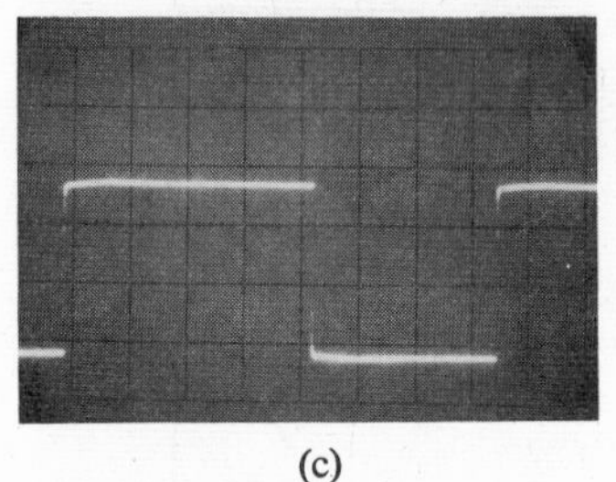

(c)

FIG. 15.50
Actual main and delayed sweep scope displays: (a) main sweep; (b) intensified main sweep; (c) delayed sweep.

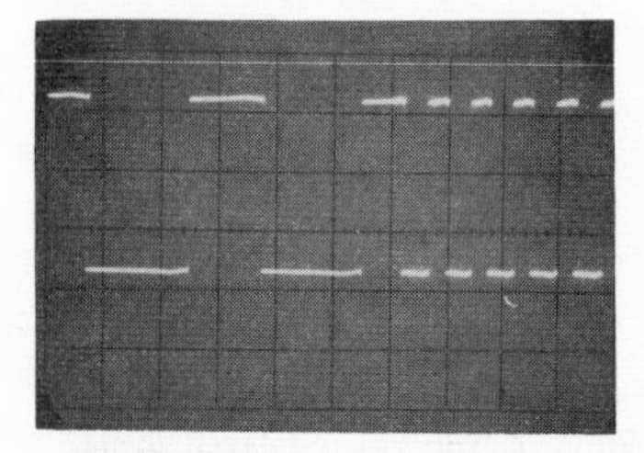

FIG. 15.51
Actual mixed sweep display.

PROBLEMS

§15.2

1. A scope CRT has a rated deflection factor of 50 V/in. How much deflection is obtained on the CRT screen for plate deflection voltages of (a) + 40 V, (b) −75 V
2. What is the vertical deflection sensitivity of a CRT that has a screen deflection of 10 mm when a voltage of 50 V is applied to the vertical deflection plates?
3. A plate deflection voltage of 80 V results in a screen deflection of

2 cm. What is the tube deflection factor and how much deflection would result if 120 V were applied to the deflection plates?

4. A CRT having 0.3 mm/V sensitivity is used. How much deflection, in inches, results from a plate deflection voltage of 150 V?
5. A CRT with accelerating potential of 10,000 V has a deflection sensitivity of 0.45 mm/V. If an applied voltage causes a deflection of 6 cm, how much deflection would result with an accelerating potential of 15,000 V? What is the amount of the applied voltage?

§ 15.4

6. A 1-kHz sinusoidal signal is fed to the vertical input of an oscilloscope. Draw the scope presentation for the following horizontal time base sweep frequencies (assume sweep triggered on positive-going slope at 0-V level). (a) 1 kHz, (b) 2 kHz, (c) 500 Hz.
7. (a) A 50-kHz square wave is fed into the vertical input of a CRO. If the horizontal sweep speed is set to 2 μsec/cm, draw the CRO display for a field of 10 cm on the CRT.
   (b) Repeat for a sweep speed of 4 μsec/cm.
   (c) Repeat for a sweep speed of 1 μsec/cm.
8. What is the CRO horizontal sweep frequency if 4 cycles of a 10-kHz signal are viewed?

§ 15.6

9. A sinusoidal signal is observed on a CRO as having a peak amplitude of 4.1 cm. If the scope vertical gain setting is 0.5 V/cm, calculate the peak and rms values of the input voltage.
10. Draw the CRO display for a 4-V rms sinusoidal waveform for a vertical scale of 2 V/cm. Indicate vertical axis and scale markings clearly.
11. What is the peak-to-peak and rms voltage for the sinusoidal waveform of Fig. 15.52?

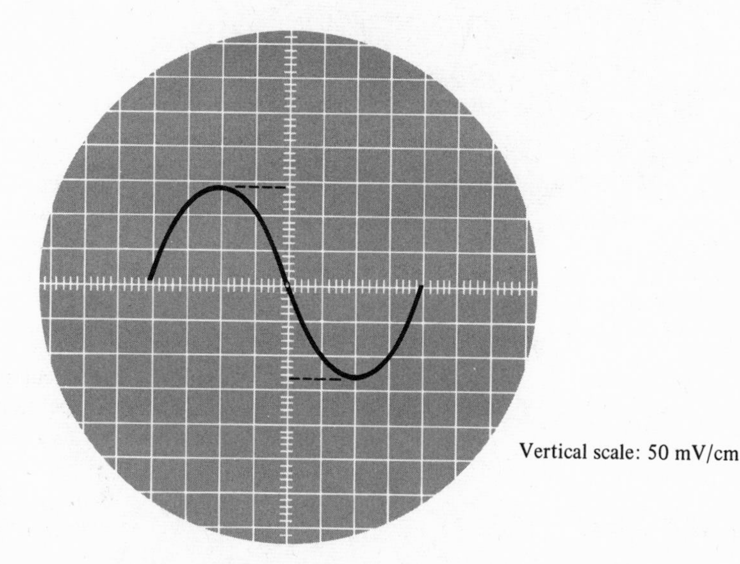

FIG. 15.52
Waveform for Problem 15.11.

12. A square-wave signal measured on a CRO has a peak amplitude of 650 mV. If the CRO scale setting was 200 mV/cm, how many centimeters of signal amplitude were observed (peak-to-peak)?
13. A pulse-type signal is observed on a CRO to have a width of 6.4 cm. Calculate the pulse width time for the following sweep scale settings. (a) 5 msec/cm, (b) 100 μsec/cm, (c) 2 μsec/cm.
14. Two pulse-type signals are observed on a CRO at a scale setting of 20 μsec/cm. If the pulse widths are measured as 1.8 and 3.2 cm, respectively, and both start at the same time, calculate the time width of each pulse and the time delay between the end of the pulses.
15. A sinusoidal signal observed on a CRO repeats 1 cycle in 6.3 cm. If the scale setting was 5 μsec/cm calculate the signal frequency.
16. Calculate the signal frequency of a square-wave signal having a width for 1 half-cycle of 10.5 cm at a scale setting of 10 μsec/cm.
17. A 400-Hz signal is observed on a CRO. How many centimeters should be observed on a CRO? How many centimeters should be observed for 3 cycles of the signal if the scale setting is 1 msec/cm?
18. For the CRO display of Fig. 15.53 calculate the following: (a) Peak-to-peak voltage (V_{p-p}) and V_{rms}, (b) Time for one complete cycle (T), (c) Frequency of waveform signal (f)

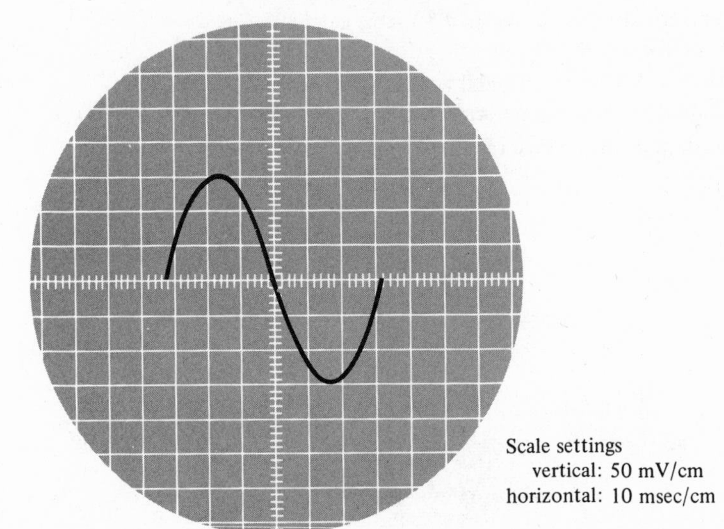

FIG. 15.53

Waveform for Problem 15.18.

19. For the CRO display of Fig. 15.54 calculate the following: (a) V_{p-p}, (b) Time for 2 cycles (T), (c) Pulse repetition rate (f)

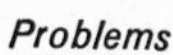

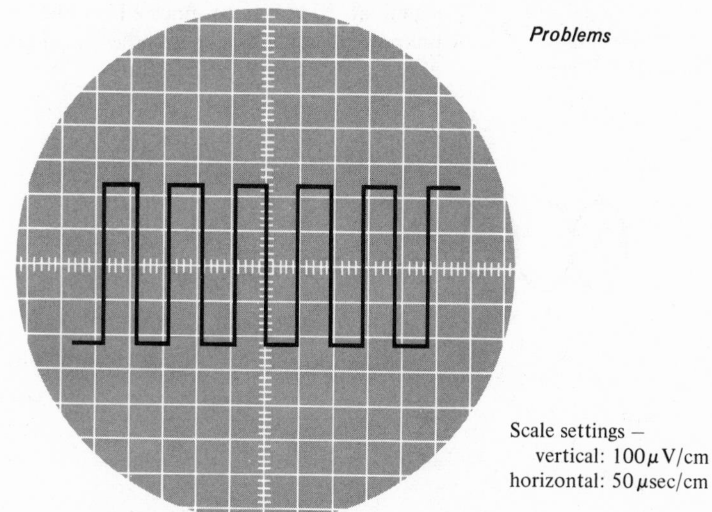

FIG. 15.54

Waveform for Problem 15.19.

20. The CRO scale is adjusted so that a sinusoidal signal takes 6 cm for one full cycle. Calculate the scale factor for this adjustment and the amount of phase shift for a reading of 1.5 cm.

21. (a) A full cycle is set to 8 cm. The phase displacement is measured as 0.5 cm. Calculate the scale factor and phase shift in degrees.
(b) Calculate the number of centimeters observed for a phase shift of 40°.

§ 15.7

22. Draw the Lissajous figures for the following phase angles: (a) $\theta = 180°$, (b) $\theta = -90°$, (c) $\theta = 45°$, (d) $\theta = 135°$

23. Calculate the phase angle for the Lissajous figures in Fig. 15.55. (Assume phase shift is less than 180°.)

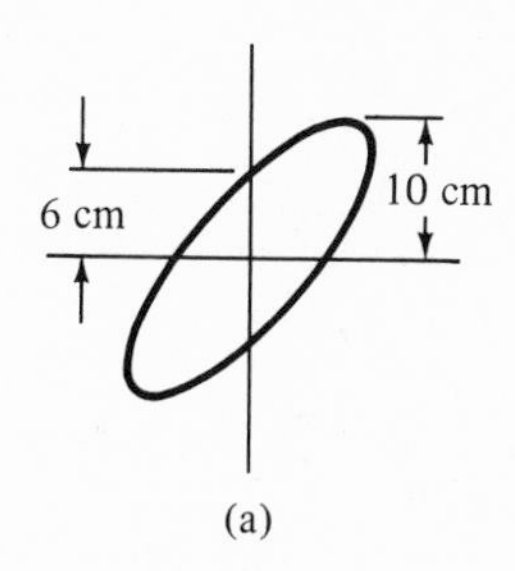

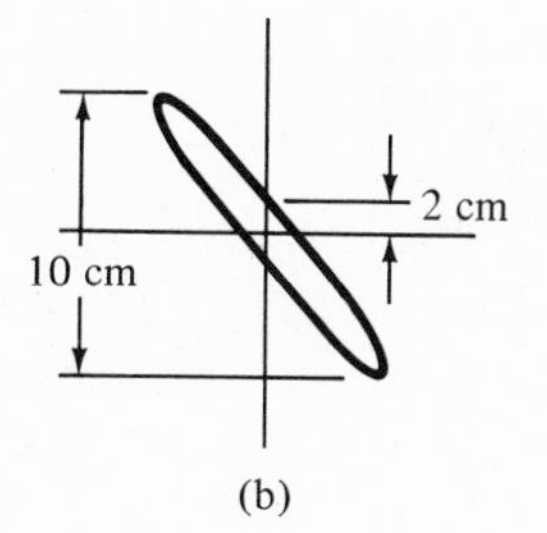

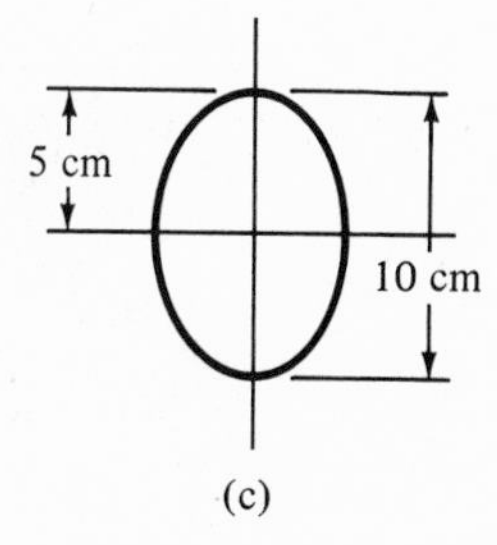

FIG. 15.55

Lissajous figures for Problem 15.23.

24. To measure the frequency of an unknown sinusoidal signal a calibrated signal at 10 kHz is connected to the vertical input. Calculate the unknown frequency (f_u) for the Lissajous figures in Fig. 15.56.

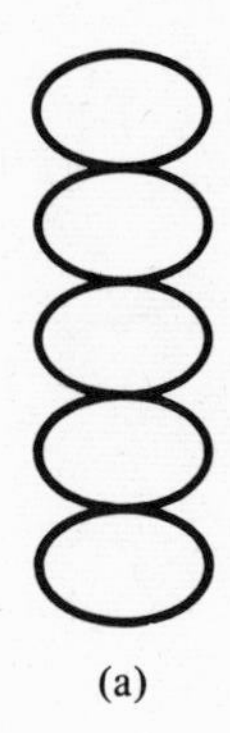

(a)

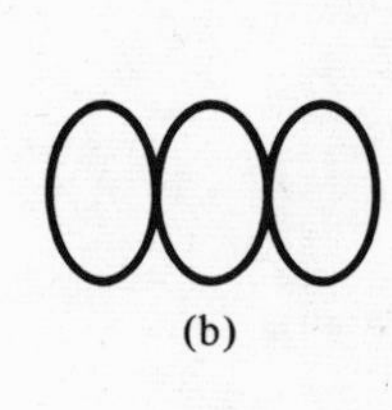

(b)

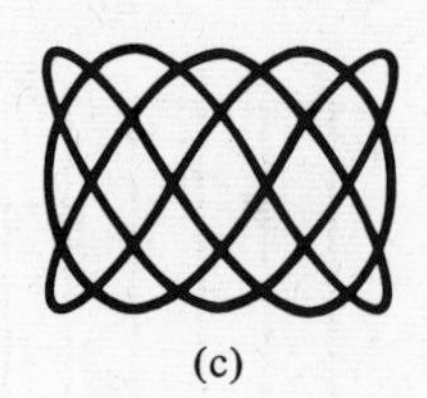

(c)

FIG. 15.56
Lissajous waveforms for Problem 15.24.

Hybrid Parameters—Conversion Equations (Exact and Approximate)

A.1 EXACT

Common-Emitter Configuration

$$h_{ie} = \frac{h_{ib}}{(1 + h_{fb})(1 - h_{rb}) + h_{ob}h_{ib}} = h_{ic}$$

$$h_{re} = \frac{h_{ib}h_{ob} - h_{rb}(1 + h_{fb})}{(1 + h_{fb})(1 - h_{rb}) + h_{ob}h_{ib}} = 1 - h_{rc}$$

$$h_{fe} = \frac{-h_{fb}(1 - h_{rb}) - h_{ob}h_{ib}}{(1 + h_{fb})(1 - h_{rb}) + h_{ob}h_{ib}} = -(1 + h_{fc})$$

$$h_{oe} = \frac{h_{ob}}{(1 + h_{fb})(1 - h_{rb}) + h_{ob}h_{ib}} = h_{oc}$$

Common-Base Configuration

$$h_{ib} = \frac{h_{ie}}{(1 + h_{fe})(1 - h_{re}) + h_{ie}h_{oe}} = \frac{h_{ic}}{h_{ic}h_{oc} - h_{fc}h_{rc}}$$

$$h_{rb} = \frac{h_{ie}h_{oe} - h_{re}(1 + h_{fe})}{(1 + h_{fe})(1 - h_{re}) + h_{ie}h_{oe}} = \frac{h_{fc}(1 - h_{rc}) + h_{ic}h_{oc}}{h_{ic}h_{oc} - h_{fc}h_{rc}}$$

$$h_{fb} = \frac{-h_{fe}(1 - h_{re}) - h_{ie}h_{oe}}{(1 + h_{fe})(1 - h_{re}) + h_{ie}h_{oe}} = \frac{h_{rc}(1 + h_{fc}) - h_{ic}h_{oc}}{h_{ic}h_{oc} - h_{fc}h_{rc}}$$

$$h_{ob} = \frac{h_{oe}}{(1 + h_{fe})(1 - h_{re}) + h_{ie}h_{oe}} = \frac{h_{oc}}{h_{ic}h_{oc} - h_{fc}h_{rc}}$$

Common-Collector Configuration

$$h_{ic} = \frac{h_{ib}}{(1 + h_{fb})(1 - h_{rb}) + h_{ob}h_{ib}} = h_{ie}$$

$$h_{rc} = \frac{1 + h_{fb}}{(1 + h_{fb})(1 - h_{rb}) + h_{ob}h_{ib}} = 1 - h_{re}$$

$$h_{fc} = \frac{h_{rb} - 1}{(1 + h_{fb})(1 - h_{rb}) + h_{ob}h_{ib}} = -(1 + h_{fe})$$

$$h_{oc} = \frac{h_{ob}}{(1 + h_{fb})(1 - h_{rb}) + h_{ob}h_{ib}} = h_{oe}$$

A.2 APPROXIMATE

Common-Emitter Configuration

$$h_{ie} \cong \frac{h_{ib}}{1 + h_{fb}}$$

$$h_{re} \cong \frac{h_{ib}h_{ob}}{1 + h_{fb}} - h_{rb}$$

$$h_{fe} \cong \frac{-h_{fb}}{1 + h_{fb}}$$

$$h_{oe} \cong \frac{h_{ob}}{1 + h_{fb}}$$

Common-Base Configuration

$$h_{ib} \cong \frac{h_{ie}}{1 + h_{fe}} \cong \frac{-h_{ic}}{h_{fc}}$$

$$h_{rb} \cong \frac{h_{ie}h_{oe}}{1 + h_{fe}} - h_{re} \cong h_{rc} - 1 - \frac{h_{ic}h_{oc}}{h_{fc}}$$

$$h_{fb} \cong \frac{-h_{fe}}{1 + h_{fe}} \cong \frac{-(1 + h_{fc})}{h_{fc}}$$

$$h_{ob} \cong \frac{h_{oe}}{1 + h_{fe}} \cong \frac{-h_{oc}}{h_{fc}}$$

Common-Collector Configuration

$$h_{ic} \cong \frac{h_{ib}}{1 + h_{fb}}$$

$$h_{rc} \cong 1$$

$$h_{fc} \cong \frac{-1}{1 + h_{fb}}$$

$$h_{oc} \cong \frac{h_{ob}}{1 + h_{fb}}$$

Ripple Factor and Voltage Calculations

B.1 RIPPLE FACTOR OF RECTIFIER

The ripple factor of a voltage is defined by

$$r \equiv \frac{\text{rms value of ac component of signal}}{\text{average value of signal}}$$

which can be expressed as

$$r = \frac{V_r\,(\text{rms})}{V_{\text{dc}}}$$

Since the ac voltage component of a signal containing a dc level is

$$v_{\text{ac}} = v - V_{\text{dc}}$$

the rms value of the ac component is

$$\begin{aligned} V_r\,(\text{rms}) &= \left[\frac{1}{2\pi}\int_0^{2\pi} v_{\text{ac}}^2\,d\theta\right]^{1/2} = \left[\frac{1}{2\pi}\int_0^{2\pi} (v - V_{\text{dc}})^2\,d\theta\right]^{1/2} \\ &= \left[\frac{1}{2\pi}\int_0^{2\pi} (v^2 - 2vV_{\text{dc}} + V_{\text{dc}}^2)\,d\theta\right]^{1/2} \\ &= [V^2\,(\text{rms}) - 2V_{\text{dc}}^2 + V_{\text{dc}}^2]^{1/2} = [V^2\,(\text{rms}) - V_{\text{dc}}^2]^{1/2} \end{aligned}$$

where V(rms) is the rms value of the total voltage.

For the half-wave rectified signal

$$V_r\,(\text{rms}) = [V^2\,(\text{rms}) - V_{\text{dc}}^2]^{1/2}$$
$$= \left[\left(\frac{V_m}{2}\right)^2 - \left(\frac{V_m}{\pi}\right)^2\right]^{1/2}$$
$$= V_m\left[\left(\frac{1}{2}\right)^2 - \left(\frac{1}{\pi}\right)^2\right]^{1/2}$$

$$\boxed{V_r\,(\text{rms}) = 0.385\,V_m, \quad \text{half-wave}} \tag{B.1}$$

For the full-wave rectified signal

$$V_r\,(\text{rms}) = [V^2\,(\text{rms}) - V_{\text{dc}}^2]^{1/2}$$
$$= \left[\left(\frac{V_m}{\sqrt{2}}\right)^2 - \left(\frac{2V_m}{\pi}\right)^2\right]^{1/2}$$
$$= V_m\left[\frac{1}{2} - \frac{4}{\pi^2}\right]^{1/2}$$

$$\boxed{V_r\,(\text{rms}) = 0.305\,V_m, \quad \text{full-wave}} \tag{B.2}$$

B.2 RIPPLE VOLTAGE OF CAPACITOR FILTER

Assuming a triangular ripple waveform approximation as shown in Fig. B.1 we can write (see Fig. B.2)

$$V_{\text{dc}} = V_m - \frac{V_r(\text{p-p})}{2} \tag{B.3}$$

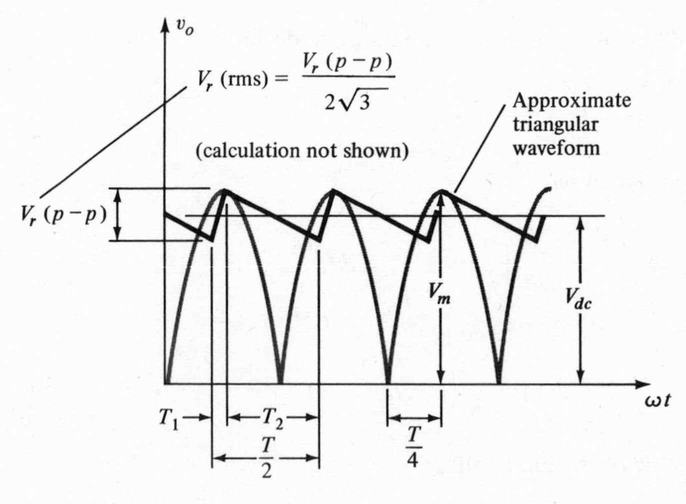

FIG. B.1

Approximate triangular ripple voltage for capacitor filter.

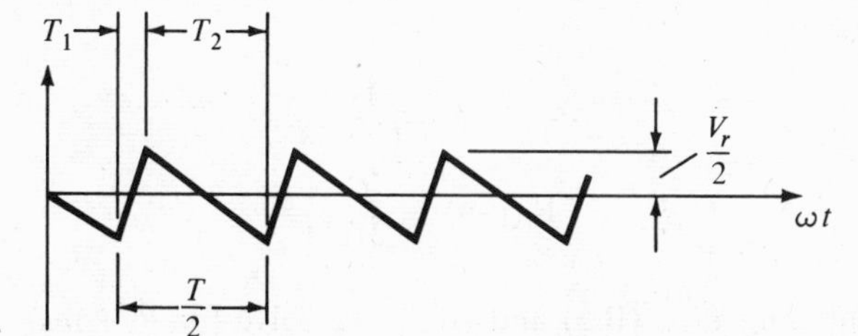

FIG. B.2
Ripple voltage.

During capacitor-discharge, voltage change across C is

$$V_r(\text{p-p}) = \frac{I_{\text{dc}} T_2}{C} \tag{B.4}$$

From the triangular waveform in Fig. B.1

$$V_r\,(\text{rms}) = \frac{V_r(\text{p-p})}{2\sqrt{3}} \tag{B.5}$$

(obtained by calculations, not shown).

Using the waveform details of Fig. B.1:

$$\frac{V_r(\text{p-p})}{T_1} = \frac{V_m}{T/4}$$

$$T_1 = \frac{V_r(\text{p-p})\,(T/4)}{V_m}$$

Also,

$$T_2 = \frac{T}{2} - T_1 = \frac{T}{2} - \frac{V_r(\text{p-p})(T/4)}{V_m} = \frac{2TV_m - V_r(\text{p-p})T}{4V_m}$$

$$T_2 = \frac{2V_m - V_r(\text{p-p})}{V_m}\,\frac{T}{4} \tag{B.6}$$

Since Eq. (B.3) can be written as

$$V_{\text{dc}} = \frac{2V_m - V_r(\text{p-p})}{2}$$

we can combine the last equation with B.6

$$T_2 = \frac{V_{\text{dc}}}{V_m}\,\frac{T}{2}$$

which, inserted into Eq. (B.4), gives

$$V_r(\text{p-p}) = \frac{I_{\text{dc}}}{C}\left(\frac{V_{\text{dc}}}{V_{\text{m}}}\,\frac{T}{2}\right)$$

Since

$$T = \frac{1}{f}$$

$$V_r(\text{p-p}) = \frac{I_{dc}}{2fC}\frac{V_{dc}}{V_m} \tag{B.7}$$

Combining Eqs. (B.5) and (B.7), we solve for V_r (rms)

$$\boxed{V_r\,(\text{rms}) = \frac{V_r(\text{p-p})}{2\sqrt{3}} = \frac{I_{dc}}{4\sqrt{3}\,fC}\frac{V_{dc}}{V_m}} \tag{B.8}$$

Charts and Tables

TABLE C.1

Greek Alphabet and Common Designations

Name	Capital	Lower case	Used to designate
alpha	A	α	Angles, area, coefficients
beta	B	β	Angles, flux density, coefficients
gamma	Γ	γ	Conductivity, specific gravity
delta	Δ	δ	Variation, density
epsilon	E	ϵ	Base of natural logarithms
zeta	Z	ζ	Impedance, coefficients, coordinates
eta	H	η	Hysteresis coefficient, efficiency
theta	Θ	θ	Temperature, phase angle
iota	I	ι	
kappa	K	κ	Dielectric constant, susceptibility
lambda	Λ	λ	Wave length
mu	M	μ	Micro, amplification factor, permeability
nu	N	ν	Reluctivity
xi	Ξ	ξ	
omicron	O	o	
pi	Π	π	Ratio of circumference to diameter = 3.1416
rho	P	ρ	Resistivity
sigma	Σ	σ	Sign of summation
tau	T	τ	Time constant, time phase displacement
upsilon	Υ	υ	
phi	Φ	ϕ	Magnetic flux, angles
chi	X	χ	
psi	Ψ	ψ	Dielectric flux, phase difference
omega	Ω	ω	Capital: ohms; lower case: angular velocity

Logarithms

Formulae: $\log ab = \log a + \log b$

$\log \frac{a}{b} = \log a - \log b$

$\log a^n = n \log a$

TABLE C.2

Common Logarithms

no.	0	1	2	3	4	5	6	7	8	9
0	....	0000	3010	4771	6021	6990	7782	8451	9031	9542
1	0000	0414	0792	1139	1461	1761	2041	2304	2553	2788
2	3010	3222	3424	3617	3802	3979	4150	4314	4472	4624
3	4771	4914	5051	5185	5315	5441	5563	5682	5798	5911
4	6021	6128	6232	6335	6435	6532	6628	6721	6812	6902
5	6990	7076	7160	7243	7324	7404	7482	7559	7634	7709
6	7782	7853	7924	7993	8062	8129	8195	8261	8325	8388
7	8451	8513	8573	8633	8692	8751	8808	8865	8921	8976
8	9031	9085	9138	9191	9243	9294	9345	9395	9445	9494
9	9542	9590	9638	9685	9731	9777	9823	9868	9912	9956
10	0000	0043	0086	0128	0170	0212	0253	0294	0334	0374
11	0414	0453	0492	0531	0569	0607	0645	0682	0719	0755
12	0792	0828	0864	0899	0934	0969	1004	1038	1072	1106
13	1139	1173	1206	1239	1271	1303	1335	1367	1399	1430
14	1461	1492	1523	1553	1584	1614	1644	1673	1703	1732
15	1761	1790	1818	1847	1875	1903	1931	1959	1987	2014
16	2041	2068	2095	2122	2148	2175	2201	2227	2253	2279
17	2304	2330	2355	2380	2405	2430	2455	2480	2504	2529
18	2553	2577	2601	2625	2648	2672	2695	2718	2742	2765
19	2788	2810	2833	2856	2878	2900	2923	2945	2967	2989
20	3010	3032	3054	3075	3096	3118	3139	3160	3181	3201
21	3222	3243	3263	3284	3304	3324	3345	3365	3385	3404
22	3424	3444	3464	3483	3502	3522	3541	3560	3579	3598
23	3617	3636	3655	3674	3692	3711	3729	3747	3766	3784
24	3802	3820	3838	3856	3874	3892	3909	3927	3945	3962
25	3979	3997	4014	4031	4048	4065	4082	4099	4116	4133
26	4150	4166	4183	4200	4216	4232	4249	4265	4281	4298
27	4314	4330	4346	4362	4378	4393	4409	4425	4440	4456
28	4472	4487	4502	4518	4533	4548	4564	4579	4594	4609
29	4624	4639	4654	4669	4683	4698	4713	4728	4742	4757
30	4771	4786	4800	4814	4829	4843	4857	4871	4886	4900
31	4914	4928	4942	4955	4969	4983	4997	5011	5024	5038
32	5051	5065	5079	5092	5105	5119	5132	5145	5159	5172
33	5185	5198	5211	5224	5237	5250	5263	5276	5289	5302
34	5315	5328	5340	5353	5366	5378	5391	5403	5416	5428
35	5441	5453	5465	5478	5490	5502	5514	5527	5539	5551
36	5563	5575	5587	5599	5611	5623	5635	5647	5658	5670
37	5682	5694	5705	5717	5729	5740	5752	5763	5775	5786
38	5798	5809	5821	5832	5843	5855	5866	5877	5888	5899
39	5911	5922	5933	5944	5955	5966	5977	5988	5999	6010
40	6021	6031	6042	6053	6064	6075	6085	6096	6107	6117
41	6128	6138	6149	6160	6170	6180	6191	6201	6212	6222
42	6232	6243	6253	6263	6274	6284	6294	6304	6314	6325
43	6335	6345	6355	6365	6375	6385	6395	6405	6415	6425
44	6435	6444	6454	6464	6474	6494	6493	6503	6513	6522
45	6532	6542	6551	6561	6571	6580	6590	6599	6609	6618
46	6628	6637	6646	6656	6665	6675	6684	6693	6702	6712
47	6721	6730	6739	6749	6758	6767	6776	6785	6794	6803
48	6812	6821	6830	6839	6848	6857	6866	6875	6884	6893
49	6902	6911	6920	6928	6937	6946	6955	6964	6972	6981
50	6990	6998	7007	7016	7024	7033	7042	7050	7059	7067
no.	0	1	2	3	4	5	6	7	8	9

TABLE C.2

Common Logarithms (continued)

no.	0	1	2	3	4	5	6	7	8	9
50	6990	6998	7007	7016	7024	7033	7042	7050	7059	7067
51	7076	7084	7093	7101	7110	7118	7126	7135	7143	7152
52	7160	7168	7177	7185	7193	7202	7210	7218	7226	7235
53	7243	7251	7259	7267	7275	7284	7292	7300	7308	7316
54	7324	7332	7340	7348	7356	7364	7372	7380	7388	7396
55	7404	7412	7419	7427	7435	7443	7451	7459	7466	7474
56	7482	7490	7497	7505	7513	7520	7528	7536	7543	7551
57	7559	7566	7574	7582	7589	7597	7604	7612	7619	7627
58	7634	7642	7649	7657	7664	7672	7679	7686	7694	7701
59	7709	7716	7723	7731	7738	7745	7752	7760	7767	7774
60	7782	7789	7796	7803	7810	7818	7825	7832	7839	7846
61	7853	7860	7868	7875	7882	7889	7895	7903	7910	7917
62	7924	7931	7938	7945	7952	7959	7966	7973	7980	7987
63	7993	8000	8007	8014	8021	8028	8035	8041	8048	8055
64	8062	8069	8075	8082	8089	8096	8102	8109	8116	8122
65	8129	8136	8142	8149	8156	8162	8169	8176	8182	8189
66	8195	8202	8209	8215	8222	8228	8235	8241	8248	8254
67	8261	8267	8274	8280	8287	8293	8299	8306	8312	8319
68	8325	8331	8338	8344	8351	8357	8363	8370	8376	8382
69	8388	8395	8401	8407	8414	8420	8426	8432	8439	8445
70	8451	8457	8463	8470	8476	8482	8488	8494	8500	8506
71	8513	8519	8525	8531	8537	8543	8549	8555	8561	8567
72	8573	8579	8585	8591	8597	8603	8609	8615	8621	8627
73	8633	8639	8645	8651	8657	8663	8669	8675	8681	8686
74	8692	8698	8704	8710	8716	8722	8727	8733	8739	8745
75	8751	8756	8762	8768	8774	8779	8785	8791	8797	8802
76	8808	8814	8820	8825	8831	8837	8842	8848	8854	8859
77	8865	8871	8876	8882	8887	8893	8899	8904	8910	8915
78	8921	8927	8932	8938	8943	8949	8954	8960	8965	8971
79	8976	8982	8987	8993	8998	9004	9009	9015	9020	9025
80	9031	9036	9042	9047	9053	9058	9063	9069	9074	9079
81	9085	9090	9096	9101	9106	9112	9117	9122	9128	9133
82	9138	9143	9149	9154	9159	9165	9170	9175	9180	9186
83	9191	9196	9201	9206	9212	9217	9222	9227	9232	9238
84	9243	9248	9253	9258	9263	9269	9274	9279	9284	9289
85	9294	9299	9304	9309	9315	9320	9235	9330	9335	9340
86	9345	9350	9355	9360	9365	9370	9375	9380	9385	9390
87	9395	9400	9405	9410	9415	9420	9425	9430	9435	9440
88	9445	9450	9455	9460	9465	9469	9474	9479	9484	9489
89	9494	9499	9504	9509	9513	9518	9523	9528	9533	9538
90	9542	9547	9552	9557	9562	9566	9571	9576	9581	9586
91	9590	9595	9600	9605	9609	9614	9619	9624	9628	9633
92	9638	9643	9647	9652	9657	9661	9666	9671	9675	9680
93	9685	9689	9694	9699	9703	9708	9713	9717	9722	9727
94	9731	9736	9741	9745	9750	9754	9759	9763	9768	9773
95	9777	9782	9786	9791	9795	9800	9805	9809	9814	9818
96	9823	9827	9832	9836	9841	9845	9850	9854	9859	9863
97	9868	9872	9877	9881	9886	9890	9894	9899	9903	9908
98	9912	9917	9921	9926	9930	9934	9939	9943	9948	9952
99	9956	9961	9965	9969	9974	9978	9983	9987	9991	9996
100	0000	0004	0009	0013	0017	0022	0026	0030	0035	0039
no.	0	1	2	3	4	5	6	7	8	9

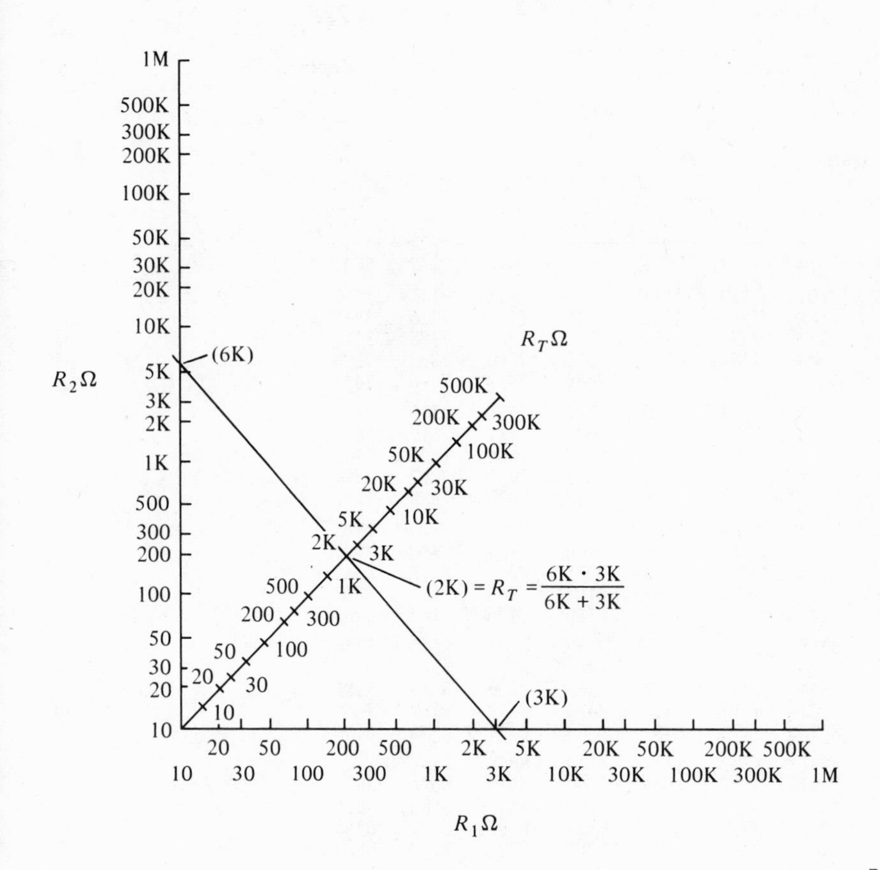

$$R_T = \frac{R_1 R_2}{R_1 + R_2}$$

Parallel Resistance Nomograph: Graph for determining the total resistance of two parallel resistors.

Solutions to Odd-Numbered Problems

CHAPTER 1

5. Yes. **9.** For $V_i = 10$ V, $V_o = 9.1$ V; For $V_i = 15$ V, $V_o = 10.4$ V; For $V_i = 20$ V, $V_o = 10.4$ V. **15.** Gas: $R_{dc} \cong 7.7$ K; Vacuum: $R_{dc} \cong 90$ K; Semiconductor $R_{dc} \cong 400\ \Omega$. **17.** $r_d \cong 0\ \Omega$ (gas); $r_d \cong 28$ K (vacuum); $r_d \cong 250\ \Omega$ (semiconductor). **29.** $V_i = -10$ V: $I_R = 16$ mA, $I_L = 8.4$ mA, $I_z =$ 7.6mA; $V_i = -14$ V: $I_R = 56$ mA, $I_L = 8.4$ mA, $I_z = 47.6$ mA.

CHAPTER 2

1. 47.7 V. **3.** 566 V, PIV. **5.** $V_{dc} = 50$ V; PIV $= 157$ V. **7.** $V_{dc} = 95.4$ V. **9.** PIV $= 566$ V. **13.** (a) Transformer $V_m = 314$ V, diode PIV $= 314$ V, turns ratio $= 1.87:1$; (b) transformer $V_m = 314$ V, diode PIV $= 314$ V, turns ratio $= 0.93:1$. **15.** 0.143. **17.** 38.4 V. **19.** 0.2. **21** 0.0686, 6.86%. **23.** 3.75 μF. **25.** 16.2%. **27.** 1.6 V. **29.** 16 mA. **31.** 42.8%. **33.** $Z_i = 6$ K, $X_{C2} \| R_L = 50\ \Omega$. **35.** 0.885 V. **37.** $C_2 = 22.5\mu$F, $L = 2.78$ H. **39.** 0.25. **41.** 32.5 V. **43.** (a) 41.5 μF, (b) 61.3 V. **45.** $L = 0.8$ H, $C = 83\ \mu$F.

CHAPTER 3

3. 7.92 mA. **5.** (a) $I_C \cong I_E = 4.8$mA; (b) $I_E = 3.3$ mA $\cong I_C$; (c) $V_{EB} = 250$ mV. **7.** (a) $I_C = 2.2$mA, (b) $V_{CE} = -3.5$ V, $V_{BE} = -380$ mV. **9.** $A_v = 1$. **11.** (a) $r_p = 10$ K; (b) $r_p = 18$ K. **13.** 4.5 V. **17.** 0.5 mA.

CHAPTER 4

1. $I_E = 1.81$ mA, $I_C = 1.78$ mA, $V_{CB} = -2.05$ V. **3.** $I_B = 55.3$ μA, $I_C = 2.5$ mA, $V_{CE} = 3.77$ V. **5.** 46; 71. **7.** $V_{CE} = -3.3$ V. **9.** $R_E = 2.06$ K. **11.** $V_C = -5.7$ V. **13.** $I_B = 23.5$ μA, $I_C =$ 1.41 mA, $V_{CE} = 2.95$ V. **15.** 15.2; 18.2. **17.** $I_B = 21.1$ μA; $I_C = 1.79$ mA $\cong I_E$; $V_E = 3.22$ V, $V_{CE} = 5.78$ V. **19.** $V_E = -1.38$ V. **21.** $V_{CE_Q} = -8$ V, $I_{C_Q} = 6$ mA. **23.** $V_{GK} = V_{GG} = -3.5$ V. **25.** $R_p = 28.6$ K. **27.** $R_C = 6$ K, $R_B = 630$ K. **29** $R_E =$ 5.3 K. **31.** $V_C = -11.5$ V. **33.** $I_C = 3.67$ mA.

CHAPTER 5

1. (a) $A_i = 57.1$; (b) $A_v = -160$; (c) $A_p = 9271$. **3.** (a) $A_i = 59$; (b) $A_{v_1} = -58.7$; (c) $A_{v_2} = -45.2$. **5.** (a) $A_v = -144$; (b) $A_i =$ 19.9. **7.** (a) $A_i = -0.23$; (b) $A_{v_1} = 228$; (c) $A_{v_2} = 45.6$ **9.** (a) $A_i = 57.8$;(b) $A_v = -161$; (c) $A_p = 9{,}273.7$;(d) $Z_i = 1.8$ K; (e) $Z_0 =$ 4.44 K. **11.** (a) $A_i = 21.6$; (b) $A_v = -169$; (c) $Z_i = 1.7$ K; (d) $Z_o = 4.3$ K. **13.** $A_v = -99$. **15.** (a) $A_v = 160$; (b) $A_i =$ -0.21; (c) $Z_i = 36.4\ \Omega$; (d) $Z_o = 8$ K. **17.** $V_o = 32.5(V_2 - V_1)$. **19** $A_v = -22.2$. **21** $A_v = -240$.

CHAPTER 6

1. $I_{DQ} = 2.5$ mA, $V_{DS_Q} = 7$ V. **3.** $V_{GS} = V_{GG} = -1.3$ V. **5.** $R_D = 8.8$ K, $R_S = 1$ K, $R_2 = 5$ K, $R_1 = 57.5$ K. **7.** $A_v = -100$. **9.** $R_D = 4.3$ K. **11.** $R_D = 66.7$ K. **13.** (a) $Z_i = 218$ K; (b) $Z_i = 4.46$ K. **15.** (a) $Z_o = 13.3$ K; (b) $Z_o = 320\ \Omega$. **17.** $A_v =$ 0.79, $Z_i = 1$ M, $Z_o = 200\ \Omega$. **19.** $R_D = 7.1$ K. **21.** $R_D =$ 3 K, $R_G = 10$ M, $A_v = -12.6$.

CHAPTER 7

1. $Z_i = 1.67$ K, $Z_o = 2.5$ K. **3.** $A_i = 1040$. **5.** $A_v = 1775$, $A_i = 1482$. **7.** $Z_i = 500\ \Omega$. **9.** (a) $A_{v_1} = -760$, $A_{v_2} = -845$, $A_v = 12{,}216$; (b) $A_v = 6108$. **11.** $A_i = 22.2$ **13.** $Z_i = 1.4$ M, $Z_{o_2} = 31\ \Omega$. **15.** (a) 13; (b) 13; (c) 7. **17.** 68. **19.** f_2 $= 1$ MHz, $f_\beta = 13.3$ MHz, $f_t = 400$ MHz. **21.** 1.1 MHz. **23.** -50. **25.** 4.6 MHz.

CHAPTER 8

1. 2.5 K. **3.** 44.7: 1. **5.** 37%. **7.** (a) 42.3%; (b) 72.9%. **15.** 4.35 W.

CHAPTER 11

3. $V_E = 0.7$ V, $I_E = 0.955$ mA, $I_{E_1} = I_{E_2} = 0.48$ mA, $V_{C_1} = -7.8$ V. **5.** $R_i = 3.6$ K, $R_o = 13.3$ K. **7.** 2.4 mA, 10.7 V. **9.** 10.7 K. **13.** 0 V. **15.** -12 V.

CHAPTER 12

1. 10. **3.** $A_f = 14.3$, $R_{if} = 31.5$ K, $R_{of} = 2.38$ K. **5.** $A = 33.3$, $A_f = 4.35$. **7.** $R_i = 2.1$ K, $R_o = 6.8$ K, $A = 3600$, $R_{of} = 20.7\ \Omega$, $R_{if} = 690$ K, $A_f = 11$. **9.** $C = 21.67$ pF, $R_D = 9.5$ K. **11.** $f_o = 50.4$ kHz, $g_m = 1100\ \mu$mhos. **13.** (b) 76 Ω. **15.** $V_B = 4$ V, $V_C = 11.5$ V, $V_{CE} = 7.66$ V. **17.** $f_o = 290$ kHz, $g_m = 333\ \mu$mhos. **19.** $h_{fe} \geq 0.5$. **23.** (a) $C_T = 0.15\ \mu$F; (b) $C_T = 1000$ pF.

CHAPTER 13

3. (a) 0 V; (b) +2 V; (c) -5 V. **5.** $V_{\text{OFF}} = -1.5$ V. **7.** 4.75. **9.** 1.1 V. **21.** 346 kHz.

CHAPTER 14

1. 5.27%. **3.** $V_L = 30$ V, $V_{R_3} = 20$ V, $I_{R_3} = 20$ mA, $I_{B_1} = 10$ mA, $I_{C_2} = 10$ mA, $I_{R_4} = 3$ mA, $I_Z = 13$ mA, $I_{B_2} = 0.1$ mA, $V_B = 20.7$ V.

CHAPTER 15

1. (a) $D = 0.8$ in.; (b) $D = -1.5$ in. **3.** $G = 40$ V/cm, $D = 3$ cm. **5.** $D_2 = 4$ cm, $V_d = 133.3$ V. **9.** $V_p = 2.05$ V, $V_{\text{rms}} = 1.45$ V. **11.** $V_{p-p} = 270$ mV, $V_{\text{rms}} = 95.6$ mV. **13.** (a) $T_{pw} = 32$ msec; (b) $T_{pw} = 6.4$ msec; (c) $T_{pw} = 12.8\ \mu$sec. **15.** $f = 31.8$ kHz. **17.** 7.5 cm. **19.** (a) $V_{p-p} = 460\ \mu$V; (b) 190 μsec; (c) $f = 10.5$ kHz. **21.** (a) scale factor = 45°/cm, phase shift = 22.5°; (b) 0.89 cm. **23.** (a) 36.9°; (b) 156.4°; (c) ± 90°.

Index

D

E

F

G

H

I

J

L

M

N

O

P

Q

R

S

U

V

W

Z